Yoshiki Oshida, Takashi Miyazaki
Biomaterials and Engineering for Implantology

Also of interest

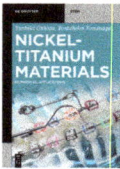

Nickel-Titanium Materials.
Biomedical Applications
Oshida, Tominaga, 2020
ISBN 978-3-11-066603-8, e-ISBN 978-3-11-066611-3

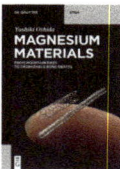

Magnesium Materials.
From Mountain Bikes to Degradable Bone Grafts
Oshida, 2021
ISBN 978-3-11-067692-1, e-ISBN 978-3-11-067694-5

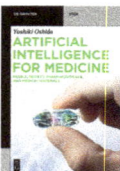

Artificial Intelligence for Medicine.
People, Society, Pharmaceuticals, and Medical Materials
Oshida, 2021
ISBN 978-3-11-071779-2, e-ISBN 978-3-11-071785-3

Materials for Medical Application
Heimann (Ed.), 2020
ISBN 978-3-11-061919-5, e-ISBN 978-3-11-061924-9

Yoshiki Oshida, Takashi Miyazaki

Biomaterials and Engineering for Implantology

In Medicine and Dentistry

DE GRUYTER

Authors
Prof. Yoshiki Oshida
School of Dentistry
University of California San Francisco
513 Parnassus Ave
San Francisco
CA 94153-0340
USA

Dr. Takashi Miyazaki
1-5-1 Nishimami Plaza
3001 Kashiba City
639-0222
Japan

ISBN 978-3-11-074011-0
e-ISBN (PDF) 978-3-11-074013-4
e-ISBN (EPUB) 978-3-11-074023-3

Library of Congress Control Number: 2021950671

Bibliographic information published by the Deutsche Nationalbibliothek
The Deutsche Nationalbibliothek lists this publication in the Deutsche Nationalbibliografie;
detailed bibliographic data are available on the Internet at http://dnb.dnb.de.

© 2022 Walter de Gruyter GmbH, Berlin/Boston
Cover image: Peterschreiber.media/iStock/Getty Images Plus
Typesetting: Integra Software Services Pvt. Ltd.
Printing and binding: CPI books GmbH, Leck

www.degruyter.com

Contents

List of abbreviations

ACL	Anterior cruciate ligament
ACWs	Antral contraction waves
ADA	American Dental Association
ADMET	Absorption, distribution, metabolism, excretion, and toxicity
AE	Acoustic emission
AES	Auger electron spectroscopy
AFM	Atomic force microscopy
AI	Artificial intelligence
aka	Also known as
ALD	Artificial liver assist device
ALP	Alkaline phosphatase
ALVAL	Aseptic lymphocyte–dominated vasculitis-associated lesions
AM	Additive manufacturing
ANN	Artificial neural network
ANS	American National Standard
ANSI	American National Standards Institute
ANT	Anterior nucleus of the thalamus
AO	Artificial organs
AP	Anterior–posterior or anteroposterior
ARDS	Acute respiratory distress syndrome
ARMD	Adverse reaction to metal debris
ARRIVE guidelines	Animal Research: Reporting of In Vivo Experiment guidelines
ASA	American Society of Anesthesiologists
ASD	Adult spinal deformity
ASTM	American Society for Testing and Materials
ATR-FTIR	Attenuated total reflectance-FTIR
AUC	Area under the receiving operating characteristic (curve)
BAF	Bone area fraction
BAL	Bioartificial liver
BCS	Bovine calf serum
BHR	Birmingham Hip Resurfacing
BIC	Bone-to-implant contact (ratio)
BII	Bone–implant interface
Bis-GMA	Bisphenol-A-glycidyldimethacrylate
BL	Bone level
BMI	Body mass index
BMP	Bone morphogenetic protein
BMPR	Bone morphogenetic protein receptor
BOPT	Biologically oriented preparation technique
BVD	Bone volume density
CAD	Computer-aided design
CAGR	Compound annual growth rate
CAL	Clinical attachment level
CAM	Computer-aided manufacturing
CAM	Class activation mapping
CAPs	Calcium phosphates
CBCT	Cone-beam computed tomography

https://doi.org/10.1515/9783110740134-203

CFRP	Carbon-fiber-reinforced thermoplastics
CHAP	Carbonate hydroxyapatite
CKD	Chronic kidney disease
CMC	Ceramic–matrix composite
CNNs	Convolutional neural networks
CNTs	Carbon nanotubes
CoC	Ceramic-on-ceramic
CoM	Ceramic-on-metal
CoP	Ceramic-on-plastic
COPD	Chronic obstructive pulmonary diseases
cpTi	Commercially pure titanium
CT	Computed tomography (μCT – micro-CT)
CVD	Chemical vapor deposition
DASH	Disabilities of the arm, shoulder, and hand
DCB	Decellularized bone
DCNN	Deep convolutional neural networks
DEXA	Dual-energy X-ray absorptiometry
D_F	Fractal dimension
DFPP	Double filtration plasmapheresis
DFR	Distal femoral replacement
DKK1	Dickkopf-related protein 1
DLC	Diamond-like carbon
DLP	Digital light processing
DLS	Deep learning system
DM	Diabetes mellitus
DMLS	Direct metal laser sintering
DVT	Deep vein thrombosis
EBM	Electron beam melting
ECF	Extracellular fluid (compartment)
ECLS	Extracorporeal lung support
ECM	Extracellular matrices
ECMO	Extracorporeal membrane oxygenation
EDM	Electric discharge machining
EDTA	Ethylenediaminetetraacetic acid
EDX	Energy-dispersive X-ray analyzer
EIS	Electrochemical impedance spectroscopy
ELSO	Extracorporeal Life Support Organization
EOS	Early-onset scoliosis
EOSQ	EOS questionnaire
e-PTFE	Expanded polytetrafluoroethylene
Er:YAG	Erbium-doped yttrium aluminum garnet (laser)
ESRD	End-stage renal disease
ETO	Ethylene oxide
FDA	U.S. Food and Drug Administration
FDI	Fédération Dentaire Internationale
FDM	Fused deposition modeling
FEM	Finite element method (analysis)
FGF	Fibroblast growth factor
FGG	Free gingival grafting

FGMs	Functionally graded materials
FMR	Full mouth reconstruction (or rehabilitation)
FTIR	Fourier-transform infrared spectroscopy
FTIR-RAS	Fourier-transform infrared reflection absorption spectroscopy
GBR	Guided bone regeneration
GDM	Gestational diabetes mellitus
GFM	Gradient functional material
GLC	Graphite-like carbon
GMP	Good manufacturing practices
GOHAI	Geriatric Oral Health Assessment Index
GP	General practice
GPLA	Graphene polylactic acid
HA	Hydroxyapatite
HAZ	Heat-affected zone
HD	Hemodialysis
HDPE	High-density polyethylene
HEMA	Hydroxyethyl methacrylate
HHP	Hydrothermal hot pressing
HIP	Hot isostatic pressing
HIPAA	Health Insurance Portability and Accountability Act
HOOS	Hip disability and osteoarthritis outcome score
hPSCs	Human pluripotent stem cells
HRQoL	Health-related quality of life
IAN	Inferior alveolar nerve
ICF	Intracellular fluid (compartment)
ICP-OES	Inductively coupled plasma optical emission spectrometry
IDIP	Initial depth of implant placement
IgA	Immunoglobulin A
IHDs	Implantable hearing devices
IOL	Intraocular lens
ISO	International Organization for Standardization
ISQ	Implant stability quotient
Laser	Light amplification by stimulated emission of radiation
LCVD	Laser chemical vapor deposition
LDL	Low-density lipoprotein
LES	Lower esophageal sphincter
LRP	LDL-receptor-related protein
LSA	Laser surface alloying
LVADs	Left ventricular assist devices
MAO	Micro-arc oxidation (or oxidization)
MARS	Molecular adsorbent recirculation system
MBL	Marginal bone loss
MCGR	Magnetic-controlled growing rods
MCL	Medical collateral ligament
MIC	Microbiology-induced corrosion
MIM	Metal injection molding
MIS	Minimally invasive surgery
MIS-C	Multisystem inflammatory syndrome in children
MMC	Metal–matrix composite

MOE	Modulus of elasticity
MoM	Metal-on-metal
MoP	Metal-on-plastic
MPC	2-Methacryloyloxyethyl phosphorylcholine
mpy	Milli-inch per year
MRI	Magnetic resonance imaging
MRSA	Methicillin-resistant *Staphylococcus aureus*
MSC	Mesenchymal stem cell
MSIA	Musculoskeletal infection society
MTT	3-(4,5-Dimethylthiazol-2-yl)-2,5-diphenyl-2H-tetrazolium bromide (assay)
MZCs	Monolithic zirconia crowns
NBS	National Bureau of Standards
NHANES	National Health and Nutrition Examination Survey
NICE	National Institute of Health and Care Excellence
NINDS	National Institute of Neurological Disorders and Stoke
NIST	National Institute for Standards and Technology
NNS	Near-net shaping
NRC	Nonrigid connector
OA	Osteoarthritis
OD	Osseodensification
OHIP	Oral Health Impact Profile
OHRQoL	Oral health-related quality of life
OHS	Oxford hip score
ON	Osteonecrosis
ONJ	Osteonecrosis of the jaw
OoC	Organ-on-a-chip
PAH	Pulmonary hypertension
PBF	Powder bed fusion
PBT	Polybutylene terephthalate
PBS	Phosphate-buffered solution
PCL	Polycaprolactone
PCL	Posterior cruciate ligament
PCU	Poly(carbonate-urethane)
PD	Peritoneal dialysis
PDGF	Platelet-derived growth factor
PDL	Periodontal ligament
PDO	Poly(*p*-dioxanone)
PE	Polyethylene
PEEK	Polyetheretherketone
PEGT	Polyethylene glycol terephthalate
PEO	Polyethylene oxide polymers
PES	Pink aesthetic score
PET	Polyethylene terephthalate
PGA	Poly(glycolic acid)
PGE2	Prostaglandin E2
PGS	Poly(glycerol sebacate)
PH	Precipitation-hardening type
PI^3 or PIII	Plasma immersion ion implantation
PIII&D	Plasma immersion ion implantation and deposition

PIM	Powder injection molding
PJI	Periprosthetic joint infection
PLA	Poly(lactic acid)
PLGA	Poly(lactic-co-glycolic acid)
PLLA	Poly(L-lactic acid)
PMA	Premarket approval
PMC	Plastic–matrix composite
PMMA	Polymethyl methacrylate
PMN	Polymorphonuclear leukocyte
PMN	Polymorphonuclear neutrophils
PMPC	Poly(MPC)
POSS	Polyhedral oligomeric silsesquioxanes
PPP	Platelet-poor plasma
PROs	Patient-reported outcomes
PROMs	Patient-reported outcome measures
PRP	Platelet-rich plasma
PTFE	Polytetrafluoroethylene
PTH	Parathyroid hormone
PTHrP	Parathyroid-hormone-related protein
PTI	Porous titanium
PTMC	Poly(trimethylene carbonate)
PU, PUR	Polyurethane
PVA	Poly(vinyl acetate)
PVD	Physical vapor deposition
PVP	Poly(vinyl pyrrolidone)
QOL	Quality of life (or living)
QSAR	Quantitative structure–activity relationship
QSPR	Quantitative structure–property relationship
RA	Rheumatoid arthritis
RANKL	Receptor activator of nuclear factor κB ligand
RFA	Resonance frequency analysis
RFGD	Radio frequency glow discharge
RGD	Arginyl-glycyl-aspartic acid
RHA	Resurfacing hip arthroplasty
ROC	Receiver operating characteristic
RPM	Remote patient monitoring
RR	(Relative) risk of revision
RSA	Radiostereometry analyses
rTSA	Reverse TSA
rTSR	Reverse TSR
SBF	Simulated body fluid
SCC	Stress corrosion cracking
SCTG	Subepithelial connective tissue graft
SE	Substantially equivalent
SEM	Scanning electron microscope
SIE	Selective infiltration etching
SLA	Stereolithography
SLAed	Sand blasted with large alumina grits, followed by acid etched
SLM	Selective laser melting (technology)

SLS	Selective laser sintering
SMA	Shape memory alloy
SQ	Sleep quality
SS	Stainless steel
STL	STereoLithography
TCP	Tricalcium phosphate
TEA CO_2 laser	Transversely excited atmospheric pressure (CO_2 laser)
TEGDMA	Triethylene glycol dimethacrylate
T_g	Glass transition temperature
TGFbeta1	Transforming growth factor (beta 1)
TGR	Traditional growing rod
THA	Total hip arthroplasty
THR	Total hip replacement
TIFP	Tooth implant-supported fixed prosthesis
TKA	Total knee arthroplasty
TKR	Total knee replacement
TL	Tissue level
T_m	Melting temperature
TMJ	Temporomandibular joint
TNT	Titania nanotube
TP	True positive
TSA	Thermographic stress analysis
TSA	Total shoulder arthroplasty
TSR	Total shoulder replacement
UES	Upper esophageal sphincter
UHMWPE	Ultra-high-molecular-weight polyethylene
UKA	Unicompartmental knee arthroplasty
UNOS	The United Network for Organ Sharing
UV	Ultraviolet
VEGF	Vascular endothelial growth factor
VRSA	Vancomycin-resistant *Staphylococcus aureus*
WES	White aesthetic score
WHO	World Health Organization
WOMAC	Western Ontario and McMaster Universities Osteoarthritis Index
XPS	X-ray photoelectron spectroscopy
XRD	X-ray diffraction
Y-TZP	Yttrium-stabilized tetragonal zirconia

Prologue

Implantation is a medical treatment for restoration of missing teeth (for dental implant) or replacing injured/damaged knee or hip (for medical or orthopedic implant). Implantation can be defined as the insertion of any object or material (as a foreign material such as an alloplastic substance or other tissue), either partially or completely, into the host's vital part of a body for therapeutic, diagnostic, prosthetic, or experimental purposes. The interface between foreign material (of placed or replaced implantable device) and receiving vital hard or soft tissue plays a dominant role to control subsequent success, survival, and longevity of placed implants. As will be discussed in Chapter 1, the implantation should be differentiated from two other similar procedures: namely, replantation and transplantation. Replantation refers to the reinsertion of a vital hard tissue such as tooth back into its jaw socket after accidental or intentional removal, whereas transplantation is the transfer of a body part from one site to another.

Implantation is a typical example of transdisciplinary treatment, in both medical implant and dental implant, which is mainly composed of biomaterials and bioengineering, bioscience, biotechnology and manufacturing engineering, surface engineering, biotribology, biomechanics, biomechanical behavior, anatomy, and basic relevant knowledge of medicine and dentistry. Each discipline has more detailed parameters. For example, biotribology contains friction, wear, and wear debris toxicity. These individual contributive elements are interrelated as shown in a figure attached below, which authors would call "implantology map." The implantology map will be shown at the top of each chapter, demonstrating that the gray-colored key words will be discussed mainly in that chapter. In this prologue, we will discuss the prevalence of medical and oral implants and their relation to quality of life (QOL) of implant-treated patients.

There are six major synovial joint systems in our body: the shoulder, elbow, wrist, hip, knee, and ankle joints. Replacement surgery is commonly performed on shoulder, hip, and knee joints, for example, hip prosthesis or hip replacement surgery becomes necessary when the hip joint has been badly damaged from causes such as arthritis, congenital malformation or abnormal development, and damage from injury. The orthopedic community and public are well aware that hip and knee replacement operations are among the most commonly performed operations in the U.S., and the important measure of the impact of joint arthroplasty on public health is known as prevalence. An estimated 4.7 million Americans have undergone total knee arthroplasty (TKA) and 2.5 million have undergone total hip arthroplasty (THA) and are living with implants. Prevalence is higher in women than in men: 3 million women and 1.7 million men are living with TKA, and 1.4 million women and 1.1 million men are living with THA [1]. Maradit-Kremers et al. [1] illustrated prevalence (%) as a function of age groups, as shown in Figure 2. It was observed that prevalence increases with age. In adults aged 80 to 89 years, about 6% (hip) and 10% (knee) have a history of total hip and knee replacement, respectively.

https://doi.org/10.1515/9783110740134-001

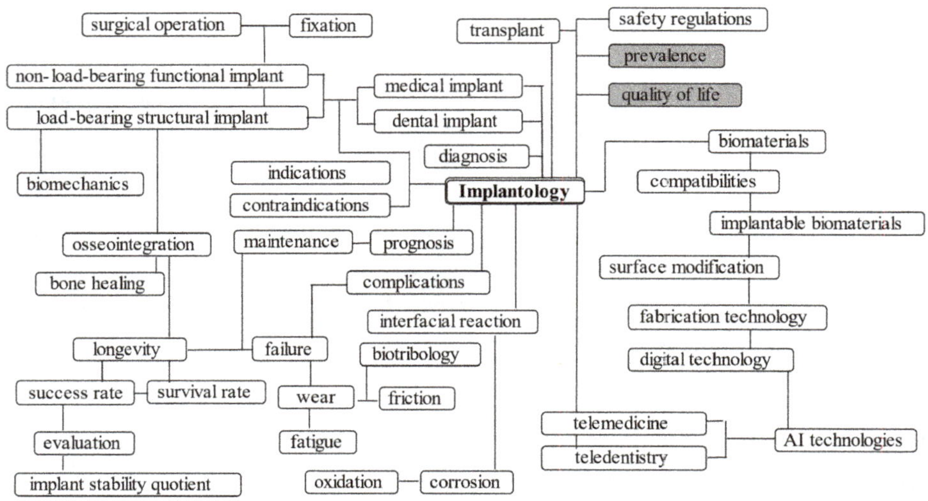

Figure 1: Implantology map.

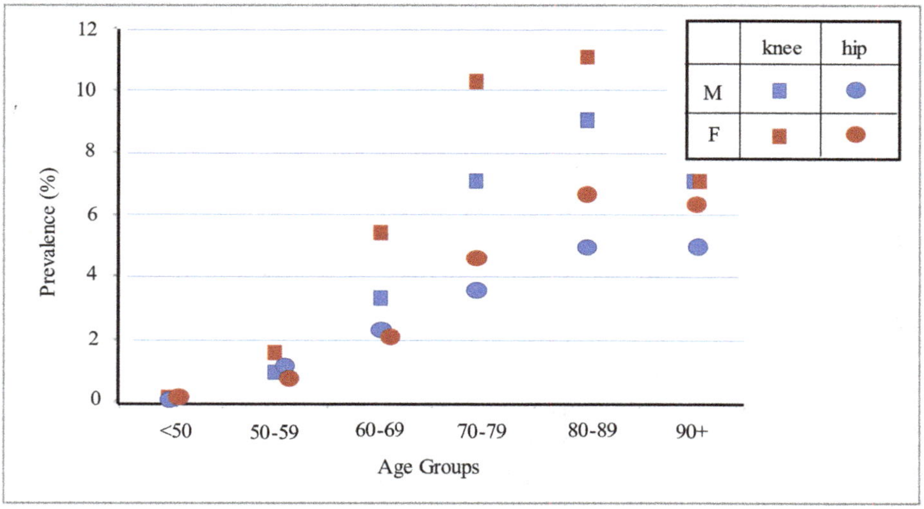

Figure 2: Prevalence of total knee and total hip replacement as a function of age groups [1].

Osteoarthritis (OA) is a serious public health issue with symptomatic disease prevalent in 9% of men and 11% of women, causing mainly pain and disability [2, 3]. Age is one of the largest risk factors for developing OA [4]. Total hip replacement (THR) is one of the most successful surgical procedures. Greater than one million operations are performed every year worldwide and this is anticipated to double within the next decade [5]. In the USA alone, the number of operations is projected to rise to

572,000 per year by 2030 [6]. An estimated 93% of operations are performed for severe OA with intractable pain and functional limitations [5]. For these patients who are refractory to conservative measures, THR is currently the recommended and most effective treatment [7].

There is wide confusion about the use of terms such as QOL, health-related quality of life (HRQoL), functional status and well-being [8]. QOL (or living) is an evaluation of all aspects of our lives, while HRQoL indicates personal health status and usually refers to aspects of our lives that are dominated or significantly influenced by our mental or physical well-being. At the same time, there are other terms such as functional status and well-being as well. Functional status describes an individual's effective performance of, or ability to perform those roles, tasks, or activities that are valued (e.g., going to work, playing sports, or maintaining the house) and well-being means subjective bodily and emotional states; how an individual feels; a state of mind distinct from functioning that pertains to behaviors and activities [8–10]. Although these terms are often used interchangeably, despite clear and distinct definitions, HRQoL is an overarching concept which has many domains including psychological and social well-being, in addition to pain and function.

QOL or HRQoL cannot be measured with an absolute number and should be evaluated by results of questionnaire. Among the many patient-reported outcome measures (PROMs) used for post-treatment outcomes, there are the two most frequently employed questionnaires on HRQoL, which include EuroQol five dimension (EQ-5D) questionnaire and SF-36/SF-12 questionnaire. Besides these, there is Western Ontario and McMaster Universities Osteoarthritis Index (WOMAC) (questionnaire) index. EQ-5D is a standardized measure of HRQoL developed by the EuroQol Group to provide a simple, generic questionnaire for use in clinical and economic appraisal and population health surveys [11]. EQ-5D assesses health status in terms of five dimensions of health and is considered a generic questionnaire because these dimensions are not specific to any one patient group or health condition. EQ-5D can also be referred to as a PROM, because patients can complete the questionnaire themselves to provide information about their current health status and how this varies over time. Five dimensions contain (1) mobility, (2) self-care, (3) usual activities, (4) pain/discomfort, and (5) anxiety/depression. The EQ-5D questionnaire also includes a visual analog scale, by which respondents can report their perceived health status with a grade ranging from 0 (the worst possible health status) to 100 (the best possible health status) [11, 12].

The SF-36 is a multi-purpose, short-form health survey with only 36 questions. It yields an 8-scale profile of functional health and well-being scores, as well as psychometrically based physical and mental health summary measures and a preference-based health utility index [13]. It is a generic measure of perceived health status, as opposed to one that targets a specific age, disease, or treatment group. It is used to measure observable and tangible limitations due to poor health and/or bodily pain in physical, social, and role activities. The SF-36, contains 35 questions, which load onto 8 scales: physical function (10 items), role limitations due to physical health problems (4 items), body pain (2 items), general health (5 items), vitality (4 items), social functioning (2 items), role limitations due to emotional problems (3 items), and emotional well-being (5 items). Some items use a three-point rating scale, while others use a five-point scale; some items call for frequency ratings, while others require judgments of quality of performance. The scales can be combined into two summary measures: physical component summary and mental component summary scores. The 36th question asks about health change and does not contribute to the scales or summary measures. Accordingly, the SF-36 has proven useful in surveys of general and specific populations, comparing the relative burden of diseases, and in differentiating the health benefits produced by a wide range of different treatments. Administration time is about 10–12 min. The 12-Item Short Form Health Survey (SF-12) was developed for the Medical Outcomes Study, a multiyear study of patients with chronic conditions. The SF-12 is an even shorter survey form that has been shown to yield summary physical and mental health outcome scores that are interchangeable with those from the SF-36. The resulting short-form survey instrument provides a solution to the problem faced by many investigators who must restrict survey length. The instrument was designed to reduce respondent burden while achieving minimum standards of precision for purposes of group comparisons involving multiple health dimensions [13, 14].

The WOMAC measures five items for pain (score range 0–20), two for stiffness (score range 0–8), and 17 for functional limitation (score range 0–68). Physical functioning questions cover everyday activities such as stair use, standing up from a sitting or lying position, standing, bending, walking, getting in and out of a car, shopping, putting on or taking off socks, lying in bed, getting in or out of a bath, sitting, and heavy and light household duties. The questions on the WOMAC are a subset of the questions of the hip disability and osteoarthritis outcome score (HOOS). Thus, a HOOS survey may also be used to determine a WOMAC score [15, 16].

Using predetermined questionnaire forms such as EQ-5D, SF-36/SF-12 or WOMAC, PROMs per se is subjective, so that there should be pros and cons on effects and analysis of the results. Siviero et al. [17] using physical and mental endpoints, investigated patient QOL at the time of knee replacement surgery and three months later. A prospective observational study was designed to evaluate patients (n = 132) scheduled for unicompartmental or TKR surgery who were assessed at baseline (preoperatively) and three months after. Physical and mental endpoints based on the component scores of the SF-12 and on the WOMAC index were used to investigate patients' QOL. Generalized estimating equation methodology was used to assess patients' baseline characteristics (age, sex, education, body mass index (BMI), comorbidity, depressive symptoms, cognitive impairment, smoking/alcohol and type of surgery), the study endpoints and their changes over a three-month post-surgery period. It was found that (i) longitudinal data analysis showed that the baseline factors associated with improvement in general QOL at the three-month post-surgery assessment were higher BMI, a high comorbidity, total (as opposed to unicompartmental) knee replacement, and low education level; (ii) data analysis of the patients who underwent rehabilitation after discharge revealed that the current smokers' physical QOL worsened over time; and (iii) the general QOL was unchanged over time in the presence of depressive symptomatology. It was, hence, concluded that (iv) these findings underline the importance of using comprehensive assessment methods to identify factors affecting functionality, QOL, and developing interventions to improve the health/wellbeing of patients after knee replacement [17]. Literatures (which used SF-36/SF-12 and WOMAC most frequently) were reviewed to evaluate QOL along patients who underwent TKA and to assess the impact of various associated factors [18]. It was mentioned that the studies made it possible to define that TKA is capable of making an overall improvement in patients' QOL. Pain and function are among the most important predictors of improvement in QOL, even when function remains inferior to that of healthy patients. It was concluded that the factors associated negatively were obesity, advanced age, comorbidities, persistence of pain after the procedure, and a lengthy wait for surgery. Canovas et al. [19] pointed out that up to 30% of patients are dissatisfied and this dissatisfaction is directly related to the patients' QOL, which they deem insufficient. Their QOL depends on various physical, behavioral, social, and psychological factors that are not taken into account by functional outcome scores.

Shan et al. [20] investigated the mid-term HRQoL after patients received THR treatment with OA. A systematic review of clinical studies published after January 2000 was performed using strict eligibility criteria. It was reported that (i) HRQoL is at least as good as reference populations in the first few years and subsequently plateaus or declines, although patient satisfaction and functional status was favorable, and (ii) the mid-term HRQoL following THR is superior to pre-operative levels in a broad range of HRQoL domains. Shan et al. [21] also conducted a systematic review meta-analysis of all studies published from January 2000 onward to evaluate HRQoL after primary TKR for OA in patients with at least three years of follow-up. It was mentioned that the TKR confers significant intermediate and long-term benefits with respect to both disease-specific and generic HRQOL (especially pain and function), leading to positive patient satisfaction.

HRQoL is affected by not only personal characteristic parameters as mentioned above, but also by material type of placed implants. Peña et al. [22] analyzed whether there are differences in QOL and functional capacity among patients undergoing TKA with conventional implants (made of Cr–Co alloy) compared to those treated with hypoallergenic heavily surface-oxidized Zr alloy implants (Oxinium: Zr 97.5% + Nb 2.5%). A pragmatic clinical study was carried out that included patients who underwent TKA between January 2013 and December 2015. During this period, 245 knees in 228 patients were treated. Eleven patients were excluded, leaving a sample of 161 conventionally treated knees, 72 knees treated with hypoallergenic implants, and 1 patient who received both implant types. It was found that (i) patients who receive hypoallergenic TKA have lower scores on the QOL and functional capacity question-naires (WOMAC, SF-12, and EQ-5D L-VAS) and worse psychological distress compared with those who receive conventional Cr–Co implants, (ii) patients with hypersensitiv-ity to metal present greater psychological distress, and this worsens QOL parameters; concluding that (iii) it is necessary to evaluate the psychological aspect in patients with knee OA and metal allergies who are going to undergo TKA [22].

Now moving to the prevalence of dental implants and how its treatment affect the HRQoL of patients (namely OHRQoL: oral health related QoL). Dental implants are an ideal option for people in good general oral health who have lost a tooth (or teeth) due to periodontal disease, an injury, or some other reason. Dental im-plants (as an artificial tooth root and usually made from titanium materials) are bio-compatible metal anchors surgically positioned in the jawbone underneath the gums to support an artificial crown where natural teeth are missing. Unlike the orthopedic implants, there are several different treatments for missing teeth, including implants, tooth-supported bridge, or removable partial denture. Hence, dental implant treat-ment is one of the many effective dental treatments for missing teeth (or tooth). Den-tal implants are long-term replacements preserving adjacent teeth. Implanting a tooth is equivalent to receiving a new tooth (teeth). In addition, it is considered as the only restorative technique that preserves and stimulates natural bone. They also give steady support to dentures (prosthetics). Furthermore, dental implants provide convenience, comfort, and improve patient's appearance (unlike removable den-tures) [23]. Luo et al. [24] assessed the trends in tooth loss among adults with and without diabetes mellitus in the United States and racial/ethnic disparities in tooth loss patterns, and to evaluate trends in tooth loss by age, birth cohorts, and survey periods. Data came from nine waves of the National Health and Nutrition Examina-tion Survey (NHANES) from 1971 through 2012. The trends in the estimated tooth loss in people with and without diabetes were assessed by age groups, survey periods, and birth cohorts. The analytical sample was 37,609 dentate (i.e., with at least one permanent tooth) adults aged 25 years or older. It was found that the estimated num-ber of teeth lost among non-Hispanic blacks with diabetes increased more with age than that among non-Hispanic whites with diabetes or Mexican Americans with dia-betes. During 1971–2012, there was a significant decreasing trend in the number of teeth lost among non-Hispanic whites with diabetes and non-Hispanic blacks with diabetes. However, adults with diabetes had about twice the tooth loss as did those without diabetes. It was, accordingly, concluded that substantial differences in tooth loss between adults with and without diabetes and across racial/ethnic groups per-sisted overtime, indicating that appropriate dental care and tooth retention need to be further promoted among adults with diabetes. Figure 3 shows estimated number of teeth lost by age groups. Japan – Ministry of Health, Labor and Welfare along with the Japan Dental association has been promoting the dental campaign "80–20," to keep 20 natural teeth by age of 80 +, which is marked with read circle in Figure 3.

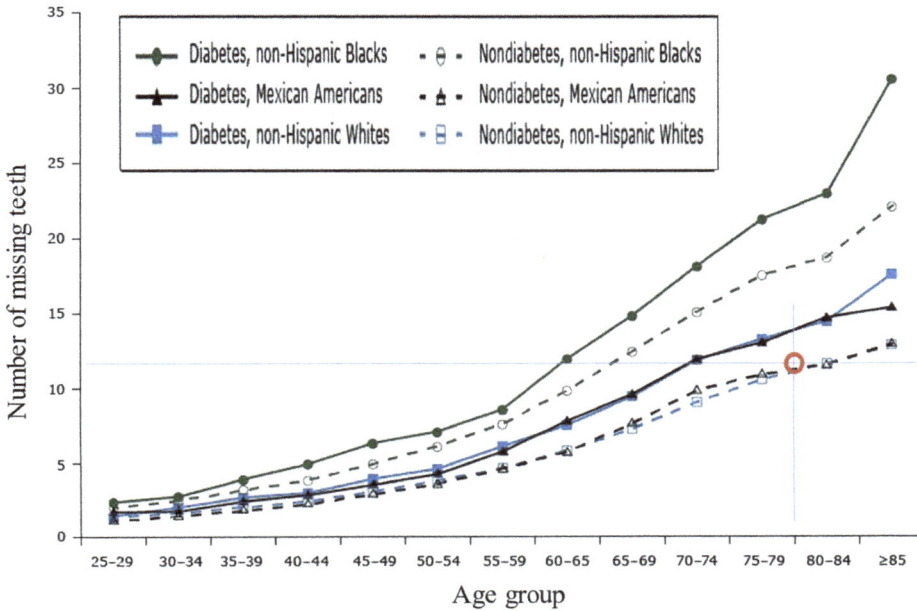

Figure 3: Estimated number of teeth lost by age groups.
National Health and Nutrition Examination Survey (NHANES) 1971–2012 [24].

Dental implants have become an increasingly popular treatment choice for replacing missing teeth. Yet, little is known about the prevalence and sociodemographic distribution of dental implant use in the United States. To address this knowledge gap, Elani et al. [25] analyzed data from seven NHANES from 1999 to 2016 and estimated dental implant prevalence among adults missing any teeth for each survey period overall as stratified by sociodemographic characteristics. It was observed that (i) there has been a large increase in the prevalence of dental implants, from 0.7% (in 1999–2000) to 5.7% (in 2015–2016), (ii) the largest absolute increase in prevalence (12.9%) was among individuals of ages ranging from 65 to 74, whereas the largest relative increase was ~1,000% among those from 55 to 64, (iii) there was an average covariate-adjusted increase in dental implant prevalence of 14% per year, and (iv) dental implant prevalence projected to 2026 ranged from 5.7% in the most conservative scenario to 23% in the least. Trend is illustrated in Figure 4 [25], where trend stops: implant prevalence is estimated to be the same average probability estimated by the regression line in 2015–2016; trend continues at the same pace: the slope of the regression line included all years 2000–2016; trend slows: the slope of the regression line included all years excluding 2015–2016; and trend steepens: the slope of the regression line included all years excluding 1999–2000 and 2001–2002. Solid lines represent the estimated prevalence (for 2016 and earlier) and projected prevalence (after 2016), and dashed lines represent the 95% prediction intervals for those estimates.

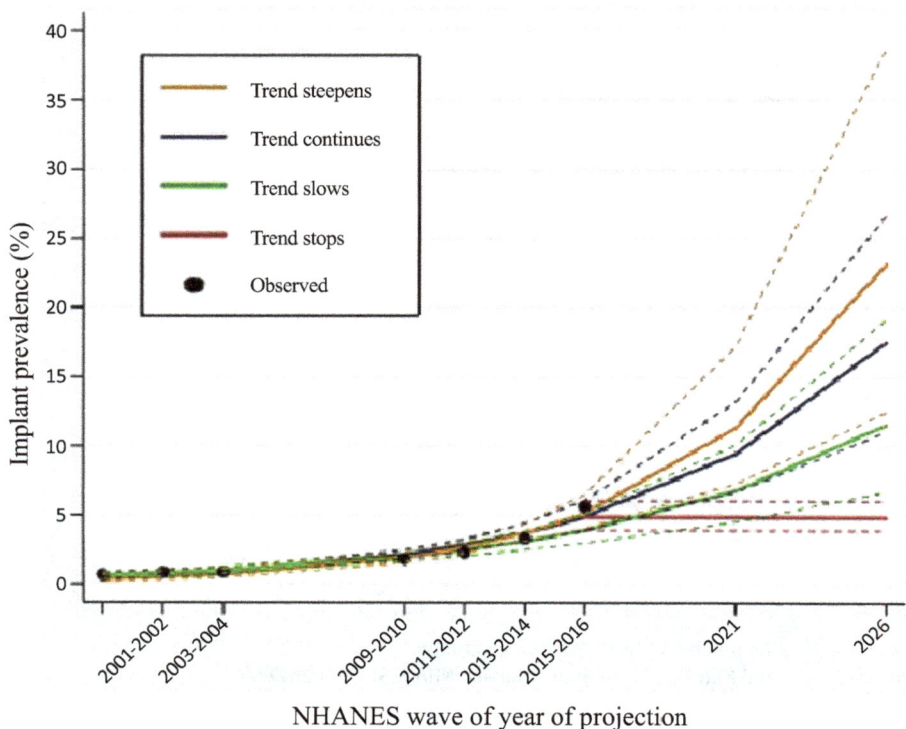

Figure 4: Trend of dental implant prevalence among adults with missing teeth in the NHANES from 1999 to 2016 and projected to 2021 to 2026 [25].

QOL includes conditions that enable good living, such that a person is able to perform everyday activities in a good physical, mental, and social state and the patient is satisfied with therapeutic efficacy, disease control, or rehabilitation. The World Health Organization (WHO) provides the following definition for health: complete physical, mental and social well-being, which not only means the absence of disease and disability but also includes body, mind, and society axes. Thus, any disability and damage to any of these three axes disrupts the individual's balance and leads to a lack of health. Following this definition, the international community's attention to the concept of QOL has increased [26, 27]. Orofacial changes, such as diseases and pain, have significant effects on individual and social aspects and can affect daily activities [28]. Tooth loss is a serious life event. Based on WHO criteria, tooth loss is a physical disorder that impairs two important functions, including eating and speaking [29]. Sargolzaie et al. [27] compared the QOL of patients requesting dental implants before and after implant in 2015. It was concluded that (i) the QOL is associated with eating; speaking clearly; cleaning teeth or dentures; performing light physical activity; being outside of the home for work or meeting others; smiling, laughing, or showing teeth without discomfort and shame; emotional conditions, such as becoming upset quicker than usual; enjoying communication with other people, such as friends, relatives, and neighbors; and performing their job significantly increased after surgery, however, the QOL associated with sleeping or resting did not improve and (ii) no significant association was noted between QOL after implant with place of residence, educational level, and sex.

Since little is known about OHRQoL for partially edentulous patients seeking implant therapy, Nickenig et al. [30] investigated preoperative and postoperative OHRQoL to determine the impact of prosthodontic treatments. Total 343 patients (219 in the partially edentulous, "study group"; 124 in the fully dentate, "control group") were assessed for OHRQoL with a German version of the Oral Health Impact Profile (OHIP) (OHIP-G21, range 0–84). Assessments were made at the preoperative evaluation and after prosthodontic treatments. It was found that (i) median OHIP scores were 17.1 and 3.4 for the study and control groups, respectively, (ii) after prosthodontic treatment, the median OHIP score of the study group decreased from 17.1 to 5.4 and (iii) the types of problems reported pre- and post-treatment were substantially different. It was, hence, concluded that (iv) preoperative assessment of OHRQoL identified clear differences for partially edentulous compared to fully dentate patients and (v) the implant therapy had a positive effect on the OHRQoL [30]. Similarly, Fillion et al. [31] studied the impact of implant therapy on OHRQoL in partially edentulous patients and analyzed the improvement of OHRQoL of patients who underwent dental implant treatment using the "functional," "psychosocial," and "pain and discomfort" categories of the Geriatric Oral Health Assessment Index (GOHAI). Within a prospective cohort of patients rehabilitated with Straumann dental implants, the OHRQoL of 176 patients (104 women and 72 men) was assessed using the GOHAI questionnaire, at two different times, before and after implant placement. It was found that (i) before treatment, the GOHAI score was lower for participants with fewer teeth, while after treatment, no difference was observed between participants, (ii) the best improvement was observed in patients who needed complete treatment, (iii) the presence of preliminary periodontal treatment, tobacco habits, age and gender of the participants did not have a significant impact on OHRQoL, and (iv) changing the time between the two evaluations (before and after treatment) had no impact on the changes in the GOHAI score; suggesting that (vi) implants enhanced the OHRQoL of participants that needed oral treatment. Yoshida et al. [32] evaluated the changes in OHRQoL during implant treatment for partially edentulous patients, and studied the influence of the type of partially edentulous arch. 20 patients with a small number of lost teeth (fewer than four teeth) who underwent implant treatment were selected. Chronological QOL change during implant treatment was measured. The subjects completed the shortened Japanese version of the OHIP (OHIP-J14) before the surgery, 1 week after the surgery, 1 week after interim prosthesis placement, and 1 week after definitive prosthesis placement. It was concluded that (i) overall OHRQoL improvement was observed after the definitive prosthesis placement and (ii) implant treatment was more effective in the unilateral free-end edentulous space.

Patel et al. [33] evaluated changes in OHRQoL in patients receiving dental implant treatment. 150 patients attending a primary care dental practice specializing in dental implant treatment were given an OHIP questionnaire to complete before their implant treatment and 6 months post-operatively. 107 patients successfully completed the study. It was obtained that (i) results of the analysis demonstrated statistically significant decreases in frequency of oral health problems post treatment and (ii) for every OHIP statement, the t-test revealed a statistically significant improvement in oral health QOL, indicating that dental implant therapy has a positive effect on oral health QOL as determined by the OHIP and long term follow up is required to provide an understanding of continuing benefits from dental implant therapy. Nemli et al. [34] reported patient-based outcomes of implant-retained maxillofacial prostheses and evaluated the effect of implant-retained maxillofacial prostheses on QOL of participants in a prospective study. 82 participants were treated with implant-retained maxillofacial prostheses. Participants were divided into two groups: a retrospective group (participants treated and under care) and a prospective group (participants willing to be treated). It was concluded that (i) implant-retained prostheses were considered highly satisfactory, indicating good QOL for patients with maxillofacial defects and (ii) a comparison of pretreatment and posttreatment assessments revealed that implant-retained maxillofacial prostheses increased patient QOL.

Kuoppala et al. [35] evaluated the OHRQoL of patients treated with implant-supported mandibular overdentures and to compare the attachment systems used. Altogether 112 patients treated with implant-supported mandibular overdentures in 1985–2004 were invited to the follow-up; 58 of them attended and replied to the OHIP (OHIP-14) -questionnaire. There were 48 overdentures with a bar connection and 10 with a ball connection, the total number of implants installed and still in use was 197. The mean follow-up time was 13.7 years. The associations between the OHIP-14 variables and the patient's age, gender, as well as the number of implants supporting the overdenture and the type of attachment used were assessed. It was found that (i) patients with implant-supported mandibular overdentures were satisfied with their OHRQoL, (ii) older patients were more satisfied than younger ones in both genders and (iii) neither the implant connection type nor the number of supporting implants seemed to have a significant influence on the OHRQoL. It was, therefore, concluded that (iv) especially older patients with mandibular implant-supported overdentures were satisfied with their OHRQoL and (v) attachment type or the number of supporting implants did not have a significant influence on the OHRQoL [35]. DeBaz et al. [36] compared the QoL in partially edentulous osteoporotic women who have missing teeth restored with dental implant retained restorations to patients who did not and reported the rate of osteonecrosis. Total 237 participants completed the Utian QoL survey, a 23-question document measuring across psychosocial domains of well-being including occupational, health, emotional, and sexual domains which together contribute to an overall score. The subset of participants having dental implant supported prosthesis (64) was compared to the subset having nonimplant supported fixed restorations (47), the subset having nonimplant supported removable restorations (60), and the subset having no restoration of missing teeth (66). It was demonstrated that implant-retained oral rehabilitation has a statistically significant impact over nonimplant and traditional fixed restorations and removable restorations.

Closing this section, important journals on orthopedic implants and dental implants are listed as follows. Important journals on transplantation are also listed.

Journals on orthopedic implants:
Acta Orthopaedica
American Journal of Sports Medicine
Arthroscopy
Clinical Orthopaedics and Related Research
Knee Surgery Sports Traumatology Arthroscopy
Journal of Bone and Joint Surgery – American
Journal of Orthopaedic & Sports Physical therapy
Journal of Physiotherapy
Journal of Shoulder and Elbow Surgery (American)
Journal of the American Academy of Orthopaedic Surgeons
Osteoarthritis and Cartilage
Physical Therapy
The Bone and Joint Journal (British)
The Journal of Arthroplasty
The Spine Journal

Journals on dental implants:
Achieves of Oral Biology
American Journal of Orthodontics and Dentofacial Orthopedics
BioMed Research International
Brazilian Dental Journal
British Dental Journal
Clinical Orthopaedics and Related Research
Dental Materials
European Journal of Prosthodontics and Restorative Dentistry
Expert Review of Medical Devices
International Journal of Orla and Maxillofacial Surgery
Journal of Biomedical Materials Research
Journal of Cranio-Maxillofacial Surgery
Journal of Clinical Periodontology
Journal of Dental Research
Journal of Dentistry
Journal of Endodontics
Journal of Indian Society of Periodontology
Journal of Materials Science: Materials in Medicine
Journal of Oral and Maxillofacial Surgery

Journal of Periodontology
Journal of Prosthetic Dentistry
Journal of Prosthodontics
Journal of the American Dental Association
Oral Surgery, Oral Medicine, Oral Pathology, and Oral Radiology
The Angle Orthodontist

Journal on transplantation:
Liver Transplantation
Transplantation
Biology of Blood and Marrow Transplantation
American Journal of Transplantation
Journal of Heart and Lung Transplantation
Nephrology Dialysis Transplantation
Bone Marrow Transplantation
Transplant International
Xenotransplantation
Transplantation Reviews

References

[1] Maradit-Kremers H, Crowson CS, Larson D, Jiranek WA, Berry DJ. Prevalence of Total Hip (THA) and Total Knee (TKA) Arthroplasty in the United States. 2014; https://www.mayoclinic.org/medical-professionals/orthopedic-surgery/news/first-nationwide-prevalence-study-of-hip-and-knee-arthroplasty-shows-7-2-million-americans-living-with-implants/mac–20431170.

[2] Zhang Y, Jordan JM. Epidemiology of osteoarthritis. Rheum Dis Clin North Am. 2008, 34, 515–29.

[3] Cushnaghan J, Coggon D, Reading I, Croft P, Byng P, Cox K, Dieppe P, Cooper C. Long-term outcome following total hip arthroplasty: A controlled longitudinal study. Arthritis Rheum. 2007, 57, 1375–80.

[4] Arden N, Nevitt MC. Osteoarthritis: Epidemiology. Best Pract Res Clin Rheumatol. 2006, 20, 3–25.

[5] Pivec R, Johnson AJ, Mears SC, Mont MA. Hip arthroplasty. Lancet. 2012, 380, 1768–77.

[6] Kurtz S, Ong K, Lau E, Mowat F, Halpern M. Projections of primary and revision hip and knee arthroplasty in the United States from 2005 to 2030. J Bone Joint Surg Am. 2007, 89, 780–85.

[7] Recommendations for the medical management of osteoarthritis of the hip and knee: 2000 update. American College of rheumatology subcommittee on osteoarthritis guidelines. Arthritis Rheum. 2000, 43, 1905–15.

[8] Hossain FS, Konan S, Patel S, Rodriguez-Merchan EC, Haddad FS. The assessment of outcome after total knee arthroplasty. J Bone Joint Surg Br. 2015, 97, 3–9.

[9] Parsons JT, Snyder AR. Health-related quality of life as a primary clinical outcome in sport rehabilitation. J Sport Rehabil. 2011, 20, 17–36.

[10] Whitcomb JJ. Functional status versus quality of life: Where does the evidence lead us?. ANS Adv Nurs Sci. 2011, 34, 97–105.

[11] Balestroni G, Bertolotti G. EuroQol-5D (EQ-5D): An instrument for measuring quality of life. Monaldi Arch Chest Dis. 2012, 78, 155–59.

[12] Brooks R, Boye KS, Slaap B. EQ-5D: A plea for accurate nomenclature. J Patient Rep Outcomes. 2020, 4, 52, doi: 10.1186/s41687-020-00222-9.

[13] Bushnik T. SF-36/SF-12. Encyclopedia of Clinical Neuropsychology. 2011; https://link.
      springer.com/referenceworkentry/10.1007%2F978-0-387-79948-3_1831.

[14] NIH. SF-36 and SF-12 Health Status Questionnaire. 2016; https://datashare.nida.nih.gov/
      instrument/sf36-and-sf12-health-status-questionnaire.

[15] American College of Rheumatology. Western Ontario and McMaster Universities
      Osteoarthritis Index (WOMAC). 2013; http://www.rheumatology.org/practice/clinical/clini
      cianresearchers/outcomes-instrumentation/WOMAC.asp.

[16] Ackerman IN, Tacey NA, Ademi Z, Bohensky MA, Liew D, Brand CA. Using WOMAC Index
      scores and personal characteristics to estimate assessment of quality of life utility scores in
      people with hip and knee joint disease. Qual Life Res. 2014, 23, 2365–74.

[17] Siviero P, Marseglia A, Biz C, Rovini A, Ruggieri P, Nardacchione R, Maggi S. Quality of life
      outcomes in patients undergoing knee replacement surgery: Longitudinal findings from the
      QPro-Gin study. BMC Musculoskelet Disord. 2020, 21, 436, doi: https://doi.org/10.1186/
      s12891-020-03456-2.

[18] Rocha da Silva R, Santos AAM, de Sampaio Carvalho J Júnior, Matos MA. Quality of life after
      total knee arthroplasty: Systematic review. Rev Bras Ortop. 2014, 49, 520–27, doi: 10.1016/
      j.rboe.2014.09.007.

[19] Canovas F, Dagneaux L. Quality of life after total knee arthroplasty. Orthop Traumatol Surg
      Res. 2018, 104, S41–6.

[20] Shan L, Shan B, Graham D, Saxena A. Total hip replacement: A systematic review and meta-
      analysis on mid-term quality of life. Osteoarthr Cartil. 2014, 22, 389–406.

[21] Shan L, Shan B, Suzuki A, Nouh F, Saxena A. Intermediate and long-term quality of life after
      total knee replacement: A systematic review and meta-analysis. J Bone Jt Surg. 2015,
      97, 156–68.

[22] Peña P, Ortega MA, Buján J, De la Torre B. Decrease of quality of life, functional assessment
      and associated psychological distress in patients with hypoallergenic total knee
      arthroplasty. J Clin Med. 2020, 9, 3270, doi: 10.3390/jcm9103270.

[23] Agarwal V. Factors Impacting Dental Implant Market Growth. 2018; https://getreferralmd.
      com/2018/11/factors-impacting-dental-implants-market-growth/.

[24] Luo H, Pan W, Sloan F, Feinglos M, Wu B. Forty-year trends in tooth loss among american
      adults with and without diabetes mellitus: An age-period-cohort analysis. CDC Prev Chronic
      Dis. 2015, 12, https://www.cdc.gov/pcd/issues/2015/15_0309a.htm, doi: 10.5888/
      pcd12.150309.

[25] Elani H, Starr J, Da Silva JD, Gallucci G. Trends in dental implant use in the U.S., 1999–2016,
      and projections to 2026. J Dent Res. 2018, doi: 10.1177/0022034518792567.

[26] Mohebbi S, Sheikhzadeh S, Bayanzadeh M, Batebizadeh A. Oral impact on daily performance
      (OIDP) index in patients attending patients clinic at dentistry school of Tehran university of
      medical sciences. J Dent Med. 2012, 25, 135–41.

[27] Sargolzaie N, Moeintaghavi A, Shojaie H. Comparing the quality of life of patients requesting
      dental implants before and after implant. Open Dent J. 2017, 11, 485–91.

[28] Berretin-Felix G, Nary Filho H, Padovani CR, Machado WM. A longitudinal study of quality of
      life of elderly with mandibular implant-supported fixed prostheses. Clin Oral Implants Res.
      2008, 19, 704–08.

[29] Enright S. Treatment of edentulous patients using implant supported mandibular
      overdentures improves quality of life. TSMJ. 2007, 8, 65–69.

[30] Nickenig H-J, Wichmann M, Andreas SK, Eitner S. Oral health-related quality of life in partially
      edentulous patients: Assessments before and after implant therapy. J Craniomaxillofac Surg.
      2008, 36, 477–80.

[31] Fillion M, Aubazac D, Bessadet M, Allègre M, Nicolas E. The impact of implant treatment on oral health related quality of life in a private dental practice: A prospective cohort study. Health Qual Life Outcomes. 2013, 11, 197, doi: https://doi.org/10.1186/1477-7525-11-197.

[32] Yoshida T, Mssaki C, Komai H, Misumi S, Mukaibo T, Kondo Y, Nakamoto T, Hosokawa R. Changes in oral health-related quality of life during implant treatment in partially edentulous patients: A prospective study. J Prosthodont Res. 2016, 60, 258–64.

[33] Patel N, Vijayanarayanan RP, Pachter D. Oral health-related quality of life: Pre- and post-dental implant treatment. Oral Surg. 2014, 8, doi: 10.1111/ors.12106.

[34] Nemli SK, Aydin C, Yilmaz H, Bal BT, Arici YK. Quality of life of patients with implant-retained maxillofacial prostheses: A prospective and retrospective study. J Prosthet Dent. 2013, 109, 44–52.

[35] Kuoppala R, Näpänkangas R, Raustia A. Quality of life of patients treated with implant-supported mandibular overdentures evaluated with the oral health impact profile (OHIP-14): A survey of 58 patients. J Oral Maxillofac Res. 2013, 4, http://www.ejomr.org/JOMR/archives/2013/2/e4/v4n2e4ht.pdf, doi: 10.5037/jomr.2013.4204.

[36] DeBaz C, Hahn J, Lang L, Palomo L. Dental implant supported restorations improve quality of life in osteoporotic women. Int J Dent. 2015, doi: https://doi.org/10.1155/2015/451923.

Chapter 1
Introduction

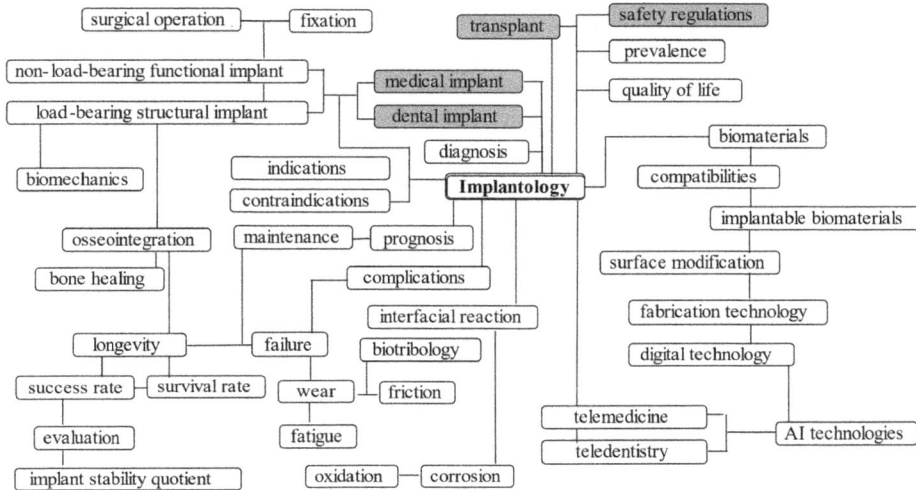

In this chapter, we will discuss the difference between transplants and implants and their associated safety regulations.

1.1 Implant versus transplant

Due to the ever-increasing risk of aging population, there are a variety of unforeseen risks jeopardizing individual's normal quality of life, including a higher incidence of coronary artery disease, diabetes, systemic disease, missing teeth, and injury-related frailty. Hence, there is a need for more reliable and safer treatments to recover original biofunctionality or enhance rehabilitation therapy. To this end, there are basically organ transplantations (of, typically, heart, intestine, kidney, liver, lung, and spleen) and medical and dental implantations to improve health-related quality of life (HRQoL) as well as oral OHRQoL.

There are significant differences between transplants and implants. Transplant is a biological tissue used as an exchange of tissues or organs of the human body but implant is a nonbiological material. Transplant requires an organ donor immunosuppression, while implant does not require it. Transplant is an active tissue in the human body but implant is the mechanical support that assists organ function. Implant is a foreign body against receiving vital tissue so that there is a risk of infection, while transplant is at risk of rejection from the human body. Compared to implants, transplantation

https://doi.org/10.1515/9783110740134-002

requires more strict consideration of ethical examination. Transplant can be used for life unless rejected, while implant might be removed if necessary.

Figure 1.1 illustrates typical implants and transplants [1–5].

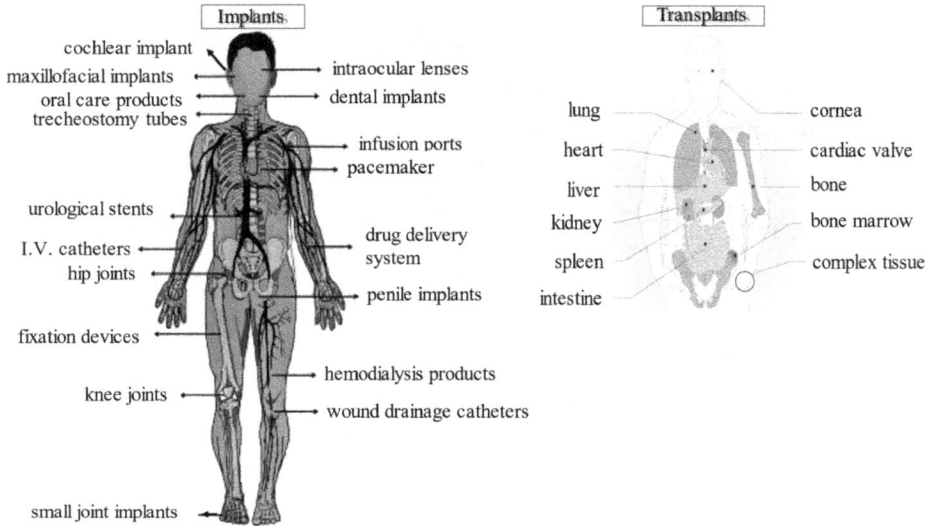

Implants

cochlear implant
maxillofacial implants
oral care products
trecheostomy tubes
intraocular lenses
dental implants
infusion ports
pacemaker
urological stents
I.V. catheters
hip joints
drug delivery system
penile implants
fixation devices
hemodialysis products
knee joints
wound drainage catheters
small joint implants

Transplants

lung — cornea
heart — cardiac valve
liver — bone
kidney — bone marrow
spleen — complex tissue
intestine

Figure 1.1: Typical implants and transplants [1–5].

Organ transplantation is the surgical implantation of an organ or section of an organ into a person with an incapacitated organ. Summarizing data on organ transplantation [6–9], 113,668 candidates waited for organ and tissue donation in 2018, 113,000 in 2019, and 110,000 in 2020; whereas 39,357 transplant operations were performed in 2018, 39,718 in 2019, and 40,000 in 2020. The United Network for Organ Sharing, administering the organ donation network, estimates that more than 120,000 Americans – more than 100,000 of whom require a kidney – are on the waiting list for life-saving organs. The average prospective kidney recipient has a 3.6-year wait, and at least 20 people waiting for an organ die each day [10]. Matching between donor and recipient is a crucial issue. Besides, success of organ transplantation depends on a number of factors: how old and how healthy the donor is, how old and how healthy the recipient is, how good a biological match can be found, and how ready the patient is to receive it [11]. More than 120,000 Americans are on the waiting list for life-saving organs. Today, despite remarkable advances in transplantation, the importance of artificial organs has not diminished. If anything, the long waiting list and the wait duration necessitate effective and immediate alternatives to organ transplant [12].

The tissue engineering evolved from the field of biomaterials development and refers to the practice of combining bioabsorbable scaffolds, cells, and biologically active molecules into functional tissues. Tissue engineering techniques require expertise in growth factor biology, a cell culture facility designed for human application, and personnel who have mastered the techniques of cell harvest, culture, and expansion. The goal of tissue engineering is to assemble functional constructs that restore, maintain, or improve damaged tissues or whole organs. Artificial skin and cartilage are examples of engineered tissues that have been approved by the Food and Drug Administration (FDA); however, currently they have limited use in human patients [13, 14]. With the advent of 3D printing [15, 16] and tissue engineering, one can think beyond electromechanical pumps that can serve as hearts to visualize an artificial one in flesh and blood. The race is on to develop a functional, tissue-based artificial organ that would mimic organs in physical and physiological functions, such as secretion of hormones, nurturing vasculature, and growth and modeling as the individual grows [12].

1.2 Types of implants

1.2.1 Medical implant

In general, medical implants can be divided into two groups: (1) load-bearing structural implants and (2) nonload-bearing functional implants. Human has six joints, including angle, knee, hip, wrist, elbow, and shoulder. Partial or total replacements of these synovial joint portions should involve complicated biotribological action, high-cycle low-stress fatigue, and biocorrosion deterioration. Hence, the component design and appropriate material selection become crucial for longevity, resulting in affecting HRQoL of the recipients. Figure 1.2 shows typical views of total hip replacement [17] and total knee replacement [18].

Besides these joint replacement orthopedic implants, there are other musculoskeletal implants such as bone plates and spina prosthesis. The Ilizarov apparatus (named after the pioneer of the technique, orthopedic surgeon Gavriil A. Ilizarov) is a type of external fixation used in orthopedic surgery to lengthen or reshape limb bones, and as a limb-sparing technique to treat complex and/or open bone fractures, as shown in Figure 1.3 [19]. This technique is sometimes also known as corticotomy [20], which is also employed in dental orthodontic treatment to shorten the treatment period and improve the efficacy of tooth mobility.

The rest of majority of medical implants are nonload-bearing functional devices or implants. They should include cardiovascular implant systems such as pacemaker, stents, artificial arteries or veins, or heart valves; organ implants such as skin, breast prostheses, kidney dialysis, artificial heart, or urinary stoma; and sensors such as cochlear, intraocular, or contact lenses [21].

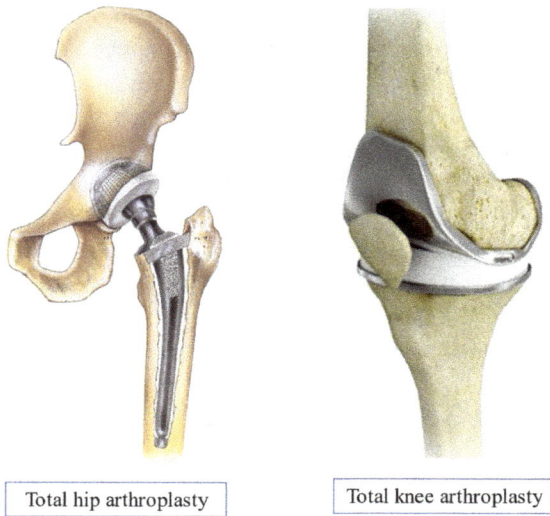

Total hip arthroplasty   Total knee arthroplasty

Figure 1.2: Typical pictures of total hip replacement [17] and total knee replacement [18].

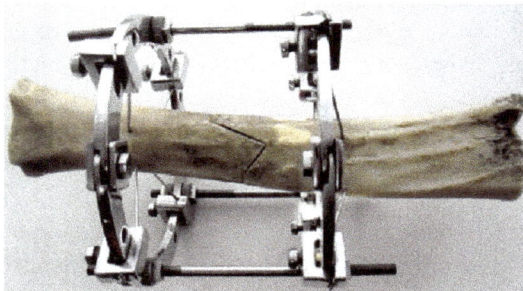

Figure 1.3: Ilizarov apparatus [19].

1.2.2 Dental implant

Oral implantology has developed and progressed into a central core of the art and science of dentistry, has revolutionized oral rehabilitation, dental prosthesis, and maxillary reconstructions, and has become a predictable method for clinicians as well as patients [22]. The restoration of missing teeth is an important aspect of modern dentistry. As teeth are lost to decay or periodontal disease, there is a demand for replacement of aesthetics and/or function. Conventional methods of restoration include a removable complete denture, a removable partial denture, or a fixed prosthesis. Each method has its own indications and its share of advantages and disadvantages. For centuries, people have attempted to replace missing teeth using implantation. It is generally known that there are 250 different implant industries in world manufacturers and markets about 500 different dental implant products with variety of diameters, lengths, surfaces, platforms, interfaces, and fixture body shapes.

Basically, there are three typical procedures for dental implantations: subperiosteal type, transosteal type, and endosteal (endosseous) type, as compared in Figure 1.4 [23–25]. *Subperiosteal implants* consist of a metallic framework that is placed/fitted onto the surface of the mandible or maxilla beneath the subperiosteum (gum tissue). As the gums heal, the frame becomes fixed to the jawbone. Vertical posts, which are part of this framework, provide through the mucosa and act as abutment for prosthesis. Unfortunately, subperiosteal implants are not as stable as endosteal implants. This is because subperiosteal implants do not go into the jawbone. Instead, they rest on top of the bone and are held in place only by the soft tissue. Although this type possesses the longest history of clinical trials, its long-term success rate is relatively low: 54%/15 years, due to poor retention force. *Transosteal implants* are surgically implanted/penetrated directly and completely through the mandible. Once the surrounding gum tissue has healed, a second surgery is needed to connect a post to the original implant. Finally, an artificial tooth (or teeth) is attached to the post – individually or grouped on a bridge or denture. This type claims the highest success rate of 90%/8–16-year period, but is limited to the mandible. *Endosteal (endosseous) implants* are the most common type of implants and are sometimes used as an alternative to a removable denture or bridge. The endosteal implants include bladed types, cylinder types (smooth), or screw types (threaded). Endosteal implants are partially submerged and anchored within the mandible or maxilla. These kinds of implants are appreciated for having one of the most natural-feeling and stable results, reporting the very high success rate over 15-year period [23–26].

For endosteal implants, there are several variations, depending on the detailed treatment plans: a single tooth dental implant, an implant-supported bridge, and implant-retained denture, as shown in Figure 1.5 [27]. Each one possesses pros and cons.

A single (unit) tooth dental implant restoration replaces, unlike other restorations, entire missing tooth from root to crown. If a patient has one missing tooth or multiple that are not adjacent to each other, then a single tooth dental implant may be the best option. However, if there are multiple missing teeth adjacent to each other, then this may not be the best option. Additionally, the next type of dental implant may save money for a case of multiple missing teeth. *An implant-supported bridge* has crowns that connect to dental implants, while typically a bridge consists of two crowns on either side of missing tooth gap with an artificial tooth held by those crowns in between. The process is similar to a single tooth dental implant; however, the teeth missing in the middle of the gap will not receive a dental implant. The benefits of an implant-supported bridge are that patient can securely replace multiple missing teeth in a row – without the cost of replacing each tooth, while the downside is that not all teeth will receive an implant, and therefore there is a risk of losing some bone mass. The problem

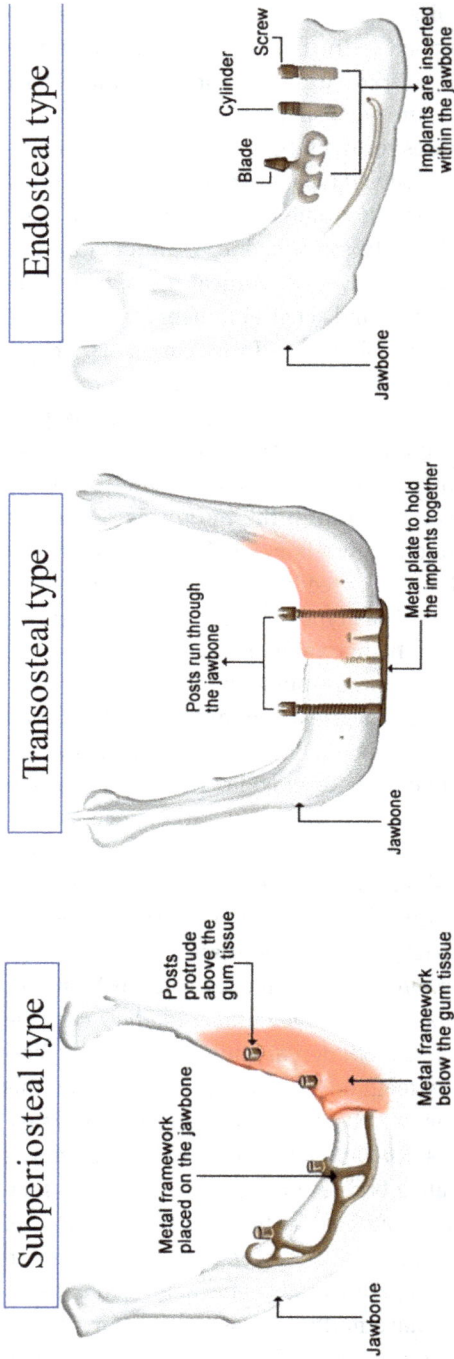

Figure 1.4: Three typical types of dental implants [23–25].

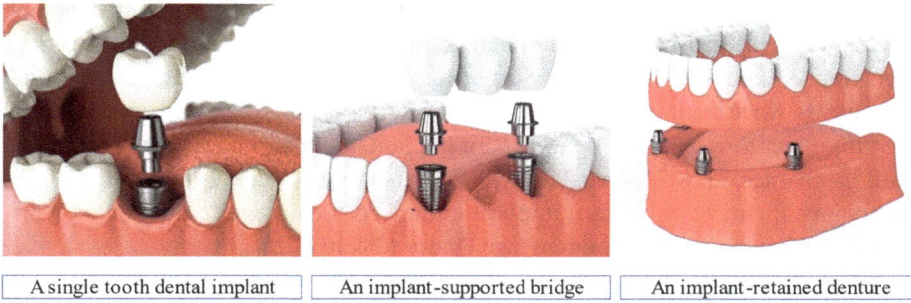

| A single tooth dental implant | An implant-supported bridge | An implant-retained denture |

Figure 1.5: Three typical treatment-related dental implant procedures [27].

associated with traditional dentures is that they are removable, which means they can slip, slide, click, fall out, and make daily tasks uncomfortable like eating and talking, and these problems can be solved by permanently securing a denture with dental implants by the *implant-retained denture treatment*. Once placed implants heal, and the current denture may be modified, so it can be worn without disrupting the healing process. Once healed, a new, custom denture is fixed to the dental implants. The result is a permanent, secure denture custom-designed to fit the facial aesthetics [27].

A relatively new procedure called "All-on-4 dental implant treatment" is available for avoiding wearing a denture. As shown in Figure 1.6 [28], a small titanium screw is placed into the jawbone, which replaces the root of the missing tooth, requiring a small surgery. Once that is done, a crown is connected, with the end result being a very real looking and functional tooth. If six implants instead of four were placed, it would be called "all-on-6 implants" [28, 29].

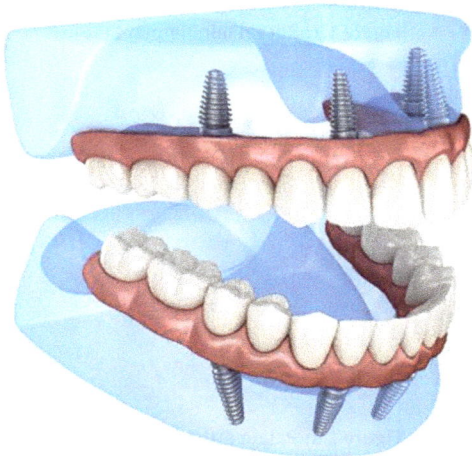

Figure 1.6: All-on-4 dental implant system [28].

Tooth movements required during the orthodontic mechanotherapy include combinations of vertical intrusion or extrusion, horizontal translation, and 3D rotation. Hence, some clinical cases cannot be fully accomplished by conventional bracket/archwire treatment. Mini-implants are small bone anchors which can be used for direct forces in tight spaces. Mini-implants offer easy insertion, straightforward mechanics, as well as versatility. When treatment is done, the mini-implants are removed. Orthodontic mini-implant-enhanced anchorage is defined as resistance to undesired tooth movement (see Figure 1.7) [30, 31].

Besides the mini-implant (as shown in Figure 1.7), there is another type of orthodontic implant, called a palatal implant (short with 4–6 mm implants placed in the roof of the mouth). A thick wire runs from the palatal implant to the selected teeth. The absolute anchorage value of the implant is transmitted to these selected teeth [32]. The palatal implant is also designed to relieve snoring and other disturbing symptoms of obstructive sleep apnea [33, 34]. Figure 1.8 shows a dental palatal implant placed in the midsagittal area of the maxillary hard palate for use as anchorage in an orthodontic treatment [34].

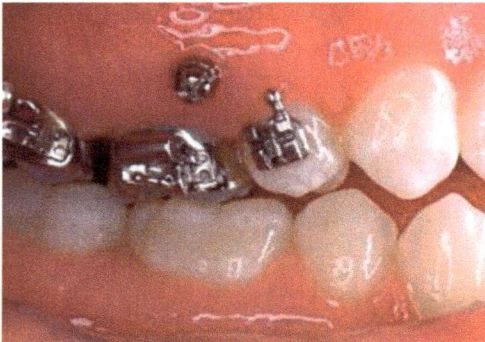

Figure 1.7: Buccal mini-implant [30].

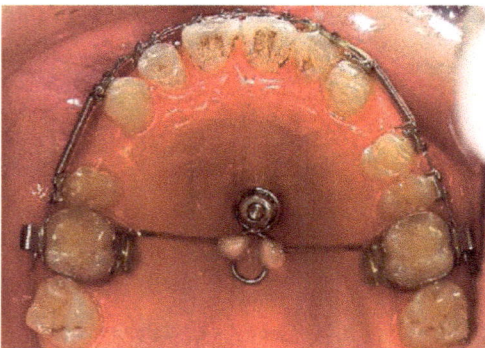

Figure 1.8: Palate orthodontic implant [34].

1.3 FDA regulations and safety issues

Until the passage in 1976 of the Medical Device Amendments to the Food and Drug Act, medical and dental materials and devices for use in the human body were not regulated by any agency of the US government. The only exception was materials for which therapeutic claims were made, which allowed the FDA to consider them as a drug. Long before 1976, the dental profession had realized a need for the development of criteria to help ensure the safety and efficacy of the dental material. The first efforts were initiated by the U.S. Army, which commissioned the National Bureau of Standards (NBS), now the National Institute for Standards and Technology, to develop specifications for dental amalgam alloys used in federal service. In 1928, the American Dental Association (ADA) assumed sponsorship of this program at NBS to develop standards for dental materials. The ADA adopted these standards as specifications in its Certification Program for Dental Materials. Each specification contained requirements for physical, mechanical, and chemical properties of a material that were felt to be relevant to the clinical application of the material and would ensure safety and efficacy. Materials that met the relevant specifications were granted certification by the ADA. Lists of certified materials were published in the *Journal of the American Dental Association* (JADA), and the manufacturer was permitted to display the ADA Seal of Certification on the product. Compliance with this program was voluntary. The ADA Council on Dental Materials, Instruments, and Equipment has the administrative responsibility for the certification program. This council also acts as an administrative sponsor of the American National Standards Institute's (ANSI) Accredited Standards Committee MD156. Committee MD156 appoints subcommittees to formulate and revise specifications. These are submitted to ANSI and, if accepted, become American National Standard (ANS). The ADA Council then has the option of adopting the ANS as a specification under the ADA certification program. There are now 51 specifications for various dental materials. A similar situation exists at the international level. The Fédération Dentaire Internationale (FDI) represents organized dentistry and acts in cooperation with the International Standards Organization (ISO) through the ISO committee TC106 – Dentistry. As of January 1992, there were 64 ISO standards under TC106, many of which have also been accepted as FDI specifications. The dental materials industry has rapidly developed into a worldwide industry with a world market. As a result, the ISO standards are becoming increasingly important [35].

The US FDA [36] defines a medical device as: (1) an instrument, apparatus, implement, machine, contrivance, implant, in vitro reagent, or other similar or related articles, including a component part or accessory which is recognized in the official National Formulary, or the United States Pharmacopoeia, or any supplement to them; (2) intended for use in the diagnosis of disease or other conditions, or in the cure, mitigation, treatment, or prevention of disease, in humans or other animals; or (3) intended to affect the structure or any function of the body of humans or other animals, and which does not achieve its primary intended purposes through chemical action within or on the body of man or other animals and which is not dependent upon being metabolized for the achievement of any of its primary intended purposes.

The Medical Device Amendments of 1976 gave the FDA jurisdiction over all materials, devises, and instruments used in the diagnosis, cure, mitigation, treatment, or prevention of disease in human. This includes materials used professionally and the over-the-counter products sold directly to the public. The FDA classifies approximately 1,700 medical devices into 16 medical specialty panels, including dentistry, cardiology, radiology, and immunology, and is further classified into 3 classes (class I, II, and III) depending on the degree of risk that may affect patients and equipment use. As the numbers increase from class I to III, the risk of life, health, and others increases. At the same time, as the risk increases, the types and amount of materials/documentations submitted to the FDA increase and become more complex, and accordingly, costs, review periods, and manufacturers' efforts increase [36–41].

Class I (general controls): class I devices have minimal contact with patients and low impact on a patient's overall health, leading to a minimum risk. In general, class I devices do not come into contact with a patient's internal organs, the central nervous system, nor cardiovascular system. These devices are subject to the fewest regulatory requirements. These include nonsterile gloves, electric toothbrush, tongue depressor, oxygen mask, reusable surgical scalpel, hospital beds, surgical instruments, cast material, crutches, bandages, and other products that do not harm patients or users even if the product is faulty. These products simplify the medical device application process and are exempt from premarket notification. Manufacturers are required to properly manage the quality of their products, comply with good manufacturing practices, and comply with labeling (labeling) rules.

Class II (special controls): Products that are more complex than class I and harm patients or users if there is a product defect are classified as class II. Devices in class II should have already been marketed for similar purposes and efficacy and are limited to those that are some degree reliable in both safety and efficacy. By showing that performance and mechanism of action are comparable to the real equivalence of medical devices that have already been approved (substantially equivalent), the FDA is possible to shorten the review. New technologies that do not meet this criterion are automatically classified as class III, no matter how simple the product is (even temporarily). Class II medical devices are more complicated than class I devices and present a higher category of risk because they are more likely to come into sustained contact with a patient. Examples include products that are implanted in the body that are not electronically controlled (implant artificial tooth root, coronary stent, bone screws, cemented Total Hip Arthroplasty (THA) system, and contact lenses), those that are used externally but directly on the skin, blood transfusion pump, and that are electronically controlled (electric muscle stimulation), complexly configured, and requiring a certain level of knowledge for safe use (such as rehabilitation instruments or electric wheelchair). This can include devices that come into contact with a patient's cardiovascular system or internal organs, and diagnostic tools. Besides these, catheters, blood pressure cuffs, pregnancy test kits, syringes, blood transfusion kits, or absorbable sutures should be included in class II devices.

Class III (premarket approval: PMA): This applies to products that have a constant or higher risk to the patient because of the content of new technologies and usages. In addition to special management, products that fall under class III require pre-commercial approval procedures (PMA). Just 10% of the devices regulated by the US FDA fall into class III. This classification is generally extended to permanent implants, smart medical devices, and life support systems. While class III is generally reserved for the most innovative and cutting-edge medical devices, there are other devices that can fall into class III for different reasons. Some devices that are categorized initially as class II may be bumped up to class III if the manufacturer is unable to demonstrate substantial equivalence to a predicate (existing product) during the PMA (510(k)) filing process. There are many products involved in the maintenance and rescue of life, and in terms of high invasiveness and required reliability, PMA applications require long-term detailed evaluation of the equipment. These include artificial joints and pacemakers embedded deep in the body, implant-type drug pumps, implant cerebellar stimulator, heart valve for transplantation, ligament replacements, defibrillators, breast implants, pacemakers, defibrillators, high-frequency ventilators, cochlear implants, fetal blood sampling monitors, implanted prosthetics, and dialysis devices. In most cases, clinical trials are required, but this process is complex and time-intensive, so labor and costly, and comprehensive evidence indicating that clearly the safety and efficacy set by the FDA is furthermore required.

Similarly, in dental field, there are three classifications. Nineteen panels were established representing different areas of medicine and dentistry for classifying dental materials. The dental panel places an item into one of three classes [35].

Class I: materials posing minimum risk. These are subject only to good manufacturing and recordkeeping procedures.

Class II: materials for which safety and efficacy need to be demonstrated and for which performance standards are available (established by the FDA or other authoritative body such as the ADA in its certification specifications). Materials must be shown to meet the performance standard.

Class III: materials that pose significant risk and materials for which performance standards have not formulated, and materials of this class are subject to PMA by the FDA for safety and efficacy, in much the same manner as a new drug.

The regulatory functions of the FDA have made the activities of the ADA and ANSI even more important. The time required to develop and approve a new federal regulation is such that the FDA would have been faced with an almost impossible task to develop new specifications for the large number of dental products that have been placed in class II [35].

References

[1]	Orthopedic Trauma Implants. GPC Medical Ltd. 2016; https://orthopedicimplantsindia.word press.com/2016/05/18/implants-in-orthopedic-trauma/.
[2]	Dutta RC, Dutta AK, Basu B. Engineering implants for fractured bones-metals to tissue constructs. J Mater Eng Appl. 2017, 1, 9–13.
[3]	Park J, Lakes RS. Biomaterials: An Introduction. New York NY USA: Springer, 2007, 564.
[4]	https://jadasingleton.weebly.com/organ-donation-overview.html.
[5]	https://www.healthdirect.gov.au/organ-transplants.
[6]	More deceased-donor organ transplants than ever. UNOS. 2020; https://unos.org/data/trans plant-trends/.
[7]	Facts: Did you know? American Transplant Foundation. https://www.americantransplantfoun dation.org/about-transplant/facts-and-myths/.
[8]	WHO. GKT1 Activity and Practices. https://www.who.int/transplantation/gkt/statistics/en/.
[9]	WHO. WHO Task Force on Donation and Transplantation of Human Organs and Tissues. https://www.who.int/transplantation/donation/taskforce-transplantation/en/.
[10]	UNOS. Transplant trends. 2020; https://unos.org/data/transplant-trends/.
[11]	Heinrich J. Organ donation – a new frontier for AI? 2017; https://phys.org/news/2017-04-don ationa-frontier-ai.html.
[12]	Innovations in Artificial Organs. Alliance of Advanced BioMedical Engineering. https:// aabme.asme.org/posts/innovations-in-artificial-organs.
[13]	NIH. Tissue engineering and regenerative medicine. National Institute of Biomedical Imaging and Bioengineering. https://www.nibib.nih.gov/science-education/science-topics/tissue-en gineering-and-regenerative-medicine.
[14]	Atala A. Tissue engineering of artificial organs. J Endourol. 2000, 14, 49–57.
[15]	Scudellari M. 3D printing bone directly into the body: A novel ink enables bioengineers to print bone-like material at room temperature with living cells. IEEE Spectrum. 2021; https://spec trum.ieee.org/the-human-os/biomedical/devices/3d-printing-bone-directly-into-the-body.
[16]	Papadopoulos L. High-speed 3D printing method takes us one step closer to printing organs. Interesting Engineering. 2021; https://interestingengineering.com/high-speed-3d-printing-takes-us-closer-to-printing-organs.
[17]	Rubin A. History of the Total Hip Replacement. 2017; https://www.flushinghospital.org/news letter/history-of-the-total-hip-replacement/.
[18]	Chia A. Total Knee Replacement; https://arthrohealth.ypo.pw/total-knee-replacement/.
[19]	Gubin A, Borzunov D, Malkova T. The Ilizarov paradigm: Thirty years with the Ilizarov method, current concerns and future research. Int Orthop. 2013, doi: 10.1007/s00264-013-1935-0.
[20]	Brutscher R, Rahn BA, Rüter A, Perren SM. The role of corticotomy and osteotomy in the treatment of bone defects using the Ilizarov technique. J Orthop Trauma. 1993, 7, 261–69.
[21]	Van Humbeek J. When does a material become a biomaterial? http://www.biotinet.eu/down loads/SummerSchool-Barcelona/SummerSchool-Humbeeck.pdf.

[22] Ong CTT, Ivanovski S, Needleman IG, Retzepi M, Moles DR, Tonetti MS, Donos N. Systematic review of implant outcomes in treated periodontitis subjects. J Clin Periodontol. 2008, 35, 438–62.

[23] Choosing the best types of dental implants: What you need to know. Renew Institute. 2020; https://www.renewinstitute.com/types-dental-implants/.

[24] What is Dental Implant concept?; http://nhakhoakaiyen.com/en/service/detail/what-is-dental-implant-concept-119.

[25] Endosteal Implants. 2019; https://sandiegoinvisaligndentist.org/endosteal-implants/.

[26] Taylor A. The Types of Dental Implants Explained: A Detailed Guide. 2021; https://hammburg.com/the-types-of-dental-implants-explained-a-detailed-guide/.

[27] Raval N, Kohnen S, Lee E. 3 Types of Dental Implants (Which One Is Best for You?). 2020; https://www.implantandperiodonticspecialists.com/blog/dental-implant-types/.

[28] All on 4 dental implants. Cutting Edge. https://cuttingedgeperiodontist.com/services/all-on-4-dental-implants/.

[29] John A. When is All-on-4 the best option? 5 factors to consider when creating your dental treatment plan. Dentistry IQ. 2017; https://www.dentistryiq.com/dentistry/implantology/article/16365915/when-is-allon4-the-best-option-5-factors-to-consider-when-creating-your-dental-treatment-plan.

[30] Baumgaertel S, Razavi MR, Hans MG. Mini-implant anchorage for the orthodontic practitioner. American Association of Orthodontists. Techno Bytes. 2008, 133, 621–27.

[31] Sorake A, Kumar J. Implants in Orthodontics. Revista Latinoamericana de Ortodoncia y Odontopediatría. 2013; https://www.ortodoncia.ws/publicaciones/2013/art–12/.

[32] Zheng X, Sun Y, Zhnag Y, Cai T, Sun F, Lin J. Implants for orthodontic anchorage. Medicine (Baltimore). 2018, 97, e0232, doi: 10.1097/MD.0000000000010232.

[33] Implant Anchorage for Orthodontics; https://www.drengen.com/orthodontics/implant-anchorage-for-orthodontics/.

[34] Crismani AG, Bernhart T, Bantleon HP, Kucher G. An innovative adhesive procedure for connecting transpalatal arches with palatal implants. Eur J Orthodont. 2005, 27, 226–30.

[35] Moore BK, Oshida Y. Materials science and technology in dentistry. In: Wise DL, ed. Encyclopedic Handbook of Biomaterials and Bioengineering – Part B. Vol. 2, Marcel Dekker, Co, 1995, 1325–430.

[36] US FDA. Medical Device Overview. 2018; https://www.fda.gov/industry/regulated-products/medical-device-overview#:~:text=The%20FDA%20defines%20a%20medical%20device%20as%3A%20%22an,United%20States%20Pharmacopoeia%2C%20or%20any%20supplement%20to%20them%2C.

[37] US FDA. List of Medical Devices, by Product Code, that FDA classifies as Implantable, Life-Saving, and Life-Sustaining Devices for purposes of Section 614 of FDASIA amending Section 519(f) of the FDC Act. 2015; https://www.fda.gov/media/87739/download.

[38] Standards and Regulatory Considerations for Orthopaedic Implants; http://user.engineering.uiowa.edu/~bme_158/lecture/Standards%20&%20Regulatory.pdf.

[39] Sheth U, Nguyen N-A, Gaines S, Bhandari M, Mehlman CT, Klein G. New orthopedic devices and the FDA. J Long Term Eff Med Implants. 2009, 19, 173–84, https://pubmed.ncbi.nlm.nih.gov/20939777/.

[40] US FDA. Classify Your Medical Device. https://www.fda.gov/medical-devices/overview-device-regulation/classify-your-medical-device.

[41] Fenton R. What are the differences in the FDA medical device classes? Qualio. 2021; https://www.qualio.com/blog/fda-medical-device-classes-differences.

Chapter 2
Biofunctionality and biomaterials

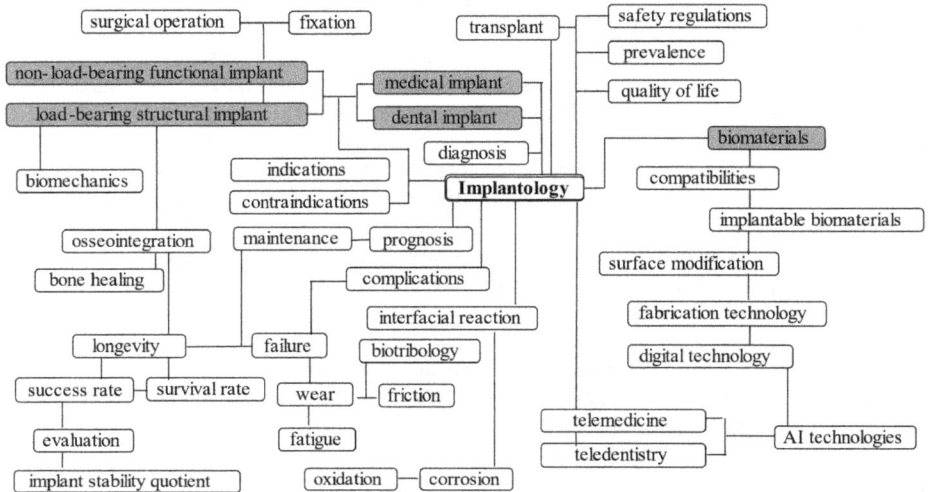

2.1 Biofunctionality

Biofunctional biomaterials are materials whose function is dependent on a biological content; hence, they play as key materials that support medical engineering and biological engineering, covering biospecific materials and biosimulating materials. Typical examples of biospecific materials are the biocompatible materials used for artificial organs. Biofunctional materials are synthesized from the viewpoint of functionality design. In order to create innovative biofunctional materials, an interdisciplinary field without boundary lines between inorganic, organic, and polymer material fields is necessary [1–5]. At the same time, as shown in Figure 2.1 [6], biofunctionality is one of supportive information and character toward the developing successful biomaterials.

On the other hand, the functionality of biomaterials in biological environment contains slightly different context and it would be much emphasized when implants made of biomaterials are placed in the human body. In engineering field, there are two types of materials (structural materials and functional materials); similarly, biomaterials can be divided into two distinctive groups as follows.

https://doi.org/10.1515/9783110740134-003

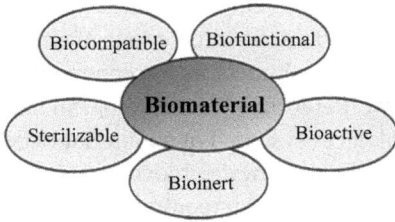

Figure 2.1: Development of successful biomaterials supported by various important parameters [6].

2.1.1 Biostructural materials

As for orthopedic implants, natural synovial joints, e.g., hip, knee, or shoulder joints, are complex and delicate structures capable of biomechanistic functioning under critical conditions. Accordingly, implants for one of six joints (angle, knee, hip, wrist, elbow, and shoulder) should exhibit satisfactory level of biomechanistic characteristics, which should include at least basic mechanical properties, biotribological resistance, and high-cycle low-stress fatigue strength. Such joint implants should also possess good bio-corrosion resistance and antiwear debris toxicity during the tribological actions.

Normally, dental implant treatments are performed with twofold purposes: recovery of mastication function and aesthetic appearance. Basically, implant treatment at the posterior area is conducted to recover the mastication/occlusal force (in range from 100 to 250 N, depending on age and gender). Accordingly, implants that are placed in the posterior area should show sufficient structural integrity. On the other hand, implants at the anterior should assist to recover the aesthetic appearance (namely, aesthetic dentistry accompanied with orthodontic therapy if needed).

2.1.2 Biofunctional materials

Opposing to the previous biostructural implants, implants such as cochlear implants, intraocular lenses, stents, pacemaker, and maxillofacial implants should be considered as biofunctional implants.

Implants made of bioabsorbable Mg materials should be also included in this category. Obviously, biofunctional materials are not expected to perform a load-bearing property.

2.2 Biomaterials

It is always not true that all materials can serve as biomaterials, rather there are several strict conditions to meet to be qualified as biomaterials. A biomaterial is a substance that has been engineered to interact with biological systems for a medical purpose, either a therapeutic (treat, augment, repair, or replace a tissue function of the body) or a diagnostic procedure to regulate the interactions of single or multiple components of living systems when applied alone or as a component of a complex device. Biomaterial should be carefully differentiated from biological materials such as the bone that is produced by a biological system. For a material to be qualified as a biomaterial, it should possess two basic criteria: biocompatibility and functional performance. The material must satisfy its design requirements in service: (i) load transmission and stress distribution (for bone replacement applications), (ii) articulation to allow movement (for in artificial joint), (iii) control of blood and fluid flow (for implanted artificial heart), (iv) electrical stimuli (for in pacemaker used for pumping blood at a normal rate), (v) light transmission in implanted lenses, or (vi) sound transmission in cochlear implant. Additionally, it should be nontoxic and noncarcinogenic. These requirements eliminate many engineering materials that are available. Since biomaterials are application-specific, some required properties should not apply to other applications. For example, in tissue engineering (TE) of the bone, the polymeric scaffold needs to be biodegradable so that as the cells generate their own extracellular matrices (ECM), the polymeric biomaterial will be completely replaced over time with the patient's own tissue. In the case of mechanical heart valves, we need materials that are biostable, are wear-resistant, and do not degrade with time. A requirement for surface energy (or wettability) is another example. Surface of dental implants is mechanically and/or chemically treated to enhance the surface wettability (in other words, hydrophilicity) to improve the subsequent osseointegration, while the inner surface of artificial vein tube should be hydrophobic [7–9].

About 20 pure elements can be used as base metals or as a major alloying element(s) in combination with an appropriate element(s) from a selection of 22 nonmetallic elements. Out of the approximately 3,500 binary (two-element) alloy systems, about 800 can be considered practical for use. From the 100,000 possible combinations for three-element (tertiary) alloy systems, only about 350 can serve as useful alloys. Hence, approximately 1,200 metallic materials (in pure or alloy form) are available. Besides metallic materials, we still have polymeric, ceramic, and composite materials. Among polymeric materials, there are about 2,000–5,000 types of polymers, depending on the molecular weight. In the realm of ceramic materials, about 10,000 different kinds exist. Additionally, there are composite materials of MMC (metal–matrix composite), PMC (plastic–matrix composite), or CMC (ceramic–matrix composite). All told, the materials available to us are extraordinarily large number and they can serve as structural and functional materials. The materials used in medical or dental applications may be referred to as biomaterials [10]. Because of demanded limitations on biomaterial candidates, as will be seen in the followings, there are quite a small number of materials serving as effective and safe biomaterials.

2.2.1 Classification of biomaterials

Instead of classification of biomaterials in terms of material kinds, there is another classification of biomaterials in terms of interaction with biological systems, including as being biotolerant, bioinert, bioactive, and biodegradable [11].

Biotolerant materials such as polymethyl methacrylate (PMMA) are characterized by a thin fibrous tissue interface, and the fibrous tissue (scar tissue) layer develops as a result of the chemical products from leaching processes (including release of ions, corrosion products, and chemical compounds from the placed implant), leading to irritation of the surrounding tissues [11, 12].

Bioinert materials are materials that do not initiate a response or interact when introduced to biological tissue and are stable in the human body, not interacting with body fluids or tissues. Generally, bioinert materials are encapsulated by fibrous tissues to isolate them from the surrounding bone, similar to biotolerant materials; however, under certain conditions, bioinert materials can have direct structural and functional connection with the adjacent bone tissue without being separated from the host tissue [13]. Samples of these are titanium and its alloys, alumina (Al_2O_3), partially stabilized zirconia (ZrO_2), and ultra-high-molecular-weight polyethylene (UHMWPE), phosphoryl choline (for contact lenses), polyethylene oxide polymers [11, 14, 15]. Generally, a fibrous capsule might form around bioinert implants; hence, its biofunctionality relies on tissue integration through the implant. These materials are capable of being in touch with bodily fluids and tissues for prolonged periods of lifetime, while eliciting little, if any, adverse reactions. Biologically inert or bioinert materials are ones that do not initiate a response or interact when introduced to the biological tissue. In other words, introducing the fabric to the body would not cause a reaction with the host. Originally, these materials were used for vascular surgery due to the need for surfaces, which do not cause clotting of the blood. For this reason, bioinert materials may sometimes even be called hemocompatible [16, 17].

Bioactive biomaterials: Bioactive refers to a material, which interacts with the surrounding bone and, in some cases, even soft tissue. This occurs through a time-dependent kinetic modification of the surface, triggered by their implantation within the living bone. An ion-exchange reaction between the bioactive implant and surrounding body fluids results in the formation of a biologically active carbonate hydroxyapatite (HA) layer on the implant that is chemically and crystallographically equivalent to the mineral phase in the bone. Hence, a bioactive material in the bone tissue environment can create an environment compatible with osteogenesis by making chemical bonds with the bone tissue [11, 18]. Typical examples of these materials are synthetic HA [$Ca_{10}(PO_4)_6(OH)_2$], glass ceramic, and bioglass.

Bioactive materials can be divided into two classes: osteoconductive and osteoinductive materials [19, 20]. Osteoconductive materials allow bone growth along the surface of the bioactive material. Ceramics such as HA and tricalcium phosphate (TCP) are examples of such osteoconductive materials [21]. Osteoconduction (meaning that bone grows on a surface) and osteoinduction are dependent on each other. Osteoconduction not only involves the action of bone-forming cells but also inculcates the role of blood vessels and various amounts of small proteins that act as growth factors during bone formation. The conduction of bone formation through these means occurs as a natural body mechanism. However, in procedures of implants and grafting, it is important for the dental surgeon to select an appropriate material that is compatible with the body tissues so that the process of bone healing and formation does not get hindered. Hence, the selection of a proper biomaterial will facilitate osteoblasts to form the new bone over the graft or around the implant. Accordingly, the graft material or implant should act as a bridge that links the existing bone and the new bone. The result of osteoconduction is a bony surface consisting of various channels, pores, and a matrix that acts as the scaffold for the new bone [22–24].

Osteoinduction is the process by which osteogenesis is induced. It is a phenomenon regularly seen in any type of bone healing process. Osteoinduction implies the recruitment of immature cells and the stimulation of these cells to develop into preosteoblasts. In a bone healing situation such as a fracture or traumatized alveolar bone in dental implant site preparation, the majority of bone healing is dependent on osteoinduction [23]. The bone-forming cells are called osteoblasts. Even though they are bone-forming cells, an injury to the bony tissue might not be completely healed by these osteoblasts alone. They need a helping hand, which is provided by cells from the tissues surrounding the site of bone injury. These sites contain cells known as undifferentiated cells that have the ability to convert into preosteoblasts, i.e., a stage before converting to the bone-forming cells. The conversion of the undifferentiated cells into preosteoblasts needs a stimulus. This stimulus that induces the conversion of the undifferentiated cells is initiated by agents that are forms of proteins. Hence, while selecting a bone graft (a material that is placed to restore the lost bony tissue), it is important to see that those inductive protein agents are present in the graft material. This entire process, where a stimulus (inductive agent) is utilized to convert a certain type of cells into preosteoblasts from bone (osteo) formation, is called osteoinduction. Apart from these inductive agents, osteoinduction can also be enhanced by certain compounds that contribute to the bone growth. These are called promoters and the process is called osteopromotion [22, 25].

When a bioactive material is implanted into the human body, it stimulates a biological response from the body, which leads to a series of biophysical and biochemical reactions between the implant and tissue that eventually led to a mechanically strong chemical bonding (i.e., osseointegration) [18].

Biodegradable materials are materials that dissolve in contact with body fluids. The dissolution products are usually secreted via the kidneys, without causing serious effects to the environment. Biodegradable materials are used for medical goods such as surgical sutures, tissues in growth materials, and controlled drug release [26–28]. The most common biodegradable materials are polymers such as polyglycolic acid (PGA) and polylactic acids (PLA), and their copolymers [29]. Examples of biodegradable ceramics are calcium phosphates (CAPs) [30], and magnesium is an example of a bio-degradable metal [28, 31, 32].

Figure 2.2 illustrates the aforementioned four different types of biomaterials [11], where (a) shows an example of a biotolerant material, indicating that a fibrous layer surrounds the screw. A bioinert material is shown in (b), where there is a direct contact between the bone tissue and the implant screw; while (c) illustrates a bioactive material, causing a chemical reaction between the implant screw and the bone tissue. Figure 2.2(d) illustrates a biodegradable material, where the material has been degraded, and the degradation products were released into the bone tissue [11].

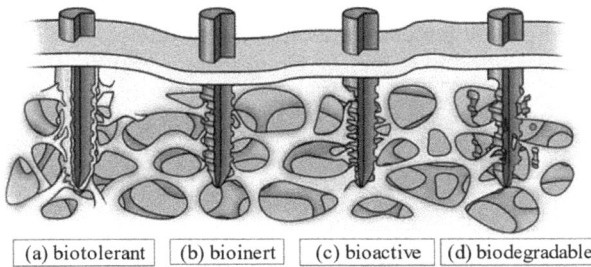

(a) biotolerant   (b) bioinert   (c) bioactive   (d) biodegradable

Figure 2.2: Schematic representation of biomaterial classifications explained using the case of a bone screw [11].

2.2.2 Metals and alloys

Metals and alloys are characterized by their high strength, high stiffness (high modulus of elasticity: E value), wear resistance, ductility (plasticity), and thermal and electrical conductivity. Metallic materials are commonly used for orthopedic implant, dental fixture, orthodontic implants, bone fixators, artificial joints, and external fixators, in addition to medical devices and related accessories. Most biocompatible metals are classified as biotolerant, and the exception is titanium and its alloys, which are classified as bioinert [33]. The most used biocompatible metals are cpTi (commercially pure titanium, grades I, II, III, and IV) and its alloys (Ti–6Al–4V and Ti–6Al–7Nb), 316 L stainless steel, and cobalt–chromium alloys [34–36]. Ti–6Al–4V is a common material for orthopedic implants due to its high strength-to-weight ratio (or specific strength) and corrosion resistance, in addition to its unique biocompatibility [37]. In 1952, Brånemark accidentally discovered the osseointegration capability of using cpTi, where the living bone formed a direct structural and functional connection with the implant surface [38]. Ti–6Al–4V is classified as inert, meaning that the material can, in certain conditions, have direct contact with the adjacent bone tissue without introducing any chemical reactions between the implant and the host tissue. Once the material is implanted into the body, it passivates itself by the formation of an adhesive oxide layer (like TiO_2). This oxide layer prevents an extensive flow of electrons and ions in the body fluid or surrounding tissue [35, 39]. Along with Ti–6Al–4V, stainless steel is a common material for biomedical implants and is a low-cost and readily available material. Out of the numerous grades of stainless steels, the austenitic 300 series (Fe–Cr–Ni) is utilized in medical applications, usually the medical grades 304 and 316L. Stainless steels do not exhibit the same biocompatibility as Ti–6Al–4V does, and it is considered a biotolerant material. With surface treatments and coatings, it is possible to increase its biocompatibility and corrosion behavior. Because of its high strength and corrosion resistance, stainless steel is often used in bone plates, screws, spinal fixation, hip and knee components, and medical devices [40, 41]. Another class of biotolerant materials is Co–Cr and Co–Cr–Mo alloys. Co–(27–30)Cr and Co–28Cr–6Mo alloys are used mostly under conditions of wear due to their high wear resistance [32, 42, 43].

2.2.3 Polymeric materials

In general, polymers can be characterized by (i) chain structures with covalent bindings (especially C), (ii) Van der Waals and H-bridges between the chains, (iii) amorphous or semicrystalline, (iv) low E value, and (v) enormous variability also within each class. Polymers have been widely studied for medical applications such as drug delivery and bone TE [44, 45]. From a biomedical perspective, polymers and copolymers can be divided into two classes: biodegradable and biotolerant. The role of polymeric implants for permanent load-bearing applications is rare [45], rather functionality of biodegradation. As Nair et al. [46] pointed out, degradable polymeric biomaterials are preferred candidates for developing therapeutic devices such as temporary prostheses, three-dimensional porous structures as scaffolds for TE, and as controlled/sustained-release drug delivery vehicles. Each of these applications demands materials with specific physical, chemical, biological, biomechanical, and degradation properties to provide efficient therapy. Other applications of degradable polymers should be surgical sutures and implants [47]. Consequently, a wide range of natural or synthetic polymers capable of undergoing degradation by hydrolytic or enzymatic route are being investigated for biomedical applications. The degradation of polymers can be classified into hydrolytical and enzymatical degradation. Most natural polymers, such as chitosan, alginate, collagen, gelatin, cellulose, and fibrinogen, undergo enzymatic degradation, meaning that the polymeric material is degraded by the enzymes that are secreted by tissues, immune system, or microorganisms that are present in the biological environment [11, 48]. The rate of enzymatic degradation varies significantly with the implantation site, depending on the availability and concentration of respective enzymes [11]. Examples of hydrolytically degradable polymers are poly(α-esters), such as PGA, PLA, poly(lactic-co-glycolic acid), and polycaprolactone [47].

2.2.4 Ceramic materials

Ceramics are generally characterized by (i) inorganic components (e.g., oxides, nitrides, or carbides) with a combination of ionic and covalent binding; (ii) complex crystal lattice or amorphous; (iii) in general, high E value; (iv) brittle nature, especially under tensile loading; and (v) relatively inert or very biodegradable. Bioceramics exhibit excellent biocompatibility and are often used as implants within bones, joints, and teeth. Bioceramics can be categorized as either bioinert or bioactive, and bioactive ceramics may be degradable or nondegradable [49]. Bioinert ceramics have high chemical stability and high mechanical strength in vivo. In addition to chemical inertness, bioinert ceramics have a significantly lower coefficient of friction and wear rate than metals so that bioinert ceramics are often used for femoral heads of Total Hip Replacement (THR) implants [50]. Examples of bioinert ceramics are aluminum oxide and zirconium oxide [51, 52]. Aluminum oxide (Al_2O_3), aka alumina, has high hardness, low friction, and excellent wear and corrosion resistance [53]. Zirconium dioxide (ZrO_2), aka zirconia, exhibits high strength and high resistance to wear so that it is often used for fixed partial dentures and other prosthetic devices [54]. Zirconia has also been used as a coating on titanium in dental implants [55]. Furthermore, zirconia implants have shown to accumulate less bacteria than cpTi implants in vivo [56], although zirconia implants possess some biomechanistic concern [57]. Bioactive ceramics are integrated with bone tissues via chemical reactions that lead to the formation of hydroxycarbonate apatite without causing inflammation [58]. The bond is usually stronger than the bone itself [59]. Bioactive ceramics include glasses, glass – ceramics, and other ceramics. Some CAPs are based on HA and TCP ($Ca_3(PO_4)_2$), which have been employed as bone replacement materials [60]. HA – $Ca_{10}(PO_4)_6(OH)_2$ – is a bioactive and nondegradable bioceramic with chemical and structural similarity to bone minerals [61]. The degradable bioactive ceramics are designed to gradually degrade in a predetermined time frame to assist as scaffolds or replace the host tissue [11, 28]. There is a difference in resorbability among CaP, TCP, and HA, influencing to maintain strength and stability integrities throughout their lifetime due to the degradation of the material. In this manner, TCP has a chemical composition similar to the bone tissue mineral and has a good resorbability and bioactivity with higher rates of biodegradation than HA under in vivo conditions. Rojbani et al. [21] evaluated the osteoconductivity of three different bone substitute materials: α-TCP, β-TCP, and HA, combined with or without simvastatin, which is a cholesterol synthesis inhibitor stimulating BMP-2 expression in osteoblasts. It was found that α-TCP, β-TCP, and HA are osteoconductive materials acting as a space maintainer for bone formation and that combining these materials with simvastatin stimulates bone regeneration and it also affects degradability of α-TCP and β-TCP, concluding that α-TCP has the advantage of higher rate of degradation allowing the more bone formation and combining α-TCP with simvastatin enhances this property.

2.2.5 Composite materials

The word "composite" refers to a heterogeneous combination, on a macroscopic scale, of two or more materials, differing in composition, morphology, and usually physical properties, made to produce specific physical, chemical, and mechanical characteristics. The combination of different elements results in a material that maximizes specific properties. The advantage of composites is that they show the best qualities of their constituents and often exhibit some properties that the single constituents do not have. Moreover, composite materials allow a flexible design, since their structure and properties can be optimized and tailored to specific applications [62]. In engineering, there are basically three types of composites: MMC, PMC, and CMC. Properties can be adapted by appropriate volume fractions and distribution of different materials, and normally the rule of mixture [63] should be applied to estimate properties (particularly mechanical properties and some physical properties) of composites. There is confusion and misleading in terminology of "biocomposite" and "composite for medical/dental applications." Biometallic materials are considered as metallic materials that can be safely used in medical and/or dental field. However, "biocomposite" is not defined by the same token, rather it is defined as a composite material composed of two or more distinct constituent materials (one being naturally derived) which are combined to yield a new material with improved performance over individual constituent materials. Normally, a polymeric resin matrix is reinforced with natural fibers [64, 65]. Accordingly, in this book, the term "composite biomaterials" is used.

Most human tissues such as bones, tendons, skin, ligaments, and teeth, are composites made up of single constituents whose amount, distribution, morphology, and properties determine the final behavior of the resulting tissue or organ. For example, bone is composed of collagen (which plays a foundational role for flexibility) and calcium (which, along with other minerals, is adhered to make the bone strong). Bone is morphologically constructed of outer hard surface layer (called compacta, aka cortical bone) and inner porous spongiosa, serving as a bone trabeculae. The tooth structure consists of mainly HA, several organics, and water. Tooth is morphologically layer-structured with the outer enamel layer (which is the hardest tissue in the entire body), dentin (which is the main body of the hard tooth tissue), cement layer (which covers entirely root surface), and the most inner dentinal pulpa in which nerves and blood vessels run. Man-made composites can, to some extent, be used to make prostheses able to mimic these biological tissues, to match their mechanical behavior, and to restore the mechanical functions of the damaged tissue. There are medical applications of composite biomaterials. Ramakrishna et al. [66] studied porous polyethylene terephthalate, polytetrafluoroethylene (PTFE), or polyurethane impregnated with collagen or gelatin for cardiovascular graft application. In orthopedic applications, most studies deal with the combination of high-density polyethylene and HA particles [62, 66–73]. For bone fracture internal fixation devices, totally bioresorbable internal fixation devices made of poly(L-lactic acid) reinforced with raw HA particles have been manufactured and characterized [74, 75]. The composite has ultra-high strength, excellent processability, and an elastic modulus close to that of natural bone. Moreover, it can retain its high strength during bone healing, shows optimal degradation and resorption behavior, and maintains osteoconductivity and bone bonding capability. This composite has been tested in vitro, to analyze the degradation process, and in vivo, to verify the biocompatibility and the bioactivity, showing good results. Kasuga et al. [76] studied composites made of PLA reinforced with HA fibers, with almost stoichiometric composition. For joint prostheses, hip replacement system composed of UHMWPE and Ti–6Al–4V alloy [77], knee replacement system consisted of UHMWPE and carbon fibers [66], and bone cement made of PMMA and HA [78] have been developed.

The use of composite biomaterials in dentistry, mainly PMCs, has considerably grown for substituting in some cases more traditional materials. PMCs are used in restorative dentistry to fill cavities, to restore fractured teeth, and to replace missing teeth [62]. In most applications, dental composite consists of a polymeric acrylic or methacrylic matrix reinforced with ceramic particles. The commercial formulations of matrices are mainly based on bisphenol-A-glycidyldimethacrylate; triethyleneglycol dimethacrylate is added to reduce the viscosity [79–83]. Some studies have addressed the combination of Ti with HA to optimize and satisfy both mechanical requirements and biological response of the host tissue to the dental implant. Graded composites combining Ti and HA have been developed [84, 85]. Research and clinical practice are devoted to reach a total bone regeneration and to inhibit both migration of epithelial cells to the implant and fibrous encapsulation. Composite materials have been studied and developed mainly for application as endosseus element. Composites made of SiC and carbon or carbon fiber reinforced carbon have been proposed [66]. These materials have some advantages over ceramics and metal alloys: they combine sufficient strength with a low elastic modulus, similar to that of natural teeth, minimizing the alteration of the stress field affecting the long-term success and host response. Moreover, they have far superior fatigue properties. The graded structure in the longitudinal direction contains more Ti in the upper section and more HA in the lower section. In fact, in the upper section, the occlusal force is directly applied and Ti offers the required mechanical performance; in the lower part, which is implanted inside the bone, the HA confers the bioactive and osteoconductive properties to the material. Ti alloy substrate has been combined also with HA granules spread over the surface, but further studies are needed to evaluate the biofunctionality and biocompatibility of this material [86]. Besides this compositional gradation, Oshida [35, 58] proposed the structural gradation of titanium substrate. A composite membrane comprising polyethylene glycol terephthalate (PEGT) and polybutylene terephthalate (PBT) copolymer reinforced with 30 wt% HA grains has been proposed [86]. This composite has excellent mechanical properties, and in vivo experimental results reveal good biocompatibility and a gradual degradation, mainly influenced by the PEGT/PBT ratio.

2.2.6 Hybrid materials

Hybrid biomaterials are another class of biomaterials and include hybrid biopolymers and biocomposites. Biopolymer gels are widely used in biomedical applications [87, 88]. Such medical/dental applications include TE [89–90] and dentistry. In particular, the combination of biopolymers with nanoparticles to fabricate membranes has been applied for periodontal regeneration [91–93]. Today, biopolymer hybrid particles are receiving more attention as novel materials for several applications in dentistry, such as drug delivery systems, bone repair, and periodontal regeneration surgery [94]. The association of such biopolymers with hybrid particles (including HA, CAP, and bioactive glass) is important for many biomedical functions, such as antimicrobial applications [95], drug delivery [96, 97], and guided bone regeneration [98–100].

Hybrid biocomposite is based on different types of fibers into a single matrix. The fibers can be synthetic or natural and can be randomly combined to generate the hybrid composites [101]. Its functionality depends directly on the balance between the good and bad properties of each individual material used. Besides, with the use of a composite that has two more types of fibers in the hybrid composite, one fiber can stand on the other one when it is blocked. The properties of this biocomposite depend directly on the fibers counting their content, length, arrangement, and also the bonding to the matrix. In particular, the strength of the hybrid composite depends on the failure strain of the individual fibers [102]. TE has evolved from the use of biomaterials for bone substitution that fulfill the clinical demands of biocompatibility, biodegradability, nonimmunogenicity, structural strength, and porosity [103]. TE has emerged as a promising alternative approach in the treatment of malfunctioning or lost organs [104]. In this approach, a temporary scaffold is needed to serve as an adhesive substrate for the implanted cells and a physical support to guide the formation of the new organs. Transplanted cells adhere to the scaffold, proliferate, secrete their own ECM, and stimulate new tissue formation. During this process, the scaffold gradually degrades and is eventually eliminated. Therefore, in addition to facilitating cell adhesion, promoting cell growth, and allowing the retention of differentiated cell functions, the scaffold should be biocompatible, biodegradable, highly porous with a large surface/volume ratio, mechanically strong, and malleable into desired shapes [105]. The most commonly used 3D porous scaffolds are constructed from two classes of biomaterials; (i) one class consists of synthetic biodegradable polymers such as aliphatic polyesters, PGA, PLA, and their copolymer of poly-(DL-lactic-co-glycolic acid), and (ii) the other class consists of naturally derived polymers such as collagen [106–110].

These biomaterials are further classified into subgroups depending on the manner of interaction with soft/hard tissue when these materials are used as implants – implantable materials that will be discussed in the next chapter. Summarizing this chapter, various biomaterials that we have been discussing in the above are involved in the human body as shown in Figure 2.3 [111].

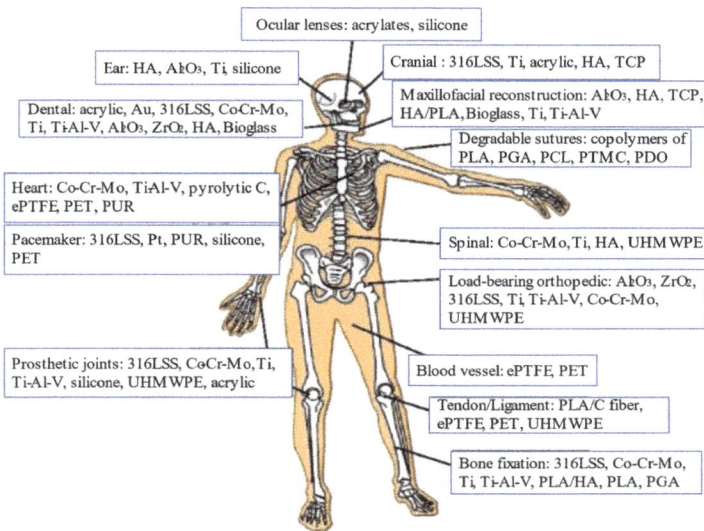

Figure 2.3: Biomaterials involved in human body [111]. PLA, polylactide; PGA, polyglycolide; PTMC, polytrimethylenecarbonate; PDO, poly(*p*-dioxanone); PUR, polyurethane; ePTFE, expanded polytetrafluoroethylene; UHMWPE, ultra-high-molecular-weight polyethylene; PET, polyethylene terephthalate; HA, hydroxyapatite; and SS, stainless steel. Ti refers to cpTi (commercially pure titanium and includes grades I, II, III, and IV) and Ti–Al–V refers to Ti–6Al–4V alloy (Ti-6-4) which is recently gradually replaced by Ti-6Al-7Nb (Ti-7-6) due to potential toxicity issues of V-element. Co–Cr–Mo alloy normally contains 60–70 wt% of Co, 27–30 wt% of Cr, and 5–7 wt% of Mo element, respectively.

References

[1]   Williams DF. Biofunctionality and biocompatibility. In: Cahn RW, et al., ed. Materials Science and Technology. A Comprehensive Treatment., Medical and Dental Materials. Vol. 14, 1991, doi: https://doi.org/10.1002/9783527603978.mst0160.

[2]   Baghdadchi J, Fatehi F. Biofunctionality: A novel learning method for intelligent agents. International Joint Conference on Neural Networks. Proceedings. 1999, 2, 1329–32; doi: 10.1109/IJCNN.1999.831155.

[3]   Cook SD, Dalton JE. Biocompatibility and biofunctionality of implanted materials. Alpha Omegan. 1992, 85, 41–47.

[4]   Rehman M, Madni A, Webster TJ. The era of biofunctional biomaterials in orthopedics: What does the future hold?. Expert Rev Med Dev. 2018, 15, 193–204.

[5]   Zeugolis DI, Pandit A. Biofunctional biomaterials – the next frontier. Bioconjugate Chem. 2015, 26, 1157, doi: https://doi.org/10.1021/acs.bioconjchem.5b00342.

[6]   Dos Santos V, Brandalise RN, Savaris M. Biomaterials: Characteristics and properties. In: Engineering of Biomaterials. Topics in Mining, Metallurgy and Materials Engineering. Cham: Springer, 2017, 5–15, doi: https://doi.org/10.1007/978-3-319-58607-6_2.

[7]   General requirements of biomaterials. https://www.britannica.com/technology/materials-science/General-requirements-of-biomaterials.

[8]   Requirements of Biomaterials; http://www.uobabylon.edu.iq/eprints/publication_12_29537_775.pdf.

[9]   Biomaterial; https://www.sciencedirect.com/topics/chemistry/biomaterial.

[10]   Oshida Y, Guven Y. Biocompatible coatings for metallic biomaterials. In: Wen C, ed. Surface Coating and Modification of Metallic Biomaterials. Woodhead Publishing, 2015, 287–343.

[11]   Ødegaard KS, Torgersen J, Elverum CW. Structural and biomedical properties of common additively manufactured biomaterials: A concise review. Metals. 2020, 10, 1677, https://www.researchgate.net/publication/347644971_Structural_and_Biomedical_Properties_of_Common_Additively_Manufactured_Biomaterials_A_Concise_Review.

[12]   Plenk H. The role of materials biocompatibility for functional electrical stimulation applications. Artif Organs. 2011, 35, 237–41.

[13]   Bergmann CP, Stumpf A. Biomaterials. In: Dental Ceramics: Microstructure, Properties and Degradation; Topics in Mining, Metallurgy and Materials Engineering. Berlin/Heidelberg, Germany: Springer, 2013, 9–13, doi: https://doi.org/10.1007/978-3-642-38224-6_2.

[14]   Bioinert materials. J Biochem Res. https://www.openaccessjournals.com/peer-reviewed-articles/bioinert-materials-1632.html.

[15]   Blokhuis TJ, Termaat MF, Den Boer FC, Patka P, Bakker FC, Haarman HJTM. Properties of calcium phosphate ceramics in relation to their in vivo behavior. J Trauma Inj Infect Crit Care. 2000, 48, 179–86.

[16]   Allan B. Closer to nature: New biomaterials and tissue engineering in ophthalmology. Br J Ophthalmol. 1999, 83, 1235–40.

[17]   Bioinert materials; https://www.omicsonline.com/bioinert-materials/leading-journals.php.

[18]   Zhao X, Courtney JM, Qian H. Bioactive Materials in Medicine: Design and Applications. Woodhead Publishing, 2011.

[19]   Jones JR. Review of bioactive glass: From Hench to hybrids. Acta Biomater. 2013, 9, 4457–86.

[20]   Barradas AMC, Yuan H, van Blitterswijk CA, Habibovic P. Osteoinductive biomaterials: Current knowledge of properties, experimental models and biological mechanisms. Eur Cell Mater. 2011, 21, 407–29, doi: 10.22203/eCM.v021a31.

[21]   Rojbani H, Nyan M, Ohya K, Kasugai S. Evaluation of the osteoconductivity of a-tricalcium phosphate, B-tricalcium phosphate, and hydroxyapatite combined with or without simvastatin in rat calvarial defect. J Biomed Mater Res Part A. 2011, 98A, 488–98.

[22]   What is the difference between osteoinduction and osteoconduction? 2020; https://dentagama.com/news/what-is-the-difference-between-osteoinduction-and-osteoconduction.

[23]   Albrektsson T, Johansson C. Osteoinduction, osteoconduction and osseointegration. Eur Spine J. 2001, 2, S96–101.

[24]   Tran PA, Sarin L, Hurt RH, Webster TJ. Opportunities for nanotechnology-enabled bioactive bone implants. J Mater Chem. 2009, 19, 2653–59.

[25]   Popa AC, Stan GE, Enculescu M, Tanase C, Tulyaganov DU, Ferreira JMF. Superior biofunctionality of dental implant fixtures uniformly coated with durable bioglass films by magnetron sputtering. J Mech Behav Biomed Mater. 2015, 51, 313–27.

[26]   Tian H, Tang Z, Zhuang X, Chen X, Jing X. Biodegradable synthetic polymers: Preparation, functionalization and biomedical application. Prog Polym Sci. 2012, 37, 237–80.

[27]   Prajapati SK, Jain A, Jain A, Jain S. Biodegradable polymers and constructs: A novel approach in drug delivery. Eur Polym J. 2019, 120, 109191, doi: https://doi.org/10.1016/j.eurpolymj.2019.08.018.

[28]   Oshida Y. Magnesium Materials. De Gruyter Pub., 2021.

[29] Godavitarne C, Robertson A, Peters J, Rogers B. Biodegradable materials. Orthop Trauma. 2017, 31, 316–20.
[30] Xie Y, Chen Y, Sun M, Ping Q. A mini review of biodegradable calcium phosphate nanoparticles for gene delivery. Curr Pharm Biotechnol. 2013, 14, 918–25.
[31] Peron M, Berto F, Torgersen J. Magnesium and Its Alloys as Implant Materials: Corrosion, Mechanical and Biological Performances. Boca Raton, FL, USA: CRC Press. 2020, https://books.google.com/books?hl=en&id=pGXXDwAAQBAJ&oi=fnd&pg=PP6&ots=OHlwBVxL0l&sig=itLZyz8MRpeShNDr5cFfb5E6EPU#v=onepage&q&f=false.
[32] Zartner P, Cesnjevar R, Singer H, Weyand M. First successful implantation of a biodegradable metal stent into the left pulmonary artery of a preterm baby. Catheter Cardiovasc Interv. 2005, 66, 590–94.
[33] Prasad K, Bazaka O, Chua M, Rochford M, Fedrick L, Spoor J, Symes R, Tieppo M, Collins C, Cao A, Markwell D, Ostrikov K, Bazaka K. Metallic biomaterials: Current challenges and opportunities. Materials (Basel). 2017, 10, 884, doi: 10.3390/ma10080884.
[34] Mavrogenis AF, Papagelopoulos PJ, Babis GC. Osseointegration of cobalt-chrome alloy implants. J Long Term Eff Med Implants. 2011, 21, 349–58.
[35] Oshida Y. Bioscience and Bioengineering of Titanium Materials. Elsevier Pub., 2007.
[36] Oshida Y, Tominaga T. Nickel-Titanium Materials. De Gruyter Pub., 2020.
[37] Venkatesh BD, Chen DL, Bhole SD. Effect of heat treatment on mechanical properties of Ti-6Al-4V ELI alloy. Mater Sci Eng. 2009, 506, 117–24.
[38] Brånemark PI, Zarb GAGA, Albrektsson T. Tissue-Integrated Prostheses: Osseointegration in Clinical Dentistry. Chicago, IL, USA: Quintessence Publishing, 1985.
[39] Biehl V, Breme J. Metallic biomaterials. Materialwiss Werkstofftech (Mater Sci Eng Technol). 2001, 32, 137–41.
[40] Oshida Y. Surface Engineering and Technology for Biomedical Implants. Momentum Press, 2014.
[41] Desai S, Bidanda B, Bártolo P. Metallic and ceramic biomaterials: Current and future developments. In: Bártolo P, et al., ed. Bio-Materials and Prototyping Applications in Medicine. Boston, MA, USA: Springer, 2008, 1–14.
[42] Poh CK, Shi Z, Tan XW, Liang ZC, Foo XM, Tan HC, Neoh KG, Wang W. Cobalt chromium alloy with immobilized BMP peptide for enhanced bone growth. J Orthop Res. 2011, 29, 1424–30.
[43] ASTM. ASTM F75 – 18: Standard Specification for Cobalt-28 Chromium-6 Molybdenum Alloy Castings and Casting Alloy for Surgical Implants (UNS R30075). https://www.astm.org/Standards/F75.
[44] Liechty WB, Kryscio DR, Slaughter BV, Peppas NA. Polymers for drug delivery systems. Annu Rev Chem Biomol Eng. 2010, 1, 149–73.
[45] Liu X, Ma PX. Polymeric Scaffolds for bone tissue engineering. Ann Biomed Eng. 2004, 32, 477–86.
[46] Nair LS, Laurencin CT. Biodegradable polymers as biomaterials. Prog Polym Sci. 2007, 32, 762–98.
[47] Ulery BD, Nair LS, Laurencin CT. Biomedical applications of biodegradable polymers. J Polym Sci Part B Polym Phys. 2011, 49, 832–64.
[48] Banerjee A, Chatterjee K, Madras G. Enzymatic degradation of polymers: A brief review. Mater Sci Technol. 2014, 30, 567–73.
[49] Rahaman MN. Bioactive ceramics and glasses for tissue engineering. In: Boccaccini AR, et al., ed. Tissue Engineering Using Ceramics and Polymers. Second ed. Cambridge, UK: Woodhead Publishing, 2014, 67–114.
[50] Chevalier J, Gremillard L. Ceramics for medical applications: A picture for the next 20 years. J Eur Ceram Soc. 2009, 29, 1245–55.

[51] Wittenbrink I, Hausmann A, Schickle K, Lauria I, Davtalab R, Foss M, Keller A, Fischer H. Low-aspect ratio nanopatterns on bioinert alumina influence the response and morphology of osteoblast-like cells. Biomaterials. 2015, 62, 58–65.

[52] Bashir M, Riaz S, Kayani ZN, Naseem S. Effects of the organic additives on dental zirconia ceramics: Structural and mechanical properties. J Sol Gel Sci Technol. 2015, 74, 290–98.

[53] Gonzalez J, Mireles J, Lin Y, Wicker R. Characterization of ceramic components fabricated using binder jetting additive manufacturing technology. Ceram Int. 2016, 42, 10559–64.

[54] Gautam C, Joyner J, Gautam A, Rao J, Vajtai R. Zirconia based dental ceramics: Structure, mechanical properties, biocompatibility and applications. Dalton Trans. 2016, 45, 19194–215.

[55] Sollazzo V, Pezzetti F, Scarano A, Piattelli A, Bignozzi CA, Massari L, Brunelli G, Carinci F. Zirconium oxide coating improves implant osseointegration in vivo. Dent Mater. 2008, 24, 357–61.

[56] Rimondini L, Cerroni L, Carrassi A, Torricelli P. Bacterial colonization of zirconia ceramic surfaces: An in vitro and in vivo study. Int J Oral Maxillofac Implant. 2002, 17, 793–98.

[57] Oshida Y, Miyazaki T, Tominaga T. Some biomechanistic concerns on newly developed implantable materials. J Dent Oral Health. 2018, 4, 1–5.

[58] Oshida Y. Hydroxyapatite: Synthesis and Applications. Momentum Press, 2015.

[59] Farid SB. Bioceramics: For Materials Science and Engineering. Amsterdam, The Netherlands: Elsevier, 2019.

[60] Dutta SR, Passi D, Singh P, Bhuibhar A. Ceramic and non-ceramic hydroxyapatite as a bone graft material: A brief review. Ir J Med Sci. 2015, 184, 101–06.

[61] Fathi MH, Hanifi A, Mortazavi V. Preparation and bioactivity evaluation of bone-like hydroxyapatite nanopowder. J Mater Process Technol. 2008, 202, 536–42.

[62] Salernitano E, Migliaresi C. Composite materials for biomedical applications: A review. J Appl Biomater Biomech. 2003, 1, 3–18.

[63] Yerbolat G, Amangeldi S, Ali MH, Badanova N, Ashirbekov A, Islam G. Composite materials property determination by rule of mixture and Monte Carlo simulation. 2018 IEEE International Conference on Advanced Manufacturing (ICAM), 2018; doi: 10.1109/ AMCON.2018.8615034.

[64] Yıldızhan Ş, Çalık A, Ozcanli M, Serin H. Bio-composite materials: A short review of recent trends, mechanical and chemical properties, and applications. Eur Mech Sci. 2018, 2, 83–91, doi: 10.26701/ems.369005.

[65] Biocomposite; https://en.wikipedia.org/wiki/Biocomposite.

[66] Ramakrishna S, Mayer J, Wintermantel E, Leong KW. Biomedical applications of polymer-composite materials: A review. Compos Sci Technol. 2001, 61, 1189–224.

[67] Ensaff H, O'Doherty DM, Jacobsen PH. The influence of the restoration-tooth interface in light cured composite restorations: A finite element analysis. Biomaterials. 2001, 22, 3097–103.

[68] Lovell LG, Lu H, Elliott JE, Stansbury JW, Bowman CN. The effect of cure rate on the mechanical properties of dental resins. Dent Mater. 2001, 17, 504–11.

[69] Schedle A, Franz A, Rausch-Fan X, Spittler A, Lucas T, Samorapoompichit P, Sperr W, Boltz-Nitilescu G. Cytotoxic effects of dental composites, adhesive substance, compomers and cements. Dent Mater. 1998, 14, 429–40.

[70] Ferracane JL, Condon JR. In vitro evaluation of the marginal degradation of dental composites under simulated occlusal loading. Dent Mater. 1999, 15, 262–67.

[71] Sarrett DC, Coletti DP, Peluso AR. The effects of alcoholic beverages on composite wear. Dent Mater. 2000, 16, 62–67.

[72] Bonfield W, Wang M, Tanner KE. Interfaces in analogue biomaterials. Acta Mater. 1998, 46, 2509–18, 38. Di Silvio L, Dalby MJ, Bonfield W. Osteoblast behaviour on HA/PE composite surfaces with different HA volumes. Biomaterials. 2002, 23, 101–7.

[73]   Wang M, Bonfield W. Chemically coupled hydroxyapatite-polyethylene composites: Structure and properties. Biomaterials. 2001, 22, 1311–20.

[74]   Shikinami Y, Okuno M. Bioresorbable devices made of forged composites of hydroxyapatite (HA) particles and poly-L-lactide (PLLA): Part I Basic characteristics. Biomaterials. 1999, 20, 859–77.

[75]   Furukawa T, Matsusue Y, Yasunaga T, Shikinami Y, Okuno M, Nakamura T. Biodegradation behavior of ultra-high-strength hydroxyapatite/poly(L-lactide) composite rods for internal fixation of bone fractures. Biomaterials. 2000, 21, 889–98.

[76]   Kasuga T, Ota Y, Nogami M, Abe Y. Preparation and mechanical properties of polylactic acid composites containing hydroxyapatite fibers. Biomaterials. 2001, 22, 19–23.

[77]   Verné E, Vitale Brovarone C, Milanese D. Glass-matrix biocomposites. J Biomed Mater Res. 2000, 53, 408–13.

[78]   Dalby MJ, Di Silvio L, Harper EJ, Bonfield W. Increasing hydroxyapatite incorporation into poly- (methylmethacrylate) cement increases osteoblast adhesion and response. Biomaterials. 2002, 23, 569–76.

[79]   Sandner B, Baudach S, Davy KWM, Braden M, Clarke RL. Synthesis of BISGMA derivatives, properties of their polymers and composites. J Mater Sci Mater Med. 1997, 8, 39–44.

[80]   Davy KWM, Kalachandra S, Pandain MS, Braden M. Relationship between composite matrix molecular structure and properties. Biomaterials. 1998, 19, 2007–14.

[81]   McCabe JF, Wassell RW. Hardness of model dental composites – the effect of filler volume fraction and silanation. J Mater Sci Mater Med. 1999, 10, 291–94.

[82]   Park C, Robertson RE. Mechanical properties of resin composites with filler particles aligned by an electric field. Dent Mater. 1998, 14, 385–93.

[83]   Imazato S, Tarumi H, Kato S, Ebisu S. Water sorption and colour stability of composites containing the antibacterial monomer MDPB. J Dent. 1999, 27, 279–83.

[84]   Watari F, Yokoyama A, Saso F, Uo M, Kawasaki T. Fabrication and properties of functionally graded dental implant. Compos Part B. 1997, 28, 5–11.

[85]   Nonami T, Kamiya A, Naganuma K, Kameyana T. Implantation of hydroxyapatite granules into superplastic titanium alloy for biomaterials. Mater Sci Eng. 1998, 6, 281–84.

[86]   Jansen JA, De Ruijter JE, Janssen PTM, Paquay YGCJ. Histological evaluation of a biodegradable polyactive®/hydroxyapatite membrane. Biomaterials. 1995, 6, 819–27.

[87]   Kartik A, Akhil D, Lakshmi D, Panchamoorthy G, Arun J, Sivaramakrishnan R, Pugazhendhi A. A critical review on production of biopolymers from algae biomass and their applications. Bioresour. Technol. 2021, 124868, doi: 10.1016/j.biortech.2021.124868.

[88]   Mallakpour S, Sirous F, Hussain CM. A journey to the world of fascinating ZnO nanocomposites made of chitosan, starch, cellulose, and other biopolymers: Progress in recent achievements in eco-friendly food packaging, biomedical, and water remediation technologies. Int J Biol Macromol. 2021, 170, 701–16.

[89]   Biswal T. Biopolymers for tissue engineering applications: A review. Mater Today Proc. 2021, 41, 397–401.

[90]   Dutta P, Giri S, Giri TK. Xyloglucan as green renewable biopolymer used in drug delivery and tissue engineering. Int J Biol Macromol. 2020, 160, 55–68.

[91]   Chang B, Ahuja N, Ma C, Liu X. Injectable scaffolds: Preparation and application in dental and craniofacial regeneration. Mater Sci Eng R Rep. 2017, 111, 1–26.

[92]   Elango J, Selvaganapathy PR, Lazzari G, Bao B, Wenhui W. Biomimetic collagen-sodium alginate-titanium oxide (TiO2) 3D matrix supports differentiated periodontal ligament fibroblasts growth for periodontal tissue regeneration. Int J Biol Macromol. 2020, 163, 9–18.

[93]   Sowmya S, Bumgardener JD, Chennazhi KP, Nair SV, Jayakumar R. Role of nanostructured biopolymers and bioceramics in enamel, dentin and periodontal tissue regeneration. Prog Polym Sci. 2013, 38, 1748–72.

[94] Chen I-H, Lee T-M, Huang C-L. Biopolymers hybrid particles used in dentistry. Gels. 2021, 7, 31, doi: https://doi.org/10.3390/gels7010031.

[95] Negut I, Floroian L, Ristoscu C, Mihailescu CN, Mirza Rosca JC, Tozar T, Badea M, Grumezescu V, Hapenciuc C, Mihailescu IN. Functional bioglass – Biopolymer double nanostructure for natural antimicrobial drug extracts delivery. Nanomaterials. 2020, 10, 385, doi: https://doi.org/10.3390/nano10020385385.

[96] Sah AK, Dewangan M, Suresh PK. Potential of chitosan-based carrier for periodontal drug delivery. Colloids Surf B Biointerfaces. 2019, 178, 185–98.

[97] Basim P, Gorityala S, Kurakula M. Advances in functionalized hybrid biopolymer augmented lipid-based systems: A spotlight on their role in design of gastro retentive delivery systems. Arch Gastroenterol Res. 2021, 2, 35–47.

[98] Niu X, with 11 co-authors. Electrospun polyamide-6/chitosan nanofibers reinforced nano-hydroxyapatite/polyamide-6 composite bilayered membranes for guided bone regeneration. Carbohydr Polym. 2021, 260, 117769, doi: https://doi.org/10.1016/j.carbpol.2021.117769.

[99] Öz UC, Toptaş M, Küçüktürkmen B, Devrim B, Saka OM, Deveci MS, Bilgili H, Ünsal E, Bozkır A. Guided bone regeneration by the development of alendronate sodium loaded in-situ gel and membrane formulations. Eur J Pharm Sci. 2020, 155, 105561, doi: https://doi.org/10.1016/j.ejps.2020.105561.

[100] Lei B, Guo B, Rambhia KJ, Ma PX. Hybrid polymer biomaterials for bone tissue regeneration. Front Med. 2019, 13, 189–201.

[101] Fazeli M, Florez J, Simão R. Improvement in adhesion of cellulose fibers to the thermoplastic starch matrix by plasma treatment modification. Comp Part B: Eng. 2018, 163, 207–16.

[102] Composite; https://en.wikipedia.org/wiki/Biocomposite.

[103] Costa VC, Costa HS, Vasconcelos WL, de Magalhães Pereira M, Oréfice RL, Mansur HS. Preparation of hybrid biomaterials for bone tissue engineering. Mater Res. 2007, 10, 21–26.

[104] Flynn LE, Kimberly A. Woodhouse, burn dressing biomaterials and tissue engineering. In: Naraya R, ed. Biomedical Materials. Springer, 2021, 537–80.

[105] Chen G, Ushida T, Tateishi T. Hybrid biomaterials for tissue engineering: A preparative method for PLA or PLGA-collagen hybrid sponges. Adv Mater. 2000, 12, 455–57.

[106] Chen Y, Lee K, Kawazoe N, Yang Y, Chen G. PLGA-collagen-ECM hybrid scaffolds functionalized with biomimetic extracellular matrices secreted by mesenchymal stem cells during stepwise osteogenesis- co -adipogenesis. J Mater Chem B. 2019, doi: https://doi.org/10.1039/C9TB01959F.

[107] de Pinho ARG, Odila I, Leferink A, van Blitterswijk C, Camarero-Espinosa S, Moroni L. Hybrid polyester-hydrogel electrospun scaffolds for tissue engineering applications. Front Bioeng Biotechnol. 2017, 7, doi: https://doi.org/10.1002/adhm.201700506.

[108] Chen Y, Lee K, Chen Y, Yang Y, Kawazoe N, Chen G. Preparation of stepwise adipogenesis-mimicking ECMs-deposited PLGA-collagen hybrid meshes and their influence on adipogenic differentiation of hMSCs. ACS Biomater Sci & Eng. 2019, 5, 6099–108.

[109] Chen G, Kawazoe N. Porous Scaffolds for regeneration of cartilage, bone and osteochondral tissue. Osteochondral Tissue Eng. 2018, 8, 171–91.

[110] Ahadian S, Civitarese R, Bannerman D, Mohammadi MH, Lu R, Wang E, Davenport-Huyer L, Lai B, Zhang B, Zhao Y, Mandla S, Korolj A, Radisic M. Organ-on-a-chip platforms: A convergence of advanced materials, cells, and microscale technologies. Adv Healthcare Mater. 2017, 7, doi: https://doi.org/10.1002/adhm.201700506.

[111] Khandelwal S. Biomaterials and its Applications; https://www.slideshare.net/khsaransh/biomaterial-and-its-applications.

Chapter 3
Interfacial reaction

3.1 Surface and interface

There are still some important issues to be considered before discussing implantable biomaterials. Metals and alloys have been commonly used for a host of dental and medical implant applications, and most of these applications are regulated by the Food and Drug Administration as class II (moderate-risk) devices and cleared for marketing through the premarket notification (519(k)) pathway after demonstration of the substantial equivalence to a legally marketed class II device or granted a de novo request, or as class III (higher risk) devices approved through the premarket approval application process after demonstration of a reasonable assurance of safety and effectiveness [1]. Safety and effectiveness should be directly or indirectly related to interactions of implant biomaterials and host vital physiological environments.

The longevity, safety, reliability, functionality, and structural integrity of biomedical materials are governed by surface and interface phenomena. In both medical and dental implants, the surface of the implant receiving vital soft or hard (host) tissue should be in contact with nonvital (foreign) implant biomaterial. The interaction between these two materials takes place at their interface and in complicated biological environment as well as biomechanistic conditions. The surface of a material defines the boundary of an object and is not just a free end of a substance, but it is a contact and boundary zone with other substances (either in gaseous,

https://doi.org/10.1515/9783110740134-004

liquid, or solid). A physical system that comprises a homogeneous component such as solid, liquid, or gas and is clearly distinguishable from each other is called a phase, and a boundary at which two or three of these individual phases are in contact is called an interface. Surface and interface reactions include reactions with organic or inorganic materials, vital or nonvital species, and hostile or friendly environments. Surface activities may vary from mechanical actions (fatigue crack initiation and propagation, stress intensification, etc.), chemical action (discoloration, tarnishing, contamination, corrosion, oxidation, etc.), mechanochemical action (corrosion fatigue, stress corrosion cracking, etc.), thermomechanical action (thermal fatigue and creep), tribological and biotribological actions (wear and wear debris toxicity, friction, etc.) to physical and biophysical actions (surface contact and adhesion, adsorption, absorption, diffusion, cellular attachment, cell proliferation and differentiation, etc.). Consequently, the properties of the surface determine the interaction of the second material [2, 3]. The surface of an object determines various properties and reactivity (especially chemical); for example, in large objects with a small surface-area-to-volume ratio, the physical and chemical properties are primarily defined by the bulk, while in small objects with a large surface-area-to-volume ratio, the properties are strongly influenced by the surface. As Patterson et al. [4] pointed out that under macroscopic observation, human tissue may appear to be chemically inert; however, at the molecular level, human tissue is a dynamic environment for immersed metals, and the metal in the living tissue is prone to corrosion. The interaction of the foreign body with the tissue involves the redox reaction (an electron exchange) at the interface, the hydrolysis (a proton exchange) of oxidehydrates as products of corrosion, and the formation of metal–organic complexes in the electrolyte [5].

3.2 Environmental chemistry and its corrosiveness

Any material applicable for surgical implantation gives rise to a wide spectrum of biochemical reactions within the body. This spectrum can be roughly divided into three categories: (i) almost inert materials, with minimal chemical reactivity; (ii) totally resorbable materials, with possible dissolution into metabolic constituents; and (iii) materials with controlled-activity surfaces. Biomaterials (particularly metallic biomaterials), in either temporary contact to a human body or semipermanently placed inside the body, should face the hostile electrolytic environment of the human body. Such corrosive environment can include blood and other constituents of the body fluid which encompass several constituents like water, sodium, chlorine, proteins, plasma, and amino acids along with mucin in the case of saliva [6]. As we will see later, the aqueous solution in the human body consists of various anions such as chloride, phosphate, and bicarbonate ions; cations like Na^+, K^+, Ca^{2+}, and Mg^{2+}; organic substances of low-molecular-weight species as well as relatively high-molecular-weight polymeric components; and dissolved oxygen. Such biological molecules upset the equilibrium of the corrosion reactions of the implant by consuming the products due to anodic or cathodic reaction [6]. As one of the prognostic studies of placed metallic implants, Mohanty et al. [7] reported that the tolerable corrosion rate for metallic implant systems should be about 0.01 milli-inch per year: mpy (or 2.5×10^{-4} mm/year). Orthodontic bracket material (2205 duplex stainless steel (SS)) was subjected to corrosion test in 37 °C, 0.9 wt% sodium chloride solution [8], and it was reported that the corrosion rate of 2205 was 0.416 mpy, whereas 316L exhibited 0.647 mpy; both mpy values are much higher than the acceptable levels, although it can be simply compared since (i) the former value is under the in vivo complicated conditions with unknown factors while the latter data were obtained in controlled in vitro corrosion test. For reference, on the other hand, most standard corrosion tables consider corrosion rates lower than 20 mpy acceptable on the premise that most corroding components have been designed with a logical corrosion allowance that will provide the desired service life [9]. Changes in the pH values also influence the corrosion. Though the pH value of the human body is normally maintained at 7.0, this value changes from 3 to 9 due to several causes such as accidents, imbalance in the biological system due to diseases, infections, and other factors; and after surgery, the pH value near the implant varies typically from 5.3 to 5.6 [6]. Dental implants can experience a wide range of pH change because of a patient's diet. Variation in pH value can also be noticed by daily intake of foods and beverages [10] (see Figure 3.1).

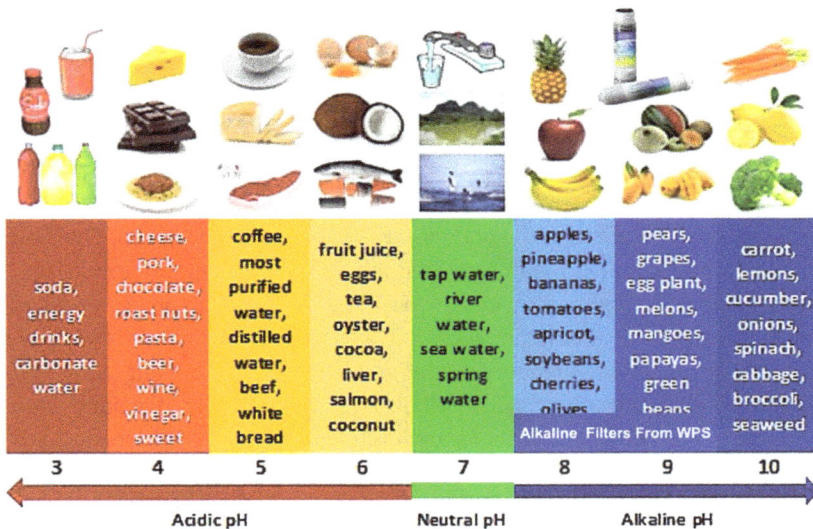

Figure 3.1: Various acidity and alkalinity of daily intakes of foods and beverages [10].

3.2.1 Saliva

Saliva is an extracellular fluid (ECF) and a thick, colorless, opalescent, complex bodily fluid that is constantly present in the mouth of humans and other vertebrates. Saliva is an ocean of 90% of water, ions, nonelectrolytes, amino acids, proteins, carbohydrates, and lipids which are flowing in waves against and into the dental surfaces. As saliva, which is produced and secreted by salivary glands – lips, tongue, and palate – 1–2 L daily into the human mouth, circulates in the mouth cavity, it picks up food debris, bacterial cells, and white blood cells [11–14].

There are five major functions of saliva: lubrication and protection, buffering, maintaining tooth integrity, antibacterial activity, and taste and digestion [11, 13–16], as shown in Figure 3.2 [17].

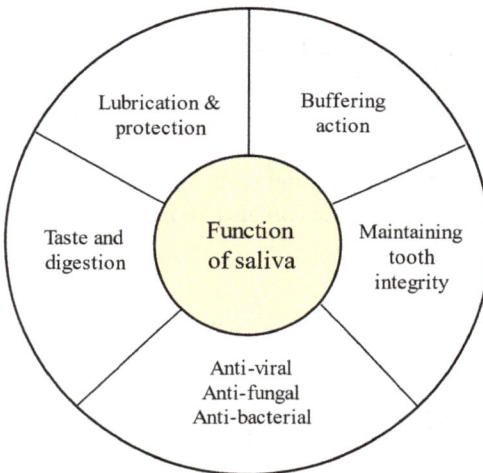

Figure 3.2: Five major biofunctions of saliva [17].

Lubrication and protection preserve the oral tissues against irritating agents. The main lubricating compounds in saliva are mucins that are complex protein molecules [18]. The lubricating effects also help mastication, speech, and swallowing processes. In tribological action during the orthodontic mechanotherapy, it was reported that friction between brackets and archwires reduced under an existence of lubricant saliva [19]. Buffering action is the second role of saliva. Whole saliva contains three major components contributing to the buffer capacity including bicarbonate, phosphate, and protein buffer [20]. Enzymes such as carbonic anhydrase also participate in controlling the pH of saliva and the buffer capacity allows saliva to maintain a relatively constant pH (normal pH range 6.7–7.3) against the acids produced by microorganisms and consumed through the diet [21, 22]. The maintaining tooth integrity balances a demineralization and a remineralization process. Demineralized processes happen when acids diffuse into the enamel resulting in crystalline dissolution and occur in a critical pH range (pH 5–5.5) for the development of caries. Dissolved mineral diffuses from the tooth into the saliva. The mineral loss may be recovered to the tooth structure from ions dissolved in the saliva by a remineralization process [23, 24]. The antibacterial activity is another important function of saliva. Saliva contains several immunologic and nonimmunologic proteins with antibacterial properties. Among the immunologic components of saliva, secretory immunoglobulin A (IgA) is the largest proportion which can neutralize viruses and bacterial, and enzyme toxins [25, 26]. The fifth function of saliva is taste and digestion. For a tasting function, the main role of saliva includes the transport of taste substances to and the protection of the taste receptors [27]. The salivary enzyme amylase has an early role in the total digestion process by breaking down starch into sugars [28].

Often head and neck radiotherapy has serious and detrimental side effects on the oral cavity including the loss of salivary gland function and a persistent complaint of a dry mouth (xerostomia) [13]. Sjögren's syndrome is salivary gland dysfunction and directly causes dry mouth and dry eyes as well [29]. Thus, saliva possesses important and beneficial biofunctions that are essential to our well-being.

3.2.2 Body fluid

Body fluids (or biofluids) are liquids within the human body [30, 31]. Human beings are mostly water, ranging from about 75% of body mass in infants to about 50–60% in adult men and women, to as low as 45% in old age [32]. The percent of body water changes with development because the proportions of the body given over to each organ and to muscles, fat, bone, and other tissues change from infancy to adulthood. As shown in Figure 3.3 [32], brain and kidneys have the highest proportions of water, which composes 80–85% of their masses. In contrast, teeth have the lowest proportion of water, at 8–10%.

In the body, water moves through semipermeable membranes of cells and from one compartment of the body to another by a process called osmosis. Osmosis is basically the diffusion of water from regions of higher concentration to regions of lower concentration, along an osmotic gradient across a semipermeable membrane. As a result, water will move into and out of cells and tissues, depending on the relative concentrations of the water and solutes found at cells or tissues. An appropriate balance of solutes inside and outside of cells must be maintained to ensure normal function [32]. The composition of tissue fluid depends upon the exchanges between the cells in the biological tissue and the blood. This means that fluid composition varies between body compartments. The total body of water is divided into two fluid compartments: the intracellular fluid (ICF) compartment and the ECF compartment in a 2:1 ratio, namely, 28 (28–32) L are inside cells and 14 (14–15) L are outside cells.

The ICF of the cytosol (or cytoplasm) is the fluid found inside cells and consists mostly of water, dissolved ions, small molecules, and large, water-soluble molecules

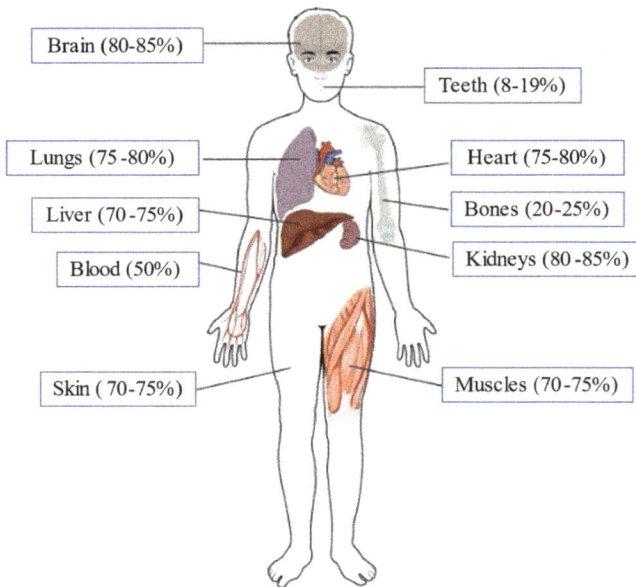

Figure 3.3: Water content of the body's organs and tissues [32].

(such as proteins). This mixture of small molecules is extraordinarily complex, as the variety of enzymes that are involved in cellular metabolism is immense. These enzymes are involved in the biochemical processes that sustain cells and activate or deactivate toxins. Most of the cytosol is water, which makes up about 70% of the total volume of a typical cell. The pH of the ICF is 7.4. The cell membrane separates cytosol from ECF but can pass through the membrane via specialized channels and pumps during passive and active transport. The concentrations of the other ions in cytosol or ICF are quite different from those in ECF. The cytosol also contains much higher amounts of charged macromolecules, such as proteins and nucleic acids, than the outside of the cell. In contrast to ECF, cytosol has a high concentration of potassium (K) ions and a low concentration of sodium (Na) ions [30–32].

The local physiochemical environment plays an important role on the corrosion performance of an implant. Local physiological factors such as pH, concentrations of ions and other small molecules, proteins, and cellular activity can influence the corrosion susceptibility of the metal [33]. Proteins at the implant site first adsorb onto the metal surface and aid in cell–metal interaction. Neutrophils and other cells of the immune system can secrete reactive oxygen species as a part of the foreign body reaction and cause a drop in local pH, which might increase corrosion susceptibility [34, 35]. Over time, as the implant integrates with the body, the local physiological environment will evolve as a part of the wound healing response [33].

The ECF usually denotes all the body fluid that is outside of the cells. The ECF has three subcompartments: interstitial fluid (or tissue fluid), transcellular fluid (which makes up only about 2.5% of the ECF), and blood plasma. The ECF compartment is divided into the interstitial fluid volume – the fluid outside both the cells and the blood vessels – and the intravascular volume (also called the vascular volume and blood plasma volume) – the fluid inside the blood vessels – in a 3:1 ratio: the interstitial fluid volume is about 12 L and the vascular volume is about 4 L. The ECF contains extracellular matrices that act as fluids of suspension for cells and molecules inside the ECF. The ECF fluid contains mainly cations and anions. The cations include Na^+, K^+, and Ca^{2+} while anions include chloride and hydrogen carbonate (HCO_3). These ions are important for water transport throughout the body [32].

3.2.3 Blood

Blood is slightly denser and is approximately three to four times more viscous than water [36–39]. Blood volume is variable but tends to be about 8% of body weight. The average adult has about 5 L of blood. Blood consists of cells that are suspended in a liquid. As with other suspensions, the components of blood can be separated by filtration. However, the most common method of separating blood is to centrifuge (spin) it. Three layers are visible in centrifuged blood. The straw-colored liquid portion, called plasma, forms at the top (~55%). The buffy coat, a thin cream-colored layer consisting of white blood cells and platelets, forms below the plasma, while red blood cells comprise the heavy bottom portion of the separated mixture (~45%). Blood consists of cellular material (99% red blood cells, with white blood cells and platelets making up the remainder), water, amino acids, proteins, carbohydrates, lipids, hormones, vitamins, electrolytes, dissolved gasses, and cellular wastes. Each red blood cell is about one-third hemoglobin, by volume. Plasma is about 92% water, with plasma proteins as the most abundant solutes. The main plasma protein groups are albumins, globulins, and fibrinogens. The primary blood gasses are oxygen, carbon dioxide, and nitrogen [40]. As shown in Figure 3.4 [40], blood consists of plasma and blood cells.

The long-term clinical success of dental implants is related to their early osseointegration. Immediately following implantation, implants are in contact with proteins and platelets from blood. The differentiation of mesenchymal stem cells will then condition the peri-implant tissue healing. Direct bone-to-implant contact is desired for a biomechanical anchoring of implants to the bone rather than the fibrous tissue encapsulation [40].

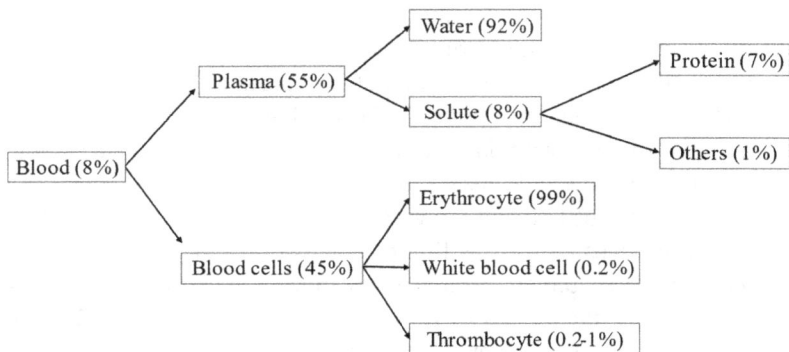

Figure 3.4: Blood components that primarily interact with the surface of dental implants [40].

3.2.4 Microbiologically induced corrosion

Proteins can bind themselves to metal ions and transport them away from the implant surface upsetting the equilibrium across the surface double layer that is formed by the electrons on the surface and excess cations in the solution. In addition, proteins that are absorbed on the surface are also found to reduce the diffusion of oxygen at certain regions and cause preferential corrosion at those regions. Hydrogen which is formed by the cathodic reaction acts as a corrosion inhibitor; however, the presence of bacteria seems to change this behavior and enhance corrosion by absorbing the hydrogen present in the vicinity of the implant [32].

Microbiology-induced corrosion (MIC, aka biocorrosion) is defined as the deterioration and degradation of metallic structures and devices by corrosion processes that occur directly or indirectly as a result of the activity of living microorganisms, which either produce aggressive metabolites, or are able to participate directly in the electrochemical reactions occurring on the metal surface. MIC occurs in aquatic habitats varying in nutrient content, temperature, stress, and pH. The oral environment of organisms, including humans, should be one of the most hospitable for MIC. Corrosion has long been thought to occur by one of three means: oxidation, dissolution, or electrochemical interaction. To this list, the newly discovered microbiological corrosion must be added [41, 42]. MIC is due to the presence of microorganisms on a metal surface which leads to changes in the rates, and sometimes also the types of the electrochemical reactions which are involved in the corrosion processes. Chang et al. [43] investigated electrochemical behaviors of commercially pure titanium (cpTi), Ti–6Al–4V, Ti–Ni, Co–Cr–Mo, 316L SS, 17–4 PH SS, and Ni–Cr in (1) sterilized Ringer's solution; (2) *S. mutans* mixed with sterilized Ringer's solution; (3) sterilized tryptic soy broth; and (4) byproduct of *S. mutans* mixed with the sterilized tryptic soy broth. It was concluded that (i) among the four electrolytes, the byproduct mixed with sterilized tryptic soy broth was the most corrosive media, leading to an increase in I_{CORR} and reduction in E_{CORR}; (ii) cpTi, Co–Cr–Mo, and SS increased their I_{CORR} significantly, but Ti–6Al–4V and Ti–Ni did not show a remarkable increase in I_{CORR}. Koh et al. [44] studied the effect of surface area ratios on bacteria galvanic corrosion of cpTi coupled with other dental alloys. cpTi was coupled with a more noble metal (type IV dental gold alloy) and a less noble metal (Ni–Cr alloy) with different surface area ratios (ranging from 4:0 to 0:4; namely, 4:0 uncouple indicates that entire surface area of a noble metal was masked, while 0:4 uncouple indicates that only cpTi surface was masked, and area ratios between 4:0 and 0:4 indicate that both surfaces were partially masked to produce different surface areas). Galvanic couples were then electrochemically tested in bacteria culture media and culture media containing bacteria byproducts. It was found that I_{CORR} (and hence, corrosion rate) profiles as a function of surface area ratio showed a straight-line trend (meaning surface area ratio independence) for Ti/Au couples, whereas the representative curve for the Ti/Ni–Cr alloy was bowl-shaped. The latter curve indicates further higher corrosion rates at both ends (larger area of either cpTi or Ni–Cr), meaning that lowest corrosion rate was exhibited at the bottom of bowl (at equal surface area ratio), supporting results obtained by Al-Ali et al. [45]. In the study [46], an inductively coupled plasma-optical emission spectrometer revealed elution of Fe, Cr, and Ni from SS appliances incubated with oral bacteria. It was reported that (i) three-dimensional laser confocal microscopy also revealed that oral bacterial culture promoted increased surface roughness and corrosion pits in SS appliances; (ii) the pH of the supernatant was lowered after coculture of appliances and oral bacteria in any combinations, but not reached at the level of depassivation pH of their metallic materials; and (iii) *Streptococcus mutans* and *Streptococcus sanguinis* which easily created biofilm on the surfaces of teeth and appliances did corrode orthodontic SS appliances. Rocher et al. [47] examined five alloys by electrochemical assays and cell culture tests with different cell types and mentioned that all tests show the high biological and electrochemical performances of Ti–6Al–4V and NiTi, and in particular a significant influence of living cells on corrosion. Bahije et al. [48] investigated the electrochemical behavior of NiTi orthodontic wires in a solution containing *S. mutans* oral bacteria. The electrochemical behavior of the alloy (NiTi) was analyzed electrochemically in Ringer's sterile artificial saliva and in artificial saliva enriched with a sterile broth and modified by addition of bacteria. It was found that (i) colonization of the metal surface by bacteria triggered a drop in the free corrosion potential; (ii) the electrochemical impedance (EIS) findings revealed no significant difference in NiTi behavior between the two media; and (iii) there was a slight difference between the two corrosion currents in favor of the bacteria-enriched solution, in which the NiTi underwent greater corrosion.

3.3 Corrosion and oxidation

3.3.1 Corrosion

As described previously, saliva and body fluid are corrosive. Once implant was placed, one of the most fundamentally important is the interaction between the surrounding physiological environment and the surface of the implant fixture body. This interaction can lead to either the failure of the implant to perform expected biofunctionality or an adverse effect on the patient resulting in the rejection of the implant by the surrounding tissue, causing the unwanted removal. Saliva is an ocean of water, ions, nonelectrolytes, amino acids, proteins, carbohydrates, and lipids. Saliva contains chloride ions and sulfurated compounds; the latter are easily metabolized into more corrosive agents by microorganisms. In some cases, blood is another aggressor. Internal fluids have a concentration of chloride ions seven times higher than that of oral fluids. Hence, the human body is not an environment that one would consider hospitable for an implanted metal alloy [49]. The human body depends on a large number of chemical reactions occurring continuously to sustain its viability. These chemical reactions produce an abundance of oxidizing agents, which create an unfriendly environment for metals and alloys. Metals implanted into this saline milieu inevitably undergo corrosion. The degradation of these metals can produce detrimental effects both locally and systemically within the human body [4, 50, 51]. As discussed later, biotribological reaction products (which are recognized as the wear debris) might cause additional corrosive environment as well as potential toxicity. In biotribological environment, an adversely synergistic effect of corrosion and tribological action is called a tribocorrosion [52]. Although titanium dental implants show very good properties, Ossowska et al. [53] pointed out that unfortunately there are still issues regarding material wear accompanied with corrosion, implant loosening, as well as biological factors – allergic reactions and inflammation leading to rejection of the implanted material. Even the most corrosion-resistant materials are not immune to the forces of nature and undergo some degree of corrosion [54].

There are two types of corrosion appearance: general corrosion and localized corrosion. With the former condition, it would be much easier to calculate the corrosion rate (in terms of mpy), resulting in easy estimation of corrosion margin. On the other hand, when the corrosion phenomenon takes place at a localized location and manner, the corrosive severity should increase (in other words, corrosion current density increases). Such localized corrosion can include pitting corrosion, crevice corrosion, galvanic corrosion, fretting corrosion, intergranular corrosion, hydrogen embrittlement, stress corrosion cracking, corrosion fatigue, and fatigue creep interaction.

3.3.2 Oxidation

Most of the metallic biomaterials (particularly, implantable materials such as Ti and its alloys, 316L-type steels, and Co–Cr alloy) are already covered with surface oxide film, which protects the substrate from further oxidation. Such protective films are known as passive films.

Passive films can be formed by in-air oxidation or electrochemically anodic treatment. Passive oxide films are TiO_2 for titanium materials, a mixture of $(Fe,Ni)O$ and $(Fe,Cr)_2O_3$ for SS and Cr_2O_3 for Co–Cr alloy [2, 3].

3.4 Biotribology

3.4.1 Biotribology

The longevity, safety, reliability, functionality, and structural integrity of biomedical materials are governed by surface and interface phenomena. Surface is not just a free end of the solid or liquid substance but is a contact zone with other substances (either in gaseous, liquid, or solid state). Surface and interface reactions include reactions with organic or inorganic materials, with vital or nonvital species, and with hostile or friendly environments. Surface activities vary from mechanical actions (fatigue crack initiation and propagation, stress intensification, etc.), chemical action (discoloration, tarnishing, contamination, corrosion, and oxidation), tribological action (wear, friction and lubricant, and wear debris toxicity) to physical action (surface contact, surface tension, diffusion, absorption and desorption, etc.). The term biotribology is a compound world of biology and tribology (tribology in biological environment). The term tribology[1] was introduced with an origin of Greek τριβω (to rub). Tribology was defined as a science and technology dealing with the friction, wear, and lubricant [2].

Tribology is the science dealing with the interaction of surfaces in tangential motion. Hence, tribology includes the nature of surfaces from both a chemical and physical point of view, including topography, the interaction of surfaces under load, and the changes in the interaction when tangential motion is introduced. Macroscopically, the interactions are manifested in the phenomena of friction and wear. Modification of the interaction through the interposition of liquid, gaseous, or solid films is known as the lubrication process. Hence, from a macroscopic point of view, tribology includes lubrication, friction, and wear. There are four major clinical

---

1 The prefix "tri" normally refers to "three" in words such as tripot, trigonometry, trilingual, and trifocal. At the same time, we just know that tribology is a science on interrelated three phenomena (friction, wear, and lubrication), so that it is easily misinterpreted that the tribology is a science dealing with the above three phenomena. But this is not true. The Greek τριβοσ does not have a prefix meaning "three." This is just a matter of coincidence.

reasons to remove the implants: (1) fracture, (2) infection, (3) wear, and (4) loosening. Among these, the removal due to the infection generally occurs in relatively early stages after implantation, while the other three incidents typically increase gradually by years because they are related to biotribological reactions. Human joints are one example of natural joints and show low wear and exceedingly low friction through efficient lubrication. Disease or accident can impair the function at a joint, and this can lead to the necessity for joint replacement [55]. Sivasankar et al. [56] mentioned that the hip is one of the largest weight-bearing joints in a body. It consists of two parts, namely, a ball (femoral head) at the top of our thighbone (femur) and it fits into a rounded socket (acetabulum) in the pelvis. A band of tissues called ligaments connect the ball to the socket and provide stability to the joint. The hip joint may get damaged due to diseases like rheumatoid arthritis, osteoarthritis, fractures, and dislocations and sometimes due to accidents too. This may cause the fracture of hip and will give the permanent handicapping to the person. There are several types of hip fractures, including (1) femoral neck fracture: pins (surgical screws) are used if the person is younger and more active, and if the broken bone is not removed much out of place. If the person is older and less active, a high strength metal device that fits into the hip socket, replacing the head of the femur (hemiarthroplasty) is needed; (2) intertrochanteric fracture: a metallic device (compression screw and side plate) holds the broken bone in place while it lets the head of the femur move normally in the hip socket. Medical implants, such as prostheses, should demonstrate very good tribological properties, as well as biological inertness. Their longevity depends significantly on the wear of the rubbing elements and must be extended as much as possible. The developments of total replacement of hip, knee, ankle, elbow, shoulder, and hand joints (and the appropriate surgical techniques) have been the major success of orthopedic surgery and would not have been possible without extensive in vitro and in vivo studies of the tribological problems, especially wear, associated with such artificial joints [56].

3.4.2 Wear debris toxicity

Toxicity can be measured by the effects on the target (organism, organ, or tissue). Because individuals typically have different levels of response to the same dose of a toxin, a population-level measure of toxicity is often used, which relates the probability of an outcome for a given individual in a population [57]. Toxicity can refer to the effect on a whole organism (such as a human, a bacterium, or a plant), or to a substructure (such as the liver, kidney, and spleen). Toxicity of a substance can be affected by many different factors, such as the route of administration, the time of exposure, the number of exposures, physical form of the toxin (solid, liquid, and gaseous), the genetic makeup of an individual, and an individual's overall health [57]. There are generally three types of toxic entities: chemical, biological, and physical.

(1) *Chemicals* include both inorganic substances, such as lead, hydrofluoric acid, and chlorine gas, as well as organic compounds such as ethyl alcohol, most medications, and poisons from living things [58, 59]. (2) *Biological* toxicity can be more complicated to measure. In a host with an intact immune system, the inherent toxicity of the organism is balanced by the host's ability to fight back; the effective toxicity is then a combination of both parts of the relationship. A similar situation is also present with other types of toxic agents. In particular, toxicity of cancer-causing agents is problematic, since for many such substances it is not certain if there is a minimal effective dose or whether the risk is just too small to see. Here, too, the possibility exists that a single cell transformed into a cancer cell is all it takes to develop the full effect [60]. (3) *Physically* toxic entities include unusual things: direct blows, concussion, sound and vibration, heat and cold, nonionizing electromagnetic radiation such as infrared and visible light, ionizing nonparticulate radiation such as X-rays and gamma rays, and particulate radiation such as alpha rays, beta rays, and cosmic rays [61].

Unlike dental implants, most orthopedic implants (and particularly, natural joint replacements such as TSR (Total Shoulder Replacement), THR (Total Hip Replacement), or TKR (Total Knee Replacement)) will be subjected to biotribological actions. Metal ions released from the implant surface are suspected of playing some contributing role in the loosening of these joint prostheses, which are substantially subjected to biotribological environments. The fresh surface will be revealed due to the formation of fraction/wear products, and there may be a chemical reaction taking place between the freshly revealed surface of implants, and the surrounding environments, and the selective dissolution of the alloy constituents into the surrounding tissues. Besides, it is generally believed that the solid powder, such as wear/friction products, particles are allergic reaction to the living tissue, so that the biological effects of wear products (debris) should be considered separately from the biocompatibility of the implants. Particulate wear debris is detected in histiocytes/macrophages of granulomatous tissues adjacent to become loose joint prostheses. Such cell–particle interactions have been simulated in vitro by challenging macrophages with particle doses according to weight percent, volume percent, and number of particles. Macrophages stimulated by wear particles are expected to synthesize numerous factors affecting events in the bone–implant interface. Wear debris from orthopedic joint implants initiates a cascade of complex cellular events that can result in aseptic loosening of the prosthesis [62]. Macrophases are cells that are mainly involved in phagocytosis and signaling and maintaining inflammation, which leads to cell damage on soft tissues and bone resorption [63, 64]. The presence of large particles provides the major stimulus for cell recruitment and granuloma formation, whereas the small particles are likely to be the main stimulus for the activation of cells which release proinflammatory products. Apoptosis can be morphologically recognized by a number of features, such as loss of specialized membrane structures, blebbing, condensation of the cytoplasm, condensation of nuclear chromatin, and splitting of the cell into a cluster of membrane-bound bodies [64, 65]. The long-term effects of metal-on-metal (MoM) arthroplasty are currently under scrutiny because of the potential biological effects of metal wear debris [66]. Information regarding metal-induced toxicity is based on a limited amount of epidemiological and experimental studies involving in vitro and in-vivo models. There are data available on the systemic effects of metal in arthroplasty patients as follows: the blood [67], the immune system [68], the liver [69], the kidney [70], the respiratory system [71], the nervous system [67, 72], the heart and vascular systems [73, 74], the musculoskeletal system [75], the endocrine system [76], the visual and auditory systems [77], the skin [78], and the reproductive system [79].

Total hip arthroplasty (THA) with MoM bearings have shown problems of biotribological wear linked to metal ion release at the local level causing an adverse reaction to metal debris and at a systemic level [80]. Figure 3.5 illustrates local and systemic toxicities, caused by wear debris products under the metal-to-metal biotribological action [80]. Metallic debris particles, released from the implant by corrosion and/or wear, cause local periprosthetic alterations called adverse reaction to metal debris (ARMD), mediated by different types of cells including macrophages, osteoclasts, giant cells, dendritic cells and lymphocytes. Tissue alterations can cause osteolysis (induced by osteoclasts and macrophages) at the bone-level, or pseudotumor formation and aseptic lymphocyte–dominated vasculitis-associated lesions (ALVAL) at the soft tissue level. Periarticular metal debris can be disseminated to different organs through circulatory and lymphatic systems causing systemic toxicity [80].

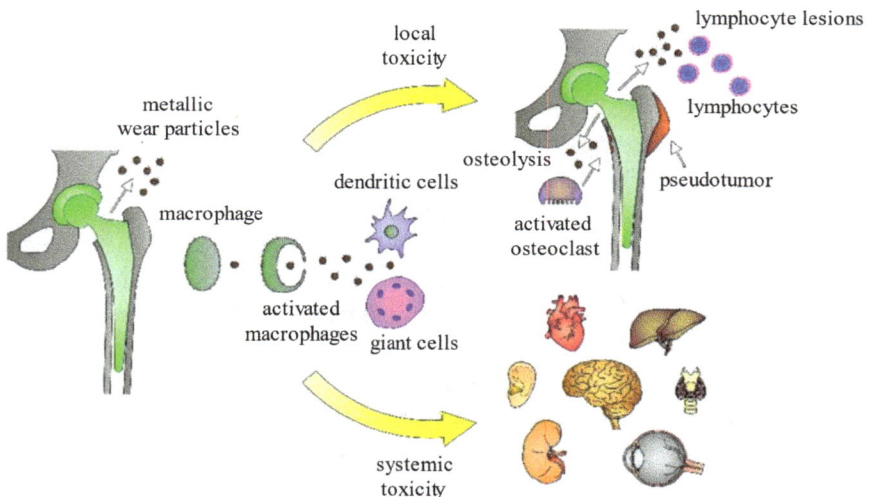

Figure 3.5: Local and systemic toxicities caused by wear debris products under the metal-on-metal biotribological action [80].

The most common traditional metals used for THA are 316L (low-carbon 18Cr–8Ni–3Mo) SS, titanium alloys (Ti–6Al–4V or Ti–6Al–7Nb), and Co–Cr–Mo alloys. Ti–6Al–7Nb has been newly developed to search for V-free and equivalent strength to Ti–4Al–6V traditional alloy [81, 82]. Co–Cr–Mo alloys are composed of 58.9–69.5 wt% Co, 27.0–30 wt% Cr, 5.0–7.0 wt% Mo, and small amount of other elements (Mn, Si, Ni, Fe, and C). These metallic alloys can be divided into two categories: high-carbon alloys (carbon content >0.20%) and low-carbon alloys (carbon content <0.08%) [83, 84]. Hence, there are a variety of metallic elements involved in biotribological action, including mainly Fe, Cr, Ni, Mo, Co, Ti, Al, V, and Nb.

In general, there are several combinations of material-to-material tribological contact and they should include MoP (metal-on-plastic), CoP (ceramic-on-plastic), CoM (ceramic-on-metal), CoC (ceramic-on-ceramic), and MoM. Merola et al. [85] compared the overall wear rate (mm^3/million cycles). For MoP combinations, wear rate of CoCr–highly cross-linked polyethylene is 6.71 and of CoCrMo–highly cross-linked polyethylene is 4.09. For CoP systems, wear rate of alumina–highly cross-linked polyethylene is 3.35 and alumina–polyethylene is 34. In CoM systems, wear rate of alumina + zirconia–CoCrMo is 0.02–0.87. In CoC systems, the wear rate of zirconia–zirconia is 0.024, while alumina–alumina has the wear rate of 0.03–0.74, alumina–alumina has 0.14, and zirconia–zirconia shows 0.18–0.20. With MoM combinations, CoCrMo–CoCrMo exhibits its wear rate of 0.11–0.60 mm^3/million cycles.

There are some quantitative data on released metallic ions (Cr, Cr, Mo, and Ti) detected in the whole blood. It was reported that for Co, 0.28 µg/L [86], 0.56 µg/L [80], or 30 µg/L [87] were detected. For Cr element, 0.1 µg/L [80], 0.43 µg/L [86], 4.6–6.5 µg/L [88], or 21 µg/L [87] were reported. As to Mo and Ti, 0.62 µg/L [86] and 1.96 µg/L [86] were reported, respectively. According to the Medicines and Healthcare Products Regulatory Agency (UK), the cutoff for metal ion levels for risks of clinical complications related to metal ion release is 7 µg/L [89]. Therefore, some data exceeds the threshold safety level.

3.4.3 Tribocorrosion

Although the MoM hip prosthesis bearings have been employed frequently, there are still remaining concerns associated with wear debris and metal ion release causing a negative response in the surrounding tissues [90–92]. Mathew et al. [93] evaluated the tribocorrosion behavior, or interplay between corrosion and wear, of a low-carbon Co–Cr–Mo alloy as a function of loading. The tribocorrosion tests were conducted by EIS and polarization resistance in (1) a linearly reciprocating alumina ball slid against the flat metal immersed in a phosphate-buffered solution and (2) the flat end of a cylindrical metal pin was pressed against an alumina ball that oscillated rotationally, using bovine calf serum as the lubricant and electrolyte. It was found that (i) the dominant wear regime for the CoCrMo alloy subjected to sliding changes from wear corrosion to mechanical wear as the contact stress increases and (ii) the proteins in the serum lubricant assist in the generation of a protective layer against corrosion during sliding. Guo et al. [94] evaluated nitrogen ion-enriched orthopedic Co–Cr–Mo alloy in tribocorrosion environment. It was concluded that (i) nitrogen implantation reduced the friction coefficient from 0.35 to 0.15 and (ii) the nitrogen implantation induces a novel composite microstructure of the nanocrystalline CrN embedded inside amorphous Co–Cr–Mo matrix in the implanted layer, which enhances hardness, corrosion, and tribocorrosion properties.

In many biomedical applications of NiTi alloys, the tribocorrosion properties of these alloys can be of critical concern. Kosec et al. [95] studied the electrochemical and tribocorrosion properties of superelastic NiTi sheet and orthodontic archwire, taking into account their microstructures and the effect of different surface finishes. In the case of the electrochemical tests, samples were tested in artificial saliva, whereas in the tribocorrosion tests the experiments were performed in ambient air, distilled water, and artificial saliva, the latter as a corrosive medium. In these tests, the total wear rate of the alloy samples was determined, together with the corresponding chemical and tribological contributions. It was mentioned that the microstructure of the investigated alloys had a significant effect on the measured electrochemical and tribocorrosion properties. Even though NiTi and SS are the most commonly used alloys for orthodontic treatments and both are known to be resistant to corrosion, there are circumstances that can lead to undesired situations, like localized types of corrosion attack, wear during sliding of an archwire through brackets, and breakdowns due to iatrogenic causes. Močnik et al. [96] analyzed the influence of environmental effects on the corrosion and tribocorrosion properties of NiTi and SS dental alloys. The effects of pH and fluorides on the electrochemical properties were studied using the cyclic potentiodynamic technique. The migration of ions from the alloy into saliva during exposure to saliva with and without the presence of wear was analyzed using inductively coupled plasma mass spectrometry analyses. Auger spectroscopy was used to study the formation of a passive oxide layer on different dental alloys. It was reported that (i) lowering the pH preferentially affects the corrosion susceptibility of NiTi alloys, whereas SS dental archwires are prone to local types of corrosion; (ii) the NiTi alloy is not affected by smaller increases of fluoride ions up to 0.024 M, while at 0.076 M (simulating the use of toothpaste) the properties are affected; (iii) a leaching test during wear-assisted corrosion showed that the concentrations of Ni ions released into the saliva exceeded the limit value of 0.5 μg/cm^2/week; (iv) the oxide films on the NiTi and SS alloys after the tribocorrosion experiment were thicker than those exposed to saliva only. Xue et al. [97] investigated the tribocorrosion behavior of a biomedical NiTi alloy in Ringer's simulated body fluid through friction experimentation, electrochemical testing, and abrasion mapping. It was mentioned that (i) due to the wear accelerated corrosion, the corrosion potential of the NiTi alloy shifted negatively and the corrosion current density increased by an order of magnitude, compared to the static electrochemical corrosion; (ii) the applied load effect on the wear rates of the NiTi alloy was higher than that of the particle concentration; (iii) the wear rates were decreased with the applied load increased, whereas the wear rates were decreased with the increase of abrasive particle concentrations; (iv) when the applied load was 1.5 N, the particle concentration was 0.03 g/cm^3, while the lowest wear rate of 9.47×10^{-5} mm^3/Nm was acquired under the corrosion wear conditions; and (v) the scanning electron microscopic wear morphology showed that the two-body abrasive wear is the dominant wear mechanism, suggesting that the wear contribution on the material loss exceeded the corrosion contribution, signifying the wear responsibility for the material loss and the corrosion wear regime was mechanical abrasion dominant with corrosion.

Oral rehabilitation devices are susceptible to biotribocorrosion phenomena in the oral environment due to the synergism of wear, chemical, biochemical, and microbiological processes [98]. Titanium-based implants are exposed to wear and corrosion challenges in the oral environment since the implantation and along the lifetime service. In addition, wear corrosion factors such as cyclic loads, micromovements, oral biofilm, and decontamination methods are also associated with dental implant degradation [2]. These environmental conditions to which dental implants are submitted lead to the release of Ti particles and ions with cytotoxic and harmful effects on peri-implant surrounding tissues [98]. Rocha et al. [99] investigated the tribocorrosion behavior of cpTi in contact with artificial saliva solutions under a reciprocating sliding geometry with movement amplitudes ranging from 200 m (fretting) to 6 mm (sliding wear) and normal loads between 2 and 10 Nm and varying pH value between 4 and 7. It was mentioned that material behavior is strongly influenced by the pH of the solution, and the acidification of the solution improves the electrochemical response of the material. Ribeiro et al. [100] investigated the tribocorrosion behavior of titanium grade 2 in reciprocating sliding conditions in contact with artificial saliva solutions. To reproduce the oral environment around the dental implant, some additives (citric acid, anodic, cathodic, and organic inhibitors) were added to simple artificial saliva constituted mainly by NaCl and KCl and with a pH between 5 and 7. It was found that titanium grade 2 in artificial saliva solution with citric acid had the highest weight loss. Although it is not directly related to tribocorrosion, Oshida [101] conducted electrochemical corrosion tests on all four grades (I, II, III, and IV) of cpTi in the 1% physiological saline water at 37 °C. All four grades of Ti materials were collected from six different material suppliers. As shown in Figure 3.6, it was found that cpTi grade III showed the highest corrosion rate, independent of material suppliers. Although the use of dental implants is growing rapidly for the last few decades and Ti-based dental implants are a commonly used prosthetic structure in dentistry, Barão et al. [102] felt that there should be a need to investigate how the acoustic emission (AE) technique can predict tribocorrosion processes in cpTi and TiZr alloys. Tribocorrosion tests were performed under potentiostatic conditions and AE detection system associated with it captures AE data. Current evolution and friction coefficient data obtained from the potentiostatic evaluations were compared with AE absolute energy showcased the same data interpretation of tribocorrosion characteristics. Other AE data such as duration, count, and amplitude matched more closely with other potentiostatic corrosion evaluations and delivered more promising results in the detection of tribocorrosion. Hence, AE can be considered as a tool for predicting tribocorrosion in dental implants. It was mentioned that experimental results revealed Ti_5Zr as one of the most appropriate dental implant materials while exposing $Ti_{10}Zr$'s lower effectiveness to withstand in the simulated oral environment.

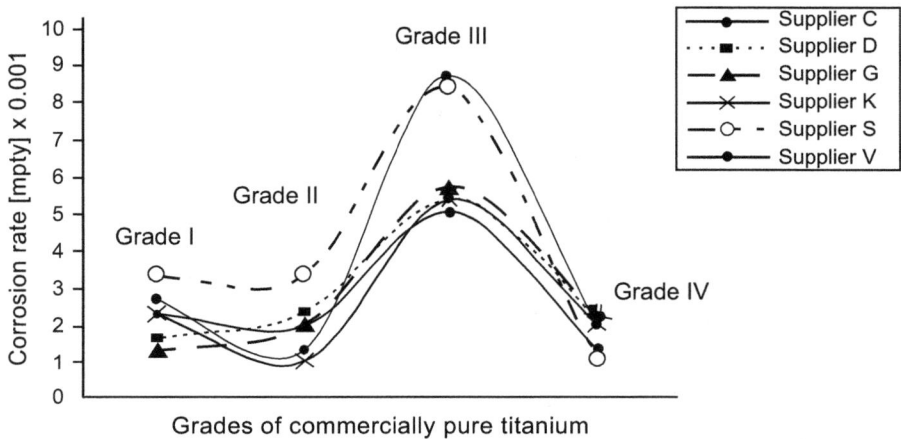

Figure 3.6: Comparison of corrosion rates among four cpTi grades [101].

As Dini et al. [98] pointed out, the mechanical (applied load, frequency, stroke distance, and number of cycles) and electrochemical (solution composition, concentration of anions, and pH) test conditions are determinants for materials' tribocorrosion performance so that there are various parameters involved in tribocorrosion phenomenon. Accordingly, appropriate standardization of testing method and evaluation methodology should be established.

References

[1]   FDA. Biological responses to metal implants. 2019; https://www.fda.gov/media/131150/ download.
[2]   Oshida Y. Bioscience and Bioengineering of Titanium Materials, 1st ed. 2nd ed. (2013), London, UK: Elsevier, 2007.
[3]   Oshida Y. Surface Engineering and Technology for Biomedical Implants. USA: Momentum Press, 2014.
[4]   Patterson SP, Daffner RH, Gallo RA. Electrochemical corrosion of metal implants. Am J Roentgenel. 2005, 184, 1219–22.

[5]   Steinemann SG. Metal implants and surface reactions. Injury. 1996, 27, S/C16–22.

[6]   Manivasagam G, Dhinasekaran D, Rajamanickam A. Biomedical implants: Corrosion and its prevention – A review. Recent Pat Corros Sci. 2010, 2, 40–54.

[7]   Mohanty M, Baby S, Menon KV. Spinal fixation device: A 6-year postimplantation study. J Biomater Appl. 2003, 18, 109–21.

[8]   Platt JA, Guzman A, Zuccari A, Thornburg DW, Rhodes BF, Oshida Y, Moore BK. Corrosion behavior of 2205 duplex stainless steel. Am J Orthod Dentofacial Orthop. 1997, 112, 69–79.

[9]   ENG-TIPS. Standard Corrosion Rates. https://www.eng-tips.com/viewthread.cfm?qid=2303.

[10] How to raise pH and alkalinity? Alkaline Water Filters. https://www.h2opurificationsystems.com/pure-water/how-to-raise-ph-and-alkalinity-alkaline-water-filters/.

[11] Maddu N. Functions of Saliva, Saliva and Salivary Diagnostics, Sridharan Gokul, IntechOpen. 2019; https://www.intechopen.com/books/saliva-and-salivary-diagnostics/functions-of-saliva.

[12] Saliva. https://en.wikipedia.org/wiki/Saliva.

[13] Tiwari M. Science behind human saliva. J Nat Sci Biol Med. 2011, 2, 53–58.

[14] Ngamchuea K, Chaisiwamongkhol K, Batchelor-McAuley C, Compton RG. Chemical analysis in saliva and the search for salivary biomarkers – A tutorial review. Analyst. 2018, 143, 81–99.

[15] Dawes C, with 14 co-authors. The functions of human saliva: A review sponsored by the World Workshop on Oral Medicine VI. Arch Oral Biol. 2015, 60, 863–74.

[16] Somaiya S. Functions of Saliva/Human Physiology. https://www.biologydiscussion.com/human-physiology/digestive-system/functions-of-saliva-human-physiology/62576.

[17] Kaplan MD, Baum BJ. The functions of saliva. Dysphagia. 1993, 8, 225–29.

[18] Slomiany BL, Murty VLN, Piotrowski J, Slomiany A. Salivary mucins in oral mucosal defense. Gen Pharmacol. 1996, 27, 761–71.

[19] Almeida FAC, Almeida APCPSC, Amaral FLB, Basting RT, França FMG, Turssi CP. Lubricating conditions: Effects on friction between orthodontic brackets and archwires with different cross-sections. Dental Press J Orthod. 2019, 24, 66–72.

[20] Bardow A, Moe D, Nyvad B, Nauntofte B. The buffer capacity and buffer systems of human whole saliva measured without loss of CO_2. Arch Oral Biol. 2000, 45, 1–12.

[21] Kivelä J, Parkkila S, Parkkila A-K, Leinonen J, Rajaniemi H. Salivary carbonic anhydrase isoenzyme VI. J Physiol. 1999, 520, 315–20.

[22] Islas-Granillo H, Borges-Yañez SA, Medina-Solís CE, Galan-Vidal CA, Navarrete-Hernández JJ, Escoffié-Ramirez M, Maupomé G. Salivary parameters (salivary flow, pH and buffering capacity) in stimulated saliva of Mexican elders 60 years old and older. West Indian Med J. 2014, 63, 758–65.

[23] Farooq I, Bugshan A. The role of salivary contents and modern technologies in the remineralization of dental enamel: A narrative review. Version 3. F1000Res. 2021, 9, 71, doi: 10.12688/f1000research.22499.3.

[24] Furgeson D, Pitts EI. Saliva, remineralization and dental caries. Decis Dent. 2019, https://decisionsindentistry.com/article/saliva-remineralization-and-dental-caries/.

[25] Tenovuo J. Antimicrobial function of human saliva – How important is it for oral health?. Acta Odontol Scand. 1998, 56, 250–56.

[26] Brandtzaeg P. Secretory immunity with special reference to the oral cavity. J Oral Microbiol. 2013, 5, doi: 10.3402/jom.v5i0.20401.

[27] Mese H, Matsuo R. Salivary secretion, taste and hyposalivation. J Oral Rehabil. 2007, 34, 711–23.

[28] Moss SJ. Clinical implications of recent advances in salivary research. J Esthet Restor Dent. 1995, 7, 197–203.

[29]   NIH. Sjögren's syndrome. National Institute of Dental and Craniofacial Research. 2018; https://www.nidcr.nih.gov/health-info/sjogrens-syndrome/more-info.

[30]   Body fluid. https://en.wikipedia.org/wiki/Body_fluid.

[31]   Body Fluids. https://courses.lumenlearning.com/boundless-ap/chapter/body-fluids /https://courses.lumenlearning.com/boundless-ap/chapter/body-fluids/.

[32]   Body Fluids and Fluid Compartments. https://open.oregonstate.education/aandp/chapter/ 26-1-body-fluids-and-fluid-compartments/.

[33]   Pound BG. The use of electrochemical techniques to evaluate the corrosion performance of metallic biomedical materials and devices. 2019, 107, 1189–98; https://onlinelibrary.wiley. com/doi/abs/10.1002/jbm.b.34212.

[34]   Asawang K, Mcbridge T, John V, Oshida Y, Kowolik MJ. Clinical instrumentation of titanium surfaces alters neutrophil oxidative responses. Jour Dent Res. 2002, 81, A–274 Abstract #2142.

[35]   De Poi RP, Kowolik M, Oshida Y, El Kholy K. The oxidative response of human monocytes to surface modified commercially pure titanium. Front Immunol. 2021, 12, 618002, doi: 10.3389/fimmu.2021.618002.

[36]   Helmenstine AM. What Is the Chemical Composition of Blood? 2019; https://www.thoughtco. com/volume-chemical-composition-of-blood-601962.

[37]   Cooke S. What elements make up the chemical compound of blood? 2016; https://socratic. org/questions/what-elements-make-up-the-chemical-compound-of-blood.

[38]   The Chemistry of Blood Types; https://www.chemistryislife.com/the-chemistry-of-blood-types.

[39]   Blood; https://en.wikipedia.org/wiki/Blood.

[40]   Lavenus S, Louarn G, Layrolle P. Nanotechnology and dental implants. Int J Biomater. 2010, 915327, doi: 10.1155/2010/915327.

[41]   Tiller AK. Aspects of microbial corrosion. In: Parkins RN, ed. Corrosion Processes. London: Applied Science Pub, 1982, 115–19.

[42]   Matasa CG. Stainless steels and direct-bonding brackets, III: Microbiological properties. Inf Orthod Kierferorthop. 1993, 25, 269–71.

[43]   Chang J-C, Oshida Y, Gregory RL, Andres CJ, Barco TM, Brown DT. Electro-chemical study on microbiology-related corrosion of metallic dental materials. J Biomed Mater Eng. 2003, 13, 281–95.

[44]   Koh I-W. Effects of bacteria-induced corrosion on galvanic couples of commercially pure titanium with other dental alloys. Indiana University Mater Degree Thesis. 2003.

[45]   Al-Ali S, Oshida Y, Andres CJ, Barco MT, Brown DT, Hovijitra S, Ito M, Nagasawa S, Yoshida T. Effects of coupling methods on galvanic corrosion behavior of commercially pure titanium with dental precious alloys. J Biomed Mater Eng. 2005, 15, 307–16.

[46]   Kameda T, Oda H, Ohkuma K, Sano N, Batbayar N, Terashima Y, Sato S, Terada K. Microbiologically influenced corrosion of orthodontic metallic appliances. Dent Mater J. 2014, 33, 187–95.

[47]   Rocher P, Medawar L, Hornez J-C, Traisnel M, Breme J, Hildebrand HF. Biocorrosion and cytocompatibility assessment of NiTi shape memory alloys. Scr Mater. 2004, 50, 255–60.

[48]   Bahije L, Benyahia H, El Hamzaoui S, Ebn Touhami M, Bengueddour R, Rerhrhaye W, Abdallaoui F, Zaoui F. Behavior of NiTi in the presence of oral bacteria: Corrosion by Streptococcus mutans. Int Orthod. 2011, 9, 110–19.

[49]   Hansen DC. Metal corrosion in the human body: The ultimate bio-corrosion scenario. The Electrochemical Society Interface. 2008; https://www.electrochem.org/dl/interface/sum/ sum08/su08_p31-34.pdf.

[50] Kruger J. Passivity of metals: A materials science perspective. Int Mater Rev. 1988, 33, 113–30, doi: https://doi.org/10.1179/imr.1988.33.1.113.

[51] Jacobs JJ, Gilbert JL, Urban RM. Current concepts review: Corrosion of metal orthopaedic implants. J Bone Joint Surg. 1998, 80, 268–82.

[52] Oshida Y. Magnesium Materials. De Gruyter Pub., 2021.

[53] Ossowska A, Zieliński A. The mechanisms of degradation of titanium dental implants. Coatings. 2020, 10, 836, doi: 10.3390/coatings10090836.

[54] Kamachimudali U, Sridhar TM, Raj B. Corrosion of bio implants. Sadhana. 2003, 28, 601–37.

[55] Dumbleton JH. Tribology Series 3: Tribology of Natural and Artificial Joints. New York: Elsevier, 1981.

[56] Sivasankar M, Arunkumar S, Bakkiyaraj V, Muruganandam A, Sathishkumar S. A review on total hip replacement. Int Res J Adv Eng Technol. 2016, 9, 589–642.

[57] Venugopal B, Luckey TD. Metal Toxicity in Mammals. NY: Plenum Press, 1978.

[58] Lemons JE, Lucas LC, Johansson B. Intraoral corrosion resulting from coupling dental implants and restorative metallic systems. Implant Dent. 1992, 1, 107–12.

[59] Zitter H, Plenk H. The electrochemical behavior of metallic implant materials as an indicator of their biocompatibility. J Biomed Mater Res. 1987, 21, 881–96.

[60] Yang RS, Tsai KS, Liu SH. Titanium implants enhance pulmonary nitric oxide production and lung injury in rats exposed to endotoxin. J Biomed Mater Res. 2004, 69A, 561–66.

[61] Toxicity; http://en.wikipedia.org/wiki/Toxicity.

[62] Stea S, Visentin M, Granchi D, Cenni E, Ciapetti G, Sudanese A, Toni A. Apoptosis in peri-implant tissue. Biomaterials. 2000, 21, 1393–98.

[63] Chiba J, Rubash H, Kim KJ, Iwaki Y. The characterization of cytokines in the interface tissue obtained from failed cementless total hip arthroplasty with and without femoral osteolysis. Clin Orthop Rel Res. 1994, 300, 304–12.

[64] Perry M, Murtuza FY, Ponsford FM, Elson CJ, Atkins RM, Learmonth D. Properties of tissue from around cemented joint implants with erosive and/or linear osteolysis. Arthroplasty. 1997, 12, 670–76.

[65] Ingham E, Fisher J. The role of macrophages in osteolysis of total hip joint replacement. Biomaterials. 2005, 26, 1271–86.

[66] Keegan GM, Learmonth ID, Case CP. Orthopaedic metals and their potential toxicity in the arthroplasty patient: A review of current knowledge and future strategies. J Bone Joint Surg. 2007, 89B, doi: https://doi.org/10.1302/0301-620X.89B5.18903.

[67] Nayak P. Aluminium: Impacts and disease. Environ Res. 2002, 89, 101–15.

[68] Hart AJ, Hester T, Sinclair K, Powell JJ, Goodship AE, Pele L, Fersht NL, Skinner J. The association between metal ions from his resurfacing and reduced T-cell counts. J Bone Joint Surg, Br. 2006, 88B, 449–54.

[69] Kametani K, Nagata T. Quantitative elemental analysis on aluminium accumulation by HVTEM-EDX in liver tissues of mice orally administered with aluminium chloride. Med Mol Morphol. 2006, 39, 97–105.

[70] Oliveira H, Santos TM, Ramalho-Santos J, De Lourdes Pereira M. Histopathological effects of hexavalent chromium in mouse kidney. Bull Environ Contam Toxicol. 2006, 76, 977–83.

[71] Antonini J, Lewis AB, Roberts JR, Whaley DA. Pulmonary effects of welding fumes: Review of worker and experimental animal studies. Am J Indust Med. 2003, 43, 350–60.

[72] Garcia GB, Biancardi M, Quiroga A. Vanadium (V)-induced neurotoxicity in the rat central nervous system: A histo-immunohistochemical study. Drug Chem Toxicol. 2005, 28, 329–44.

[73] Linna A, Oksa P, Groundstroem K, Halkosaari M, Palmroos P, Huikko S, Uitti J. Exposure to cobalt in the production of cobalt and cobalt compounds and its effects on the heart. Occup Environ Med. 2004, 61, 877–85.

[74] Lippmann M, Ito K, Hwang JS, Maciejczyk P, Chen LC. Cardiovascular effects of nickel in ambient air. Environ Health Perspect. 2006, 114, 1662–69.
[75] Vermes C, Glant TT, Hallab NJ, Fritz EA, Roebuck KA, Jacobs JJ. The potential role of the osteoblast in the development of periprosthetic osteolysis: Review of in vitro osteoblast responses to wear debris, corrosion products, and cytokines and growth factors. J Arthroplasty. 2001, 16, 95–100.
[76] Darbre PD. Metalloestrogens: An emerging class of inorganic xenoestrogens with potential to add to the oestrogenic burden of the human breast. J Appl Toxicol. 2006, 26, 191–97.
[77] Steens W, von Foerster G, Katzer A. Severe cobalt poisoning with loss of sight after ceramic-metal pairing in a hip: A case report. Acta Orthop. 2006, 77, 830–32.
[78] Hallab N, Mikecz K, Jacobs J. Metal sensitivity in patients with orthopaedic implants. J Bone Joint Surg, Am. 2001, 83A, 428–36.
[79] Aruldhas M, Subramaniam S, Sekar P, Vengatesh G, Chandrahasan G, Govindarajulu P, Akbasha MA. Chronic chromium exposure-induced changes in testicular histoarchitecture are associated with oxidative stress: Study in a non-human primate (*Macaca radiata* Geoffroy). Human Reprod. 2005, 20, 2801–13.
[80] Pozzuoli A, Berizzi A, Crimi A, Belluzzi E, Frigo AC, De Conti G, Nicolli A, Trevisan A, Biz C, Ruggieri P. Metal ion release, clinical and radiological outcomes in large diameter metal-on-metal total hip arthroplasty at long-term follow-up. Diagnostics (Basel). 2020, 10, 941, doi: 10.3390/diagnostics10110941.
[81] Chlebus E, Kuźnicka B, Kurzynowski T, Dybała B. Microstructure and mechanical behaviour of Ti-6Al-7Nb alloy produced by selective laser melting. Mater Charact. 2011, 62, 488–95.
[82] Ti-6Al-7Nb; https://en.wikipedia.org/wiki/Ti-6Al-7Nb#:~:text=Ti%2D6Al%2D7Nb%20(UNS,a%20material%20for%20hip%20prostheses.
[83] Affatato S, Traina F, Ruggeri O, Toni A. Wear of metal-on-metal hip bearings: Metallurgical considerations after hip simulator studies. Int J Artif Organs. 2011, 34, 1155–64.
[84] Ihaddadene R, Affatato S, Zavalloni M, Bouzid S, Viceconti M. Carbon composition effects on wear behaviour and wear mechanisms of metal-on-metal hip prosthesis. Comput Methods Biomech Biomed Eng. 2011, 14, 33–34.
[85] Merola M, Affatato S. Materials for hip prostheses: A review of wear and loading considerations. Materials (Basel). 2019, 12, 495, doi: 10.3390/ma12030495.
[86] Reiner T, Sorbi R, Müller M, Nees T, Kretzer JP, Rickert M, Moradi B. Blood metal ion release after primary total knee arthroplasty: A prospective study. Orthop Surg. 2020, 12, 396–403.
[87] Nicolli A, Trevisan A, Bortoletti I, Pozzuoli A, Ruggieri P, Martinelli A, Gambalunga A, Carrieri M. Biological monitoring of metal ions released from hip prostheses. Int J Environ Res Public Health. 2020, 17, 3223, doi: 10.3390/ijerph17093223.
[88] Lhotka C, Szekeres T, Steffan I, Zhuber K, Zweymüller K. Four-year study of cobalt and chromium blood levels in patients managed with two different metal-on-metal total hip replacements. J Orthop Res. 2003, 21, 189–95.
[89] MHRA Medical Device Alert: MDA/2017/018: All Metal-on-Metal (MoM) hip Replacements – Updated Advice for Follow-up of Patients. 2020; https://www.gov.uk/drug-device-alerts/all-metal-on-metal-mom-hip-replacements-updated-advice-for-follow-up-of-patients.
[90] Landolt D, Mischler S, Stemp M. Electrochemical methods in tribocorrosion: A critical appraisal. Electrochim Acta. 2001, 46, 3913–29.
[91] Mathew MT, Pai PS, Pourzal R, Fischer A, Wimmer MA. Significance of tribocorrosion in biomedical applications: Overview and current status. Adv Tribol. 2009, doi: https://doi.org/10.1155/2009/250986.
[92] Oshida Y, Tomizaga T. NiTi Materials. De Gruyter Pub., 2020.

[93] Mathew MT, Runa MJ, Laurent M, Jacobs JJ, Rocha LA, Wimmer MA. Tribocorrosion behavior of CoCrMo alloy for hip prosthesis as a function of loads: A comparison between two testing systems. Wear. 2011, 271, 1210–19.

[94] Guo Z, Pang X, Yan Y, Gao K, Volinsky AA, Zhang T-Y. CoCrMo alloy for orthopedic implant application enhanced corrosion and tribocorrosion properties by nitrogen ion implantation. Appl Surf Sci. 2015, 347, 23–34.

[95] Kosec T, Močnik P, Legat A. The tribocorrosion behaviour of NiTi alloy. Appl Surf Sci. 2013, 288, 727–35.

[96] Močnik P, Kosec T, Kovač J, Bizjak M. The effect of pH, fluoride and tribocorrosion on the surface properties of dental archwires. Mater Sci Eng C Mater Biol Appl. 2017, 78, 682–89.

[97] Xue Y, Hu Y, Wang Z. Tribocorrosion behavior of NiTi alloy as orthopedic implants in Ringer's simulated body fluid. Biomed Phys Eng Express. 2019, 5, 045002, https://iopscience.iop.org/article/10.1088/2057-1976/ab1db0/pdf.

[98] Dini C, Costa RC, Sukotjo C, Takoudis CG, Mathew MT, Barão VAR. Progression of bio-tribocorrosion in implant dentistry. Front Mech Eng. 2020, doi: https://doi.org/10.3389/fmech.2020.00001.

[99] Rocha LA, Ribeiro AR, Vieira AC, Ariza E, Gomes JR, Celis J-P. Tribocorrosion studies on commercially pure titanium for dental applications. Proceedings of the European Corrosion Congress (EUROCORR '05), Lisbon, Portugal, September 2005; https://repositorium.sdum.uminho.pt/handle/1822/2271.

[100] Ribeiro RL, Rocha LA, Ariza E, Gomes JR, Celis J-P. Tribocorrosion behaviour of titanium grade 2 in alternative linear regime of sliding in artificial saliva solutions. Proceedings of the European Corrosion Congress (EUROCORR '05), pp. 1–10, Lisbon, Portugal, September 2005; https://repositorium.sdum.uminho.pt/handle/1822/5818.

[101] Oshida Y. Unpublished data, 2019.

[102] Barão VAR, Ramachandran RA, Matos AO, Badhe RV, Grandini CR, Sukotjo C, Ozevin D, Mathew M. Prediction of tribocorrosion processes in titanium-based dental implants using acoustic emission technique: Initial outcome. Mater Sci Eng C. 2021, 123, 112000, doi: 10.1016/j.msec.2021.112000.

Chapter 4
Implantability and compatibility

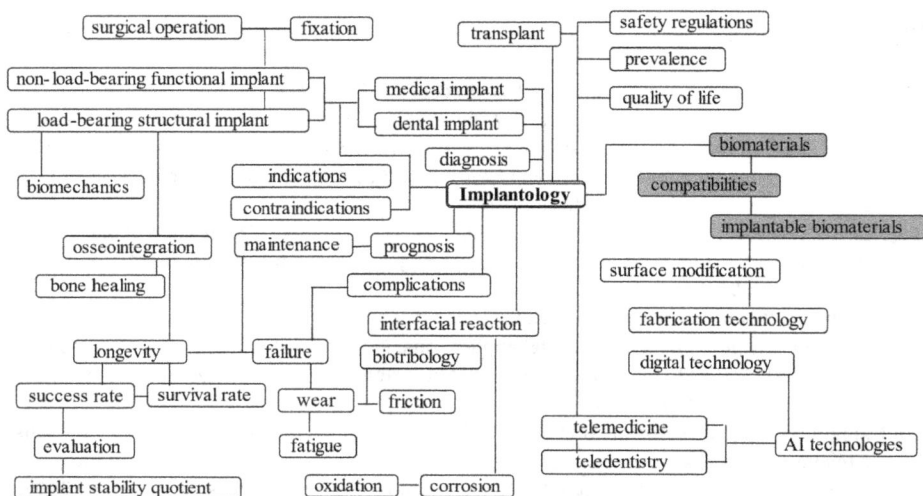

In previous chapters, we have discussed the type of biomaterials and reactions (physiological, chemical, and tribological) to the vital soft/hard tissues and body fluids. There are still hurdles for biomaterials to overcome to be qualified for their implantability.

4.1 Miniaturization and implantability

Global market of biomaterials has been ever growing by various driving factors, including the increasing aging population, increasing demand for minimally invasive procedures, increasing research and development investment, growing demand for orthopedic implants and plastic surgery, miniaturization of implant devices, and advanced technologies such as 3D printing of biomaterials [1, 2]. Most of the medical devices and implants are categorized in FDA class II. A certain type of devices is demanded to be fabricated as small size as possible, for the particularly preferred purpose of operating the minimum invasive surgery (MIS). On the other hand, since dental implants and orthopedic implants (such as TSR, THR, and TKR devices) are required to sustain their original strength or stronger and tougher, instead of size miniaturization, materials design and fabrication have been subject to R&D to enhance their mechanical properties as well as structural integrity by nanotechnology or other advanced technologies which will be discussed in later chapters.

https://doi.org/10.1515/9783110740134-005

Recently, with an ever-prevalence growth and advent of MIS, combined with the demand for both portable and wearable devices, the need for ever smaller miniaturized medical components has been noticeable. From temperature sensors, integrated circuits, and miniature valves to micromotors and drive systems, miniaturized components are enabling a new generation of medical devices and applications, including digital healthcare, remote monitoring, and wearable diagnostics. In addition, robotic surgery seemingly leveled the playing field, coordinating precision suturing, six degrees of surgical freedom, camera motion, and retraction, all with 3D vision [3].

After the first laparoscopic nephrectomy in 1991, urologists charted a minimally invasive surgical quest to limit morbidity without sacrificing success during complex operations. Proficient laparoscopy required years to develop with expertise limited to fewer surgeons, especially with such challenging procedures as radical prostatectomy, partial nephrectomy, and pyeloplasty [1]. Management of pediatric stone disease is challenging, with standard percutaneous nephrolithotomy (PCNL) having a good stone-free rate (SFR), but with associated high complication rates. Miniaturization of this technique has led to the rise of minimally invasive PCNL techniques such as micro- (<10 F) and ultra-mini (<15 F) PCNL procedures. Jones et al. [4] conducted a systematic review of the literature to evaluate the success and complication rates of minimally invasive PCNL techniques in the pediatric age group (<18 years). It was concluded that miniaturized PCNL techniques can deliver high SFRs with a small risk of Clavien I/II complications, and the size of tract seems to influence the nature of complications, with higher hematuria and renal extravasation with increasing tract size. Usui et al. [5] compared the treatment outcomes between minimally invasive ECIRS (mini-ECIRS: endoscopic combined intrarenal surgery) using 16.5 Fr percutaneous access sheath and standard ECIRS using 24 Fr access sheath for renal stones. It was concluded that (i) when compared to standard ECIRS, mini-ECIRS maintained SFR without increasing perioperative complications tended to reduce postoperative pain and had a potential to reduce bleeding-related complications, and (ii) the advantages of ECIRS miniaturization for renal stones were indicated.

The use of mechanical circulatory support to treat patients with congestive heart failure has grown enormously, recently surpassing the number of annual heart transplants worldwide. The latest generation of left ventricular assist devices (LVADs) is characterized by improved technologies. In addition, the size of new LVAD systems is considerably reduced when compared to older generation devices [6, 7]. The result is that MIS is now possible for the implantation, explantation, and exchange of LVADs. Minimally invasive procedures improve surgical outcome; for example, they lower the rates of operative complications (such as bleeding or wound infection) [6, 7]. For manipulators used in MIS, variable stiffness and miniaturization are very important characteristics. In order to improve the traditional devices with insufficient variable stiffness characteristics, Park et al. [8] proposed a manipulator that can be magnetically steered by a permanent magnet at the end and can have variable stiffness characteristics by a phase transition of graphene polylactic acid so that the proposed manipulator is easy to fabricate and miniaturize as a magnetic steering MIS manipulator. The growing adoption of MIS is driving surgical robotics development and increased miniaturization of surgery instrumentation. Several upcoming surgical robotic technologies promise to accelerate growth in this field. Enriquez [9] introduced two newly developed technologies: robotically controlled forceps that can pass through a hole about 3 mm wide – about the thickness of two pennies held together [10], and designing novel surgical robotic systems that perform MIS through an incision no wider than 3 cm for various abdominal or pelvic surgical procedures [11].

4.2 Compatibility

At the final check point before placing implants (in either dental or medical) in the body, there are groups of various compatibilities. Without overcoming these hurdles, biomaterials should not be considered as implantable biomaterials. Some compatibility requires proper surface manipulations of biomaterial candidates through thermal, chemical/electrochemical, or mechanical techniques.

As described previously, an implantation generates an intricate situation between vital host soft/hard tissue and nonvital foreign material. Since, among various types of implants, dental implantation can represent the above situation clearer, it would be worthy to compare various properties and behaviors between natural tooth and artificial tooth root structure and their surrounding tissues. As shown in Figure 4.1 [12–17], the most obvious difference between natural dentin and implant is the presence of the periodontal ligament (PDL), which possesses unique biological and biomechanical functions. We will revisit this figure when discussing the biomechanical compatibility.

In the field of dental and/or orthopedic implantology, there are many compatibilities which essentially determine acceptance, rejection, and long-term survival. These compatibilities should include (1) biological compatibility, (2) biomechanical compatibility, (3) morphological compatibility, (4) hemocompatibility, (5) cytocompatibility, and (6) magnetic resonance imaging (MRI) compatibility. In this chapter, we review three major compatibilities and additional two important compatibilities against blood (hemocompatibility) and cell (cytocompatibility), since these molecules need to be adhered, proliferated, and differentiated properly right after placing implants. These compatibilities, especially the surface energy condition, are controlling factors, and certain types of surface modifications can change the surface energy.

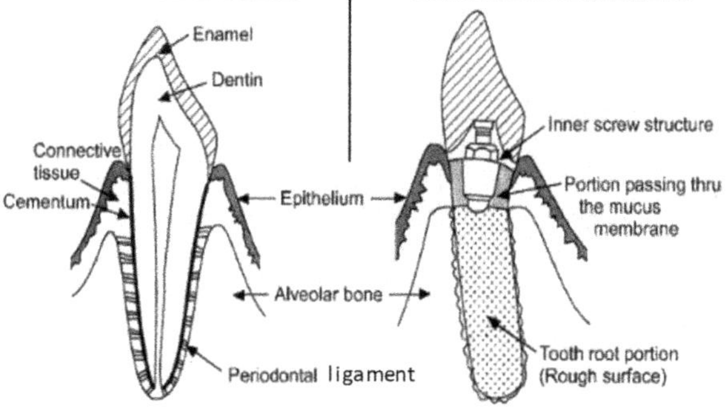

	Natural tooth	Implant
Schematic diagram		
Junction	PDL (periodontal ligament)	Osseointegration and functional ankylosis
Junction Epithel (JE)	Hemidemosomesve basal lamia	Hemidemosomes and basal lamia
Connect Tissue (CT)	13 groups: Vertical surfaces and tooth surface	2 groups: parallel vertical fibers No attachment on implant and bone surface
Biological Width (BW)	JE : 0.97 – 1.14mm CT: 0.77 – 1.07 mm BW: 2.04 – 2.91 mm	JE : 1.88 mm CT: 1.05 mm BW: 3.08 mm
Blood Supply	High	Low
Probing Depth	3 mm in healthy tissue	2.5 – 5.0 mm, according to soft tissue depth

Figure 4.1: Comparison of various properties between natural tooth and implant [12–17].

	High	Low
Pressure Sensitivity	High	Low
Axial Movability	25 – 100 nm	3 – 5 nm
Movement Type	Two phased. Primary: complex and non-linear movement Secondary: linear and elastic movement 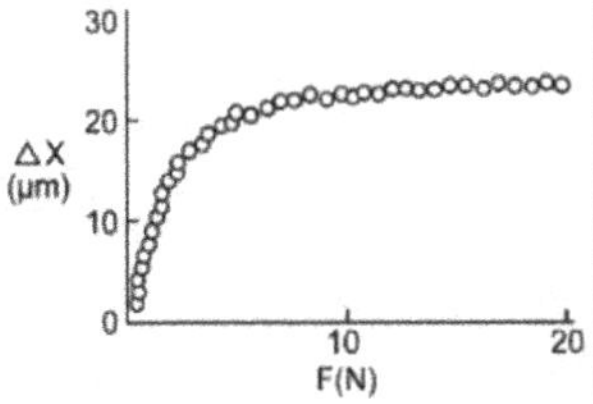 (graph: ΔX (µm) vs $F(N)$; curve rising and plateauing near 20, axes marked 0, 10, 20, 30 and 10, 20)	Linear and elastic movement 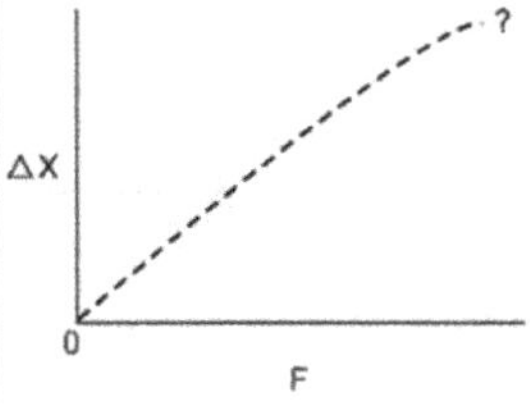 (graph: ΔX vs F; dashed rising curve ending with "?")
Movement Forms	Primary: Urgent movement Secondary: Progressive move	Gradual movement
Hinge point in lateral movements	1/3 apex region of the root	Crestal Bone
Property of frightening	Shock absorption mechanism and stress distribution	Concentration and stress increase in crestal bone
Overload Findings	Widening in periodontal ligament, movement, abrasion surface, fremitus, pain	Loss of screw or fracture, fracture in abutment or prosthesis, bone loss, implant fracture

Figure 4.1 (continued)

4.2.1 Biological compatibility (or biocompatibility)

Corrosion is one of the major processes that cause problems when metals are used as implants in the body. Their proper application to minimize such problems requires that one has an understanding of principles underlying the important degradative process of corrosion. To have such an understanding will result in proper application, better design, choice of appropriate test methods to develop better designs, and the possibility of determining the origin of failures encountered in practice [18, 19]. The service conditions in the mouth are hostile, both corrosively and mechanically. All intraorally placed parts are continuously bathed in saliva, an aerated aqueous solution of about 0.1 N chlorides, with varying amounts of Na, K, Ca, PO_4, CO_2, sulfur compounds, and mucin. The pH value is normally in the range of 5.5–7.5, but under plaque deposits it can be as low as 2. Temperatures can vary ±36.5 °C, and a variety of food and drink concentrations apply for short periods. Loads may go up to 1,000 N (with normal masticatory force ranging from 150 to 250 N), sometimes at impact speeds. Trapped food debris may decompose to create sulfur compounds, causing discoloration of placed devices [20]. With such hostile conditions, biocompatibility (biological compatibility) of metallic materials essentially equates to corrosion resistance because it is thought that alloying elements can only enter the surrounding organic system and develop toxic effects by conversion to ions through chemical or electrochemical process. After implant placement, initial healing of the bony compartment is characterized by formation of blood clots at the traumatized wound site, protein adsorption, and adherence of polymorphonuclear leukocyte. Then approximately 2 days after placement of the implant, fibroblasts proliferate into the blood clot, organization begins, and an extracellular matrix is produced. Approximately 1 week after the implant is placed, appearance of osteoblast-like cells and new bone is seen. New bone reaches the implant surface by osseoconduction (through growth of bone over the surface and migration of bone cells over the implant surfaces) [21]. During healing process, metallic ions (e.g., Ti, Co, Cr, Al, V, and Fe) release corrosion products (which is mainly oxides or hydroxides) into the surrounding tissue and fluids even though it is covered by a thermodynamically stable oxide film [22–24]. An increase in oxide thickness, as well as incorporation of elements from the extracellular fluid (P, Ca, and S) into the oxide, has been observed as a function of implantation time [25]. Moreover, changes in the oxide stoichiometry, composition, and thickness have been associated with the release of products [26]. Properties of the oxide, such as stoichiometry, defect density, crystal structure and orientation, surface defects, and impurities were suggested as factors determining biological performance [27–29]. Using Auger electron spectroscopy to study the change in the composition of the titanium surface during implantation in the human bone, it is observed that the oxide formed on titanium implants grows and takes up minerals during the implantation [25]. The growth and uptake occur even though the adsorbed layer of protein is present on the oxide, indicating that mineral ions pass through the adsorbed protein. It was shown that, using Fourier transform infrared reflection absorption spectroscopy, phosphate ions are adsorbed by the titanium surface after the protein has been adsorbed. Using X-ray photoelectron spectroscopy (XPS) [30], it was demonstrated that oxides on commercially pure titanium (cpTi) and titanium alloy (Ti–6Al–4V) change into complex phosphates of titanium and calcium containing hydroxyl groups which bind water on immersion in artificial saliva (pH 5.2) [31]. All these studies indicate that the surface oxide on titanium materials reacts with mineral ions, water, and other constituents of biofluids and that these reactions in turn cause a remodeling of the surface.

It was shown that titanium is in almost direct contact to the bone tissue, separated only by an extremely thin cell-free noncalcified tissue layer. Transmission electron microscopy revealed an interfacial hierarchy that consisted of a 20–40 nm thick proteoglycan layer within 4 nm of the titanium oxide, followed by collagen bundles as close as 100 nm and Ca deposits within 5 nm of the surface [32]. To reach the steady-state interface, both the oxide on titanium and the adjacent tissue undergo various reactions. The physiochemical properties of titanium have been associated with the unique tissue response to the materials: these include the biochemistry of released corrosion products, kinetics of release and the oxide stoichiometry, crystal defect density, thickness, and surface chemistry [33]. The performance of titanium and its alloys in surgical implant applications can be evaluated with respect to their biocompatibility and capability to withstand the corrosive species involved in fluids within the human body. This may be considered as an electrolyte in an electrochemical reaction. It is well documented that the excellent corrosion resistance of titanium materials is due to the formation of a dense, protective, and strongly adhered film, which is called a passive film. Such a surface situation is referred to passivity or a passivation state. As shown in Figure 4.2 [34], corrosion resistance, in general, is linearly related to the extent of biocompatibility.

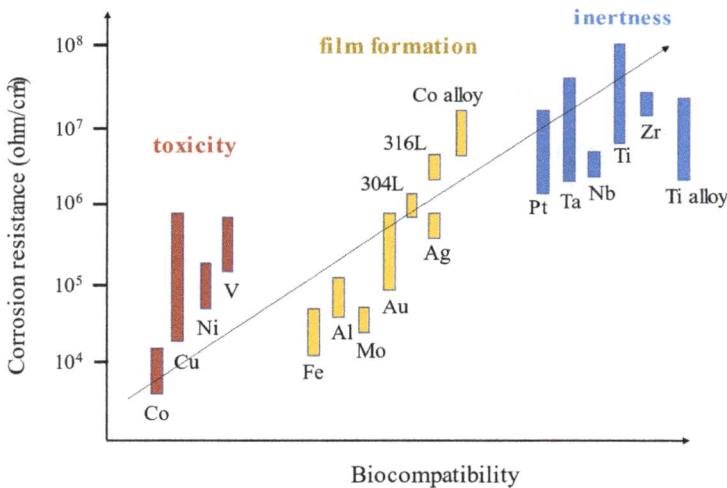

Figure 4.2: Linear relationship between corrosion resistance and biocompatibility [34].

Pourbaix [pH–E] diagram indicates thermodynamically stable phases (at chemical equilibrium) of pure metallic element on aqueous electrochemical system and normally it consists of three zones: immune zone, corrosive zone, and passive zone. In the following QA session, reader is welcome to exercise how to use the Pourbaix diagram.

Question
Referring to the Pourbaix's [pH–E] diagram for Ti–O system (see Figure 4.3a [35]), prove that Ti implant is an acceptable material in the intraoral environment. Hint: you need to search for data in reasonable range of pH value and intraoral potential range as well.

Answer (as an example)
In general, Pourbaix diagram is composed mainly of three zones: immune, corrosive, and passive. If we can identify the information on intraoral conditions in the passive zone of Ti, it can be concluded that the surface of Ti material is protected by the passive film so that Ti is an acceptable biomaterial in the intraoral environment.

Step 1: Range of pH value. Although pH value ranges very widely (from 3 to 10, as shown in Figure 3.1, depending on the intake dietary types), it is normally believed that the reasonable range of pH would be 5–8.

Step 2: Intraoral potential. For this information, we need to do some literature research. Not surprisingly, there are not enough data to summarize the appropriate value of intraoral potential range, Corso et al. [36] reported that the intraoral potential is in a range from −300 to + 300 mV. Reclaru et al. [37] reported a range from 0 to +300 mV. Ewers et al. [38] reported that it ranges between − 380 and + 50 mV. Ewers et al. [39] reported that it ranged from −2 to + 200 mV. Hence, the range of overlapping data is a very narrow window of potential zone from 0 to + 200 mV.

Step 3: If we add two line-range (pH and intraoral potential) into a given [pH–E] diagram, we will have Figure 4.3b.

Answer: From Figure 4.3b, it can be said that Ti can be found within the passivity zone, indicating that the surface of Ti is protected by a passive film in the intraoral environment and can exhibit excellent biocompatibility.

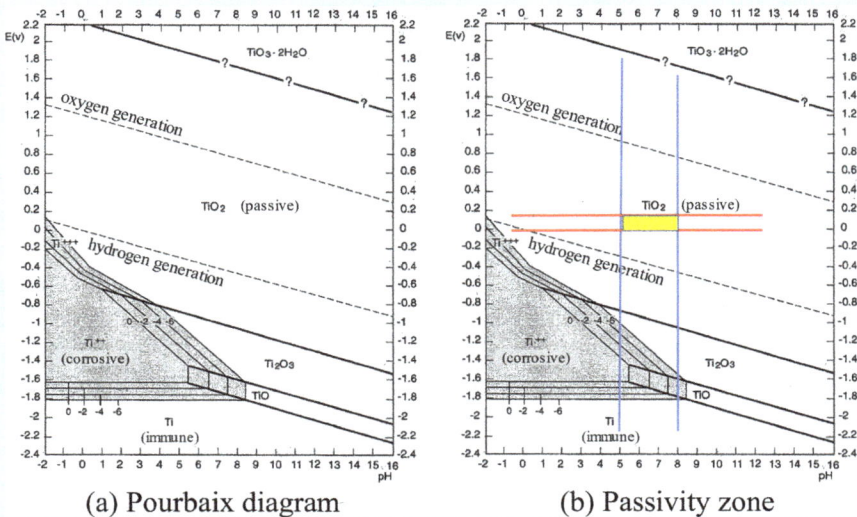

(a) Pourbaix diagram (b) Passivity zone

Figure 4.3: (a) Pourbaix [pH–E] diagram for Ti–O. (b) With limitations of pH value range (blue lines) and intraoral range (red lines), it is found that Ti is within the passivity zone, indicating that the surface of Ti can be covered by the protective passive film.

The concept of biocompatibility has a certain limitation in explaining the phenomena involved in biomaterial-based tissue repair, although biocompatibility is the basic requirement of biomaterials for tissue repair. Since new materials, particularly those for tissue engineering and regeneration, have been developed with common characteristics, i.e., they participate deeply into important chemical and biological processes in the human body, and the interaction between the biomaterials and tissues is far more complex. Understanding the interplay between these biomaterials and tissues is vital for their development and functionalization. Based on this background, Wang [40] introduced the concept of "bioadaptability" of biomaterials. The proposed bioadaptability describes the three most important aspects that can determine the performance of biomaterials in tissue repair: 1) the adaptability of the microenvironment created by biomaterials to the native microenvironment in situ; 2) the adaptability of the mechanical properties of biomaterials to the native tissue; 3) the adaptability of the degradation properties of biomaterials to the new tissue formation, emphasizing both the material's characteristics and biological aspects within a certain microenvironment and molecular mechanism.

4.2.2 Biomechanical compatibility

Biomechanics involved in implantology should include at least (1) the nature of the biting forces on the implants, (2) transferring of the biting forces to the interfacial tissues, and (3) the interfacial tissue reaction, biologically, to stress transfer conditions. Interfacial stress transfer and interfacial biology represent more difficult, interrelated problems. While many engineering studies have shown that variables such as implant shape, elastic modulus, and the extent of bonding between implant and bone can affect the stress transfer conditions, the unresolved question is whether there is any biological significance to such differences. The successful clinical results achieved with osseointegrated dental implants underscore the fact that such implants easily withstand considerable masticatory loads. In fact, one study showed that bite forces in patients with these implants were comparable to those in patients with natural dentitions [41]. A critical aspect affecting the success or failure of an implant is the manner in which mechanical stresses are transferred from the implant to the bone smoothly. It is essential that neither implant nor bone be stressed beyond the long-term fatigue capacity. It is also necessary to avoid any relative motion that can produce abrasion of the bone or progressive loosening of the implants. An osseointegrated implant provides a direct and relatively rigid connection of the implant to the bone. This is an advantage because it provides a durable interface without any substantial change in form or duration. There is a mismatch of the mechanical properties and mechanical impedance at the interface of the implant material and the bone. It is interesting to observe that from a mechanical standpoint, the shock-absorbing action would be the same if the soft layer were between the metal implant and the bone. In the natural tooth, the periodontum aka PDL, which forms a shock-absorbing layer, is in this position between the tooth and the jawbone [42, figure 4.1]. Natural teeth and implants have different force transmission characteristics of the bone. Compressive strains were induced around natural teeth and implants as a result of static axial loading, whereas combinations of compressive and tensile strains were observed during lateral dynamic loading [12]. Magnitude of strain around the natural tooth is significantly lower than the opposing implant and occluding implants in the contra lateral side for most regions under all loading conditions. It was reported that there was a general tendency for increased strains around the implant opposing natural tooth under higher loads and particularly under lateral dynamic loads [41]. By means of finite element analysis, stress distribution in the bone around implants was calculated with and without stress-absorbing element [43]. A freestanding implant and an implant connected with a natural tooth were simulated. For the freestanding implant, it was concluded that the variation in the modulus of elasticity of the stress-absorbing element had no effect on the stresses in the bone. Changing the shape of the stress-absorbing element had little effect on the stresses in the cortical bone. For the implant connected with a natural tooth, it was concluded that a more uniform stress was obtained around the implant with a low modulus of elasticity of the stress-absorbing element. It was also concluded that the bone surrounding the natural tooth showed a decrease in the height of the peak stresses.

The dental or orthopedic prostheses, particularly the surface zone thereof, should respond to the load transmitting function. The placed implant and receiving tissues establish a unique stress–strain field. Between them, there should be an interfacial layer. During the loading with implant/bone couple, the strain-field continuity should be held (if not, it should indicate that implant is not fused to the vital bone), although the stress field is obviously in a discrete manner due to different values of modulus of elasticity (E) between the host tissue and the foreign implant material. Namely, stress at the bone $\sigma_B = E_B \varepsilon_B$ and stress at the implant $\sigma_I = E_I \varepsilon_I$. Under the continuous strain field, $\varepsilon_B = \varepsilon_I$. However, $E_B \neq E_I$ is due to the dissimilar material couple condition. If the magnitude of the difference in modulus of elasticity is large, then the interfacial stress, accordingly, could become so large that the placed implant system will face a risky failure or detachment situation. In other words, if interfacial stress due to stress difference $\Delta\sigma = \Delta(\sigma_I - \sigma_B)$ is larger than the retention strength of the osseointegrated implant, the placed implant will be failed. Therefore, materials for implant or surface zone of implants should be mechanically compatible to mechanical properties (especially, modulus of elasticity) of receiving tissues to minimize the interfacial discrete stress. This is the second compatibility and is called as the biomechanical compatibility [12, 44].

Figure 4.4 [45] compares yield strengths and modulus of elasticity of various biomaterials in log–log plot.

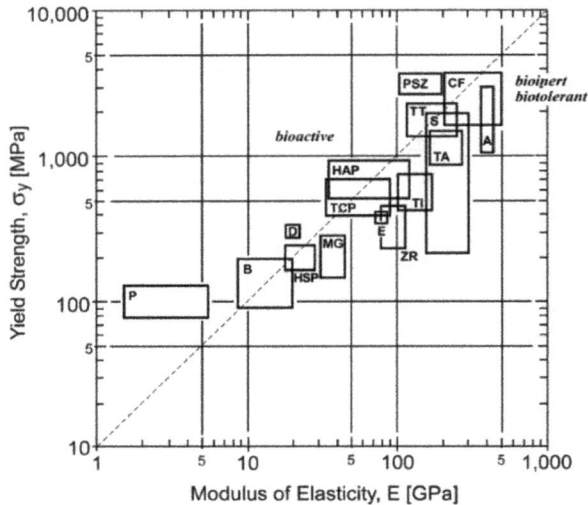

Figure 4.4: A relationship between yield strength and modulus of elasticity (or rigidity) of various types of biomaterials and bones that receive the vital tissue for placed implants [45]. P, polymeric materials; B, bone; HSP, high-strength polymers (such as Kevlar, Kapton, and PEEK); D, dentin; TCP, tricalcium phosphate; HAP, hydroxyapatite; E, enamel; MG, magnesium; TI, commercially pure titanium (all unalloyed grades); TA, titanium alloys (e.g., Ti–6Al–4V and Ti–6Al–7Nb); S, steels (e.g., 304L and 316L stainless steels); A, alumina; PSZ, partially stabilized zirconia; CF, carbon fiber.

As shown clearly, strength and rigidity of all biomaterials concerned are related linearly on both log scales. From the point of strain-continuity viewpoint, it is ideal to choose any implantable materials that have both strength and rigidity values close to those of the receiving bone. Hydroxyapatite (HA) coating onto the titanium implant has been widely adopted since both HA and receiving vital bone possess similar chemical compositions; hence, early adaptation can be highly expected. At the same time, E_{HA} is positioned in between the values of E_B and E_I; as a result, HA coating will have a second function for mechanical compatibility to make the stress a smooth transfer (or to minimize the interfacial stress). This is one of the typical hindsight, because HA coating is originally and still now performing due to its similarity of its chemical composition to the receiving bone.

Question
Recently, two new implantable materials have been introduced: TiZr alloy [46–49] and zirconia ceramic [50–54]. Assuming that these new materials are already evaluated as biomaterials and exhibit excellent biocompatibility, discuss if they are accepted in terms of biomechanical compatibility.

Answer
The typical chemical composition of TiZr implantable alloy is Ti–15Zr–4Nb–4Ta [31] and exhibits mechanical property range as shown in Figure 4.5, which is marked with red TZ. Although mechanical strength of TZ appears to be similar to that of TA, TZ shows higher MOE value than TA, possibly resulting in creating higher level of interfacial stress than the case of bone and TA as discussed previously. Another new implantable material of zirconia is also shown in Figure 4.5, marked with red ZO. As shown clearly, newly developed zirconia ceramic possesses higher strength as well as rigidity, indicating that the risk of stress discrete situation between the bone and zirconia should be the highest among any combinations foreseen from Figure 4.5. It should not be ignored that an implant recipient is ever-aging so that the bone's strength and toughness are also reduced, as shown with blue arrow mark. In addition, ceramic material is brittle so that the surface modification is not easily modified such as HA coating or foam-structure texturing [55].

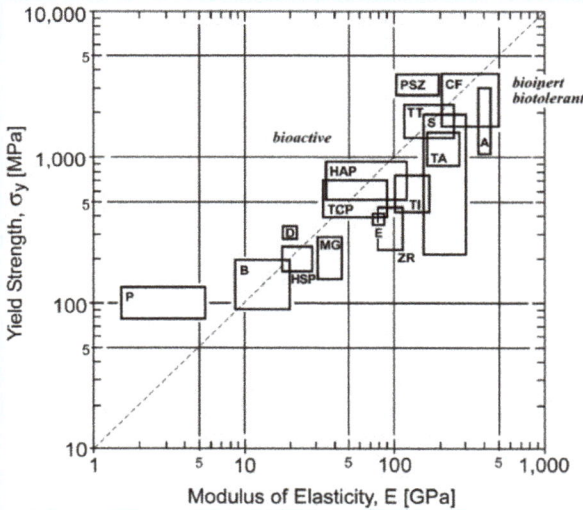

Figure 4.4: [E–σ_Y] diagram.

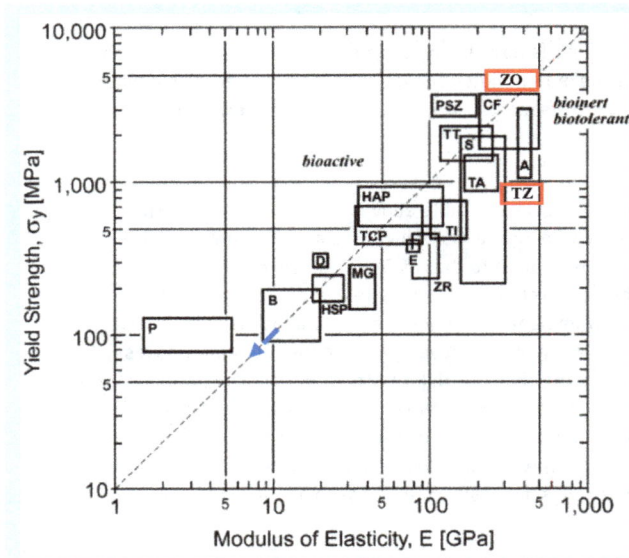

Figure 4.5: $[E-\sigma_Y]$ diagram with new materials.

4.2.3 Morphological compatibility

Surface plays a crucial role in biological interactions for four reasons. First, the surface of a biomaterial is the only part in contact with the bioenvironment. Second, the surface region of a biomaterial is almost always different in morphology and composition from the bulk. Differences arise from molecular rearrangement, surface reaction, and contamination. Third, for biomaterials that do not release nor leak biologically active or toxic substances, the characteristics of the surface govern the biological response. And fourth, some surface properties, such as topography, affect the mechanical stability of the implant/tissue interface [56, 57]. In a scientific article [44], it was found that surface morphology of successful implants has an upper and lower limitations in average roughness (1–50 μm) and average particle size (10–500 μm), regardless of the types of implant materials (either metallic, ceramics, or polymeric materials) and independent of surface modifications. If a particle size is smaller than 10 μm, the surface will be more toxic to fibroblastic cells and have an adverse influence on cells due to their physical presence independent of any chemical toxic effects. If the pore is larger than 500 μm, the surface zone does not maintain sufficient structural integrity because it is too coarse. This is the third compatibility – morphological compatibility [44, 58]. Figure 4.6 illustrates appropriate ranges in surface roughness of clinically successful implants [59].

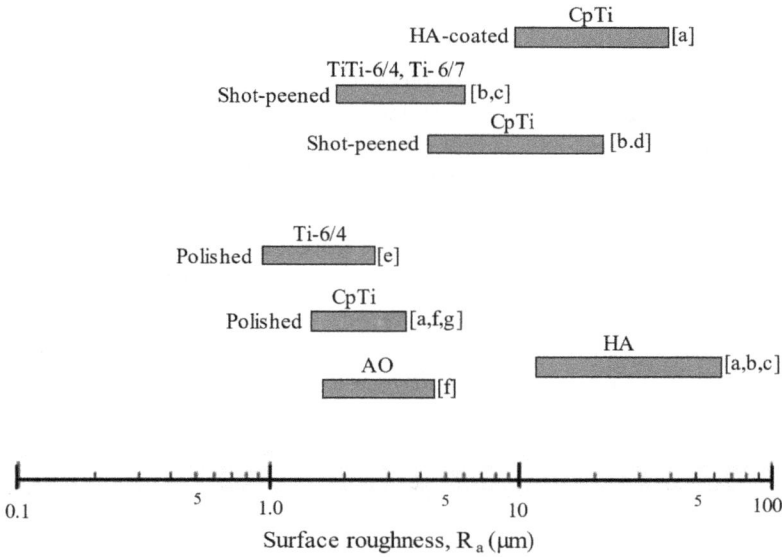

Figure 4.6: Distribution of surface roughness of successfully implanted biomaterials [59].
References:
[a] Buser D, Schenk RK, Steinemann JP, Fox CH, Stich H. Influence of surface characteristics on bone integration of titanium implants. A histomorphometric study in miniature pigs. J Biomed Mater Res. 1991, 25, 889–902.
[b] Jansen JA, van der Waerden JP, Wolke JG. Histologic investigation of the biologic behavior of different hydroxyapatite plasma-sprayed coatings in rabbits. J Biomed Mater Res. 1993, 27, 603–10.
[c] Wang BC, Lee TM, Chang E, Yang CY. The shear strength and the failure mode of plasma-sprayed hydroxyapatite coating to bone: the effect of coating thickness. J Biomed Mater Res. 1993, 27, 1315–27.
[d] Steinemann JP, Eulenberger J, Maeusli PA, Schoeder A. Biological and biomechanical performances of biomaterials. Amsterdam: Elsevier Science, 1986, 409–14.
[e] Hayashi K, Inadome T, Mashima T, Sugioka Y. Comparison of bone-implant interface shear strength of solid hydroxyapatite and hydroxyapatite-coated titanium implants. J Biomed Mater Res. 1993, 27, 557–63.
[f] Thomas KA, Cook SD. An evaluation of variables influencing implant fixation by direct bone apposition. J Biomed Mater Res. 1985, 19, 875–901.
[g] Li J. Behaviour of titanium and titania-based ceramics in vitro and in vivo. Biomaterials. 1993, 14, 229–32.

It has been shown that preparation methods of implant surface can significantly affect the resultant properties of the surface and subsequently the biological responses that occur at the surface [60–62]. Recent efforts have shown that the success or failure of dental implants can be related not only to the chemical properties of the implant surface but also its macromorphologic nature [63–65]. From an in vitro standpoint, the response of cells and tissues at implant interfaces can be affected by surface topography or geometry on a macroscopic basis [64, 66], as well as by surface morphology or roughness on a microscopic level [64, 67]. These characteristics undoubtedly affect how cells and tissues respond to various types of biomaterials. Of all the cellular responses, it has been suggested that cellular adhesion is considered the most important response necessary for developing a rigid structural and functional integrity at the bone/implant interface [68]. Cellular adhesion alters the entire tissue response to biomaterials [69]. The effect of surface roughness (Ra: 0.320, 0.490, and 0.874 μm) of the titanium alloy Ti–6Al–4V on the short- and long-term response of human bone marrow cells in vitro and on protein adsorption was investigated [70]. Cell attachment, cell proliferation, and differentiation (alkaline phosphatase-specific activity) were determined. The protein adsorption of bovine serum albumin and fibronectin from single protein solutions on rough and smooth Ti–6Al–4V surfaces was examined with XPS and radio labeling. It was found that (i) cell attachment and proliferation were surface roughness sensitive and increased as the roughness of Ti–6Al–4V increased, (ii) human albumin was adsorbed preferentially onto the smooth substratum, and (iii) the rough substratum bound a higher amount of total protein (from culture medium supplied with 15% serum) and fibronectin (10-fold) than did the smooth one [70], suggesting an importance of the rugophilicity.

Events lead to integration of an implant into the bone; hence, determining the long-time performance of the device take place largely at the interface formed between the tissue and the implant [71]. The development of this interface is complex and is influenced by numerous factors, including surface chemistry and surface topography of the foreign material [28, 72–75]. For example, Oshida et al. [76] treated NiTi by acid pickling in $HF–HNO_3–H_2O$ (1:1:5 by volume) at room temperature for 30 s to control the surface topology and selectively dissolve Ni, resulting in the Ti-enriched surface layer, demonstrating that surface topology can be managed.

The role of surface roughness on the interaction of cells with titanium model surfaces of well-defined topography was investigated using human bone–derived cells (MG63 cells). The early phase of interactions was studied using a kinetic morphological analysis of adhesion, spreading, and proliferation of the cells. Scanning electron microscopy (SEM) and double immunofluorescent labeling of vinculin and actin revealed that the cells responded to nanoscale roughness with a higher cell thickness and a delayed apparition of the focal contacts. A singular behavior was observed on nanoporous oxide surfaces, where the cells were more spread and displayed longer and more numerous filopods. On electrochemically microstructured surfaces, the MG63 cells were able to penetrate inside, adhere, and proliferate in cavities of 30 or 100 μm in diameter, whereas they did not recognize the 10 μm diameter cavities. Cells adopted a 3D shape when attaching inside the 30 μm diameter cavities. It was concluded that nanotopography on surfaces with 30 μm diameter cavities had little effect on cell morphology compared to flat surfaces with the same nanostructure, but cell proliferation exhibited a marked synergistic effect of microscale and nanoscale topography [77]. On a macroscopic level (roughness > 10 μm), roughness influences the mechanical properties of the titanium/bone interface, the mechanical interlocking of the interface, and the biocompatibility of the material [78, 79]. Surface roughness in the range from 10 nm to 10 μm may also influence the interfacial biology, since it is the same order as the size of the cells and large biomolecules [80]. Micro-roughness at this level includes material defects, such as grain boundaries, dislocation steps and kinks, and vacancies that are active sites for adsorption, and therefore influence the bonding of biomolecules to the implant surface [81]. Microrough sur-faces promote significantly better bone apposition than smooth surfaces, resulting in a higher percentage of bone in contact with the implant. Microrough surfaces may influence the mechanical properties of the interface, stress distribution, and bone remodeling [82]. Increased contact area and mechanical inter-locking of bone to a microrough surface can decrease stress concentrations resulting in decreased bone resorption. Bone resorption takes place shortly after loading smooth surfaced implants [83], resulting in a fibrous connective tissue layer, whereas remodeling oc-curs on rough surfaces [84].

Recently developed clinical oral implants have been focused on topographical changes of implant surfaces, rather than alterations of chemical properties [65, 85–88]. These attempts may have been based on the concept that mechanical interlocking be-tween tissue and implant materials relies on surface irregularities in the nanometer to micron level. Recently published in vivo investigations have shown significantly im-proved bone tissue reactions by modification of the surface oxide properties of Ti im-plants [89–96]. It was found that in animal studies, bone tissue reactions were strongly reinforced with oxidized titanium implants, characterized by a titanium oxide layer thicker than 600 nm, a porous surface structure, and an anatase type of Ti oxide with large surface roughness compared with turned implants [94, 95]. This was later sup-ported by the work done by Lim et al. [97]. Oshida [98] and Elias et al. [99] found that the alkali-treated cpTi surface was covered mainly with anatase-type TiO_2 and exhib-ited hydrophilicity, whereas the acid-treated CpTi was covered with rutile-type TiO_2 with hydrophobicity. Besides this characteristic crystalline structure of TiO_2, it was mentioned that good osseointegration, bony apposition, and cell attachment of Ti implant systems [100, 101] are partially due to the fact that the oxide layer, with unusually high dielectric constant of 50–170, depending on the TiO_2 concentration, may be the responsible feature [82, 102, 103].

Based on what we learn in the above about optimum three major compatibil-ities (biological, biomechanical, and morphological), Oshida [45, 60] proposed a conceptually integrated implant accompanied with gradated biofunctionality, as shown in Figure 4.7.

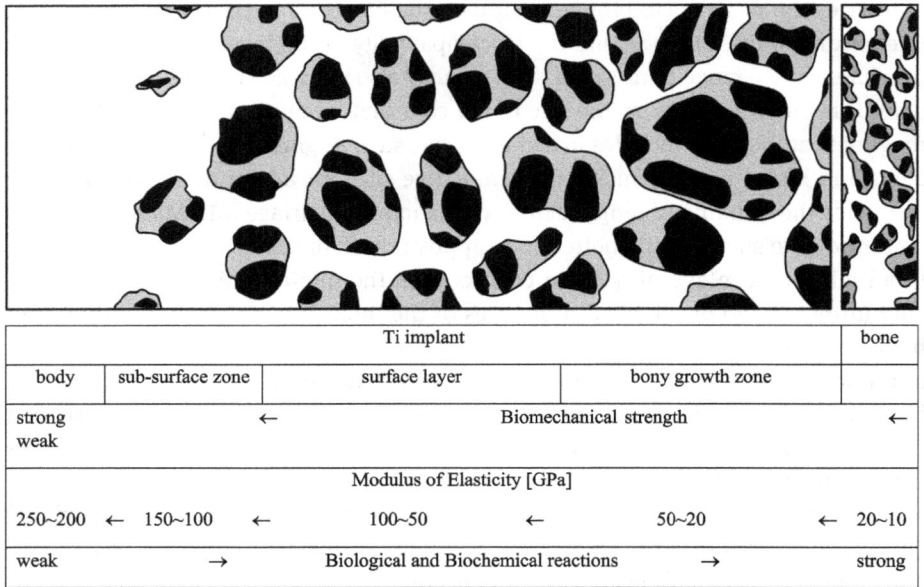

Ti implant				bone
body	sub-surface zone	surface layer	bony growth zone	
strong weak	←	Biomechanical strength		←
		Modulus of Elasticity [GPa]		
250~200 ← 150~100 ←		100~50 ←	50~20 ← 20~10	
weak	→	Biological and Biochemical reactions	→	strong

Figure 4.7: Schematic and conceptual Ti implant with gradated functions, accompanied with biological, biomechanical, and morphological compatibilities.

4.2.4 Hemocompatibility

Due to the increasing importance of biomaterials in our society, there are numerous new developments of blood contacting materials. A big percentage of the industrialized society owes their QOL to medical products like stents, hemodialyzers, central venous catheters, artificial heart valves, orthopedic knee, or hip implants and dental implants. During the surgical procedure of implantation, the biomaterial most likely will encounter blood. Almost instantly following contact with blood, the implant surface will be covered with plasma proteins that become adsorbed to the surface [104, 105]. Hence, the hemocompatibility (blood compatibility) is the most important aspect and a key property for biomaterials that come in contact with blood. Hemocompatibility of biomaterials is mainly dependent on its surface characteristics, including composition, roughness, wettability (or surface energy), and morphology. There are two opposing required properties in terms of blood compatibility, that is, they are hydrophobic hemocompatibility and hydrophilic hemocompatibility and these are controlled and managed by appropriate surface modifications.

Hydrophobic hemocompatibility

Catheters, guidewires, dialyzer, oxygenators (artificial lungs), heart-supporting systems, cardiac pacemaker, vascular grafts, stents, heart valves, microparticles, and nanoparticles are widely used medical devices and materials coming in direct contact with blood [106, 107]. Hemocompatibility is one of the major criteria, which limit the clinical applicability of blood-contacting biomaterials. These materials come in close contact with blood, which is a complex "organ," comprising 55% plasma, 44% erythrocytes, and 1% leukocytes and platelets. Thus, adverse interactions between newly developed materials and blood should be extensively analyzed to prevent activation and destruction of blood components [107]. The initially adsorbed protein layer on the biomaterial surface mainly triggers the adverse reactions, such as the activation of coagulation via intrinsic pathway, the activation of leukocytes, which results in inflammation, and the adhesion and activation of platelets [108]. As a result, the number of blood cells can decrease and a thrombus can be formed. Thus, the applied blood-contacting biomaterials should not adversely interact with any blood components and activate or destruct blood components [107]. Prior to clinical application, the hemocompatibility of blood-contacting medical materials have to be analyzed, and therefore, a guidance is developed by the ISO (International Organization for Standardization: ISO 10993-4) [109]. According to this guideline, five different categories, thrombosis, coagulation, platelets, hematology, and immunology (complement system and leukocytes), are indicated for hemocompatibility evaluation. The devices are divided into three categories concerning blood contact: (1) externally communicating devices with indirect blood contact, for example, cannulas and blood collection sets; (2) externally communicating devices with direct blood contact, for example, catheters and hemodialysis equipment; (3) implant devices, for example, heart valves, stents, and vascular grafts.

Natural heart valves can become diseased or damaged, which means they either do not open properly (stenosis) or do not close properly (regurgitation), indicating that the artificial heart valve is demanded for appropriate opening and closing biofunctions. Artificial durable heart valves, which must replicate their cyclic biofunction over an entire life with an estimated total demand of at least 3×10^9 cycles, are cardiac structures whose physiological function is to ensure directed blood flow through the heart over the cardiac cycle. The primarily passive structures are driven by forces exerted by the surrounding blood and heart [110]. Hence, valvular function should be considered in terms of mechanical and mechanobiological behaviors of the constituent biological materials (e.g., extracellular matrix proteins and cells). Biomaterial surfaces initiate blood coagulation. This poses a significant challenge to the development of blood-contacting devices used in human beings and especially for the implementation of cardiovascular devices. The performance of medical products such as catheters, blood vessel grafts, vascular stents, extracorporeal oxygenator membranes, and LVAD are significantly impaired by the thrombosis problem [111]. A thrombosis is composed of crosslinked fibrin clots and aggregated platelets connected by fibrinogen. When thrombi come off the site as emboli and travel through the blood stream, they may block small vessels downstream leading to stroke or tissue death. With the increasing demand for cardiovascular healthcare worldwide, it is imperative to understand the mechanisms leading to thrombus formation on biomaterial surfaces in order to develop strategies for improved hemocompatibility [111]. Blood coagulation can be potentiated by plasma-phase coagulation and/or platelet-mediated reactions [112]. When a biomaterial is implanted and/or vessel is injured, a variety of biological responses are initiated triggering processes of platelet activation/aggregation and plasma coagulation. The mechanism of platelet aggregation involved the adhesion, activation, and aggregation of platelets at the site of injured wall or on implanted biomaterial surfaces [113].

The hemocompatibility of the titanium oxide films was characterized in terms of clotting, time, blood platelet adhesion and activation, thrombin time, and thrombinogen time. Huang et al. [114] studied the behavior of protein adsorption by irradiation labeling with 125I. Ti-O thin films were prepared by plasma immersion ion implantation (PI^3 or PIII) and deposition, and by sputtering. Systematic evaluation of hemocompatibility, including in vitro clotting time, thrombin time, prethrombin time, blood platelet adhesion and in vivo implantation into a dog's ventral aorta or right auricle from 17 to 90 days, proved that Ti-O films have excellent hemocompatibility. It is, hence, suggested that the significantly lower interface tension between Ti-O films, blood, and plasma proteins, and the semiconducting nature of Ti-O films give them their improved hemocompatibility [114]. Already widely used in the orthopedic field as biomaterials, Ti-6Al-4V could be selected as the construction material for blood-contacting devices, such as LVAD [115]. Several authors have discussed the biocompatibility and hemocompatibility of pure titanium [115–119]. Protein adsorption studies on titanium oxide (TiO$_2$) have been reported [120, 121]. Ti-6Al-4V alloy behavior toward bovine serum albumin and bovine gamma-globulin has been studied using an enzyme-linked immunosorbent assay and has demonstrated a greater protein binding for Ti-6Al-4V alloy than for stainless steel [122]. These results indicate that the Ti-6Al-4V alloy is well tolerated by blood [116]. It was also suggested that the Ti-6Al-4V alloy is well tolerated by the blood and, therefore, is a promising biocompatible material for construction of biomedical devices, either in the orthopedic field or in the cardiovascular one [116].

Muramatsu et al. [123] evaluated the fibrin deposition and blood platelet adhesion onto alkali- and heat-treated CpTi, alkali- and water-treated CpTi, and alkali- and heat-treated CpTi formed with apatite in simulated body fluid. They were evaluated by exposure to anticoagulated blood or washed platelet suspension under static conditions and subsequent observation with SEM. It was reported that (i) thrombus formation on alkali-treated CpTi surfaces, which were exposed to heparinized whole blood for 1 h, was significantly less than that on untreated control CpTi, on which pronounced depositions of fibrin-erythrocytes and lymphocytes were observed; (ii) no thrombus was observed on the apatite-like formed CpTi surface; and (iii) the apatite-like formed CpTi surface exhibits stronger antithrombogenic characteristics than CpTi and other materials examined in heparinized blood [123]. Untreated Ti–6Al–4V and glow-discharge-treated Ti–6Al–4V samples were tested on human peripheral blood mononuclear cells. It was found that (i) apoptosis, undetectable after 24 h contact of peripheral blood mononuclear cells with the two sample types, is induced after 48 h by treated samples, and, after 48 h, but in the presence of 1.5 µg/mL PHA, by both sample types, and (ii) although plasma-treated titanium alloy shows a better biocompatibility in comparison with the untreated one, attention must be paid to the careful control of the first signs of inflammation [124]. Thor et al. [125] investigated the thrombogenic responses of whole blood, platelet-rich plasma, and platelet-poor plasma in contact with a highly thrombogenic surface as titanium. The thrombogenic responses of clinically used surfaces as HA, machined titanium, titania-blasted titanium, and fluoride ion-modified grit-blasted titanium were also investigated. It was concluded that (i) the whole blood is necessary for sufficient thrombin generation and platelet activation during placement of implants, and (ii) a fluoride ion modification seems to segment the thrombogenic properties of titanium [125]. Maitz et al. [126] treated superelastic NiTi by oxygen or helium plasma-immersion ion implantation to deplete surface Ni, leading to the formation of a nickel-poor titanium oxide surface with a nanoporous structure. Fibrinogen adsorption and conformation changes, blood platelet adhesion, and contact activation of the blood clotting cascade have been checked as in vitro parameters of blood compatibility; metabolic activity and release of cytokines IL-6 and IL-8 from cultured endothelial cells on these surfaces give information about the reaction of the blood vessel wall. It was found that the oxygen-ion-implanted NiTi surface adsorbed less fibrinogen on its surface and activated the contact system less than the untreated nitinol surface, but conformation changes of fibrinogen were higher on the oxygen-implanted nitinol. No difference between initial and oxygen-implanted NiTi was found for the platelet adherence, endothelial cell activity, or cytokine release, and oxygen-ion implantation is seen as a useful method to decrease the nickel concentration in the surface of nitinol for cardiovascular applications [126].

There is a demand for generating surfaces repelling various liquids that would have broad technological implications for areas ranging from biomedical devices and fuel transport to architecture [127]. Inspirations from natural nonwetting structures [128, 129], particularly the leaves of the lotus, have led to the development of liquid-repellent microtextured surfaces that rely on the formation of a stable air–liquid interface [130]. Despite over a decade of intense research, these surfaces are, however, still plagued with problems that restrict their practical applications: limited oleophobicity with high contact angle hysteresis [130, 133], failure under pressure [131–133], and upon physical damage [132, 134, 135]. Wong et al. [127] developed slippery surfaces that will be useful in fluid handling and transportation, optical sensing, medicine, and as self-cleaning and antifouling materials operating in extreme environments.

Hydrophilic hemocompatibility

As Albrektsson et al. [101] and Tominaga et al. [136] among many other researchers pointed out, the favorable hydrophilic property in vital environment is crucial for the placed implants to exhibit successful osseointegration. To this end, there are various surface modifications proposed to enhance the wettability. Hence, this hydrophilic hemocompatibility is asking for the exact opposing surface physics from the previous hydrophobic hemocompatibility.

Sartoretto et al. [137] evaluated the early osseointegration of two different implant surfaces, a sandblasted and acid-etched surface (TN) compared with the same geometry and surface roughness modified to be hydrophilic/wettable by conditioning in an isotonic solution of 0.9% sodium chloride (TA) through histological and histomorphometric analysis after sheep tibia implantation. Forty dental implants, divided into two groups (TN and TA), were placed in the left tibia of 20 healthy, skeletally mature Santa Ines sheep ($n = 5$/experimental period). After 7, 14, 21, and 28 days postimplantation, the implants were removed and the sheep were kept alive. It was reported that the (i) analysis of resin sections (30 µm) allowed the quantification of bone area (BA) and bone-to-implant contact (BIC), and TA group presented nearly 50% increase in BA at 14 days compared with 7 days; (ii) the TA presented higher values than the TN for BA and BIC at 14, 21, and 28 days after placement, stabilizing bone healing; and (iii) TA hydrophilic surface promoted early osseointegration at 14 and 21 days compared to TN, accelerating bone healing period post-implant placement in sheep tibia. Hou et al. [138] investigated the surface characteristic, biomechanical behavior, hemocompatibility, bone tissue response, and osseointegration of the optimal micro-arc oxidation surface-treated titanium (MST-Ti) dental implant. The surface characteristic, biomechanical behavior and hemocompatibility of the MST-Ti dental implant were performed using scanning electron microscope, finite element method, blood dripping, and immersion tests. The mini-pig model was utilized to evaluate the bone tissue response and osseointegration of the MST-Ti dental implant in vivo. It was found that (i) the hybrid volcano-like micro/nanoporous structure was formed on the surface of the MST-Ti dental implant; (ii) the hybrid volcano-like micro/nanoporous surface played an important role to improve the stress transfer between fixture, cortical bone, and cancellous bone for the MST-Ti dental implant; (iii) the MST-Ti implant was considered to have the outstanding hemocompatibility; (iv) the in vivo testing results showed that the BIC ratio significantly altered as the implant with micro/nanoporous surface; (v) after 12 weeks of implantation, the MST-Ti dental implant group exhibited significantly higher BIC ratio than the untreated dental implant group; and (vi) the MST-Ti dental implant group also presented an enhancing osseointegration, particularly in the early stages of bone healing. Based on these findings, it was concluded that the micro-arc oxidation approach induced the formation of micro/nanoporous surface is a promising and reliable alternative surface modification for Ti dental implant applications [138].

For successful osseointegration during the bone healing process, favorable wettability to blood flow is just one element among many others, and a detailed discussion will be continued in the later chapters on osseointegration.

4.2.5 Cytocompatibility

Cytology (or cell biology) is a branch of biology dealing with the structure, function, multiplication, pathology, and life history of cells, and its ability to be compatible to human body is called as cytological compatibility (in short, cytocompatibility). While cytotoxicity involves negative criteria such as cellular alterations, cell death, and hampered growth, then cytocompatibility suggests positive criteria. A material will be considered as cytocompatible if both structure and functions of the tissue in direct contact with it remain unchanged. The maintenance of these functions depends directly upon the quality of the material surface [139].

It is widely accepted that implant surface factors affect the quality of the bone-to-implant interface via enhancement of osteoblast cell adhesion. Cell characteristics (including cell adhesion strength, cell viability, cell proliferation, and cell stretching) all should affect directly the osseointegration along with the surface configurations of placed implant biomaterials, which will be discussed in the section of osseointegration. When a biological system encounters an implant, reactions are induced at the implant–tissue interface. Such interfacial reactions include that (i) the body wants to keep the implant isolated; (ii) a protective layer of macrophages, monocytes, and giant cells is formed; and (iii) a wall of collagen fibers (capsule) is established, and the type of material may influence the fibrous capsule thickness [12, 140]. Host response to biomaterials should be considered first as an inflammatory reaction and the host defenses will attempt to eliminate it. The aforementioned inflammatory processes should include four main stages: initial events (redness, swelling, and pain), cellular invasion (white cells invade tissues), tissue remodeling, repair (being orchestrated by macrophages; occurs differently in different tissues; bone may completely remodel; usually complete within 3–4 weeks), and resolution (extrusion, resorption, integration, and encapsulation). Furthermore, cellular invasion has neutrophil action (main function is phagocytes; die at tissue site; release further inflammatory medicators; prostaglandins and leukotrinets) and macrophages (phagocytes; removal of dead cells; number and activity depend on the pressure of a particulate material). Moreover, resolution is an attempt to return to the original condition: extrusion and resorption of implant material for reestablishment of homeostasis, and integration and encapsulation with a layer of fibrous tissue to establish a steady state. These responses resemble normal wound healing regarding cell recruitment and persistence [141, 142].

Hou et al. [143] used the rat oral implant model to assess histological changes in the mechanical environment surrounding loaded and unloaded dental implants.

The maxillary left first molar from rats was extracted, and the site was allowed to heal for 1 month. A titanium miniscrew implant was then placed into the site and allowed to heal for 21 days. The mandibular left first molars in one group of rats were extracted to create an unloaded condition. In a second group of rats, the mandibular left first molars were left in occlusion with the opposing screw head to simulate loading. Radiographs were taken on the day of placement and at 7, 14, and 21 days after postimplantation and were used to estimate the BIC ratio. It was reported that (i) areas of high shear stress adjacent to the helical threads of loaded implants were associated with osteocalcin localization and bone formation but only minimal localization of matrix metalloproteinase 13, and (ii) the bone adjacent to unloaded implants showed fibrous tissue and extensive matrix metalloproteinase 13 localization surrounding the apical two-thirds of each implant [143]. In terms of wettability, hydrophilicity has been gaining increasing interest as a factor that might influence the osseointegration of dental implants. Using nine screw-type implant systems from eight manufacturers, dynamic water contact angles were measured using the tensiometry by Rupp et al. [144]. Wettability was quantified by first advancing contact angles. It was concluded that (i) the first advancing mean contact angles of all implants ranged from 0° to 138°, demonstrating statistically significant differences among implant systems, and (ii) because of kinetic hysteresis, initially hydrophobic implants become hydrophilic by following immersion loops [144]. Normally, measuring the surface contact angle is performed by placing one liquid droplet onto the target substrate, followed by measuring the height (h) of the droplet and length (d) of the droplet after wetting. Using the sphere approximation, the contact angle (θ) can be obtained by the formula $\theta = 2 \tan^{-1}(2\,h/d)$. And it is the common practice to measure the advancing angle instead of receding angle which is the much lower angle since the wetting front is traveling on already wet area. The contact angle (if the used liquid does not show any corrosiveness against the substrate surface material) normally spreads; hence, the contact angle (θ) can be partially differentiated with respect to time (t) as the formula $\delta\theta/\delta t = [-4\,h\,(\delta d/\delta t) + 4\,d\,(\delta h/\delta t)] / (d^2 + 4\,h^2)$ [145].

Park et al. [146] evaluated compatibility of 9 (nine) types of pure metal (Ag, Al, Cu, Mn, Mo, Nb, V, Zr) and 36 experimental titanium (Ti) alloys containing 5, 10, 15, and 20 wt% of each alloying element. The cell viabilities for each test group were compared with that of cpTi using the WST-1 test and agar overlay test. It was found that (i) the ranking of pure metal cytotoxicity from most potent to least potent was as follows: Cu > Al > Ag > V > Mn > Cr > Zr > Nb > Mo > Ti; (ii) the mean cell viabilities for pure Cu, Al, Ag, V, and Mn were 21.6%, 25.3%, 31.7%, 31.7%, and 32.7%, respectively, which were significantly lower than that for the control group; (iii) the mean cell viabilities for pure Zr and Cr were 74.1% and 60.6%, respectively; (iv) pure Mo and Nb demonstrated good biocompatibility with mean cell viabilities of 93.3% and 93.0%, respectively; (v) the mean cell viabilities for all the Ti-based alloy groups were higher than 80% except for Ti–20 Nb (79.6%) and Ti–10V (66.9%); and (vii) the Ti–10 Nb alloy exhibited the highest cell viability (124.8%), which was higher than that of cpTi, suggesting that these data can serve as a guide for the development of new Ti-based alloy implant systems.

The potential for producing medical components via selective laser melting technology (SLM) was assessed [147]. The material tested consisted of the biocompatible titanium alloy Ti–6Al–4V. The research involved the testing of laboratory specimens produced using SLM technology both in vitro and for surface roughness. The aim of the research was to clarify whether SLM technology affects the cytocompatibility of implants and, thus, whether SLM implants provide suitable candidates for medical use following zero or minimum postfabrication treatment. The specimens were tested with an osteoblast cell line and, subsequently, two posttreatment processes were compared: nontreated (as-fabricated) and glass-blasted. Interactions with MG-63 cells were evaluated by means of metabolic MTT assay and microscopic techniques (SEM and fluorescence microscopy). Surface roughness was observed on both the nontreated and glass-blasted SLM specimens. It was concluded that (i) the glass blasting of SLM Ti–6Al–4V significantly reduces the surface roughness; (ii) the arithmetic mean roughness Ra was calculated at 3.4 μm for the glass-blasted and 13.3 μm for the nontreated surfaces; however, the results of in vitro testing revealed that the nontreated surface was better suited to cell growth. Ou et al. [148] fabricated various Ti–xZr (x = 5, 15, 25, 35, 45 wt%) alloys with low elastic modulus and high mechanical strength as a novel implant material. The biocompatibility of the Ti–xZr alloys was evaluated by osteoblast-like cell line (MG63) in terms of cytotoxicity, proliferation, adhesion, and osteogenic induction using CCK-8 and live/dead cell assays, electron microscopy, and real-time polymeric chain reaction. It was reported that (i) the Ti–xZr alloys were nontoxic and showed superior biomechanics compared to cpTi and (ii) Ti–45Zr had the optimum strength/elastic modulus ratio and osteogenic activity; thus, this is a promising dental implantable Ti-based alloy.

Mikulewicz et al. [149] performed an extensive review on the cytocompatibility of medical biomaterials containing nickel, as assessed by cell culture of human and animal osteoblasts or osteoblast-like cells, based on a survey of 21 studies on biomaterials including stainless steel, NiTi alloys, pure Ni, Ti, and other pure metals. It was mentioned that (i) the observation that the layers significantly reduced the initial release of metal ions, and increased cytocompatibility was confirmed in cell culture experiments; (ii) physical and chemical characterization of the materials was performed, including surface characterization (roughness, wettability, corrosion behavior, quantity of released ions, microhardness, and characterization of passivation layer); (iii) cytocompatibility tests of the materials were conducted in the cultures of human or animal osteoblasts and osteoblast-like cells in terms of cell proliferation and viability test, adhesion test, morphology (by fluorescent microscopy or SEM); (iv) in the majority of works, it was found that the most cytocompatible materials were stainless steel and NiTi alloy, while pure Ni was rendered and less cytocompatible; and (v) all papers confirmed that the consequence of the formation of protective layers was in significant increase of cytocompatibility of materials, indicating the possible further modifications of the manufacturing process (formation of the passivation layer). Nickel–titanium shape memory alloys are increasingly being used in orthopedic applications. However, there is a concern that Ni is harmful to the human body. NiTi archwires were studied as-received and after conditioning for 24 h or 35 days in a cell culture medium under static conditions [150]. It was mentioned that (i) all of the tested archwires, including their conditioned medium (CM), were noncytotoxic for L929 cells, but Rematitan SE (both as received and conditioned) induced the apoptosis of rat thymocytes in a direct contact; (ii) in contrast, TruFlex SE and Equire TA increased the proliferation of thymocytes; and (iii) the cytotoxic effect of Rematitan SE correlated with the higher release of Ni ions in CM, higher concentration of surface Ni, and an increased oxygen layer thickness after the conditioning. Based on these findings, it was concluded that the apoptosis assay on rat thymocytes, in contrast to the less sensitive standard assay on L929 cells, revealed that Rematitan SE was less cytocompatible compared to other archwires and the effect was most probably associated with a higher exposition of the cells to Ni on the surface of the archwire, due to the formation of unstable oxide layer. Hang et al. [151] fabricated NiTiO (nickel–titanium–oxygen) nanopores with different lengths (0.55–114 μm), which were anodically grown on nearly equiatomic NiTi alloy. Length-dependent corrosion behavior, nickel ion (Ni^{2+}) release, cytocompatibility, and antibacterial ability were investigated by electrochemical, analytical chemistry, and biological methods. It was reported that (i) constructing nanoporous structure on the NiTi alloy improved its corrosion resistance; however, the anodized samples release more Ni^{2+} than that of the bare NiTi alloy, suggesting chemical dissolution of the nanopores rather than electrochemical corrosion governs the Ni^{2+} release; (ii) in addition, the Ni^{2+} release amount increases with nanopore length; (iii) the anodized samples show good cytocompatibility when the nanopore length is <11 μm; and (iv) encouragingly, the length scale covers the one (1–11 μm) that the nanopores showing favorable antibacterial ability, concluding that the nanopores with length in the range of 1–11 μm are promising as coatings of biomedical NiTi alloy for anti-infection, drug delivery, and other desirable applications.

The titanium dioxide layer is composed mainly of anatase and rutile. This layer is prone to break (mainly due to its brittle ceramic nature), releasing particles to the milieu. Based on the previous finding that commercial anatase TiO_2 was deposited in organs with macrophagic activity, transported in the blood by phagocytic mononuclear cells, and induced an increase in the production of reactive oxygen species, Olmedo et al. [152] further evaluated the effects of rutile TiO_2, using male Wistar rats being injected with a suspension of rutile TiO_2 powder. It was reported that titanium was found in phagocytic mononuclear cells, serum, and in the parenchyma of all the organs tested. Although rutile TiO_2 provoked an augmentation of reactive oxygen species, it failed to induce damage to membrane lipids, possibly due to an adaptive response. Rutile TiO_2 is less bioactive than the anatase TiO_2 [146]. Antibacterial effect and cytocompatibility of a nanostructured TiO_2 (titania) film containing Cl on CpTi surface was investigated [153]. The nanostructured TiO_2 was prepared by anodization with hydrofluoric acid, followed by secondary anodization with NaCl solution of different concentrations (0.4, 1, and 2 M). The antibacterial effect was evaluated by film adhesion method, and cytocompatibility was tested by the viability of MG-63 cells in an MTT assay. It was found that (i) the contact angle (as a function of surface wettability) showed anodized titanium as a hydrophilic nature, and (ii) TiO_2 nanostructured film anodized in 1 M NaCl exhibited the best antibacterial effect and cell cytocompatibility [153]. Yang et al. [154] investigated the in vitro cytocompatibility with osteogenic cells and the in vivo anti-infection activity of titanium implants with HACC-loaded nanotubes (NT-H). The titanium implant (Ti), nanotubes without polymer loading (NT), and nanotubes loaded with chitosan (NT-C) were fabricated and served as controls. Firstly, we evaluated the cytocompatibility of these specimens with human bone marrow-derived mesenchymal stem cells in vitro. The observation of cell attachment, proliferation, spreading, and viability in vitro showed that NT-H has improved osteogenic activity compared with Ti and NT-C. A prophylaxis rat model with implantation in the femoral medullary cavity and inoculation with methicillin-resistant *Staphylococcus aureus* was established and evaluated by radiographical, microbiological, and histopathological assessments. It was reported that the in vivo study demonstrated that NT-H coatings exhibited significant anti-infection capability compared with the Ti and NT-C groups, suggesting that HACC-loaded nanotubes fabricated on a titanium substrate show good compatibility with osteogenic cells and enhanced anti-infection ability in vivo, providing a good foundation for clinical application to combat orthopedic implant-associated infections [154]. The positive cell response to the implant material is reflected by the capacity of cells to divide, which leads to the tissue regeneration and osseointegration. Prachár et al. [155] studied the properties of titanium nitride (TiN) and zirconium nitride (ZrN) coatings deposited by PVD (physical vapor deposition), which were applied to substrates of pure titanium, Ti-based alloys (Ti–6Al–4V and Ti–35Nb–6Ta), and CoCrMo dental alloy. Different treatments of substrate surfaces were used: polishing, etching, and grit blasting. Cytocompatibility tests assessed the cell colonization and their adherence to substrates. It was found that (i) TiN layers deposited by PVD are suitable for coating all substrates studied, (ii) the polished samples and those with TiN coating exhibited a higher cell colonization, and (iii) this coating technique meets the requirements for the biocompatibility of the implanted materials.

4.2.6 MRI compatibility

MRI is a technology developed in medical imaging that is probably the most innovative and revolutionary other than the computed tomography. MRI is a 3D imaging technique used to image the protons of the body by employing magnetic fields, radio frequencies, electromagnetic detectors, and computers [156]. For millions of patients worldwide, MRI examinations provide essential and potentially life-saving information. Some devices, such as pacemakers and neurostimulators, have limitations related to MRI safety and may be contraindicated for use with MRI. Even more devices, such as stents, vena cava filters, and some types of catheters and guidewires, are safe for use with MRI but have limited MRI image compatibility. Some of these devices are simply not well-imaged under MRI. Others have properties that interfere with the MRI image by causing an image artifact (distortion) in the area in and around the device, limiting the effectiveness of MRI for assisting placement or diagnostic follow-up on these implants. It may be contraindicated in certain situations because the magnetic field present in the MRI environment may, under certain circumstances, result in movement or heating of a metallic orthopedic implant device. Metals that exhibit magnetic attraction in the MRI setting may be subject to movement (deflection) during the procedure. Both magnetic and nonmagnetic metallic devices of certain geometries may also be subjected to heating caused by interactions with the magnetic field. Of secondary concern is the possibility of image artifacts that can compromise the procedure and image quality.

There are different types of contraindications that would prevent a person from being examined with an MRI scanner. MRI systems use strong magnetic fields that attract any ferromagnetic objects with enormous force. Caused by the potential risk of heating, produced from the radio frequency pulses during the MRI procedure, metallic objects like wires, foreign bodies, and other implants need to be checked for compatibility. High-field MRI requires particular safety precautions. In addition, any device or MRI equipment that enters the magnet room has to be MRI compatible. MRI examinations are safe and harmless, if these MRI risks are observed. Safety concerns in MRI include the magnetic field strength, possible missile effects caused by magnetic forces, the potential for heating of body tissue due to the application of radio frequency energy, the effects on implanted active devices such as cardiac pacemakers or insulin pumps, magnetic torque effects on the indwelling metal (clips), the audible acoustic noise, danger due to cryogenic liquids, and the application of contrast medium. MRI use is contraindicated for the following devices [157, 158]: cardiac pacemakers are absolutely contraindicated; however, MRI-compatible pacemakers are being developed. A few pacemaker patients have been scanned for life-threatening situations or inadvertently. Subsequently, their pacemakers must be meticulously checked for function. Some deaths have been reported; intracranial aneurysm clips are contraindicated, unless the specific type of MRI-compatible clip can be absolutely documented; and neurostimulators/spinal-fusion stimulators are generally contraindicated. There are two manufacturers that seek FDA approval for usage of their products in MRI but under strict guidelines, and drug fusion pumps are generally contraindicated. Synchro Med and Synchro Med EL are the two models that are currently FDA approved with specific guidelines, and metallic foreign bodies may or may not preclude MRI scanning depending on the type, size, and location. These patients must be screened by X-ray. Detection should also target possible orbital metallic foreign bodies, cochlear implants (contraindicated), and other otologic implants (generally, these are nonferromagnetic and MRI compatible). The exception is the McGee stapedectomy piston prosthesis, which is not MRI compatible. Dental implants are generally compatible, except for those that contain magnetically activated components; ocular implants are MRI compatible; and intravascular stents, filters, and coils are compatible 6–8 weeks following placement, unless they are made of nonferrous metal, for example, titanium, in which case they can be imaged right after the placement. Drug-eluting stents must be cleared by the implanting physician if they have been in less than 3 months; vascular access ports and catheters are compatible, excluding the Swan Ganz catheter; penile implants are of mixed compatibility and should be checked; orthopedic implants/prosthesis are compatible, though MRI may cause local heating and MRI will cause local image artifact; heart valve prosthesis is compatible, but there is a prototypic electromagnetically controlled heart valve that is being developed, which is contraindicated; pacer wires without pacemaker having MRI compatibility are controversial but are probably fine at low MRI fields and questionable at high-field imaging. All patients must have a chest X-ray prior to having an MRI to ensure that the wires are not looped or crossed, Holter monitor (contraindicated), ventricular-peritoneal shunts (compatible, except SOPHY-adjustable pressure valve type), Swan Ganz catheter (contraindicated; the thermal dilutor may melt), dermal patches (can cause burns during MRI study and should be removed), pessary and diaphragm (compatible, but it will produce local image artifact), hearing aids (must be removed), permanent eyeliner/tattoos (can cause local burns), body rings/spikes (can cause local burns, dislodgment, or artifacts; it is suggested to remove these devices under normal MRI or the use under low-field MRI), breast tissue expanders (not compatible), and bullets, shrapnel, and pellets (depends on the foreign body's location and duration of the MRI; an X-ray is required before an MRI can be done) [157, 158].

MRI is widely used as an important diagnostic tool, especially for orthopedic and brain surgery. This method has remarkable advantages for obtaining various cross-sectional views and for diagnosis of the human body with no invasion and no exposure of the human body to X-ray radiation. However, MRI diagnosis is inhibited when metals are implanted in the body, since metallic implants such as stainless steels, Co–Cr alloys, and Ti alloys become magnetized in the intense magnetic field of the MRI instrument, and artifacts occur in the image [159–162]. Ernstberger et al. [163] determined whether magnesium as a lightweight and biocompatible metal is suitable as a biomaterial for spinal implants based on its MRI artifacting behavior. To compare artifacting behaviors, different test spacers made of magnesium, titanium, and carbon-fiber-reinforced thermoplastics (CFRP) were implanted into one cadaveric spine, as shown in Figure 4.8. All test spacers were scanned using two T1-TSE MRI sequences. The artifact dimensions were traced on all scans and statistically analyzed. MRI was performed with a 1.5-T MRI (Magnetom Symphony, Siemens AG Medical Solutions, Erlangen, Germany). The T1w-TSE sequences were used to acquire a slice thickness of 3 mm (see Figure 4.8), which included a first sequence (TR 600; TE 14; flip angle 15; band width 150) and a second sequence (TR 2,260, TE 14, flip angle 15, band width 150). A matrix of 512 × 512 pixels combined with a field of view of 500 mm was chosen for the study. It was reported that the total artifact volume and median artifact area of the titanium spacers were statistically significantly larger than magnesium spacers, while magnesium and CFRP spacers produced almost identical artifacting behaviors, suggesting that spinal implants made with magnesium alloys will behave more like CFRP devices in MRI scans, and given its osseoconductive potential as a metal, implant alloys made with magnesium would combine the advantages of the two principal spacer materials currently used but without their limitations, at least in terms of MRI artifacting [163].

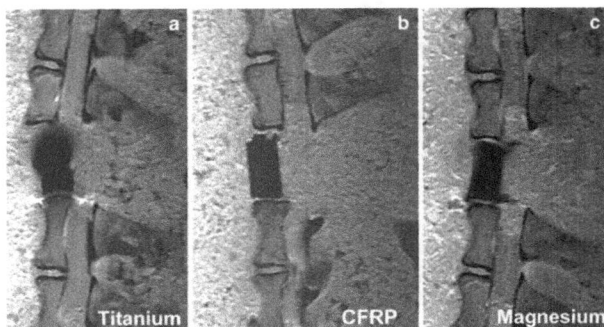

Figure 4.8: Median MRI artifact range depicted in a selection of three large test implants: (a) Ti–6Al–4V, (b) CFRP (carbon-fiber-reinforced thermoplastic), and (c) AM50 (Mg–5Al–0.5Mn–0.2Zn) [163].

There are currently several researchers as well as an ASTM committee exploring methods for accurately assessing the MRI compatibility of implant devices. The primary focus of the research has been the measurement of implant movement in response to a magnetic field. Shellock and coworkers [164–167] conducted several studies in which the movement/deflection of various orthopedic implants was measured in the high magnetic field (0.3–1.5 Tesla) region of MRI units. The results of these studies show no measurable movement of implants fabricated from cobalt, titanium, and stainless steel alloys. The movement/deflection of selected orthopedic implants in a 3.0 Tesla MRI unit was also examined and it was found that devices fabricated from cobalt, titanium, and stainless steel exhibited little or no movement/deflection [166]. Ferromagnetic metal will cause a magnetic field inhomogeneity, which, in turn, causes a local signal void, often accompanied by an area of high signal intensity, as well as a distortion of the image. They create their own magnetic field and dramatically alter precession frequencies of protons in the adjacent tissues. Tissues adjacent to ferromagnetic components become influenced by the induced magnetic field of the metal hardware rather than the parent field and, therefore, either fail to process or do so at a different frequency and hence do not generate useful signal. Two components contribute to susceptibility artifact, induced magnetism in the ferromagnetic component itself, and induced magnetism in protons adjacent to the component. Artifacts from metal may have varied appearances on MRI scans due to different types of metal or configuration of the piece of metal. In relation to imaging titanium alloys are less ferromagnetic than both cobalt and stainless steel, induce less susceptibility artifact, and result in less marked image degradation [167].

Medical biomaterials are being used to replace or support severely damaged or completely missing tissue and organs at the human body. Current usage of these products in medicine has become quite prevalent, thanks to developments of biomaterial technology. However, in case a medical biomaterial is exposed to an environment including extremely strong energy that MRI device generates, undesirable results in terms of human health may occur. Hence, evaluation of MRI compatibility as well as safety is essential [168]. ASTM's medical device and implant standards are instrumental in specifying and evaluating the design and performance requirements of a number of biomedical materials, tools, and equipment. These apparatuses are used in surgical procedures that involve the placement of such devices to the specified parts and structures of the body (both humans and animals) for the purpose of enhancement or as an aid in a disability. These medical devices and implant standards allow material and product manufacturers, medical laboratories, and other concerned institutions to inspect and assess such instruments to ensure proper quality and workmanship [169]. List of medical device standards and implant standards developed by ASTM include 42 designations for arthroplasty, 7 for assessment for temps, 26 for biocompatibility test methods, 20 for biomaterials and biomolecules for temps, 17 for cardiovascular standards, 2 for cell signaling, 11 for cells and tissue-engineered constructs for temps, 10 for ceramic materials, 3 for classification and terminology for temps, 2 for computer-assisted orthopedic surgical systems, 1 for consumer rubber products, 1 for GI applications, 2 for human clinical trials, 1 for implantable hearing devices, 56 for material test methods, 18 for medical/surgical instruments, 48 for metallurgical materials, 4 for neurosurgical standards, 15 for osteosynthesis, 4 for plastic and reconstructive surgery, 24 for polymeric materials, 16 for spinal devices, and 1 for urological materials and devices. ASTM F2503 – 20 designates the "Standard Practice for Marking Medical Devices and Other Items for Safety in the Magnetic Resonance Environment" [170].

References

[1]   Implantable Biomaterials Global Market: Forecast to 2023 – Growing Demand for Plastic Surgery – Research and Markets. 2016; https://www.businesswire.com/news/home/ 20161220005897/en/Implantable-Biomaterials-Global-Market-Forecast-to-2023–Growing-De mand-for-Plastic-Surgery–Research-and-Markets.

[2] $21.12 Billion Implantable Biomaterials Market: Global Forecasts to 2023 – Miniaturization of Implant Devices and Increasing Minimally Invasive Surgery. Research & Markets. 2017;

https://www.globenewswire.com/fr/news-release/2017/03/10/934363/28124/en/21-12-Bil
lion-Implantable-Biomaterials-Market-Global-Forecasts-to-2023-Miniaturization-of-Implant-
Devices-and-Increasing-Minimally-Invasive-Surgery.html.

[3] Badal J, Canvasser N. Robotic miniaturization: Reducing the surgical footprint. Urology Times. 2019; https://www.urologytimes.com/view/robotic-miniaturization-reducing-surgical-footprint.

[4] Jones P, Bennett G, Aboumarzouk OM, Griffin S, Somani BK. Role of minimally invasive percutaneous nephrolithotomy techniques-micro and ultra-mini PCNL (<15F) in the pediatric population: A systematic review. J Endourol. 2017, 31, 816–24.

[5] Usui K, Komeya M, Taguri M, Kataoka K, Asai T, Ogawa T, Yao M, Matsuzaki J. Minimally invasive versus standard endoscopic combined intrarenal surgery for renal stones: A retrospective pilot study analysis. Intl Urol Nephrol. 2020, 52, 1219–25.

[6] Rojas SV, Avsar M, Uribarri A, Hanke JS, Haverich A, Schmitto JD. A new era of ventricular assist device surgery: Less invasive procedures. Minerva Chir. 2015, 70, 63–68.

[7] Rojas SV, Avsar M, Hanke JS, Khalpey Z, Maltais S, Haverich A, Schmitto JD. Minimally invasive ventricular assist device surgery. Artif Organs. 2015, 39, 473–79.

[8] Park J, Lee H, Kee H, Park S. Magnetically steerable manipulator with variable stiffness using graphene polylactic acid for minimally invasive surgery. Sens Actuators A Phys. 2020, 309, 112032, doi: 10.1016/j.sna.2020.112032.

[9] Enriquez J. Demand For Minimally Invasive Surgery Driving Surgical Robotics Development, Miniaturization. Med Device Online. 2016; https://www.meddeviceonline.com/doc/demand-for-minimally-invasive-surgery-driving-surgical-robotics-development-miniaturization-0001.

[10] BYU. Surgical tools made smaller with origami to make surgery less invasive. 2016; https://www.eurekalert.org/pub_releases/2016-03/byu-stm030716.php.

[11] HKPloyU. The world's first internally motorized minimally invasive surgical robotic system. 2016; https://www.eurekalert.org/pub_releases/2016-03/thkp-twf030216.php.

[12] Oshida Y. Bioscience and Bioengineering of Titanium Materials, Elsevier Pub., 2007.

[13] Panagiotopoulou O, Kupczik K, Cobb SN. The mechanical function of the periodontal ligament in the macaque mandible: A validation and sensitivity study using finite element analysis. J Anat. 2011, 218, 75–86.

[14] Sours CL. Implant-supported fixed bridgework; http://www.charleslsoursdds.com/library/8036/Implant-SupportedFixedBridgework.html.

[15] Smith VS. Dental implant maintenance. Dear Doctor, Dentistry and Oral Health. https://www.deardoctor.com/inside-the-magazine/issue-21/dental-implant-maintenance/.

[16] McCormack SW, Witzel U, Watson PJ, Fagan MJ, Gröning F. The biomechanical function of periodontal ligament fibres in orthodontic tooth movement. PLos One. 2014, 9, e102387, doi: 10.1371/journal.pone.0102387.

[17] Pavasant P, Yongchaitrakul T. Role of mechanical stress on the function of periodontal ligament cells. Periodontology. 2000, 2011, 56, 154–65.

[18] Kruger J. Fundamental aspects of the corrosion of metallic implants. In: ASTM STP 684. Syrett BC, et al., eds. American Society for Testing and Materials, Corrosion and Degradation of Implant Materials, 1979, 107–27.

[19] Greene ND. Corrosion of surgical implant alloys: A few basic ideas. In" ASTM STP 859. Fraker AC, et al., eds. Corrosion and Degradation of Implant Materials: Second Symposium. 1983, 5–10.

[20] Brockhurst PJ. Dental materials: New territories for materials science. Metals Forum Australian Inst Metals. 1980, 3, 200–10.

[21] Steinmann SG. Corrosion of surgical implants – In vivo and in vitro tests. In: Winter GD, Leray JL, de Groot K, eds. Evaluation of Biomaterials. New York: John Wiley & Sons, 1980, 1–34.

[22] Ferguson AB, Akahoshi Y, Laing PG, Hodge ES. Characteristics of trace ions released from embedded metal implants in the rabbit. J Bone Joint Surg. 1962, 44-A, 323–36.

[23] Meachim G, Williams DF. Changes in nonosseous tissue adjacent to titanium implants. J Biomed Mater Res. 1973, 7, 555–72.

[24] Ducheyne P, Williams G, Martens M, Helsen J. In vivo metal-ion release from porous titanium-silver material. J Biomed Mater Res. 1984, 18, 293–308.

[25] Sundgren J-E, Bodö P, Lundström I. Auger electron spectroscopic studies of the interface between human tissue and implants of titanium and stainless steel. J Colloid Interf Sci. 1986, 110, 9–20.

[26] Ducheyne P, Healy KE. Surface spectroscopy of calcium phosphate ceramic and titanium implant materials. In: Ratner BD, ed. Surface Characterization of Biomaterials. Amsterdam: Elsevier Science, 1988, 175–92.

[27] Fraker AC, Ruff AW. Corrosion of titanium alloys in physiological solutions. In: Titanium Science and Technology, Vol. 4. New York: Plenum Press, 1973, 2447–57.

[28] Albrektsson T. Direct bone anchorage of dental implants. J Prosth Dent. 1983, 50, 255–61.

[29] Albreksson T, Hansson HA. An ultrastructural characterization of the interface between bone and sputtered titanium or stainless steel surfaces. Biomaterials. 1986, 7, 201–05.

[30] Liedberg B, Ivarsson B, Lundstrom I. Fourier transform infrared reflection absorption spectroscopy (FTIR-RAS) of fibrinogen adsorbed on metal and metal oxide surfaces. J Biochem Biophys Method. 1984, 9, 233–43.

[31] Hanawa T. Titanium and its oxide film: A substrate for formation of apatite. In: Davies JE, ed. The Bone-biomaterial Interface. Toronto: Univ. of Toronto Press, 1991, 49–61.

[32] Healy KE, Ducheyne P. Hydration and preferential molecular adsorption on titanium in vitro. Biomaterials. 1992, 13, 553–61.

[33] Healy KE, Ducheyne P. The mechanisms of passive dissolution of titanium in a model physiological environment. J Biomed Mater Res. 1992, 26, 319–38.

[34] Oshida Y, Tominaga T. Nickel-Titanium Materials, De Gruyter Pub., 2020.

[35] Pourbaix MJN. Atlas of Electrochemical Equilibria in Aqueous Solutions. NY: Pergamon Press, 1966.

[36] Corso PP, German RM, Simmons HD. Corrosion evaluation of gold-based dental alloys. J Dent Res. 1985, 64, 854–59.

[37] Reclaru L, Meyer JM. Zonal coulometric analysis of the corrosion resistance of dental alloys. J Dent Res. 1995, 23, 301–11.

[38] Ewers GJ, Thornber MR. The effect of a simulated environment on dental alloys. J Dent Res. 1983, 62, 330, Abstract No. 330.

[39] Ewers GJ, Greener EH. The electrochemical activity of the oral cavity – A new approach. J Oral Rehabil. 1985, 12, 469–76.

[40] Wang Y. Bioadaptability: An innovative concept for biomaterials. J Mater Sci Technol. 2016, 32, 801–09.

[41] Skalak R. Biomechanical considerations in osseointegrated prostheses. J Prosthet Dent. 1983, 49, 843–48.

[42] Hekimoglu C, Anil N, Cehreli MC. Analysis of strain around endosseous implants opposing natural teeth or implants. J Prosthet Dent. 2004, 92, 441–46.

[43] van Rossen IP, Braak LH, de Putter C, de Groot K. Stress-absorbing elements in dental implants. J Prosthet Dent. 1990, 64, 198–205.

[44] Oshida Y, Hashem A, Nishihara T, Yapchulay MV. Fractal dimension analysis of mandibular bones: Toward a morphological compatibility of implants. J BioMed Mater Eng. 1994, 4, 97–107.

[45] Oshida Y. Magnesium Materials, De Gruyter Pub., 2021.

[46] Ikarashi Y, Toyoda K, Kobayashi E, Doi H, Yoneyama T, Hamanaka H, Tsuchiya T. Improved biocompatibility of titanium–zirconium (Ti–Zr) Alloy: Tissue reaction and sensitization to Ti–Zr alloy compared with pure Ti and Zr in rat implantation study. Mater Trans. 2005, 46, 2260–67.

[47] Grandin HM, Berner S, Dard M. A review of titanium zirconium (TiZr) alloys for use in endosseous dental implants. Materials. 2012, 5, 1348–60.

[48] Cordeiro JM, Faveranic LP, Grandini CR. Characterization of chemically treated Ti-Zr system alloys for dental implant application. Mater Sci Eng C. 2018, 92, 849–61.

[49] Chiapasco M, Casentini P, Zaniboni M, Corsi E, Anello T. Titanium-zirconium alloy narrow-diameter implants (Straumann Roxolid(®)) for the rehabilitation of horizontally deficient edentulous ridges: Prospective study on 18 consecutive patients. Clin Oral Implants Res. 2012, 23, 1136–41.

[50] Hoffmann O, Angelov N, Gallez F, Jung RE, Weber FE. The zirconia implant-bone interface: A preliminary histologic evaluation in rabbits. Int J Oral Maxillofac Implants. 2008, 23, 691–95.

[51] Özkurt Z, Kazazoğlu E. Zirconia dental implants: A literature review. J Oral Implantol. 2011, 37, 367–76.

[52] Adatia D, Bayne SC, Cooper LF, Thompson JY. Fracture resistance of yttria-stabilized zirconia dental implant abutments. J Prosthodont. 2009, 18, 17–22.

[53] Karapataki S. From titanium to zirconia implants. Ceramic Implants. 2017, 1, 6–12.

[54] Donaca R, Rausch P. Shifting of dental implants through ISO standards. Ceramic Implants. 2017, 1, 34–39.

[55] Oshida Y, Miyazaki T, Tominaga T. Some biomechanistic concerns on newly developed implantable materials. J Dent Oral Health. 2018, 4, https://scientonline.org/open-access/some-biomechanistic-concerns-on-newly-developed-implantable-materials.pdf

[56] Eriksson C, Lausmaa J, Nygren H. Interactions between human whole blood and modified TiO_2-surfaces: Influence of surface topography and oxide thickness on leukocyte adhesion and activation. Biomaterials. 2001, 22, 1987–96.

[57] Wen X, Wang X, Zhang N. Microsurface of metallic biomaterials: A literature review. J BioMed Mater Eng. 1996, 6, 173–89.

[58] Oshida Y. Requirements for successful biofunctional implants. The 2nd Symposium International of Advanced BioMaterials. Montreal, Canada. 2000, 5–10.

[59] Oshida Y, Tuna EB. Science and technology integrated titanium dental implant systems. In: Basu B, et al., eds. Advanced Biomaterials. John Wiley & Sons, 2009, 143–77.

[60] Keller JC, Dougherty WJ, Grotendorst GR, Wrightman JP. In vitro cell attachment to characterized cpTi surfaces. Adhesion. 1989, 28, 115–33.

[61] Keller JC, Wrightman JP, Dougherty WJ. Characterization of acid passivated cpTi surfaces. J Dent Res. 1989, 68, 872, Abstract.

[62] Keller JC, Draughn RA, Wrightman JP, Dougherty WJ. Characterization of sterilized CP titanium implant surfaces. Int J Oral Maxillofac Implants. 1990, 5, 360–69.

[63] Schroeder A, van der Zypen E, Stich H, Sutter F. The reactions of bone, connective tissue and epithelium to endosteal implants with titanium sprayed surfaces. J Maxillofac Surg. 1981, 9, 15–25.

[64] Rich A, Harris AK. Anomalous preferences of cultured macrophages for hydrophobic and roughened substrata. J Cell Sci. 1981, 50, 1–7.

[65] Buser D, Schenk RK, Steinemann S, Fiorellinni JP, Fox CH, Stich H. Influence of surface characteristics on bone integration of titanium implants. A histomorphometric study in miniature pigs. J Biomed Mater Res. 1991, 25, 889–902.

[66] Buser D, Nydegger T, Oxland T, Cochran DL, Schenk RK, Hirt HP, Sneitivy D, Nolte LP. Interface shear strength of titanium implants with a sandblasted and acid-etched surface: A biomechanically study in the maxilla of miniature pigs. J Biomed Mater Res. 1999, 45, 75–83.

[67] Murray DW, Rae T, Rushton N. The influence of the surface energy and roughness of implants on bone resorption. J Bone Joint Surg. 1989, 71B, 632–37.

[68] Cherhoudi B, Gould TR, Brunette DM. Effects of a grooved epoxy substratum on epithelial behavior in vivo and in vitro. J Biomed Mater Res. 1988, 22, 459–77.

[69] von Recum AF. In: Heimke G, et al., eds. Clinical Implant Materials. Amsterdam, The Netherlands: Elsevier Science Pub., 1990, 297–302.

[70] Deligianni DD, Katsala N, Ladas S, Sotiropoulou D, Amedee J, Missirlis YF. Effect of surface roughness of the titanium alloy Ti-6Al-4V on human bone marrow cell response and on protein adsorption. Biomaterials. 2001, 22, 1241–51.

[71] Yang Y, Cavin R, Ong LJ. Protein adsorption on titanium surfaces and their effect on osteoblast attachment. J BioMed Mater Res A. 2003, 67, 344–49.

[72] Kasemo B, Lausmaa J. Surfaces science aspects on inorganic. Biomater CRC Crit Rev Biocomp. 1986, 2, 335–80.

[73] Schenk RK, Buser D. Osseointegration: A reality. Periodontology. 1998, 17, 22–35.

[74] Masuda T, Yliheikkaila PK, Fleton DA, Cooper LF. Generalizations regarding the process and phenomenon of osseointegration. Part I. *In vivo* studies. Int J Oral Maxillofac Implants. 1998, 13, 17–29.

[75] Larsson C, Esposito M, Liao H, Thomsen P. In: Titanium in Medicine: Materials Science, Surface Science, Engineering, Biological Responses, and Medical Applications. Brunette DM, et al., eds. New York, NY, USA: Springer, 2001, 587–648.

[76] Oshida Y, Sachdeva R, Miyazaki S. Microanalytical characterization and surface modification of NiTi orthodontic archwires. J BioMed Mater Eng. 1991, 2, 51–69.

[77] Zinger O, Anselme K, Denzer A, Habersetzer P, Wieland M, Jeanfils J, Hardouin P, Landolt D. Time-dependent morphology and adhesion of osteoblastic cells on titanium model surfaces featuring scale-resolved topography. Biomaterials. 2004, 25, 2695–711.

[78] Ratner BD. Surface characterization of biomaterials by electron spectroscopy for chemical analysis. Ann Biomed Eng. 1983, 11, 313–36.

[79] Baro AM, Garcia N, Miranda R, Vázquez L, Aparicio C, Olivé J, Lausmaa J. Characterization of surface roughness in titanium dental implants measured with scanning tunneling microscopy at atmospheric pressure. Biomaterials. 1986, 7, 463–66.

[80] Kasemo B. Biocompatibility of titanium implants: Surface science aspects. J Pros Dent. 1983, 49, 832–37.

[81] Moroni A, Caja VL, Egger EL, Trinchese L, Chao EY. Histomorphometry of hydroxyapatite coated and uncoated porous titanium bone implants. Biomaterials. 1994, 15, 926–30.

[82] Keller JC, Young FA, Natiella JR. Quantitative bone remolding resulting from the use of porous dental implants. J Biomed Mater Res. 1987, 21, 305–19.

[83] Pilliar RM, Deporter DA, Watson PA, Valiquette N. Dental implant design-effect on bone remodeling. J Biomed Mater Res. 1991, 25, 467–83.

[84] Gilbert JL, Berkery CA. Electrochemical reaction to mechanical disruption of titanium oxide films. J Dent Res. 1995, 74, 92–96.

[85] Deporter D, Watson P, Pharoah M, Levy D, Todescan R. Five-to six-year results of a prospective clinical trial using the ENDOPORE dental implant and a mandibular overdenture. Clin Oral Implants Res. 1999, 10, 95–102.

[86] Palmer RM, Plamer PJ, Smith BJ. A 5-year prospective study of Astra single tooth implants. Clin Oral Implants Res. 2000, 11, 179–82.

[87] Testori T, Wiseman L, Woolfe S, Porter SS. A prospective multicenter clinical study of the Osseotite implant: Four-years interim report. Int J Oral Maxillofac Implants. 2001, 16, 193–200.

[88] Sul Y-T. The significance of the surface properties of oxidized titanium to the bone response: Special emphasis on potential biochemical bonding of oxidized titanium implant. Biomaterials. 2003, 24, 3893–907.

[89] Ishizawa H, Fujiino M, Ogino M. Mechanical and histological investigation of hydrothermally treated and untreated anodic titanium oxide films containing Ca and P. J Biomed Mater Res. 1995, 29, 1459–68.

[90] Larsson C, Emanuelsson L, Thomsen P, Ericson L, Aronsson B, Rodahl M, Kasemo B, Lausmaa J. Bone response to surface-modified titanium implants: Studies on the tissue response after one year to machined and electropolished implants with different oxide thicknesses. J Mater Sci : Mater Med. 1997, 8, 721–29.

[91] Skripitz R, Aspenberg P. Tensile bond between bone and titanium. Acta Orthop Scand. 1998, 6, 2–6.

[92] Fini M, Cigada A, Rondelli G, Chiesa R, Giardino R, Giavaresi G, Aldini N, Torricelli P, Vicentini B. In vitro and vivo behaviour of Ca and P-enriched anodized titanium. Biomaterials. 1999, 20, 1587–94.

[93] Henry P, Tan AE, Allan BP. Removal torque comparison of TiUnite and turned implants in the Greyhound dog mandible. Appl Osseointegr Res. 2000, 1, 15–17.

[94] Sul Y-T, Johansson CB, Jeong Y, Röser K, Wennerberg A, Albreksson T. Oxidized implants and their influence on the bone response. J Mater Sci : Mater Med. 2001, 12, 1025–31.

[95] Sul Y-T, Johansson CB, Jeong Y, Wennerberg A, Albrektson T. Resonance frequency and removal torque analysis of implants with turned and anodized surface oxide. Clin Oral Implants Res. 2002, 13, 252–59.

[96] Sul Y-T, Johansson CB, Albreksson T. Oxidized titanium screws coated with calcium ions and their performance in rabbit bone. Int J Oral Maxillofac Implants. 2002, 17, 625–34.

[97] Lim YJ, Oshida Y, Andres CJ, Barco MT. Surface characterizations of variously treated titanium materials. Int J Oral Maxillofac Implants. 2001, 16, 333–42.

[98] Oshida Y. Surface science and technology – Titanium dental implant systems. J Soc Titanium Alloys Dent. 2007, 5, 52–53.

[99] Elias CN, Oshida Y, Lima JHC, Muller CA. Relationship between surface properties (roughness, wettability and morphology) of titanium and dental implant torque. J Mech Behav Biomed Mater. 2008, 1, 234–42.

[100] Lausmaa GJ. Chemical composition and morphology of titanium surface oxides. Mat Res Soc Symp Proc. 1986, 55, 351–59.

[101] Albrektsson T, Jacobsson M. Bone-metal interface in osseointegration. J Prosthet Dent. 1987, 57, 5–10.

[102] Brånemark PI. Tissue-Integrated Prostheses. In: Brånemark PI, ed. Chicago, IL, USA: Quintessence Pub., 1985, 63–70.

[103] Kasemo B, Lausmaa J. In: Tissue-Integrated Prostheses. Brånemark PI, ed. Chicago, IL, USA: Quintessence, 1985, 108–15.

[104] Chehroudi B, McDonnel D, Brunette DM. The effects of micromachined surfaces on formation of bonelike tissue on subcutaneous implants as assessed by radiography and computer image processing. J Biomed Mater Res. 1997, 34, 279–90.

[105] Larsson C, Thomsen P, Lausmaa J, Rodahl M, Kasemo B, Ericson LE. Bone response to surface modified implants: Studies on electropolished implants with different oxide thickness and morphology. Biomaterials. 1994, 15, 1062–64.

[106] Simon-Walker R, Cavicchia J, Prawel DA, Dasi LP, James SP, Popat KC. Hemocompatibility of hyaluronan enhanced linear low density polyethylene for blood contacting applications. J Biomed Mater Res B Appl Biomater. 2018, 106, 1964–75.

[107] Weber M, Steinle H, Golombek S, Hann L, Schlensak C, Wendel HP, Avci-Adali M. Blood-contacting biomaterials: In vitro evaluation of the hemocompatibility. Front Bioeng Biotechnol. 2018, 6, 99, doi: 10.3389/fbioe.2018.00099.

[108] Liu X, Yuan L, Li D, Tang Z, Wang Y, Chen G, Chen H, Brash JL. Blood compatible materials: State of the art. J Mater Chem B. 2014, 2, 5718–38.

[109] International Organization for Standardization. Biological Evaluation of Medical Devices. 2000; https://www.iso.org/standard/63448.html.

[110] Sacks MS, Yoganathan AP. Heart valve function: A biomechanical perspective. Philos Trans R Soc Lond B Biol Sci. 2007, 362, 1369–91.

[111] Ratner BD. The catastrophe revisited: Blood compatibility in the 21st century. Biomaterials. 2007, 28, 5144–47.

[112] Vogler EA, Siedlecki CA. Contact activation of blood-plasma coagulation. Biomaterials. 2009, 30, 1857–69.

[113] Kuharsky AL, Fogelson AL. Surface-medicated control of blood coagulation: The role of binding site densities and platelet deposition. Biophys J. 2001, 80, 1050–74.

[114] Huang N, Yang P, Leng YX, Chen JY, Sun H, Wang J, Wang JG, Ding PD, Xi TF, Leng Y. Hemocompatibility of titanium oxide films. Biomaterials. 2003, 24, 2177–87.

[115] Dion I, Baquey C, Monties JR, Havlik P. Harmocompatibility of Ti6Al4V alloy. Biomaterials. 1993, 14, 122–6.

[116] Eriksson AS, Bjursten LM, Ericson LE, Thomsen P. Hollow implants in soft tissue allowing quantitative studies of cells and fluid at the implant interface. Biomaterials. 1988, 9, 86–90.

[117] Eriksson AS, Thomsen P. Leukotriene B4, interleukin 1 and leucocyte accumulation in titanium and PTFE chambers after implantation in the rat abdominal wall. Biomaterials. 1991, 12, 827–30.

[118] Elwing H, Tengvall P, Askendal A, Lunstroem I. Lens on surface: A versatile method for the investigation of plasma protein exchange reactions on solid surfaces. J Biomater Sci Polmer Edu. 1991, 3, 7–16.

[119] Graham TR, Dasse K, Coumbe A, Salih V, Marrinan MT, Frazier OH, Lewis CT. Neo-intimal development on textured biomaterial surfaces during clinical use of an implantable left ventricular assist device. Eur J Cardio-Thorac Surg. 1990, 4, 182–90.

[120] Ellingen JE. A study on the mechanism of protein adsorption to TiO_2. Biomaterials. 1991, 12, 593–96.

[121] Sunny MC, Sharma CP. Titanium-protein interaction changes with oxide layer thickness. J Biomater Appl. 1991, 6, 89–98.

[122] Merritt K, Edwards CR, Brown SA. Use of an enzyme linked immunosorbent assay (ELISA) for quantification of proteins on the surface of materials. J Biomed Mater Res. 1988, 22, 99–109.

[123] Muramatsu K, Uchida M, Kim H-M, Fujisawa A, Kokubo T. Thrombo-resistance of alkali- and heat-treated titanium metal formed with apatite. J Biomed Mater Res. 2003, 65A, 409–16.

[124] Martinesi M, Bruni S, Stio M, Treves C, Borgioli F. In vitro interaction between surface-treated Ti-6Al-4V titanium alloy and human peripheral blood mononuclear cells. J Biomed Mater Res. 2005, 74A, 197–207.

[125] Thor A, Rasmusson L, Wennerberg A, Thomsen P, Hirsch JM, Nilsson B, Hong J. The role of whole blood in thrombin in contact with various titanium surfaces. Biomaterials. 2007, 28, 966–74.

[126] Maitz MF, Shevchenko N. Plasma-immersion ion-implanted nitinol surface with depressed nickel concentration for implants in blood. J Biomed Mater Res. 2006, 76A, 356–65.

[127] Wong T-S, Kang S-H, Tang SKY, Smythe EJ, Hatton BD, Grinthal A, Aizenberg J. Bioinspired self-repairing slippery surfaces with pressure-stable omniphobicity. Nature. 2011, 477, 443–47.

[128] Hansen WR, Autumn K. Evidence for self-cleaning in gecko setae. Proc Natl Acad Sci USA. 2005, 102, 385–89.

[129] Gao X, Yan X, Yan X, Zhang K, Zhang J, Yang B, Jiang L. The dry-style antifogging properties of mosquito compound eyes and artificial analogues prepared by soft lithography. Adv Mater. 2007, 19, 2213–17.

[130] Tuteja A, Choi W, Mabry JM, McKinley GH, Cohen RE. Robust omniphobic surfaces. Proc Natl Acad Sci USA. 2008, 105, 18200–05.

[131] Nguyen TPN, Brunet P, Coffinier Y, Boukherroub R. Quantitative testing of robustness on superomniphobic surfaces by drop impact. Langmuir. 2010, 26, 18369–73.

[132] Bocquet L, Lauga E. A smooth future?. Nat Mater. 2011, 10, 334–37.

[133] Poetes R, Holtzmann K, Franze K, Steiner U. Metastable underwater superhydrophobility. Phys Rev Lett. 2010, 105, 166104, doi: https://doi.org/10.1103/PhysRevLett.105.166104.

[134] Quéré D. Non-sticking drops. Rep Prog Phys. 2005, 68, 2495–32.

[135] Quéré D. Wetting and roughness. Annu Rev Mater Res. 2008, 38, 71–99.

[136] Tominaga T, Miyazaki T, Burr DB, Oshida Y. Osseointegration evaluation on dental implants retrieved from a cadaver mandible. J Dent Oral Disorders. 2018, 4, 1105–11.

[137] Sartoretto SC, Calasans-Maia JA, da Costa YO, Louro RS, Granjeiro JM, Calasans-Maia MD. Accelerated healing period with hydrophilic implant placed in sheep tibia. Braz Dent J. 2017, 28, doi: https://doi.org/10.1590/0103-6440201601559.

[138] Hou P-J, Ou K-L, Wang -C-C, Hunag C-F, Ruslin M, Sugiatno E, Yang T-S, Chou -H-H. Hybrid micro/nanostructural surface offering improved stress distribution and enhanced osseointegration properties of the biomedical titanium implant. J Mech Behav Biomed Mater. 2018, 79, 173–80.

[139] Sigot-Luizard MF, Warocquier-Clerout R. In vitro cytocompatibility tests. In: Dawids S, ed. Test Procedures for the Blood Compatibility of Biomaterials. Springer, 1993, 569–94, doi: https://doi.org/10.1007/978-94-011-1640-4_48.

[140] Kasemo B, Lausmaa J. Biomaterials and implant surfaces: A surface science approach. Int J Oral Maxillofac Implants. 1988, 3, 247–59.

[141] Masuda T, Salvi GE, Offenbacher S, Fleton D, Cooper LF. Cell and matrix reactions at titanium implants in surgically prepared rat tibiae. Int J Oral Maxillofac Implants. 1997, 1, 472–85.

[142] Rosengren A, Johansson BR, Danielsen N, Thomsen P, Ericson LE. Immuno-histochemical studies on the distribution of albumin, fibrigen, fibronectin, IgG and collagen around PTFE and titanium implants. Biomaterials. 1996, 17, 1779–86.

[143] Hou X, Weiler A, Winger JN, Morris JR, Borke JL. Rat model for studying tissue changes induced by the mechanical environment surrounding titanium implants. Int J Oral Maxillofac Implants. 2009, 24, 800–07.

[144] Rupp F, Scheideler L, Eichler M, Geis-Gerstorfer J. Wetting behavior of dental implants. Int J Oral Maxillofac Implants. 2011, 26, 1256–66.

[145] Oshida Y, Sachdeva R, Miyazaki S. Changes in contact angles as a function of time on some pre-oxidized bio-materials. J Mater Sci: Mater Med. 1992, 3, 306–12.

[146] Park Y-J, Song Y-H, An J-H, Song H-J, Anusavice KJ. Cytocompatibility of pure metals and experimental binary titanium alloys for implant materials. J Dent. 2013, 41, 1251–58.

[147] Matouskova L, Ackermann M, Horakova M, Capek L, Henys P, Safka J. How does the surface treatment change the cytocompatibility of implants made by selective laser melting?. Expert Rev Med Devices. 2018, 15, 313–21.

[148] Ou P, Hao C, Liu J, He R, Wang B, Ruan J. Cytocompatibility of Ti–xZr alloys as dental implant materials. J Mater Sci Mater Med. 2021, 32, 50, doi: https://doi.org/10.1007/s10856-021-06522-w.

[149] Mikulewicz M, Chojnacka K. Cytocompatibility of medical biomaterials containing nickel by osteoblasts: A systematic literature review. Biol Trace Elem Res. 2011, 142, 865–89.

[150] Čolić M, Tomić S, Rudolf R, Marković E, Šćepan I. Differences in cytocompatibility, dynamics of the oxide layers' formation, and nickel release between superelastic and thermo-activated nickel–titanium archwires. J Mater Sci Mater Med. 2016, 27, doi: 10.1007/s10856-016-5742-1.

[151] Hang R, Liu Y, Bai L, Zhang X, Huang X, Jia H, Tang B. Length-dependent corrosion behavior, Ni^{2+} release, cytocompatibility, and antibacterial ability of Ni-Ti-O nanopores anodically grown on biomedical NiTi alloy. Mater Sci Eng C Mater Biol Appl. 2018, 89, 1–7.

[152] Olmedo DG, Tasat DR, Guglielmotti EP, Cabrini RL. Biological response of tissues with macrophagic activity to titanium dioxide. J Biomed Mater Res A. 2008, 84A, 1087–93.

[153] Kang MK, Moon SK, Kim KM, Kim KN. Antibacterial effect and cytocompatibility of nano-structured TiO_2 film containing Cl. Dent Mater J. 2011, 39, 790–98.

[154] Yang Y, Ao H, Wang Y, Lin W, Yang S, Zhang S, Yu Z, Tang T. Cytocompatibility with osteogenic cells and enhanced in vivo anti-infection potential of quaternized chitosan-loaded titania nanotubes. Bone Res. 2016, 4, 16027, doi: https://doi.org/10.1038/boneres.2016.27.

[155] Prachár P, Bartáková S, Březina V, Cvrček L, Vanek J. Cytocompatibility of implants coated with titanium nitride and zirconium nitride. Bratisl Lek Listy. 2015, 116, 154–56.

[156] Lauterbur PC. Image formation by induced local interactions: Example employing nuclear magnetic resonance. Nature. 1973, 242, 190–92.

[157] Oshida Y. Surface Engineering and Technology for Biomedical Implants, Momentum Press, 2014.

[158] www.stmri.com/ht_docs/safety-2.html; www.mri.tju.edu/Policies-contraindications.html.

[159] Abbaszadeh K, Heffez LB, Mafee MF. Effect of interference of metallic objects on interpretation of T1-weighted magnetic resonance images in the maxillofacial region. Oral Surg Oràl Med Oral Pathol Oral Radiol Endod. 2000, 89, 759–65.

[160] Hopper TAJ, Vasilić B, Pope JM, Jones CE, Epstein C, Song HK, Wehrli FW. Experimental and computational analyses of the effects of slice distortion from a metallic sphere in an MRI phantom. Magn Reson Imaging. 2006, 24, 1077–85.

[161] Elison JM, Leggitt VL, Thomson M, Oyoyo U, Wycliffe ND. Influence of common orthodontic appliances on the diagnostic quality of cranial magnetic resonance images. Am J Orthod Dentofacial Orthop. 2008, 134, 563–72.

[162] Costa ALF, Appenzeller S, Yasuda CL, Pereira FR, Zanardi VA, Cendes F. Artifacts in brain magnetic resonance imaging due to metallic dental objects. Med Oral Patol Oral Cir Bucal. 2009, 14, E278–82.

[163] Ernstberger T, Buchhorn G, Heidrich G. Artifacts in spine magnetic resonance imaging due to different intervertebral test spacers: An in vitro evaluation of magnesium versus titanium and carbon-fiber-reinforced polymers as biomaterials. Neuroradiology. 2009, 51, 525–29.

[164] Shellock FG, Morisoli S, Kanal E. MR procedures and biomedical implants, materials, and devices: Update. Radiology. 1993, 189, 587–99.

[165] Shellock FG, Mink JH, Curtin S, Friesman MJ. MR imaging and metallic implants for anterior cruciate ligament reconstruction: Assessment of ferromagnetism and artifact. J Magn Reson Imaging. 1992, 2, 225–28.

[166] Shellock FG. Biomedical implants and devices: Assessment of magnetic field interactions with a 3.0 Tesla MR system. J Magn Reson Imaging. 2002, 16, 721–32.

[167] Shellock FG, Fieno DS, Thompson LJ, Talavage TM, Berman DS. Cardiac pacemaker: In vitro assessment at 1.5T. Am Heart J. 2006, 151, 436–43.

[168] Istanbullu O, Akdoğan G. Evaluation of MRI compatibility and safety risks for biomaterials. Mater Sci Med Technol Natl Conf (TIPTEKNO). 2015, doi: 10.1109/TIPTEKNO.2015.7374579.

[169] Medical Device Standards and Implant Standards. https://www.astm.org/Standards/medical-device-and-implant-standards.html.

[170] MRI Safety Information and Labeling (ASTM F2503). https://medinstitute.com/services/mri-safety/mri-safety-information-and-labeling-astm-f2503/.

Chapter 5
Surface modification for implantable materials

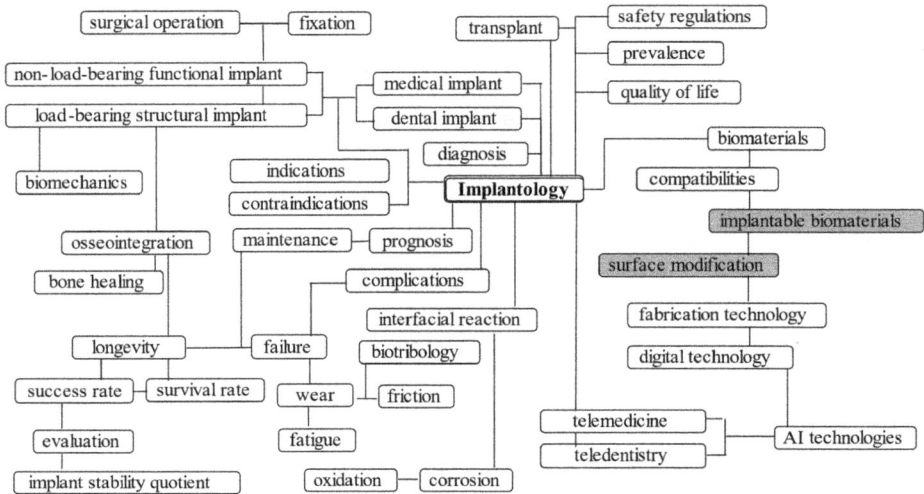

We have a tremendous number of artificial organs or devices that help to maintain pa-tients' quality of life; they include hydrocephalus shunts, ocular and contact lenses, orbital floors, artificial ears, cochlear implants, nasal implants, artificial chins, mandib-ular mesh, artificial skin, blood substitutes, artificial hearts and heart valves, pace-makers, breast prostheses, pectus implants, glucose biosensors, dialysis shunts and catheters, absorbable pins, temporary tendons, artificial kidneys, birth control im-plants, vascular grafts, artificial live, spinal fixations, finger joints, cartilage replace-ments, artificial legs, bone plates and bars (including sacroiliac joint), and Harington spinal bars. There are also a variety of dental prostheses, including dental implants, bridges and crowns, endodontic devices, and orthodontic archwires and brackets. The development of dental and orthopedic implants (particularly, TSAs, THAs, and TKAs) is important and the most challenging because it involves the collective knowledge and synthesis of several diverse disciplines such as biomaterials science and engineer-ing, and surface modification and technology in order to obtain its ultimate goal: that their surfaces are biologically accepted by a host's vital hard and soft tissues and bio-logically function in body. In previous chapters, we have discussed requirements for successful implant materials and systems. In this chapter, to these ends, surface modi-fication and alternations will be described.

https://doi.org/10.1515/9783110740134-006

5.1 Nature of surface

Before discussing surface coating and additive modification in detail, it is worth to review the nature of surface and interface. The surface of a material defines the boundary of an object and is not just a free end side of a substance, but it is a contact and boundary zone with other substances (either in gaseous, liquid, or solid). A physical system which comprises of a homogeneous component such as solid, liquid, or gas and is clearly distinguishable from each other is called a phase, and a boundary at which two or three of these individual phases are in contact is called an interface. Surface and interface reactions include reactions with organic or inorganic materials, vital or nonvital species, hostile or friendly environments, and so on. Surface activities may vary from mechanical actions (fatigue crack initiation and propagation, stress intensification, dislocation movement, etc.), chemical action (discoloration, tarnishing, contamination, corrosion, oxidation, etc.), mechano-chemical action (corrosion fatigue, stress-corrosion cracking, etc.), thermo-mechanical action (thermal fatigue and creep), tribological and biotribological actions (wear and wear debris toxicity, friction, etc.) to physical and biophysical actions (surface contact and adhesion, adsorption, absorption, diffusion, cellular attachment, cell proliferation and differentiation, etc.). Consequently, the properties of the surface determine the interaction of the second material [1].

Biological survival, particularly that of biological adhesive joints, is often dependent on thin surface films. Surfaces and interfaces behave completely different from bulk properties. The characteristics of a biomaterial surface governs the processes involved in biological response. Surface properties such as surface chemistry, surface energy, and surface morphology may be studied in order to understand the surface region of biomaterials. The surface plays a crucial role in biological interactions for four reasons: (1) the surface of a biomaterial is the only part contacting with the bio-environment, (2) the surface region of a biomaterial is almost always different in morphology and composition from the bulk, (3) for biomaterials that do not release or leak biologically active or toxic substance, the characteristics of the surface governs the biological response (foreign material vs. host tissue), and (4) some surface properties such as topography affect the mechanical stability of the implant–tissue interface [2–6]. The surface has a certain characteristic thickness: (1) In cases where interatomic reactions such as wetting or adhesion are dominant, it is important to analyze atoms up **to a depth of 100 nm,** (2) when mechanical interactions such as tribology and surface hardening come into play, a thickness of about **0.1 ~ 10 μm** will be important since the elasticity due to surface contact and the plastically deformed layer will be a governing area, and (3) for cases when mass transfer or corrosion is involved, the effective layer for preventing the diffusion will be within **1 ~ 100 μm** [7]. Since the thickness of the desired coating layer is accordingly selected, it becomes essential to control the thickness of coated or surface-modified film/layer. Coating is just a small portion of surface science and engineering. Structural elements beneath the surface function principally to support the surface and its characteristics determine not only esthetic appearance, but also interaction with the environment, including mechanical interactions, chemical interactions, optical and thermal interaction and interactions [8], and of course biological interactions. Accordingly, the longevity, safety, reliability, and structural integrity of dental and medical materials and devices are greatly governed by these surface phenomena, which can be detected, observed, characterized, and analyzed by virtue of various means of analytical equipment and technologies [9–11].

As demonstrated in Figure 4.6 of Chapter 4, successful implant surfaces should possess a certain range of roughness. To achieve satisfactory level of surface roughness by surface engineering, there are, in general, two categories from aspect of topological appearances: (1) surface concave texturing and (2) surface convex texturing. Surface concave textures can be achieved by either material removal (or subtractive technique) by chemical/electrochemical etching, or mechanical indentations (caused by sandblasting, shot peening, or laser peening). Chemical/electrochemical treatments are normally accompanied with a formation of reaction product (in other words, surface oxides or hydroxides) which can be expected to exhibit additional biocompatibility). By blasting or peening technique, surface layer will be locally forged with high-speed bombardment of blasting/peening media carrying high kinetic energy, resulting in generating compressive residual stress on surface layer which is benefit in terms of enhancing fatigue life and strength. On the other hand, surface convex textured surfaces can be formed by depositing certain types of particles by one of several physical or chemical depositing techniques (like CVD, PVD, plasma-spraying) or diffusion bonding. These are recognized as additive techniques.

Surface modification of dental implants is a key process in the production of these medical devices, and especially titanium implants used in the dental practice are commonly subjected to surface modification processes before their clinical use. A wide range of treatments, such as sand blasting, acid etching, plasma etching, plasma spray deposition, sputtering deposition, and cathodic arc deposition, have been studied over the years in order to improve the performance of dental implants. There are basically three major methods for surface modifications, namely, mechanical surface engineering, chemical/electrochemical modification, and physical or thermal treatment. Besides these, there are complicated methods combined by the above basic technique(s) and chemical/physical coating of biomaterials.

5.2 Mechanical modification

Sandblasting, as well as shot peening, possesses three purposes: (1) cleaning surface contaminants, (2) roughening surfaces to increase effective surface area, and (3) producing beneficial surface compressive residual stress [12]. As a result, such treated surfaces exhibit higher surface energy, indicating higher surface chemical and physical activities, and enhancing fatigue strength as well as fatigue life [13,14]. Aluminum oxide (Al_2O_3) particles are commonly used as a sandblasting medium, particularly in dentistry, for multiple purposes including divesting the casting investment materials and increasing effective surface area for enhancing the mechanical retention strengths of subsequently applied fired porcelain or luting cements. Fine aluminum oxide particles are normally recycled within the sandblasting machine. Ceramics such as aluminum oxides are brittle in nature, therefore, some portions of recycling aluminum oxide particles might be brittle-fractured. If fractured sandblasting particles are involved in the recycling media, it might result in creating an irregularity on metallic materials surface as well as the recycling sandblasting media itself be contaminated. Using fractal dimension analysis [15], a sample plate surface was weekly analyzed in terms of topographic changes, as well as chemical analysis of sampled recycled Al_2O_3 particles. It was found that the fractal dimension remained a constant value of about 1.4, prior to that it continuously increased from 1.25 to 1.4. By the electron probe microanalysis on collected blasting particles, it was found that unused Al_2O_3 contains 100% Al, whereas used (accumulated usage time was about 2,400 s) particles contained Al (83.32 wt%), Ti (5.48), Ca (1.68), Ni (1.36), Mo (1.31), S (1.02), Si (0.65), P (0.55), Mn (0.49), K (0.29), Cl (0.26), and V (0.08), strongly indicating that used alumina powder was heavily contaminated and was a high risk for the next material surface to be contaminated. Such contaminants are from previously blasted materials having various chemical compositions, materials used for tube and nozzles, and investing materials as well [13].

There are several data on surface contamination due to mechanical abrasive actions. As a metallographic preparation, the surface needs to be mechanically polished with a metallographic paper (which is normally SiC-adhered paper) under running water [16]. It is worth mentioning here that polishing paper should be changed between different types of materials, and particularly when a dissimilar metal couple is used for galvanic corrosion tests, such couple should not be polished prior to corrosion testing because both materials could become cross-contaminated. Hence, there are attempts to use TiO_2 powder for blasting onto titanium material surfaces. It was reported that titanium surfaces were sandblasted using TiO_2 powder (particle size ranging from 45 µm, 45–63 µm, and 63–90 µm) to produce the different surface textures prior to fibroblast cell attachment [17]. In the rabbit tibia, CpTi implants, which were sandblasted with 25 µm Al_2O_3 and TiO_2 particles, were inserted in the rabbit tibia for 12 weeks. Even though the amount of Al on the implant surface was higher than for the Al_2O_3-blasted implants compared to implants not blasted with Al_2O_3, any negative effects of the Al element were not detected [18], which is in contrast to those reported by Johansson et al. [19], who reported that Al release from Ti–6Al–4V implants was found to coincide with a poorer bone-to-implant over a three-month period. It is possible that the lack of differences between TiO_2-blasted and the Al_2O_3-blasted implants depends on lower surface concentrations of toxic Al ions than those reported by Johansson et al. [19]. Wang [20] investigated the effects of various surface modifications on porcelain bond strengths. Such modifications included Al_2O_3 blasting, TiO_2 blasting, $HNO_3 + HF + H_2O$ treatment, H_2O_2 treatment, and preoxidation in air at 600 °C for 10 min. Ti–porcelain couples were subjected to three-point bending tests. It was concluded that TiO_2 air abrasion showed the highest bond strength, which was significantly different from other surface treatments. Recently, it was reported that sandblasting using alumina as a media caused a remarkable distortion on a Co–Cr alloy and a noble alloy [21, 22]. It was estimated that the stress causing the deflection exceeded the yield strength of tested materials. It was also suggested that the sandblasting should be done using the lowest air pressure, duration of blasting period, and particle size alumina in order to minimize distortion of crowns and frameworks. To measure distortion, Co–Cr alloy plates (25 mm long, 5 mm wide, and 0.7 mm thick) were sandblasted with Al_2O_3 of 125 µm. Distortion was determined as the deflection of the plates at a distance of 20 mm from the surface. It was reported that (i) the mean deflections varied between 0.37 mm and 1.72 mm, and (ii) deflection increased by an increase in duration of the blasting, pressure, particle size, and by a decrease in plate thickness [21]. All these parameters are related to cumulative input energy of the media.

It is well documented that blasting or shot peening cause enlarge the effective surface area by concave-shaped indentations. As illustrated in Figure 5.1 [23], when a blasting (or peening) media particle (with red mark) with an average diameter D is bombarded onto substrate (gray-colored surface), a remarkable indentation is created with height h and diameter d. Although there are varieties of media for blasting such as aluminum oxide, garnet, silicon carbide, crushed glass grit, glass bead, steel shot, steel grit, plastic pumice, corn cob, or walnut shell, there is an empirical equation among these three important parameters: $h = [D - (D^2 - d^2)^{1/2}]/2$ [24]. It can be roughly estimated that the original flat surface can be doubled by blasting, resulting in (1) increase in surface roughness and (2) increase in effective surface area, which can be calculated knowing h (indent height), d (indent diameter), and d/D (indent ratio) under an assumption of 100% coverage.

The effect of a dual treatment of titanium implants and the subsequent bone response after implantation were investigated [25]. The implant, which was dually blasted with TiO_2 particles of two different sizes, was compared with implants that were blasted with only one of these particle sizes. Implants in group 1 were grit blasted with small particles, 22–28 µm in size, and group 2 with coarser particles, 180–220 µm size. Group 3 implants were blasted first with the 180–220 µm particles and subsequently with the 22–28 µm particles. Group 2 implants, which were blasted with only the coarse particles, showed a significantly better functional attachment than the other two groups. Group 1, which was blasted with only small particles, showed the lowest retention in bone. Kamal [26] conducted a series of experimental studies to investigate cellular responses to surface modifications of titanium implant material. With machined (turned) surfaces as controls, the topography was altered by blasting with TiO_2 particles with varying size. It was reported that (i) surface roughness increased with

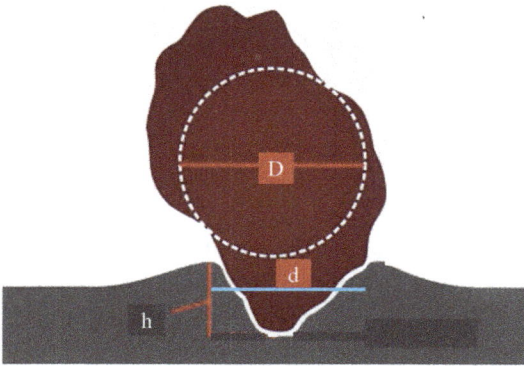

Figure 5.1: Indentation created by sandblasting or shot peening [23].

increasing particle size, up to 106–180 μm, (ii) further increasing particle size to 300 μm did not significantly increase surface roughness, and (iii) surface roughness, achieved by blasting titanium surfaces with various sizes of TiO_2 particles, significantly favored proliferation, differentiation, and production of TGF beta1 (transforming growth factor) and PGE2 (prostaglandin E2). Increasing surface roughness to this range may modulate the activity of cells interacting with an implant, thereby enhancing tissue healing and implant success.

Shot peening (which is a process similar to sandblasting, but with more controlled peening power, intensity, coverage, and direction) is a cold-working process in which the surface of a part is bombarded with small spherical media called shot. Each piece of shot striking the material acts as a tiny hammer, imparting to the surface small indentations or dimples. In order for the dimple to be created, the surface fibers of the material must be yielded in tension. Below the surface, the fibers try to restore the surface to its original shape, thereby producing below the dimple a hemisphere of cold-worked material highly stressed in compression. Overlapping dimples (which are sometimes called forged dimples) develop an even layer of metal in residual compressive stress. It is well known that cracks will not initiate or propagate in a compressively stressed zone due to a tendency of crack-closure phenomenon. Since nearly all fatigue and stress corrosion cracking (SCC) failures originate at the surface of a part, compressive surface residual stresses induced by shot peening provide considerable increases in part life, since advancing crack-opening is suppressed by preexisting compressive residual stress. The maximum compressive residual stress produced at or under the surface of a part by shot peening is at least as great as half the yield strength of the material being peened. Many materials will also increase in surface hardness due to the cold-working effect of shot peening [12, 27]. Both compressive stresses and cold-working effects are used in the application of shot peening in forming metal parts, called "shot forming" [27]. Oshida investigated the compressive residual stress distribution in depth underneath the surface of Ti–6Al–4V alloy using the X-ray diffraction (XRD). It was found that the compressive residual stress at the surface was −50 MPa, followed by the maximum compressive residual stress of −80 MPa at 20 μm below the surface. Then, the compressive residual stress starts to decrease to show almost zero stress at about 50 μm underneath the surface [28]. It was also shown that shot peening resulted in better bonding strengths between titanium substrate and porcelain than sandblasting [28]. Recent advances in biomaterials' research suggest that electrical charges on a dental implant surface significantly improve its osseointegration to living bone, as a result of selective osteoblast activation and fibroblast inhibition. Guo et al. [29] investigated the possibility of using sandblasting to modify the electrical charges on the surface of titanium materials, using Al_2O_3 grits on CpTi (grade II) plates, for durations between 3 and 30 s. After sandblasting, Ti surfaces were measured for their electrostatic voltage. The results indicate a novel finding, that is, negative static charges are generated on the titanium surface, which may stimulate osteoblast activity to promote osseointegration around dental implant surface. This finding may at least partially explain the good osseointegration results of sandblasted titanium dental implants, in addition to other known reasons, such as topological changes on the implant's surface. However, the static charges accumulated on the titanium surface during sandblasting decayed to a lower level with time. It remains a challenging task to seek ways to retain these charges after quantification of desired level of negative charges needed to promote osteoblast activity for osseointegration around dental implants [29].

The laser peening technology is recently developed, claiming noncontact, no-media, and contamination-free peening method. Before treatment, the workpiece is covered with a protective ablative layer (paint or tape) and a thin layer of water. High-intensity (5–15 GW/cm^2) nanosecond pulses (10–30 ns) of laser light beam (3–5 mm width) striking the ablative layer generate a short-lived plasma which causes a shock wave to travel into the workpiece. The shock wave induces compressive residual stress that penetrates beneath the surface and strengthens the workpiece [30–33], resulting in improvements in fatigue life and retarding in SCC occurrence. Cho et al. [34] laser-treated CpTi screws and inserted in right tibia metaphysics of white rabbits for 8 weeks. It was reported that (i) SEM of laser-treated implants demonstrated a deep and regular honeycomb pattern with small pores, and (ii) 8 weeks implantation, the removal torque was 23.58 N-cm for control machined and 62.57 N-cm for laser-treated implants. Gaggl et al. [35] reported that (i) surfaces of laser-treated Ti implants showed a high purity with enough roughness for good osseointegration, and (ii) the laser-treated Ti had regular patterns of micropore with interval of 10–12 μm, diameter of 25 μm, and depth of 20 μm. Jindal et al. [36] studied the effect of a novel process of surface modification, surface nanostructuring by ultrasonic shot peening, on osteoblast proliferation and corrosion behavior of CpTi, in simulated body fluid (SBF). A mechanically polished disc of CpTi was subjected to ultrasonic shot peening with stainless steel (SS) balls to create nanostructure at the surface. A nanostructure (<20 nm), with inhomogeneous distribution, was revealed by atomic

force and scanning electron microscopy. There was an increase of ~10% in cell proliferation, but there was drastic fall in corrosion resistance. Corrosion rate was increased
by 327% in the shot-peened condition. In order to examine the role of residual stresses,
a part of the shot-peened specimen was annealed at 400 °C for 1 h. It was found that
(i) surface nanostructure was much prominent, with increased number density and
sharper grain boundaries, (ii) cell proliferation was enhanced to ~50%, and corrosion
rate was reduced by 86.2% and 41%, as compared with that of the shot peened and
the as received conditions treated, respectively, (iii) the highly significant improvement in cell proliferation, resulting from annealing of the shot-peened specimen, was
attributed to increased volume fraction of stabilized nanostructure, stress recovery,
and crystallization of the oxide film, and (iv) increase in corrosion resistance, from annealing of shot peened material, was related to more effective passivation. Thus,
the surface of CpTi, modified by this novel process, possessed unique quality of enhancing cell proliferation as well as the corrosion resistance. It could be highly effective in reducing treatment time of patients adopting dental and orthopedic implants of
titanium and its alloys [36]. Electric discharge machining (EDM) can alter the surface
energy status [37, 38]. Hercuba et al. [37] investigated the properties of Ti–6Al–4V
alloy after surface treatment by the EDM process. The EDM process with high peak currents proved to induce surface macroroughness and to cause chemical changes to the
surface. Evaluations were made of the mechanical properties by means of tensile tests,
and of surface roughness for different peak currents of the EDM process. The EDM process with peak current of 29 A was found to induce sufficient surface roughness, and
to have a low adverse effect on tensile properties. The chemical changes were studied
by scanning electron microscopy equipped with an energy dispersive X-ray (EDX) analyzer. The surface of the benchmark samples was obtained by plasma-spraying a titanium dioxide coating. An investigation of the biocompatibility of the surface-treated
Ti–6Al–4V samples in cultures of human osteoblast-like MG 63 cells revealed that the
samples modified by EDM provided better substrates for the adhesion, growth, and viability of MG 63 (human osteoblast-like) cells than the TiO_2-coated surface. Thus, it was
reported that EDM treatment can be considered as a promising surface modification to
orthopedic implants, in which good integration with the surrounding bone tissue is required [37]. Plastic deformation can also change the surface conditions [39, 40].
The processes of equal channel angular pressing and high-pressure torsion are now
established for the fabrication of ultrafine-grained metals having superior properties
by comparison with their coarse-grained counterparts. Although titanium and zirconium alloys are considered to be promising materials for orthopedics because of their
biocompatibility with tissues, their main drawbacks for application as implants have
generally been considered to be insufficient levels of mechanical and tribological
properties. The influence of equal channel angular pressing and nitrogen ion implantation on the structure and properties of Ti and Zr alloys has been investigated to ensure the optimum combination of the bulk material and surface layer properties [40].
The data obtained showed that the equal channel angular pressing and nitrogen ion

implantation can be efficiently used to improve bulk and surface properties of Ti- and Zr-based implants [40].

There are various techniques to measure and characterize the surface roughness. They include that (1) surface roughness can be measured using a profilometer with sharp edge stylus, which is a contact method, (2) atomic force microscopy (AFM) can provide noncontact surface topography from which the surface roughness can be indirectly measured, and (3) fractal dimension analysis can be used to present the surface roughness in non-Euclidian dimension [13, 15, 41, 42]. Recently, Hansson et al. [43] employed computer simulations to measure surface roughness. The lateral resolution was defined as the pixel size of a profiling system. A surface roughness was simulated by a trigonometric function with random periodicity and amplitude. The function was divided into an array of pixels simulating the pixels of the profiling system. The mean height value for each pixel was used to calculate the surface roughness parameters. It was found that the accuracy of all the surface roughness parameters investigated decreased with increasing pixel size. This tendency was most pronounced for mean slope and developed length ratio, amounting to about 80% of their true values for a pixel size of 20% of the true mean high-spot spacing. It was concluded that the lateral resolution of an instrument/method severely compromises the precision of surface roughness parameters which are measured for roughness features with a mean high-spot spacing less than five times the lateral resolution [43].

5.3 Chemical and electrochemical modifications

In general, chemical method has been adopted for (i) forming a stable oxide film and (ii) resulting surface having an appropriate level of surface roughness. On the other hand, electrochemical method also possesses two-fold purposes; (i) forming a stable oxide film, and (ii) through an electrochemical reaction to deposit target substance(s) onto substrate surface [44]. In either method, enhanced biocompatibility of implants (in particular, titanium implants) highly depends on the possibility of achieving high degrees of surface functionalization for an enhanced mineralization of bioactive minerals (e.g., hydroxyapatite (HA)) [45].

In chemical treatment, a variety of acid and alkali agents were found. Variola et al. [46] used a mixture of H_2SO_4 and H_2O_2 on Ti–6Al–4V surfaces having both a microtexture and a nanotexture to promote cell activity at the surface of implants. It was reported that the resulting new surfaces selectively promote the growth of osteoblasts while inhibiting the growth of fibroblasts, making them promising tools for regulating the activities of cells in biological environments [46]. A mixture of H_2SO_4 and H_2O_2 [45, 47], a mixture of H_2SO_4 and HCl [48], and 3% H_2O_2 in hydrothermal [49], in common, can promote easy deposition bioactive substance and further adhesion of biomimetic substances. Grit-blasted (with alumina particles) titanium was treated in hydrofluoric acid (HF) to evaluate the osseointegration in ovariectomized rats [50].

It was found that, 12 weeks after implant insertion, (i) the fluoride-modified implants showed improved osseointegration compared to control, with the bone area ratio and bone-to-implant contact (BIC) increased by 0.9- and 1.4-fold in histomorphometry, (ii) the bone volume ratio and percentage of osseointegration by 0.8- and 1.3-fold, and (iii) the maximal push-out force and ultimate shear strength by 1.2- and 2.0-fold in biomechanical test. It was then concluded that HF treatment of Ti surface improved implant osseointegration in rats and suggested the feasibility of using fluoride modification to improve Ti implant osseointegration in osteoporotic bone [50]. To enhance the bone bonding and osteoblast response, H_3PO_4 was used for titanium surface treatment [51–53]. Acid etching is a popular method to texture the surface of dental implants. During etching, the titanium oxide protective layer is dissolved and small native hydrogen ions diffuse into the unprotected implant surface. They enrich the implant surface with hydrogen and precipitate into titanium hydride (TiH). Szmukler-Moncler et al. [54] measured the concentration of TiH at the implant surface and the total concentration of hydrogen at five commercially available implant systems, made of either CpTi or titanium alloy. XRD was conducted on each implant system to determine the compounds present at the implant surface. Following a TiH_2/Ti calibration curve, the concentration of TiH was determined. Concentration of hydrogen in the implants was measured by the inert gas fusion thermal conductivity/infrared detection method. It was found that (i) XRD data showed that TiH was present on all CpTi implants but not on the alloyed implants, (ii) TiH concentration varied between 5% and 37%, and hydrogen concentration varied between 43 and 108 ppm, no difference in uptake was found between CpTi and alloyed implants, (iii) low solubility of hydrogen in α-titanium is responsible for precipitation into TiH, and (iv) stronger etching conditions led to higher concentration of TiH_{2-x}. Based on these results, it was concluded that high solubility of hydrogen in the β-phase of the alloy is preventing hydrogen from precipitating into TiH. All implants, even those lacking TiH at the surface, were enriched with hydrogen, and in all implants, hydrogen concentration was within the normative limit of 130 ppm [54]. Sodium hydroxide (NaOH) was equally employed as pretreatment for subsequent depositing of bioactive materials [47, 53, 55, 56]. Alkali treatment of the Ti−6Al−7Nb alloys with subsequent heat treatment has been adopted as an important surface treatment procedure for apatite formation in dental implants. Park et al. [57] examined the effects of alkali treatment on the precipitation of apatite on a Ti−6Al−7Nb alloy. All samples were immersed in a Hanks' balanced salts solution (SBF) at pH 7.4 and 36.5 °C for 15 days. The surface structural changes of samples due to the alkali treatment and immersing in SBF were analyzed by XRD, SEM, and XPS. The cell toxicity was evaluated based on the optical density of the surviving cells. The samples were implanted into the abdominal connective tissue of mice for 4 weeks. A sodium titanate hydrogel layer was formed after immersion in a NaOH solution. A dense and uniform bone-like apatite layer was precipitated on the alkali and heat-treated Ti−6Al−7Nb alloy in the SBF. There was a significant difference in cell toxicity between the treated and untreated Ti−6Al−7Nb.

The thickness of the fibrous capsule formed around the implant body was decreased significantly by the alkali and heat treatment. It was, hence, concluded that the alkali treatment samples showed a better biocompatibility than the untreated commercial metal samples [57].

Among the many methods of titanium surface modification, electrochemical techniques are simple. Anodic oxidation is the anodic electrochemical technique while electrophoretic and cathodic depositions are the cathodic electrochemical techniques. By anodic oxidation, it is possible to obtain desired roughness, porosity, and chemical composition of the oxide. Electrochemistry deals with charge transfer across interfaces. One of the first applications of electrochemistry was for coating conducting surfaces with either metals, for example, electroplating, or by inorganic or organic substances, such as oxides and polymers. Coatings ranging from a monoatomic layers (e.g., via under potential deposition, to microns thick have been carried out), which affected the physical and chemical properties of the coated surface. Hence, electrochemically deposited films have been used to inhibit corrosion, accommodate enzymes and biosensors, control optical properties and in numerous other applications [58]. Anodizing in electrochemical reaction can control the thickness of oxide films formed on substrate depending on used electrolyte [59, 60]. Xiao et al. fabricated titania nanotubes by anodization method using an electrolyte dimethyl sulfoxide electrolyte containing 1 wt% HF solution at above 50 °C and reported that the formed oxide was anatase type TiO_2 [61]. Major purpose of HA coating can be found for better implant–bone fixation in dental and orthopedic implants under load-bearing conditions. In recent works, it has been demonstrated that an in vitro, chemically deposited, bone-like apatite layer with bone-bonding ability could be induced on a titanium surface. By reproducing that chemical procedure, a dense bone-like apatite layer can be formed on the surface of the titanium in SBF. For the last decade, different surface modifications have been investigated to provide these implants with bone-bonding ability.

Most of the problems that have been associated with HA-coated implants are caused by the electrochemical deposition method. Due to the high temperatures that occur during the plasm spraying process, the resulting coatings do not have an exactly defined phase composition. Additionally, layers produced by this method have a rather high thickness, resulting in insufficient adhesion strength. When the selected electrolyte contains Ca and P ions as major constituents, it is highly expected to see bone-like compounds being deposited onto substrates [62–64]. For example, Simka et al. used mixed electrolytes NaH_2PO_2 in 4.3 M H_3PO_4 and $Ca(H_2PO_2)_2$ in 4.3 M H_3PO_4 [65], and Ca $(H_2PO_2)_2$, H_3PO_4, or $(HCOO)_2Ca$ [66]. Sharma et al. [67] investigated the influence of synthesis parameters like pH of suspension and current density. It was found that (i) the addition of chitosan increased the adhesive strength of the composite coating, (ii) modulus of elasticity of the coating was found to be 9.23 GPa, and (iii) in presence of chitosan, dense negatively charged surface with homogenous nucleation was the primary factor for sheet-like evolution of apatite layer. Based on these findings, it was suggested that incorporation of chitosan with apatite–

wollastonite in composite coating could provide excellent in vitro bioactivity with enhanced mechanical properties [67]. Similarly, using electrochemical reaction, deposition of biocompatible platinum foil (Pt) onto TiN substrate [68], conductive polymer, poly(3,4-ethylenedioxythiophene) film onto biodegradable Mg substrate [69], and (Ti,Al)N onto Ni–Cr dental base alloy [70] have been reported.

Anodic oxidation at high voltages can improve the crystallinity of the oxide. The chief advantage of this technique is doping of the coating of the bath constituents and incorporation of these elements improves the properties of the oxide. Electrophoretic deposition uses HA powders dispersed in a suitable solvent at a particular pH value. Under these operating conditions, these particles acquire positive charge, and coatings are obtained on the cathodic titanium by applying an external electric field. These coatings require a post-sintering treatment to improve the coating properties. Cathodic deposition is another type of electrochemical method where HA is formed in situ from an electrolyte containing calcium and phosphate ions. It is also possible to alter structure and/or chemistry of the obtained deposit. Nanograined HA has higher surface energy and greater biological activity and therefore emphasis is being laid to produce these coatings by cathodic deposition [71]. Titanium surface characteristics determine the degree of success of permanent implants. The topography, morphology of the surface in micro- and nanoscales, the impurities are present and other characteristics are main concerns, and therefore a multitechnique approach is required in order to evaluate modification process effects on the surface. Surface modification of titanium in the nanometrical range was performed by means of anodization in phosphoric with the aim of improving both the biocompatibility and the corrosion resistance in the biological media. Sanchez et al. [72] investigated biocompatible characteristics of the modified titanium surface, as the presence of anatase in the oxide film and the incorporation of phosphate to the surface. It was reported that the increase in the film thickness from 3 to 42 nm was estimated from electrochemical impedance spectroscopy results when anodizing potentials from 0 to 30 V were applied, whereas a bilayer structure of the protective oxides formed was determined.

It may be worth reviewing the types of oxides formed on titanium. TiO_2 possesses three crystalline structures: anatase, rutile, and brookite. Anatase-type TiO_2 is a tetragonal crystalline system with $a_o = 3.78$ Å and $c_o = 9.50$ Å; the rutile-type TiO_2 is also a tetragonal structure, but the lattice constants are quite different from those of anatase type (i.e., $a_o = 4.58$ Å and $c_o = 2.98$ Å). The third type is brookite type and has an orthorhombic crystalline structure with $a_o = 9.17$ Å, $b_o = 5.43$ Å, and $c_o = 5.13$ Å. Among these oxides, rutile is known to be the most stable phase and the crystalline structure (between rutile and anatase) appears to be controlled by oxidation conditions, as shown in Figure 5.2 [73]. Lee et al. [74] examined the effect of electrolyte temperature on the surface characteristics of anodized pure titanium in DL-α-glycerophosphate and calcium acetate. It was reported that (i) the anodic oxide films contained a large proportion of anatase with some rutile, (ii) the relative intensity of the anatase peaks and the surface roughness increased with increasing electrolyte temperature, and (iii) as the electrolyte temperature was increased, the pore size increased to 1–4 μm, and the apatite crystals became coarser and denser.

Condition	Type of crystallibe structure of formed Ti oxde		
Physical deposition			**Anatase**
Dry oxidation	**Rutile**		
Wet oxidation	**Rutile**	**R + A**	**Anatase**
pH value	acid - - - - - - - - - - - - - - - - - - - → alkaline		

Figure 5.2: Types and stabilities of oxides formed on titanium surface [73].

5.4 Physical modification

5.4.1 Hydrophilicity and hydrophobicity

The interaction between an implant's metal surface and the host's cells/tissues are important parameters for the biocompatibility, in particular for materials used in either dental or orthopedic implants. Hydrophilicity, or surface wettability, is a function of surface energy. It plays an important role in the biological response to an implant, as it controls the adsorption of proteins. These proteins promote and greatly influence whether cells attach to the implant's surface and the subsequent cell spreading [18, 75]. Spreading ($\delta\theta/\delta t$) phenomenon is a function of time (t) and interaction rate of wetting substance on substrate surface.

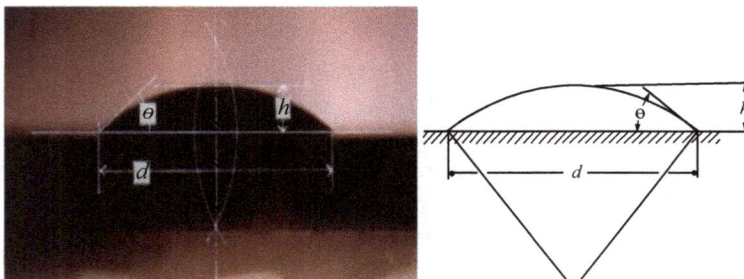

Figure 5.3: Typical drop profile for measuring surface contact angle [76].

Referring to Figure 5.3, the surface contact angle (in other words, wettability in terms of θ can be calculated by: wettability (θ) = $2\tan^{-1}(2\,h/d)$, where h is the highest height of the droplet and d is the width of the droplet. The spreading is everchanging as a function of time (t), hence the spreadability ($\delta\theta/\delta t$) can be calculated by the following equation [76, 77]: ($\delta\theta/\delta t$) = $[-4\,h(\delta d\delta t) + 4d(\delta h/\delta t)]/(d^2 + 4h^2)$.

Many techniques and methods have been reported and proposed to increase surface hydrophilicity. Polymer nanofibers were surface-functionalized by sputter coating with TiO$_2$. It was reported that the surface wettability of TiO$_2$ sputter-coated nanofibers was significantly improved after UV irradiation [78]. Similarly, TiO$_2$ coating resulted in increased surface wettability [79, 80]. Surface hydrophilicity can also be improved by surface texturing [81–84]. Fleming et al. [82] treated a titanium surface by sandblasting and dip coating it with colloidal silica nanoparticles in order to produce superhydrophilicity. It was reported that the treated surface was suitable for the potential use in a variety of applications, such as prosthetic dentistry and other biomedical fields [82]. Ti–6Al–4V surface was modified using a combination of various techniques such as cold spray, selective laser sintering, pulsed electroerosion treatment, and magnetron sputtering to control surface topography (roughness and blind porosity), surface chemistry, and wettability, i.e., the characteristics which affect osseointegration. It was concluded that the combination of high surface roughness, blind porosity, hydrophilicity, and biocompatibility makes fabricated metal-ceramic materials promising candidates for applications involving tissue regeneration [84]. Nano- or microstructured materials composed of TiO$_2$ [85], Y$_2$O$_3$ [86], and Ti [87, 88] showed improving surface wettability. An HA coating increases the surface hydrophilicity of Ti–6Al–4V [89, 90]. Another study that demonstrated an increase in surface hydrophilicity showed that when TiO$_2$ was incorporated with Sr and P elements, the treated surfaces showed significantly higher wettability and surface energy. Also, after immersion in Hank's balanced salt solution, considerable apatite deposition was observed on the Sr- and P-doped surfaces, suggesting that Sr- and P-incorporated Ti oxide surfaces may improve implant bone healing by enhancing attachment, viability, and differentiation of osteoblastic cells [91]. Tsuji et al. [92] employed an Ar-ion implantation technique on a polystyrene surface and reported that atomic bonds of C–O, C=O, and O=C–O endowed hydrophilicity to the polystyrene surface. It was also observed that human umbilical vein endothelial cells (HUVEC) only exhibited cell growth and attachment on Ag-implanted surfaces.

Lin et al. deposited a TiO$_x$ film by electron-beam evaporation system and then treated it with ultraviolet or visible light irradiation [93]. It was reported that ultraviolet irradiation treated TiO$_x$ film became highly hydrophilic and returned to its original hydrophobic state by visible-light irradiation. This demonstrates the responsiveness and plasticity of TiO$_x$: its wettability is dependent on exposure to light irradiation. Wu et al. deposited TiO$_2$ film on unheated glass substrates by using a twin dc magnetron sputtering system [94]. After irradiating the oxide film surface with UV, the water contact angle of the fresh TiO$_2$ film was found to be within 100–112°. However, within 60 min of UV irradiation, the film became highly hydrophilic; the water contact angle was nearly zero. Also, the surface was considerably a rougher surface of the films with 12.6–14.5 nm. Hence, it was concluded that high hydrophobicity and photo-induced hydrophilicity can be attributed to fully oxidized chemical composition and higher roughness on the film surface [94]. In some tribological orthopedic implants, high hydrophobicity is required. There are reports about the superhydrophobic effect perfluoropolyether (PFPE) [95] and perfluorooctyl trichlorosilane [96] when deposited on Ti surfaces.

Increased wettability may be an important factor for reosseointegration. Duske et al. [97] applied cold atmospheric pressure gas-discharge plasma to titanium discs with different surface topography in order to reduce water contact angles and to improve the spreading of osteoblastic cells. It was reported that the contact angle of the titanium discs (baseline values: 68°–117°) were significantly reduced, close to 0°, irrespective of surface topography after the application of argon plasma with 1.0% oxygen admixture for 60 s or 120 s. Also, the cell size of osteoblastic cells grown on argon–oxygen plasma-treated titanium discs was significantly larger than on nontreated surfaces irrespective of surface topography; indicating that the application of cold plasma may be supportive in the treatment of peri-implant lesions and may improve the process of reosseointegration [97]. Functionalizing implant surfaces implies the increasing surface energy (with improved wettability). Schliephake et al. [98] hypothesized that organically coating titanium screw implants provide binding sites for integrin receptors that can enhance peri-implant bone formation. Ten adult female foxhounds received experimental titanium screw implants in the mandible 3 months after removal of all premolar teeth. Four types of implants were evaluated in each animal: (1) implants with machined titanium surfaces, (2) implants coated with collagen I, (3) implants with collagen I and cyclic RGD (arginylglycalaspartic acid) peptide coating (Arg–Gly–Asp) with low RGD concentrations (100 µmol/mL), and (4) implants with collagen I and RGD coating with high RGD concentrations (1,000 µmol/mL). Peri-implant bone regeneration was assessed histomorphometrically after 1 and 3 months, in five dogs per interval, by measuring BIC and the bone volume density (BVD) of the newly formed peri-implant bone. It was found that after 1 month, BIC was significantly enhanced only in the group of implants coated with the higher concentration of RGD peptides and that the volume density of the newly formed peri-implant bone was significantly higher in all implants with organic coatings – no significant difference was found between collagen coating and RGD coatings. Furthermore, after 3 months, BIC was significantly higher in all implants with organic coating than in implants with machined surfaces. The peri-implant BVD was significantly increased in all coated implants in comparison to machined surfaces. It was therefore concluded that the organic coating of machined screw implant surfaces provided binding sites for integrin receptors and enhanced both BIC contact and peri-implant bone formation [98].

5.4.2 Surface modification

The microarc oxidation (MAO), ion implantation including plasma immersion ion implantation (PIII), other plasma treatments, and laser technologies are typical methods to modify surface of biomaterials. Wu et al. [99], using Ti–24Nb–4Zr–7.9Sn alloy (which is a newly developed β-titanium alloy having a low value of modulus of elasticity), evaluated the effects of MAO treatment on biological performance and characterized treated surface and reported that MAO treatment helps to form a porous surface with a biologically active bone-like apatite layer on Ti alloy. Since TiO_2 surface possesses high apatite-forming ability, porous Ti was MAO processed to form TiO_2 thin layer [100], and CpTi was subjected to MAO treatment to form porous TiO_2, indicating that the cortex-like coating was conducive to cell retention and implant stabilization [101]. Most of MAO application can be found in depositing bioactive or biocompatible substances onto biomaterials. The Ca- and P-containing coating electrolyte (α-$Ca(PO_3)_2$, or $CaTiO_3$) was coated on Ti–6Al–4V [102], and on Mg [103]. HA was MAO processed on titanium substrate and a mixture of HA and TiO_2 was detected [104]. Song et al. [105] treated MAO processed CpTi surface (which has porous TiO_2) with an electrolyte containing Ca and P ions by hydrothermal treatment. A composite of potassium titanate ($K_2Ti_6O_{13}$) was also coated on titanium substrate and it was indicated that the potassium/TiO_2 bioceramic coatings possess excellent capability of including bone-like apatite to deposit [106]. A new technique combining MAO and electrophoresis was introduced to develop a biocompatible oxide layer on CpTi implant surface. Originally developed alkaline electrolyte containing nanoscale HA powder suspension was used in the new technique. In the electric field, nanoscale HA powder was electrophoretically moved and sintered into the gradually formed oxide layer on titanium anode. Physiochemical properties and in vitro biological performance of the newly formed surface were examined and evaluated. An 8.5-μm thick oxide layer with high surface energy and roughness, which was composed of titanium dioxide and calcium phosphates as well as HA, was formed on titanium surface by the modified MAO technique. Osteoblasts cultured on the modified MAO titanium surface showed significantly increased alkaline phosphatase activity comparing to machined titanium and MAO titanium surface. Natural oxide surface of titanium could be transformed into a hybrid oxide layer by modified MAO treatment. The modified titanium surface, which is rough and porous, contains calcium phosphates and proved to be more biocompatible in vitro [107].

Plasma treatment is an interesting method to modify the implant surface. Not only can this treatment alter the surface charge, but this treatment can also alter the chemistry and the topography [108, 109]. Thermal plasma treatment has been traditionally used as a method to utilize HA coatings on implant surfaces (plasma spraying) [110, 111]. Another form of plasma treatment, the atmospheric pressure (cold) plasmas, has shown to alter the surface energy and the chemistry due to the generation of high concentration of reactive species that are generated [108]. This has been reported to be beneficial for the enhancement of osteogenic responses, since surfaces treated with atmospheric plasma significantly enhanced the wettability and improved the initial cellular interaction. Plasma surface modification has become a popular method to modify the surface structure and biological properties of biomaterials. By modifying selective surface mechanical and biological properties, conventional materials can be redesigned with their favorable bulk attributes retained. Plasma surface modification can enhance the multifunctionality, mechanical properties, as well as biocompatibility of artificial biomaterials and medical implants [112]. In order to improve the properties of titanium and expand its clinical application, many methods have been applied to modify the surface of titanium, which should include the application of plasma surface modification technologies (such as plasma spraying and PIII and deposition), to improve the bioactivity of titanium [113]. Plasma technology can be applied not only to metallic materials, but also to polymeric materials [114, 115].

Plasma-based ion implantation and deposition techniques are widely used in many applications, including microelectronics, biomedical engineering, and nanotechnology [116–118]. Ion implantation has been successful in biomaterials modification, such as in improving the wear resistance of artificial joint components, in improving wettability, anticoagulability, anticalcific behavior of biomedical polymers, in minimizing biofouling of medical devices, and depositing HA or Ca–P compounds [119–121]. Application of PIII technique to oxidation can be applied on different types of biomaterials: CpTi [122–127], Ti–6Al–4V [128–131], Ti–5Al–2.5Fe [130], NiTi [123, 125, 126], and Zr [132]. Xie et al. [133] investigated the effects of ($H_2O + H_2$) implanted titanium by PIII technique on human osteoblast cells culture, and concluded that ($H_2O + H_2$) implanted titanium exhibited good adhesion and growth, suggesting that a practical means to improve the surface bioactivity and cytocompatibility of medical implants made of titanium.

Among plasma technology, the most popular application is a plasma-spraying HA on titanium surface [134–139]. Montazeri et al. [137], using the plasma electrolytic oxidation, which is one of the effective techniques, coated titania and HA onto Ti–6Al–4V. It was reported that by increasing the operation time, the amount of the formed HA increased, and the sample coated at 500 V and 15 min showed the best corrosion resistance in Ringer's solution. Ti was subjected to plasma oxidation and showed that improved wetting was obtained, which is potentially associated with shorter osseointegration periods [137]. Huan et al. [139] modified the surface of NiTi shape memory alloy (SMA) by plasma electrolytic oxidation in Na_3PO_4 with an aim to produce porous NiTi surfaces for biomedical applications. It was concluded that the wettability and surface free energy of NiTi increased significantly, indicating that the plasma oxidation process shows potential for expanding the biofunctionality of NiTi [139]. Similarly, plasma nitrization can be achieved. Ti–6Al–4V, Ti–4Al–2.5Fe [140], and Ti–24Al–10Nb intermetallic alloys [141] were plasma nitrided and TiN film was detected formed on these alloy surfaces. Fouquet et al. plasma treated Ti–6Al–4V using a N_2–H_2 gas mixture under 10 Pa [142]. It was reported that (i) the plasma treatment enables the formation of a δ-TiN layer which can act as a diffusion barrier for nitrogen, and (ii) the underneath previously formed ε-Ti_2N grains undergo a nitrogen rearrangement with the remaining α-Ti grains leading to the formation of a nitrogen poor phase that was identified as the α-$TiN_{0.26}$ phase [142]. The application of atmospheric plasma is increasing in numerous situations especially in the biomedical field due to their practical capability to low temperature providing plasma that are not spatially bound or confined by electrodes [143, 144].

Light amplification by stimulated emission of radiation (laser) is a coherent and monochromatic source of electromagnetic radiation that can propagate in a straight line with negligible divergence. As a result, laser finds diverse applications ranging from mere mundane to most sophisticated uses either for totally commercial or purely scientific purposes, and from lifesaving to life-threatening causes. High-power lasers can produce intense heating and perform various manufacturing operations or material processing. The manufacturing processes covered have been broadly divided into four major categories, namely, laser-assisted forming, joining, machining, and surface engineering. Surface engineering means the design and modification of a surface to enhance hardness, wear resistance, heat resistance, or some other property. Although most laser surface engineering processes appear expensive compared to other technologies, in many cases they are actually more cost-effective because of their precision and speed. Due to the intrinsic properties of high coherence and directionality, laser beam can be focused onto metallic surface to perform a broad range of treatments such as remelting, alloying, and cladding, which are used to improve the wear and corrosion resistance of titanium alloys [145–149]. Romanos et al. [149] evaluated the use of different laser systems in implant dentistry. Although many case studies indicate extensive use of lasers and promising results in dental implantology, lasers may be used for uncovering submerged implants atraumatically to prevent crestal bone loss, recontouring peri-implant soft tissues and sculpting emergence profile for prosthetic components, raising surgical flaps, osseous recontouring, and creating parabolic tissue architecture. Additionally, bone harvesting of block grafts, window preparation in sinus lift procedures, ridge splitting, and debridement of extraction sockets for immediate implant placement can be included. It was concluded that aside from the many benefits associated with the use of lasers in implant-related procedures, there are also risks to consider from the laser irradiation on the implant surface and the peri-implant tissues. Therefore, an appropriate training on laser use is mandatory to increase the clinical outcome and to control the potential of complications [149]. Laser surface modification has been extensively employed [150–156].

5.5 Thermal modification

Titanium surface can be hydrothermally treated on CpTi [157–161], Ti–6Al–4V [162], Ti–13Nb–13Zr [163], surgical SS [164]. Polymeric materials can also be thermal sprayed, although thermally sprayed polymer coatings are gaining significant attention from many industries, including potential applications that are limited in petrochemical, automotive, and aircraft industries [165]. HA ceramics have been prepared from α-tricalcium phosphate (α-TCP) with addition of water or ammonia water by hydrothermal hot-pressing (HHP) method at 300 °C under pressure of 40 MPa for 2 h.

When the additive is water, it appears that the HA grains produced show whisker-like and plate-like features. The bend strength of the HA ceramic obtained is measured to be 24.5 MPa. When the additive is ammonia water, the obtained HA ceramic consists of HA-whiskers only. The bending strength and fracture toughness reach 56 MPa and 1.52 MPa $\cdot$ m$^{1/2}$, respectively. It was concluded that treating α-TCP with addition of ammonia water by the HHP method is thus a useful method for in situ fabrication of HA-whisker ceramics [166].

Tang et al. [167] reported a new method of preparation of uniform porous HA biomaterials. In order to obtain uniform porous biomaterials, disk samples were formed by the mixture of HA powders and monodispersed polystyrene microspheres, and then HA uniform porous materials with different diameter and different porosity (diameter: 436 ± 25 nm, 892 ± 20 nm and 1,890 ± 20 nm, porosity: 46.5%, 41.3% and 34.7%, respectively) were prepared by sintering these disk samples at 1,250 °C for 5 h. A single electrodischarge-sintering (EDS) pulse (1.0 kJ/0.7 g), from a 150 °C capacitor, was applied to atomized spherical Ti–6Al–4V powder in air to produce microporous compact. A solid core surrounded by a porous layer was self-assembled by a discharge in the middle of the compact. X-ray photoelectron spectroscopy (XPS) was used to study the surface characteristics of the compact material. C, N, O and Ti were the main constituents, with smaller amounts of Al and V. The surface was lightly oxidized and was primarily in the form of TiO$_2$. A lightly etched EDS sample showed the surface form of metallic Ti, indicating that EDS breaks down the oxide film of the as-received Ti–6Al–4V powder during the discharge process. The EDS Ti–6Al–4V compact surface also contained small amounts of TiN in addition to TiO$_2$, resulting in the reaction between nitrogen in air and the Ti substrate in times as short as 125 μs [168]. Corrosion tests were performed on sintered and unsintered titanium porous compacts of various porosities, as well as on solid titanium which acted as a control material. The tests were conducted in 0.9% aqueous NaCl solution maintained at 37 °C to simulate the physiological environment encountered by surgical implants in the human body. The results confirm that solid titanium and sintered porous titanium both possess a distinct passivation range. Unsintered porous titanium does not seem to passivate at all and suffers greater corrosion than solid Ti. Unsintered specimens compacted at higher pressures experienced more corrosion than those compacted at lower pressures, although sintered specimens behaved in the exact opposite fashion [169]. Dewidar et al. [170] fabricated porous surface with solid core Ti–6Al–4V implant compacts by traditional powder metallurgy. Powder metallurgy technique was used to produce three different porous surfaced implant compacts 30%, 50%, and 70% in vacuum atmosphere. The solid core formed in the center of the compact possesses similar microstructure of near full density of Ti–6Al–4V. The compressive yield strength was up to 270 MPa and significantly depended on the surface porosity, core size, and temperature of sintering. Porous-surfaced Ti–6Al–4V implant compacts with a solid core have much higher compressive strengths compared to the human bone. The in-growth of bone tissue into the outer porous surface layer results in part fixation, while the solid inner core region provides the necessary mechanical strength for a device used for the replacement of heavy load bearing joint regions such as the hip and knee. Two-step thermal oxidation of CpTi was investigated with a focus on the formation of anatase type TiO_2. A first-step treatment was conducted in Ar–(0.1–20)%CO atmosphere at a temperature range of 500–900 °C for a holding time of 0 or 1 h, and a subsequent second-step treatment was conducted in air at 200–600 °C for 1 h. It was reported that (i) titanium oxides and titanium oxycarbide were obtained in the first step, with relative amounts depending on heating temperature, holding time, and CO partial pressure, (ii) an anatase-rich layer on Ti was obtained after second-step treatment in air at 300–350 °C in cases where single-phase titanium oxycarbide formed in the first step, and (iii) the bonding strength of an anatase-rich layer with a thickness of 0.5 um was calculated to be around 90 MPa [171].

Besides the above popular coating thermal methods, there are still several important ways for coatings. Rodionov et al. [172] employed the steam-thermal oxidation of medical titanium implants. A porous-coated Ti–6Al–4V implant was fabricated by electrical resistance sintering [173]. Shenhar et al. [174] developed power immersion reaction assisted coating nitriding method, which is suitable for surface modification of large complex shaped orthopedic implants. Ti–6Al–4V alloy samples were annealed at 850–1,100 °C in sealed SS containers that allow selective diffusion of nitrogen atoms from the atmosphere and it was reported that (i) Ti_3Al intermetallic phase was detected at the Ti_2N/α–Ti interface acting as a barrier for nitrogen diffusion and (ii) importantly for biomedical applications, no toxic Al or V was detected in the surface layer nitrided Ti–6Al–4V alloy.

5.6 Combined technology

There are research combining several methods mentioned above; some of them are conducted simultaneously while the others are done consecutively. To improve the bioactivity of TiO_2 coatings, Zhang et al. [175] conducted MAO in an electrolyte containing Ca and P ions, and MAO in a 1M NaOH electrolyte and reported that alkaline-treated TiO_2 surface was found to be responsible for the rapid formation of HA during a SBF soaking, and the previous introduction of Ca and P can increase the opportunity to form HA. To increase osseoinductivity and corrosive-wear resistance, a hybrid treatment of microarc discharge oxidation and electrophoretic deposition was conducted by forming a double layer HA-TiO_2 coating on titanium alloys with HA as the top layer and a dense TiO_2 film (which is believed to be effective as chemical barriers against the in vivo release of metal ions from the implants) as the inner layer, so that good combination of bioactivity, chemical stability and mechanical integrity can be expected [176]. It was reported that a hybrid combination of MAO and electrophoretic deposition provided a phase-pure HA top layer and anticorrosive TiO_2 interlayer, which should show good mechanical and biochemical stability in the corrosive environment of the human body. MAO was combined with hydrothermal treatment in electrolytes containing calcium glycerphosphate and calcium acetate on Ti–6Al–4V, and it was found that (i) the film with a Ca/P ratio equivalent to HA was obtained when final voltage and current density were 350 V and 200 A/m^2 in the electrolyte containing 0.06 M calcium glycerphosphate and 0.25 M calcium acetate, and (ii) after hydrothermal treatment, the HA was precipitated on the surface of the film obtained by MAO with a thickness of about 4 μm [177]. Martines et al. [178] employed a sandblasted/acid-etched (SLAed) implant versus a smooth-surface implant to compare implant mobility (MO) and clinical reactions of peri-implant tissues to experimentally induced peri-implantitis in Beagle dogs. The right and left mandibular premolars were extracted from five Beagle dogs, and two smooth-surface implants and two SLAed implants were placed in each animal. The following were the findings. After 120 days, healing abutments were connected. Fifteen days later, the prosthetic abutments were connected, the hygiene regimen was suspended, and peri-implantitis was induced by the insertion of cotton ligatures into the soft tissue around the implants. At baseline and 30, 60, and 90 days later, clinical attachment level (CAL), probing depth (PD), and MO were measured. PD increased significantly in the SLA group alone when baseline PD was compared with 30-, 60-, and 90-day evaluations. The loss in CAL was significant in both groups when the baseline value was compared with 30-, 60-, and 90-day evaluations. Comparison between the two implant groups revealed a greater loss in CAL in the SLA group at the 90-day evaluation period. A significant increase in MO was seen in both groups when baseline values and 90-day evaluations were compared. Experimentally induced peri-implantitis results in a greater loss of CAL in SLA implants than in smooth-surface implants in dogs; however, no differences in MO or in PD have been noted between the two implant groups [178]. Wei et al. [179] heat treated NiTi alloy in air at 600 °C to form TiO$_2$ film and alkali-treated in NaOH solutions (1M, 2M, 3M, 5M) to improve their bioactivity, followed by soaking SBF to evaluate their in vitro performance, and was reported that (i) the 3 M NaOH treatment is the most appropriate method, (ii) a large amount of apatite formed within 1 day's soaking in SBF, and (iii) after 7 day's soaking TiO$_2$/HA composite layer formed on the NiTi surface [179]. Nishiguchi et al. [180] investigated the effects of the alkali and heat treatments on the bone-bonding behavior of porous titanium implants. Porous titanium implants had a 4.6 mm solid core and a porous 0.7 mm-thick outer layer using pure titanium plasma-spray technique. Three types of porous implants were prepared from these pieces: (1) control implant (CL implant) as manufactured, (2) AW-glass ceramic bottom-coated implant (AW implant) in which AW-glass ceramic was coated on only the bottom of the pore of the implant, and (3) alkali- and heat-treated implant (AH implant), where implants were immersed in 5 mol/L NaOH solution at 60 °C for 24 h and subsequently heated at 600 °C for 1 h. The implants were inserted into bilateral femora of six dogs hemitranscortically in a random manner. At 4 weeks, push-out tests revealed that the mean shear strengths of the CL, AW, and AH implants were about 10.8, 12.7, and 15.0 MPa, respectively. At 12 weeks, there was no significant difference between the bonding strengths of the three types of the porous implants (16.0–16.7 MPa). It was also reported that, histologically and histomorphologically, direct bone contact with the implant surface was significantly higher in the AH implants than the CL and AW implants both at 4 and 12 weeks. Thus, the higher bonding strength between bone and alkali- and heat-treated titanium implants was attributed to the direct bonding between bone and titanium surface. It was concluded that alkali and heat treatments can provide porous titanium implants with earlier stable fixation [180].

Juodzbalys et al. [181] investigated to create an acid-etched implant surface that is similar to that created by sandblasting combined with acid etching and to compare it with the surfaces of various commercially available screw-type implants. CpTi (grade V) disks were machined in preparation for acid etching. Tests were carried out using different acids and combinations of them with varying time exposures. It was found that the surface similar to the SLAed surface was best obtained with a combination of sulfuric and hydrochloric acids. It was concluded that the new experimental acid-etched titanium surface had the features of a roughened titanium surface, with glossily microroughness and large waviness. In general, the experimental surface was significantly rougher than the selected commercially available implants and similar to a SLAed surface (top Sa: 2.08 ± 0.36 μm, Sdr: 1.34 ± 0.3 μm, valleys: 1.16 ± 0.1 μm and 0.68 ± 0.1 μm, flanks: 2.24 ± 0.8 μm and 1.27 ± 0.1 μm, respectively) [181]. Increased surface roughness of dental implants has demonstrated greater bone apposition; however, the effect of modifying surface chemistry remains unknown. Buser et al. [108] evaluated bone apposition to a modified SLAed titanium surface, as compared with a standard SLA surface, during early stages of bone regeneration. Experimental implants were placed in miniature pigs, creating two circular bone defects. Test and control implants had the same topography but differed in surface chemistry. Test surface was prepared by submerging the implant in an isotonic NaCl solution following acid-etching to avoid contamination with molecules from the atmosphere. It was shown that (i) test implants demonstrated a significantly greater mean percentage of BIC contact as compared with controls at 2 (49.30% vs. 29.42%) and 4 weeks (81.91% vs. 66.57%) of healing, and (ii) at 8 wks, similar results were observed. It was, hence concluded that a modified SLAed titanium surface promoted enhanced bone apposition during early stages of bone regeneration [108].

Plasma-coated HA possesses a few crucial problems including easy dissolution or resorption in living tissue, and a potentially weak interfacial bond between the coating and a metal substrate, so that the low stability of the HA coating may lead to the failure of the implant. Besides, HA cannot be deposited uniformly due to geometrical limitations such as on inner or undercut surfaces of a porous structure. Ishizawa et al. [182] employed combined methods of electrochemical and hydrothermal reactions to precipitate HA crystals on porous Ti. HA ceramic has been applied on Ti implants (which was previously modified by laser beam irradiation) to increase the corrosion resistance, gain better biocompatibility and implant–bone tissue biological adhesion, by coating polyvinyl dense fluoride/HA composite which was obtained by casting method, and it was reported that uniform coating with a small thickness variation along the coated surface was successfully obtained [183]. Moreover, to control surface topography (roughness and blind porosity), surface chemistry, and wettability affecting the osseointegration, Shtansky et al. [84] employed a combination of various techniques such as cold spray, selective laser sintering, pulsed electroerosion treatment, and magnetron sputtering on Ti materials. In particular, for modifying surface chemistry, multifunctional bioactive nanostructured TiCaPCON films were deposited on cold sprayed, laser-sintered and electroerosion treated using a composite $TiC_{0.5}$ + $Ca_3(PO_4)_2$ target. It was reported that (i) the wettability measurements showed that the cold sprayed and electroerosion treated surfaces exhibit high values of water contact angle, and (ii) ion etching in vacuum and TiCaPCON film deposition made the samples highly hydrophilic; concluding that the combination of high surface roughness and blind porosity with hydrophilicity and biocompatibility makes the fabricated metal-ceramic materials promising candidates for applications involving tissue regeneration [84]. In order to improve the tribological properties of aluminum alloys, Sun [184] applied a process including the fabrication of a rutile-type TiO_2 coating on an aluminum alloy by the combination of sputter deposition of a pure titanium coating on the substrate and subsequent thermal oxidation. It was found that (i) the coating has a layered structure, comprising a rutile-TiO_2 layer at the top, an oxygen and nitrogen dissolved α-Ti layer in the middle, and a titanium aluminide layer in the interfacial region, and (ii) such a hybrid coating system has good adhesion with the substrate, can significantly enhance the surface hardness and tribological properties of the aluminum alloy in terms of much reduced friction coefficient and increased wear resistance [184].

To make metals bioactive for orthopedic applications, apatite/TiO_2 composite coatings were formed on Ti and NiTi SMA using a H_2O_2 oxidation and hot water aging technique and the subsequent accelerated biomimetic process. Nanoindentation testing conducted on cross sections of composite coatings indicated that there was no significant difference in nanohardness and elastic modulus between apatite/TiO_2 composite coatings formed on Ti and NiTi SMA samples. It was reported that (i) the enhancement of the adhesion between the apatite layer and the metal substrates arose from the TiO_2-intermediate layer in the composite coating, (ii) the highest values of coating adhesion strength for Ti and NiTi SMA samples, as measured by scratch tests, were 22.58 N and 19.07 N, respectively, and (iii) however, compared to corresponding Ti samples, NiTi SMA samples had better tribological properties [185]. To enhance the biological performance of titanium coating, hierarchical structuring was investigated to produce a hybrid structure of nano-, micro-, and macroscale layers which each of the substructure possesses at different biological functions. Such multifunction and multiscale structure was generated by sandblasting, acid etching, and alkali treating the vacuum plasma sprayed titanium coating following hot water immersion [186, 187]. The effects of different thickness of HA coatings on bone stress distribution near the dental implant–bone interface are very important factors for the HA-coated dental implant design and clinical application. By means of finite element analysis, the bone stress distributions near the dental implant coated with different thicknesses from 0 to 200 µm were calcu-lated and analyzed under the 200 N chewing load. In all cases, the maximal von Mises stresses in the bone are at the positions near the neck of dental implant on the lingual side and decrease with the increase of the HA coatings' thickness. The HA coatings weaken the stress concentration and improve the biomechanical property in the bone, however, in HA coatings thickness range of 60–120 µm, the distinctions of that benefit are not obvious. In addition, considering the technical reason of HA coatings, it was concluded that thickness of HA coatings ranging from 60 to 120 µm would be the bet-ter choice for clinical application [188].

5.7 Coating biomaterials

5.7.1 Metallic materials

Metal elements as well as alloys are chosen and used extensively as coating materi-als. A nanotopographic noble metal (Ag, Au, Pd) coating has been applied on com-mercial urinary catheters and used in more than 80,000 patients and was indicated that, by varying the noble metal ratio at implant surfaces, it is possible to modulate inflammation and fibrosis in soft tissue [189]. Use of silver element was due to the excellent antibacterial property [190]. Bioactive coatings are in high demand to in-crease the functions of cells for numerous medical devices. Metal element (Au, Ti) was coated on several potential orthopedic polymeric materials (specifically, ultra-high-molecular-weight polyethylene (UHMWPE), polyetheretherketone (PEEK), and polytetrafluoroethylene (PTFE) by ionic plasma deposition process which creates a surface-engineered nanostructure (with features below 100 nm) and their osteoblast (bone-forming cell) adhesion was improved [191].

Solid Ti metal was widely selected as coating elements. Ti was coated on surgi-cal 316 SS substrate and it was reported that cell proliferation, alkaline phosphatase activity, migration, and adhesion were increased, and osteoblast-like cells on the grit-blasted, titanium-coated, microarc-oxidated surface were strongly adhered, and prolif-erated well [192]. Nanoscale titanium was coated on silicon surfaces and increased cell spreading and proliferation rates were reported on surfaces with 50-nm-thick Ti coatings [193]. Nanostructures entail a high potential for improving implant surfaces, for instance, in stent applications. Ti nanoparticles were coated on an electropolished NiTi stent surface [194]. The solid solution phase of β-(Ti, Nb) [195] and Ti–6Al–4V film [196] were also selected as coating materials. Although carbon fiber-reinforced PEEK (CF/PEEK) exhibits properties suitable for load-bearing orthopedic implants, its hydrophobicity induces the deposition of a peri-implant fibrous tissue capsule pre-venting bone apposition. Devine et al. coated CF/PEEK screws coated with titanium by vacuum plasma spraying and physical vapor deposition (PVD) and found that Ti coating of CF/PEEK screws significantly improve bone apposition and removal torque compared with uncoated CF/PEEK screws [197].

Tantalum (Ta) metal has successfully been used for implants due to excellent biocompatibility [198, 199]. It has been found that tantalum coatings show great potential within both industrial and medical applications. Tantalum coating with a very ductile nature is suitable, where a product can be improved by combining material characteristics of the substrate with an exceptional corrosion-resistant surface [200]. As a result, tantalum is an excellent candidate for using as coatings for NiTi alloys to improve its anticorrosion property and radiopacity [201]. Ta was coated on Co–Cr alloy with excellent outcomes of good mechanical properties (high elastic modulus, high hardness value, low friction coefficient, high wear resistance) and chemical properties (high wettability, good corrosion resistance), so that it is of interest for joints presenting good biocompatibility and great longevity [202, 203]. Porous tantalum, a low modulus metal with a characteristic appearance similar to cancellous bone, is currently available for use in several orthopedic applications (hip and knee arthroplasty, spine surgery, and bone graft substitute). It was reported that this transition metal maintains several interesting biomaterial properties, including: a high volumetric porosity (70–80%), low modulus of elasticity (3 MPa), and high frictional characteristics along with an excellent biocompatibility and tantalum–chondrocyte composites have yielded successful early results in vitro and may afford an option for joint resurfacing in the future, suggesting promising applications in orthopedic surgery [204–206]. Ta along with TaO was coated on surgical 316 L SS and it was mentioned that (i) the mechanical stability and repassivation properties of the material are guaranteed up to the onset of substrate plastic deformation, (ii) biocompatibility is obtained by tantalum oxide, and (iii) the ductility is achieved by the tantalum interface which, at the same time, ensures continued film adhesion, even after plastic deformation of the steel substrate [207].

Olivares-Navarrete et al. [208] deposited niobium (Nb) by magnetron sputtering as a possible surface modification for SS substrates in biomedical implants and the biocompatibility of the coatings was evaluated using human alveolar bone derived cells, in terms of cellular adhesion, proliferation, and viability. It was observed that (i) no toxic response was observed for any of the surfaces, indicating that the Nb coatings act as a biocompatible, bioinert material, (ii) cell morphology test confirmed the healthy state of the cells on the Nb surface, and (iii) water contact angle measurements showed that the Nb surface is more hydrophobic than the SS substrate [209]. Nb thin film and NbN were deposited on SS to evaluate the biocompatibility of the surfaces by testing the cellular adhesion and viability/proliferation of human cementoblasts during different culture times, up to 7 days [209]. It was reported that (i) the niobium oxide films were amorphous and of stoichiometric Nb_2O_5, while the niobium nitride films were crystalline in the FCC phase (c-NbN) and were also stoichiometric with a Nb to N ratio of one, (ii) the biocompatibility of the SS could be improved by any of the two films, but neither was better than the Ti–6Al–4V alloy, and (iii) on the other hand, comparing the two films, the c-NbN seemed to be a better surface than the oxide in terms of the adhesion and proliferation of human cementoblasts [209].

NiTi films [210] or porous NiTi [211] and NiTi foam [212] were prepared. Liu et al. [211] evaluated bone ingrowth under actual load-bearing conditions and interfacial bonding strength by histological analysis and push-out test. It was reported that the porous NiTi materials bond very well with newly formed bone tissues and the highest average strength of 357 N and best ductility are achieved from the porous NiTi materials [211].

High-purity biodegradable magnesium (Mg) is believed to exhibit excellent biocompatibility, and coating of Mg element is expected to be appropriate for implant applications and to improve the interaction between the implant and the biological environment [213]. In recent years, research on magnesium alloys had increased significantly for hard tissue replacement and stent application due to their outstanding advantages, because (i) Mg alloys have mechanical properties similar to bone which avoid stress shielding, (ii) they are biocompatible and essential to the human metabolism as a factor for many enzymes, (iii) main degradation product Mg is an essential trace element for human enzymes, and (iv) the most important reason is they are perfectly biodegradable in the body fluid [213–215]. Mg–Y–Nd was evaluated for its feasibility as bone implant applications mainly focused on biocompatibility and corrosion resistance [216]. It was concluded that the biodegradable Mg–Y–Nd implant is superior to the controlled Ti–6Al–4V with respect to both bone–implant interface strength and osseointegration. Mg–Zr–Sr alloys have recently been developed as biodegradable implant materials [217]. It was reported that (i) both Zr and Sr are evaluated as excellent alloying elements in manufacturing biodegradable Mg alloy implants, (ii) Zr additionally refined the grain size, improved the ductility, smoothed the grain boundaries, and enhanced the corrosion resistance of Mg alloys, (iii) Sr, in addition, led to an increase in compressive strength, better in vitro biocompatibility, and significantly higher bone formation in vivo, and (iv) Mg–xZr–ySr alloys with x and y ⩽5 wt% would make excellent biodegradable implant materials for load-bearing applications [217]. Biodegradable Mg–Zn–Zr was also developed as an orthopedic implant material due to its biodegradable feature and suitable mechanical properties [218]. Amorphous Si film (which was prepared by plasma-enhanced chemical vapor deposition of SiH_4) was coated on Mg–Y–Nd alloy for biomedical application [300]. The hemolysis test and blood platelets adhesion test were conducted, and it was reported that the hemolysis of Mg–Y–Nd alloy decreased after being coated by Si and the platelets attached on the Si film were at the inactivated stage with a round shape [219].

Functionally graded, hard and wear-resistant Co–Cr–Mo alloy was coated on Ti–6Al–4V alloy and it was shown that the addition of the Co–Cr–Mo alloy significantly increased the surface hardness without any intermetallic phases in the transition region [220]. Selenium (in the form of sodium selenite) coating on Ti material was evaluated and it was found that there was no inhibitory effect of the selenium coating on the osteoblastic cell growth, and Se coating is a promising method to reduce bacterial attachment on prosthetic material [221].

5.7.2 Polymeric materials

Biodegradable polymers (which perform a structural application and are designed to be completely resorbed and become weaker over time) form a unique class of materials that created an entirely new concept when originally proposed as biomaterials. Recently, in tissue engineering, a biodegradable scaffold seeded with an appropriate cell type provides a substitute for damaged human tissue while the natural process of regeneration is completed [222, 223]. An adhesive biodegradable polymer, poly(vinyl acetate) (PVA), was coated on Mg-based alloys by the dip-coating technique in order to control the degradation rate and enhance the biocompatibility of magnesium alloys. Since the cytocompatibility of osteoblast cells (MC3T3) revealed high adherence, proliferation, and survival on the porous structure of PVA-coated Mg alloy, which was not observed for the uncoated samples, it was concluded that the PVA coating is a promising candidate for biodegradable implant materials, which might widen the use of Mg-based implants [224]. A biodegradable polymeric membrane fabricated by polycaprolactone and dichloromethane was coated on Mg-based alloy to control high corrosion rate, and accumulation of hydrogen gas upon degradation hinders its clinical application [225]. Polyurethane (PU) [226] and polyhedral oligomeric silsesquioxanes (POSS) and poly(carbonate-urea)urethane [227] were coated onto NiTi stent to control the thrombogenicity. When Ti–6Al–4V alloy is used in several ventricular assist devices, the thrombogenicity resistance is highly required. A plasma-induced graft polymerization of 2-methacryloyloxyethyl phosphorylcholine (MPC) was carried out and poly(MPC) (PMPC) chains were covalently attached onto the Ti–6Al–4V surface [228]. It was reported that platelet deposition was markedly reduced on the PMPC grafted surfaces and platelet activation in blood that contacted the PMPC-grafted samples was significantly reduced relative to the unmodified Ti–6Al–4V and polystyrene control surfaces [228]. For bone and teeth repairs, biomaterials are demanded of both good biocompatibility and reliable mechanical properties for long periods. Due to the nature of the HA of low fracture toughness and fracture energy, there are several ways to overcome these disadvantages of HA materials, including incorporating with ZrO_2 or Al_2O_3. Nakahara et al. [229] infiltrated porous HA with nylon to improve fracture toughness of HA with 1.65 MPa√m.

5.7.3 Ceramics – metallic oxides, nitrides, and carbides

Titanium oxides (TiO_2) have been used in different forms such as thick layer, thin film, porous, and foam, and in nanotube, titanium nitrides, and carbides along with other metals' oxides. Titanium dioxide (TiO_2, titania) is a widely abundant and inexpensive material. In bulk form, it is produced as a white powder and it is the most widely used white pigment because of its brightness and very high refractive index. Applications include filler pigment in paints, cosmetics, pharmaceuticals, food products, tooth whitening gel, and toothpaste [230]. Bioactive TiO_2 was deposited on surface-modified polyethylene terephthalate (PET) plates by using a sol–gel method and was indicated that this modification possesses a mechanically stable surface-modified layer, excellent bone-bonding ability, osteoconductive ability, and biocompatibility in bone [231]. Ti–O film was deposited on surgical 316 L SS by PIII and deposit method [232].

NiTi was surface-modified by plasma electrolytic oxidation in aqueous solutions of sodium sulfate and sodium hydroxide (Na_2SO_4–NaOH) using an AC power supply. It was found that a thick and porous oxide layer with micron-sized pores was formed on the NiTi substrate, with the thickness of the oxide layer ranging from a few μm to over 10 μm, depending on the processing time [233]. Titanium oxide coatings were conducted by MAO in galvanostatic regime on biomedical NiTi alloy in H_3PO_4 electrolyte using DC power supply [234]. Titanium oxide (TiO_2) films were deposited onto the polymer substrates of PTFE, polyethylene (PE), and PET, which were premodified with polydopamine coating, by a simple liquid-phase deposition process. It was indicated that the fabricated TiO_2 films could markedly improve the in vitro cytocompatibility [235]. TiO_2 was coated on Ti–6Al–4V substrate by air plasma spraying method. It was reported that, according to the osteoblast adhesion morphology and proliferation tests, osteoblast-like cell morphology was not influenced by process parameters, but cell proliferation was affected to some extent by surface roughness and porosity among TiO_2 [235]. Titania has three distinct types of crystalline structures: brookite, anatase, and rutile. Although most of the works on TiO_2 coating did not identify the crystalline structure of the coating titania, there are several works which identified crystalline structures [236]. Anatase phase titanium dioxide (TiO_2) thin films were grown by pulsed laser deposition on SiO_2 substrates and anatase-type TiO_2 coatings were grown on Ti and Si by atomic layer deposition [237]. On the other hand, a rutile-type titania film was deposited on a silicon substrate by RF-magnetron sputtering using a sintered oxide target in an argon-gas atmosphere [238] and an amorphous titania was formed by treating titanium plates by Ar and oxygen plasma treatment, and furthermore, nano-(α + rutile-TiO_2) were formed on the amorphous-like oxide layer following glow-discharging.

Thin porous titania film was also coated differently. Porous titania film is prepared by alkali (in NaOH) treatment of NiTi alloy followed by soaking treatment in HCl solution. The benefit of this porous titania film as an interlayer to improve adhesion and integrity of the sol–gel titania coating on NiTi alloy substrate is evidenced by surface morphological observations [239]. Effective immobilization of bioactive substances such as adhesive proteins, synthetic peptides, and growth factors on metallic substrates is required for a number of medical applications. It was reported that (i) an alkoxy-derived nanoporous titanium oxide coating (which was synthesized electrochemically on titanium in methanolic electrolytes) may act as an effective interface for functionalizing a titanium surface, and (ii) nanoporous oxide coatings could facilitate fast diffusion of small organic molecules within the oxide network and form strong chemical bonds with the functional groups of these molecules at room temperature [240]. Nanotubes of titania was coated. Oriented aligned TiO_2 nanotube arrays were fabricated by anodizing titanium foil in 0.5% HF electrolyte solution [241]. A net-like structured TiO_2 film was produced with a low-temperature hydrothermal process, followed by acid washing and calcination in air, and was deposited on surgical 316 L SS substrate [242]. For improving cell adhesion and differentiation and as a result of these improvements, nanofibrous modification of dental and bone implants might enhance osseointegration [243]. TiO_2 nanofibers were fabricated by the electrospinning method using a mixture of Ti isopropoxide and poly(vinyl pyrrolidone) in acidic alcohol solution. It was reported that the diameter of the TiO_2 nanofibers was controlled within the range of 20–350 nm [243].

Titania composites were also chosen as coating materials. To improve the bioactivity of titanium, CpTi was hydrothermally treated in 0.1 mol/L $MgCl_2$ solutions with different pH values (5.5, 7.5, and 9.5) at 200 °C for 24 h. It was reported that (i) nanosized anatase precipitations (including $MgTiO_3$ at low pH and $Mg(OH)_2$ at high pH) were obtained and (ii) hydrothermal treatment in $MgCl_2$ solution can improve the bioactivity of titanium by immobilizing Mg into oxide layer, however, the chemical state and amount of Mg should be well controlled [244]. Mg-incorporated titania was produced by hydrothermally treating Ti in an alkaline Mg-containing solution. It was concluded that the Mg-incorporated sub-microporous Ti oxide surface produced by hydrothermal treatment may improve implant osseointegration by enhancing the attachment, spreading, and differentiation of osteoblastic cells [245]. Ni_2O_3-doped TiO_2 nanotubes were deposited on NiTi alloy using electrochemical anodization method. It was reported that (i) forming of Ni_2O_3-doped TiO_2 film on the surface of NiTi alloy increased its wettability, especially for 600 °C annealed sample, so the biological response is expected to be improved, and (ii) annealing the Ni_2O_3-doped TiO_2 film coated NiTi alloy at 450 °C is desirable for cardiovascular applications [246]. There are still several different types of oxides which are ready for coating processing. A mixture of TiO_2 and ZnO was sputter-coated on glass substrates [247]. To improve corrosion resistance and biocompatibility, Fe–O film was coated on iron stents [248]. It was reported that the results of HUVECs culture showed that HUVECs had good adhesion and proliferation behavior on the Fe–O film, and, after depositing Fe–O thin film by PIII and deposition technique under the low oxygen flux, the corrosion resistance and biocompatibility of pure iron were effectively improved [248].

Due to excellent mechanical stability and repassivation properties of the tantalum (Ta) are guaranteed up to the onset of substrate plastic deformation. Macionczyk et al. [249] coated TaO onto SS implants and obtained results indicating that, while biocompatibility is obtained by tantalum oxide, the ductility is achieved by the tantalum interface which at the same time ensures continued film adhesion, even after plastic deformation of the steel substrate. Thin TaO film was also coated on Ti material by layer-by-layer sol–gel deposition technique [250]. A composite of TaO and carbon was coated on Ti surface to improve its propensity to a long-term degradation in physiological conditions and its weak osseointegrative capacities [251].

Zirconia (ZrO_2) coating was evaluated by the histologic analysis, indicating that (i) bone growth is more evident around coating fixtures than in controls and (ii) a more mature bone is present in the peri-implant coated surface than in controls, hence zirconium oxide coating can enhance implant osseointegration [252]. Yan et al. [253] prepared macroporous and nanocrystalline zirconia coating film by MAO, followed by chemical treatment in H_2SO_4 or NaOH aqueous solution to evaluate the apatite-forming ability. A zirconia (ZrO_2) was coated on Ti and Zr to exhibit lower friction and superior wear properties, suggesting that they could be used in hip and knee implants [254].

Nano-Mg_2SiO_4 (forsterite) was coated onto porous HA to enhance the compressive strength of porousHA scaffold (porosity: ~83%, mean pore size: ~740 μm) and found that (i) the coating microstructure consisted of the grains with the range between 35 and 80 nm, (ii) the compressive strength of highly porous HA was improved from 0.12 to 1.61 MPa, indicating that the scaffolds obtained provided a good mechanical support while maintaining bioactivity, and hence they could be used as tissue engineering scaffolds for low-load-bearing applications [255]. For controlling the porosity of bioglasses at the nanometric scale, a composite of silica-based glasses with organic components (in a form of new organic–inorganic hybrid material) was prepared by the sol–gel process [256]. Alumina (Al_2O_3) was deposited on Fe–Cr–Al–Y alloy by the thermal oxidation technique [257].

Now moving to metal nitrides, TiN film was coated on mild steel by a new combined laser/sol–gel processing technique, indicating that nanohardness measurements revealed a hardness value of the order of 22–27 GPa [258]. Titanium and Ti–6Al–4V alloy samples were coated using a powder immersion reaction assisted coating nitriding method in order to modify their surface properties, and it was reported that strongly adherent single (TiN)- or double (Ti_2N/TiN)-layer coatings were obtained on both substrates, suggesting that titanium nitride coatings can provide surgical titanium alloys with the enhanced fretting wear and corrosion resistant surface thereby minimizing the ion- and particulate-generating potential of modular orthopedic implants [259]. Nanostructured TiN film was prepared by reactive plasma spraying in the air. The coated film was identified as mainly composed of TiN (max. 86.3%) and a small quantity of Ti_3O [260]. Sinter-coating combines sintering and TiN coating in one step by sintering the green specimens in N_2 atmosphere was conducted on titanium foam to produce TiN film layer [261]. It was reported that, after alkali treatment by soaking the foams in NaOH solution and then heating at 600 °C, treated foams were immersed into SBF to induce a bone-like apatite layer on foam surface [261]. TiN-based composite coatings with and without the addition of Cr were prepared and deposited by reactive plasma spraying in air. Both sintered and mixed powder of Ti and B_4C were used for this process. It was found that the coating deposited using sintered Ti and B_4C powder is composed of two main phases (TiN and TiN$_{0.3}$), two minor phases (Ti$_2$O$_3$ and TiB$_2$), and a small fraction of TiC phase, and the composition of the coating deposited using the mixed powder with Cr added is predominantly in the TiN and TiB$_2$ phases, a smaller phase fraction of Ti$_2$O$_3$ and TiO$_2$, and some unreacted Cr [262]. TiN/Ti multilayer was deposited on 304 SS by magnetron sputtering to improve pitting corrosion resistance [263]. TiN was also coated on cellulosic fiber substrates by atomic layer deposition to produce biocompatible cellulose and other implant materials [264].

In TiN group, there are several composites prepared for coating purposes. To improve cytotoxicity, cell adherency, and tribological performance, TiN particle reinforced Ti–6Al–4V composite coatings were prepared. It was found that (i) the composite coatings contain distinct TiN particles embedded in α + β phase matrix, (ii) the average top surface hardness of Ti–6Al–4V alloy increased from 394 ± 8 HV to 1,138 ± 61 HV with 40 wt% TiN reinforcement, and (iii) among the composite coatings, the coatings reinforced with 40 wt% TiN exhibited the highest wear resistance of 3.74×10^{-6} mm^3/Nm, which is lower than the wear rate, 1.04×10^{-5} mm^3/Nm, of laser processed Co–Cr–Mo alloy tested under identical experimental conditions [265]. Türkan et al. [266] coated a medical grade Co–Cr–Mo alloy with TiN by means of PVD technique at 550 °C for 6 h. Static immersion test was conducted to investigate the effectiveness of TiN coating in preventing the dissolution of metal ions into the SBF from the substrate by atomic absorption spectrometry (AAS) and inductively coupled plasma optical emission spectrometry (ICP-OES). It was found that (i) the SEM analysis indicated quite uniform and highly dense TiN coated layer (about 3 μm thick) with a columnar growth mode reaching from substrate to coating surface, (ii) the AAS and ICP-OES results showed that the presence of the TiN coating prevented the release of cobalt and chromium metal ions from the substrate Co–Cr–Mo alloy whereas cobalt was preferentially dissolved from the as-polished material, and (iii) calcium phosphate precipitation was observed on the surface of the as-polished material, indicating a degree of bioactivity of the as-polished surface which is absent in the TiN coated substrate alloy. Titanium nitride oxide (TiNO$_x$) coating was performed on roughened Ti and SS by plasma/PVD deposition technique and on cellulose fiber substrates by atomic layer deposition [267]. It was reported that TiNO$_x$ coatings yield similar proliferation and differentiation rates when applied onto roughened Ti and SS, suggesting that a more effective osseointegration of endosseous implants can be produced [267].

Titanium aluminum nitride (Ti,Al)N was coated on 304 SS substrate. After thus coated implants were placed into the mid-shaft of the femur of Wistar rats, it was found that (i) all implants exhibited a favorable response in bone with no evidence of fibrous encapsulation, (ii) there was no significant difference in the amount of new bone formed around the different rods (osseoconduction), and (iii) however, there was a greater degree of shrinkage separation of bone from the coated rods than from the plain rods, suggesting that titanium aluminum nitride coating may result in reduced osseointegration between bone and implant [268]. Mukherjee et al. [269] coated the hard Ti-based coating TiAlN on SS by PIII-assisted deposition to improve wear resistance in orthopedic conditions. Furthermore, titanium aluminum nitride oxide (TiAlNO) coatings were deposited on the Corning glass by means of the cosputtering technique. It was indicated that TiAlNO is versatile material that could be used for both mechanical and electronic applications; however, it is important to control fabrication conditions [270]. Shtansky et al. [271] studied complicate composite of TiCaPCON films which were prepared by DC magnetron sputtering of composite comprising $TiC_{0.5}$ and $Ca_3(PO_4)_2$ target produced by self-propagating high-temperature synthesis. It was reported that (i) the in vitro studies showed that human fibroblasts well adhered and spread on the surface of the PTFE sample coated with TiCaPCON films, and (ii) the in vivo studies using rat hip and rabbit calvarial defect models demonstrated a high osseointegration potential of the TiCaPCON/PTFE implants [271].

Titanium carbide (TiC) was coated on titanium substrate by pulsed laser deposition method to improve implant hardness, biocompatibility through surface stability, and osseointegration through improved bone growth [272] and TiC and TiB (titanium boride) were coated by laser surface alloying of Ti–6Al–4V with graphite and boron mixed powders to enhance wear resistance [273]. Levashov et al. [274] coated composite of TiCCaPCON on NiTi by rapid plastic deformation technology, thermomechanical treatment, and magnetron sputtering.

5.7.4 Ceramics – nonmetallic compounds

In recent years, implants coated with bioactive layers to promote fixation through osseointegration mechanism has become increasingly common. Fabrication routes that provide control over features such as biocompatibility and bioreactivity through management of coating chemistry and structure are still being developed. Current coating fabrication technology includes plasma spraying, rf suppering, CVD, laser ablation, and sol–gel processing, although plasma spraying is used to produce most commonly available bioceramic coatings for orthopedic and dental implants [275]. The term apatite denotes a family of crystals with the formula $M_{10}(RO_4)_6X_2$, where M is usually calcium, R is usually phosphorus, and X is hydroxide or a halogen such as fluorine. Bone mineral was found to be quite complex and included various types of hydrated calcium phosphates, the most common being calcium HA – $Ca_{10}(PO_4)_6$ $(OH)_2$ [276]. Although no standard manufacturing guideline exists for depositing HA – $(Ca_{10}(PO_4)_6(OH)_2)$ – on implant surfaces, HA depositions have been widely applied to generate bioactive surfaces in simulated biological environments [277, 278]. HA was coated on titanium plates [279–281]. In order to improve biocompatibility of Ti substrates, Hahn et al. [281] coated 1-μm-thick nanostructured HA, deposited on the substrates through aerosol deposition, which sprays HA powder with an average particle size of 3.2 μm at room temperature in vacuum. Ti–6Al–4V is also a popular biomedical metallic substrate for HA coating [282, 283]. Gopi et al. [284] passivated surgical grade SS (316 L) in borate through poly-*ortho*-phenylenediamine, and subsequently HA was coated by a dip-coating method to minimize the release of metallic ions. Nonmetallic substrates are also subjected to HA coatings. In order to improve bioactivity and to enhance osseointegration of titanium implant materials, nanotubular TiO_2 structure was subject to HA coating, and the bond strength was further improved by annealing the HA-coated nanoporous titania in an argon atmosphere [285]. Cao et al. [286] coated the fusion-cage-like carbon fiber reinforced carbon composite implants with HA by the plasma spraying technique for bone tissue reconstruction. After autogenously bone filled fusion-cage-like implants were grafted in hybrid goats' tibia for 328 days, it was reported that the coating can significantly speed up the bone defect healing process and improve the surface bioactivity of carbon fiber reinforced carbon composites [286]. Nanocomposite ZrO_2–Al_2O_3 was HA-coated for improving biocompatibility, and from the mechanical and biological evaluations, it was concluded that a mixture of coated HA with 30 vol% and 70 vol% of ZrO_2–Al_2O_3 composite was found to be the optimal composition for load-bearing biological applications [287].

HA material as a coating species is not limited in form of relatively thick solid layer, but it can be formed in various shapes. HA thin films for applications in the biomedical field were grown by pulsed laser deposition and radio frequency magnetron sputtering techniques [288–291]. To achieve excellent biocompatibility of metal-based implants, HA thin film was coated on nanotube-formed Ti–35Nb–10Zr alloys after femtosecond laser texturing [291]. The thus treated surface was found to be a composite with TiO_2, Nb_2O_5, and ZrO_2. It was furthermore reported that (i) the HA-coated surface on the nanotubes showed the lowest contact angle compared with the other surfaces, and (ii) from FE-SEM observations, cell attachment and spreading of MG 63 cells showed significantly higher tendency for surfaces covered by HA coating and nanotubes [291]. Not only thickness but the size is also controlled for coating HA material [292, 293]. A highly crystalline nano-HA was coated on CpTi using inductively coupled radio frequency plasma spray [293]. It was characterized that (i) depending on the plasma-processing conditions, a coating thickness between 300 and 400 μm was achieved where the adhesive bond strengths were found to be between 4.8 and 24 MPa, (ii) the coating was made of multigrain HA particles of ~200 nm in size, which consisted of recrystallized HA grains in the size range of 15–20 nm [293]. In most cases, HA has been prepared as a synthetic material, but Duta et al. [294] prepared HA as the synthesis of novel ovine and bovine derived HA thin films and coated it on titanium substrates by pulsed laser deposition. It was reported that (i) the micrographs of the films showed a uniform distribution of spheroidal particulates with a mean diameter of ~2 μm, (ii) pull-off measurements demonstrated excellent bonding strength values between the HA films and the titanium substrates, and (iii) because of their physiochemical properties and low-cost fabrication from renewable resources, these new coating materials could be considered as a prospective competitor to synthetic HA used for implantology applications [294]. Based on this variation possibility, there are several ion substitutions inside the HA structure (which is hexagonal closed-packed crystalline structure). Among proposed variations, fluoridated HA seems to be major one [295–297]. Zahrani et al. [295] prepared nano-crystalline fluoridated HA powders with a chemical composition of $Ca_{10}(PO_4)_6OH_{2-x}F_x$ (where x values were selected equal to 0.0, 0.5, 1.0, 1.5, and 2.0) through a modified simple sol–gel technique. Similarly, the sol–gel technique was applied to produce fluoridated HA [296, 297]. Carbon can be incorporated in HA structure to show bio-function of excellent mechanical property of carbon in form of nanotube [298] or carbonate [299, 300] with biocompatible HA. Li et al. [301] coated Na-doped HA (which was Ca-deficient) directly prepared onto carbon–carbon composites using electro-chemical deposition. It was reported that (i) the Na/P molar ratios of the coating formed on carbon–carbon substrate was 0.097, (ii) the mean thickness of the coating was approximately 10 ± 2 µm, (iii) the average shear bonding strength of Na–HA coating on carbon/carbon was 5.55 ± 0.77 MPa and (vi), after soaking the samples in a SBF for 14 days, the Na–HA coated carbon–carbon composites can rapidly induce bone-like apatite nucleation and growth on its surface in SBF; suggesting that the Na–HA coating might be an effective method to improve the surface bioactivity and biocompatibility of carbon/carbon composites [302]. Silicon-substituted HA was pre-pared and was coated on Ti substrate via a magnetron cosputtering technique and coated surfaces were subjected to an in vitro study using primary human osteoblast cells to evaluate their biological property [302]. It was reported that (i) although human osteoblast cells showed initial poor adhesion and spreading on hydrophobic Ti surface, Si-incorporated HA coatings on Ti substrates renders the surface more hy-drophilic, with water contact angles between 30° and 40°, (ii) human osteoblast cells attached, spread, and proliferated well on these coatings, (iii) enhanced calcification (formation of calcium phosphate nodules across the collagenous matrices) was ob-served on Si–HA coatings with increasing Si content; suggesting that Si-incorpo-ration into HA structure enhances the surface wettability, and promotes cell proliferation and calcification [304]. Guo et al. [303] investigated the physicochemical properties and biocompatibilities of La-containing apatites with the formula La_2Ca_8 $(PO_4)_6O_2$ as typical one. It was reported that the sintered La-incorporated apatite block achieves a maximal flexure strength of 66.69 ± 0.98 MPa at 5% La content with 320% increase when compared with the La-free apatite, suggesting that the La-incor-porated apatite possesses application potential in developing a new type of bioactive coating material for metal implants and also as a promising La carrier for further ex-ploring the beneficial functions of La in the human body [303]. Sr (strontium)-substi-tuted HA was prepared with 10 mol% Ca^{2+} replaced by Sr^{2+} and the implant fixation in ovariectomized rats was investigated after Sr–HA was coated on Ti implant by the sol–gel dip methods [304]. It was reported that (i) Sr–HA-coated implants revealed im-proved osseointegration compared to HA, with the bone area ratio and BIC increased by 70.9% and 49.9% in histomorphometry, (ii) the bone volume ratio and percent os-seointegration by 73.7% and 45.2% in micro-CT evaluation, and (iii) the maximal push-out force and ultimate shear strength by 107.2% and 132.9% in push out test;

suggesting that the feasibility of using Sr–HA coatings to improve implant fixation in osteoporotic bone [304]. Similarly, Sr–HA was coated on Mg–Sr (with Sr in the range of 0.3–2.5%) [305]. HA-based composites are also prepared. Ji et al. [308] prepared HA-carbon nanotube/titania double layer and coated it on titanium substrates intended for biomedical applications, reporting that the composite HA coating promoted the adhesion of preosteoblasts and the unique surfaces combined with the osteoconductive properties of HA exhibited excellent mechanical properties [306]. A double coating layer of the first thin precoating made with polyalkoxysilanes was prepared to promote the adhesion of HA/polyalkoxysilanes composite second layer [307]. Furthermore, HA was admixed with TCP with a Ca/P molar ratio from 1.56 to 1.77 [308], with TiO_2 with bonding strength between HA coating and substrate of 27.2 ± 1.6 MPa [309], and ZrO_2 [310].

Aside from being known for its excellent mechanical properties and aesthetic effect, zirconia has recently attracted attention as a new dental implant material. Many studies have focused on HA coating for obtaining improved biocompatibility, however the coating stability was reduced by a by-product produced during the high-temperature sintering process. Kim et al. [311], in order to overcome the aforementioned problem, coated the zirconia surface with a sol–gel-derived HA layer, followed by sintering. It was reported that (i) in vitro cell experiments using a preosteoblast cell line revealed that the HA-coated zirconia surface acts as a preferable surface for cell attachment and proliferation than bare zirconia surface, (ii) in vivo animal experiments also demonstrated that the osteoconductivity of zirconia were dramatically enhanced by HA coating, which was comparable to that of Ti implant, suggesting that (iii) the sol–gel-based HA-coated zirconia (ZrO_2) has a great potential for use as a dental implant material.

Biocompatible calcium–phosphate (Ca–P) has been widely employed as coating material for improving the substrate's biocompatibility. Ca–P was coated on CpTi substrates [312–316], and various Ti-based alloys including Ti–6Al–4V [317–319], Ti–7.5Mo alloy [320] and Ti-3Zr-2Sn-3Mo-25Nb [321]. Uniquely, Wang et al. [322] prepared functionally graded bioceramic coating composed of essentially calcium phosphate compounds. It was expected that the coating was graded in accordance to adhesive strength, bioactivity, and bioresorbability; i.e., the bond coat on the Ti–6Al–4V substrate is deposited with a particle range of the HA that provides a high adhesive strength and bioactivity but has poor bioresorption properties; while the top coat is composed predominantly of α-TCP that is highly bioresorbable [322]. Chen et al. [323] used electrochemically deposited dicalcium phosphate dihydrate coating of a porous substrate on CpTi implant surface and reported that Ca_2P coatings improve the extent of cancellous bone ingrowth in the early post-operative phase following uncemented arthroplasty [323]. Titanium substrate has been also coated with another type of Ca-containing species. Load-bearing Ti implants were coated with TCP [324, 325]. Ti was also coated with composite of Ag–HA composite [326] or TCP doped by Mg element in form of $Ca_{2.8}Mn_{0.2}(PO_4)_2$ [327]. Ti–6Al–4V alloy was coated with bioactive calcium alkali phosphate [319], Ca_2SiO_4 [328], $CaTiSiO_5$ (sphene) [329], TiO_2 + CaP compound [330], corrosion-resistant zeolite (aluminosilicate) for eliminating the release of cytotoxic Al and V ions [331], or ceramics treated in K_4ZrF_6–H_3PO_4 and $NaAlO_2$–Na_3PO_4 solutions [332]. Bioglass should be included in this group as biocompatible coating material. CpTi as well as Ti-based alloys were coated with bioglass; two phase ($\alpha + \beta$) Ti–6Al–7Nb titanium alloy was coated with a composite PEEK + Bioglass [333], and Ti–6Al–4V was coated with bioactive borate glass comprised of Na_2O–CaO–B_2O_3 system, modified by additions of SiO_2, Al_2O_3, and P_2O_5 with adhesive strengths of 36 ± 2 MPa [334], or alkali-free bioglass thin films with the highest mean value of pull-out adherence (60.3 ± 4.6 MPa) [335].

Coatings such as diamond-like carbon (DLC) and titanium nitride (TiN) are employed in joint implants due to their excellent tribological properties. Recently, graphite-like carbon (GLC) and tantalum (Ta) have been proven to have good potential as coating since they possess mechanical properties similar to bones – high hardness and high flexibility. Hee et al. [336] conducted systematic literature review to summarize the coating techniques of these four materials in order to compare their mechanical properties and tribological outcomes. It was mentioned that although experiment conditions varied, Ta has the lowest wear rate compared to DLC, GLC, and TiN because it has a lower wear rate with high contact pressure as well as higher hardness to elasticity ratio.

5.7.5 Composites, hybrids, and biomimetic materials

There are a huge number of papers published on HA-based biocompatible composites. HA–TiO_2 composites [337–340] and HA–ZrO_2 composites [341] are prepared. Moreover, HA–glass composites [342], HA–carbon composites [343], and HA–wollastonite ($CaSiO_3$) composites [344] are produced. In order to enhance the hard tissue biocompatibility and an excellent osteoconductivity, HA was admixed with chitosan [345, 346]. In order to improve bioactivity, the initial cell proliferation, and finally the osseointegration, HA has been also incorporated with collagen [347–349]. For promoting implant osseointegration or bone tissue engineering, biomimetic coating materials were introduced, for example, HA–amelogenin composite [350] and HA–fibronection [351].

Titania (TiO_2) has been admixed/incorporated with Zn element [352], Ag element [353], carbon [354], $CaTiSiO_5$ (sphene) [355], SiO_2 [354], and $CaSiO_3$ [356]. Similarly, zirconia (ZrO_2) is mixed with Ca-P [357, 358] or polypyrrole [359] to improve corrosion resistance and bio-compatibility. As nitride composites, there are reports on AlN–SiO_2 [360] and TiN–carbon composites [361]. Other metallic oxide composites are also prepared. Gurappa [362] deposited a mixture of alumina, magnesia-stabilized zirconia, and yttria-stabilized zirconia with different thicknesses on 316 L SS substrate by the air plasma spray technique to improve corrosion resistance [362]. Dense and ultrafine alumina–zirconia composites (Al_2O_3–16 wt%ZrO_2 and ZrO_2–20 wt%Al_2O_3) were prepared to enhance the bioactivity [363]. Nelson et al. [364] deposited a mixture of powders of titanium alloy (Ti–6Al–4V) and bioactive glass by flame spraying to fabricate composite porous coatings for potential use in bone fixation implants. After immersing samples in SBF, it was reported that HA was found on the bioactive glass–alloy composite coatings after 7 days of immersion, while no HA was observed after 14 days on the pure titanium alloy control coating [364]. Balasubramanian et al. [365] prepared nanocomposite Ti–Si–N coatings by reactive dc magnetron sputtering in a mixture of Ar and N_2 gases and deposited the composite on implantable 316 L SS substrates. It was reported that theta-Si–N nanocomposite coatings exhibited superior corrosion resistance compared with the Si_3N_4, TiN single layer, and the bare substrate in SBF solution.

With an attempt at achieving faster osseointegration to hasten the overall treatment process, the use of biomimetic agents represents a growing area of research in implant dentistry. Major biomimetic agents should include (1) biocompatible ceramics, (2) bioactive agents, (3) ions, and (4) polymers, and their respective importance in the early stages of osseointegration. The potential bioactive agents include bone morphogenetic proteins, growth factors, type I collagen, RGD peptide, fluoride, or chitosan, among others. The ideal characteristics that biomimetic agents should uphold and factors that may influence their effectiveness are reviewed [366]. It was pointed out that (i) some of these agents, such as bioceramics (calcium phosphate salts) or ions (fluoride), are already commercially available and have shown clinical success, (ii) others such as bone morphogenetic proteins are very promising, with an excellent therapeutic potential, and (iii) a specific implant surface coating may enhance the percentage of BIC as well as speed of osseointegration that allows clinicians to overcome many challenging clinical scenarios [369]. In implantolgy, particularly dental implant treatment, the increased demand on implant performance and the broadened treatment indications have led to the development of new moderately rough surfaces and alterations in both the surface chemistry and topography for chemical influences on bone tissue (or bioactivity) [367–370]. Furthermore, surface alterations should include manipulations for surface energy, surface wettability, cellular maturation state, nutrition status, and microstresses that alter the degree of bioactivity [368, 369]. It was reported that implant roughness promotes bacterial colonization [371]. In order to enhance bone formation, implants have been coated with bone specific biomolecules [372, 373]. Based on the above background, there have been many ideas as well as coating materials proposed toward the promising biomimetic materials for surface engineering for biomaterials. For the particular case of artificial bone grafts, synthetic materials which are to be used in biological environments must display an adequacy of both their surface and bulk characteristics in order to fulfill the dual requirements of biocompatibility and suitable mechanical properties for the given application [374]. Otherwise, due to a poor biocompatibility of improper compounds, fibrous tissues always encapsulate the implants made from such materials, which prolong the healing time. In the surgical disciplines of both dental and medical field, where bones have to be repaired, augmented, or improved, an interest has dramatically increased in application of synthetic bone grafts with excellent mechanical stability [374].

The surface microtopography of implant distinctly influences the rate of bone formation and the ratio of BIC. However, it is indicated that the surface of titanium implant existed contamination after different surface treatments. Chitosan is one of the most abundant natural polymers for its adsorption property for metal ions. He et al. [375] reported on the pretreatment of the metal surface and comparison of the efficacy of three treatments for the binding ability of collagen and chitosan. The compound of collagen and chitosan was immobilized on the titanium oxide, and the morphology and chemical composition were used to characterize the titanium surfaces by scanning electron microscope, and X-ray photoelectron spectrometry. The results showed that the surface displayed irregularities after roughness treatments and the rough surface was beneficial to the adsorption and attachment of collagen and chitosan [375]. Collagen is a highly versatile material, extensively used in the medical, dental, and pharmacological fields. Resorbable forms of collagen have been used to dress oral wounds, for closure of graft and extraction sites, and to promote healing. Collagen-based membranes also have been used in periodontal and implant therapy as barriers to prevent epithelial migration and allow cells with regenerative capacity to repopulate the defect area [376]. In particular, collagen type I is a major component of the extracellular matrix of most tissue and it is increasingly utilized for surface engineering of biomaterials to accelerate receptor-mediated cell adhesion [377]. Collagen was coated on Ti-based [378], Co-based biomaterials [377], or SS [379], suggesting that collagen coating can enhance the rate of bone healing and increase new bone formation at the implant surface. Collagen was even coated on poly(DL-lactide-co-glycolide) fibers [380] and TiO_x surface oxide layer [381].

Fibronectin functions in cell adhesion, migration, survival, proliferation, and differentiation as well as tissue organization and interacts with other biomolecules, such as collagen, proteoglycan, heparin, hyaluronic acid, fibrin/fibrinogen, plasmin, gangliosides, complement components, and also integral proteins of cell plasma membrane-integrins, as well as with itself [382]. Fibronection was coated on Ti surface [382–384] and the porous TiO_2/perlite composite [385], reporting that fibronection coating improved cellular spreading and enhanced the tissue ingrowth. Immobilization of bone morphogenetic proteins as growth factor onto material surfaces is promisingsince these proteins can accelerate and/or enhance the quality of osseointegration [386]. Protein was coated on Ti substrate [387, 388] and on Ti–6Al–4V [387]. Fe–Pd based ferromagnetic shape memory actuators for medical applications were also coated with protein [389]. Covalently bound recombinant human tropoelastin as a major regulator of vascular cells in vivo was prepared to enhance endothelial cell interactions; suggesting that such coating for metal alloys with multifaceted biocompatibility can resist delamination and is nonthrombogenic, with implications for improving coronary stent efficacy [390]. Prevention of bacterial colonization and formation of a bacterial biofilm on implant surfaces has been a challenge in orthopedic surgery [391]. The antimicrobial peptide was coated on Ti biomaterials [392–394], suggesting that the integration of soft tissue on titanium dental implants was improved by significantly protecting implants from peri-implant inflammation and enhancement long-term implant stabilization. Similarly, for the prevention of the peri-implant infection in orthopedics, Ca–P coated titanium surface was coated with the peptide [395]. To prevent the unwanted infection on and around titanium implants, antibacterial coatings were performed on titanium implants [396]. Reyes et al. [397] coated titanium implant surface with glycine-phenylalanine-hydroxyproline-glycine-glutamate-arginine collagen-mimetic peptide to improve osseointegration and reported that a biologically active and clinically relevant implant coating that enhances bone repair and orthopedic implant integration was developed [397]. Chitosan is the deacetylated derivative of the natural polysaccharide, chitin. Chitosan has been shown to be biocompatible, biodegradable, osteoconductive, and to accelerate wound healing [398, 399]. Ti was coated with chitosan to promote osseointegration [400]. Frosch et al. [401] coated titanium implant surfaces with stem cell to evaluate the partial surface replacement of the knee and to provide a basis for a successful treatment of large osteochondral defects, indicating that a partial joint resurfacing of the knee with stem cell-coated titanium implants occur. Similarly, stem cell was coated to functionalized titanium films, reporting that stem-cell coated titanium films are beneficial for sustained in situ inducing osteoprogenitor cells to differentiate into mature osteoblasts over long time [402]. Achneck et al. [403] evaluated the viability and adherence of peripheral blood-derived porcine endothelial progenitor cells on thin Ti layers and reported that (i) peripheral blood-derived cells adhere and function normally on Ti surfaces, and (ii) the cells may be used to seed cardiovascular devices prior to implantation to ameliorate platelet activation and thrombus formation [403]. Moreover, electrochemically functionalized Ti-alloy surfaces were coated with a cell specific aptamer in order to enhance the cell adhesion and by aptamer-based capture molecules on cell adhesion [404]. Methicillin-resistant *Staphylococcus aureus* (MRSA) and vancomycin-resistant *S. aureus* (VRSA) can seriously jeopardize bone implants. Zarghami et al. [405] examined the potential synergistic effects of melittin and vancomycin in preventing MRSA- and VRSA-associated bone implant infections. Chitosan/bioactive glass nanoparticles/vancomycin composites were coated on hydrothermally etched titanium substrates by casting method. The composite coatings were coated by Melittin through drop-casting technique. It was reported that (i) melittin raised the proliferation of MC3T3 cells, making it an appropriate option as osseoinductive and antibacterial substance in coatings of orthopedic implants, (ii) composite coatings having combined vancomycin and melittin eliminated both planktonic and adherent MRSA and VRSA bacteria, whereas coatings containing one of them failed to kill the whole VRSA bacteria, indicating that (iii) chitosan/bioactive glass/vancomycin/melittin coating can be used as a bone implant coating because of its antiinfective properties.

Dental implant surface modification has been developed over the years on both commercial and research levels in order to promote osseointegration, faster healing time, higher BIC ratio, and longevity of titanium implants. There are excellent reviews covering various surface modification techniques for dental implants [406, 407].

5.8 Sterilization

Because there are so many different types of implant materials currently in use (including metals, polymers, and diverse biological materials), the response of tissue to these different materials varies dramatically in both pre- and post-sterilization treatment [408]. Material specificity in implant–tissue interactions derives primarily from the surface properties (chemical composition, microstructure, etc.) of the implant. Such surface features can include cleanness, contamination, or sterilization. Kasemo et al. [409] pointed out that (i) the surface status of a particular implant material may vary widely depending on its preparation and handling history, (ii) the surface status of implants is expected to be important for in vivo function and should thus be controlled and standardized and (iii) it is usually not possible to predict how a change in surface status will affect the long-term *in vivo* function of an implant. In order to enhance osseointegration of dental and orthopedic implants, many surface modification strategies have been pursued focusing on the important role of the biomaterial surface properties, as we have been discussing so far [410–412].

An appropriate sterilization is performed for various types of dental devices and medical tools. Among these devices, without any exceptions, endodontic instruments (such as files or reamers) are the most frequently subjected to the routine sterilization. Resterilization of instruments used on one patient for reuse on another has been common practice in dentistry. Resterilization (conducted between reuse occasions) is the repeated application of the terminal process designed to remove or destroy all viable forms of microbial life, including bacterial spores, to an acceptable sterility assurance level [413, 414]. These devices are class I instruments as defined by the US FDA and can be reused if sterility can be guaranteed [415]; however, there is now evidence that the sterilization process is complex and that if strict adherence to an effective protocol is not followed, contamination of instruments may result, as will be seen later in this section. Endodontics is the aspect of dentistry involved in the treatment or precautions taken to maintain the vital pulp, moribund tooth, or nonvital tooth in the dental arch [416]. Reuse of instruments in dentistry is common and endodontic treatment involves the use of instruments which are usually reused. During endodontic instrumentation, vital tissue, dentin shavings, necrotic tissue, bacteria, blood, blood by-products, and other potential irritants are encountered with accumulation of the debris on the flutes of the instruments [417]. Transfer of these debris from patient to patient and to dental staff is highly undesirable, as these debris can act as antigens and infecting agents capable of transmitting diseases such as Creutzfeldt–

Jakob disease [418, 419]. The presence of debris has been reported to interfere with sterilization by forming a protective barrier that prevents the complete sterilization of the surface beneath [420]. If this debris is not removed, sterilization procedure may be pointless [421]. In the absence of adequate infection control procedures, there is a high probability of transmitting pathogenic microorganism through endodontic instruments [421]. Sterilization is a process to render an object free from viable microorganisms including bacterial spores and viruses. Sterilization of instruments in dentistry is required to protect patients and oral health care staff from cross contamination through instruments [422].

Depending on the biomedical device or material, secondary cleaning or sterilization (and including resterilization) may be done in a clinical setting. Therefore, it is possible for biomaterials to undergo several uncounted cleaning and sterilization steps in an uncontrolled manner. In situations where implant devices are approved for reuse, cleaning and sterilization are key steps in the reconditioning of the implant to its initial state but may also contribute to modification from initial surface properties [423]. Cleaning and sterilization can be distinguished in terms of function. The purpose of cleaning is to remove or reduce visible soils including blood, protein and debris on the surface of substrata. Sterilization serves to eliminate or stop reproduction of microorganisms including bacteria, spores, and fungi and usually, cleaning is done first, followed by sterilization [412, 424]. Sterilization can be achieved through various means, including heat, chemicals, irradiation, high pressure, and filtration and there are many different sterilization techniques available such as autoclaving, carbon dioxide laser/plasma sterilization, chemical sterilization (with glutaraldehyde), glass-bead sterilization, and gamma irradiation sterilization, each with different modes of action and resultants [416, 425].

Although they can share the common and specific aim among these different sterilization methods, there are studies on various influences (in both advantage and disadvantage, or no remarkable affects) on surface morphology, surface structure, surface chemistry, tissue/cell response, and osseointegration. It is well understood that the biological, chemical, and mechanical properties of biomaterials are susceptible to change by diverse sterilization methods [426–428]. However, the contribution of the outermost molecular layer of an implant to its in vivo function is less well-established [412]. This layer plays an important role in determining the surface properties including surface energy, chemistry, and wettability. These properties directly affect the interactions with surrounding host tissue in vitro and in vivo [429, 430]. Cleaning and sterilization can cause changes in the surface properties of biomaterials [431], but how these changes alter biological response is less understood.

Physical and chemical properties of materials are sensitive to small variations in their surface chemical composition and morphology [432–434]. Sterilization is required for using any material or device in contact with the human body. Zuleta et al. [435] investigated the effect of four sterilization methods (steam autoclave, hydrogen peroxide plasma, ethylene oxide (ETO), and gamma sterilization) on the surface chemistry and in vitro bioactivity of pseudowollastonite (psW) coatings on Ti–6Al–4V substrates. The psW coatings in Ti–6Al–4V substrates obtained by laser ablation technique were sterilized and immersed in Kokubo's SBF up to 30 days. It was found that (i) no changes in the chemical composition were noted after sterilization, (ii) however, a Ca/P layer of different thickness, identified as similar to HA was developed on all the samples after soaking, although, the ETO sterilized samples present a nonhomogeneous and ~55.9% thinner HA-like layer. The influence of conventional sterilization methods such as gamma irradiation, ETO treatment and steam exposure on the physical properties of poly(glycerol sebacate) (PGS) was investigated [436]. It was reported that (i) SEM observations, attenuated total reflectance-Fourier transform infrared spectroscopy and Raman spectra demonstrated that the applied sterilization methods do not induce adverse surface modifications, (ii) the sterilized samples also maintained their thermal and mechanical properties post-sterilization, and (iii) gamma-sterilized PGS films did not induce any toxicity when films were in contact with murine fibroblast 3T3 cells.

There are few suitable techniques available to sterilize biodegradable polyester three-dimensional (3D) tissue engineering scaffolds because they are susceptible to degradation and/or morphological degeneration by high temperature and pressure. Holy et al. [437] used a novel poly(lactide-co-glycolide) scaffold (Osteofoam™) to determine the optimal sterilization procedure, namely a sterile product with minimal degradation and deformation. It was found that, an argon plasma created at 100 W for 4 min was optimal for sterilizing Osteofoam™ scaffolds without affecting their morphology. The radio frequency glow discharge (RFGD) plasma sterilization method was compared to two well-established techniques – ETO and y-irradiation (y) – which were in turn compared to disinfection in 70% ethanol. Disinfection in 70% ethanol serves as a useful control because it affects neither the morphology nor the molecular weight of the polymer, yet, ethanol is unsuitable as a sterilization method because it does not adequately eliminate hydrophilic viruses and bacterial spores. The three sterilization techniques, ETO, y, and RFGD plasma, were compared in terms of their immediate and long-term effects on the dimensions, morphology, molecular weight and degradation profile of the scaffolds. It was observed that (i) scaffolds shrank to ~60% of their initial volume after ETO sterilization whereas their molecular weight (Mw) decreased by ~50% after y-irradiation, so that both ETO and y-irradiation posed immediate problems as sterilization techniques for 3-D biodegradable polyester scaffolds, (ii) during the in vitro degradation study, all sterilized samples showed advanced morphological and volume changes over time relative to ethanol (EtOH) disinfected samples, with the greatest changes observed for y-irradiated samples, (iii) ETO, RFGD plasma sterilized and EtOH disinfected samples showed similar changes in Mw and mass over the 8-week time frame, and (iv), overall, of the three sterilization techniques studied, RFGD plasma was the best [437].

Plasma-based sterilization is a promising alternative to the use of pure ETO, for low-temperature clinical sterilization of medical instruments and devices, although few studies have been published that evaluate its safety in terms of possible damage to materials, particularly polymers. Lerouge et al. [438] evaluated polymer surface modifications induced by commercial plasma-based sterilizers, in comparison with pure ETO. It was reported that (i) surface oxidation and wettability changes were observed on all samples sterilized by plasma-based techniques, the degree of modifications depending on the sterilizer (Sterrad, Plazlyte) and the type of polymer, (ii) drastic changes in surface appearance were also observed by SEM on PVC samples sterilized by Plazlyte and by pure ETO. Thierry et al. [439] studied the effect of dry heat, steam autoclaving, ETO, peracetic acid, and plasma-based sterilization techniques on the surface properties of NiTi. After processing electropolished NiTi disks with these techniques, surface analyses were performed by Auger electron spectroscopy (AES), AFM, and contact angle measurements. It was mentioned that (i) AES analyses revealed a higher Ni concentration (6–7% vs. 1%) and a slightly thicker oxide layer on the surface for heat and ETO processed materials, (ii) studies of surface topography by AFM showed up to a threefold increase of the surface roughness when disks were dry heat sterilized, and (iii) an increase of the surface energy of up to 100% was calculated for plasma treated surfaces, indicating that some surface modifications are induced by sterilization procedures.

The surgical template enables a predictable and a safe minimally invasive surgery. The main objective of surgical template is to direct the implant drilling system and provide an accurate placement of the implant according to the surgical treatment plan. A surgical guide is the union of two components, that is, the guiding cylinders and the contact surface. The contact surface fits either on an element of a patient's gums or on the patient's jaw (i.e., the bone, the teeth). Cylinders within the drill guides help in transferring the plan by guiding the drill in the exact location and orientation [440]. The virtually planned implant position is transferred to the surgical area by using an implant drill guide template and resulting greater control of the implant procedure [441, 442]. Modern 3D imaging technologies and software can be great assistance in the pre-operative planning of dental implant surgery [443, 444]. Implant surgery is an invasive procedure, during which the surgical templates come in contact with blood, injured mucous membranes, and bone. If the drill template is not sterilized properly, microorganisms can easily enter to the surgical wound and negatively affect the success of the surgery and the lifespan of the implant. Therefore, like all other instruments used in the implant surgery, drill templates should also be sterilized to avoid infection [445]. Based on the above background, Török et al. [446] investigated the effects of disinfection and three different sterilization methods on the dimensional changes and mechanical properties of 3D-printed surgical guide for implant therapy, in order to assess the effects of sterilization procedures in 3D printed drill guide templates with destructive and nondestructive material testing. Sterilizations were disinfection (4% Gigasept®, 60 min), plasma sterilization, and steam autoclave sterilization (121 °C × 20 min and 134 °C × 10 min). It was reported that (i) evaluation of the hardness measurements of the various specimens shows that the

hardness of the material was not changed by the plasma sterilization, steam sterilization on 121 °C or disinfection process; the statistical analysis revealed significant difference in hardness strength of the autoclave sterilized (134 °C) specimens, (ii) there was
no significant difference between the groups regarding the scanning electron microscopic and stereomicroscopic examinations, (iii) there was no significant difference regarding the X-ray visibility of the templates to the effect of the disinfection, plasma
sterilization and steam sterilization on 121 °C and steam sterilization on 134 °C, and (iv)
the effect of the sterilization was the same in case of both flexural and compressive
strength of the material, concluding that (v) plasma sterilization and steam sterilization
at 121 °C were both suitable for sterilizing the tested 3D printed surgical guides [446].

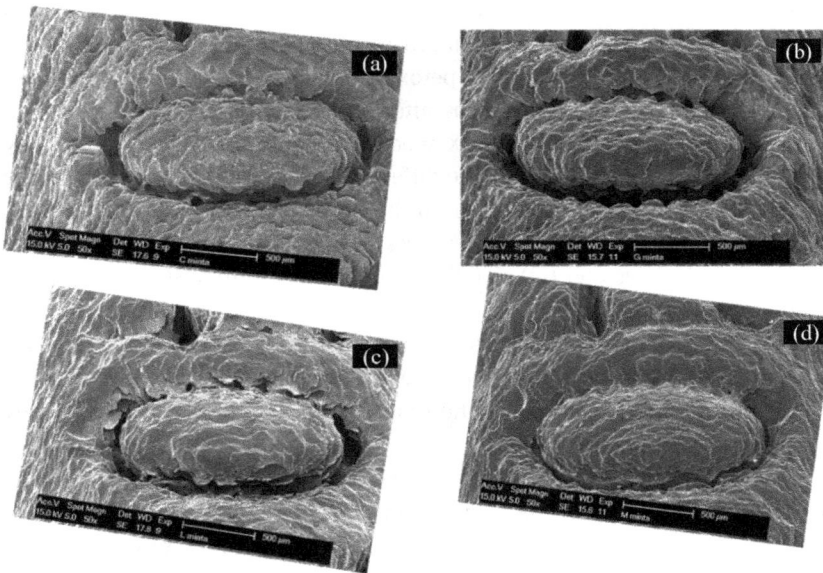

Figure 5.4: SEM images of the bridging element of (a) control, (b) autoclave-sterilized at 121 °C,
(c) plasma-sterilized, and (d) autoclave-sterilized at 134 °C specimen [446].

Figure 5.4 compares SEM images of the bridging element of template among different sterilization methods [446]. It was reported that (i) SEM images of specimens
soaked in disinfectant solution shows no morphologic changes, (ii) the plasma-
sterilized specimens also show no surface changes, (iii) comparison of images
taken after autoclave sterilization with the baseline control images indicates that
the specimens did not suffer any damage or morphologic changes, and (iv) despite the high-temperature treatment, the heat-sensitive polymer did not melt,
and the layered structure of the specimens is still clearly visible. Material was PolyJet materials (MED610 Syratasys Polyjet systems), which is biocompatible and is

a flexible, transparent material, enabling direct printing of indirect bonding trays as well as soft gingival masks for implantology cases [447].

Implant drills are reusable devices and various sizes of step drills are used for final sizing of the osteotomy when placing tapered implants. These drills are designed to accommodate the varying lengths of tapered implants without the need for length specific tapered drills. Each drill has two diameters of straight walled design. This promotes maximum implant engagement with the bone. When considering drill geometry in general, twist drills and taps are used to prepare sites for screw-shaped implants, while triflute drills are used to prepare sites for cylindrical implants. Drills are normally made out of SS or Ti-based alloys. Recently, ceramic drills are made of zirconia (ZrO_2). Scarano et al. [448] evaluated effects on SS and zirconia implant drills of 50 cycles of sterilization through different processes. A total of 24 SS and 24 zirconia drills were treated with three different sterilization processes: 50 cycles of immersion in glutaraldehyde 2%, 50 cycles in 6% hydrogen peroxide, and 50 cycles of heat. It was obtained that (i) zirconia drills seem not to be affected by the different treatments; no significant differences were found with EDX spectroscopy nor through thermography controls; while SS drills were affected by the different treatments, as confirmed by the increased roughness of the SS samples after all the cycles of sterilization/disinfection, measured at SEM, (ii) the zirconia drills roughness was not particularly affected by the chemical and thermal cycles and (iii) significant differences were observed regarding the temperature, between steel and zirconia drills, concluding that (iv) the disinfection agents had a weak impact on the temperature changes during implant bone preparation, while heat sterilization processes had no effect on either of the drills evaluated and (v) the disinfection agents increased the roughness of the steel drills, while they had no effect on the zirconia drills.

There are several studies on effects of sterilization on mechanical properties. The research on the development and characterization of potential magnesium biomaterials is a steadily expanding [449]. Commonly, implants present a high risk of infection for their recipients. For this reason, a preoperative sterilizing process is required. Due to the temperatures and media which are used while sterilizing, effects may occur which cause a change in the mechanical strength of certain magnesium alloys [450]. Seitz et al. [450] investigated four commonly used sterilization methods (autoclave sterilization, dry heat sterilization, gamma sterilization and ETO sterilization) to gain information about their influences on the quasi-static mechanical behavior of LAE442 (Mg-4Li-4Al-2RE by mol%), $MgCa_{0.8}$ (Mg–0.8Ca by mol%) alloys, as well as pure magnesium. It was found that (i) the mechanical properties exhibited by the sterilized and nonsterilized alloys refer to susceptibilities of the mechanical strengths to the investigated sterilization methods; such susceptibilities appear to be dependent on the combination of alloy and method of sterilization, (ii) however, the maximum changes in mechanical strength appear in the range of $\pm 10\%$ and (iii) ETO sterilization caused the least changes in the mechanical strength of the alloys and appears to be the best performer.

As mentioned previously, endodontic files are the most frequently instruments required for appropriate sterilization between patients [425]. Surface roughness of NiTi files increases by dry heat sterilization [451], autoclave sterilization [452, 453], or irrigation [454]. It is, in general, believed that fatigue crack is initiated at very localized surface irregular site, so that roughening surface might be harmful for fatigue behavior. Most of research results indicates that sterilization exhibits insignificant negative influence on fatigue life and strength [455–459], although Viana et al. [460] evaluated the effect of repeated sterilization cycles in dry oven or autoclave, on the mechanical behaviors and fatigue resistance of rotary endodontic NiTi instruments and mentioned that the sterilization procedures are safe as they produced a significant increase in the fatigue resistance of the instruments. Özyürek et al. [461] compared the cyclic fatigue resistances of ProTaper Universal, ProTaper Next (PTN), and ProTaper Gold (PTG) and the effects of sterilization by autoclave on the cyclic fatigue life of NiTi instruments and concluded that autoclaving increased significantly the cyclic fatigue resistances of PTN and PTG.

Torsional property of rotary NiTi files is another important factor to evaluate efficacy and safety of endodontic treatments. There are reports on no adversely effects of sterilization on torsional properties [462, 463]. In terms of risk for fracture and corrosion-assisted fracture (in other words, SCC), it was suggested that NiTi files can be cleaned up to 10 times without affecting fracture susceptibility or corrosion but should not be immersed in NaOCl overnight [464]. It was also indicated that NiTI alloy was susceptible to irradiation-assisted SCC when in the presence of a constraining stress, fluorinated polymers, and gamma irradiation [465, 466].

There are still numerous studies on polymeric materials which are subjected to various sterilization methods. Wear of UHMWPE acetabular cups in hip prostheses produces billions of fine wear particles (as known as wear debris) annually that can cause osteolysis and loosening of the components. Thus, substantial improvement of the wear resistance of UHMWPE could extend the clinical life span of total hip prostheses. McKellp et al. [467] examined the conditions, under which UHMWPE cups have been sterilized, that can markedly affect their long-term wear properties, and new sterilization methods and other modifications have been developed to minimize the negative effects. It was reported that (i) the cross-linking induced by gamma irradiation improves the wear resistance of UHMWPE, while oxidation reduces it, (ii) without thermal aging, the two types of cups that were sterilized with gamma irradiation while in low-oxygen packaging exhibited about a 50% lower rate of wear than did either the nonsterilized cups or the nonirradiated cups sterilized with gas plasma, and (iii) Hylamer cups (that is, those that were sterilized with gas plasma) exhibited wear properties very close to those of the nonsterilized UHMWPE cups (the controls) with or without aging. Based on these findings, it was concluded that (iv) sterilizing an UHMWPE acetabular cup without radiation (e.g., with ETO or gas plasma) avoids immediate and long-term oxidative degradation of the implant but does not improve the inherent wear resistance of the PE, (v) sterilizing with use of gamma irradiation with the implant packaged in a low-oxygen atmosphere avoids immediate oxidation and cross-links the PE, thereby increasing its wear resistance, but long-term oxidation of the residual free radicals may markedly reduce the wear resistance, indicating that ideally, cross linking with gamma irradiation to reduce wear should be done in a manner that avoids both immediate and long-term oxidation [467]. The effects of gamma radiation and low temperature hydrogen peroxide gas plasma sterilization on structure and cyclic mechanical properties were examined for orthopedic grade UHMWPE and compared to each other as well as to no sterilization (control) [468]. It was found that (i) density was monitored with a density gradient column and was found to be directly influenced by the sterilization method employed: gamma radiation led to an increase, while plasma did not, (ii) oxidation of the polymer was studied by observing changes in the carbonyl peak with Fourier transform infrared spectrometry and was found to be strongly affected by both gamma radiation and subsequent aging, while plasma sterilization had little effect, (iii) gamma radiation resulted in embrittlement of the polymer and a decreased resistance to fatigue crack propagation. This mechanical degradation was a direct consequence of post-radiation oxidation and molecular evolution of the polymer and was not observed in the plasma-sterilized polymer. Both gamma radiation and plasma sterilization led to improved wear performance of the UHMWPE compared to the nonsterile control material [468].

As Linkow et al. [469] stated that plasma glow discharge treatment is an ideal technique to increase the wettability of the metal and to ensure a sound bone-implant interface, sterilization could alter the surface physiochemistry. The effects of cleaning and sterilization on surface characteristics were studied using contaminated and pure Ti substrata [412]. Two different surface structures were prepared: pretreated titanium (PT, Ra: 0.4 µm) (i.e., surfaces that were not modified by sandblasting and/or acid etching); (SLA, Ra: 3.4 µm). Cleaned specimens were sterilized with autoclave, gamma irradiation, oxygen plasma, or ultraviolet light. XPS, contact angle measurements, profilometry, and scanning electron microscopy were used to examine surface chemical components, hydrophilicity, roughness, and morphology, respectively. It was mentioned that (i) small organic molecules present on contaminated Ti surfaces were removed with cleaning, (ii) XPS analysis confirmed that surface chemistry was altered by both cleaning and sterilization, (iii) cleaning and sterilization affected hydrophobicity and roughness, (iv) these modified surface properties affected osteogenic differentiation of human MG63 osteoblast-like cells and (v) specifically, autoclaved SLA surfaces lost the characteristic increase in osteoblast differentiation seen on starting SLA surfaces, which was correlated with altered surface wettability and roughness. It was then indicated that (vi) recleaned and resterilized Ti implant surfaces cannot be considered the same as the first surfaces in terms of surface properties and cell responses; therefore, the reuse of Ti implants after resterilization may not result in the same tissue responses as found with never-before-implanted specimens. Serro et al. [470] investigated the effect of the sterilization processes on the mineralization of titanium implants induced by incubation in various biological model fluids. Titanium samples were submitted to the following sterilization processes used for implant materials: steam autoclaving, glow discharge Ar plasma treatment and y-irradiation. It was reported that the most significant modifications were detected on the wettability: while the samples treated with Ar plasma became highly hydrophilic (water contact angle ~0°), y-irradiation and steam sterilization induced an increase in the hydrophobicity.

References

[1]   Oshida Y. Bioscience and Bioengineering of Titanium Materials. 1st edition. London, UK: Elsevier, 2007.

[2]   Kasemo B, Lausmaa J. Biomaterials and implant surfaces: A surface science approach. Int J Oral Maxillofac Implants. 1988, 3, 247–59.

[3]   Baier RE, Meyer AE. Implant surface preparation. Int J Oral Maxillofac Implants. 1988, 3, 9–20.

[4]   Reitz WE. The eighth international conference on surface modification technology. J Metals. 1995, 47, 14–16.

[5]   Smith DC, Pilliar RM, Chernecky R. Dental implant materials. I. Some effects of preparative procedures on surface topography. J Biomed Mater Res. 1991, 25, 1045–68.

[6]   Buddy D, Ratner B, Thomas JL. Biomaterial surfaces. J Biomed Mater Res. 1987, 21, 59–89.

[7]   Shabalovskaya SA, Tian H, Anderegg JW, Schryvers DU, Carroll WU, Van Humbeeck J. The influence of surface oxides on the distribution and release of nickel from Nitinol wires. Biomaterials. 2009, 30, 468–77.

[8]   Tucker RC. Surface engineering. J Metals. 2002, 160, 1–3.

[9]   Allen P. Titanium alloy development. Adv Mater Processes. 1996, 154, 35–37.

[10] Oshida Y, Hashem A, Nishihara T, Yapchulay MV. Fractal dimension analysis of mandibular bones – Toward a morphological compatibility of implants. J Bio-Med Mater Eng. 1994, 4, 397–407.

[11] Oshida Y, Güven Y. Biocompatible coatings for metallic biomaterials. In: Wen C, ed. Surface Coating and Modification of Metallic Biomaterials. Woodhead Pub., 2015, 287–343.

[12] Oshida Y, Daly J. Fatigue damage evaluation of shot-peened high strength aluminum alloy. In: Meguid SA, ed. Surface Engineering. London: Elsevier Applied Science, 1990, 404–16.

[13] Oshida Y, Munoz CA, Winkler MM, Hashem A, Ito M. Fractal dimension analysis of aluminum oxide particle for sandblasting dental use. J Biomed Mater Eng. 1993, 3, 117–26.

[14] Oshida Y, Tuna EB, Aktören O, Gençay K. Dental implant systems. Intl J Mol Sci. 2010, 11, 1580–78.

[15] Chesters S, Wen HY, Lundin M, Kasper G. Fractal-based characterization of surface texture. Appl Surf Sci. 1989, 40, 185–92.

[16] Miyakawa O, Okawa S, Kobayashi M, Uematsu K. Surface contamination of titanium by abrading treatment. Dent In Jpn. 1998, 34, 90–96.

[17] Mustafa K, Lopez BS, Hultenby K, Arvison K. Attachment of HGF of titanium surfaces blasted with TiO_2 particles. J Dent Res. 1997, 76, 85.

[18] Wennerberg A, Albrektsson T, Johnsson C, Andersson B. Experimental study of turned and grit-blasted screw-shaped implants with special emphasis on effects of blasting material and surface topography. Biomaterials. 9996, 17, 15–22.

[19]   Johansson CB, Albrektsson T, Thomsen P, Snnerby I, Lodding A, Odelius H. Tissue reactions to titanium-6aluminum-4vanadium alloy. Eur J Exp Musculoskel Res. 1992, 1, 161–69.

[20]   Wang C-S. Surface modification of titanium for enhancing titanium-porcelain bond strength. Indiana University Master Degree Thesis. 2004.

[21]   Peutzfeld A, Asmussen E. Distortion of alloy by sandblasting. Am J Dent. 1996, 9, 65–66.

[22]   Peutzfeld A, Asmussen E. Distortion of alloys caused by sandblasting. J Dent Res. 1996, 75, 259.

[23]   Quatman C. Evaluating the Performance Characteristics of Abrasive Media. 2018; https://kta.com/kta-university/abrasive-media-evaluation/.

[24]   Kirk D. Peening indent dimensions. Shot Peener. 2010; https://www.shotpeener.com/library/pdf/2010020.pdf.

[25]   Rønold HJ, Lyngstadaas SP, Ellingsen JE. A study on the effect of dual blasting with TiO2 on titanium implant surfaces on functional attachment in bone. J Biomed Mater Res. 2003, 67A, 524–30.

[26]   Kamal M. Cellular Responses to Titanium Surfaces Blasted with TiO2 Particles. Karolinska Institutet, 2001, https://openarchive.ki.se/xmlui/handle/10616/42693.

[27]   Metal Improvement Company, Inc. Shot Peening Applications. 7th Edition, 1990.

[28]   Oshida Y. unpublished data 2018.

[29]   Guo CY, Matinlinna JP, Tang ATH. A novel effect of sandblasting on titanium surface: Static charge generation. J Adhesion Sci Tech. 2012, 26, 2603–13.

[30]   DeWald AT, Rankin JE, Hill MR, Lee MJ, Chen H-L. Assessment of tensile residual stress mitigation in Alloy 22 welds due to laser peening. J Eng Mater Tech, Trans ASME. 2004, 126, 81–89.

[31]   Fairland BP, Wilcox BA, Gallagher WJ, Williams DN. Laser shock-induced microstructural and mechanical property changes in 7075 aluminum. J Appl Phys. 1972, 43, 3893–95.

[32]   Fairland BP, Clauer AH. Laser generation of high amplitude stress waves in materials. J Appl Phys. 1979, 50, 1497–502.

[33]   Dane CB, Hackel LA, Daly J, Harrison J. High power laser for peening of metals enabling production technology. Mater Manuf Process. 2000, 15, 81–96.

[34]   Cho S-A, Jung S-K. A removal torque of the laser-treated titanium implants in rabbit tibia. Biomaterials. 2003, 24, 4859–63.

[35]   Gaggl A, Schultes G, Muller WD, Karcher H. Scanning electron microscopical analysis of laser-treated titanium implants surfaces – A comparative study. Biomaterials. 2000, 21, 1067–73.

[36]   Jindal S, Bansal R, Singh BP, Pandey R, Narayanan S, Wani MR, Singh V. Enhanced osteoblast proliferation and corrosion resistance of cp-Ti through surface nanostructuring by ultrasonic shot peening and stress relieving. J Oral Implant. 2012, 2014, 40, doi: 10.1563/AAID-JOI-D-12-00006.

[37]   Harcuba P, Bačáková L, Stráský J, Bačáková M, Novotná K, Janeček M. Surface treatment by electric discharge machining of Ti-6Al-4V alloy for potential application in orthopaedics. J Mech Behav Biomed Mat. 2012, 7, 96–105.

[38]   Otsuka F, Kataoka Y, Miyazaki T. Enhanced osteoblast response to electrical discharge machining surface. Den Mat J. 2012, 31, 309–15.

[39]   Valiev RZ, Langdon TG. Achieving exceptional grain refinement through severe plastic deformation: New approaches for improving the processing technology. Metall Mater Trans A. 2011, 42, 2942–51.

[40]   Byeli AV, Kukareko VA, Kononov AG. Titanium and zirconium based alloys modified by intensive plastic deformation and nitrogen ion implantation for biocompatible implants. J Mech Behav Biomed Mat. 2012, 6, 89–94.

[41]   Mandelbrot BB. The Fractal Geometry of Nature. NY: Freeman, 1983, 34.

[42] Sayles RS, Thomas TR. Surface topography as a non-stationary random process. Nature. 1978, 271, 431–34.

[43] Hansson S, Hansson KN. The effect of limited lateral resolution in the measurement of implant surface roughness: A computer simulation. J Biomed Mater Res. 2005, 75A, 472–77.

[44] Elias CN, Oshida Y, Lima JHC, Muller CA. Relationship between surface properties (roughness, wettability and morphology) of titanium and dental implant removal torque. J Mech Behav Biomed Mat. 2008, 1, 234–42.

[45] Ajami E, Aguey-Zinsou K-F. Formation of OTS self-assembled monolayers at chemically treated titanium surfaces. J Mat Sci Mat Med. 2011, 22, 1813–24.

[46] Variola F, Yi J-H, Richert L, Wuest JD, Rosei F, Nanci A. Tailoring the surface properties of Ti6Al4V by controlled chemical oxidation. Biomaterials. 2008, 29, 1285–98.

[47] Pisarek M, Roguska A, Andrzejczuk M, Marcon L, Szunerits S, Lewandowska M, Janik-Czachorm M. Effect of two-step functionalization of Ti by chemical processes on protein adsorption. App Sur Sci. 2011, 257, 8196–204.

[48] Kawai T, Takemoto M, Fujibayashi S, Akiyama H, Yamaguchi H, Pattanayak DP, Doi K, Matsushita T, Nakamura T, Kokubo T. Osteoconduction of porous Ti metal enhanced by acid and heat treatments. J Mater Sci Mater Med. 2013, 24, 1707–15.

[49] Yoneyama Y, Matsuno T, Hashonoto Y, Satoh T. In vitro evaluation of H2O2 hydrothermal treatment of aged titanium surface to enhance biofucntional activity. Dent Mat. 2013, 32, 115–21.

[50] Li Y, Zou S, Wang D, Feng G, Bao C, Hu J. The effect of hydrofluoric acid treatment on titanium implant osseointegration in ovariectomized rats. Biomaterials. 2010, 31, 3266–73.

[51] Viornery C, Chevolot Y, Léonard D, Aronsson B-O, Péchy P, Mathieu HJ, Descouts P, Grätzel M. Surface modification of titanium with phosphonic acid to improve bone bonding: Characterization by XPS and ToF-SIMS. Langmuir. 2002, 18, 2582–89.

[52] Park J-W, Kim Y-J, Jang J-H, Kwon T-G, Bae Y-C, Suh JY. Effects of phosphoric acid treatment of titanium surfaces on surface properties, osteoblast response and removal of torque forces. Acta Biomat. 2010, 6, 1661–70.

[53] Hsu H-C, Wu S-C, Fu C-E, Ho W-F. Formation of calcium phosphates on low-moldulus Ti-7.5 Mo alloy by acid and alkali treatments. J Mater Sci. 2010, 45, 3661–70.

[54] Szmukler-Moncler S, Bischof M, Nedir R, Ermrich M. Titanium hydride and hydrogen concentration in acid-etched commercially pure titanium and titanium alloy implants: A comparative analysis of five implant systems. Clin Oral Implants Res. 2010, 21, 944–50.

[55] De Souza GB, Lepienski CM, Foerster DE, Kuromoto NK, Soares P, De Araújo Ponte H. Nanomechanical and nanotribological properties of bioactive titanium surfaces prepared by alkali treatment. J Mech Behav BiomedMat. 2011, 4, 756–65.

[56] Xie J, Juan BL. Formation of hydroxyapatite coating using novel chemo-biomimetic method. J Mater Sci Mater Med. 2008, 19, 3211–20.

[57] Park IS, Choi UJ, Yi HK, Park BK, Lee MH, Bae TS. Biomimetic apatite formation and biocompatibility on chemically treated Ti-6Al-7Nb alloy. Surf Interf Anal. 2008, 40, 37–42.

[58] Guslitzer-Okner R, Mandler D. Applications of electrochemistry and nanotechnology in biology and medicine, I: Electrochemical coating of medical implants. Modern Aspects Electrochem. 2011, 52, 291–342.

[59] Gao Y, Gao B, Wang R, Wu J, Zhang LJ, Hao YL, Tao XJ. Improved biological performance of low modulus Ti–24Nb–4Zr–7.9Sn implants due to surface modification by anodic oxidation. App Sur Sci. 2009, 255, 5009–15.

[60] Minagar S, Berndt CC, Wang J, Ivanova E, Wen C. A review of the application of anodization for the fabrication of nanotubes on metal implant surfaces. Acta Biomater. 2012, 8, 2875–88.

[61] Xiao X, Ouyang K, Liu R, Liang J. Anatase type titania nanotube arrays direct fabricated by anodization without annealing. Appl Sur Sci. 2009, 255, 3659–63.

[62] Necula BS, Apachitei I, Fratila-Apachitei LE, Van Langelaan EJ, Duszczyk J. Titanium bone implants with superimposed micro/nano-scale porosity and antibacterial capability. App Sur Sci. 2013, 273, 310–14.

[63] Xie J, Luan BL, Wang J, Liu XY, Rorabeck C, Bourne R. Novel hydroxyapatite coating on new porous titanium and titanium-HDPE composite for hip implant. Surf Coatings Tech. 2008, 201, 2960–68.

[64] Iwaniak A, Nawrat G, Maciej A, Michalska J, Radwański K, Gazdowicz J, Simka W. Modification of titanium oxide layer by calcium and phosphorus. Electrochim Acta. 2009, 54, 6983–88.

[65] Simka W, Krząkała A, Korotin DM, Zhidkov IS, Kurmaev EZ, Cholakh SO, Kuna K, Dercz G, Michalska J, Suchanek K, Gorewoda T. Modification of a Ti–Mo alloy surface via plasma electrolytic oxidation in a solution containing calcium and phosphorus. Electrochim Acta. 2013, 96, 180–90.

[66] Chávez-Valdez A, Herrmann H, Boccaccini AR. Alternating current electrophoretic deposition (EPD) of TiO2 nanoparticles in aqueous suspensions. J Colloid and Interface Sci. 2012, 375, 102–05.

[67] Sharma S, Soni VP, Bellare JR. Chitosan reinforced apatite-wollastonite coating by electrophoretic deposition on titanium implants. J Mater Sci. 2009, 20, 1427–36.

[68] Morcos BM, O'Callaghan JM, Amira MF, Van Hoof C, De Beeck MO. Electrodeposition of platinum thin films as interconnects material for implantable medical applicationselectrochemical/electroless deposition. J Electrochem Soc. 2013, 160, D300–6.

[69] Sebaa MA, Dhillon S, Liu H. Electrochemical deposition and evaluation of electrically conductive polymer coating on biodegradable magnesium implants for neural applications. J Mater Sci Mater Med. 2013, 24, 307–16.

[70] Chung KH, Liu GT, Duh JG, Wang JH. Biocompatibility of a titanium–aluminum nitride film coating on a dental alloy. Surface Coatings Tech. 2004, 188/189, 745–49.

[71] Kim K-H, Ramaswamy N. Electrochemical surface modification of titanium in dentistry. Den Mat J. 2009, 28, 20–36.

[72] Sanchez AG, Schreiner W, Duffó G, Ceré S. Surface modification of titanium by anodic oxidation in phosphoric acid at low potentials. Part 1. Structure, electronic properties and thickness of the anodic films. Surf Interf Anal. 2013, 45, 1037–46.

[73] Oshida Y, Tuna EB. Science and technology integrated titanium dental implant systems. In: Basu B, et al., ed. Advanced Biomaterials. Wiley & Sons, 2009, 143–77.

[74] Lee IG, Kim YK, Park IS, Park JM, Lee MH, Bae TS, Park CW. Influence of electrolyte temperature on pure titanium modified by electrochemical treatment for implant. Surf Interf Anal. 2008, 40, 1538–44.

[75] Lim YJ, Oshida Y, Barco T, Andres CJ. Surface characterization of variously treated titanium materials. Int J Oral Maxillofac Implants. 2001, 16, 333–42.

[76] Oshida Y, Sachdeva R, Miyazaki S. Changes in contact angles as a function of time on some pre-oxidized bio-materials. J Mater Sci Mater Med. 1992, 3, 306–12.

[77] Oshida Y, Sachdeva R, Miyazaki S. Effects of shot peening on surface contact angles of biomaterials. J Mater Sci Mater Med. 1994, 4, 443–47.

[78] Wei QF, Huang FL, Hou DY, Wang YY. Surface functionalisation of polymer nanofibres by sputter coating of titanium dioxide. Appl Surf Sci. 2006, 252, 7874–77.

[79] Tian Y, Ding S, Peng H, Lu S, Wang G, Xia L, Wang P. Osteoblast growth behavior on porous-structure titanium surface. Appl Surf Sci. 2012, 261, 25–30.

[80]   Lin Z, Lee G-H, Liu C-M, Lee I-S. Controls in wettability of TiOx films for biomedical applications. Surf Coat Technol. 2010, 205, S391–7.
[81]   Park JH, Schwartz Z, Olivares-Navarrete R, Boyan BD, Tannenbaum R. Enhancement of surface wettability via the modification of microtextured titanium implant surfaces with polyelectrolytes. Langmuir. 2011, 27, 5976–85.
[82]   Fleming RA, Zou M. Fabrication of stable superhydrophilic surfaces on titanium substrates. J Adhes Sci Technol. 2012, doi: 10.1080/01694243.2012.697754.
[83]   Khosroshahi ME, Mahmoodi M, Tavakoli J. Characterization of Ti6Al4V implant surface treated by Nd: YAG laser and emery paper for orthopedic applications. Appl Surf Sci. 2007, 253, 8772–81.
[84]   Shtansky DV, Batenina IV, Yadroitsev IA, Ryashin NS, Kiryukhantsev-Korneev PV, Kudryashov AE, Sheveyko AN, Zhitnyak IY, Gloushankova NA, Smurov IY, Levashov EA. A new combined approach to metal-ceramic implants with controllable surface topography, chemistry, blind porosity, and wettability. Surf Coat Technol. 2012, 208, 14–23.
[85]   Vasilev K, Poh Z, Kant K, Chan J, Michelmore A, Losic D. Tailoring the surface functionalities of titania nanotube arrays. Biomaterials. 2010, 31, 532–40.
[86]   Barshilia HC, Chaudhary A, Kumar P, Manikandanath NT. Wettability of Y2O3: A relative analysis of thermally oxidized, reactively sputtered and template assisted nanostructured coatings. Nanomater. 2012, 2, 65–78.
[87]   Park JH, Wasilewski CE, Almodovar N, Olivares-Navarrete R, Boyan BD, Tannenbaum R, Schwartz Z. The responses to surface wettability gradients induced by chitosan nanofilms on microtextured titanium mediated by specific integrin receptors. Biomaterials. 2012, 33, 7386–93.
[88]   Park JH, Olivares-Navarrete R, Wasilewski CE, Boyan BD, Tannenbaum R, Schwartz Z. Use of polyelectrolyte thin films to modulate Osteoblast response to microstructured titanium surfaces. Biomaterials. 2012, 33, 5267–77.
[89]   Mekayarajjananonth T, Winkler S. Contact angle measurement on dental implant biomaterials. J Oral Implantol. 1999, 25, 230–36.
[90]   Paital SR, He W, Dahotre NB. Laser pulse dependent micro textured calcium phosphate coatings for improved wettability and cell compatibility. J Mater Sci Mater Med. 2010, 21, 2187–200.
[91]   Park J-W, Kim Y-J, Jang J-H. Enhanced osteoblast response to hydrophilic strontium and/or phosphate ions-incorporated titanium oxide surfaces. Clin Oral Implants Res. 2010, 21, 398–408.
[92]   Tsuji H, Satoh H, Ikeda S, Gotoh Y, Ishikawa J. Contact angle lowering of polystyrene surface by silver-negative-ion implantation for improving biocompatibility and introduced atomic bond evaluation by XPS. Nucl Instrum Meth Phys Res B. 1998, 141, 197–201.
[93]   Lin Z, Lee G-H, Liu C-M, Lee I-S. Controls in wettability of TiOx films for biomedical applications. Surf Coat Technol. 2006, 205, S391–7.
[94]   Wu K-R, Wang -J-J, Liu W-C, Chen Z-S, Wu J-K. Deposition of graded TiO2 films featured both hydrophobic and photo-induced hydrophilic properties. Appl Surf Sci. 2006, 252, 5829–38.
[95]   Panjwani B, Sujeet K, Sinha SK. Tribology and hydrophobicity of a biocompatible GPTMS/PFPE coating on Ti6Al4V surfaces. J Mech Behav Biomed Mater. 2012, 15, 103–11.
[96]   Ou J, Liu M, Li W, Wang F, Xue M, Li C. Corrosion behavior of superhydrophobic surfaces of Ti alloys in NaCl solutions. Appl Surf Sci. 2012, 258, 4724–28.
[97]   Duske K, Koban I, Kindel E, Schröder K, Nebe B, Holtfreter B, Jablonowski L, Weltmann KD, Kocher T. Atmospheric plasma enhance wettability and cell spreading on dental implant metals. J Cin Periodontol. 2012, 39, 400–07.

[98] Schliephake H, Scharnweber D, Dard M, Sewing A, Aref A, Roessler S. Functionalization of dental implant surfaces using adhesion molecules. J Biomed Mater Res B. 2005, 73B, 88–96.

[99] Wu J, Liu Z-M, Zhao X-H, Gao Y, Hu J, Gao B. Improved biological performance of microarc-oxidized low-modulus Ti-24Nb-4Zr-7.9Sn alloy. J Biomed Mater Res B. 2010, 92B, 298–306.

[100] Fan X, Feng B, Di Y, Lu X, Duan K, Wang J, Weng J. Preparation of bioactive TiO film on porous titanium by micro-arc oxidation. Appl Surf Sci. 2012, 258, 7584–88.

[101] Liu Z, Wang W, Liu H, Wang T, Qi M. Formation and characterization of titania coatings with cortex-like slots formed on Ti by micro-arc oxidation treatment. Appl Surf Sci. 2013, 266, 250–55.

[102] Yu S, Yang X, Yang L, Liu Y, Yu Y. Novel technique for preparing Ca- and P-containing ceramic coating on Ti-6Al-4V by micro-arc oxidation. J Biomed Mater Res B. 2007, 83B, 623–27.

[103] Gan J, Tan L, Yang K, Hu Z, Zhang Q, Fan X, Li Y, Li W. Bioactive Ca–P coating with self-sealing structure on pure magnesium. J Mater Sci Mater Med. 2013, 24, 889–901.

[104] Abbasi S, Golestani-Fard F, Rezaie HR, Mirhosseini SMM. MAO-derived hydroxyapatite/ TiO2 nanostructured multi-layer coatings on titanium substrate. Appl Surf Sci. 2012, 261, 37–42.

[105] Song H-J, Kim J-W, Kook M-S, Moon W-J, Park Y-J. Fabrication of hydroxyapatite and TiO2 nanorods on microarc-oxidized titanium surface using hydrothermal treatment. Appl Surf Sci. 2010, 256, 7056–61.

[106] Zhao Z, Chen X, Chen A, Huo G, Li H. Preparation of K2Ti6O13/TiO2 bio-ceramic on titanium substrate by micro-arc oxidation. J Mater Sci. 2009, 44, 6310–16.

[107] Ma PW, Wang S-H, Wu G-F, Liu B-L, Wei J-H, Xie C, Li D-H. Preparation and in vitro biocompatibility of hybrid oxide layer on titanium surface. Surf Coat Tech. 2010, 205, 1736–42.

[108] Buser D, Broggini N, Wieland M. Enhanced bone apposition to a chemically modified SLA titanium surface. J Den Res. 2004, 83, 529–33.

[109] Foest R, Schmidt M, Becker K. Microplasmas, an emerging field of low-temperature plasma science and technology. Int J Mass Spectrometry. 2006, 248, 87–102.

[110] Quaranta A, Iezzi G, Scarano A, Soelho PG, Vozza I, Marincola M, Piattelli A. A histomorphometric study of nanothickness and plasma-sprayed calcium-phosphorous-coated implant surfaces in rabbit bone. J Periodont. 2010, 81, 556–1.

[111] Laroussi M, Akan T. Arc-free atmospheric pressure cold plasma jets: A review. Plasma Process Polym. 2009, 9, 777–88.

[112] Chu PK. Plasma surface treatment of artificial orthopedic and cardiovascular biomaterials. Surf Coatings Tech. 2009, 201, 5601–06.

[113] Liu X, Poon RWY, Kwoka SCH, Chu PK, Ding C. Plasma surface modification of titanium for hard tissue replacements. Surf Coatings Tech. 2004, 186, 227–33.

[114] Jacobs T, Morent R, De Geyter N, Dubruel P, Leys C. Plasma surface modification of biomedical polymers: influence on cell-material interaction. Plasma Chem Plasma Process. 2012, 32, 1039–73.

[115] Hu X, Shen H, Shuai K, Zhang E, Bai Y, Cheng Y, Xiong X, Wang S, Fang J, Shicheng W. Surface bioactivity modification of titanium by CO2 plasma treatment and induction of hydroxyapatite: In vitro and in vivo studies. App Sur Sci. 2011, 257, 1813–23.

[116] Chu PK. Recent applications of plasma-based ion implantation and deposition to microelectronic, nano-structured, and biomedical materials. Surf Coatings Tech. 2010, 204, 2853–63.

[117] Chu PK. Applications of plasma-based technology to microelectronics and biomedical engineering. Surf Coatings Tech. 2009, 203, 2793–98.

[118] Jagielski J, Piatkowska A, Aubert P, Thomé L, Turos A, Kader A. Ion implantation for surface modification of biomaterials. Surf Coatings Tech. 2006, 200, 6355–61.

[119] Mändl S, Krause D, Thorwarth G, Sader R, Zeilhofer F, Horch HH, Rauschenbach B. Plasma immersion ion implantation treatment of medical implants. Surf Coat Tech. 2001, 142/144, 1046–50.

[120] Lee I-S, Zhao B, Lee G-H, Choi S-H, Chung S-M. Industrial application of ion beam assisted deposition on medical implants. Surf Coatings Tech. 2007, 201, 5132–37.

[121] Sawase T, Wennerberg A, Baba K, Tsuboi Y, Sennerby L, Johansson CB, Albrektsson T. Application of oxygen ion implantation to titanium surfaces: Effects on surface characteristics, corrosion resistance, and bone response. Clin Implant Dent Related Res. 2001, 3, 221–29.

[122] Liu F, Li B, Sun J, Li H, Wang B, Zhang S. Proliferation and differentiation of osteoblastic cells on titanium modified by ammonia plasma immersion ion implantation. Appl Surf Sci. 2012, 258, 4322–27.

[123] Mändl S. PIII treatment of Ti alloys and NiTi for medical applications. Surf Coat Tech. 2007, 201, 6833–38.

[124] Yankov RA, Shevchenko N, Rogozin A, Maitz MF, Richter E, Möller W, Donchev A, Schütze M. Reactive plasma immersion ion implantation for surface passivation. Surf Coatings Tech. 2004, 201, 6752–58.

[125] Shevchenko N, Pham M-T, Maitz MF. Studies of surface modified NiTi alloy. Appl Surf Sci. 2004, 235, 126–31.

[126] Qi H, Wu HY. Effect of surface modification of pure Ti on tribological and biological properties of bone tissue. Surf Eng. 2013, 29, 300–05.

[127] Firouzi-Arani M, Savaloni H, Ghoranneviss M. Dependence of surface nano-structural modifications of Ti implanted by N+ ions on temperature. Appl Surf Sci. 2010, 256, 4502–11.

[128] Berberich F, Matz W, Kreissig U, Richter E, Schell N, Möller W. Structural characterisation of hardening of Ti-Al-V alloys after nitridation by plasma immersion ion implantation. Appl Surf Sci. 2001, 179, 13–19.

[129] Gordin DM, Gloriant T, Chane-Pane V, Busardo D, Mitran V, Höche D, Vasilescu C, Drob SI, Cimpean AJ. Surface characterization and biocompatibility of titanium alloys implanted with nitrogen by Hardion+ technology. J Mater Sci Mater Med. 2012, 23, 2953–66.

[130] Leitão E, Silva RA, Barbosa MA. Electrochemical and surface modifications on N+-ION implanted Ti-5Al-2.5Fe immersed in HBSS. Corros Sci. 1997, 39, 377–83.

[131] Pye D. Ion Nitriding – Basics. Ad Mater Process. 2004, 162, 41–43.

[132] Byeli AV, Kukareko VA, Kononov AG. Titanium and zirconium-based alloys modified by intensive plastic deformation and nitrogen ion implantation for biocompatible implants. J Mech Behav Biomed Mater. 2012, 6, 89–94.

[133] Xie Y, Liu X, Huang A, Ding C, Chu PK. Improvement of surface bioactivity on titanium by water and hydrogen plasma immersion ion implantation. Biomaterials. 2005, 26, 6129–35.

[134] Azarmi F. Vacuum plasma spraying. Ad Mater Process. 2005, 163, 37–39.

[135] Fomin AA, Steinhauer AB, Lyasnikov VN, Wenig SB, Zakharevich AM. Nanocrystalline structure of the surface layer of plasma-sprayed hydroxyapatite coatings obtained upon preliminary induction heat treatment of metal base. Tech Phys Lett. 2012, 38, 481–83.

[136] Latka L, Goryachev SB, Kozerski S, Pawlowski L. Sintering of fine particles in suspension plasma sprayed coatings. Materials. 2010, 3, 3845–66.

[137] Montazeri M, Dehghanian C, Shokouhfar M, Baradaran A. Investigation of the voltage and time effects on the formation of hydroxyapatite-containing titania prepared by plasma electrolytic oxidation on Ti-6Al-4V alloy and its corrosion behavior. Appl Surf Sci. 2011, 257, 7268–75.

[138] Silva MAM, Martinelli AE, Alves C, Nascimento RM, Távora MP, Vilar CD. Surface modification of Ti implants by plasma oxidation in hollow cathode discharge. Surf Coatings Tech. 2006, 200, 2618–26.

[139] Huan Z, Fratila-Apachitei LE, Apachitei I, Duszczyk J. Porous NiTi surfaces for biomedical applications. Appl Surf Sci. 2012, 258, 5244–49.
[140] Rie K-T, Stucky T, Silva RA, Leitao E, Bordji K, Jouzeau J-Y, Mainard D. Plasma surface treatment and PACVD on Ti alloys for surgical implants. Surf Coatings Tech. 1995, 74/75, 973–80.
[141] Abd El-Rahman AM, Maitz MF, Kassem MA, El-Hossary FM, Prokert F, Reuther H, Pham MT, Richter E. Surface improvement and biocompatibility of TiAl24Nb10 intermetallic alloy using rf plasma nitriding. Appl Surf Sci. 2007, 253, 9067–72.
[142] Fouquet V, Pichon L, Drouet M, Sraboni A. Plasma assisted nitridation of Ti-6Al-4V. Appl Surf Sci. 2004, 221, 248–58.
[143] Sawase T, Jimbo R, Wennerberg A, Suketa N, Tanaka Y, Atsuta M. A novel characteristic of porous titanium oxide implants. Clin Oral Implants Res. 2007, 18, 680–85.
[144] Stachowski MJ, Medige J, Baier RE. Methodology for testing the mechanical properties of the bone/titanium implant interface. Environ Degrad Eng Mater. 1987, 3, 493–500.
[145] Dutta Majumdar JD, Manna I. Laser material processing. Int Mater Rev. 2011, 56, 341–88.
[146] Dahotre NB. Laser surface engineering. Ad Mater Process. 2002, 160, 35–39.
[147] Narayan RJ, Jin C, Patz T, Doraiswamy A, Modi R, Chrisey DB, Su -Y-Y, Lin SJ, Ovsianikov A, Chichkov B. Laser processing of advanced biomaterials. Ad Mater Process. 2005, 163, 39–42.
[148] Tian YS, Chen CZ, Li ST, Huo QH. Research progress on laser surface modification of titanium alloys. Appl Surf Sci. 2005, 242, 177–84.
[149] Romanos GE, Gupta B, Yunker M, Romanos EB, Malmstrom H. Lasers use in dental implantology. Implant Dent. 2013, 22, 282–88.
[150] Bandyopadhyay A, Balla VK, Roy M, Bose S. Laser surface modification of metallic biomaterials. J Metals. 2011, 63, 94–99.
[151] Liu X-B, Wang H-M. Modification of tribology and high-temperature behavior of Ti-8Al-2Cr-2Nb intermetallic alloy by laser cladding. Appl Surf Sci. 2006, 252, 5735–44.
[152] Trtica M, Gakovic B, Batani D, Desai T, Panjan P, Radak B. Surface modifications of a titanium implant by a picosecond Nd: YAGlaser operating at 1064 and 532 nm. Appl Surf Sci. 2006, 253, 2551–56.
[153] Pető G, Karacs A, Pászti Z, Guczi L, Divinyi T, Joób A. Surface treatment of screw shaped titanium dental implants by high intensity laser pulses. Appl Surf Sci. 2002, 186, 7–13.
[154] Roy M, Bandyopadhyay A, Bose S. Laser surface modification of electrophoretically deposited hydroxyapatite coating on titanium. J Amer Ceramic Soc. 2008, 91, 3517–21.
[155] Ciganovic J, Stasic J, Gakovic B, Momcilovic M, Milovanovic D, Bokorov M, Trtica M. Surface modification of the titanium implant using TEA CO2 laser pulses in controllable gas atmospheres – Comparative study. Appl Surf Sci. 2012, 258, 2741–48.
[156] Stübinger S, Etter C, Miskiewicz M, Homann F, Saldamli B, Wieland M, Sader R. Surface alterations of polished and sandblasted and acid-etched titanium implants after Er: YAG, carbon dioxide, and diode laser irradiation. Int J Oral Maxillofac Implants. 2010, 25, 104–11.
[157] Sultana R, Kon M, Hirakata LM, Fujihara E, Asaoka K, Ichikawa T. Surface modification of titanium with hydrothermal treatment at high pressure. Dent Mater J. 2006, 25, 470–79.
[158] Drnovšek N, Daneu N, Rečnik A, Mazaj M, Kovač J, Novak S. Hydrothermal synthesis of a nanocrystalline anatase layer on Ti6A4V implants. Surf Coatings Tech. 2009, 203, 1462–68.
[159] Valanezhad A, Tsuru K, Maruta M, Kawachi G, Matsuya S, Ishikawa K. A new biocompatible coating layer applied on titanium substrates using a modified zinc phosphatizing method. Surf Coatings Tech. 2012, 206, 2207–12.
[160] Puerta DG. Thermal spray coating – Characetrization and evaluation. Ad Mater Process. 2013, 171, 16–19.
[161] Haman JD, Boulware AA, Lucas LC, Crawmer DE. High-velocity oxyfuel thermal spray coatings for biomedical applications. J Thermal Spray Tech. 1995, 4, 179–84.

[162] Lima RS, Dimitrievska S, Bureau MN, Marple BR, Petit A, Mwale F, Antoniou J. HVOF-sprayed nano TiO2-HA coatings exhibiting enhanced biocompatibility. J Thermal Spray Tech. 2010, 19, 336–43.

[163] Park J-W, Tsutsumi Y, Lee CS, Park CH, Kim Y-J, Jang J-H, Khang D, Im Y-M, Doi H, Nomura N, Hanawa T. Surface structures and osteoblast response of hydrothermally produced CaTiO3 thin film on Ti–13Nb–13Zr alloy. Appl Surf Sci. 2011, 257, 7856–63.

[164] Siva D, Krishna R, Sun Y. Thermally oxidised rutile-TiO2 coating on stainless steel for tribological properties and corrosion resistance enhancement. Appl Surf Sci. 2005, 252, 1107–16.

[165] Petrovicova E, Schadler LS. Thermal spraying of polymer. Int Mater Rev. 2002, 47, 169–90.

[166] Li J-G, Hashida T. In situ formation of hydroxyapatite-whisker ceramics by hydro-thermal hot-pressing method. J Amer Ceramic Soc. 2006, 89, 3544–46.

[167] Tang YJ, Tang YF, Lv CT, Zhou ZH. Preparation of uniform porous hydroxyapatite biomaterials by a new method. Appl Surf Sci. 2008, 254, 5359–62.

[168] Lee WH, Hyun CY. Surface characteristics of self-assembled microporous Ti-6Al-4V compacts fabricated by electro-discharge-sintering in air. Appl Surf Sci. 2007, 253, 4649–51.

[169] Seah KHW, Thampuran R, Chen X, Teoh SH. A comparison between the corrosion behaviour of sintered and unsintered porous titanium. Corros Sci. 1995, 37, 1333–40.

[170] Dewidar MM, Lim JK. Properties of solid core and porous surface Ti–6Al–4V implants manufactured by powder metallurgy. J Alloys Comp. 2008, 454, 442–46.

[171] Okazumi T, Ueda K, Tajima K, Umetsu N, Narushima T. Anatase formation on titanium by two-step thermal treatment. J Mater Sci. 2011, 46, 2998–3005.

[172] Rodionov IV. Steam-thermal oxide coatings for titanium medical implants. Biomed Eng. 2012, 46, 58–61.

[173] Lee WH, Kim SJ, Lee WJ, Byun CS, Kim DK, Kim JY, Hyun CY, Lee JG, Park JW. Mechanism of surface modification of a porous-coated Ti-6Al-4V implant fabricated by electrical resistance sintering. J Mater Sci. 2001, 36, 3573–77.

[174] Shenhar A, Gotman I, Gutmanas EY, Ducheyne P. Surface modification of titanium alloy orthopaedic implants via novel powder immersion reaction assisted coating nitriding method. Mater Sci Eng A. 1999, 268, 40–46.

[175] Zhang P, Zhang Z, Li W, Zhu M. Effect of Ti-OH groups on microstructure and bioactivity of TiO2coating prepared by micro-arc oxidation. Appl Surf Sci. 2013, 268, 381–86.

[176] Nie X, Leyland A, Matthews A. Deposition of layered bioceramic hydroxyapatite/ TiO2 coatings on titanium alloys using a hybrid technique of micro-arc oxidation and electrophoresis. Surf Coatings Tech. 2000, 125, 407–14.

[177] Liu F, Wang F, Shimizu T, Igarashi K, Zhao L. Formation of hydroxyapatite on Ti-6Al-4V alloy by microarc oxidation and hydrothermal treatment. Surf Coatings Tech. 2005, 199, 220–24.

[178] Martines RT, Sendyk WR, Gromatzky A, Cury PR. Sandblasted/acid-etched vs smooth-surface implants: Implant mobility and clinical reaction to experimentally induced peri-implantitis in beagle dogs. J Oral Implant. 2008, 34, 185–89.

[179] Wei Q, Cui Z-D, Yang X-J, Shi J. Improving the bioactivity of NiTi shape memory alloy by heat and alkali treatment. Appl Surf Sci. 2008, 255, 462–65.

[180] Nishiguchi S, Kato H, Neo M, Oka M, Kim H-M, Kokubo T, Nakamura T. Alkali- and heat-treated porous titanium for orthopedic implants. J Biomed Mat Res. 2001, 54, 198–208.

[181] Juodzbalys G, Sapragoniene M, Wennerberg A, Baltrukonis T. Titanium dental implant surface micromorphology optimization. J Oral Implant. 2007, 33, 177–85.

[182] Ishizawa H, Ogino M. Thin hydroxyapatite layers formed on porous titanium using electrochemical and hydrothermal reaction. J Mater Sci. 1996, 31, 6279–84.

[183] Ribeiro AA, Marques RFC, Guastaldi AC, Sinézio JCC. Hydroxyapatite deposition study through polymeric process on commercially pure Ti surfaces modified by laser beam irradiation. J Mater Sci. 2009, 44, 4056–62.

[184] Sun Y. Tribological rutile-TiO2 coating on aluminum alloy. Appl Surf Sci. 2004, 233, 328–35.

[185] Sun T, Wang M. Mechanical performance of apatite/TiO2 composite coatings formed on Ti and NiTi shape memory alloy. Appl Surf Sci. 2008, 255, 404–08.

[186] Xie Y, Zheng X, Huang L, Ding C. Influence of hierarchical hybrid micro/nano-structured surface on biological performance of titanium coating. J Mater Sci. 2011, 47, 1411–17.

[187] Zheng CY, Nie FL, Zheng YF, Cheng Y, Wei SC, Valiev RZ. Enhanced in vitro biocompatibility of ultrafine-grained titanium with hierarchical porous surface. Appl Surf Sci. 2011, 257, 5634–40.

[188] Jiang W, Wang WD, Shi XH, Chen HZ, Zou W, Guo Z, Luo JM, Gu ZW, Zhang XD. The effects of hydroxyapatite coatings on stress distribution near the dental implant–bone interface. Appl Surf Sci. 2008, 255, 273–75.

[189] Suska F, Svensson S, Johansson A, Emanuelsson L, Karlholm H, Ohrlander M, Thomsen P. In vivo evaluation of noble metal coatings. J Biomed Mater Res B. 2010, 92B, 86–94.

[190] Chen Y, Zheng X, Xie Y, Ji H, Ding C. Antibacterial properties of vacuum plasma sprayed titanium coatings after chemical treatment. Surf Coat Technol. 2009, 204, 685–90.

[191] Yao C, Storey D, Webster T. Nanostructured metal coatings on polymers increase osteo-blast attachment. Int J Manomed. 2007, 2, 487–92.

[192] Young L, Soon K, Doo S, Yong K. Enhanced biocompatibility of stainless steel implants by titanium coating and microarc oxidation. Clin Orthopaedics Related Res. 2011, 469, 330–38.

[193] Mwenifumbo S, Li M, Chen J, Beye A, Soboyejo W. Cell/surface interactions on laser micro-textured titanium-coated silicon surfaces. J Mater Sci Mater Med. 2007, 18, 9–23.

[194] Neumeister A, Bartke D, Bärsch N, Weingärtner T, Guetaz L, Montani A, Compagnini G, Barcikowski S. Interface of nanoparticle-coated electropolished stents. Langmuir. 2012, 28, 12060–66.

[195] Fallah V, Corbin SF, Khajepour A. Process optimization of Ti–Nb alloy coatings on a Ti–6Al–4V plate using a fiber laser and blended elemental powders. J Mater Process Tech. 2010, 210, 2081–87.

[196] Sonoda T, Saito T, Watazu A, Katou K, Asahina T. Coating of Ti–6Al–4V alloy with pure titanium film by sputter-deposition for improving biocompatibility. Surf Interf Anal. 2006, 38, 797–800.

[197] Devine DM, Hahn J, Richards RG, Gruner H, Wieling R, Pearce SG. Coating of carbon fiber-reinforced polyetheretherketone implants with titanium to improve bone apposition. J Biomed Mater Res B. 2-13, 101B, 591–98.

[198] Matsuno H, Yokoyama A, Watari F, Uno M, Kawasaki T. Biocompatibility and osteogenesis of refractory metal implants, titanium, hafnium, niobium, tantalum and rhenium. Biomaterials. 2001, 22, 1253–62.

[199] Dolatshahi-Pirouz A, Jensen T, Kraft DC, Foss M, Kingshott P, Hansen JL, Larsen AN, Chevallier J, Besenbacher F. Fibronectin adsorption, cell adhesion, and proliferation on nanostructured tantalum surfaces. ACS Nano. 2010, 4, 2874–82.

[200] Macionczyk R, Gerold B, Thull R. Repassivating tantalum/tantalum oxide surface modification on stainless steel implants. Surf Coat Technol. 2001, 142/144, 1084–87.

[201] Cheng Y, Cai W, Li HT, Zheng YF. Surface modification of NiTi alloy with tantalum to improve its biocompatibility and radiopacity. J Mater Sci. 2006, 41, 4961–64.

[202] Spriano S, Vernè E, Guala C, Faga MG, Eitel F. Surface Modification of cobalt alloys for articular prostheses. J Bone Joint Surg Br. 2009, 91-B, 265–66.

[203] Pham V-H, Lee S-H, Li Y, Kim H-E, Shin K-H, Koh Y-H. Utility of tantalum (Ta) coating to improve surface hardness in vitro bioactivity and biocompatibility of Co–Cr. Thin Solid Films. 2012, 536, 269–74.

[204] Levine BR, Sporer S, Poggie RA, Della Valle CJ, Jacobs JJ. Experimental and clinical performance of porous tantalum in orthopedic surgery. Biomaterials. 2006, 27, 4671–81.

[205] Macheras GA, Papagelopoulos PJ, Kateros K, Kostakos AT, Baltas D, Karachalios TS. Radiological evaluation of the metal-bone interface of a porous tantalum monoblock acetabular component. J Bone Joint Surg Br. 2006, 88, 304–09.

[206] Balla VK, Bodhak S, Bose S, Bandyopadhyay A. Porous tantalum structures for bone implants: Fabrication, mechanical and in vitro biological properties. Acta Biomater. 2010, 6, 3349–59.

[207] Macionczyk F, Gerold B, Thull R. Repassivating tantalum/tantalum oxide surface modification on stainless steel implants. Surf Coat Technol. 2001, 142/144, 1084–87.

[208] Olivares-Navarrete R, Olaya JJ, Ramírez C, Rodil SE. Biocompatibility of niobium coatings. Coatings. 2011, 1, 72–87.

[209] Ramírez G, Rodil SE, Arzate H, Muhl S, Olaya JJ. Niobium based coatings for dental implants. Appl Surf Sci. 2011, 257, 2555–59.

[210] Liu BT, Yan XB, Zhang X, Zhou Y, Guo YN, Bian F, Zhang XY. Investigation of oxidation resistance of Ni–Ti film used as oxygen diffusion barrier layer. Appl Surf Sci. 2009, 255, 6179–82.

[211] Liu X, Wu S, Yeung KWK, Chan YL, Hu T, Xu Z, Liu X, Chung JCY, Cheung KMC, Chu PK. Relationship between osseointegration and superelastic biomechanics in porous NiTi scaffolds. Biomaterials. 2011, 32, 330–38.

[212] Bansiddhi A, Sargeant TD, Stupp SI, Dunand DC. Porous NiTi for bone implants: A review. Acta Biomater. 2008, 4, 773–82.

[213] Salunke P, Shanov V, Witte F. High purity biodegradable magnesium coating for implant application. Mater Sci Eng B. 2011, 176, 1711–17.

[214] Yang J, Cui F, Lee IS. Surface Modifications of Magnesium Alloys for Biomedical Applications. Annals Biomed Eng. 2011, 39, 1857–71.

[215] Kirkland NT, Birbilis N, Staiger MP. Assessing the corrosion of biodegradable magnesium implants: A critical review of current methodologies and their limitations. Acta Biomater. 2012, 8, 925–36.

[216] Yuen CK, Ip WY. Theoretical risk assessment of magnesium alloys as degradable biomedical implants. Acta Biomater. 2010, 6, 1808–12.

[217] Li Y, Wen C, Mushahary D, Sravanthi R, Harishankar N, Pande G, Hodgson P. Mg–Zr–Sr alloys as biodegradable implant materials. Acta Biomater. 2012, 8, 3177–88.

[218] Yang X, Li M, Lin X, Tan L, Lan G, Li L, Yin Q, Xia H, Zhang Y, Yang K. Enhanced in vitro biocompatibility/bioactivity of biodegradable Mg–Zn–Zr alloy by micro-arc oxidation coating contained Mg2SiO4. Surf Coat Technol. 2013, 233, 65–73.

[219] Li M, Cheng Y, Zheng YF, Zhang X, Xi TF, Wei SC. Plasma enhanced chemical vapor deposited silicon coatings on Mg alloy for biomedical application. Surf Coat Technol. 2012, 228, S262–5.

[220] Krishna BV, Xue W, Bose S, Bandyopadhyay A. Functionally graded Co-Cr-Mo coating on Ti-6Al-4V alloy structure. Acta Biomater. 2008, 4, 697–706.

[221] Holinka J, Pilz M, Kubista B, Presterl E, Windhager R. Effects of selenium coating of orthopaedic implant surfaces on bacterial adherence and osteoblastic cell growth. Bone Joint J. 2013, 95, 678–82.

[222] Gomes ME, Reis RL. Biodegradable polymers and composites in biomedical applications: From catgut to tissue engineering. Part 1: Available systems and their properties. Int Mater Rev. 2004, 49, 261–73.

[223] Gomes ME, Reis RL. Biodegradable polymers and composites in biomedical applications: From catgut to tissue engineering. Part 2: Systems for temporary replacement and advanced tissue regeneration. Int Mater Rev. 2004, 49, 274–85.

[224] Abdal-hay A, Dewidar M, Lim JK. Biocorrosion behavior and cell viability of adhesive polymer coated magnesium-based alloys for medical implants. Appl Surf Sci. 2012, 261, 536–46.

[225] Wong HM, Yeung KWK, Lam KO, Tam V, Chu PK, Luk KDK, Cheung KMC. A biodegradable polymer-based coating to control the performance of magnesium alloy orthopaedic implants. Biomaterials. 2010, 31, 2084–96.

[226] Tepe G, Schmehl J, Wendel HP, Schaffner S, Heller S, Gianotti M, Claussen CD, Duda SH. Reduced thrombogenicity of nitinol stents – In vitro evaluation of different surface modifications and coatings. Biomaterials. 2006, 27, 643–50.

[227] Bakhshi R, Darbyshire A, Evans JE, You Z, Lu J, Seifalian AM. Polymeric coating of surface modified nitinol stent with POSS-nanocomposite polymer. Colloids Surf B Biointerfaces. 2011, 86, 93–105.

[228] Ye SH, Johnson CA, Woolley R, Oh H-I, Gamble LJ, Ishihara K, Wagner WR. Surface modification of a titanium alloy with a phospholipid polymer prepared by a plasma-induced grafting technique to improve surface thromboresistance. Colloids Surf B Biointerfaces. 2009, 74, 96–102.

[229] Nakahira A, Tamai M, Miki S, Pezzotti G. Fracture behavior and biocompatibility evaluation of nylon-infiltrated porous hydroxyapatite. J Mater Sci. 2002, 37, 4425–30.

[230] Oshida Y. Bioscience and Bioengineering of Titanium Materials. 2nd edition. London UK: Elsevier, 2011.

[231] Saito T, Takemoto M, Fukuda A, Kuroda Y, Fujibayashi S, Neo M, Honjoh D, Hiraide T, Kizuki T, Kokubo T, Nakamura T. Effect of titania-based surface modification of polyethylene terephthalate on bone–implant bonding and peri-implant tissue reaction. Acta Biomater. 2011, 7, 1558–69.

[232] Xie D, Wan G, Maitz MF, Sun H, Huang N. Deformation and corrosion behaviors of Ti–O film deposited 316L stainless steel by plasma immersion ion implantation and deposition. Surf Coat Technol. 2013, 214, 117–23.

[233] Siu HT, Man HC. Fabrication of bioactive titania coating on nitinol by plasma electrolytic oxidation. Appl Surf Sci. 2013, 274, 181–87.

[234] Wang HR, Liu F, Zhang YP, Yu DZ, Wang FP. Preparation and properties of titanium oxide film on NiTi alloy by micro-arc oxidation. Appl Surf Sci. 2011, 257, 5576–80.

[235] Ou J, Wang J, Zhang D, Zhang P, Liu S, Yan P, Liu B, Yang S. Fabrication and bio- compatibility investigation of TiO2 films on the polymer substrates obtained via a novel and versatile route. Colloids Surf B Biointerfaces. 2010, 76, 123–27.

[236] Kim H-K, Jang J-W, Lee C-H. Surface modification of implant materials and its effect on attachment and proliferation of bone cells. J Mater Sci Mater Med. 2004, 15, 825–30.

[237] György E, Socol G, Axente E, Mihailescu IN, Ducu C, Ciuca S. Anatase phase TiO2 thin films obtained by pulsed laser deposition for gas sensing applications. Appl Surf Sci. 2005, 247, 429–33.

[238] Grigal IP, Markeev AM, Gudkova SA, Chernikova AG, Mityaev AS, Alekhin AP. Correlation between bioactivity and structural properties of titanium dioxide coatings grown by atomic layer deposition. Appl Surf Sci. 2012, 258, 3415–19.

[239] Advincula M, Fan X, Lemons J, Advincula R. Surface modification of surface sol–gel derived titanium oxide films by self-assembled monolayers (SAMs) and non-specific protein adsorption studies. Colloids Surf B Biointerfaces. 2005, 42, 29–43.

[240] Fu T, Liu BG, Zhou YM, Wu XM. Sol-gel titania coating on NiTi alloy with a porous titania film as interlayer. J Sol-Gel Sci Technol. 2011, 58, 307–11.

[241] Lai YK, Sun L, Chen C, Nie CG, Zuo J, Lin CJ. Optical and electrical characterization of TiO2 nanotube arrays on titanium substrate. Appl Surf Sci. 2005, 252, 1101–06.

[242] Yun H, Lin C, Li J, Wang J, Chen H. Low-temperature hydrothermal formation of a net-like structured TiO2 film and its performance of photogenerated cathode protection. Appl Surf Sci. 2008, 255, 2113–17.

[243] Lim JI, Yu B, Woo KM, Lee Y-K. Immobilization of TiO2 nanofibers on titanium plates for implant applications. Appl Surf Sci. 2008, 255, 2456–60.

[244] Shi X, Tsuru K, Xu L, Kawachi G, Ishikawa K. Effects of solution pH on the structure and biocompatibility of Mg-containing TiO2 layer fabricated on titanium by hydrothermal treatment. Appl Surf Sci. 2013, 270, 445–51.

[245] Park J-W, Kim Y-J, Jang J-H, An C-H. In vitro biocompatibility of magnesium- incorporated submicro-porous titanium oxide surface produced by hydrothermal treatment. Appl Surf Sci. 2010, 257, 925–31.

[246] Huang F, Xie B, Wu B, Shao L, Li M, Wang H, Jiang Y, Song Y. Enhancing the crystallinity and surface roughness of sputtered TiO2 thin film by ZnO underlayer. Appl Surf Sci. 2009, 255, 6781–85.

[247] Hang R, Huang X, Tian L, He Z, Tang B. Preparation, characterization, corrosion behavior and bioactivity of Ni2O3-doped TiO2 nanotubes on NiTi alloy. Electrochim Acta. 2012, 70, 382–93.

[248] Zhu S, Huang N, Xu L, Zhang Y, Liu H, Lei Y, Sun H, Yao Y. Biocompatibility of Fe–O films synthesized by plasma immersion ion implantation and deposition. Surf Coat Technol. 2009, 203, 1523–29.

[249] Macionczyk F, Gerold B, Thull R. Repassivating tantalum/tantalum oxide surface modification on stainless steel implants. Surf Coat Technol. 2001, 142, 1084–87.

[250] Arnould C, Volcke C, Lamarque C, Thiry PA, Delhalle J, Mekhalif Z. Titanium modified with layer-by-layer sol–gel tantalum oxide and an organodiphosphonic acid: A coating for hydroxyapatite growth. J Colloid Interf Sci. 2009, 336, 497–503.

[251] Maho A, Linden S, Arnould C, Detriche S, Delhalle J, Mekhalif Z. Tantalum oxide/carbon nanotubes composite coatings on titanium, and their functionalization with organophosphonic molecular films: A high quality scaffold for hydroxyapatite growth. J Colloid Interf Sci. 2012, 371, 150–58.

[252] Sollazzo V, Pezzetti F, Scarano A, Piattelli A, Bignozzi CA, Massari L, Brunelli G, Carinci F. Zirconium oxide coating improves implant osseointegration in vivo. Dent Mater. 2008, 24, 357–61.

[253] Yan Y, Han Y, Lu C. The effect of chemical treatment on apatite-forming ability of the macroporous zirconia films formed by micro-arc oxidation. Appl Surf Sci. 2008, 254, 4833–39.

[254] Balla VK, Xue W, Bose S, Bandyopadhyay A. Laser-assisted Zr/ZrO2 coating on Ti for load-bearing implants. Acta Biomater. 2009, 5, 2800–09.

[255] Emadi R, Tavangarian F, Esfahani SIR, Sheikhhosseini A, Kharaziha M. Nanostructured Forsterite Coating Strengthens Porous Hydroxyapatite for Bone Tissue Engineering. J Amer Ceramic Soc. 2010, 93, 2679–83.

[256] Arcos D, Vallet-Regí M. Sol-gel silica-based biomaterials and bone tissue regeneration. Acta Biomater. 2010, 6, 2874–88.

[257] Pérez P, Haanappel VAC, Stroosnijder MF. Formation of an alumina layer on A FeCrAlY alloy by thermal oxidation for potential medical implant applications. Surf Coat Technol. 2001, 139, 207–15.

[258] Ezz T, Crouse P, Li L, Liu Z. Synthesis of TiN thin films by a new combined laser/sol–gel processing technique. Appl Surf Sci. 2007, 253, 7903–07.

[259] Shenhar A, Gotman I, Radin S, Ducheyne P, Gutmanas EY. Titanium nitride coatings on surgical titanium alloys produced by a powder immersion reaction assisted coating method: Residual stresses and fretting behavior. Surf Coat Technol. 2000, 126, 210–18.

[260] Xiao L, Yan D, He J, Zhu L, Dong Y, Zhang J, Li X. Nanostructured TiN coating prepared by reactive plasma spraying in atmosphere. Appl Surf Sci. 2007, 253, 7535–39.

[261] Mutlu I. Sinter-Coating Method for Production of TiN Coated Titanium Foam for Biomedical Implant Applications. Surf Coat Technol. 2013, 232, 396–402.

[262] Mao Z, Ma J, Wang J, Sun B. The effect of powder preparation method on the corrosion and mechanical properties of TiN-based coatings by reactive plasma spraying. Appl Surf Sci. 2007, 255, 3784–88.

[263] Flores M, Huerta L, Escamilla R, Andrade E, Muhl S. Effect of substrate bias voltage on corrosion of TiN/Ti multilayers deposited by magnetron sputtering. Appl Surf Sci. 2007, 253, 7192–96.

[264] Hyde GK, McCullen SD, Jeon S, Stewart SM, Jeon H, Loboa EG, Parsons GN. Atomic layer deposition and biocompatibility of titanium nitride nano-coatings on cellulose fiber substrates. Biomed Mater. 2009, 4, doi: 10.1088/1748-6041/4/2/025001.

[265] Balla VK, Bhat A, Bose S, Bandyopadhyay A. Laser processed TiN reinforced Ti6Al4V composite coatings. J Mech Behav Biomed Mater. 2012, 6, 9–20.

[266] Türkan U, Öztürk O, Eroğlu AE. Metal ion release from TiN coated CoCrMo orthopedic implant material. Surf Coat Technol. 2006, 200, 16/17, 5020–27.

[267] Rieder P, Scherrer S, Filieri A, Wiskott HWA, Durual S. TiNOx coatings increase human primary osteoblasts proliferation independency of the substrate – A short report. Bio-Med Mater Eng. 2012, 22, 277–81.

[268] Freemanm CO, Brook IM. Bone response to a titanium aluminium nitride coating on metallic implants. J Mater Sci Mater Med. 2006, 17, 465–70.

[269] Mukherjee S, Maitz MF, Pham MT, Richter E, Prokert F, Moeller W. Development and biocompatibility of hard Ti-based coatings using plasma immersion ion implantation-assisted deposition. Surf Coat Technol. 2005, 196, 312–16.

[270] García GL, Hernández TJ, Flores RN, Argumedo MP, López VA, Araujo LDJ, Courrech AAM. Influence of nitrogen in TiAlNO coatings fabricated by co-sputtering. IOP Conf Ser: Mater Sci Eng. 2013, 45, doi: 10.1088/1757-899X/45/1/012024.

[271] Shtansky DV, Grigoryan AS, Toporkova AK, Arkhipov AV, Sheveyko AN, Kiryukhantsev-Korneev PV. Modification of polytetrafluoroethylene implants by depositing TiCaPCON films with and without stem cells. Surf Coat Technol. 2011, 206, 1188–95.

[272] Brama M, Rhodes N, Hunt J, Ricci A, Teghil R, Migliaccio S, Rocca CD, Leccisotti S, Lioi A, Scandurra M, De Maria G, Ferro D, Pu F, Panzini G, Politi L, Scandurra R. Effect of titanium carbide coating on the osseointegration response in vitro and in vivo. Biomaterials. 2007, 28, 595–608.

[273] Tian YS, Chen CZ, Chen LB, Chen LX. Study on the microstructure and wear resistance of the composite coatings fabricated on Ti-6Al-4V under different processing conditions. Appl Surf Sci. 2006, 253, 1494–99.

[274] Levashov EA, Petrzhik MI, Kiryukhantsev-Korneev FV, Shtansky DV, Prokoshkin SD, Gunderov DV, Sheveiko AN, Korotitsky AV, Valiev RZ. Structure and mechanical behavior during indentation of biocompatible nanostructured titanium alloys and coatings. Metallurgist. 2012, 56, 395–407.

[275] Rogers KD, Etok SE, Scott R. Structural characterization of apatite coatings. J Mater Sci. 2004, 39, 5747–54.

[276] Jaffe WL, Scott DF. Current concepts review – Total hip arthroplasty with hydroxyapatite-coated prostheses. J Bone Joint Surg Am. 1996, 78, 1918–34.

[277] Ong JL, Daniel CN, Chan DCN. Hydroxyapatite and their use as coatings in dental implants: A review. Crit Rev Biomed Eng. 2000, 28, 667–707.

[278] Arias JL, Mayor MB, Pou J, León B, Pérez-Amor M. Transport of ablated material through a water vapor atmosphere in pulsed laser deposition of hydroxylapatite. Appl Surf Sci. 2002, 186, 448–52.

[279] Stiegler N, Bellucci D, Bolelli G, Cannillo V, Gadow R, Killinger A, Lusvarghi L, Sola A. High-velocity suspension flame sprayed (HVSFS) hydroxyapatite coatings for biomedical applications. J Thermal Spray Technol. 2012, 21, 275–87.

[280] Guo CY, Tang ATH, Matinlinna JP. Insights into surface treatment methods of titanium dental implants. J Adhesion Sci Technol. 2012, 26, 189–205.

[281] Hahn B-D, Park D-S, Choi -J-J, Ryu J, Yoon W-H, Kim K-H, Park C, Kim H-E. Dense nanostructured hydroxyapatite coating on titanium by aerosol deposition. J Amer Ceramic Soc. 2009, 92, 683–87.

[282] Mansur MR, Wang J, Berndt CC. Microstructure, composition and hardness of laser assisted hydroxyapatite and Ti-6Al-4V composite coatings. Surf Coat Technol. 2013, 232, 482–88.

[283] Luo R, Liu Z, Yan F, Kong Y, Zhang Y. The biocompatibility of hydroxyapatite film deposition on micro-arc oxidation Ti6Al4V alloy. Appl Surf Sci. 2013, 266, 57–61.

[284] Gopi D, Indira J, Kavitha L, Ferreira JMF. Hydroxyapatite coating on selectively passivated and sensitively polymer-protected surgical grade stainless steel. J Appl Electro-chemistry. 2013, 43, 331–45.

[285] Kar A, Raja KS, Misra M. Electrodeposition of hydroxyapatite onto nanotubular TiO2 for implant applications. Surf Coat Technol. 2006, 201, 3723–31.

[286] Cao N, Dong J, Wang Q, Ma Q, Xue C, Li M. An experimental bone defect healing with hydroxyapatite coating plasma sprayed on carbon/carbon composite implants. Surf Coat Technol. 2010, 205, 1150–56.

[287] Kong Y-M, Bae C-J, Lee S-H, Kim H-W, Kim H-E. Improvement in biocompatibility of by addition of HA. Biomaterials. 2005, 26, 509–17.

[288] Grigorescu S, Carradò A, Ulhaq C, Faerber J, Ristoscu C, Dorcioman G, Axente E, Werckmann J, Mihailescu IN. Study of the gradual interface between hydroxyapatite thin films PLD grown onto Ti-controlled sublayers. Appl Surf Sci. 2007, 254, 1150–54.

[289] Yang S, Xing W, Man HC. Pulsed laser deposition of hydroxyapatite film on laser gas nitriding NiTi substrate. Appl Surf Sci. 2009, 255, 9889–92.

[290] Lee K, Choe H-C, Kim B-H, Ko Y-M. The biocompatibility of HA thin films deposition on anodized titanium alloys. Surf Coat Technol. 2010, 205, S267–70.

[291] Jeong Y-H, Choe H-C, Brantley WA, Sohn I-B. Hydroxyapatite thin film coatings on nanotube-formed Ti-35Nb-10Zr alloys after femtosecond laser texturing. Surf Coat Technol. 2013, 217, 13–22.

[292] Khalid HM, Jauhari I, Dom AHM. Development of nanolayer hydroxyapatite (HA) on titanium alloy via superplastic deformation method. Metall Mater Trans A. 2012, 43, 3776–85.

[293] Roy M, Bandyopadhyay A, Bose S. Induction plasma sprayed nano hydroxyapatite coatings on titanium for orthopaedic and dental implants. Surf Coat Technol. 2012, 205, 2785–92.

[294] Duta L, Oktar FN, Stan GE, Popescu-Pelin G, Serban N, Luculescu C, Mihailescu IN. Novel doped hydroxyapatite thin films obtained by pulsed laser deposition. Appl Surf Sci. 2013, 265, 41–49.

[295] Zahrani ME, Fathi MH, Alfantazi AM. Sol-gel derived nanocrystalline fluoridated hydroxyapatite powders and nanostructured coatings for tissue engineering applications. Metall Mater Trans A. 2011, 42, 3291–309.

[296] Tredwin CJ, Georgiou G, Kim H-W, Knowles JC. Hydroxyapatite, fluor-hydroxyapatite and fluorapatite produced via the sol–gel method: Bonding to titanium and scanning electron microscopy. Dent Mater. 2013, 29, 521–29.

[297] Miao S, Lin N, Cheng K, Yang D, Huang X, Han G, Weng W, Ye Z. Zn-releasing FHA coating and its enhanced osseointegration ability. J Amer Ceramic Soc. 2011, 94, 255–60.

[298] Balani K, Anderson R, Laha T, Andara M, Tercero J, Crumpler E, Agarwal A. Plasma-sprayed carbon nanotube reinforced hydroxyapatite coatings and their interaction with human osteoblasts in vitro. Biomaterials. 2007, 28, 618–24.

[299] Zhou H, Lawrence JG, Touny AH, Bhaduri SB. Biomimetic coating of bisphosphonate incorporated CDHA on Ti6Al4V. J Mater Sci Mater Med. 2011, 23, 365–74.

[300] Hayakawa T, Takahashi K, Okada H, Yoshinari M, Hara H, Mochizuki C, Yamamoto H, Sato M. Effect of thin carbonate-containing apatite (CA) coating of titanium fiber mesh on trabecular bone response. J Mater Sci Mater Med. 2008, 19, 2087–96.

[301] Li H, Zhao X, Cao S, Li K, Chen M, Xu Z, Lu J, Zhang L. Na-doped hydroxyapatite coating on carbon/carbon composites: Preparation, in vitro bioactivity and biocompatibility. Appl Surf Sci. 2012, 263, 163–73.

[302] Thian ES, Huang J, Barber ZH, Best SM, Bonfield W. Surface modification of magnetron-sputtered hydroxyapatite thin films via silicon substitution for orthopaedic and dental application. Surf Coat Technol. 2011, 205, 3472–77.

[303] Guo DG, Wang AH, Han Y, Xu KW. Characterization, physicochemical properties and biocompatibility of La-incorporated apatites. Acta Biomater. 2009, 5, 3512–23.

[304] Li Y, Li Q, Zhu S, Luo E, Li J, Feng G, Liao Y, Hu J. The effect of strontium-substituted hydroxyapatite coating on implant fixation in ovariectomized rats. Biomaterials. 2010, 31, 9006–14.

[305] Bornapour M, Muja N, Shum-Tim D, Cerruti M, Pekguleryuz M. Biocompatibility and biodegradability of Mg–Sr alloys: The formation of Sr-substituted hydroxyapatite. Acta Biomater. 2013, 9, 5319–30.

[306] Ji X, Lou W, Wang Q, Ma J, Xu H, Bai Q, Liu C, Liu J. Sol-Gel- Derived Hydroxyapatite-Carbon Nanotube/Titania Coatings on Titanium Substrates. Int J Mol Sci. 2012, 13, 5242–53.

[307] Damia C, Sarda S, Deydier E, Sharrock P. Study of two hydroxyapatite/ poly(alkoxysilane) implant coatings. Surf Coat Technol. 2006, 201, 3008–15.

[308] Tkalcec E, Sauer M, Nonninger R, Schmidt H. Sol-gel-derived hydroxyapatite powders and coatings. J Mater Sci. 2001, 36, 5253–63.

[309] Wang L, Luo J. Formation of Hydroxyapatite Coating on Anodic Titanium Dioxide Nanotubes via an Efficient Dipping Treatment. Metall Mater Trans A. 2011, 42, 3255–64.

[310] Chou B-Y, Chang E. Influence of deposition temperature on mechanical properties of plasma-sprayed hydroxyapatite coating on titanium alloy with ZrO2 intermediate layer. J Thermal Spray Technol. 2003, 12, 199–207.

[311] Kim J, Knag I-G, Cheon K-H, Lee S, Park S, Kim H-E, Han C-M. Stable sol-gel hydroxyapatite coating on zirconia dental implant for improved osseointegration. J Mater Sci Mater Med. 2021, 32, 81, doi: https://doi.org/10.1007/s10856-021-06550-6.

[312] Leem JM, Lim YJ. In vitro investigation of anodization and CaP deposited titanium surface using MG63 osteoblast-like cells. Appl Surf Sci. 2010, 256, 3086–92.

[313] Junker R, Manders PJD, Wolke J, Borisov Y, Braceras I, Jansen JA. Loaded microplasma-sprayed CaP-coated implants in vivo. J Dent Res. 2010, 89, 1489–93.

[314] Zhang B, Lu J, Chen J, Zhang X, Gu Z, Yang X. Biomimetic Ca–P coating on pre-calcified Ti plates by electrodeposition method. Appl Surf Sci. 2010, 256, 2700–04.

[315] Ribeiro AA, Balestra RM, Rocha MN, Peripolli SB, Andrade MC, Pereira LC, Oliveira MV. Dense and porous titanium substrates with a biomimetic calcium phosphate coating. Appl Surf Sci. 2013, 265, 250–56.

[316] Narayanan R, Seshadri SK, Kwon TY, Kim KH. Calcium phosphate-based coatings on titanium and its alloys. J Biomed Mater Res B. 2008, 85B, 279–99.

[317] Paital SR, Bunce N, Nandwana P, Honrao C, Nag S, He W, Banerjee R, Dahotre NB. Laser surface modification for synthesis of textured bioactive and biocompatible Ca–P coatings on Ti-6Al-4V. J Mater Sci Mater Med. 2011, 22, 1393–406.

[318] Durdu S, Deniz ÖF, Kutbay I, Usta M. Characterization and formation of hydroxyapatite on Ti6Al4V coated by plasma electrolytic oxidation. J Alloys Comp. 2013, 551, 422–29.

[319] Symietz C, Lehmann E, Gildenhaar R, Koter R, Berger G, Krüger J. Fixation of bioactive calcium alkali phosphate on Ti6Al4V implant material with femtosecond laser pulses. Appl Surf Sci. 2011, 257, 5208–12.

[320] Hsu H-C, Wu S-C, Fu C-É, Ho W-F. Formation of calcium phosphates on low-modulus Ti-7.5 Mo alloy by acid and alkali treatments. J Mater Sci. 2010, 45, 3661–70.

[321] Yu S, Yu Z, Wang G, Han J, Ma X, Dargusch MS. Biocompatibility and osteoconduction of active porous calcium–phosphate films on a novel Ti–3Zr–2Sn–3Mo–25Nb biomedical alloy. Colloids Surf B. 2011, 85, 103–15.

[322] Wang Y, Khor KA, Cheang P. Thermal spraying of functionally graded calcium phosphate coatings for biomedical implants. J Thermal Spray Technol. 1998, 7, 50–57.

[323] Chen D, Bertollo N, Lau A, Taki N, Nishino T, Mishima H, Kawamura H, Walsh WR. Osseointegration of porous titanium implants with and without electrochemically deposited DCPD coating in an ovine model. J Orthopaedic Surg Res. 2011, 6, 56, doi: 10.1186/1749-799X-6-56.

[324] Lin K, Yuan W, Wang L, Lu J, Chen L, Jiang ZW. Evaluation of host inflammatory responses of β-tricalcium phosphate bioceramics caused by calcium pyrophosphate impurity using a subcutaneous model. J Biomed Mater Res B. 2011, 99B, 350–58.

[325] Roy M, Krishna BV, Bandyopadhyay A, Bose S. Laser processing of bioactive tricalcium phosphate coating on titanium for load-bearing implants. Acta Biomater. 2008, 4, 324–33.

[326] Mo A, Liao J, Xu W, Xian S, Li Y, Bai S. Preparation and antibacterial effect of silver-hydroxyapatite/titania nanocomposite thin film on titanium. Appl Surf Sci. 2008, 255, 435–38.

[327] Sima F, Socol G, Axente E, Mihailescu IN, Zdrentu L, Petrescu SM, Mayer I. Bio-compatible and bioactive coatings of Mn2+-doped β-tricalcium phosphate synthesized by pulsed laser deposition. Appl Surf Sci. 2007, 254, 1155–59.

[328] Mirhosseini N, Crouse PL, Li L, Garrod D. Combined laser/sol–gel synthesis of calcium silicate coating on Ti–6Al–4V substrates for improved cell integration. Appl Surf Sci. 2007, 253, 7998–8002.

[329] Wu C, Ramaswamy Y, Gale D, Yang W, Xiao K, Zhang L, Yin Y, Zreiqat H. Novel sphene coatings on Ti–6Al–4V for orthopedic implants using sol–gel method. Acta Biomater. 2008, 4, 569–76.

[330] Ng BS, Annergren I, Soutar AM, Khor KA, Jarfors AEW. Characterisation of a duplex TiO2/CaP coating on Ti6Al4V for hard tissue replacement. Biomaterials. 2005, 26, 1087–95.

[331] Bedi RS, Beving DE, Zanello LP, Yan Y. Biocompatibility of corrosion-resistant zeolite coatings for titanium alloy biomedical implants. Acta Biomater. 2009, 5, 3265–71.

[332] Yao Z, Jiang Y, Jia F, Jiang Z, Wang F. Growth characteristics of plasma electrolytic oxidation ceramic coatings on Ti-6Al-4V alloy. Appl Surf Sci. 2008, 254, 4084–91.

[333] Moskalewicz T, Seuss S, Boccaccini AR. Microstructure and properties of composite polyetheretherketone/Bioglass® coatings deposited on Ti-6Al-7Nb alloy for medical applications. Appl Surf Sci. 2013, 273, 62–67.

[334] Peddi L, Brow RK, Brown RF. Bioactive borate glass coatings for titanium alloys. J Mater Sci Mater Med. 2008, 19, 3145–52.

[335] Stan GE, Popa AC, Galca AC, Aldica G, Ferreira JMF. Strong bonding between sputtered bioglass-ceramic films and Ti-substrate implants induced by atomic inter-diffusion post-deposition heat-treatments. Appl Surf Sci. 2013, 280, 530–38.

[336] Hee AC, Choudhury D, Nine MJ, Osman NAA. Effects of surface coating on reducing friction and wear of orthopaedic implants. Sci Technol Adv Mater. 2014, 15, 1–21.

[337] Lee C-K. Fabrication, characterization and wear corrosion testing of bioactive hydroxyapatite/nano-TiO2 composite coatings on anodic Ti-6Al-4V substrate for biomedical applications. Mater Sci Eng B. 2012, 177, 810–18.

[338] Han JY, Yu ZT, Zhou L. Hydroxyapatite/titania composite bioactivity coating processed by sol–gel method. Appl Surf Sci. 2008, 255, 455–58.

[339] Wei D, Zhou Y, Jia D, Wang Y. Chemical treatment of TiO2-based coatings formed by plasma electrolytic oxidation in electrolyte containing nano-HA, calcium salts and phosphates for biomedical applications. Appl Surf Sci. 2008, 254, 1775–82.

[340] Lima RS, Dimitrievska S, Bureau MN, Marple BR, Petit A, Mwale F, Antoniou J. HVOF-Sprayed Nano TiO2-HA Coatings Exhibiting Enhanced Biocompatibility. J Thermal Spray Technol. 2010, 19, 336–43.

[341] Evis Z, Doremus RH. Hot-pressed hydroxyapatite/monoclinic zirconia comnposites with improved mechanical properties. J Mater Sci. 2007, 42, 2426–31.

[342] Tanaskovic D, Jokic B, Socol G, Popescu A, Mihailescu IN, Petrovic R, Janackovic D. Synthesis of functionally graded bioactive glass-apatite multistructures on Ti substrates by pulsed laser deposition. Appl Surf Sci. 2007, 254, 1279–82.

[343] Pei X, Wang J, Wan Q, Kang L, Xiao M, Bao H. Functionally graded carbon nanotubes/ hydroxyapatite composite coating by laser cladding. Surf Coat Technol. 2011, 205, 4380–87.

[344] Lin K, Zhang M, Zhai W, Qu H, Chang J. Fabrication and Characterization of Hydroxy- apatite/ Wollastonite Composite Bioceramics with Controllable Properties for Hard Tissue Repair. J Amer Ceramic Soc. 2011, 94, 99–105.

[345] Pang SX, Zhitomirsky I. Electrophoretic deposition of composite hydroxyapatite– Chitosan– heparin coatings. J Mater Process Technol. 2009, 209, 1597–606.

[346] Rath PC, Singh BP, Besra L, Bhattacharjee S. Multiwalled Carbon Nanotubes Reinforced Hydroxyapatite-Chitosan Composite Coating on Ti Metal: Corrosion and Mechanical Properties. J Amer Ceramic Soc. 2012, 95, 2725–31.

[347] Yang -C-C, Huang C-Y, Lin -C-C, Yen S-K. Electrolytic Deposition of Collagen/HA Composite on Post HA/TiO2 Coated Ti6Al4V Implant AlloyElectrochemical Synthesis and Engineering. J Electrochem Soc. 2011, 158, E13–20.

[348] Xia Z, Yu X, Wei M. Biomimetic collagen/apatite coating formation on Ti6Al4V substrates. J Biomed Mater Res B. 2012, 100B, 871–81.

[349] Sugata Y, Sotome S, Yuasa M, Hirano M, Shinomiya K, Okawa A. Effects of the systemic administration of alendronate on bone formation in a porous hydroxyapatite/collagen composite and resorption by osteoclasts in a bone defect model in rabbits. J Bone Joint Surg. 2011, 93-B, 510–16.

[350] Du C, Schneider GB, Zaharias R, Abbott C, Seabold D, Stanford C, Moradian-Oldak J. Apatite/ amelogenin coating on titanium promotes osteogenic gene expression. J Dent Res. 2005, 84, 1070–74.

[351] Wang G, Yang H, Li M, Lu S, Chen X, Cai X. The use of silk fibroin/ hydroxyapatite composite co-cultured with rabbit bone-marrow stromal cells in the healing of a segmental bone defect. J Bone Joint Surg. 2010, 92-B, 320–25.

[352] Zhang Z, Gu B, Zhang W, Kan G, Sun J. The enhanced characteristics of osteoblast adhesion to porous Zinc–TiO2 coating prepared by plasma electrolytic oxidation. Appl Surf Sci. 2012, 258, 6504–11.

[353] Dastjerdi R, Montazer M, Shahsavan S. A novel technique for producing durable multifunctional textiles using nanocomposite coating. Colloid Surf B. 2010, 81, 32–41.

[354] Cho J, Schaab S, Roether JA, Boccaccini AR. Nanostructured carbon nanotube/TiO2 composite coatings using electrophoretic deposition (EPD). J Nanopart Res. 2008, 10, 99–105.

[355] Cheng S, Wei D, Zhou Y. Formation and structure of sphene/titania composite coatings on titanium formed by a hybrid technique of microarc oxidation and heat-treatment. Appl Surf Sci. 2011, 257, 3404–11.

[356] Wang X, Wu G, Zhou B, Shen J. Thermal annealing effect on optical properties of binary TiO2-SiO2 Sol-Gel coatings. Materials. 2013, 6, 76–84.

[357] Hu H, Qiao Y, Meng F, Liu X, Ding C. Enhanced apatite-forming ability and cytocompatibility of porous and nanostructured TiO2/CaSiO3 coating on titanium. Colloid Surf B. 2013, 101, 83–90.

[358] Pardun K, Treccani L, Volkmann E, Li Destri G, Marletta G, Streckbein P, Heiss C, Rezwan K. Characterization of wet powder-sprayed zirconia/calcium phosphate coating for dental implants. Clin Implant Dent Relat Res. 2015, 17, 186–98.

[359] Kumar AM, Rajendran N. Influence of zirconia nanoparticles on the surface and electrochemical behaviour of polypyrrole nanocomposite coated 316L SS in simulated body fluid. Surf Coat Technol. 2012, 213, 155–66.

[360] Gazia R, Mandracci P, Mussano F, Carossa S. AlNx and a-SiOx coatings with corrosion resistance properties for dental implants. Surf Coat Technol. 2011, 206, 1109–15.

[361] Sokołowska A, Rudnicki J, Niedzielski P, Boczkowska A, Bogusławski G, Wierzchoń T, Mitura S. TiN-NCD composite coating produced on the Ti6Al4V alloy for medical applications. Surf Coat Technol. 2005, 200, 87–89.

[362] Gurappa I. Development of appropriate thickness ceramic coatings on 316 L stainless steel for biomedical applications. Surf Coat Technol. 2002, 161, 70–78.

[363] Martorana S, Fedele A, Mazzocchi M, Bellosi A. Surface coatings of bioactive glasses on high strength ceramic composites. Appl Surf Sci. 2009, 255, 6679–85.

[364] Nelson GM, Nychka JA, McDonald AG. Flame spray deposition of titanium alloy-bioactive glass composite coatings. J Thermal Spray Technol. 2011, 20, 1339–51.

[365] Balasubramanian S, Ramadoss A, Kobayashi A, Muthirulandi J. Nanocomposite Ti-Si-MN coatings deposited by reactive dc magnetron sputtering for biomedical applications. J Am Ceramic Soc. 2012, 95, 2746–52.

[366] Avila G, Misch K, Galindo-Moreno P, Wang H-L. Implant surface treatment using biomimetic agents. Implant Dent. 2009, 18, 17–26.

[367] Bougas K, Jimbo R, Xue Y, Mustafa K, Wennerberg A. Novel implant coating agent promotes gene expression of osteogenic markers in rats during early osseointegration. Int J Biomater. 2012, doi: 10.1155/2012/579274.

[368] Jemt T, Johansson J. Implant treatment in the edentulous maxillae: A 15-year follow-up study on 76 consecutive patients provided with fixed prostheses. Clinical Implant Dent Related Res. 2006, 8, 61–69.

[369] Jemt T. Single implants in the anterior maxilla after 15 years of follow-up: Comparison with central implants in the edentulous maxilla. Int J Prosthodont. 2008, 21, 400–08.

[370] Åstrand P, Ahlqvist J, Gunne J, Nilson H. Implant treatment of patients with edentulous jaws: A 20-year follow-up. Clinical Implant Dent Related Res. 2008, 10, 207–17.

[371] Fröjd V, De Paz LC, Andersson M, Wennerberg A, Davies JR, Svensäter G. In situ analysis of multispecies biofilm formation on customized titanium surfaces. Molec Oral Microbiol. 2011, 26, 241–52.

[372] Thorey F, Menzel H, Lorenz C, Gross G, Hoffmann A, Windhagen H. Osseointegration by bone morphogenetic protein-2 and transforming growth factor beta2 coated titanium implants in femora of New Zealand white rabbits. Indian J Orthopaedics. 2011, 45, 57–62.

[373] Abtahi J, Tengvall P, Aspenberg P. A bisphosphonate-coating improves the fixation of metal implants in human bone: A randomized trial of dental implants. Bone. 2012, 50, 1148–51.

[374] Dorozhkin SV. Calcium orthophosphate coatings, films and layers. Progress in Biomat. 2012, doi: 10.1186/2194-0517-1-1.

[375] He HW, Liu ML, Zhu ZL, Yang MZ, Li QL, Chen ZQ. Influence of surface morphology of cpTi on the adsorption and attachment of collagen/chitosan. Appl Surf Sci. 2008, 255, 509–11.

[376] Patino MG, Neiders ME, Andreana S, Noble B, Cohen RE. Collagen as an implantable material in medicine and dentistry. I Oral Implant. 2002, 28, 220–25.

[377] Müller R, Abke J, Schnell E, Scharnweber D, Kujat R, Englert C, Taheri D, Nerlich M, Angele P. Influence of surface pretreatment of titanium- and cobalt-based biomaterials on covalent immobilization of fibrillar collagen. Biomaterials. 2006, 27, 4059–68.

[378] Rammelt S, Illert T, Bierbaum S, Scharnweber D, Zwipp H, Schneiders W. Coating of titanium implants with collagen, RGD peptide and chondroitin sulfate. Biomaterials. 2006, 27, 5561–71.

[379] Müller R, with 10 co-authors. Surface engineering of stainless steel materials by covalent collagen immobilization to improve implant biocompatibility. Biomaterials. 2005, 26, 6962–72.

[380] Glattauer V, Briggs KL, Zappe S, Ramshaw JAM, Truong YB. Collagen-based layer-by-layer coating on electrospun polymer scaffolds. Biomaterials. 2012, 33, 9198–204.

[381] Yang X, Jiang B, Huang Y, Tian Y, Chen H, Chen J, Yang B. Collagen nanofilm immobilized on at surfaces by electrodeposition method. J Biomed Mater Res B. 2009, 90-B, 608–13.

[382] Elias CN, Gravina PA, De Filho CS, De Paula Nascente PA. Preparation of bioactive titanium surfaces via fluoride and fibronectin retention. Int J Biomater. 2012, doi: 10.1155/2012/290179.

[383] Petrie TA, Raynor JE, Reyes CD, Burns KL, Collard DM, García AJ. The effect of integrin-specific bioactive coatings on tissue healing and implant osseointegration. Biomaterials. 2008, 29, 2849–57.

[384] Ku Y, Chung CP, Jang J-H. The effect of the surface modification of titanium using a recombinant fragment of fibronectin and vitronectin on cell behavior. Biomaterials. 2005, 26, 5153–57.

[385] Erli H-J, Rüger M, Ragoß C, Jahnen-Dechent W, Hollander DA, Paar O, Von Walter M. The effect of surface modification of a porous TiO2/perlite composite on the ingrowth of bone tissue in vivo. Biomaterials. 2006, 27, 1270–76.

[386] Simon Z, Deporter DA, Pilliar RM, Clokie CM. Heterotopic bone formation around sintered porous-surfaced Ti-6Al-4V implants coated with native bone morphogenetic proteins. Implant Dent. 2006, 15, 265–74.

[387] Guillot R, Gilde F, Becquart P, Sailhan F, Lapeyrere A, Logeart-Avramoglou D, Catherine Picart C. The stability of BMP loaded polyelectrolyte multilayer coatings on titanium. Biomaterials. 2013, 34, 5737–46.

[388] Nayak S, Dey T, Naskar D, Kundu SC. The promotion of osseointegration of titanium surfaces by coating with silk protein sericin. Biomaterials. 2013, 34, 2855–64.

[389] Allenstein U, Ma Y, Arabi-Hashemi A, Zink M, Mayr SG. Fe–Pd based ferromagnetic shape memory actuators for medical applications: Biocompatibility, effect of surface roughness and protein coatings. Acta Biomater. 2013, 9, 5845–53.

[390] Waterhouse A, Yin Y, Wise SG, Bax DV, McKenzie DR, Bilek MMM, Weiss AS, Ng MKC. The immobilization of recombinant human tropoelastin on metals using a plasma-activated coating to improve the biocompatibility of coronary stents. Biomaterials. 2010, 31, 8332–40.

[391] Kazemzadeh-Narbat M, Lai BFL, Ding C, Kizhakkedathu JN, Hancock REW, Wang R. Multilayered coating on titanium for controlled release of antimicrobial peptides for the prevention of implant-associated infections. Biomaterials. 2013, 34, 5969–77.

[392] Meyers SR, Khoo X, Huang X, Walsh EB, Grinstaff MW, Kenan DJ. The development of peptide-based interfacial biomaterials for generating biological functionality on the surface of bioinert materials. Biomaterials. 2009, 30, 277–86.

[393] Yazici H, Fong H, Wilson B, Oren EE, Amos FA, Zhang H, Evans JS, Snead ML, Sarikaya M, Tamerler C. Biological response on a titanium implant-grade surface functionalized with modular peptides. Acta Biomater. 2013, 9, 5341–52.

[394] Werner S, Huck O, Frisch B, Vautier D, Elkaim R, Voegel J-C, Brunel G, Tenenbaum H. The effect of microstructured surfaces and laminin-derived peptide coatings on soft tissue interactions with titanium dental implants. Biomaterials. 2009, 30, 2291–301.

[395] Kazemzadeh-Narbat M, Kindrachuk J, Duan K, Jenssen H, Hancock REW, Wang R. Antimicrobial peptides on calcium phosphate-coated titanium for the prevention of implant-associated infections. Biomaterials. 2010, 31, 9519–26.

[396] Busscher HJ, Rinastiti M, Siswomihardjo W, Van Der Mei HC. Biofilm formation on dental restorative and implant materials. J Dent Res. 2010, 89, 657–65.

[397] Reyes CD, Petrie TA, Burns KL, Schwartz Z, García AJ. Biomolecular surface coating to enhance orthopaedic tissue healing and integration. Biomaterials. 2007, 28, 3228–35.

[398] Leedy MR, Martin HJ, Norowski PA, Jennings JA, Haggard WO, Bumgardner JD. Use of chitosan as a bioactive implant coating for bone-implant applications; chitosan for biomaterials II. Ad in Polymer Sci. 2011, 244, 129–65.

[399] Darder M, Aranda P, Ruiz-Hitzky E. Chitosan-clay bio-nanocomposites. In: Avérous L, et al., ed. Environmental Silicate Nano-Biocomposites. London: Springer-Verlag, 2012, 365–91.

[400] Bumgardner JD, Chesnutt BM, Yuan Y, Yang Y, Appleford M, Oh S, McLaughlin R, Elder SH, Ong JL. The integration of chitosan-coated titanium in bone: An in vivo study in rabbits. Implant Dent. 2007, 16, 66–79.

[401] Frosch K-H, Drengk A, Krause P, Viereck V, Miosge N, Werner C, Schild D, Stürmer EK, Stürmer KM. Stem cell-coated titanium implants for the partial joint resurfacing of the knee. Biomaterials. 2006, 27, 2542–49.

[402] Hu Y, Cai K, Luo Z, Zhang R, Yang L, Deng L, Jandt KD. Surface mediated in situ differentiation of mesenchymal stem cells on gene-functionalized titanium films fabricated by layer-by-layer technique. Biomaterials. 2009, 30, 3626–35.

[403] Achneck HE, with 20 coauthors. The biocompatibility of titanium cardiovascular devices seeded with autologous blood-derived endothelial progenitor cells: EPC-seeded antithrombotic Ti Implants. Biomaterials. 2011, 32, 10–18.

[404] Guo K-T, Scharnweber D, Schwenzer B, Ziemer G, Wendel HP. The effect of electrochemical functionalization of Ti-alloy surfaces by aptamer-based capture molecules on cell adhesion. Biomaterials. 2007, 28, 468–74.

[405] Zarghami V, Ghorbani M, Bagheri KP, Shokgozar MA. Prevention the formation of biofilm on orthopedic implants by melittin thin layer on chitosan/bioactive glass/vancomycin coatings. J Mater Sci Mater Med. 2021, 32, 75:, doi: 10.1007/s10856-021-06551-5.

[406] Subramani K, Mathew RT. Chapter 6 – Titanium surface modification techniques for dental implants – from microscale to nanoscale. In: Emerging Nanotechnologies in Dentistry: Processes, Materials and Applications. Micro and Nano Technologies. 2012, 85–102.

[407] El-Banna A, Bissa MW, Khurshid Z, Zohaib S, Asiri FYI, Zafar MS. 4 – Surface modification techniques of dental implants. In: Dental Implants: Materials, Coatings, Surface Modifications and Interfaces with Oral Tissues. Woodhead Publishing Series in Biomaterials, 2020, 49–68.

[408] An YH, Alvi FI, Kang Q, Laberge M, Drews MJ, Zhang J, Matthews MA, Arciola CR. Effects of sterilization on implant mechanical property and biocompatibility. Int J Artif Organs. 2005, 28, 1126–37.

[409] Kasemo B, Lausmaa J. Biomaterial and implant surfaces: On the role of cleanliness, contamination, and preparation procedures. J Biomed Mater Res. 1988, 22, S145–58.

[410] Mavrogenis AF, Dimitriou R, Parvizi J, Babis GC. Biology of implant osseointegration. J Musculoskeletal Neuronal Interact. 2009, 9, 61–71.

[411] Schwarz F, with 13 coauthors. Potential of chemically modified hydrophilic surface characteristics to support tissue integration of titanium dental implants. J Biomed Mater Res B. 2009, 88B, 544–57.

[412] Park JH, Olivares-Navarrete R, Baier RE, Meyer AE, Tannenbaum R, Boyan BD, Schwartz Z. Effect of cleaning and sterilization on titanium implant surface properties and cellular response. Acta Biomater. 2012, 8, 1966–75.

[413] Kommmineni NK, Dappili SRR, Prathyusha P, Vanaja P, Reddy KVKK, Vasanthi D. Comparative evaluation of sterilization efficacy using two methods of sterilization for rotary endodontic files: An in vitro study. J Dr NTR Univ Health Sci. 2016, 5, 142–46.

[414] Dunn D. Reprocessing single-use devices – The ethical dilemma. AORN J. 2002, 75, 989–99.

[415] Dunn D. Reprocessing single-use devices – Regulatory roles. AORN J. 2002, 76, 100–08.

[416] Enabulele JE, Omo JO. Sterilization in endodontics: Knowledge, attitude, and practice of dental assistants in training in Nigeria – A cross-sectional study. Saudi Endod J. 2018, 8, 106–10.

[417] Mustafa M. Knowledge, attitude and practice of general dentists towards sterilization of endodontic files: A cross – Sectional study. Indian J Sci Tech. 2016, 9, 11–16.

[418] Limbhore M, Saraf A, Medha A, Jain D, Mattigatti S, Mahaparale R. Endodontic hand instrument sterilization procedures followed by dental practitioners. Unique J Med Dent Sci. 2014, 2, 106–11.

[419] Ferreira MM, Michelotto AL, Alexandre AR, Morgantio R, Carnillo EV. Endodontic files: Sterilize or discard. Dent Press Endod. 2012, 2, 46–51.

[420] Aslam A, Panuganti V, Nanjundarethy JK, Halappa M, Krishna VH. Knowledge and attitude of endodontic post graduate students towards sterilization of endodontic files: A cross – Sectional study. Saudi Endod J. 2016, 4, 18–22.

[421] Linsuwamont P, Parashos P, Messor HH. Cleaning of rotary nickel titanium endodontic instruments. Int Endod J. 2004, 37, 19–28.

[422] Takkar H, Kumar SA, Kumar MS, Takkar S. Contribution of endodontic field in clean India campaign by the dentists – Survey in Sri Ganganagar District, Rajanthan. J Adv Med Dent Sci Res. 2015, 3, 23–28.

[423] Martin JY, Dean DD, Chran DL, Simpson J, Boyan BD, Schwartz Z. Proliferation, differentiation, and protein synthesis of human osteoblast-like cells (MG63) cultured on previously used titanium surfaces. Clin Oral Implan Res. 1996, 7, 27–37.

[424] Kilpadi DV, Weimer JJ, Lemons JE. Effect of passivation and dry heat-sterilization on surface energy and topography of unalloyed titanium implants. Colloid Surf A. 1998, 135, 89–101.

[425] Oshida Y, Tominaga T. Nickel-Titanium Materials. De Gruyter Pub, 2020.

[426] Premnath V, Harris WH, Jasty M, Merrill EW. Gamma sterilization of UHMWPE articular implants: An analysis of the oxidation problem. Biomaterials. 1996, 17, 1741–53.

[427] Qiu QQ, Connor J. Effects of gamma-irradiation, storage and hydration on osteoinductivity of DBM and DBM/AM composite. J Biomed Mater Res A. 2008, 87A, 373–79.

[428] Pandiyaraj KN, Selvarajan V, Pavese M, Falaras P, Tsousleris D. Investigation on surface properties of TiO_2 films modified by DC glow discharge plasma. Curr Appl Phys. 2009, 9, 1032–37.

[429] Boyan BD, Lossdoerfer S, Wang L, Zhao G, Lohmann CH, Cochran DL, Schwartz Z. Osteoblasts generate an osteogenic microenvironment when grown on surfaces with rough microtopographies. Eur Cells Mater. 2003, 6, 22–27.

[430] Olivares-Navarrete R, Hyzy SL, Hutton DL, Erdman CP, Wieland M, Boyan BD, Schwartz Z. Direct and indirect effects of microstructured titanium substrates on the induction of mesenchymal stem cell differentiation towards the osteoblast lineage. Biomaterials. 2010, 31, 2728–35.

[431] Pegueroles M, Gil FJ, Planell JA, Aparicio C. The influence of blasting and sterilization on static and time-related wettability and surface-energy properties of titanium surfaces. Surf Coat Tech. 2008, 202, 3470–79.

[432] Lausmaa J, Kasemo B, Mattsson H. Surface spectroscopic characterization of titanium implant materials. Appl Surf Sci. 1990, 44, 133–46.

[433] Esposito M, Lausmaa J, Hirsch JM, Thomsen P. Surface analysis of failed oval titanium implants. J Biomed Mater Res. 1999, 48, 559–68.

[434] Zhao G, Raines AL, Wieland M, Schwartz Z, Boyan BD. Requirement for both micron- and submicron scale structure for synergistic responses of osteoblasts to substrate surface energy and topography. Biomaterials. 2007, 28, 2821–29.

[435] Zuleta FA, Velasquez P, De Aza PN. Effect of various sterilization methods on the bioactivity of laser ablation pseudowollastonite coating. J Biomed Mater Res B Appl Biomater. 2010, 94B, 399–405.

[436] Rai R, Tallawi M, Roether JA, Detsch R, Barbani N, Rosellini E, Kaschta J, Schubert DW, Boccaccini AR. Sterilization effects on the physical properties and cytotoxicity of poly (glycerol sebacate). Mater Lett. 2013, 105, 32–35.

[437] Holy CE, Cheng C, Davies JE, Shoichet M. Optimizing the sterilization of PLGA scaffolds for use in tissue engineering. Biomaterials. 2000, 22, 25–31.

[438] Lerouge S, Tabrizian M, Wertheimer MR, Marchand R, Yahia L. Safety of plasma-based sterilization: Surface modifications of polymeric medical devices induced by Sterrad and Plazlyte processes. Biomed Mater Eng. 2002, 12, 3–13.

[439] Thierry B, Tabrizian M, Savadogo O, Yahia L. Effects of sterilization processes on NiTi alloy: Surface characterization. J Biomed Mater Res. 2000, 49, 88–98.

[440] Scherer U, Stoetzer M, Ruecker M, Gellrich N-C, Von See C. Template-guided vs. non-guided drilling in site preparation of dental implants. Clin Oral Investig. 2015, 19, 1339–46.

[441] Fortin T, Isidori M, Blanchet E, Perriat M, Bouchet H, Coudert JL. An image-guided system—drilled surgical template and trephine guide pin to make treatment of completely edentulous patients easier: A clinical report on immediate loading. Clin Implant Dent Relat Res. 2004, 6, 111–19.

[442] D'haese J, Van De Velde T, Komiyama A, Hultin M, De Bruyn H. Accuracy and complications using computer-designed Stereolithographic surgical guides for Oral rehabilitation by means of dental implants: A review of the literature. Clin Implant Dent Relat Res. 2012, 14, 321–35.

[443] Moslehifard E, Nokar S. Designing a custom made gauge device for application in the access hole correction in the dental implant surgical guide. Indian Prosthodont Soc. 2012, 12, 123–29.

[444] De Vico G, Spinelli D, Bonino M, Schiavetti R, Pozzi A, Ottria L. Computer-assisted virtual treatment planning combined with flapless surgery and immediate loading in the Rehabilitaion of partial Edentulie. Oral Implantol. 2012, 5, 3–10.

[445] Sennhenn-Kirchner S, Weustermann S, Mergeryan H, Jacobs HG, Von Zepelin MB, Kirchner B. Preoperative sterilization and disinfection of drill guide templates. Clin Oral Investig. 2008, 12, 179–87.

[446] Török G, Gombocz P, Bognár E, Nagy P, Dinya E, Kispélyi B, Hermann P. Effects of disinfection and sterilization on the dimensional changes and mechanical properties of 3D printed surgical guides for implant therapy – Pilot study. BMC Oral Health. 2020, 20, 19, doi.org, doi: 10.1186/s12903-020-1005-0.

[447] https://www.stratasys.com/materials/search/biocompatible.

[448] Scarano A, Petrini M, Mastrangelo F, Noumbissi S, Lorusso F. The effects of liquid disinfection and heat sterilization processes on implant drill roughness: Energy dispersion X-ray microanalysis and infrared thermography. J Clin Med. 2020, 9, 1019, doi: 10.3390/jcm9041019.

[449] Oshida Y. Magnesium Materials. De Gruyter Pub., 2021.

[450] Seitz J-M, Collier K, Wulf E, Bormann D, Angrisani N, Meyer-Lindenberg A, Bach F-W. The effect of different sterilization methods on the mechanical strength of magnesium based implant materials. Adv Eng Mater. 2011, 13, 1146–51.

[451] Thierry B, Tabrizian M, Savadogo O, Yahia LH. Effects of sterilization processes on NiTi alloy: Surface characterization. J Biomed Mater Res. 2000, 49, 88–98.

[452] Spagnuolo G, Ametrano G, D'Antò V, Rengo C, Simeone M, Riccitiello F, Amato M. Effect of autoclaving on the surfaces of TiN -coated and conventional nickel-titanium rotary instruments. Int Endod J. 2012, 45, 1148–55.

[453] Yılmaz K, Uslu G, Özyürek T. Effect of multiple autoclave cycles on the surface roughness of HyFlex CM and HyFlex EDM files: An atomic force microscopy study. Clin Oral Investig. 2018, 22, 2975–80.

[454] Cai JJ, Tang XN, Ge JY. Effect of irrigation on surface roughness and fatigue resistance of controlled memory wire nickel-titanium instruments. Int Endod J. 2017, 50, 718–24.

[455] Mize SB, Clement DJ, Pruett JP, Carnes DL Jr. Effect of sterilization on cyclic fatigue of rotary nickel-titanium endodontic instruments. J Endod. 1998, 24, 843–47.

[456] Bergeron BE, Mayerchak MJ, Roberts MJ, Jeansonne BG. Multiple Autoclave Cycle Effects on Cyclic Fatigue of Nickel-Titanium Rotary Files Produced by New Manufacturing Methods. J Endodontics. 2011, 37, 72–74.

[457] Plotino G, Costanzo A, Grande NM, Petrovic R, Testarelli L, Gambarini G. Experimental Evaluation on the Influence of Autoclave Sterilization on the Cyclic Fatigue of New Nickel-Titanium Rotary Instruments. J Endodontics. 2012, 38, 222–25.

[458] Janardhanan S, Kanisseri M, John MK. Influence of sterilization on mechanical properties and fatigue resistance of nickel-titanium rotary endodontic instruments: An *in vitro* study. Int J Oral Care Res. 2018, 6, 5–11.

[459] Pedullà E, Benites A, La Rosa GM, Plotino G, Grande NM, Rapisarda E, Generali L. Cyclic fatigue resistance of heat-treated nickel-titanium instruments after immersion in sodium hypochlorite and/or sterilization. J Endod. 2018, 44, 648–53.

[460] Viana AC, Gonzalez BM, Buono VT, Bahia MG. Influence of sterilization on mechanical properties and fatigue resistance of nickel-titanium rotary endodontic instruments. Int Endod J. 2006, 39, 709–15.

[461] Özyürek T, Yılmaz K, Uslu G. The effects of autoclave sterilization on the cyclic fatigue resistance of ProTaper Universal, ProTaper Next, and ProTaper Gold nickel-titanium instruments. Restor Dent Endod. 2017, 42, 301–08.

[462] Canalda-Sahli C, Brau-Aguadé E, Sentís-Vilalta J. The effect of sterilization on bending and torsional properties of K-files manufactured with different metallic alloys. Int Endod J. 1998, 31, 48–52.
[463] Hilt BR, Cunningham CJ, Shen C, Richards N. Torsional properties of stainless-steel and nickel-titanium files after multiple autoclave sterilizations. J Endod. 2000, 26, 76–80.
[464] O'Hoy PYZ, Messer HH, Palamara JEA. The effect of cleaning procedures on fracture properties and corrosion of NiTi files. Int Endod J. 2003, 36, 724–32.
[465] Smith SA, Gause B, Plumley DL, Drexel MJ. Irradiation-assisted stress-corrosion cracking of nitinol during eBeam sterilization. J Mater Eng Perform. 2012, 21, 2638–42.
[466] Norwich DW. Fracture of Polymer-coated nitinol during gamma sterilization. J Mater Eng Perform. 2012, 21, 2618–21.
[467] McKellop H, Shen FW, Lu B, Campbell P, Salovey R. Effect of sterilization method and other modifications on the wear resistance of acetabular cups made of ultra-high molecular weight polyethylene. A hip-simulator study. J Bone Joint Surg Am. 2000, 82-A, 1708–25.
[468] Goldman M, Pruitt L. Comparison of the effects of gamma radiation and low temperature hydrogen peroxide gas plasma sterilization on the molecular structure, fatigue resistance, and wear behavior of UHMWPE. J Biomed Mater Res. 1998, 40, 378–84.
[469] Linkow LI, Giauque F. Introduction to plasma glow discharge treatment of dental implants. Implant Soc. 1993, 4, 15–16.
[470] Serro AP, Saramago B. Influence of sterilization on the mineralization of titanium implants induced by incubation in various biological model fluids. Biomaterials. 2003, 24, 4749–60.

Chapter 6
Advanced fabrication technologies for implantable device

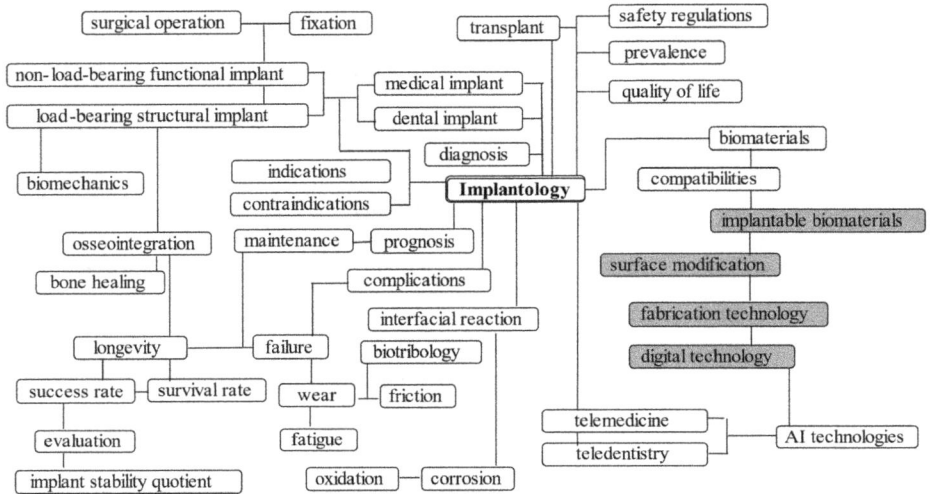

Due to the ever-increasing risk of aging population, there are a variety of unforeseen risks jeopardizing individual's quality of life, including a higher incidence of coronary artery disease, diabetes, systemic disease, missing teeth, and injury-related frailty. Hence, there is a need for more reliable and safer treatments to recover original biofunctionality or enhance rehabilitation therapy. In this chapter, the advanced technologies involved in manufacturing implantable devices are presented and reviewed; in particular, the additive electrical, chemical, and physical processes are currently being employed [1]. Biomanufacturing integrates life science and engineering fundamentals to produce biocompatible products enhancing the health-related quality of life.

6.1 In general

ASTM (American Society for Testing and Materials) subcommittee defined the term additive manufacturing (AM) as "a process of joining materials to make objects from 3D model data, usually layer upon layer, as opposed to subtractive manufacturing methodologies." Additionally, synonyms are expressed as additive fabrication, additive processes, additive techniques, additive layer manufacturing, layer manufacturing,

https://doi.org/10.1515/9783110740134-007

and freeform fabrication [2]. Opposing to structural components and devices, any functional structures and devices in either engineering or medical areas possess complex shapes and their sizes, depending on application protocols and users (or patients). Therefore, the production of, for example, biomedical implants using conventional technologies always accompanies with significant time, material, and energy costs. Despite the fact that the type of the production process or raw material characteristics may be different in AM techniques, the fabrication consists of eight similar stages of all types throughout the development process, as shown in Figure 6.1 [3], where the STL stands for STereoLithography. The AM technologies can be broadly divided into three types. The first type is sintering by which the material is heated without being liquefied to create complex objects. Direct metal laser sintering (DMLS) uses metal powder, whereas selective laser sintering (SLS) uses a laser on thermoplastic powders so that the particles stick together. The second AM technology melts the materials, including DMLS which uses a laser to melt layers of metal powder and electron beam melting (EBM) which uses electron beams to melt the powders. And the third type of AM technology is stereolithography (SLA), which uses a process called photopolymerization, whereby an ultraviolet (UV) laser is fired into a vat of photopolymer resin to create torque-resistant ceramic parts that are able to endure extreme temperatures [4, 5]. Referring to Figure 6.1, the AM is a technique for fabricating parts in precise geometry using computer-aided design (CAD) and computer-aided manufacturing (CAM) [6]. In each AM technique, the three-dimensional (3D) model designed in CAD software is converted to STL format, which is a triangular mesh of the object, and then the STL format is sliced into 2D profile layers. Each sliced layer of the model is bonded to the previous layer on the build platform until a 3D part is fabricated [7]. The principal AM technologies are SLS, SLA, fused deposition modeling (FDM), DMLS, and inkjet 3D printing (3DP) techniques [8, 9]. The most significant elements that should be considered in choosing an appropriate AM technology for a particular purpose are accuracy, time, and cost of fabrication. The parameter of accuracy refers to the thickness of the layers and the system of consolidation, and since AM techniques are tool-free fabrication methods, time of production can outweigh increased fabrication costs per item [9, 10].

Additive electrochemical and physical processes, through which physical objects are created from computer-generated models, emerged in the 1980s. The basic concept of additive fabrication is layer laminate manufacturing, in which 3D structures are formed by laminating thin layers according to the 2D slice data obtained from a 3D model. The main advantages of additive electrochemical and physical techniques are the capacity to rapidly produce very intricated 3D models and the ability to use various raw materials. When combined with clinical imaging data, these fabrication techniques can be used to produce constructs customized to the shape of the defect or injury. Some processes operate at room temperature, thus allowing cell encapsulation and biomolecule incorporation without significantly affecting viability. Basically, AM technologies can include electrospinning, SLA, laser sintering and melting processes, EBM process, extrusion-based processes, or inkjet printing processes [11].

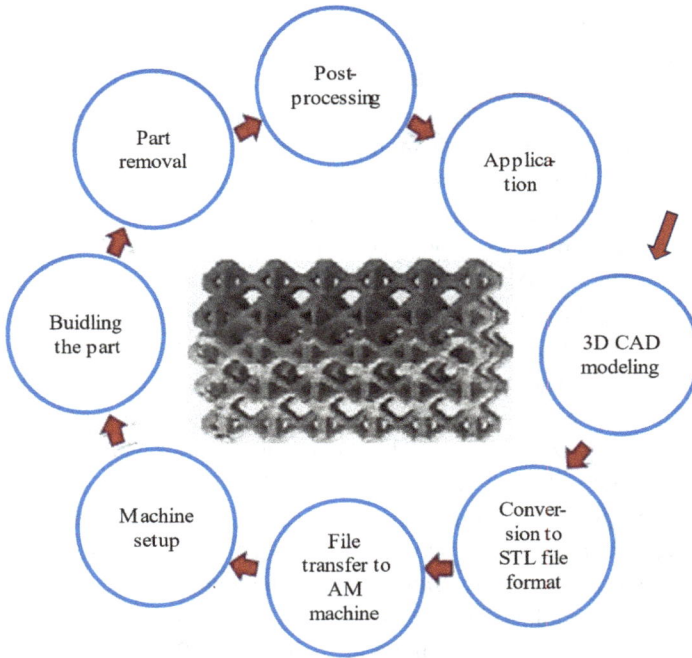

Figure 6.1: Process sequence of additive manufacturing [3].

Although numerous AM technologies are available for medical and tissue engineering purposes, some techniques involve the printing of live cells along with other materials, known as bioprinting [12]. Such load-bearing technologies can include powder bed fusion (PBF), binder jetting, material extrusion, material jetting, and vat polymerization [13–15], although some of those mentioned are overlapped. Briefly, the PBF methods use either electron beam or laser to selectively consolidate material powder. These techniques are known as EBM, selective laser melting (SLM), and SLS. Both SLM and EBM fully melt and fuse the powder material, while SLS heats it to the point that the powder can fuse together on a molecular level. All PBF techniques involve spreading material powder over the previous layers. Binder jetting: the binder jetting technique is similar to the PBF technique in that it utilizes material powder that is spread over previous layers. However, unlike PBF, which melts and fuses the powder, this technique uses a binder as an adhesive for its consolidation in layers of defined cross sections. Material extrusion: the material extrusion technique, also known as FDM, pushes the raw material in the form of polymer wires through a heated nozzle. The material is deposited as polymer roads that are arranged to define a cross section of the part. These lines are then stacked in a layer-by-layer fashion. Material jetting (MJ): the MJ technique uses a liquid photopolymer resin that is cured with UV or near-UV light. Similar to the material extrusion technique, the material is deposited from a nozzle that moves horizontally across the build platform. The material is then cured, defining a cross section of the part. Individual cross sections are consolidated in a layer-by-layer fashion as the building platform moves in the vertical direction. Vat polymerization: the vat polymerization technique is similar to the MJ technique inasmuch as it employs photopolymer resins that are cured with UV light in a layer-by-layer fashion. In contrast to MJ, the resin remains in a material vat, where the build platform is submerged. The build platform moves downward (or upward depending on the position of the light source) to create additional layers on top of the previous [13–15].

AM approaches, particularly SLS and 3DP, are simple and adaptable to using a broad range of powders to produce porous ceramics, polymers, metal-based tissues [16, 17], and scaffolds filled with a porous spacer for the ingrowth of blood vessel as well [18].

6.2 Application of laser technologies

6.2.1 Laser surface alloying (LSA)

In the field of laser fabrication, variety of applications can be found. Alloying can be achieved by not only conventional melting/casting method, but it can be also done by solid-state powder metallurgy, mechanical alloying, and laser surface alloying (LSA). LSA is an advanced material processing technology that utilizes the high power of density from a focused laser beam as the source of energy to produce an extremely dense and crack-free structure by heating and melting a surface while injecting alloy elements or compound powder onto the surface [19, 20]. Prior to the laser alloying process, a coating material is deposited onto the base material. The coating material and the base material are then targeted by the laser beam. The laser beam fuses, or alloys, these two materials together. The laser bonding results in new components with highly resistant surfaces against wear at both high and low temperatures. The material surface properties can be influenced by the addition of alloying powders, which form intermetallic compounds via a chemical reaction between the powder and the material. These compounds are a solid phase consisting of two or more metallic elements in precise proportions, generally characterized as hard materials. A thin layer on the surface of the base material is rapidly melted and mixed at the same time as the alloy to be added to form a surface melting layer having a thickness of 10–1,000 μm so that (i) the material remains cool and acts as a heat sink; (ii) the surface of the material forms a surface having a desired depth and chemical composition in a short time; (iii) there is a large temperature gradient across the boundary between the melted surface region and the underlying solid substrate, which results in rapid self-quenching and resolidification; and (iv) the low heat input that can be achieved by laser alloying reduces the size of the heat-affected zone (HAZ), which allows the base material to retain much of its original properties [19–21]. The process is attractive because a wide variety of chemical and microstructural states that can be retained because of the rapid quench from the liquid phase. This will include chemical profiles where the alloyed element is highly concentrated near the atomic surface and decreases in concentration over shallow depths. It also includes uniform profiles where the concentration is the same throughout the entire melted region [19–21]. The minimized HAZ can result in coatings with high hardness and excellent wear resistance so that applications of LSA include items that need increased wear resistance such as certain areas of tooling.

6.2.2 Laser deposition

The laser deposition is another type of solid freeform metal fabrication AM method in which instead of having a powder bed, the powder is selectively deposited and simultaneously melted by a laser beam, resulting in a molten pool on the substrate. The laser/deposition head and/or the substrate can travel in space directions, demonstrating the ability to fabricate 3D objects, directly from their computer representation, by combining the deposition process with CAD/CAM methods similar to those used in rapid prototyping [22, 23]. The most important process parameters in this method that need to be optimized are laser-substrate relative velocity (traverse speed), laser power and beam diameter, hatch spacing, powder feeding rate, and scanning strategy. Laser powder deposition has considerable advantages as compared with alternative processes, namely, the wide size range of parts that can be produced, its versatility in terms of materials that can be used, the possibility of producing monolithic multimaterial parts and functionally graded materials (FGMs), a wide processing window, low power and material consumption, low environmental and health impacts, and the facility of creating parts with excellent properties if the deposition process is adequately controlled. Laser melting and solidification of metallic materials also allow refining their microstructure, dissolving inclusions and precipitates, and forming nonequilibrium supersaturated solid solutions and quasicrystalline and amorphous materials, often with improved properties. Due to these characteristics, a wide range of applications can be found in industries such as aerospace, energy, automotive, and biomedical [22, 23].

Laser deposition was performed to produce nanostructured TiO_2 film [24] and TiO_x nanoparticles [25]. Calcium phosphate (CaP) as well as HA can be deposited by laser treatment [26]. Thin films of calcium hydroxyapatite ($Ca_{10}(PO_4)_6(OH)_2$), was deposited on polished substrates of Ti–6Al–4V [27], and CaP compounds were deposited on acid- and alkali-treated Ti substrate [28], indicating that such laser-treated surfaces enhance the biocompatibility of titanium and its alloys. Laser chemical vapor deposition is a new manufacturing process that holds great potential for the production of small and complex metal and ceramic parts [29]. A complex compound of silver nanoparticles (size 40 nm) combined with polymer blends (polyethylene glycol/poly(lactide-co-glycolide) was deposited through laser-assisted pulsed evaporation method to cover the substrate surface with an antibacterial feature [30]. Besides major surface coating and modification technology involved in physical reaction as described above, there are still various methods available. Mishra et al. [31] investigated synthesizing hydroxyapatite (HA) nanorods by microwave irradiation technique, taking ethylenediaminetetraacetic acid as a complexing reagent. After examining calcinizing in the temperature range from 450 to 900 °C, it was found that the precursor was needed above 700 °C to produce nanorods with ~200 nm. Nanocomposite Ti–Si–N coating was performed by reactive dc magnetron sputtering in a mixture of Ar and N_2 gases onto bioimplantable 316L stainless steel (SS) substrates [32]. It was found, by X-ray diffraction, that the thus-processed Ti–Si–N nanocomposite coatings are mainly composed of amorphous Si_3N_4 and TiN crystals as major constituents along with TiN, TiO_2, and Si_3N_4. For modifying titanium dental implants, Yoshinari et al. [33] employed a dry process on titanium surface, including the ion beam dynamic mixing (thin CaPs), ion implantation (Ca^+, N^+, F^+), titania spraying, ion plating (TiN, alumina), and ion beam mixing (Ag, Sn, Zn, Pt) with Ar^+ to enhance surface compatibility with the host bone tissue, subepithelial connective tissue, and epithelial tissue so that the tissue-compatible surface can be obtained. It was concluded that the dry process surface modification was useful in controlling the physicochemical nature of surfaces, including the surface energy and the surface electrical charge, and in developing tissue-compatible implants [33]. Hydroxyapatite was cold-sprayed to biodegradable Mg alloy and it was reported that biocompatible coatings of the order of 20–30 μm thickness was obtained by preheating at 400 °C prior to the cold spraying [34]. Cai et al. [35] used the layer-by-layer self-assembly technique for improving the surface biocompatibility of titanium films, based on the polyelectrolyte-mediated electrostatic adsorption of chitosan and gelatin. It was suggested that titanium films could be modified with chitosan and gelatin. The electrostatic layer-by-layer self-assembly technique was also employed for producing polymers and enzymes [36]. Since the presence of appropriate porosity is required for tissue ingrowth and for the concurrent degradation of implanted structure, formation of sufficiently large 3D nano-/macroporosity in bone scaffolds is important. Zhang et al. [37] introduced a new technique of combining the slip casting and polymer sponge methods using the slips with different powders to prepare HA scaffolds with bimodal pore sizes. It was reported that (i) the scaffolds were prepared with an open, uniform, and interconnected porous structure with a bimodal pore size of 100–300 μm and (ii) the bimodal porous HA scaffold sintered at 1,200 °C had a large flexural strength of 73.3 MPa and a porosity of 52.5 vol%.

6.2.3 Laser nitriding and oxidation

Laser nitriding and oxidation are another remarkable application of laser technology. Laser nitriding can be described as the irradiation of metal surfaces by short laser pulses in nitrogen containing atmospheres. This may lead to a strong take-up of nitrogen into the metal and nitride formation which can improve the metal's surface properties, for example, the hardness or the corrosion and wear resistance [38]. Commercially pure titanium (cpTi) was treated using Q-switched Nd:YAG laser radiation to control early stage of oxidation so that the oxygen content did not exceed 50 at% (TiO) [39]. CpTi surfaces were subjected to laser nitrization [40, 41]. It was reported that laser-treated cpTi was covered with δ-TiN phase [41] or tetragonal δ'-Ti$_2$N phase [42], and it was mentioned that films containing tetragonal Ti$_2$N phases possess better mechanical properties, corrosion, and wear resistance in comparison with cubic TiN [41]. Similarly, Ti–6Al–4V alloy surface was laser-treated under various aims [42–44]. Excimer laser was employed to improve the pitting corrosion resistance of the alloy and found that pitting corrosion resistance was improved due to TiN precipitates acting as galvanic cathodes at high corrosion potentials [42]. For improving tribological properties, laser diffusion nitriding was conducted [43, 44]. Nitriding and carburizing by laser irradiation in combination with a process gas is a common method to enhance the surface properties of various materials. Carburizing and hybrid nitrocarburizing can be applicable [45].

6.2.4 Selective laser melting (SLM)

SLM is an AM technique that utilizes laser melting to construct complex 3D geometries, which exploits the powerful capabilities of medical imaging such as X-rays, magnetic resonance imaging, and computed tomography scans to produce implants that complement or replicate the anatomy of the host tissues. One of the main advantages of SLM technology is its ability to process a wide range of metals including biomedical-grade SS (such as 304 and 316 SSs), Co–Cr alloy (such as ASTM-F75, containing mainly Co, with 27–30 wt% Cr and 5–7 wt% Mo) Ti (cpTi with four grades), and Ti-based biomaterials (such as Ti–6Al–4V, Ti–6Al–7Nb, and Ti–Zr alloys). For this reason, SLM has emerged as a promising technique for the fabrication of variety of medical devices. The complex anatomies of the human knee, hip, skull, dental, and craniofacial tissues require a precise design and high-dimensional accuracy during their fabrication to match and fit the target implantation sites. SLM has been, therefore, explored exclusively in the past two decades for fabricating tissue engineering scaffolds and implants from a wide range of metals. Recent developments in SLM technologies provide a pathway for the clinical success of orthopedic surgeries where most body parts can be reconstructed by combining AM and medical imaging techniques. These AM technologies have enabled surgeons and materials scientists to fabricate patient-specific medical devices in nearly any imaginable geometry and sometimes within 24 h [46]. SLM fully melts the powder, resulting in that it needs to reach a higher temperature than other metal 3DP technique. The build chamber is filled with an inert gas (either argon or nitrogen at oxygen levels below 500 parts per million) in order to create the perfect conditions for melting. The full melting process allows the metal to form a homogeneous block with good resistance. It fits perfectly for pure metals like titanium or aluminum. Because we need the higher temperature to fully melt the material, the cooling time will be longer than for DMLS. SLM is also distinct from EBM in that SLM uses a high-power laser while EBM uses a beam of electrons. SLM is also synonymous with direct melt laser melting and laser PBF, all of which describe the same underlying process of fusing metal powder using heat from a laser source [47–49].

As any other technologies, although SLM is a great alternative for traditional manufacturing methods, there are advantages and disadvantages associated with SLM technique when compared with traditional methods [46, 47].

Advantages:
- SLM creates full-metal, high-performance parts that are highly accurate and detailed
- Design freedom and concept validation by prototyping
- The range of materials in SLM is large, encompassing high-strength and specialty metals
- SLM can reduce part numbers by printing whole assemblies and can create highly complex geometries
- Ability to realize complex shapes or internal features (which would be incredibly difficult or expensive to achieve via traditional manufacturing)
- SLM speeds up metal manufacturing techniques, reducing delays in repairs and increasing the pace of production, resulting in cost savings and reduction of time to market
- SLM reduces material usage and waste, especially when compared with traditional manufacturing methods, leading to an environment-friendly method

Disadvantages:
- Only single-component metals and specified materials with good flow characteristics are acceptable in SLM
- SLM is a high-energy process, leading to temperature gradients that can stress/dislocate parts and compromise their structural integrity
- SLM parts need extensive support structures and SLM requires a source of inert gas
- SLM parts have a rough surface finish out of print and require a lot of postprocessing to take place
- SLM has a size restriction on parts and is very expensive, limited it to small-batch production runs

- Rough surface finish
- Specialized design and manufacturing skills and knowledge needed
- Limited currently to relatively small parts
- Lots of postprocessing required

The most common applications for this technology are in the aerospace industry, as complex parts can be made with AM, which overcomes the limitations of conventional manufacturing. It can also result in the reduction of parts weight needed. SLM also has applications in the medical field where some prosthetics are created with this technology, allowing the model to be customized to the patient's anatomy.

Saedi et al. [50–52] had extensively investigated the properties of SLMed NiTi alloys on various aspects. Using Ni-rich $Ni_{50.8}Ti_{49.2}$ (at%) alloys, effects of postfabrication heat treatment were studied [50, 51], and it was reported that (i) the SLM method and postheat treatments can be used to tailor the microstructure and shape memory response; (ii) partial superelasticity was observed after the SLM process; (iii) solution treatment of fabricated samples increased the strength and improved the superelasticity but slightly decreased the recoverable strain; (iv) samples aged at 350 °C showed better recovery in superelasticity tests, where 350 °C × 18 h aged samples exhibited almost perfect superelasticity with 95% recovery ratio with 5.5% strain in the first cycle and stabilized superelasticity with a recoverable strain of 4.2% after 10th cycle, while 450 °C × 10 h aged sample exhibited 68% recovery with a recoverable strain of 4.2% in the first cycle and stabilized recoverable strain of 3.8% after 10th cycle [50, 51]. Studying the same NiTi alloy [52], it was reported that (i) the sample fabricated with a laser power of 100 W and scanning speed of 125 mm/s exhibited almost perfect superelasticity with a recovery ratio of 96% and strain recovery of 5.77% in the first cycle and (ii) the corresponding stabilized superelastic response demonstrated full strain recovery of 5.5% after 10 cycles. For enhancing superelasticity of porous NiTi structures (with 32–58% porosity), Saedi et al. [52] heat treated the fabricated samples by solution annealing + aging at 350 °C for 15 min. It was reported that (i) SLMed NiTi with up to 58% porosity can display shape memory effect with full recovery under 100 MPa nominal stress, (ii) dense SLMed NiTi could show almost perfect superelasticity with strain recovery of 5.65 after 6% deformation at body temperatures, (iii) the strain recoveries were 3.5%, 3.6%, and 2.7% for samples with porosity levels of 32%, 45%, and 58%, respectively, and (iv) the stiffness of NiTi parts can be tuned by adjusting the porosity levels to match the properties of the bones. This last conclusion is related to the biomechanical compatibility [53] and stress shielding effect [54].

Mullen et al. [55] investigated a novel porous titanium structure fabricated by SLM technique with specific requirements, for developing a bone ingrowth structure; in particular, functionally graded structures with bone ingrowth surfaces exhibiting properties comparable to those of the human bone. It was found that porosity ranging from 10% to 95% and resultant compression strength from 0.5 to

350 MPa are comparable to the typical naturally occurring range, and optimized structures have been produced that possesses ideal qualities for bone ingrowth applications and that these structures can be applied in the production of orthopedic devices. Research group of Shishkovskii [56, 57] investigated conditions of a layer-by-layer synthesis of 3D parts made of nitinol and HA additions using SLS/SLM since optimization for SLS/SLM parameters are needed for the synthesis of NiTi + HA to be used in tissue engineering and manufacture of medical devices (pins, nails, porous implants, and drug delivery systems). It was reported that (i) no significant destruction of HA ceramics under laser treatment was observed, (ii) the amount of nickel released to the surface of 3D parts decreases owing to the additional oxidation of free titanium during SLS/SLM and the formation of a protective HA layer, and (iii) full-density 3D parts are produced from nitinol by SLM including preheating to 300 °C.

Moghaddam et al. [58] evaluated the anisotropic tensile properties of $Ni_{50.1}$ $Ti_{49.9}$ (at%) components fabricated by SLM method. It was mentioned that (i) SLMed samples in the horizontal orientation had the highest ultimate tensile strength(606 MPa) and elongation (6.8%) with the strain recovery of 3.54% after four shape memory effect cycles, and (ii) at stress levels ≤200 MPa, these samples had the actuation strain >3.8% without accumulation of noticeable residual strain. Chekotu et al. [59] reviewed the most recent publications related to the SLM processing of nitinol identify the various influential factors involved and process-related issues. It was described that (i) powder quality and material composition exhibit a significant effect on the produced microstructures and phase transformations, (ii) the effect of heat treatments after SLM fabrication on the functional and mechanical properties are noted, and (iii) optimization of several operating parameters were found to be critical in fabricating nitinol parts of high density.

6.2.5 Selective laser sintering (SLS)

Laser beam of CO_2 or Nd:YAG is normally employed for scanning successive layers of powdered materials to create a 3D object, and the scanning patterns of each layer are automatically digitized. Fabrication of the final parts using the SLS method is composed of 3D CAD design of the concept and transfer of the CAD data to the SLS machine to carry out fabrication with the desired powders [7, 60]. Each AM system has a unique binding mechanism to bind the layers. The binding mechanism of SLS technology can be classified into three main categories [61–63]: (1) solid-state sintering is a thermal process. The binding mechanism in this category occurs between $T_m/2$ and T_m, wherein lies the melting temperature of the material in question. (2) Liquid-phase-assisted sintering is commonly used for materials that are difficult to sinter. Liquid-phase-assisted sintering is the process of adding an additive to the powder that will melt before the matrix phase. This method is widely employed for fabrication of 3D parts from ceramic materials with incorporation of a small amount of polymers which will gradually decompose and completely disappear [64]. (3) Full melting is used for metallic and ceramic materials more than polymers. In this mechanism, near-full density is reached in one step by melting the powders completely by laser beam, thus avoiding lengthy postprocessing steps. Various types of materials can be used in SLS, namely, they are thermoplastics (in semicrystalline or amorphous) with different T_m (melting) and T_g (glass transition) temperatures, powder-based or slurry-based ceramics, biocompatible Ti and Ti-based alloys (e.g., Ti–6Al–4V, Ti–6Al–7Nb, or Ti–Zr), Ta, Co–Cr, and Co–Cr–Mo alloys, SSs (such as 304L and 316 SSs), and even composites with ceramics and polymers [7, 65]. The SLS has been proved to be a reliable and robust AM technology for producing different categories of biomedical devices. The SLS system has a great advantage of being tailored to a wide range of requirements through the adjustment of laser and process parameters [66, 67].

There are several characteristics recognized with SLS products [68]. In SLS, almost all process parameters are preset by the machine manufacturer. The default layer height used is 100–120 µm, which is close to the preferred particle size range of 10–150 µm [69]. A key advantage of SLS is that it needs no support structures. The unsintered powder provides the part with all the necessary support. For this reason, SLS can be used to create freeform geometries that are impossible to manufacture with any other method. Taking advantage of the whole build volume is very important when printing with SLS, especially for small batch productions. In SLS, the bond strength between the layers is excellent. This means that SLS-printed parts have almost isotropic mechanical properties. A typical SLS-printed part is about 30% porous so that porosity gives SLS parts their characteristic grainy surface finish. It also means that SLS parts can absorb water, so they can be easily dyed in a hot bath to a large range of colors but also that they require special postprocessing if they are to be used in a humid environment. SLS parts are susceptible to warping and possess 3–3.5% shrinkage.

6.2.6 3D printing (3DP)

There are various sintering and modeling techniques as sequential processes prior to the 3DP and they should include (1) FDM by material extrusion; (2) SLA-digital light processing for vat polymerization; (3) SLS by PBF of polymers; (4) MJ-drop-on-demand by jetting demanded material; (5) binder jetting; and (6) DMLS, SLM, or EBM for PBF of metals [68]. SLS 3DP is used for both prototyping of functional polymer components and for small production runs, as it offers a very high design freedom and high accuracy and produces parts with good and consistent mechanical properties, unlike FDM or SLA. The capabilities of the technology can be used to its fullest though only when the designer takes into consideration its key benefits and

limitations [69]. Freedom of design, waste minimization, and the ability to manufacture complex structures, as well as fast prototyping are the main benefits of AM
or 3DP [13, 15]. AM, aka 3DP, is driving major innovations in many areas, and recent
advances have enabled 3DP of biocompatible materials, cells, and supporting components into complex 3D functional living tissues. 3D bioprinting is being applied
to regenerative medicine to address the need for tissues and organs suitable for
transplantation. Compared with nonbiological printing, 3D bioprinting involves additional complexities, such as the choice of materials, cell types, growth and differentiation factors, and technical challenges related to the sensitivities of living cells
and the construction of tissues. Addressing these complexities requires the integration of technologies from the fields of engineering, biomaterials science, cell biology,
physics, and medicine. 3D bioprinting has already been used for the generation and
transplantation of several tissues, including multilayered skin, bone, vascular grafts,
tracheal splints, heart tissue, and cartilaginous structures. Other applications include
developing high-throughput 3D-bioprinted tissue models for research, drug discovery, and toxicology [12].

In medical and healthcare fields, 3DP promises to address the needs of personalized medicine, producing solutions that match the individual anatomical needs of
patients, and to advance tissue and organ engineering through printing of cells and
complex multimaterial scaffolds for tissue regeneration [70, 71]. The former application has been utilized clinically in the production of personalized orthopedics and
stomatology solutions, whereas the latter at this stage is mostly limited to the printing of 3D tissue scaffolds under laboratory conditions [71]. The ability to produce
implants with good matching to the anatomy of the patient is a valuable attribute
of 3DP, particularly for reconstructive surgery to treat craniofacial breaks or fractures [72], for patients suffering from dwarfism, or in cancer patients, where the implant is made to match the excised tissue and thus may reduce the pressure placed
onto the existing bone compared with a noncustomized implant [73]. Yet, it is a
combination of these two approaches that hold most promise for the future of the
personalized medicine [71].

Gaget [74] mentioned that AM is an advanced technology and is now increasing
the possibilities to get on to measure treatments for the patients. 3DP is offering
great solutions to make reconstructions, or help patients in their daily lives. One of
the most important advantages that 3DP can bring to the medical sector is mass
customization. Indeed, it is a way to get printed parts perfectly adapted to the patient that will perfectly fit his or her needs and body. For example, 3DP is well used
to create on to measure prosthesis. It is now common to 3DP prosthesis for arms or
legs. Each body and each morphology are different; by using this technology, prostheses are totally fitting the morphology of the patient and replace correctly his or
her missing member. This is the reason why it is now a major benefit; it enables
more efficient prosthesis, adapted to the mechanical constraints of the body. It was
reported that the world's first 3D-printed prosthesis has been used to recreate a new

sternum and part of rib cage for a 54-year-old man who suffered from tumor in the chest wall. Surgeons knew that reconstruction would be difficult due to the complex geometry of the chest cavity and the fact that conventional titanium implants are held together by screws, which can come undone over time, causing further complications. The Spanish patient suffering from chest wall sarcoma received a 3D-printed replacement sternum and rib cage after many of the natural bones were surgically removed. The titanium implant was designed, specifically for the unique anatomical needs of this particular patient. Post-surgery, the patient was discharged in 12 days and prognosis was reported well [75], as shown in Figure 6.2.

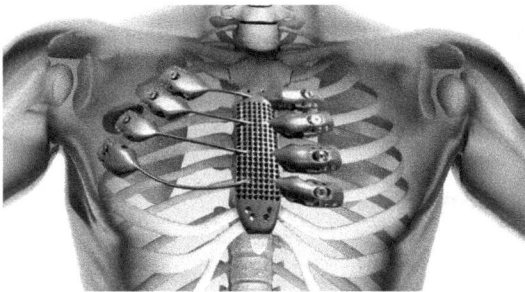

Figure 6.2: World's first 3D-printed prosthesis to recreate a new sternum and partial ribcage [75].

As 3DP continues to transform manufacturing, doctors are hoping it could also help the 30 million people worldwide in need of prosthetic limbs, braces, or other mobility devices. Slowly but surely, 3D bioprinting has been revolutionizing aspects of medicine since the start of the century. Experts have developed 3D-printed skin for burn victims, airway splints for infants, facial reconstruction parts for cancer patients, orthopedic implants for pensioners. The fast-developing technology has churned out more than 60 m customized hearing-aid shells and ear molds, while it is daily producing thousands of dental crowns and bridges from digital scans of teeth, replacing the traditional wax modeling methods used for centuries [76]. According to Amputee Coalition [77], in the United States alone, there are nearly 2 million people living with limb loss. Approximately 185,000 amputations occur in the United States each year due to causes varying from vascular disease to trauma [78]. It was also reported that around 1 in every 1,900 newborns in the United States are born with a limb reduction deficit, and there are an estimated 30 million people worldwide who need a prosthetic. Of those, <20% actually have one. Around 200,000 amputations are performed in the United States per year, leaving many with stumps who struggle to live as they used to. But 3D-printed prosthetics offer a potential solution [79].

Broken jaw is repaired with implants made form fibula. According to Johns Hopkins researchers, ever year due to birth defect, trauma, or cancer, more than 200,000 people will need replacement bones in their face or skull. Typically, doctors would remove a part of the patient's fibula and try to carve it into the required shape and implant the bone back into the patient's face. While the procedure typically results in the bone regrowing and healing the damage in the face, it is not the ideal solution. Depending on the damage being corrected, the bone fragment often cannot be shaped to fit the face very well, which leaves the patient with significant scarring. The removal of part of the fibula also creates trauma in the patient's leg which, when combined with the ongoing trauma in their face skull, can be quite stressful [80]. Tissue-engineered approaches to regenerate bone in the craniomaxillofacial region utilize biomaterial scaffolds to provide structural and biological cues to stem cells to stimulate osteogenic differentiation. Bioactive scaffolds are typically comprised of natural components but often lack the manufacturability of synthetic materials. To make a good framework for filling in the missing bone, mix at least 30% pulverized natural bone with some special man-made plastic and create the needed shape with a 3D printer. Hung et al. [81] performed 3D-printed materials comprised of decellularized bone (DCB) matrix particles combined with polycaprolactone (PCL) to create novel hybrid DCB/PCL scaffolds for bone regeneration, as shown in Figure 6.3.

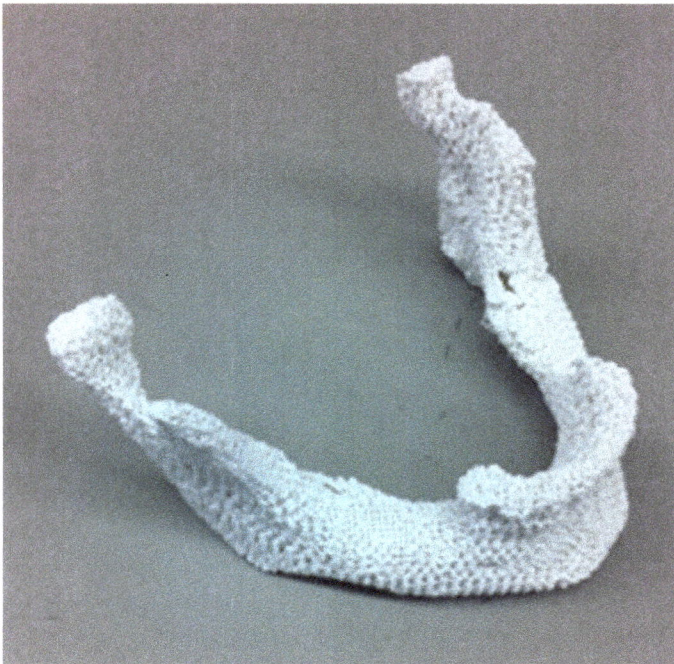

Figure 6.3: 3D-printed scaffold, matching the lower jaw of a female patient [81].

Jaw is an important part of craniofacial bone, which maintains the integrity of facial shape, undertakes masticatory function, and is closely related to swallowing, language, breathing, and other functions. After implantation, jaw prosthesis is not only pulled by surrounding muscles, but also by masticatory pressure. Therefore, the prosthesis should restore the patient's personalized appearance and have enough mechanical strength to ensure its stability and function in vivo [82]. Biocompatible titanium and titanium-based alloys can be used to make titanium plate, titanium mesh, retaining screw, and artificial prosthesis to repair maxillofacial bone defects. Because of the complex anatomical structure and contour of the jaw, traditional techniques cannot accurately prepare titanium alloy prostheses that match the morphological and biomechanical characteristics of the defect area and can complete denture restoration. The 3D bioprinting technology has the potential and unique advantages of completing complex structural design and making complex structures more accurately and quickly, especially in craniofacial defect prosthesis with complex shape and structure as well as complex internal structure. Therefore, with the help of digital medical imaging technology and CAD software, 3DP of titanium and titanium alloy is expected to achieve the purpose of personalized design, production, and repair of jaw defects, as shown in Figure 6.4 [82].

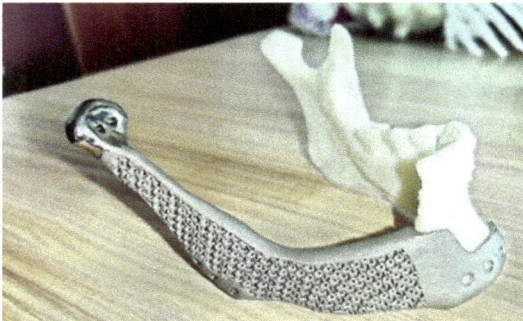

Figure 6.4: Partially repair jawbone by 3D bioprinting technique [82].

6.3 Metal injection molding (MIM)

Metal injection molding (MIM) – sometimes called powder injection molding – is an advanced metalworking process in which finely powdered metal (particle size ranging from several μm to several 10 μm) is mixed with the binder material to create a feedstock that is then shaped and solidified using injection molding machine. MIM technology as an advanced powder metallurgy offers near-net shaping and mass production of products of the same design [83, 84]. MIM has been used successfully in the industrial field, particularly in watch manufacturing, which requires a precise near-net shape forming [84]. The MIM process comprises the following sequences: (1) homogenous mixing/kneading of metal powder and binding agent, (2) feedstock formation with sufficient viscosity which is governed by the rheology, (3) injection molding, (4) binder removal (or debinding) with shape retention to densify the powders, and (5) sintering to high density with linear shrinkage as high as 15–20%. For certain applications, such as the automotive, medical, and aerospace sectors, HIP (hot isostatic pressing) can be used to completely remove any residual porosity. Figure 6.5 compares the shape and appearance at three major processes during the MIM [85].

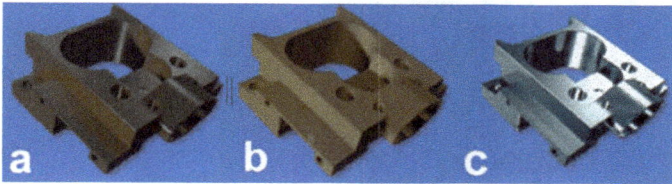

Figure 6.5: MIM part: (a) as molded, (b) debinded, and (c) sintered [85].

The binding material (adding volume % of 40–50 of organic binding substances) consists of thermoplastic resins such as polyethylene or polypropylene, wax, plasticizer, and lubricant. The principal purpose of mixing the binder is to provide sufficient flow mobility when the compound is injected into the mold. The polymer imparts viscous flow characteristics to the mixture to aid forming, die filling, and uniform packing. The molded products are then subjected to debinding by either a heating process or a dissolving process in order to remove the binding material. The atmosphere used in the debinding stage includes nitrogen, air, vacuum, wet hydrogen, and hydrogen with hydrochloric acid [86].

The MIM process can replace other metal-forming techniques such as investment casting and machining and following advantages are recognized [87]. Figure 6.6 shows various MIMed products [88]:

– Complex geometries
– Efficient use of material
– Less material waste as a result of producing near-net shape components and considered a green technology
– Repeatability
– Excellent mechanical properties
– Tailored solutions using unique materials formulated to meet component/application requirements
– MIMed products can be brazed/joined to a variety of components for complete assembly solutions

- MIM is a repeatable process for complex components made from high-temperature alloys
- MIMed parts are near fully dense, which provides excellent mechanical, magnetic, corrosion, and hermetic sealing properties, and allows secondary operations like plating, heat treating, and machining to be easily performed
- Complex shapes are achieved through innovative tooling techniques similar to those used in the plastic injection molding industry
- High volumes are attained through multicavity tooling

Figure 6.6: Various MIMed products [88].

Crawford [89] reported the similarities and dissimilarities between MIM and metal-AM. The goal of MIM and metal-AM is the same – to make durable, high-quality metal components, implants, instruments, and other devices. Both methods offer medical device manufacturers and their contract manufacturers an alternative to traditional machining. MIM is well-suited for small, complex geometries and intricately shaped components. It is also a good option for manufacturing medium- to high-volume parts for medical devices for laparoscopic surgeries and other high-volume applications such as surgical instruments. MIM is also a highly scalable and cost-effective metal manufacturing technique. For metal-AM, medical device applications continue to expand as the process becomes more widely accepted as a viable alternative to traditional machining. AM is especially popular in orthopedics because of its ability to create monolithic, porous structures that can promote bone ingrowth. In addition, metal-AM eliminates the need for tooling, which helps to reduce costs. The metal-AM is also a great solution for the production of personalized implants and instruments because of the design freedom it can provide. The most utilized metal AM processes for medical device production are PBF technologies, which weld layers of metal powder together utilizing a high-energy source, typically a laser or electron beam. Titanium is the metal of choice for most AM applications for fabricating implantable devices with integrated lattice or porous structures that encourage osteointegration. Other materials include cobalt–chrome for implants. Both MIM and 3MP have surface finishes that, in their raw states, are inferior to machined components. In addition, dimensional control can often be an issue. In the raw state, MIM surface finish and dimension control are superior to those of 3MP and many MIM, and 3MP manufacturers have secondary machining and surface finishing capabilities that can close the gap between these two technologies. Engineers often use metal-AM when they need to create specialized parts that require high-strength characteristics or have intricate design features. The metal-AM does not require molds, which provides designers with more creative possibilities for shape, tolerance, and functionality. The main advantage of PBF is its ability to produce complex lattice and porous structures. These types of complex structures cannot be achieved using MIM or other manufacturing processes. However, the disadvantages of PBF are higher cost per part and lower volume production capability. On the other hand, the advantage of MIM is that it is cost-effective and can produce high volumes. Products with complex structures are best suited to PBF, whereas simpler structures in need of high-volume production are better suited to MIM [89].

There are several reports on medical and dental applications of MIM technique. In medical field, there are surgical stapling unit made of 17–4 PH (precipitation–hardening type) SS [90], surgical instruments made from 304L SS, 316L SS, 17–4 PH SS [91], orthopedic prosthesis made of 316L SS, Co–28Cr–6Mo, Ti–6Al–4V, unalloyed Ti, and unalloyed Ta [91], total knee replacement cup made of Ti–6Al–4V alloy [90], lumbar spinal device made of C–Cr alloy [90], and a base plate for an implantable infusion pump made from Ti–6Al–4V [91]. In dental field, there are dental implants made of 316L SS [92], and orthodontic brackets made of 316L SS [93] or unalloyed Ti [94].

6.4 Foam structure

A metal foam is a cellular structure consisting of a solid metal with gas-filled pores comprising a large portion of the volume. The pores can be sealed (closed-cell foam) or interconnected (open-cell foam). The defining characteristic of metal foams is a high porosity with typically only 5–25% of the volume is the base metal (or typically 75–95% of volume of the metal). Aslan et al. [94] fabricated porous Ti–6Al–4V structures having the homogeneously distributed (with interconnected open pores) porosities at 41.08%, 52.37%, and 64.10% by adding 40%, 50%, and 60% spherical magnesium (Mg) powder with 350 µm particle sizes in average as spacers and evaporating magnesium via the atmosphere-controlled sintering.

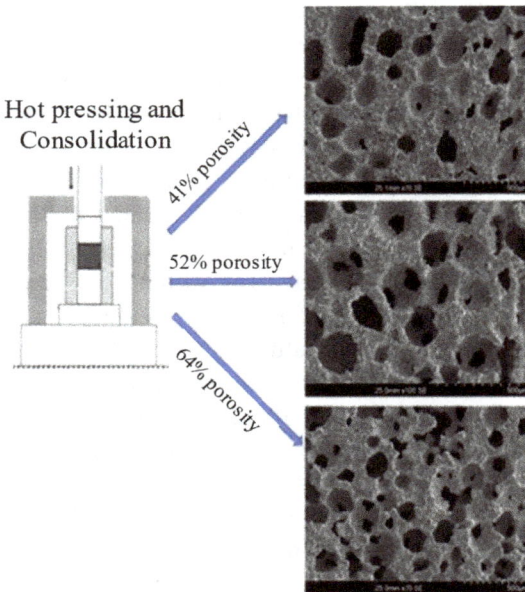

Figure 6.7: Porosity-controlled foamed Ti–6Al–4V [94].

Metal foams have great specific stiffness (ratio of stiffness to weight), and an almost reversible quasielastic zone, so they are applicable to the light structure. For metal foams with a 1/5 density, the specific stiffness is five times that of normally dense metal with the same weight. Metal foams typically retain some physical properties of their base material [95, 96].

The pores can be sealed (closed-cell foam) or interconnected (open-cell foam), as shown in Figure 6.8 [97]. Furthermore, Kränzlin et al. [98] subdivided into metal foams, metal sponges, and nanoporous metals, as shown in Figure 6.9. However, this classification possesses a sort of controversial issue since there are numerous studies, which will be cited later in this section, and utilize the term "open-cell metal foam" for tissue engineering applications. Syntheses are also listed to each of these cellular structures, although in most cases, a method suitable for a specific shape might not be the best option for controlling the microscopic features. It was further mentioned that the production strategy greatly affects the efficiency and performance of the porous metal structure in its final application, and some applications rely on more complex geometries and pore structures than others [98]. Foams are commonly fabricated by injecting a gas or mixing foam-making agent into a molten metal. Melts can be foamed by creating gas bubbles in the material. Normally, bubbles in molten metal are highly buoyant in the high-density liquid and rise quickly to the surface. This rise can be slowed by increasing the viscosity of the molten metal by adding ceramic powders or alloying elements to form stabilizing particles in the melt, or by other means. Metallic melts can be foamed in one of three ways: (1) by injecting gas into the liquid metal from an external source; (2) by causing gas formation in the liquid by admixing gas-releasing blowing agents with the molten metal; and (3) by causing the precipitation of gas that was previously dissolved in the molten metal [99]. On the other hand, open-cell foams are normally manufactured by foundry or powder metallurgy. In the powder method, the space holders are used to occupy the pore spaces and channels. In casting processes, foam is cast with an open-celled polyurethane foam skeleton.

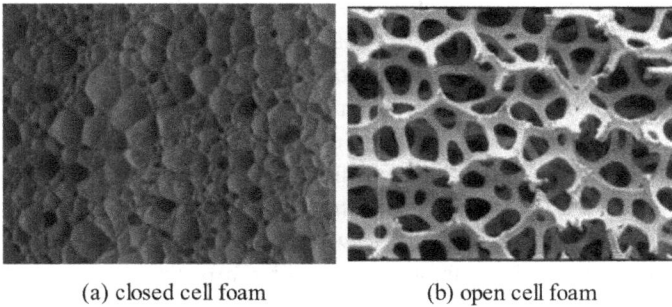

(a) closed cell foam   (b) open cell foam

Figure 6.8: Comparison of SEM images of (a) closed-cell foam and (b) open-cell foam [97].

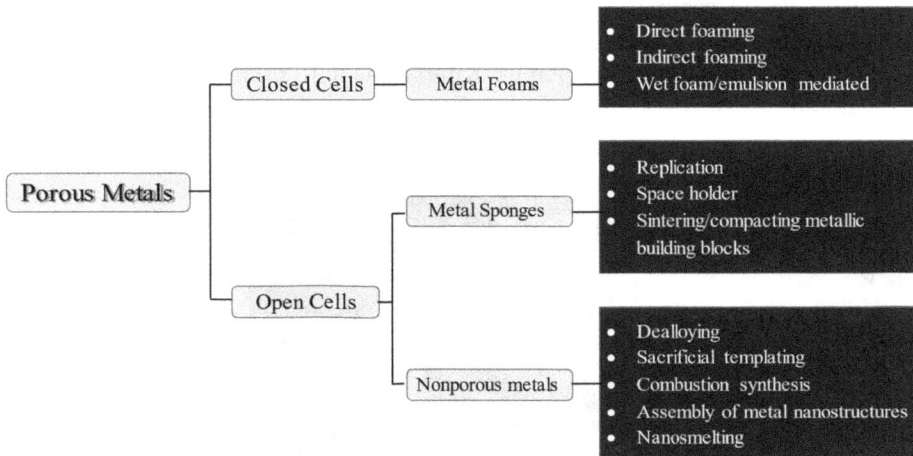

Figure 6.9: Classification of cellular structures and suggested syntheses [98].

Although metal foams typically retain some physical properties of their base material, the strength of the foam structure depends on the density (or porosity). Even if porous metals are divided into closed-cell pores and open-cell pores and these parameters control to some extent of basic properties of cellular structures, additional features like relative density, pore structure, and macroscopic shape are also critical in determining the functionality of porous metals and thus their application potential beyond structural materials [98, 100]. The applicability of porous metal structure is strongly controlled by three parameters: (1) the composition, (2) the macroscopic shape, and (3) the pore structure. The composition determines the intrinsic properties and thus decides on whether or not a specific metal or alloy fulfills the targeted physical (density, electrical/thermal conductivity, strength, elasticity, ductility, machinability) or chemical (oxidation behavior and corrosion stability) requirements. The macroscopic shape is critical for the integration into predesigned devices or for their combination with other materials. The pore structure (relative density, surface area density, mechanical stability, pore size, pore size distribution, and connectivity of pores) defines the efficiency of the material during operation. Only when all these three characteristics are carefully engineered, the material can be considered as a possible candidate for a given application [95, 101]. Functional applications of metal foam structures are therefore versatile, including medical areas (in which devices for tissue engineering, orthopedic implants, and stents should be included), automotive industry (including sound damping and catalytic converters), better heat transfer like heat sinks, energy absorption, and electromagnetic shielding, arresting flames, and catalytic engineering [102].

It is believed that open-cell foams are normally manufactured by foundry or powder metallurgy, in which the space holders are used to occupy the pore spaces and channels. In casting processes, foam is cast with an open-celled polyurethane foam skeleton. However, recently, it was mentioned that the AM technology can be applied to manufacture porous materials as a 3D-printing technology does [103], as shown in Figure 6.10 [104].

Figure 6.10: Open-cell Al foam by the additive manufacturing [104].

Gotman et al. [105] fabricated that highly porous NiTi scaffolds for bone ingrowth were fabricated by reactive conversion of commercially available Ni foams. It was mentioned that these open-cell trabecular NiTi scaffolds possess high strength and ductility and exhibit low Ni ion release; and (ii) reactive conversion deposition of a thin titanium nitride (TiN) layer further improves the corrosion characteristics of trabecular NiTi and allows for material bioactivation by alkali treatment or biomimetic Ca phosphate deposition. Barrabés et al. [106] examined NiTi foams that have been treated using a new oxidation treatment for obtaining Ni-free surfaces that could allow the ingrowth of living tissue, thereby increasing the mechanical anchorage of implants. It was reported that (i) a significant increase in the effective surface area of these materials can decrease corrosion resistance and favor the release of Ni which might induce allergic reactions or toxicity in the surrounding tissues; (ii) these foams have pores in an appropriate range of sizes and interconnectivity, and thus their morphology is similar to that of bone; and (iii) their mechanical properties are biomechanically compatible with the bone. Tissue engineering is a field that aims to regenerate damaged tissues by enhancing tissue growth through the porous architecture of the scaffolds which is desired to mimic the human cancellous bone. Mg-based scaffolds are gaining importance in the field of tissue engineering, owing to its potential application as a biomaterial [107]. There are several reasons for promising applicability of open-cell Mg foam to fabricate scaffold structures; (1) technically it is not difficult to manufacture open-cell Mg foams, (2) Mg material exhibits an excellent biodegradability, and (3) Mg shows an excellent osteoinductivity. Porous Mg and Mg alloys can be used as bone substitutes with bone-mimicking characteristics. Since the structures of bone tissue and porous Mg materials are similar, host bone cells can exhibit the bone ingrowth into the pores of an Mg scaffold. Moreover, the gradual biodegradation of porous Mg and Mg alloys after implantation, followed by the absorption of Mg ions in the body, eliminates the necessity of removing the implant via a subsequent surgical procedure, which makes Mg scaffolds more favorable than other porous metallic biomaterials [108].

6.5 Gradient functional material (GFM)

A material is composed of multilayers, with each layer having unique characteristics, yet adjacent layers having some similarity is called gradient functional material (GFM). Although such functions can include various properties, it is limited to mechanical, physical, or thermal properties, since other properties, such as chemical or electrochemical, are more likely important and susceptible to the surface layer's property, and not related to bulky or semibulky behavior [53]. FGM is another term which means the same as GFM. Figure 6.11 illustrates the differences in several properties between ordinary composite material and GFM [109].

GFMs are advanced materials that have different properties as the dimension varies. Human body is also a highly graded organization. For instance, the longitudinal cross section of a long bone resembles a sponge while the bone in the middle

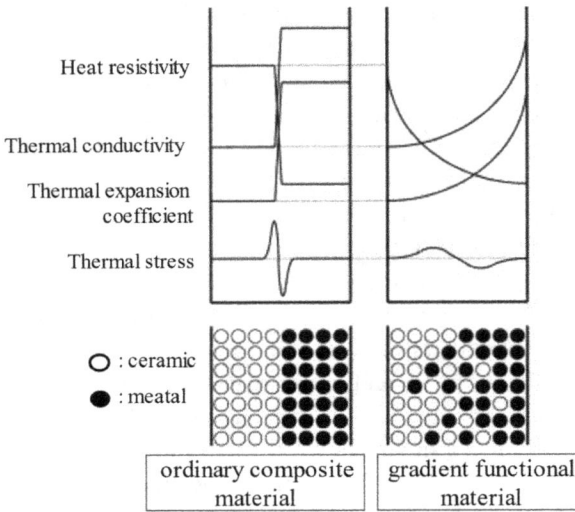

Figure 6.11: Differences in material structures and properties of cross-sectional view between ordinary composite material and gradient functional material [109].

has less porosity. Therefore, GFM is considered as a biomimetic material. The overall properties of GFM are unique and different from individual material that forms it. GFM usage is expected to increase when the cost of material processing and fabrication processes are reduced [110]. An example of such materials can be graded porous materials that have several advantages in terms of mechanical properties. Different porosities are required for different regions of implants to obtain optimum bone ingrowth that differs between tissues [111]. For example, if HA is needed to spray-coat onto CpTi, this GFM concept can be applied. Instead of applying HA powder directly onto the CpTi surface, a multilayer of HA/HA + Al_2O_3/Al_2O_3 + TiO_2/ TiO_2/CpTi can be prepared to enhance the bonding strength [112]. From the HA side to CpTi side, the mechanical properties (particularly, modulus of elasticity) and thermal properties (such as linear coefficient of thermal expansion) are gradually changing so that when this HA-coated CpTi is subjected to stressing, interfacial stress between each constituent layer can be minimized, so that the degree of discreteness in the stress field can also be minimized [53]. The National Industrial Research Institute of Nagoya of the Agency of Industrial Science and Technology has established technology for forming a functional gradient titanium-oxide film on a titanium alloy (Ti–6Al–4V) by the reactive DC sputtering vapor deposition method [113]. The method was developed for fabricating denture bases and artificial implants, and the oxygen concentration was changed continuously during sputtering to provide a gradient in the film composition, by which adhesivity to the alloy, surface hardness, and biocompatibility were improved. As a result of investigations, a method of coating the surface with titanium, a new formation of a gradient film featuring excellent biocompatibility, firm adhesivity, and great surface hardness was developed. Denture bases produced by superplastic forming are cleaned with an organic solvent, the oxygen concentration is changed continuously during sputtering, and pure titanium is vapor deposited by the reactive DC sputtering. In the initial stage, intermetallic bonding is achieved by oxygen-free vapor deposition, so the adhesion is excellent and there is no fear of exfoliation. But farther away from the metal surface, the oxygen concentration is raised gradually to form a gradient film. At the surface, some titanium oxides are formed. Titanium oxide features excellent biocompatibility, and since it is a hard material, it resists damage. The overall film thickness in the experiment was 3 μm, and the Vickers hardness of the surface was 1,500 (200–300 for pure titanium) [113]. Bogdanski et al. [114] fabricated the FGM, which was prepared through powder metallurgical processing with thoroughly mixed powders of the elements. Ten mixtures were prepared ranging from Ni:Ti of 90:10 (by atomic weight) through Ni:Ti = 80:20, . . ., to Ni:Ti = 10:90, and pure Ti. Each mixture was homogenized in a mixer for 24 h in a bottle without addition of grinding balls to keep impurities low. The compaction was done by HIP at 1,050 °C and 195 MPa for 5 h. It was reported that using cells comprised of osteoblast-like osteosarcoma cells, primary human osteoblasts, and murine fibroblasts shows good biocompatibility of Ni–Ti shape memory alloy with 50:50% Ni:Ti [114].

In some applications, to which both structural integrity and functionality are important, it is not necessary to fabricate the entire structure in porous form, rather it should possess gradated porosity from one end to the other. Although biomaterials (including bioceramics, biometals, biopolymers, and biocomposites) are important materials for the replacement and regeneration of human tissues, dense bioceramics and dense biometals pose the problem of stress shielding due to their high moduli of elasticity compared to those of bones [53]. On the other hand, porous biomaterials exhibit the potential of bone ingrowth, which will depend on porous parameters such as pore size, pore interconnectivity, and porosity. Unfortunately, a highly porous biomaterial results in poor mechanical properties. To optimize the mechanical and the biological properties, porous biomaterials with graded/gradient porosity, pore size, and/or composition have been developed. Graded/gradient porous biomaterials have many advantages over graded/gradient dense biomaterials and uniform or homogenous porous biomaterials. The internal pore surfaces of graded/gradient porous biomaterials can be modified with organic, inorganic, or biological coatings, and the internal pores themselves can also be filled with biocompatible and biodegradable materials or living cells. However, graded/gradient porous biomaterials are generally more difficult to fabricate than uniform or homogenous porous biomaterials. With the development of cost-effective processing techniques, graded/gradient porous biomaterials can find wide applications in bone defect filling, implant fixation, bone replacement, drug delivery, and tissue engineering [111]. Yang et al. [115] fabricated surface-porous Mg–Al alloys with different microstructures via a new method of electrochemical dealloying in a neutral 0.6 M NaCl solution. It was reported that (i) a bimodal porous structure with 47.57 ± 11.43 µm large pores and 265.60 ± 78.68 nm honeycomb-like fine pores was obtained by direct electrochemical dealloying of as-cast Mg–20Al alloy; (ii) by subsequent annealing, the eutectic structure disappeared and a bicontinuous, single-sized, porous structure with 7.10 ± 1.96 µm ligaments was created by an annealing-electrochemical dealloying approach; and (iii) the porous formation mechanism is governed by selective dissolution of the α-Mg phase, which leaves the $Mg_{17}Al_{12}$ phase as the porous layer framework. Oshida et al. [116] proposed a concept of gradation of controlled porosity which should correspond to the surrounding anatomical conditions when an implant is placed into the bony structure. As illustrated in Figure 6.12, surface area needs to possess a certain level of porosity to accommodate the bone ingrowth activity, so the biological reaction is high, while the inner portion of the implant should be strong enough to bear an occlusal force (if a case of dental implant) and zones in between should gradually change by both increasing the biomechanical strength and decreasing the biological activity. As well known, the main reason for the HA coating is for establishing the surface roughness to exhibit the morphological compatibility with the receiving hard tissue [53].

As mentioned earlier, AM is a promising manufacturing method of biomaterials of various kinds, particularly when it comes to patient-specific applications. In recent years, research and clinical applications of porous metallic AM scaffolds and the implantation of AM biomaterials into the human body has been actively pursued. As Ødegaard et al. pointed out [14], it is necessary to establish a proper method of predicting elastic moduli, and fatigue properties by fabricated scaffold design will have high potential for further customization and utilization of the AM benefits. A scenario can be envisioned where more active patients with higher bone density may receive an implant with a higher stiffness as opposed to patients with lower bone density. Here the design can be adapted to the X-ray image of the bone around the site of implantation. With such possibilities, the full potential of AM can be unfolded lying in the direct conversion of computer-generated architectures into physical devices. As seen for other fields with fully digital workflows enabled by AM (such as hearing aid shells), we could expect to see a transition to more clinical research with load-bearing applications [14]. All necessary clinical data should be digitized toward the successful digital medicine and digital dentistry.

6.6 Nanotechnology

Nanotechnology is science, engineering, and technology conducted at the nanoscale, which is about 1–100 nm [117]. The prefix "nano" is referred to a Greek prefix meaning "dwarf" or something very small and depicts one thousand millionth of a meter (10^{-9} m). We should distinguish between nanoscience and nanotechnology [118]. The relative comparison among macroscale, mesoscale, microscale, and nanoscale dimensions can be visualized as shown in Figure 6.13 [119].

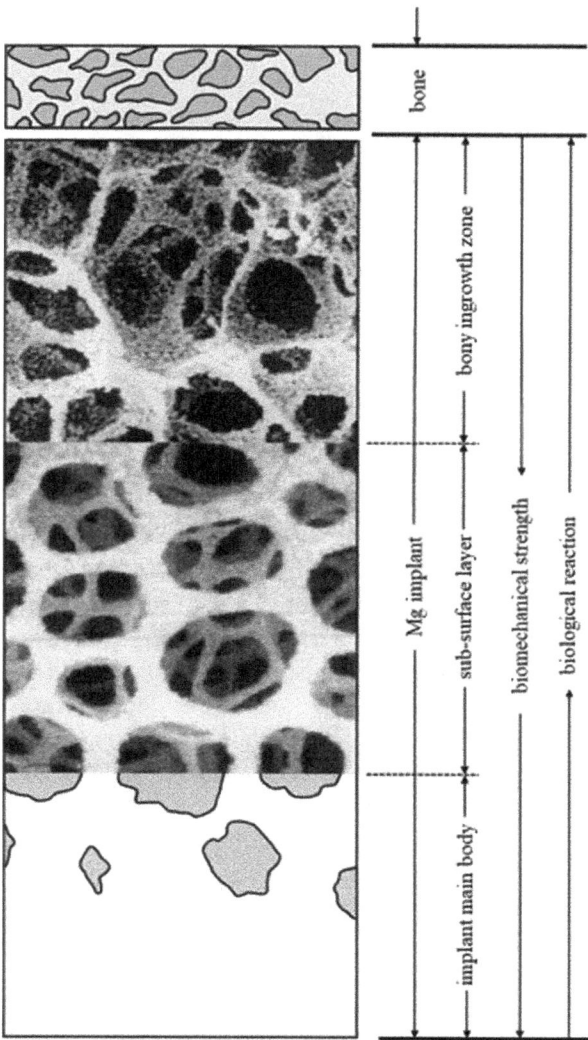

Figure 6.12: Conceptual design of implant with gradated porosity [53].

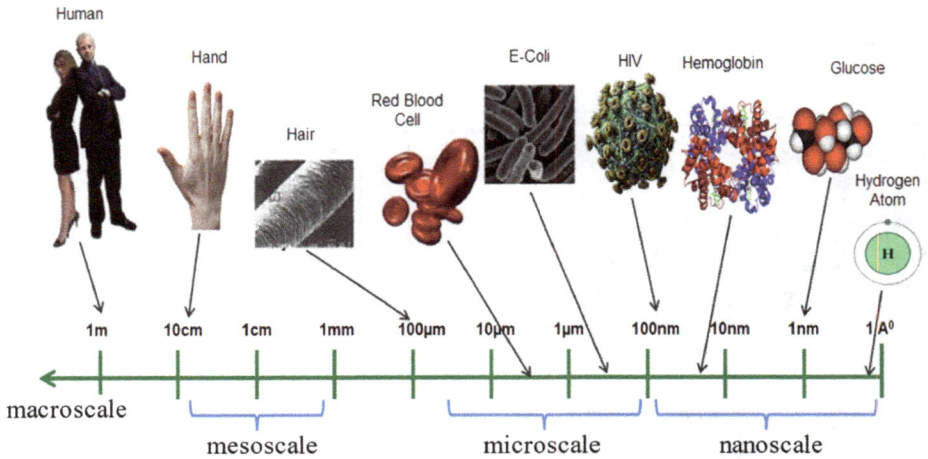

Figure 6.13: Comparison among different scales [119].

Nanoscience and nanotechnology are the study and application of extremely small things and can be used across all the other science fields as diverse as surface science, organic chemistry, molecular biology, semiconductor physics, energy storage, engineering, microfabrication, and molecular engineering [118, 120]. Bayda [118] distinguished between nanoscience and nanotechnology clearly; nanoscience is a convergence of physics, materials science, and biology, which deal with manipulation of materials at atomic and molecular scales, while nanotechnology is the ability to observe measure, manipulate, assemble, control, and manufacture matter at the nanometer scale. Nanoscience is the study of structures and materials on an ultra-small scale, and the unique and interesting properties of what these materials demonstrate. Nanoscience is a multidisciplinary research approach, indicating that scientists from a range of fields including chemistry, physics, biology, medicine, computing, materials science and engineering are studying it and using it to better understand our world. Nanotechnology (aka, molecular manufacturing) is the design, production, and application of structures, devices, and systems at the nanoscale so that, essentially, nanoscience is studying nanomaterials and their properties, and nanotechnology is using those materials and properties to create something new or different. Recently, the term "nanobiotechnology" was introduced and it is an interdisciplinary area of nanotechnology and biology, indicating the merger of biological research with various fields of nanotechnology and it is based on concepts that are enhanced through nanobiology include nanodevices (such as biological machines), nanoparticles, and nanoscale phenomena that occur within the discipline of nanotechnology [121]. Nanotechnology has been used to improve the environment and to produce more efficient and cost-effective energy, such as generating less pollution during the manufacture of materials, producing solar cells that generate electricity at a competitive cost, cleaning up organic chemicals polluting groundwater, and cleaning volatile organic compounds from air [118]. The need for computational applications at the nanoscale has given rise to the field of nanoinformatics. Powerful artificial intelligence machine learning algorithms and predictive analytics can considerably facilitate the design of more efficient nanocarriers. Such algorithms provide predictive knowledge on future data, and have been mainly applied for predicting cellular uptake, activity, and cytotoxicity of nanoparticles. Data mining, network analysis, quantitative structure–property relationship, quantitative structure–activity relationship, and "absorption, distribution, metabolism, excretion, and toxicity" predictors are some of the other prominent property evaluations being carried out in nanoinformatics [118]. Sharma et al. [122] mentioned that nanoinformatics could hasten up the advancements in anticancer nanomedicines through the use of computational tools, nanoparticle repositories, and various modeling and simulation methods.

6.6.1 Nanomedicine

Nanotechnology in medicine is a wide area that encompasses disease diagnosis, target-specific drug delivery, molecular imaging, and implantology particularly within orthopedics. Nanomedicine has transformed orthopedics through recent advances in bone tissue engineering, implantable materials, diagnosis and therapeutics, and surface adhesives [123]. The potential for nanotechnology within the field of orthopedics is vast. Nanotechnology has proven particularly successful in transforming drug delivery and manufacturing as well as medical diagnostic tools [124]. As researchers learn more about the mechanisms and characteristics of medical nanoparticles, these molecules have become increasingly known for their pharmacologic potential in improving drug synthesis and carriers as well as optimizing materials and reducing toxicity [124]. Some benefits of nanotechnology that have already become apparent in medicine include permanent implantation of small devices, development-implantable bone grating, and surface modification for promoting the osseointegration. Within cancer biology, nanotech has been used to deliver drugs such as doxorubicin in a way that shuts off cancer genes that normally allow cells to escape the drug [125, 126]. These examples are diverse but yet only represent a small fraction of the capabilities of nanomedicine, which has found a place in nearly every branch of medicine [123].

Sensors and diagnostics

Nanotechnology has been used to diagnose bone diseases, such as Paget's disease, renal osteodystrophy, and osteoporosis [123, 127]. This is often done using biosensors. Nanoelectronics makes the medical sector undergo deep changes by exploiting the traditional strengths of the semiconductor industry – miniaturization and integration. While conventional electronics have already found many applications in biomedicine –

medical monitoring of vital signals, biophysical studies of excitable tissues, implantable electrodes for brain stimulation, pacemakers, and limb stimulation – the use of nanomaterials and nanoscale applications will bring a further push toward implanted electronics in the human body [128]. The development of a nanobioelectronic system triggers enzyme activity and, in a similar vein, the electrically triggered drug release from smart nanomembranes; an artificial retina for color vision; nano material-based breathalyzers as diagnostic tools; nanogenerators to power self-sustained biosystems and implants; and future bionanotechnology might even use computer chips inside living cells. A lot of nanotechnology work is going on in the area of brain research. For instance, the use of a carbon nanotube (CNT) ropes to electrically stimulate neural stem cells; nanotechnology to repair the brain and other advances in fabricating nanomaterial-neural interfaces for signal generation. Molecular sensing and molecular electronics (nanoelectronics) is a diverse area that can include molecular conformational changes, changes in charge distribution, changes in optical absorbance and emission, or changes in electrical conductivity along or across simple or complex-shaped molecules, all in response to a target input. Each of these approaches can be integrated into a transduction system that provides a measurable and desired change in response to a specific or range of inputs. The ability to integrate such transduction mechanisms with biomolecules or to use biomolecules as the source of such materials provides, to varying extent, biocompatibility with other systems. Plasmonic nanobiosensors could ultimately become a key asset in personalized medicine by helping to diagnose diseases at an early stage. In other works, nanobiosensors that were originally designed to detect herbicides can help diagnose multiple sclerosis, and a new type of smartphone-based nanobiosensor has shown promise for early detection of tuberculosis [128]. These sensors are available in many designs and forms and can be implanted. Often, biosensors employ CNTs, as their unique properties make the sensors strong and electrically conductive. Bioapplications of CNTs have been predicted and explored ever since the discovery of these one-dimensional (1D) carbon allotropes. Indeed, CNTs have many interesting and unique properties potentially useful in a variety of biological and biomedical systems and devices. Significant progress has been made in an effort to overcome some of the fundamental and technical barriers toward bioapplications, especially on issues concerning the aqueous solubility and biocompatibility of CNTs and on the design and fabrication of prototype biosensors. In this chapter, we take a comprehensive look at the advances in this fast-moving and exciting research field. We review the current status of available methodologies for the aqueous dispersion and solubilization of CNTs, discuss the results on modifications of CNTs with various biological and bioactive species, and highlight some of the recent achievements in the fabrication and evaluation of CNT-based bioanalytical devices [129]. There is a diversity of detection products employing nanotechnology and revolutionizing the field of orthopedics. For instance, for osteoporosis, techniques for diagnosis have great importance in providing precise data detection in a timely, affordable, and noninvasive manner. Prior to techniques employing nanomaterials, there were few reasonable options for detection. However, new methods using nanotechnology allow for detection of osteoporosis with a handheld device. Specifically, research has led to the development of novel biochip that uses gold nanoparticles to detect a protein that is indicative of osteoporosis [25]. It has been shown to effectively evaluate bone conditioning and provide accurate detection and identification of the degree of damage to bones [130]. Jain [131] mentioned that nanotechnologies will extend the limits of current molecular diagnostics and enable point-of-care diagnostics, integration of diagnostics with therapeutics, and development of personalized medicine. Although the potential diagnostic applications are unlimited, the most important current applications are foreseen in the areas of biomarker discovery, cancer diagnosis, and detection of infectious microorganisms. Safety studies are needed for in vivo use. Because of its close interrelationships with other technologies, nanobiotechnology in clinical diagnosis will play an important role in the development of nanomedicine in the future.

Target-specific drug delivery
Nanotechnology has revolutionized therapeutics, allowing for greater precision of drug delivery, proving to be especially beneficial in the field of orthopedics [132]. One way this is done has been accomplished through coupling drug delivery to nanosensors. Alternatively, nanophase delivery systems can be used for drug delivery without an accompanying sensor [123]. Nanotechnology for targeted drug therapy originally began through the use of large particles (i.e., growth factors) and has more recently incorporated smaller particles (i.e., silver) into nanostructured materials [133, 134]. Using nanotechnology for drug delivery improves precision and also serves a vital purpose in reducing bacterial growth, and thus, the risk of infection [123].

The potential of nanomedicine in cancer therapy is infinitely promising due to the fact that novel developments are constantly being explored. This is particularly the case in the use of nanoparticles in both tumor diagnosis and treatment. This chapter attempts to describe some recent advances using nanoparticle drug delivery system in cancer therapy [135]. Remarkable progresses have been made also in the field of nano-oncology by improving the efficacy of traditional chemotherapy drugs for a plethora of aggressive human cancers [136]. Hence, nano-oncology is a very attractive application of nanoscience and allows for the improvement of tumor response rates in addition to a significant reduction of the systemic toxicity associated with current chemotherapy treatments [123].

Hyperthermia, an established adjuvant in cancer treatment, potentiates the effects of anticancer drugs and synergetic combination can be obtained, improving the therapeutic outcome. Mohamed et al. [137] mentioned that magnetizable nanoparticles and antineoplastic agents are able to identify and deliver drug therapies to cancerous bones in a targeted manner. A magnetite-enriched collagen/HA composite material was prepared and obtained by a coprecipitation method [138]. Hu et al. [139]

prepared 3D nanomagnetite/chitosan rod via in situ precipitation method, which was dispersed homogeneously in the chitosan matrix with magnetite crystal size of 11.5 nm. Murakami et al. [140] fabricated the magnetite/HA composite that facilitates direct bonding to bones through HA, and generation of heat from magnetite exposed to AC magnetic field is considered suitable for hyperthermia therapies of cancer in bones. The composite had micro-sized pores of about 400 μm and submicron-sized pores of about 0.2 μm in size and magnetite particle aggregates were strongly trapped in the cages of rod-shaped HA particles only when the magnetite contents were 30 mass% or less.

Nanopharmaceuticals are of two types: (1) a therapeutic agent itself acts as a own nanocarrier system, and (2) therapeutic agents with modified characteristics by coated nanoparticle carriers or nanoengineered drugs are used. They are functionalized or entrapped to develop new properties. Poorly soluble, poorly absorbed drugs acquire new properties suited for better drug delivery [141, 142]. Labile biologically active substances are converted by nanoparticulate systems into stable and ideal drugs for better delivery. Recently, newer drug carriers are developed using third-generation vectors. It consists of biodegradable core together with a polymer envelope with membrane recognition ligands. Dubin [143] and Dass et al. [144] pointed out that nanoparticles used for drug carriers for chemotherapeutics deliver medications directly to tumors, without any harm to the healthy tissues. Precision drug targeting depends on the efficiency of drug-targeting systems on the three components, namely, targeting structure, drug attachment or carrying element, and the drug delivered. Unique molecular feature for different diseases known as "unique address" makes it possible for site-specific targeting of drugs. It is a milestone in the improvement of drug delivery systems. The therapeutic target must be abundantly expressed by most diseased cells and is absent from healthy issues. The methods used for drug targeting are (1) unique molecular structures target cancer molecules of various organs and tissues or (2) antibodies to these unique molecular structures are used as therapeutic agents. Stabilization of drug molecules by nanoscale drug delivery systems is used as gene delivery vectors and prevents rapid degradation of drugs. Nanoparticles, nanowires, nanocages, and dendrimers are the different types of nanosized carriers with clear advantages: (i) drug reaches the target cell without being degraded, (ii) enhances drug absorption in cancer cells, (iii) avoids side effects, and (iv) it has no interaction with normal cells [141].

Nanoinformatics has provided a major supplementary platform for nanoparticle design and analysis to overcome such in vitro barriers. Nanoinformatics exclusively deals with the assembling, sharing, envisaging, modeling, and evaluation of significant nanoscale-level data and information. Nanoinformatics also facilitates chemotherapy by improving the nanomodeling of the tumor cells and aids detection of the drug-resistant tumors easily. Hyperthermia-based targeted drug delivery and gene therapy approaches are the latest nanoinformatic techniques proven to treat cancer with least side effects [122]. The empowering role of nanoinformatics in design and elucidation of nanoparticles for effective cancer treatment has made this field a fascinating area for researchers, inspiring them to enhance up the quality and efficacy of existing anticancer medicines. Theoretical and computational modeling is being seen as a forefront solution for problems related to surface chemistry, optimized geometry, or other properties in nanoparticle designing and drug delivery [122].

Bone grafting and osseointegration

Nanotechnology is the application of science and engineering at the nanoscale. A diverse range of applications are beginning to emerge in all areas of medicine. Nanostructured materials have been proposed as the next generation of orthopedic implant properties by creating a surface environment more conducive for osteoblast function. Bone substitute materials, whose nanoscale composition emulates the hierarchic organization of natural bone, show initiation of the desirable formation of an apatite layer. Nanotechnology has also been harnessed to improve the cutting performance and quality of surgical blades. Postoperative infection rates may be reduced by using nanofibrous membrane wound dressings containing antibacterial properties. The most notable application of nanotechnology in orthopedics may be drug delivery, including nanotherapeutics for treating bone cancer and arthritis. Nanotechnology is being used in orthopedics and will likely play a valuable role in future developments [145].

Traditional treatment methods for bone defect (such as bone allografts and autografts) are still often utilized; however, these techniques are accompanied by many risks, including infection, rejection by the immune system, and lengthy times for complete repair, especially for minor defects [123, 146]. Bone augmentation (or grafting) is required not only for orthopedic implantation but also for dental implant treatment, if the bone quality is poor (or less density) and less mass quantity. For successful osseointegration, a placed implant material must be inhabitable for bone-forming cells (osteoblasts) so that they can colonize the implant surface and synthesize new bone tissue [147]. The success of both orthopedic implants and tissue-engineered construct is highly dependent on the interactions between the selected biomaterial and the host tissue. One of the key factors identified in the failure of both types of implants was insufficient tissue regeneration around the biomaterial immediately after implantation [148]. This has been attributed to poor surface interaction of biomaterials with the host tissue. In dental implant treatment, there are several different types of bone grafting. The type of bone graft used depends on the extent of the damage and the location of the lost tooth. They can include the following grafts: (i) *a socket graft* that is ideally conducted immediately after a tooth is removed in order to conserve the alveolar ridge and stop any bone deterioration from occurring. Bone material is inserted into the vacant socket left by the displaced tooth. In most cases, modern-day socket grafts use "xenograft materials" or bone from a nonhuman source, such as animals (usually a cow). Amazingly, over time, the body adjusts to the composition of this foreign bone material until it becomes the human bone. It usually takes 3–6 months for the graft to heal before you can add a dental implant. (ii) *Block bone grafts* typically use human bone in the form of a small "block" taken from the patient's chin or lower jaw near where they once had wisdom teeth. They are selectively used where a xenograft would not render enough bone thickness to build up the already deteriorated ridge. This process is very similar to a socket graft with about the same healing timeline. (iii) *Sinus lift grafts* are needed when a patient requires an implant in the upper jaw, since this area is not typically stable enough to hold a dental implant. If the maxillary sinus cavity is too close to where a dental implant is needed, a sinus lift graft is performed. This type of grafting customarily uses the equine bone, because it does not dissolve as quickly as human bone and microscopically is more comparable to the human bone. This equine bone creates a kind of "scaffold" that promotes the growth of bone in the sinus. After a sinus lift graft, the healing process typically takes 8–12 months. Artificial bone (as bone replacement materials) is a laboratory-created bone-like material used as bone graft. Hydroxyapatite and collagen fibers are the major components of bone. Chondroitin sulfate, keratan sulfate, and lipid are also present [141]. Organic polysaccharides (chitin, chitosan, alginate) and minerals (HA) are the material types prepared in bone grafting procedures [149, 150]. Nanoclay filler-reinforced bone cement has enhanced mechanical properties. Selenium, nanoceramics, alumina, titania, carbon, nanometals, Ti–6Al–4V, cobalt chrome alloys, and nanocrystalline diamond display nanophase characteristics [151, 152]. Bone replacement materials are used to treat fractures of the bones, periprosthetic fractures during hip revision surgery, acetabular reconstruction, filling cages in spinal column surgery, osteotomies, and for filling bone defects in children [153].

There are a number of important reasons to explore the potential for the application of nanomaterials in orthopedic surgery. The use of nanotechnology has been tested on a wide range of materials (such as metals, ceramics, polymers, and composites), where either nanostructured surface features or constituent nanomaterials (including grains, fibers, or particles with at least 1D from 1 to 100 nm) have been utilized. These nanomaterials have demonstrated superior properties compared with their conventional (or micron structured) counterparts, due to their distinctive nanoscale features and the novel physical properties that ensue. The aim of this chapter is to explore how nanotechnology can really improve the future of orthopedic implants and scaffolds for bone and cartilage defects. Here we are showing the most relevant works about the use of nanotechnologies for the treatment of osteochondral defects [146]. Nanocomposites and nanostructured materials are believed to play a pivotal role in orthopedic research since the bone itself is a typical example of a nanocomposite [147]. Implants using nanobiomaterials have improved upon many of those risks, but still are not without fail. Implants derived from nanomaterials are not yet able to provide restoration of full functionality, nor do they often have longevity beyond a decade or two, at best. Complete failure of implants may occur and can be particularly challenging, requiring extensive and expensive reoperations [154]. Nevertheless, nanotechnology has proven incredibly beneficial for use in orthopedic implants, improving the treatment of many types of bone defects and orthopedic traumas. Several nanoscale biomaterials have been investigated and applied, leading to the use of a wide array of potential materials with their own unique properties and benefits. Examples of materials include gelatin, bioactive ceramics, biodegradable polymers, and polysaccharides such as agarose [146]. These nanomaterials are able to work well within the human body, as their physical properties and nanoscale features allow them to promote cell growth and tissue regeneration. The ability of these nanomaterials to mimic cellular environment is key in replicating mechanisms of cells, which also have nanometer dimensions and come together to form extracellular matrices [146]. Furthermore, implants with nanomaterials are able to form a greater surface area, which helps cultivate a healthy environment for bone growth and reduce infection rates. Nanomaterials have a wide array of uses for implants and scaffolding in orthopedics, ultimately contributing to faster recoveries, decreasing risks of surgery, and improving overall health of the affected area. However, many potential uses are yet to be investigated, and there is still a lack of clarity surrounding long-term safety and clinical benefits [146].

Nanosurgery

Nanotechnology applications in medicine are efficiently utilized to revolutionize the prevention, diagnosis, and treatment of disease including drug delivery [147, 155]. Surgical specialties, such as orthopedic surgery, are among those developing nanotechnology applications for clinical use. Orthopedic surgery addresses disorders of the musculoskeletal system including repair by both surgical and nonsurgical means of tendons, ligaments, muscles, bones, and nerves injured due to trauma or disease [155]. Medical interventions targeting orthopedic conditions are becoming increasingly important, given current epidemiologic trends in these conditions.

Nanotechnology has played a major role in all surgical specialties with fabrication of surgical implants and tissue engineering products. They have improved imaging technology, drug delivery systems, and scaffolds fabrication with improved material–cell interaction that are useful in surgical practice of wound healing. Nanotechnology applications are spread over almost all surgical specialties and have revolutionized treatment of various medical and surgical conditions. Clinically relevant applications of nanotechnology in surgical specialties include development of surgical instruments, suture materials, imaging, targeted drug therapy, visualization methods, and wound healing techniques. Management of burn wounds and scar is an important application of nanotechnology. Prevention, diagnosis, and treatment of various orthopedic conditions are crucial aspects of technology for functional recovery of patients. Improvement in standard of patient care, clinical trials, research, and development of medical devices for safe use are improved with nanotechnology [141].

As to surgical tools, there are surgical tools (plasma-polished blade, blades with diamond nanolayers), suture needles (made of nanosized SS crystals), surgical sutures with drug-eluting capacities such as sutures incorporated with drugs like aceclofenac or insulin, or nanotweezers for precise cellular level surgery possible [141]. As to smart nanobiomaterials, there are woven fabrics and textiles using CNTs for wound care [141], silver-based nanoparticles incorporated with antimicrobial agent for burn wound healing [156], or nitric oxide-delivering silica nanoparticles for wound healing [157, 158]. These surgical nanodevices and smart nanomaterials can make MIS (minimally invasive surgery) easy and enable [159, 160]. Surgery, which is supported by or assisted with nanotechnology in the form of MIS and others, can include maxillofacial surgery, thoracic surgery, vascular surgery, neurosurgery, ophthalmic surgery, or plastic and reconstructive surgery.

Although several conventional therapies have been replaced by nanotechnology-assisted medicine, Garimella et al. [123] pointed out that there are concerns to be taken into considerations and questions to be answered such as long-term clinical safety, potential risks of lung cytotoxicity and inflammation of internal organs [151, 161, 162], viability and toxicity of nanosized particles and sensors, and enhancement of longevity of nanobiomaterials for further investigations [123].

6.6.2 Nanodentistry

As nanotechnology is becoming very important technology in medicine (now recognized as nanomedicine), its application in dental field is becoming indispensable, in particular, for implant surface modification with a purpose of improving osseointegration. With the application of nanotechnology, the composition of dental implants, surface energy, and roughness and topography can be improved for better osseointegration and it can also influence the events occurring at the bone–implant interface. The cellular activities and tissue responses occurring at the bone–implant interface can be altered by nanoscale modifications and can result in better treatment outcomes [163–165]. Immediately following implantation, implants are in contact with proteins and platelets from blood. The differentiation of mesenchymal stem cells will then condition the peri-implant tissue healing. Direct bone-to-implant contact is desired for a biomechanical anchoring of implants to the bone rather than fibrous tissue encapsulation. Surface properties such as chemistry and roughness play a determinant role in these biological interactions. Physicochemical features in the nanometer range may ultimately control the adsorption of proteins as well as the adhesion and differentiation of cells [166].

The structures encountered by osteoblasts in the human body are not only in micrometer scale, since bones are made up of nanostructures. Thus, there is a need to produce better implant materials having also nanometer roughness [167]. Several studies have suggested that nanophase materials produced from various chemistries, such as metals, polymers, composites, and ceramics, improved cellular activities when compared with conventional microrough materials [168, 169]. Nanobiomaterials have an increased percentage of atoms and crystal structures, and also provide a higher surface area than the conventional ones. Thus, nanoscale surfaces possess high surface energy leading to increasing initial protein adsorption that is very important in regulating the cellular interactions on the implant surface. Webster et al. [170] suggested increased osteoblast adhesion on nanophase materials. Numerous studies have shown that osteoblasts cultured on nanophase biomaterials exhibited better osteogenic behavior, including adhesion, extracellular matrix production, and mineralization than on conventional materials [171, 172]. In recent years, several methods have also been developed to produce nanoscale structures on titanium surface [167]. While irregular nanomorphologies can be established using solution chemistry [173], the electrochemical anodization of titanium is the most popular and novel strategy to produce controlled structures (including nanotubes, pillar-like nanostructures, and nanodots) on implant surfaces for load-bearing approaches [174, 175].

Nanotechnology may involve 1D concepts (nanodots and nanowires) or the self-assembly of more complex structures (nanotubes) [176]. Materials are also classified according to their form and structure as nanostructures, nanocrystals, nanocoatings, nanoparticles, and nanofibers [154]. Nanoscale modification of the titanium endosseous implant surface may lead to alteration in the topography as well as chemistry of the surface. Therefore, the goal of nanoscale modification should be a specific chemical modification of cpTi. A distinct implication associated with nanoscale manipulation of any material is that it also leads to inherent chemical changes on the material surface. Albrektsson et al. [177] divided implant surface quality into three categories: (i) mechanical properties, (ii) topographic properties, and (iii) physicochemical properties and pointed out that these characteristics are interrelated and a change in any of these groups affects the others as well.

It is important to note that the idea of creating metallic implants with decreased surface feature dimensions (i.e., into the nanometer regime) in order to mimic the roughness of extracellular matrices in the bone has also been utilized by others [169, 178, 179]. However, in such studies, the modified synthetic materials varied in a number of properties, not just the degree of nanometer surface roughness that may have also influenced osteoblast function [178]. For example, it has been reported that Ti and Ti–6Al–4V treated with H_2SO_4 and H_2O_2 to create nanotextured surfaces promoted osteoblast osteopontin and bone sialoprotein synthesis [178]. Although this represents a novel finding due to chemistry changes that may have occurred during chemical treatment of Ti and Ti–6Al–4V, it is not clear which property (chemistry or nanometer roughness) increased functions of osteoblasts.

There are various methods to create nanoscale features at the implant surface. These methods include (i) physical methods, like self-assembly of nanolayers and compaction of nanoparticles and ion beam deposition; (ii) chemical methods, like acid etching, peroxidation, alkali treatment (NaOH), and anodization; (iii) nanoparticle deposition like sol–gel (colloidal particle deposition) and discrete crystalline deposition; and (iv) lithography and contact printing technique [169]. By self-assembly of monolayers, the exposed functional end group could be a molecule with different functions (an osseoinductive or cell adhesive molecule). The compaction of nanoparticles conserves the chemistry of the surface among different topographies. Ion beam deposition can impart nanofeatures to the surface based on the materials used. Acid etching (material subtraction technique) combined with other methods (such as sandblasting and/or peroxidation) can impart nanofeatures to the surface and remove contaminants. Peroxidation can produce a titania (TiO_2) gel layer and both chemical and topographic changes are imparted. Alkali treatment (with NaOH) produces a sodium gel layer allowing HA deposition, and both chemical and topographic changes are imparted. Anodization can impart nanofeatures to the surface creating a new oxide layer. Sol–gel colloidal particle adsorption creates a thin film of controlled chemical characteristics, and atomic-scale interactions display string physical interactions. Discrete crystalline deposition can superimpose a nanoscale surface topographic complexity on the surface. By the lithography and contact printing technique, many different shapes and materials can be applied over the surface, and approaches are labor intensive and require considerable development prior to clinical translation and application on implant surface [169].

TNT (titania nanotube)

The titania nanotube (TNT) arrays are one of the most promising candidates of titanium nanosurfaces for dental implantology. Several in vitro studies have demonstrated that cells cultured on these nanotubular surfaces showed higher adhesion, proliferation, alkaline phosphatase activity, and bone matrix deposition [179, 180]. Fabrication of TNT is achieved basically by electrochemical anodization. Zakir et al. [181] used an electrolyte consisting of a mixture of glycerol (92%) with distilled water (8%) and ammonium fluoride (0.4 M) for 60 min, at a voltage ranging from 30 to 60 V, using a two-electrode cell with the titanium sample as the anode and a platinum electrode as the counter-electrode, followed by furnace annealing at 600 °C for 2 h to enhance the crystallinity of synthesized TNT films. Sreekantan et al. [182] employed a similar anodization procedure using a mixed electrolyte of glycerol (85%, containing 15% water) containing 6 wt% ethylene glycol (EG) and 5 wt% ammonium fluoride (NH_4F), at 50 V for 60 min, followed by furnace annealing at 400 °C for 2 h in argon to obtain the crystalline phase. Li et al. [183] used the electrolyte of a mixed solution of EG + 0.3 wt% NH_4F + 2 vol%H_2O at 30–60 V, followed by annealing at 450 °C for 2 h in high-purity N_2 gas. Figure 6.14 shows typical SEM images of fabricated TNT structure [188]. Oh et al. [180] anodized pure Ti sheet in a 0.5% HF solution at 20 V for 30 min at room temperature with a platinum electrode as the counter-electrode, followed by crystallization heat treatment at 500 °C for 2 h. Figure 6.14 shows a typical SEM image of TiO_2 nanotubes (TNT) fabricated at anodization voltage of 50 V [174].

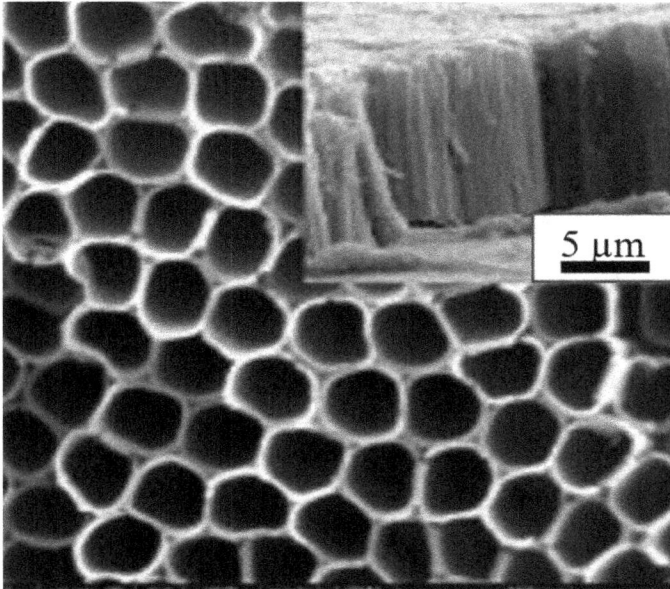

Figure 6.14: SEM image of TiO$_2$ nanotubes anodized at voltage of 50 V [182].

Gulati et al. [184] fabricated unique anisotropic titania nanopores (TNPs) on the surface of titanium (via electrochemical anodization) toward enhancing the soft tissue integration and wound healing abilities of the conventional abutments. It was reported that (i) improved cell viability was observed on TNPs as compared to controls, and (ii) cellular spreading morphology indicated cell alignment along the direction of the nanopores with strong anchoring evident by enhanced filopodia and stress fibers.

These increased in vitro cellular activities for TNTs also translated to in vivo bone bonding. Nanotubular surfaces significantly improved bone bonding strength by as much as ninefold compared with grit-blasted surfaces, and histological analysis revealed greater bone–implant contact and collagen type I expression confirming the better in vivo behavior of TNTs [185, 186]. It has also been shown that various nanomorphological features of TNTs, such as length, diameter, and wall thickness, have a major impact on the cellular responses, providing the evidence that cells are susceptible to nanoscale dimensions [187, 188]. Besides, nanotubular structures on titanium provide a suitable infrastructure for loading and subsequent releasing of antibiotics [180, 189] or for immobilizing biosignaling molecules for better osseointegration [190]. However, there is still a need for additional studies that would optimize the fabrication of nanotubes for better bioactivity.

Four material-related factors, which can impact events at the bone–implant interface, are surface roughness, implant surface material composition, surface topography, and surface energy. These properties can be modified using different methodologies such as sandblasting, acid etching, laser etching, and surface anodization. With regard to the surfaces themselves, there are three levels of surface structure, namely, nano-scale, microscale, and macroscale surface topography. At best, current surface structures are controlled at the micron level, but primarily, processes controlled at the nanoscale level determine tissue response. A critical role is played by surface profiles in the nanometer range with regard to the adsorption of proteins, osteoblastic cell adhesion, and ultimately the rate, level, and quality of osseointegration [191]. Based on the above background, Gupta et al. [191] reviewed the use of nanotechnology for the purpose of implant surface modification in order to improve the osseointegration process of ceramic implants, but also the various techniques and methods by which nanofeatures can be applied to zirconia implant surfaces. It was reported that (i) there are various methods reported for nanotexturization on zirconia surfaces, including solid-state laser, CaP coating, selective infiltration etching, Nd:YAG laser ablation + Ag/Au particles deposition, anodization, coating sol–gel-derived titania, femtosecond laser, self-assembly nanoislands, or hydrothermal treatment; (ii) nanotechnology for roughening of zirconia dental implant surfaces is still evolving; (iii) nanometer-controlled surfaces have an excellent impact on healing after implant placement, as several in vitro and animal studies have shown, positively affecting the blood clot formation, adsorption of proteins and cell division, and differentiation occurring upon implantation; and (iv) nanotechnology has opened a new range of possibilities for improvement of zirconia implants.

Nanoparticle coating
Current approaches in implant surface modification can generally be categorized into three areas: ceramic coatings, surface functionalization, and patterning on the microscale to nanoscale. The distinctions among these are imprecise, as some or all of these approaches can be combined to improve in vivo implant performance. These surface improvements have resulted in durable implants with a high percentage of success and long-term function [192].

There is a strong belief that nanoscale materials will produce a new generation of implant materials with high efficiency, low cost, and high volume. The nanoscale in materials processing is truly a new frontier and includes the fabrication of novel coatings and nanopatterning of dental implants. The ultimate goal is to produce materials and therapies that will bring state-of-the-art technology to the bedside and improve quality of life and current standards of care [165]. The combined requirements imposed by the enormous scale and overall complexity of designing new implants or complete organ regeneration are well beyond the reach of present technology in many dimensions, including nanoscale, as we do not yet have the basic knowledge required to achieve these goals. The need for a synthetic implant to address multiple physical and biological factors imposes tremendous constraints on the choice of suitable materials. There is a strong belief that nanoscale materials will produce a new generation of implant materials with high efficiency, low cost and high volume. The nanoscale in materials processing is truly a new frontier. Metallic dental implants have been successfully used for decades but they have serious shortcomings related to their osseointegration and the fact that their mechanical properties do not match those of the bone. This chapter reviews recent advances in the fabrication of novel coatings and nanopatterning of dental implants. It also provides a general summary of the state of the art in dental implant science and describes possible advantages of nanotechnology for further improvements. The ultimate goal is to produce materials and therapies that will bring state-of-the-art technology to the bedside and improve quality of life and current standards of care. Various coatings have been developed to improve an implant's ability to bond to living tissues, particularly bone. The idea is to apply a thin ceramic layer that will bond both to the implant and to the surrounding tissue while promoting bone apposition. Candidate coating materials are bioactive compounds that are able to promote cell attachment, differentiation, and bone formation. The most prevalent bioactive materials are CaPs (such as HA or tricalcium phosphate) and bioactive glasses. When implanted, these bioactive ceramics form a carbonated apatite layer on their surfaces through dissolution and precipitation. This phase is equivalent in composition and structure to the mineral phase of osseous tissue. At the same time, collagen fibrils can be incorporated into the apatite agglomerates. The sequence of events is poorly understood but appears to be adsorption of biological moieties and action of macrophages, attachment of stem cells and differentiation, formation of matrix, and, finally, complete mineralization [193–195].

Nanoparticles have been introduced as materials with good potential to be extensively used in biological and medical applications. Nanoparticles are clusters of atoms in the size range of 1–100 nm. Inorganic nanoparticles and their nanocomposites are applied as good antibacterial agents. The metallic nanoparticles are the most promising as they show good antibacterial properties due to their large surface-area-to-volume ratios, which draw growing interest from researchers due to increasing microbial resistance against metal ions, antibiotics, and the development of resistant strains. Metallic nanoparticles can be used as effective growth inhibitors in various microorganisms and are therefore are applicable to diverse medical devices [196]. It was mentioned that (i) coating of Ti surfaces with nanoparticles can improve soft tissue integration and osteogeneration that leads to improved fixation of implants, (ii) osteoconductive nanoparticles induce a chemical bond with the bone to attain good biological fixation for implants, (iii) surface modification of implants using antibacterial properties can also decrease the potential for infection, and certainly, present improved clinical outcomes, and (iv) nanosurfaced features of the dental implants improve osseointegration [197, 198].

There are studies on coating effects of CaP and applications to dental implant surfaces: thin CaP coating [199, 200], calcium ion modification [201], crystalline hydroxyapatite, and amorphous CaP compacts functionalized with the arginine–

glycine–aspartic acid peptide sequence [202], porous CaP [203], or nanodimensioned hydroxyapatite [204].

Besides CaP, there are several metallic oxide nanoparticles used: silver nanoparticle coating using antimicrobial properties [205], zinc oxide nanoparticles to inhibit bacterial adhesion and promote osteoblast growth [206], titania coating incorporated with silver nanoparticles for antibacterial ability [207], and copper oxide nanoparticles that attribute to combat infection, to control the formation of biofilms within the oral cavity, as a function of their biocidal, antiadhesive, and delivery capabilities [208].

Several anti-inflammatory agents have been employed: quercitrin nanocoating to decrease initial bacterial adhesion while increasing human gingival fibroblast attachment [209], or chlorhexidine hexametaphosphate nanoparticles to exhibit antimicrobial efficacy against oral primary colonizing bacterium *Streptococcus gordonii* [210, 211].

References

[1]   Bartolo P, Kruth J-P, Silva J, Levy G, Malsje A, Rajurkar K, Mitsuishi M, Ciurana J, Leu M. Biomedical production of implants by additive electro-chemical and physical processes. CIRP Ann Manuf Technol. 2012, 61, 635–55.

[2]   ASTM Committee F42 on Additive Manufacturing Technologies; http://www.astm.org/COM MITTEE/F42.htm.

[3]   Peduk GSA, Dilibal S, Harrysson O, Özbek S. Comparison of the production processes of nickel-titanium shape memory alloy through additive manufacturing. International Symposium on 3D Printing (Additive Manufacturing). 2017, 3/4: https://nickel-titanium.com/ wp-content/uploads/NiTi_AM_Paper_Symposion-on-3D_AM_2017.pdf.

[4]   Jahangir MN, Mamun MAH, Sealy MP. A review of additive manufacturing of magnesium alloys. AIP Conference Proceedings. 2018: https://doi.org/10.1063/1.5044305

[5]   Karunakaran R, Ortgies S, Tamayol A, Michael FB, Sealy P. Additive manufacturing of magnesium alloys. Bioact Mater. 2020, 5, 44–54.

[6]   Hutmacher DW, Sittinger M, Risbud MV. Scaffold-based tissue engineering: Rationale for computer-aided design and solid free-form fabrication systems. Trends Biotechnol. 2004, 22, 354–62.

[7]   Shirazi SFS, Gharehkhani S, Mehrali M, Yarmand H, Metselaar HSC, Kadri NA, Osman NAA. A review on powder-based additive manufacturing for tissue engineering: Selective laser sintering and inkjet 3D printing. Meta Sci Technol Adv Mater. 2015, doi: 10.1088/1468-6996/ 16/3/033502.

[8]   Pham DT, Gault RS. A comparison of rapid prototyping technologies. Int J Mach Tools Manuf. 1998, 38, 1257–87.

[9]   Wendel B, Rietzel D, Kühnlein F, Feulner R, Hülder G, Schmachtenberg E. Additive processing of polymers. Macromol Mater Eng. 2008, 293, 799–809.

[10]   Melchels FPW, Feijen J, Grijpma DW. A review on stereolithography and its applications in biomedical engineering. Biomaterials. 2010, 31, 6121–30.

[11]   Oshida Y, Tominaga T. Nickel-Titanium Materials. De Gruyter Pub., 2020.

[12]   Murphy SV, Atala A. 3D bioprinting of tissues and organs. Nat Biotechnol. 2014, 32, 773–85.

[13]   Ngo TD, Kashani A, Imbalzano G, Nguyen KTQ, Hui D. Additive manufacturing (3D printing): A review of materials, methods, applications and challenges. Compos Part B Eng. 2018, 143, 172–96.

[14]   Ødegaard KS, Torgersen J, Elverum CW. Structural and biomedical properties of common additively manufactured biomaterials: A concise review. Metals. 2020, 10, 1677, doi: https://doi.org/10.3390/met10121677.

[15]   Jasiuk I, Abueidda DW, Kozuch C, Pang S, Su FY, McKittrick J. An overview on additive manufacturing of polymers. JOM. 2018, 70, 275–83.

[16]   Warnke PH, Seitz H, Warnke F, Becker ST, Sivananthan S, Sherry E, Liu Q, Wiltfang J, Douglas T. Ceramic scaffolds produced by computer-assisted 3D printing and sintering: Characterization and biocompatibility investigations. J Biomed Mater Res B: Appl Biomater. 2010, 93B, 212–17.

[17]   Rahmati S, Abbaszadeh F, Farahmand F. An improved methodology for design of custom-made hip prostheses to be fabricated using additive manufacturing technologies. Rapid Prototyp J. 2012, 18, 389–400.

[18]   Bohner M, Van Lenthe GH, Grunenfelder S, Hirsiger W, Evison R, Muller R. Synthesis and characterization of porous beta-tricalcium phosphate blocks. Biomaterials. 2005, 26, 6099–105.

[19]   Yiming C, Guochao G, Huijun Y, Chuanzhong C. Laser surface alloying on aluminum and its alloys: A review. Opt Lasers Eng. 2018, 100, 23–37.

[20]   Taylor-Smith K. What is Laser Surface Alloying? 2019; https://www.azooptics.com/Article.aspx?ArticleID=1595.

[21]   Tian YS, Chen CZ, Wang DY, Huo QH, Lei TQ. Laser surface alloying of pure titanium with TiN-B-Si-Ni mixed powders. Appl Surf Sci. 2005, 250, 223–27.

[22]   Vilar R. Laser powder deposition. Comprehen Mater Process. 2014, 10, 163–216.

[23]   Nematollahi M, Jahadakbar A, Mahtabi MJ, Elahinia M. Additive Manufacturing (AM). Metals for Biomedical Devices. 2nd ed. Woodhead Publishing Series in Biomaterials. 2019, 331–53.

[24]   Walczak M, Papadopoulou EL, Sanz M, Manousaki A, Marco JF, Castillejo M. Structural and morphological characterization of TiO2 nanostructured films grown by nanosecond pulsed laser deposition. Appl Surf Sci. 2009, 255, 5267–70.

[25]   Kurland H-D, Stötzel C, Grabow J, Zink I, Müller E, Staupendahl G, Müller FA. Preparation of spherical titania nanoparticles by CO2 laser evaporation and process-integrated particle coating. J Amer Ceramic Soc. 2010, 93, 1282–89.

[26]   Mihailescu IN, Ristoscu C, Bigi A, Mayer I. Laser-surface interactions for new materials production. In: Advanced Biomimetic Implants Based on Nanostructured Coatings Synthesized by Pulsed Laser Technologies. Springer Series in Materials Science, Vol. 130, 2010, 235–60.

[27]   Cotell CM, Chrisey DB, Grabowski KS, Sprague JA, Gossett CR. Pulsed laser deposition of hydroxylapatite thin films on Ti-6Al-4V. J Appl Biomater. 1992, 3, 87–93.

[28]   Liang CY, Yang XJ, Wei Q, Cui ZD. Comparison of calcium phosphate coatings formed on femtosecond laser-induced and sand-blasted titanium. Appl Surf Sci. 2008, 255, 515–18.

[29]   Duty C, Jean D, Lackey WJ. Laser chemical vapour deposition: Materials, modelling, and process control. Int Mater Rev. 2001, 46, 271–87.

[30]   Paun IA, Moldovan A, Luculescu CR, Dinescu M. Antibacterial polymeric coatings grown by matrix assisted pulsed laser evaporation. Appl Phys A. 2013, 110, 895–902.

[31]   Mishra VK, Srivastava SK, Asthana BP, Kumar D. Structural and spectroscopic studies of hydroxyapatite nanorods formed via microwave-assisted synthesis route. J Amer Ceramic Soc. 2012, 95, 2709–15.

[32] Balasubramanian S, Ramadoss A, Kobayashi A, Muthirulandi J. Nanocomposite Ti–Si–N coatings deposited by reactive dc magnetron sputtering for biomedical applications. J Amer Ceramic Soc. 2012, 95, 2746–52.

[33] Yoshinari M, Oda Y, Inoue T, Shimono M. Dry-process surface modification for titanium dental implants. Metallurg Mater Trans A. 2002, 33, 511–19.

[34] Noorakma ACW, Zuhailawati H, Aishvarya V, Dhindaw BK. Hydroxyapatite-coated magnesium-based biodegradable alloy: Cold spray deposition and simulated body fluid studies. J Mater Eng Performance. 2013, doi: 10.1007/s11665-013-0589-9.

[35] Cai K, Rechtenbach A, Hao J, Bossert J, Klaus D, Jandt KD. Polysaccharide-protein surface modification of titanium via a layer-by-layer technique: Characterization and cell behaviour aspects. Biomaterials. 2005, 26, 5960–71.

[36] Ai H, Jones SA, Lvov YM. Biomedical applications of electrostatic layer-by-layer nano-assembly of polymers, enzymes, and nanoparticles. Cell Biochem Biophys. 2003, 39, 23–43.

[37] Zhang Y, Kong D, Yokogawa Y, Feng X, Tao Y, Qiu T. Fabrication of porous hydroxyapatite ceramic scaffolds with high flexural strength through the double slip-casting method using fine powders. J Amer Ceramic Soc. 2012, 95, 147–52.

[38] Schaaf P. Laser nitriding of metals. Prog I Mater Sci. 2002, 47, 1–161.

[39] Lavisse L, Grevey D, Langlade C, Vannes B. The early stage of the laser-induced oxidation of titanium substrates. Appl Surf Sci. 2002, 186, 150–55.

[40] Höche D, Rapin G, Schaaf P. FEM simulation of the laser plasma interaction during laser nitriding of titanium. Appl Surf Sci. 2007, 254, 888–92.

[41] György E, Del Pino AP, Serra P, Morenza JL. Surface nitridation of titanium by pulsed Nd: YAG laser irradiation. Appl Surf Sci. 2002, 186, 130–34.

[42] Yue TM, Yu JK, Mei Z, Man HC. Excimer laser surface treatment of Ti-6Al-4V alloy for corrosion resistance enhancement. Mater Lett. 2002, 52, 206–12.

[43] Man HC, Bai M, Cheng FT. Laser diffusion nitriding of Ti-6Al-4V for improving hardness and wear resistance. Appl Surf Sci. 2011, 258, 436–41.

[44] Dahotre SN, Vora HD, Pavani K, Banerjee R. An integrated experimental and computational approach to laser surface nitriding of Ti-6Al-4V. Appl Surf Sci. 2013, 271, 141–48.

[45] Hoeche D, Kaspar J, Schaaf P. Laser nitriding and carburization of materials. In: Lawrence J, et al., ed. Laser Surface Engineering: Processes and Applications. Woodhead, 2014, doi: 10.1016/B978-1-78242-074-3.00002-7.

[46] Munir K, Bieseikierski A, Wen C, Li Y. Selective laser melting in biomedical manufacturing. In: Metallic Biomaterials Processing and Medical Device Manufacturing. Woodhead Publishing Series in Biomaterials, 2020, 235–69, https://doi.org/10.1016/B978-0-08-102965-7.00007-2Getrightsandcontent.

[47] Murphy J. Selective Laser Melting (SLM) – 3D Printing Simply Explained. 2019; https://all3dp.com/2/selective-laser-melting-slm-3d-printing-simply-explained/.

[48] Yap CY, Chua CK, Dong ZL, Liu ZH, Zhang DQ, Loh LE, Sing SL. Review of selective laser melting: Materials and applications. Appl Phys Rev. 2015, 2, 041101, doi: https://doi.org/10.1063/1.4935926.

[49] Cavallo C. All About Selective Laser Melting 3D Printing; https://www.thomasnet.com/articles/custom-manufacturing-fabricating/selective-laser-melting-3d-printing/.

[50] Saedi S, Turabi AS, Andani MT, Haberland C, Elahinia M, Karaca H. Thermomechanical characterization of Ni-rich NiTi fabricated by selective laser melting. Smart Mater Struct. 2016, 25, Article 035005, https://iopscience.iop.org/article/10.1088/0964-1726/25/3/035005/pdf.

[51] Saedi S, Turabi AS, Andani MT, Haberland C, Karaca H, Elahinia M. The influence of heat treatment on the thermomechanical response of Ni-rich NiTi alloys manufactured by selective laser melting. J Alloys Compd. 2016, 677, 204–10.

[52] Saedi S, Saghaian SE, Jahadakbar A, Shayesteh Moghaddam N, Taheri Andani M, Saghaian SM, Lu YC, Elahinia M, Karaca HE. Shape memory response of porous NiTi shape memory alloys fabricated by selective laser melting. J Mater Sci Mater Med. 2018, 29, doi: 10.1007/s10856-018-6044-6.

[53] Oshida Y. Bioscience and Bioengineering of Titanium Materials. Elsevier Pub., 2007.

[54] Engh CA, McGovern TF, Bobyn JO, Harris WH. A quantitative evaluation of periprosthetic bone remodeling after cementless total hip arthroplasty. J Bone Joint Surg. 1992, 74-A, 1009–20.

[55] Mullen L, Stamp RC, Brooks WK, Jones E, Sutcliffe CJ. Selective laser melting: A regular unit cell approach for the manufacture of porous, titanium, bone in-growth constructs, suitable for orthopedic applications. J Biomed Mater Res Part B Appl Biomater. 2009, 89B, 325–34.

[56] Shishkovskii IV, Yadroitsev IA, Smurov IY. Powder metal. Metal Ceram. 2011, 50, 275, https://doi.org/10/1007/s11106-011-9329-6.

[57] Shishkovsky I, Yadroitsev I, Smurov I. Direct selective laser melting of nitinol powder. Phys Procedia. 2012, 39, 447–54.

[58] Moghaddam NS, Saghaian SE, Amerinatanzi A, Ibrahim H, Li P, Toker GP, Karaca HE, Elahinia M, Moghaddam NS. Anisotropic tensile and actuation properties of NiTi fabricated with selective laser melting. Mater Sci Eng A. 2018, 724, 220–30.

[59] Chekotu JC, Groarke R, O'Toole K, Brabazon D. Advances in selective laser melting of nitinol shape memory alloy part production. Materials (Basel). 2019, 12, E809, doi: 10.3390/ma12050809.

[60] Palermo E. What is Selective Laser Sintering? 2013; https://www.livescience.com/38862-selective-laser-sintering.html.

[61] Eshraghi S, Das S. Mechanical and microstructural properties of polycaprolactone scaffolds with one-dimensional, two-dimensional, and three-dimensional orthogonally oriented porous architectures produced by selective laser sintering. Acta Biomater. 2010. 2020, 6, 2467–76.

[62] Sallica-Leva E, Jardini AL, Fogagnolo JB. Microstructure and mechanical behavior of porous Ti-6Al-4V parts obtained by selective laser melting. J Mech Behav Biomed Mater. 2013, 26, 98–108.

[63] Vaezi M, Seitz H, Yang S. A review on 3D micro-additive manufacturing technologies. Int J Adv Manuf Technol. 2013, 67, 1721–54.

[64] Shuai C, Zhuang J, Peng S, Wen X. Inhibition of phase transformation from β-to & tricalcium phosphate with addition of poly (L-lactic acid) in selective laser sintering. Rapid Prototyp J. 2014, 20, 369–76.

[65] Witek L, Marin C, Granato R, Bonfante EA, Campos F, Bisinotto J, Suzuki M, Coelho PG. Characterization and in vivo evaluation of laser sintered dental endosseous implants in dogs. J Biomed Mater Res Part B. 2012, 100B, 1566–73.

[66] Riza SH, Masood SH, Rashid RAR, Chandra S. Selective laser sintering in biomedical manufacturing. In: Metallic Biomaterials Processing and Medical Device Manufacturing. Woodhead Publishing Series in Biomaterials, 2020, 193–233, https://doi.org/10.1016/B978-0-08-102965-7.00006-0Getrightsandcontent.

[67] Selective Laser Sintering. https://www.3dsystems.com/resources/information-guides/selective-laser-sintering/slsSLS Applications.

[68] Varotsis AB. Introduction to SLS 3D printing. https://www.hubs.com/knowledge-base/introduction-sls-3d-printing/.

[69] Duan B, Wang M, Zhou WY, Cheung WL, Li ZY, Lu WW. Three-dimensional nanocomposite scaffolds fabricated via selective laser sintering for bone tissue engineering. Acta Biomater. 2010, 6, 4495–505.

[70] Prasad K, Bazaka O, Chua M, Rochford M, Fedrick L, Spoor J, Symes R, Tieppo M, Collins C, Cao A, Markwell D, Ostrikov K, Bazaka K. Metallic biomaterials: Current challenges and opportunities. Materials (Basel). 2017, 10, 884, doi: 10.3390/ma10080884.

[71] Mulford JS, Babazadeh S, Mackay N. Three-dimensional printing in orthopaedic surgery: Review of current and future applications. ANZ J Surg. 2016, 86, 648–53.

[72] Gu Q, Hao J, Lu Y, Wang L, Wallace GG, Zhou Q. Three-dimensional bio-printing. Sci Chin Life Sci. 2015, 58, 411, doi: 10.1007/s11427-015-4850-3.

[73] Eltorai AE, Nguyen E, Daniels A. Three-dimensional printing in orthopedic surgery. Orthopedics. 2015, 38, 684–87.

[74] Gaget L. Medical 3D printing: Additive manufacturing for Jaw Reconstruction. 2018; https://www.sculpteo.com/blog/2018/03/28/medical-3d-printing-additive-manufacturing-for-jaw-re construction/.

[75] Sebin D. World's first 3D printed rib cage for cancer patient. 2015; https://www.dailyrounds.org/blog/worlds-first-3d-printed-rib-cage-for-cancer-patient/.

[76] Birrell I. 3D-printed prosthetic limbs: the next revolution in medicine. 2017; https://www.the guardian.com/technology/2017/feb/19/3d-printed-prosthetic-limbs-revolution-in-medicine.

[77] Limb Loss Statistics. https://www.amputee-coalition.org/resources/limb-loss-statistics/.

[78] Fuentes L. The most common 3D printed Prosthetics in 2021. 2021; https://all3dp.com/2/the-most-common-3d-printed-prosthetics/.

[79] How 3D Printed Prosthetics Can Change The Lives of Millions of Amputees. https://www.3dsourced.com/guides/3d-printed-prosthetics/.

[80] The Search for Better Bone Replacement: 3-D Printed Bone with Just the Right Mix of Ingredients. 2016; https://www.hopkinsmedicine.org/news/media/releases/the_search_for_better_bone_replacement_3_d_printed_bone_with_just_the_right_mix_of_ingredients.

[81] Hung BP, Naved BA, Nyberg EL, Dias M, Holmes CA, Elisseeff JH, Dorafshar AH, Grayson WL. Three-dimensional printing of bone extracellular matrix for craniofacial regeneration. ACS Biomater Sci Eng. 2016, 2, 1806–16.

[82] Application of 3D Printing Titanium and Titanium Alloy in Repair of Jaw Defect. 2019; https://www.hosnti.com/2019/08/28/application-of-3d-printing-titanium-and-titanium-alloy-in-re pair-of-jaw-defect/.

[83] Bulger M. Metal injection molding. Adv Mater Process. 2005, 163, 39–40.

[84] Yamagishi T, Ito M, Kou Y, Obata A, Deguchi T, Hayashi S, Igarashi Y. Mechanical properties of sintered titanium using metal injection molding. J Jpn Dent Mater. 1995, 14, 1–7.

[85] An overview of the metal injection moulding process. Powder Injection Molding Intl. https://www.pim-international.com/metal-injection-molding/an-overview-of-the-metal-injection-moulding-process/.

[86] Wei TS, German RM. Injection molded tungsten heavy alloy. Int J Powder Metallurgy. 1988, 24, 327–35.

[87] Metal Injection Molding (MIM); https://mppinnovation.com/metal-injection-molding/.

[88] Metal injection molding. https://en.wikipedia.org/wiki/Metal_injection_molding.

[89] Crawford M. Metal Injection Molding and Metal 3D Printing Examination. 2021; https://www.mpo-mag.com/issues/2021-07-01/view_features/metal-injection-molding-and-metal-3d-printing-examination/.

[90] Applications for metal injection molding: medical and orthodontic. Powder Injection Molding Intl.; https://www.pim-international.com/metal-injection-molding/applications-for-mim-i-medical-and-orthodontic/.

[91]   Johnson JL. Mass production of medical devices by metal injection molding. 2002; Medical Device & Diagnostic Industry. https://www.mddionline.com/manufacturing-processes/mass-production-medical-devices-metal-injection-molding.

[92]   Ferreira TJ, Vieira M-T F, Costa J, Silva M, Gago PT. Manufacturing dental implants using powder injection molding. J Orthod Endod. 2016, 2, https://www.researchgate.net/publica tion/297397305_Manufacturing_Dental_Implants_using_Powder_Injection_Molding.

[93]   Deguchi T, Ito M, Obata A, Koh Y, Yamagishi T, Oshida Y. Trail production of titanium orthodontic brackets fabricated by sintering metal injection molding methods. J Den Res. 1996, 75, 1491–96.

[94]   Aslan N, Aksakal B, Findik F. Fabrication of porous-Ti6Al4V alloy by using hot pressing technique and Mg space holder for hard-tissue biomedical applications. J Mater Sci: Mater Med. 2021, 32, 80, doi: https://doi.org/10.1007/s10856-021-06546-2.

[95]   Banhart J. Manufacture, characterization and application of cellular metals and metal foams. Prog Mater Sci. 2001, 46, 559–632.

[96]   Strano M. A new FEM approach for simulation of metal foam filled tubes. J Manufact Sci Eng. 2011, 133, 061003, doi: 10.1115/1.4005354.

[97]   https://www.slideshare.net/aliaahmeddiaa/foam-by-alia.

[98]   Kränzlin N, Niederberger M. Controlled fabrication of porous metals from the nanometer to the macroscopic scale. Mater Horiz. 2015, 2, 59–377.

[99]   Banhart J. Manufacturing routes for metallic foams. JOM Miner Met Mater Soc, 2000, 52, 22–27.

[100] Liu PS, Chen GF. Chapter three – application of porous metals. In: Porous Materials; Processing and Applications. 2014, 113–88.

[101] DeGroot CT, Straatman AG, Betchen LJ. Modeling forced convection in finned metal foam heat sinks. J Electron Packag. 2009, 131, doi: https://doi.org/10.1115/1.3103934.

[102] https://en.wikipedia.org/wiki/Metal_foam.

[103] Guo N, Leu MC. Additive manufacturing: Technology, applications and research needs. Frontiers Mech Eng. 2013, 8, 215–43.

[104] https://www.machinedesign.com/materials/article/21830512/additive-manufacturing-comes-to-metal-foam.

[105] Gotman I. Fabrication of load-bearing NiTi Scaffolds for bone ingrowth by Ni foam conversion. Adv Eng Mater. 2010, 12, B320–5.

[106] Barrabés M, Michiardi A, Aparicio C, Sevilla P, Planell JA, Gil FJ. Oxidized nickel–titanium foams for bone reconstructions: Chemical and mechanical characterization. J Mater Sci Mater Med. 2007, 18, 2123–29.

[107] Singh S, Vashisth P, Shrivastav A, Bhatnagar N. Synthesis and characterization of a novel open cellular Mg-based scaffold for tissue engineering application. J Mech Behav Biomed Mater. 2019, 94, 54–62.

[108] Vahidgolpayegni A, Wen C, Hodgson P, Li Y. Production methods and characterization of porous Mg and Mg alloys for biomedical applications. In: Wen C, ed. Metallic Foam Bone: Processing, Modification and Characterization and Properties. Duxford UK: Woodhead Pub, 2017, pp 25–82, doi: 10.1016/B978-0-08-101289-5.00002-0.

[109] Shinohara Y. Functionally graded materials. In: Handbook of Advanced Ceramics. 2nd ed. Materials, Applications, Processing, and Properties. 2013, 1179–87.

[110] Mahamood RM, Akinlabi ET, Shukla M, Pityana S. Functionally graded material: An overview. Proceedings of the World Congress on Engineering. International Association of Engineers. 2012, 1593–97.

[111] Miao X, Sun D. Graded/gradient porous biomaterials. Materials (Basel). 2010, 3, 26–47.

[112] Oshida Y. unpublished data, 2021.

[113] JETRO. Ti-O coating on Ti-6Al-4V alloy by DC reactive sputtering method. New Technol Japan. No. 94-06-001-01. 1994, 22, 18.
[114] Bogdanski D, Köller M, Müller D, Muhr G, Bram M, Buchkremer HP, Stöver D, Choi J, Epple M. Easy assessment of the biocompatibility of Ni-Ti alloy by in vitro cell culture experiments on a functionally graded Ni-NiTi-Ti material. Biomaterials. 2002, 23, 4549–55.
[115] Yang F, Yan Z-Y, Wei Y-H, Li Y-G, Hou L-F. Fabrication of surface-porous Mg–Al alloys with different microstructures in a neutral aqueous solution. Corros Sci. 2018, 130, 138–42.
[116] Oshida Y, Tuna EB, Aktören O, Gençay K. Dental implant systems. Int J Mol Sci. 2010, 11, 1580–678.
[117] Nanotechnology; https://en.wikipedia.org/wiki/Nanotechnology.
[118] Bayda S, Adeel M, Tuccinardi T, Cordani M, Rossolia F. The history of nanoscience and nanotechnology: From chemical – physical applications to nanomedicine. Molecules. 2020, 25, 112, doi: 10.3390/molecules25010112.
[119] Ojigblo S. Applications of nanotechnology in drug delivery and design – Part 1. Pharmanews. 2016; https://pharmanewsonline.com/applications-of-nanotechnology-in-drug-delivery-and-design-part-1/.
[120] Feynman R. What is Nanotechnology? https://www.nano.gov/nanotech-101/what/definition.
[121] Nanobiotechnology; https://en.wikipedia.org/wiki/Nanobiotechnology.
[122] Sharma N, Sharma M, Sajid Jamal QM, Kamal MA, Akhtar S. Nanoinformatics and biomolecular nanomodeling: A novel move en route for effective cancer treatment. Environ Sci Pollut Res Int. 2020, 27, 19127–41.
[123] Garimella R, Eltorai AEM. Nanotechnology in orthopedics. J Orthop. 2017, 14, 30–33.
[124] Bentley W. Nanotechnology medical applications. http://www.research.umd.edu/sites/default/files/documents/brochures/nanotechnology-medical-applications.pdf.
[125] Trafton A. One-two punch knocks out aggressive tumors. MIT News. 2013; http://news.mit.edu/2013/one-two-punch-knocks-out-aggressive-tumors-1021.
[126] Buntz B. Nanotech breakthroughs you should know about (updated). MDDI. 2014; http://www.qmed.com/mpmn/article/10-nanotech-breakthroughs-you-should-know-about-updated.
[127] Yun YH, Eteshola E, Bhattacharya A, Dong Z, Shim JS, Conforti L, Kim D, Schulz MJ, Ahn CH, Watts N. Tiny medicine: Nanomaterial-based biosensors. Sensors (Basel). 2009, 9, 9275–99.
[128] Nanobiotechnology; https://www.nanowerk.com/nanobiotechnology.php.
[129] Lin Y, Taylor S, Li H. Advances toward bioapplications of carbon nanotubes. J Mater Chem. 2004, 14, 527–41.
[130] Singh K, Kim KC. Early detection techniques for osteoporosis. In: Dionyssiotis Y, ed. Osteoporosis. IntechOpen, 2012, doi: 10.5772/29798;, https://www.intechopen.com/chapters/29542.
[131] Jain KK. Applications of nanobiotechnology in clinical diagnostics. Clin Chem. 2007, 53, 2002–09.
[132] Brenner SA, Ling JF. Nanotechnology applications in orthopedic surgery. J Nanotechnol Eng Med. 2012, 3, doi: https://doi.org/10.1115/1.4006923.
[133] Wei G, Jin Q, Giannobile WV, Ma PX. Nano-fibrous scaffold for controlled delivery of recombinant human PDGFBB. J Control Release. 2006, 112, 103–10.
[134] Xing ZC, Chae WP, Huh MW. In vitro anti-bacterial and cytotoxic properties of silver-containing poly(l-lactide-co-glycolide) nanofibrous scaffolds. J Nanosci Nanotechnol. 2011, 11, 61–65.
[135] Lee PY, Wong KKY. Nanomedicine: A new frontier in cancer therapeutics. Curr Drug Deliv. 2011, 8, 245–53.

[136] Yuan Y, Gu Z, Yao C, Luo D, Yang D. Nucleic acid–based functional nanomaterials as advanced cancer therapeutics. Small. 2019, 15, 1900172, doi: 10.1002/smll.201900172.

[137] Mohamed M, Borchard G, Jordan O. In situ forming implants for local chemotherapy and hyperthermia of bone tumors. J Drug Deliv Sci Technol. 2012, 22, 393–408.

[138] Andronescu E, Ficai M, Voicu G, Ficai D, Maganu M, Ficai A. Synthesis and characterization of collagen/hydroxyapatite: Magnetite composite material for bone cancer treatment. J Mater Sci Mater Med. 2010, 21, 2237–42.

[139] Hu Q, Chen F, Li B, Shen J. Preparation of three-dimensional nanomagnetite/chitosan rod. Mater Lett. 2006, 60, 368–70.

[140] Murakami S, Hosono T, Jeyadevan B, Kamitakahara M, Ioku K. Hydrothermal synthesis of magnetite/hydroxyapatite composite material for hyperthermia therapy for bone cancer. J Ceram Soc Jpn. 2008, 116, 950–54.

[141] Mariappan N. Recent trends in nanotechnology applications in surgical specialties and orthopedic surgery. Biomed Pharmacol J. 2019, doi: https://dx.doi.org/10.13005/bpj/1739.

[142] Qi B, Wang C, Ding J, Tao W. Editorial: Applications of nanobiotechnology in pharmacology. Front Pharmacol. 2019, 10, 1451, doi: 10.3389/fphar.2019.01451.

[143] Dubin CH. Special delivery: Pharmaceutical companies aim to target their drugs with nano precision. Mech Eng Nanotechnol. 2004, 126, 10–12.

[144] Dass CR, Su T. Particle-mediated intravascular delivery of oligonucleotides to tumors: Associated biology and lessons from genotherapy. Drug Deliv. 2001, 8, 191–213.

[145] Tasker LH, Sparey-Taylor GJ, Noles LDM. Applications of nanotechnology in orthopaedics. Clin Orthopaedics and Related Res. 2007, 456, 243–49.

[146] Parchi PD, Vittorio O, Andreani L, Piolanti N, Cirillo G, Lemma F, Hampel S, Lisanti M. How nanotechnology can really improve the future of orthopedic implants and scaffolds for bone and cartilage defects. J Nanomed Biotherap Discov. 2013, 3, 114, doi: 10.4172/2155-983X.1000114.

[147] Boccaccini AR, Keim S, Ma R, Li Y, Zhitomirsky I. Electrophoretic deposition of biomaterials. J R Soc Interface. 2010, 7, S581–S613, doi: https://doi.org/10.1098/rsif.2010.0156.focus.

[148] Marolt D, Knezevic M, Novakovic GV. Bone tissue engineering with human stem cells. Stem Cell Res Ther. 2010, 1, doi: 10.1186/scrt10.

[149] Balasundaram G, Webster TJ. Nanotechnology and biomaterials for orthopedic medical applications. Nanomedicine. 2006, 1, 169–76.

[150] Montaser MG. Applications of nanotechnology in orthopedics. Benha Med J. 2016, 33, 1–2, https://www.proquest.com/openview/634e79a774c204c190aa1bf0de2e5874/1?pq-origsite=gscholar&cbl=2042888.

[151] Sato M, Webster TJ. Nanobiotechnology: Implications for the future of nanotechnology in orthopedic applications. Expert Rev Med Devices. 2004, 1, 105–14.

[152] Durmus NG, Webster TJ. Nanostructured titanium: The ideal material for improving orthopedic implant efficacy?. Nanomedicine. 2012, 7, 791–93.

[153] Chris Arts JJ, Verdonschot N, Schreurs BW, Buma P. The use of a bioresorbable nano-crystalline hydroxyapatite paste in acetabular bone impaction grafting. Biomaterials. 2006, 27, 1110–18.

[154] Christenson EM, with 13 co-authors. Nanobiomaterial applications in orthopedics. J Orthop Res. 2007, 25, 11–22.

[155] Brenner SA, Ling JF. Nanotechnology applications in orthopedic surgery. J Nanotechnol Eng Med. 2012, 3, 024501, doi: https://doi.org/10.1115/1.4006923.

[156] Atiyeh BS, Costagliola M, Hayek SN, Dibo SA. Effect of silver on burn wound infection control and healing: Review of the literature. Burns. 2007, 33, 139–48.

[157] Miller C, McMullin B, Ghaffari A, Stenzler A, Pick N, Roscoe D, Ghahary A, Road J, Av-Gay Y. Gaseous nitric oxide bactericidal activity retained during intermittent high-dose short duration exposure. Nitric Oxide. 2009, 20, 16–23.

[158] Barraud N, Schleheck D, Klebensberger J, Webb JS, Hassett DJ, Rice SA, Kjelleberg S. Nitric oxide signaling in Pseudomonas aeruginosa biofilms mediates phosphodiesterase activity, decreased cyclic di-GMP levels, and enhanced dispersal. J Bacteriol. 2009, 191, 7333–42.

[159] Yang Y, Wang H. Perspectives of nanotechnology in minimally invasive therapy of breast cancer. J Healthcare Eng. 2013, 4, doi: https://doi.org/10.1260/2040-2295.4.1.67.

[160] Mattei TA, Rehman AA. Extremely minimally invasive: Recent advances in nanotechnology research and future applications in neurosurgery. Neurosurg Rev. 2015, 38, 27–37.

[161] Sullivan MP, McHale KJ, Parvizi J, Mehta S. Nanotechnology: Current concepts in orthopaedic surgery and future directions. Bone Joint J. 2014, 96B, 569–73.

[162] Polyzois I, Nikolopoulos D, Michos I, Patsouris E, Theocharis S. Local and systemic toxicity of nanoscale debris particles in total hip arthroplasty. J Appl Toxicol. 2012, 32, 255–69.

[163] Gupta A, Singh G, Afreen S. Application of nanotechnology in dental implants. Corpus. 2017, https://www.semanticscholar.org/paper/Application-of-Nanotechnology-In-Dental-Implants-Gupta-Singh/12086c7c18aedfecf07827844ee4985044771611.

[164] Ozak ST, Ozkan P. Nanotechnology and dentistry. Eur J Dent. 2013, 7, 145–51.

[165] Tomsia A, Launey M, Lee J, Mankani M, Wegst U, Saiz E. Nanotechnology approaches for better dental implants. Int J Oral Maxillofac Implants. 2011, 26, 25–46.

[166] Lavenus S, Louarn G, Layrolle P. Nanotechnology and dental implants. Int J Biomater. 2010, doi: https://doi.org/10.1155/2010/915327.

[167] Ramazanoglu M, Oshida Y. Osseointegration and bioscience of implant surfaces – current concepts at bone-implant interface. In: Implant Dentistry – A Rapidly Evolving Practice, Ilser Turkyilmaz. IntechOpen., 2011, https://www.intechopen.com/chapters/18415.

[168] Gutwein LG, Webster TJ. Increased viable osteoblast density in the presence of nanophase compared to conventional alumina and titania particles. Biomaterials. 2004, 25, 4175–83.

[169] Webster TJ, Ejiofor JU. Increased osteoblast adhesion on nanophase metals: Ti, Ti6Al4V, and CoCrMo. Biomaterials. 2004, 25, 4731–39.

[170] Webster TJ, Schadler LS, Siegel RW, Bizios R. Mechanisms of enhanced osteoblast adhesion on nanophase alumina involve vitronectin. Tissue Eng. 2001, 7, 291–301.

[171] Elias KL, Price RL, Webster TJ. Enhanced functions of osteoblasts on nanometer diameter carbon fibers. Biomaterials. 2002, 23, 3279–87.

[172] Price RL, Gutwein LG, Kaledin L, Tepper F, Webster TJ. Osteoblast function on nanophase alumina materials: Influence of chemistry, phase, and topography. J Biomed Mater Res A. 2003, 67, 1284–93.

[173] Mendonça G, Mendonça DB, Aragão FJ, Cooper LF. The combination of micron and nanotopography by H(2)SO(4)/H(2)O(2) treatment and its effects on osteoblast-specific gene expression of hMSCs. J Biomed Mater Res A. 2010, 94, 169–79.

[174] Oh S, Daraio C, Chen LH, Pisanic TR, Fiñones RR, Jin S. Significantly accelerated osteoblast cell growth on aligned TiO2 nanotubes. J Biomed Mater Res A. 2006, 78, 97–103.

[175] Sjöström T, Dalby MJ, Hart A, Tare R, Oreffo RO, Su B. Fabrication of pillar-like titania nanostructures on titanium and their interactions with human skeletal stem cells. Acta Biomater. 2009, 5, 1433–41.

[176] Thakral GK, Thakral R, Sharma N, Seth J, Vashisht P. Nanosurface – the future of implants. J Clin Diag Res. 2014, 8, doi: 10.7860/JCDR/2014/8764.4355.

[177] Albrektsson T, Wennerberg A. Oral implant surfaces: Part 1, Focusing on topographic and chemical properties of different surfaces and in vivo responses to them. Int J Prosth. 2004, 17, 536–43.

[178] De Oliveira PT, Nanci A. Nanotexturing of titanium-based surfaces upregulates expression of bone sialoprotein and osteopontin by cultured osteogenic cells. Biomaterials. 2004, 25, 403–13.

[179] Kawaguchi H, McKee MD, Okamoto H, Nanci A. Immunocytochemical and lectin-gold characterization of the interface between alveolar bone and implanted hydroxyapatite in the rat. Cells Mater. 1993, 3, 337, https://digitalcommons.usu.edu/cellsandmaterials/vol3/iss4/1.

[180] Popat KC, Leoni L, Grimes CA, Desai TA. Influence of engineered titania nanotubular surfaces on bone cells. Biomaterials. 2007, 28, 3188–97.

[181] Zakir O, Idouhli R, Elyaagoubi M, Khadiri M, Aityoub A, Koumya Y, Rafqah S, Abouelfida A, Outzourhit A. Fabrication of TiO2 nanotube by electrochemical anodization: Toward photocatalytic application. J Nanomater. 2020, doi: https://doi.org/10.1155/2020/4745726.

[182] Sreekantan S, Saharudin KA, Wei LC. Formation of TiO2 nanotubes via anodization and potential applications for photocatalysts, biomedical materials, and photoelectrochemical cell. Int Symp Global Multidiscip Eng Conf Series: Mater Sci Eng. 2011, 21, doi: 10.1088/1757-899X/21/1/012002.

[183] Li H, Wang G, Niu J, Wang E, Niu G, Xie C. Preparation of TiO2 nanotube arrays with efficient photocatalytic performance and super-hydrophilic properties utilizing anodized voltage method. Results Phys. 2019, 14, 102499, doi: https://doi.org/10.1016/j.rinp.2019.102499.

[184] Gulati K, Moon H-J, Kumar PTS, Han P, Ivanovski S. Anodized anisotropic titanium surfaces for enhanced guidance of gingival fibroblasts. Mater Sci and Eng C. 2020, 112, 110860, doi: https://doi.org/10.1016/j.msec.2020.110860.

[185] Bjursten LM, Rasmusson L, Oh S, Smith GC, Brammer KS, Jin S. Titanium dioxide nanotubes enhance bone bonding in vivo. J Biomed Mater Res A. 2010, 92, 1218–24.

[186] Von Wilmowsky C, Bauer S, Lutz R, Meisel M, Neukam FW, Toyoshima T, Schmuki P, Nkenke E, Schlegel KA. In vivo evaluation of anodic TiO2 nanotubes: An experimental study in the pig. J Biomed Mater Res Part B Appl Biomater. 2009, 89, 165–71.

[187] Brammer KS, Oh S, Cobb CJ, Bjursten LM, Van Der Heyde H, Jin S. Improved bone-forming functionality on diameter-controlled TiO(2) nanotube surface. Acta Biomater. 2009, 5, 3215–23.

[188] Park J, Bauer S, Schlegel KA, Neukam FW, Von Der Mark K, Schmuki P. TiO2 nanotube surfaces: 15 nm–an optimal length scale of surface topography for cell adhesion and differentiation. Small. 2009, 5, 666–71.

[189] Aninwene GE, Yao C, Webster TJ. Enhanced osteoblast adhesion to drug-coated anodized nanotubular titanium surfaces. Int J Nanomedicine. 2008, 3, 257–64.

[190] Balasundaram G, Yao C, Webster TJ. TiO2 nanotubes functionalized with regions of bone morphogenetic protein-2 increases osteoblast adhesion. J Biomed Mater Res Part A. 2008, 84, 447–53.

[191] Gupta S, Noumbissi S, Kunrath MF. Nano modified zirconia dental implants: Advances and the frontiers for rapid osseointegration. Med Devices Sens. 2020, doi: https://doi.org/10.1002/mds3.10076.

[192] Tomsia AP, Lee JS, Wegst UGK, Saiz E. Nanotechnology for dental implants. Int J Oral Maxillofac Implants. 2013, 28, e535–46.

[193] Boyan BD, Lohmann CH, Dean DD, Sylvia VL, Cochran DL, Schwartz Z. Mechanisms involved in osteoblast response to implant surface morphology. Annu Rev Mater Res. 2001, 31, 357–71.

[194] Kasemo B. Biological surface science. Surf Sci. 2002, 500, 656–77.

[195] Hench LL, Xynos ID, Polak JM. Bioactive glasses for in situ tissue regeneration. J Biomater Sci (Polymer). 2004, 15, 543–62.

[196] Hamouda IM. Current perspectives of nanoparticles in medical and dental biomaterials. J Biomed Res. 2012, 26, 143–51.

[197] Feridoun P, Javad Y, Vahid J, Solmaz MD. Overview of nanoparticle coating of dental implants for enhanced osseointegration and antimicrobial purposes. J Pharm Pharm Sci. 2017, 20, 148–60.

[198] Praveena C, Chaughule RS, Satyanarayana KV. Nanotechnology in implant dentistry. In: Chaughule RS, et al., ed. Advanced in Dental Implantology Using Nanomaterials and Allied Technology Applications. Springer, 2021, doi: https://doi.org/10.1007/978-3-030-52207-0_1.

[199] Junker R, Dimakis A, Thoneick M, Jansen JA. Effects of implant surface coatings and composition on bone integration: A systematic review. Clin Oral Implants Res. 2009, 20, 185–206.

[200] Choi AH, Ben-Nissan B, Matinlinna JP, Conway RC. Current perspectives: Calcium phosphates nanocoatings and nanocomposite coatings in dentistry. J Dent Res. 2013, 92, 853–59.

[201] Park JW, Han SH, Hanawa T. Effects of surface nanotopography and calcium chemistry of titanium bone implants on early blood platelet and macrophage cell function. Biomed Res Int. 2018, 1362958, doi: https://doi.org/10.1155/2018/1362958.

[202] Balasundaram G, Sato M, Webster TJ. Using hydroxyapatite nanoparticles and decreased crystallinity to promote osteoblast adhesion similar to functionalizing with RGD. Biomaterials. 2006, 27, 2798–805.

[203] Schildhauer TA, Seybold D, Gebmann J, Muhr G, Koller M. Fixation of porous calcium phosphate with expanded bone marrow cells using an autologous plasma clot. Mater Werkst. 2007, 38, 1012–14.

[204] Oh SH, Finones RR, Daraio C, Chen LH, Jin S. Growth of nano-scale hydroxyapatite using chemically treated titanium oxide nanotubes. Biomaterials. 2005, 26, 4938–43.

[205] Bressan E, Ferroni L, Gardin C, Rigo C, Stocchero M, Vindigni V, Cairns W, Zavan B. Silver nanoparticles and mitochondrial interaction. Int J Dent. 2013, 2013, 1–8.

[206] Memarzadeh K, Sharili AS, Huang J, Rawlinson SC, Allaker RP. Nanoparticulate zinc oxide as a coating material for orthopedic and dental implants. J Biomed Mater Res A. 2015, 103, 981–89.

[207] Zhang P, Zhang Z, Li W. Antibacterial TiO2 coating incorporating silver nanoparticles by microarc oxidation and ion implantation. J Nanomater. 2013, 2013, 2, doi: https://doi.org/10.1155/2013/542878.

[208] Anu K, Maleeka Begum SF, Rajesh G, Renuka Devi KP. Wet biochemical synthesis of copper oxide nanoparticles coated on titanium dental implants. Int J Adv Res Sci Eng Technol. 2016, 3, 1191–94.

[209] Gomez-Florit M, Pacha-Olivenza MA, Fernández-Calderón MC, Córdoba A, González-Martín ML, Monjo M, Ramis JM. Quercitrin-nanocoated titanium surfaces favour gingival cells against oral bacteria. Sci Rep. 2016, 6, 1–7.

[210] Barbour ME, Maddocks SE, Wood NJ, Collins AM. Synthesis, characterization, and efficacy of antimicrobial chlorhexidine hexametaphosphate nanoparticles for applications in biomedical materials and consumer products. Int J Nanomed. 2013, 8, 3507–19.

[211] Wood NJ, Jenkinson HF, Davis SA, Mann S, O'Sullivan DJ, Barbour ME. Chlorhexidine hexametaphosphate nanoparticles as a novel antimicrobial coating for dental implants. J Mater Sci Mater Med. 2015, 26, 1–10.

Chapter 7
Medical implants

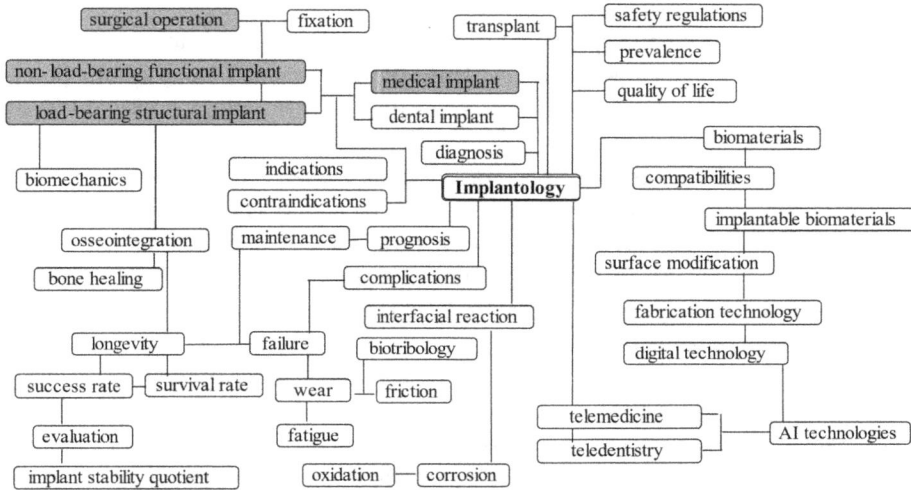

7.1 Introduction

An artificial organ implantation and an organ transplantation are two distinctive methods for treating severe organ dysfunctions that cannot be recovered by preservation treatment, trauma, or organ symptoms associated with surgery. These treatments are also called organ replacement therapy. The latter is a supplementary transplantation of organs of others, while the former is to compensate for the missing organ function with artificial developed devices. This chapter will describe the artificial organ implantation. Artificial organs (AO) are widely used as medical implants, and there are two types of implants, namely one that mainly plays a functional role and another one that is responsible for the load in terms of biomechanics.

AOs are medical devices that are implanted or connected to the human body as a functional substitute for organs so that patients can return to normal life as much as possible. There are no functional restrictions on the features that are substituted, but they are often fatal. Some believe that the characteristic of AOs is that they are not connected to fixed external power supplies, filters, chemical processing equipment, and so on, and that devices that require regular charging and replacement of consumables should not be classified as AOs; for example, a dialysis machine is a functional substitute for the kidneys and a medical device connected to the human body, but not an AO. However, in this chapter, dialysis machines are included in AOs since

https://doi.org/10.1515/9783110740134-008

they have been developed for the purpose of acting on behalf of sick organs. The world is facing the problem of a shortage of organ donors. While the number of organ donors has remained unchanged for many years, the number of transplant applicants is increasing every year. However, in recent years, there has been a dramatic development in medical technology and cell tissue engineering, and the development of AOs has also been dramatic.

There are three main purposes and roles of AOs: (1) the first is to assist patients facing organ dysfunction, especially patients who are waiting for organ transplants at the end of the disease, for instance, an artificial heart and an artificial kidney can be included; (2) the second is to allow patients to manage their own illnesses without relying on doctors or nurses. For example, diabetics can regularly measure blood sugar levels in real time through the artificial pancreas and adjust insulin secretion accordingly; and (3) the third is to improve quality of health-related quality of life (HRQoL). Cochlear implant and artificial vision devices can eliminate inconveniences in daily life, promote human-to-human communication, and improve the mental health of patients [1–3].

As mentioned briefly in Chapter 1, there are several clear differences between transplantation and implantation. Transplantable tissue is a biological tissue used as an exchange of tissues, organs, or organs of the human body, but implant is a nonbiological material. Transplant requires an organ donor immunosuppression, while implant does not require it. Transplant is an active tissue in the human body, but implant is the mechanical support that assists organ function. Implant is a foreign body against receiving vital tissue, so that there is a risk of infection. On the other hand, transplant is at risk of rejection from the human body. Compared to implantations, transplantations require stricter consideration of ethical examination. Transplant can be used for life unless rejected, while implant might be removed if necessary. Table 7.1 compares several important issues between AO implantation and organ transplantation [4–9].

Table 7.1: Comparison between artificial organ implantation and organ transplantation [4–9].

	Artificial organ implantation	Organ transplantation
Supply	Possible mass production	Absolute number shortage
Implant	Impossible, except several devices	Possible
Dimension	Large freedom	Limitations (proper selection)
Preservation	Possible for a long time	Short life (except some organs)
Preparation	Anytime be ready	Brain-death patient – informed consent
Biofunction	Incomplete	Almost complete

Table 7.1 (continued)

	Artificial organ implantation	**Organ transplantation**
Patient's reaction	Foreign-body reaction, thrombus formation, lime deposition	Rejection
Pharmacotherapy	Anticoagulants	Immunosuppressant
Social issue	Cost	Ethics (brain death, license), cost
Medical system	Support system	Organ bank, transportation means, international collaboration

7.2 HRQoL evaluations for medical implants

Changes in quality of life (QoL) for patients who require transplantation or permanent AO support begin with the onset of serious organ dysfunction. These changes, which initially are almost universally negative, greatly affect the patient as well as close family members [10].

TSA

Total shoulder arthroplasty (TSA) is considered as the standard reconstructive surgery for patients suffering from severe shoulder pain and dysfunction caused by arthrosis. Multiple patient-reported outcome measures (PROMs) have been developed and validated that can be used to evaluate TSA outcomes. When selecting an outcome measure both content and psychometric properties must be considered. Most research to date has focused on psychometric properties [11]. Lu et al. investigated the assessment of PROMs on TSA outcomes to classify the type of measure (International society for quality of life (ISOQOL) using definitions of functioning, disability, and health (FDH), QoL, and HRQoL) and to compare the content of these measures by linking them to the International Classification of Functioning, Disability and Health (ICF) framework. It was concluded that (i) there is an inconsistency and lack of clarity in conceptual frameworks of identified PROMs; despite this, common core constructs are evaluated, and (ii) decision-making about individual studies or core sets for outcome measurement for TSA would be advanced by considering our results, patient priorities, and measurement properties. Deshmukh et al. [12] examined long-term outcomes of TSA via survivorship analysis, patient questionnaires, and minimum 10-year physical examinations on 320 consecutive TSAs performed in 267 patients between 1974 and 1988. Diagnoses included rheumatoid arthritis (RA) (69%), osteoarthritis (22%), and juvenile RA (4.7%). Of the shoulders, 22 (6.9%)

required a revision, most commonly for loosening of one or both components (15 shoulders). It was reported that (i) the patients' subjective assessments of TSA were favorable in that 92% felt that their shoulder was "much better" or "better" after TSA, (ii) the long-term analysis of the Neer-type TSA revealed survivorship rates comparable to other joint replacements, and (iii) the significant improvements in relief of pain, shoulder range of motion (ROM), and strength are associated with a high degree of patient satisfaction. Leite et al. [13] evaluated the HRQoL of patients who have undergone reverse shoulder arthroplasty (RSA) for rotator cuff arthropathy (RCA) on 35 patients who underwent RSA from August 2007 to July 2015 and concluded that (i) patients who had undergone RSA for RCA had good HRQoL, (ii) longer follow-up time was associated with better HRQoL, and (iii) good results were maintained over the follow-up period.

THA

Osteoarthritis (OA), in the hip as a chronic disease impairing patients' function, causes pain and reduces HRQoL, but the total hip arthroplasty (THA) as elective surgery has been shown to relieve pain and improve physical function and HRQoL [14]. Hip arthroplasty patients also showed greater improvement in pain and function and were more satisfied with the outcomes than patients undergoing knee arthroplasty. Patients undergoing arthroplasty seem to have psychological distress, although it was not associated with self-perceived functional recovery among hip arthroplasty patients. The stronger the anxiety trait, the more probable it is that the individual will experience more state anxiety in a threatening situation. HRQoL is thought to include the elements of biological function, symptoms, functional status, and general health perceptions that are influenced by individual and environmental characteristics, post-operation conditions, and prognosis [12]. Psychosocial improvements were seen sooner than physical improvements, but after 6 months, improvements in all dimensions of HRQoL were found [15]. Various scales have been reported on HRQoL, for example [16–18].

Shan et al. [19] investigated mid-term HRQoL after THA in patients with OA and mentioned that THA confers significant mid-term HRQoL benefits across a broad range of health domains. Shi et al. [20] applied the generalized estimating equations in a large-scale prospective cohort study of predictors of HRQoL in a Taiwanese population, including all patients who had undergone primary THA performed between March 1998 and December 2002 by either of two orthopedic surgeons in two hospitals. The Short Form 36 Health Survey questionnaire (SF-36) was used in preoperative and postoperative assessments of 335 patients. The SF-36 is normally used to indicate the health status of particular populations, to help with service planning, and to measure the impact of clinical and social interventions. It was described that (i) young age, male gender, minimal comorbidity, use of epidural anesthesia, lack of readmission within the previous 30 days, and higher preoperative functional status were positively associated with HRQoL; (ii) patients should be advised that their postoperative HRQoL may depend not only on their postoperative health care but also on their preoperative functional status; and (iii) these analytical results should be applicable to other Taiwanese hospitals and to other countries with similar social and cultural practices.

There is an increasing focus on measuring patient-reported outcomes (PROs) as part of routine medical practice, particularly in fields such as joint replacement surgery where pain relief and improvement in HRQoL are primary outcomes. Between-country comparisons of PROs may present difficulties due to cultural differences and differences in the provision of health care. Gordon et al. [21] investigated factors influencing HRQoL one year after total hip replacement (THR) surgery in Sweden and in Denmark. It was concluded that (i) there are clear similarities in how basic predictors influence PROs in patients with THA in Sweden and Denmark, and (ii) these known predictors of good or poor HRQoL outcomes are not specific for each country.

To assess the HRQoL, there are several studies using the body mass index (BMI). Villalobos et al. [22] examined the contribution of patient weight and other preoperative variables to improvements in the general physical health of patients undergoing THA. Data were prospectively collected from 63 THA patients (28 males and 35 females). The primary outcome measure was the improvement in general health (SF-12) at three months post-THA. It was found that (i) patients with BMI > 28 kg/m^2 showed greater improvements in function and in the physical component of general health after THA, and (ii) stepwise regression analyses revealed that the BMI and WOMAC general index were independent and significant predictors of physical function and together explained 34.2% of the variance in physical function scores, suggesting that the BMI before surgery and improvements in hip function are relevant contributors to post-THA improvements in general health.

Poulsen et al. [23] examined patient-reported HRQoL and hip function after a completed reimplantation in a two-stage revision. All together 82 patients were identified retrospectively in the National Patient Register. Fifty-seven patients were alive and asked to complete the questionnaires EuroQol-5D (EQ-5D) and Oxford Hip Score (OHS) in November 2014. It was indicated that (i) patients who underwent the two-stage revision after a periprosthetic hip joint infection (PJI) have lower scores on HRQoL than the general population, and (ii) patients who are reinfected following revision have a lower HRQoL score than patients not reinfected.

There are several more reports on HRQoL after the two-stage revision of PJI. The survival of repeat two-stage revision hip arthroplasty was reported [24]. Brown et al. [24] retrospectively identified 19 hips (19 patients) that had undergone repeat two-stage revision THA for infection between 2000 and 2013. There were seven female patients (37%) and the mean age was 60 years (30 to 85). The patients were classified according to the Musculoskeletal Infection Society (MSIS) system, and risk factors for failure were identified. Mean follow-up was four years (2 to 11). It was found that (i) reinfection after two-stage exchange hip arthroplasty for PJI presents a challenging scenario, and (ii) repeat two-stage exchange arthroplasty has a low survival free from revision at five years (45%) and a high rate of reinfection (42%). The clinical outcomes for one-stage and two-stage exchange arthroplasty performed in patients with chronic culture-negative PJI were investigated [25].

In the above, we have been reviewing the two-stage revision, while there is the one-stage revision treatment. The prosthesis is replaced in the same operation (one-stage) or replaced at a delayed interval between 2 weeks and 12 months (two-stage). In a two-stage revision, a temporary spacer or temporary joint replacement may be fitted, but the patient has no definitive THR until it is replaced in the second operation. Poulsen et al. [26] evaluated changes in HRQoL and patient-reported hip function two years following a cementless one-stage revision for chronic PJI and concluded that patients treated with a cementless one-stage revision for chronic PJI experienced a marked increase in HRQoL and patient-reported hip function, and matched population norms on many parameters. Abdelaziz et al. [27], after one-stage revision arthroplasty for PJI of the hip, tried to find answers as to what factors were associated with an increased risk of re-revision and what factors were associated with an increased risk of reinfection. It was concluded that (i) prolonged wound drainage after the one-stage revision arthroplasty for PJI of the hip must be treated rigorously, and (ii) patients with a history of a prior surgical procedure due to hip infection should be informed about the risk of further re-revision when deciding for the one-stage exchange. In case of enterococcal isolation, surgeons may consider another treatment approach rather than the one-stage exchange, suggesting the use of dual mobility cups when performing the one-stage revision hip arthroplasty to reduce the risk of dislocation.

TKA

Total knee arthroplasty (TKA) is a highly successful and frequently performed operation. Technical outcomes of surgery are excellent, with favorable early postoperative HRQoL. Shan et al. [28] conducted a systematic review, and a meta-analysis of all studies published from January 2000 onward was performed to evaluate HRQoL after primary total knee replacement (TKR) for OA in patients with at least three years of follow-up. It was concluded that TKR confers significant intermediate and long-term benefits with respect to both disease-specific and generic HRQoL, especially pain and function, leading to positive patient satisfaction. Liebs et al. [29] compared the survivorship and HRQoL after lateral versus medial cemented mobile-bearing unicompartmental knee arthroplasties (UKAs) and determined whether there is an association of survival to modifications of surgical technique in one of three phases. A total of 558 patients who underwent mobile-bearing UKAs from 2002 to 2009 were subjected to retrospective review. It was found that (i) implant survival was 88% at 9 years; (ii) there were similar implant survival rates for medial (90%) and lateral UKAs (83%); (iii) in all HRQoL measures, patients receiving a medial UKA had better mean scores compared with patients who had a lateral UKA: WOMAC physical function (23 versus 34, respectively), pain (21 versus 34), and SF-36 PCS (41 versus 38); and (iv) there were no survival differences by surgical phase, suggesting that a medial UKA is associated with superior HRQoL when compared with a lateral UKA, although implant survival is similar.

Scoliosis

Progressive early onset scoliosis (EOS) in children may lead to surgical interventions with growth-friendly implants, which require repeated lengthening procedures in order to allow adequate growth. Hell et al. [30] studied the QoL using the validated German version of the EOS questionnaire (EOSQ-24-G) in surgically treated EOS children with different lengthening modalities. It was concluded that, using the validated EOSQ-24-G, no statistically significant differences were found between the group of children receiving repetitive surgeries and children with external lengthening procedures without surgery; however, results were influenced by the etiology, complication rate, or ambulatory ability. Doany et al. [31] compared QoL and caregiver burden in traditional growing rod (TGR) and magnetic controlled growing rod (MCGR) patients. Inclusion criteria were (1) younger than 10 years of age at index procedure, (2) major curve is larger than 30°, (3) no previous spine surgery, and (4) minimum one-year postoperative follow-up. The previously validated 24-item EOS questionnaire (EOSQ-24) was utilized to assess QoL. Statistic methods were applied to compare domain scores between TGR and MCGR patients. It was reported that HRQoL data reveal superior outcomes in overall satisfaction and financial burden domains in the MCGR group; however, the positive effects of MCGR decrease when controlled for length of follow-up, indicating that the MCGR is not yet a magic fix-all and that the TGR remains an option in the treatment of EOS.

Adult spinal deformity (ASD) is assessed radiologically with the spinopelvic parameters and clinically with HRQoL scores. The revision rate after ASD surgery is high and usually occurs during the first or second postoperative year. Bourghli et al. [32] examined clinical or radiological factors that could predict revision surgery in the second postoperative year. Inclusion criterion was that ASD patients operated on by instrumented posterior fusion with more than two years follow-up were enrolled prospectively. It was concluded that (i) the revision rate at the second-year post-surgery (13.4%) remains high and demonstrates that a two-year follow-up is mandatory, (ii) in addition to usual risk factors for mechanical complications in ASD surgery, stabilization or worsening of the HRQOL scores between the 6th and 12th month postop was highly predictive of revision rate, and (iii) these findings are beneficial for ASD patient follow-up as clinical symptoms clearly precede mechanical failure [32]. Núñez-Pereira et al. [33] investigated the impact of early reoperations within the first year on HRQoL and the impact on the likelihood of reaching the minimally clinically important difference (MCID) after ASD surgery by a retrospective analysis of prospectively collected data from consecutive surgically treated adult deformity surgery patients included in a multicenter international database. A total of 280 patients were included from a multicenter international prospective database. It was reported that early unanticipated revision surgery has a negative impact on mental health at six months and reduces the chances of reaching an MCID improvement in SRS-22, SF-36 PCS, and ODI at the two-year follow-up. Total hip and knee arthroplasty (THA/TKA) are surgical procedures with proven benefits. Although the literature reports outcomes of fusion of the lumbar spine comparable to those of THA/TKA in general HRQoL questionnaires, functional assessment is nevertheless needed for these results to be of use in clinical practice and management. Irimia et al. [34] investigated that lumbar spinal fusion has similar if not better outcomes than THA/TKA using intervention-specific HRQoL questionnaires and functional assessment questionnaires. It was concluded that lumbar spinal fusion and THA/TKA are comparable in terms of functional improvement when thoroughly studied with health, quality-of-life, and functional assessment questionnaires.

Hemodialysis

Measures of HRQoL have a significant predictive value on patient survival and hospitalizations, especially in patients with chronic kidney disease (CKD). Avramovic et al. [35] reviewed studies performed in patients with different stages of renal failure. The most used instrument for measuring HRQoL is the SF-36. It was found that (i) patients with predialysis CKD had higher SF-36 scores than a large cohort of hemodialysis (HD) or peritoneal dialysis (PD) patients, but lower scores than those reported for the adult population, and (ii) kidney transplantation offers better HRQoL than dialysis. HRQoL and sleep quality (SQ) were impaired in patients with end-stage renal disease (ESRD). The impairment of both HRQoL and SQ and being in a depressive mood were found to be associated with increased morbidity and mortality in dialysis patients. Accordingly, Turkmen at al. [36] investigated the association between SQ, HRQoL, and depression to define independent predictors of SQ and depression in PD and HD patients. It was reported that (i) HD and PD patients had similar total SQ scores. Physical and mental component scale of HRQoL were found to be significantly higher in HD patients; (ii) PD patients were found to be in much more depressive mood when compared with HD patients; (iii) independent predictors of depression in patients were mental component scale of HRQoL, gender (being female), and dialysis modality (being PD patient); and (iv) physical component scale was also found to be an independent predictor of SQ, suggesting that despite similar SQ scores between two groups HD patients had better HRQoL and less depression than PD patients [36].

7.3 Un-load-bearing implants

There are two types of AOs and implants: one is AO based on mechanical technologies and the other is based on tissue engineering [35]. AOs and implants fabricated through mechanical technology can include artificial bones made of ceramics, dental materials used in implant treatment, and mechanical auxiliary artificial hearts. Other examples include a heart pacemaker that embeds directly in a patient with a heart condition, or an artificial lens that embeds directly into the eyeballs of a patient who has had his lens removed due to worsening cataracts. In addition, AOs that have been put to practical use in recent years after clinical trials have been completed include cochlea (implanted in people who are deaf for hearing recovery) by connecting electronic technology with the auditory nerves of the brain. In addition, research and development is progressing toward the practical application of artificial visual systems (artificial retinas, artificial eyes) that give sight to blind people by electrical stimulation to the visual center using multipoint electrodes. In recent years, with the aim of assisting functional disorders caused by physical fitness decline and unforeseen accidents, development of artificial limbs, artificial arms, auxiliary devices, and so on, combining machinery and computer technology is also progressing for practical use. Until now, it was not possible to move prosthetic hands and prosthetic legs on patient's own will, but by developing sensors that can catch trace amounts of electrical signals resulting from the motor nervous system, it has enabled to move prosthetic hands and prosthetic legs at will. Of course, in the future, the development of artificial limbs arms, and so on that take into account feedback with the sensory nervous system, or the like, will also proceed. As a result, there is a possibility that lost sensations can be regained [36].

It is an AO born by forming a new tissue using living cells for the organ and the part to be substituted. The technology that creates this kind of AO is called tissue engineering, and it is a technology created not only by medical researchers but also by researchers in molecular biology dealing with cells and genes, and researchers in materials engineering dealing with plastics and polymeric materials, who conduct interdisciplinary research and development beyond the boundaries of academic disciplines. Organs and tissues such as the skin, blood vessels, bones, trachea, the eating path, vagina, cartilage, lungs, kidneys, and liver have been regenerated at least partially by artificial treatment. In addition, attempts have been made to artificially create organs using induced pluripotent stem (iPS) cells, and researchers from pharmaceutical companies and research institutes around the world. Owing to developed encapsulating cell technology, a treatment is also being tried to wrap cells in a semipermeable membrane, making them into small capsules which are transplanted. It is thought that it mainly wraps the islets of Langerhans cells in the pancreas and is used to treat diabetes [37].

7.3.1 Nerve system

Artificial brain

Rapid advancements in neurostimulation technologies are providing relief to an un-precedented number of patients affected by debilitating neurologic and psychiatric disorders, including epilepsy, which is a chronic neurological condition character-ized by recurring seizures or abnormal bursts of electrical activity in the brain that can trigger jerky movements, strange sensations or emotions, unusual behavior, and/or loss of consciousness. Neurostimulation therapies include invasive and non-invasive approaches that involve the application of electrical stimulation to drive neural function within a circuit [38, 39]. Research supported by the National Insti-tute of Neurological Disorders and Stoke (NINDS) and conducted by NINDS scien-tists figured prominently in the development of two FDA-approved devices that deliver electrical stimulation to the brain in different ways to reduce seizure fre-quency in people who do not achieve good seizure control with medication alone. Deep brain stimulation (DBS) for epilepsy, approved in 2018, delivers chronic stimula-tion to the anterior nucleus of the thalamus (ANT), a small brain structure involved in the spread of an initially localized seizure. In contrast, responsive neurostimulation (RNS®), approved in 2013, delivers stimulation directly to the source of an individu-al's seizures, but only when continuously monitored brain activity suggests a seizure may be beginning [39]. Figure 7.1 shows typical neurological stimulation devices [38].

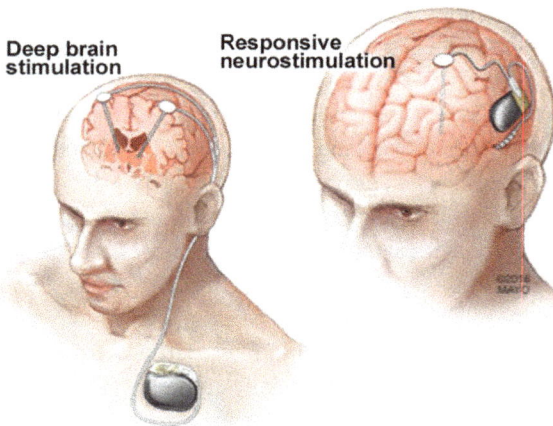

Figure 7.1: Typical devices for deep brain stimulation and responsive neurostimulation [38].

Gamillo [41] introduced the miniature brains with a set of eye-like formations called optic cups. The optic cups are precursors to the retina, and its development within the mini-organoids resembled the emergence of eye structures in human embryos. Organo-ids are small, three-dimensional tissue cultures that can replicate organs. It was reported that out of the 314 created mini-brains, 72% grew optic cups. Tan et al. [40] mentioned, in their review article, that (i) reconstruct aspects of major neurological disorders under static or dynamic conditions have been developed, and (ii) engineered human mini-brains contribute to advancing the study of the physiology and etiology of neurological disorders, and the development of personalized medicines for them.

Artificial nerve
The artificial nerve in regenerative medicine is a type of AO and one of the reconstructive techniques of nerve tissue using the resilience of the living body and mainly the peripheral nervous system (PNS) tissue if the human body has been studied. Simply cut nerves can be recovered by suturing, but difficulties are made when nerves which are more than 3 cm long in the human body are lost. Artificial neural technology does not create or culture nerve tissue or nerve cells themselves outside the body but instead uses the regenerative abilities inherent in the human body to reconstruct nerve tissue. Specifically, by connecting both ruptures of the nerve with a thin tube made of silicone, the stretch distance of the nerve axon is extended. When both nerves are connected using artificial nerves, the gap is filled by a gel containing fibrin protein that fills the silicone tube. On the central side, regeneration buds consisting of axons and Schwann cells from neurons are born and begin to grow little by little into the fibrin gel. On the peripheral side, Wallerian degeneration (the degeneration of distal axons) begins, and the scaffold that greets the axon from the central side proceeds by the alignment of Schwann cells (Schwann cell cord). Before long, both come into contact, the axon extends the passage of the Schwann cell, and the Schwann tube is extended even after it reaches the cut-off on the peripheral side; it reaches the organ and the muscle before long, and the regeneration process is completed [42].

Owing to advances in medical devices, some of the functionalities of the lost limbs can be restored by artificial arms or legs, or prostheses. People with prosthetic legs can walk, run, or dance with ease, and those with prosthetic hands can control each finger and grip in a natural, coordinated way. However, current prostheses lack one important aspect of natural limbs – the tactile senses of human skin. Sensory nerves carry information from the outside world to our spinal cord and brain. In particular, our ability to perceive touch sensation is achieved by a type of sensory nerve ending called mechanoreceptors which are located in our skin. When pressure is applied to the skin, the mechanoreceptors respond by changing their electric voltage (i.e., a measure of electrical energy) [43, 44]. Desarrollo [45] introduced the first prosthesis in the world that connects directly to the bone, nerves, and muscles, allowing users to experience sensations and offering free mobility, and is controlled via brain/ computer interface, as seen in Figure 7.2. The device becomes an extension of the human body through osseointegration, meaning that it connects directly to the bone via a titanium implant and offers a robust and intuitive control of the artificial hand through the neuronal and muscle-binding interfaces so that it is possible to move the limb just by thinking about it. The prosthesis can be controlled with the mind using an electrode system.

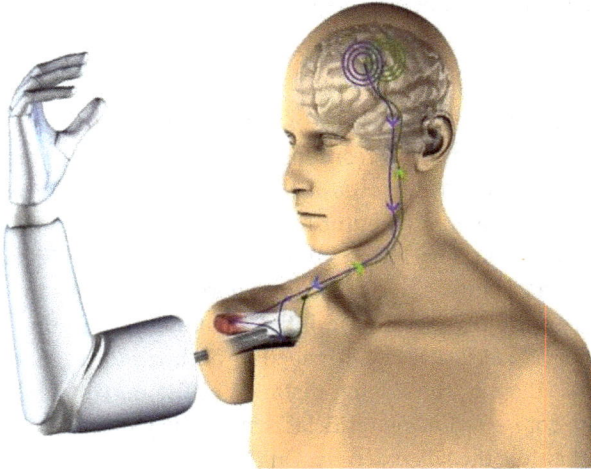

Figure 7.2: The first prosthesis with direct connection to bone, nerves, and muscles [45].

7.3.2 Head area

Artificial ear and cochlear implant
A prosthetic ear is an artificial auricle. It is used for the purpose of keeping the shape of the surrounding ear tissue in a normal state when the auricle is lost due to trauma or disease, or when hearing is lost due to atrophy, or when the shape is naturally different from the general state, such as small ear disease (aka microtia). The function of the auricular membrane is not so conscious in daily life, but the sound from the front is reflected so that it reaches the back of the ear (eardrum), and if you lose it, you will have hearing loss that will cause your hearing to drop even if other functions such as the eardrum are not impaired. It is the opposite of hearing a small sound in the distance if the auricle is assisted by the palm. The prosthetic ear is an artifact that mimics this as an alternative to this auricle, but since the face and head are also parts that usually touch the public eye, there is a side as an epithesis that adjusts the appearance with a structure closer to the body, and there are also things that are tailored to the color of the wearer's skin [46, 47].

Cochlear implant

A cochlear implant is an electronic device that can provide auditory sensation to a person with severe to profound sensorineural hearing loss. Cochlear implants are instruments that contact electrodes in the cochlea of the hearing-impaired inner ear to assist hearing. Cochlear implants are designed for people who do not receive adequate benefit from traditional amplification, those patients including mid-hearing people to hear audio to a certain extent, or maybe deaf and/or hearing-impaired infants to support the language development.

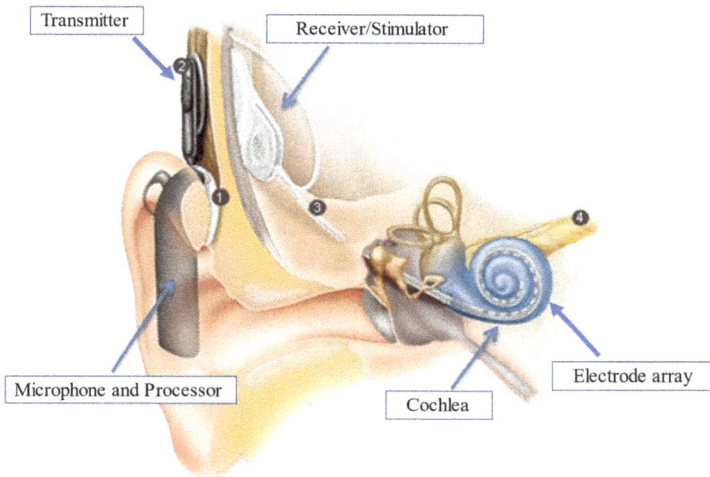

Figure 7.3: Cochlear implant [46].

Referring to Figure 7.3 [46], the cochlear implant consists of a microphone, a voice analyzer, a stimulation electrode, and a radio wave transmitter and receiver. The microphone captures the outside audio and converts the sound into an electrical signal with an out-of-body audio analyzer. The electrical signal is sent in a noncontact manner to the electrode in the inner ear, where the electrode stimulates the auditory nerve. Since the cochlea has frequency specificity by the site, several electrodes are embedded. The degree to which electrode is stimulated is determined by the processor in the speech analyzer. The hearing nerve fibers in the cochlea pick up the signals and send them to the brain, giving the sensation of sound [46, 47]. Although a cochlear implant is usually attached to one ear, wearing it in both ears allows to expect a greater effect than the case of only one ear, such as a three-dimensional effect of speech.

Visual prosthesis

A visual prosthesis (which is sometimes referred to as a bionic eye) is an experimental visual device intended to restore functional vision in those suffering from partial or total blindness. Visual prosthetics are being developed as a potentially valuable aid for individuals with visual degradation [48, 49]. The ability to give sight to a blind person via a bionic eye depends on the circumstances surrounding the loss of sight. For retinal prostheses, which are the most prevalent visual prosthetics under development (due to ease of access to the retina among other considerations), patients with vision loss due to degeneration of photoreceptors (retinitis pigmentosa, choroideremia, geographic atrophy macular degeneration) [50] are the best candidates for treatment. Candidates for visual prosthetic implants find the procedure most successful if the optic nerve was developed prior to the onset of blindness. Persons born with blindness may lack a fully developed optical nerve, which typically develops prior to birth, though neuroplasticity makes it possible for the nerve, and sight, to develop after implantation [48, 51, 52].

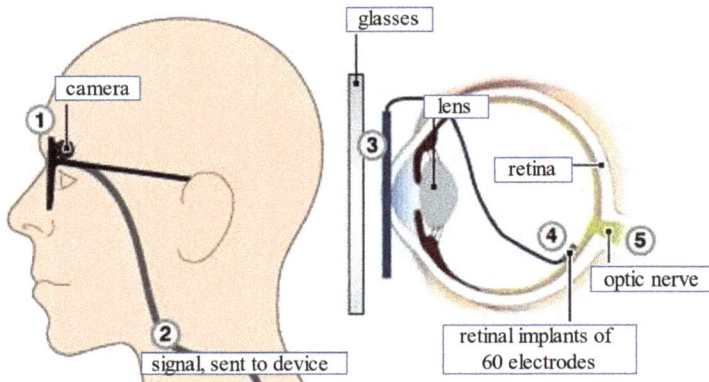

Figure 7.4: Typical bionic eye system [52].

Referring to Figure 7.4 [52], the camera installed on the glasses views the image and signals are sent to a handheld device. The processed information is sent back to the glasses and wirelessly transmitted to the receiver under surface of the eye, and the receiver sends information to electrodes in the retinal implant. And finally, electrodes stimulate the retina to send input information to brain to recognize the image.

Intraocular lens (IOL)

When the lens is removed during cataract surgery, it becomes a lens-free and has a strong farsightedness. For this case, the intraocular lens (IOL) is an artificial lens inserted. In general, monofocal intraocular lenses are used, and after surgery, they become monofocal. After the operation, the adjustment force is theoretically no more, but the residual (false adjustment) of some adjustment force is recognized, although the mechanism is unknown yet. By varying the degree of IOL, the refractive index after surgery can be changed. As a result, myopia and farsighted vision can be corrected, and it has the side of refractive surgery. Therefore, the degree is determined by various calculation formulas referring to the lifestyle of the patient himself before surgery. In addition, there are IOLs inserted with a lens for myopia correction purposes, not cataracts [53–55]. Figure 7.5 shows a typical IOL [54].

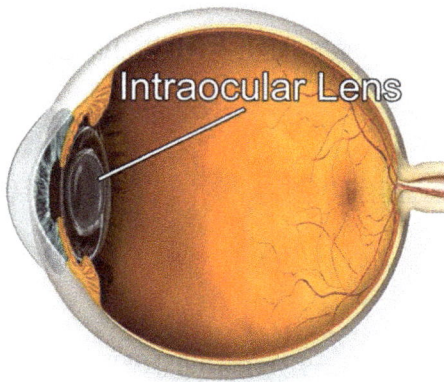

Figure 7.5: Intraocular lens [54].

Contact lens
A contact lens is a thin, curved lens placed on a film of tears that covers the surface of your eye. The lens itself is naturally clear but is often given the slightest tinge of color to make them easier for wearers to handle. Today's contact lenses are either hard or soft. The base curve radius of a contact lens is the measure of an important factor to determine the size of the contact lens and parameter of a lens in optometry. It determines the type of fit the lens must have to match the natural curvature of your eye. It is usually expressed in millimeters and may be further characterized as steep, median, or flat. Typical base curve values range between 8.0 and 10.0 mm, though it can be flatter (from 7.0 mm) if you have a rigid gas-permeable lens. A person with a higher base curve number has a flatter cornea (the clear, front surface of the eye) compared to someone with a lower base curve number, which indicates a steeper cornea. Typical values for a contact lens are from 8.0 to 10.0 mm. The base curve is the radius of the sphere of the back of the lens that the prescription describes (the lower the number, the steeper the curve of the cornea and the lens and the higher the number, the flatter the curve of the cornea and the lens). This number is important in order to allow the contact lens to fit well on the wearer's cornea for comfort, to facilitate tear exchange, and to allow oxygen transmission [56].

Prosthetic eye

The surgical procedure to remove the entire natural eye is referred to as enucleation. The surgical procedure to remove the contents of a natural eye is referred to as evisceration [57–59]. Both procedures will require prosthetic restoration with the fitting of a prosthetic eye, as shown in Figure 7.6 [60]. This process can usually begin approximately six weeks following the initial surgery. Fitting the prosthetic eye begins with taking an impression mold or other advanced fitting methods to accurately contour or manipulate the anophthalmic socket. These techniques are utilized to ensure the prosthesis has optimal cosmesis, comfort, and movement [57, 59].

Figure 7.6: Prosthetic eye [60].

Prosthetic nose

A prosthetic nose is an artificial external nose used for the purpose of keeping the shape of the nostrils in a normal state when the external nose atrophies and loses the sense of smell, or when the external nose is lost. A nasal prosthesis (artificial silicone nose) is fabricated to restore normal contour and improve function for patients who have experienced partial or total loss of their nose to traumatic injury, disease, or due to surgical removal of the nose (rhinectomy). Rhinectomy (surgical removal of the nose) often occurs as a necessary treatment in the eradication of malignant neoplasms (skin cancers) such as basal cell or squamous cell carcinomas or malignant melanomas. Malignant tumors or mucormycosis (fungal infection) may also arise in the bone or the cavities (sinuses) near the nose, requiring surgery that involves the nose [61, 62]. Nasal prostheses are attached using either medical adhesive attachment, anatomical retention – directly to the anatomy – or craniofacial bone-anchored implants [64]. Nasal prosthetics is a learned specialty like any other that begins with basic materials and skills. New scanners and 3D printers now impressively capture soft tissue contour and create working models that represent a very accurate snapshot of nasal form [63].

7.3.3 Pharynx area

Electrolarynx
The electric artificial pharynx or artificial vocal cords are devices that allow a person who has lost their vocal cords and unable to utter words on their own to utter words electrically. An electrolarynx, sometimes referred to as a throat back, is a medical device about the size of a small electric razor used to produce clearer speech by those people who have lost their voice box, usually due to cancer of the larynx. The most common device is a handheld, battery-operated device pressed against the skin under the mandible, which produces vibrations to allow speech; other variations include a device similar to the talk box electronic music device, which delivers the basis of the speech sound via a tube placed in the mouth [64]. There are two main kinds of electrolarynx: one that you hold to your neck and one that has a small tube (oral tube) that you place in your mouth. Both types are small, handheld, and battery-operated [65].

The definition of human voice is a sound produced by means of the lungs and larynx or the faculty of utterance. Voice production with an anatomically normal vocal tract requires three distinct elements: (1) air generator – the lungs generate an air stream that flows through the larynx; (2) vibrating apparatus – apposition of the paired vocal folds when supplied by the air stream generates undulating vibrations to create speech sound; and (3) articulating tract: sound is modulated by the pharynx and oral cavity to produce phonetic voice. On the other hand, total laryngectomy removes the vibrating apparatus and redirects the air stream, significantly impacting vocal abilities [65, 66], as seen in Figure 7.7. Patients who have undergone total laryngectomy have had their vibrating apparatus removed and although the air generator and articulating tract remain, the air stream is diverted and does not pass through the articulating tract. It is, by this mechanism, that they lose the ability to produce sound (as seen in Figure 7.7). Furthermore, the articulating tract may be modified as well during the surgical excision depending on the extent of concomitant pharyngeal or tongue base disease involvement. Voice restoration aims to artificially create a sound source by reintroducing a vibrating air column that is then modified by the articulating apparatus [67, 68].

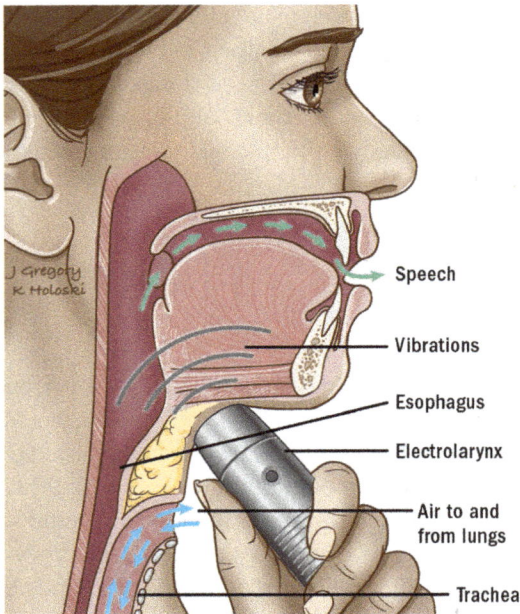

Figure 7.7: Typical electrolarynx [65].

Artificial trachea

The trachea (aka windpipe) and bronchus play important roles in the human body, providing a pathway that leads air to reach the lungs. It is made up of cartilage, muscle, and connective tissue. Tracheas are often around 4 inches long and are comprised of approximately 20 cartilage rings. On the inside lining of the trachea is a mucus membrane that is able to catch harmful airborne particles and bacteria before it reaches your lungs. Swallowing and sneezing are ways of removing these particles [69–72]. Breathing involves two main steps: oxygen enters the body when you breathe in and carbon dioxide leaves the body when you exhale. When someone takes a breath, oxygen goes through the nose and mouth down their trachea. From here, the trachea splits up into two bronchi tubes which lead to the left and right lung. Next, the oxygen travels through the smaller branches, bronchioles, to millions of small alveoli located throughout the lungs. Alveoli are small sacs in the lungs that allow for the exchange of gases with the blood stream. Here oxygen is exchanged with carbon dioxide. The red blood cells deliver the oxygen to the rest of the body while the carbon dioxide is expelled out the same way the oxygen came in. To avoid food and drink from entering the lungs, a flap known as the epiglottis closes while swallowing in order to prevent them from entering the trachea and lungs [70]. Diseases of the trachea that can benefit from a trachea replacement are: (1) tracheal cancer, (2) tuberculosis, (3) tracheal stricture (stenosis), and (4) tracheomalacia. Surgery represents the main form of treatment for these diseases. However, resecting over 50% of the tracheal length is currently unfeasible because of the need for remnant trachea. To overcome this limitation and improve curability rates, alternatives are needed. Many investigators in both research fields and clinical practice have reported on artificial trachea. Autologous tissue, synthetic materials, and allograft have all been used as artificial trachea [72, 73]. Etienne et al. [74] classified tracheal substitutes into five types: (1) synthetic prosthesis implantation, (2) allografts, (3) cadaver tracheal transplantation, (4) tissue engineering, and (5) autologous tissue composite. The ideal tracheal substitute is still unclear, but some techniques have shown promising clinical results.

In order to fabricate an artificial trachea, the patient is first imaged by a CT scan. This is so the tissue engineers can make a scaffold that fits perfectly within the patient's body. The scaffold is then made from nano-sized polyethylene terephthalate (PET) fibers using these specifications. Next, it is seeded with stem cells from the patient's bone marrow by placing it in a bioreactor with transcription factors, which may help force the stem cells to differentiate into trachea-specific cells. These cells then grow and divide to produce cartilage around the scaffold after 2–3 days. Finally, it is surgically implanted by suturing it into the patient's throat and lungs [75, 76]. Figure 7.8 shows a typical artificial trachea with differentiated stem cells [75].

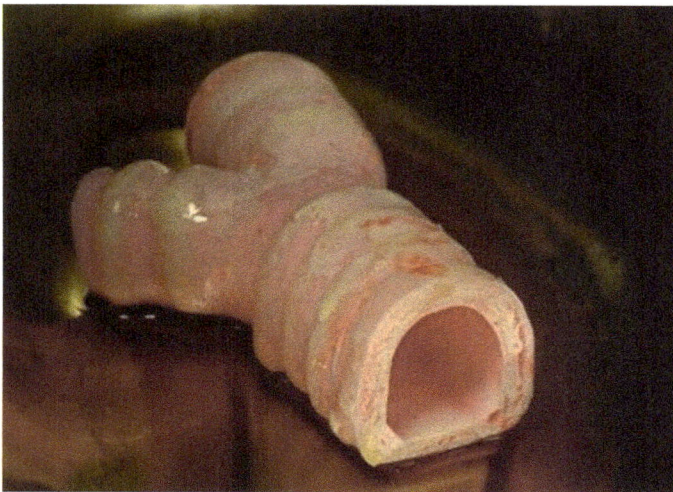

Figure 7.8: Artificial trachea with differentiated stem cells [75].

Among various technologies to manufacture effective artificial trachea systems, a tissue engineering method using 3D bioprinting has been developed [70, 77]. Park et al. [77] fabricated a multilayered scaffold with polycaprolactone (PCL) and hydrogel with nasal epithelial and auricular cartilage cells in the printing process. Three-dimensional

(3D) bioprinting serves as an additional manufacturing method that uses a 3D bio-printer and enables 3D stacking of certain biomaterials layer by layer according to the intended design. Recent advances in tissue engineering enabled 3D bioprinting using various biocompatible materials including living cells, thereby making the product clinically applicable. Additionally, mesenchymal stem cells (MSC), originally isolated from the bone marrow, are a promising cell type for tissue engineering applications because of their proliferation ability and their potential to differentiate in vitro and in vivo into multiple functional tissue-specific cell types, such as adipocytes, chondro-cytes, osteoblasts, and skeletal myocytes. Based on this background, Bae et al. [70] manufactured dual cell–containing artificial trachea using a 3D bioprinting method with epithelial cells and undifferentiated bone marrow-derived MSC (bMSC) or epi-thelial cells and chondrogenic-differentiated bMSC. Figure 7.9 shows 3D bioprinted artificial trachea, depicting that the SEM image revealed micropores in the inner-most and outermost PCL layers and showed that the middle non-pore PCL layer sep-arates the two alginate hydrogel layers [70]. Figures 7.9A and B show general appearance with 15 mm length and 5 mm inner diameter with five-layered structure. In SEM image (Figure 7.9C), a-, c-, and e-layers are composed of PCL, while b-layer is alginate layer with MSC or d-MSC, and d-layer is alginate layer with epithelial cells. The innermost (e) and outermost (a) layers showed a microporous feature. The non-porous third layer (c) separates the epithelial cell layer (d) and the MSC or d-MSC layer (b) [70].

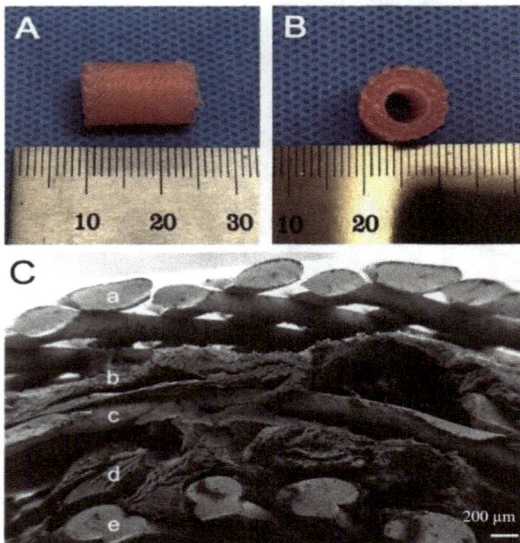

Figure 7.9: 3D bioprinted artificial trachea [70].

Artificial esophagus

The current treatments for esophageal diseases, such as carcinomas, trauma, or congenital malformations, require surgical intervention and esophageal reconstruction using redundant parts of the gastrointestinal tract [78, 79]. However, the use of gastrointestinal segments can cause various surgical morbidities and mortality because additional abdominal surgery may be required at the expense of other anatomic structures. Therefore, tissue engineering using various biomaterial or cell sources has emerged as an alternative strategy of biomimicking the native esophageal tissue that could be implanted as an artificial graft. Although tissue engineering techniques have promise as an effective regenerative strategy, no functional solution currently exists for esophageal reconstruction [78].

Figure 7.10 illustrates a typical esophagus system [80]. The esophagus is a muscular tube connecting the throat (pharynx) to the stomach. The esophagus is about 8 inches long and is lined by moist pink tissue called mucosa. The esophagus runs behind the windpipe (trachea) and heart, and in front of the spine. Just before entering the stomach, the esophagus passes through the diaphragm. The upper esophageal sphincter (UES) is a bundle of muscles at the top of the esophagus. The muscles of the UES are under conscious control, used when breathing, eating, belching, and vomiting. They keep food and secretions from going down the windpipe. The lower esophageal sphincter (LES) is a bundle of muscles at the low end of the esophagus, where it meets the stomach. When the LES is closed, it prevents acid and stomach contents from traveling backwards from the stomach. The LES muscles are not under voluntary control [80].

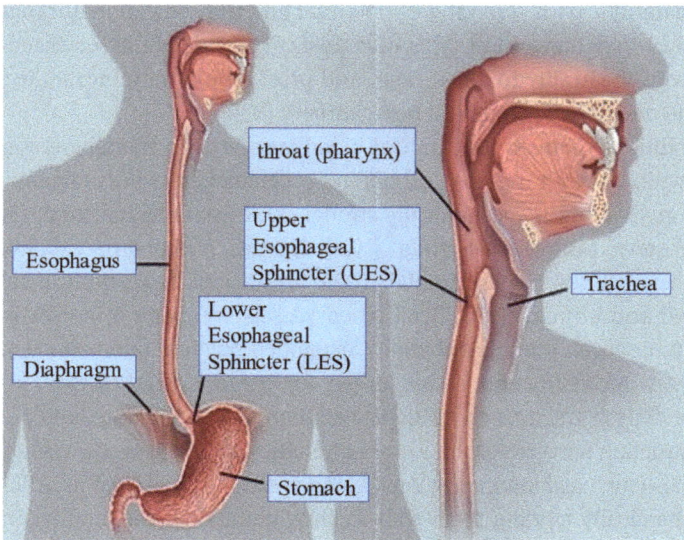

Figure 7.10: A typical esophagus system, consisting of upper esophageal sphincter (UES), esophagus main tube, and lower esophageal sphincter (LES) [80].

7.3.4 Cardiovascular system

Artificial blood vessels

Cardiovascular diseases, especially ones involving narrowed or blocked blood vessels with diameters smaller than 6 mm, are the leading cause of death globally. Vascular grafts have been used in bypass surgery to replace damaged native blood vessels for treating severe cardiovascular and peripheral vascular diseases. However, autologous replacement grafts are not often available due to prior harvesting or the patient's health. Furthermore, autologous harvesting causes secondary injury to the patient at the harvest site. Therefore, artificial blood vessels have been widely investigated in the last several decades [81]. About 300,000 people a year in the United States receive regular kidney dialysis, which removes and filters a patient's blood before returning it. To speed the procedure, doctors typically implant a small blood vessel between a vein and an artery in the patient's arm. Blood is then removed and reinserted through an intravenous line inserted into this bypass vessel. When possible, doctors typically harvest a piece of a vein from a patient to make this bypass, called a shunt. But over time, these shunts often fail, forcing doctors to use shunts made with plastics and other synthetic materials that can trigger immune reactions or blood-flow problems downstream [82]. The incidence of surgical correction of congenital heart defects has increased dramatically over the last several decades. These defects, considered fatal just 30 years ago, can often be corrected successfully with overall operative mortality of < 2%. Reconstruction or replacement of blood vessels, valves, and cardiac chambers is frequently required to repair or reform the appropriate anatomic configuration. The use of synthetic materials, with zero growth potential and unpredictable durability, is often the only way to achieve these operative goals. The availability of tissue-engineered material, with the ability to grow, heal, and provide long-term durability, would revolutionize the practice of congenital heart surgery [83].

The preferable conditions required for artificial blood vessels are generally as follows: (1) good biocompatibility with the host, (2) nontoxic, carcinogenic, nonantigenic, (3) hard to deteriorate in vivo and is mechanically strong, (4) resistant to infection, (5) easy to sterilize and preserve, (6) available grafts of various sizes, (7) good operability, (8) not leaking blood, and (9) nonthrombotic [84]. There are various types of artificial blood vessels developed and in market. Artificial blood vessels are made of polyester cloth in either knit form or woven form. Hydrophobic expanded polytetrafluoroethylene (ePTFE) is used for venous reconstruction, shunt surgery for cyanosis heart disease, internal shunt surgery for dialysis treatment in patients with chronic renal failure, and peripheral arterial reconstruction surgery for the lower extremities. Artificial blood vessels derived from biomaterials (such as human or animal blood vessels) can be used as artificial blood vessels by chemically treating them with alcohol, glutaraldehyde, and so on to reduce antigenicity. When used in clinical practice, it is said that there are disadvantages, that is, they are easy to clog with blood clots and deteriorate easily. Artificial blood vessels made of synthetic polymeric materials do not attach blood clots to the inner surface and are difficult to attach to endothelial cells in living organisms such as polyester [81, 85]. In order to remain open for a long time, it is important to have the stability of pannus extending from the biological side in the transition part (anastomosis part) between artificial blood vessels and living organisms. Hybrid artificial blood vessels are a combination of materials or structures in their composition and artificial blood vessels made by mixing artificial and biological materials. Furthermore, artificial blood vessels are fabricated extensively using tissue engineering [86], including (1) cell seeding type (an artificial blood vessel covered with the inner surface of a patient's endothelial cells for providing antithromboticity), (2) an artificial blood vessel in which a patient's cells and tissues are seeded in an artificial blood vessel (such as endogenous cytokine-active artificial blood vessels) [82], (3) regenerated blood vessels (made by culturing a patient's cells), (4) cultured artificial blood vessels (using human-derived materials, artificial blood vessels with histological three-layer structures such as the inner, middle, and outer membranes which were created outside the body, close to normal arteries. Artificially assembling the components of a blood vessel wall from the cellular level is time-consuming and costly, but technically possible), [87] or (5) genetically introduced artificial blood vessels, providing a place for cell transplantation for gene therapy (as a function to incorporate into endothelial cells, genes can be selected for the purpose of preventing arteriosclerosis, antithrombotic properties, suppressing endothelial growth, and vasodilation) [84].

Figure 7.11 shows a vascular implant made of thermoplastic polyurethane (PU), offering a new and desirable form of biodegradable vascular implant, exhibiting equivalent long-term performance characteristics compared to the clinically used, nondegradable material with improvements in intimal hyperplasia and ingrowth of host cells [88]. The blood vessels were printed from a bioink containing human smooth muscle cells (harvested from an aorta) and endothelial (lining) cells from an umbilical vein, having the same dual-layer architecture of natural blood vessels and outperforming existing engineered tissues, as seen in Figure 7.12 [86].

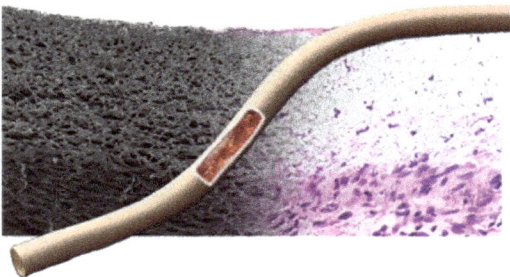

Figure 7.11: Biodegradable thermoplastic polyurethane vascular implant [88].

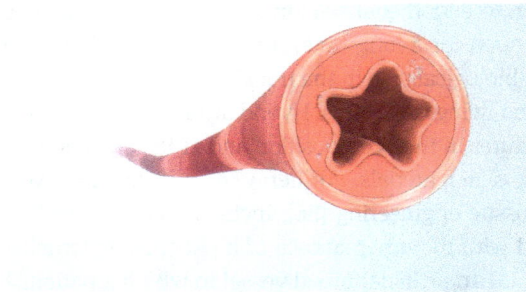

Figure 7.12: Blood vessel fabricated by 3D bioprinting [88].

Syedain et al. [89] created a new lab-grown blood vessel replacement that is composed completely of biological materials, but surprisingly does not contain any living cells at implantation. The vessel, which could be used as an "off the shelf" graft for kidney dialysis patients, performed well in a recent study with nonhuman primates. It is the first-of-its-kind nonsynthetic, decellularized graft that becomes repopulated with cells by the recipient's own cells when implanted. The discovery could help tens of thousands of kidney dialysis patients each year. The grafts could also be adapted in the future for use as coronary and peripheral bypass blood vessels and tubular heart valves [89, 90], as seen in Figure 7.13 [89].

Figure 7.13: Bioengineered artificial blood vessels, being capable of growth within the recipient [89].

Electrically conductive artificial blood vessels that may serve as implants to replace diseased native vessels have been developed [91]. The flexible and biodegradable constructs consist of a metal-polymer conductive membrane, and an electric current can be passed through the vessel when it is implanted in the body. The electric stimulation appears to encourage the proliferation and migration of endothelial cells and may improve the integration of the implanted vessel with the surrounding tissue. It can also be used to deliver gene therapy or drugs into those tissues. The electronic blood vessel would be an ideal platform to enable diagnostics and treatments in the cardiovascular system and can greatly empower personalized medicine by creating a direct link of the vascular tissue-machine interface [91].

Artificial heart

The artificial heart is typically used to bridge the time to heart transplantation, or to permanently replace the heart in case heart transplantation is impossible. Our heart mainly serves as a pump that controls blood circulation. The blood that has returned to the heart from the whole body is sent out to the lung circulation, then back to the heart again, and finally to the whole body. The function of a heart is to pump. The artificial heart is a mechanical act on behalf of the pump's function. The development of artificial hearts was caused by high mortality rates due to heart failure in the United States. In 1964, the FDA launched an artificial heart program, and the development of artificial hearts (blood pumps) began in earnest. Meanwhile, in 1967, the world's first heart transplant was performed in South Africa, and two major activities were available as a treatment for severe terminal heart failure [92]. There are two types of artificial hearts in clinical applications. One is to remove one's heart (ventricular part) and replace it with two blood pumps, called a completely substituted artificial heart. The other is called the left ventricle assisted heart, which removes blood from a place called the left ventricle and returns blood to the aorta. Although there are still two main issues associated with the artificial heart system, including quality of battery and miniaturization of system body, the following are recognized as basic purposes and functions of the artificial heart. One is semipermanently dependent on the artificial heart to maintain blood circulation (permanent use), and the other is for temporary use until a heart transplant donor (heart donor) is found (bridge use) [92, 93]. Figure 7.14 illustrates a typical ventricular assist device (VAD), designed to assist left ventricle in pumping oxygenated blood through the aorta and to the body's tissues. The pump is placed inside the chest cavity, while the power source and system controller are carried on a harness outside the body [94].

Birla et al. [95] mentioned that 3D printing technologies are emerging as a disruptive innovation for the treatment of patients in cardiac failure. The ability to create custom devices, at the point of care, will affect both the diagnosis and treatment of cardiac diseases. The introduction of bioinks containing cells and biomaterials and the development of new computer-assisted design and computer-assisted manufacturing systems have ushered in a new technology known as 3D bioprinting. Small-scale 3D bioprinting has successfully created cardiac tissue microphysiological systems. 3D bioprinting provides an opportunity to evaluate the assembly of specific parts of the heart and most

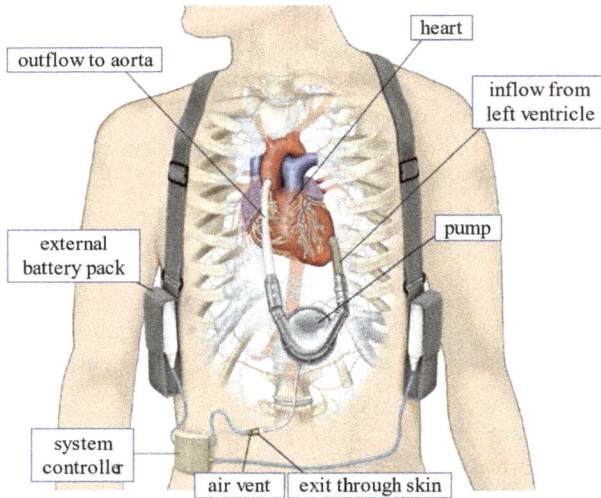

Figure 7.14: Typical ventricular assist device [94].

notably heart valves. With the continuous development of instrumentation and bioinks and a complete understanding of cardiac tissue development, it is proposed that 3D bioprinting may permit the assembly of a heart described as a total biofabricated heart. Smith [96] introduced a new technique of using donor's tissue as bioink sources. As Goodwin [97] introduced a 3D bioprinted artificial heart (see Figure 7.15).

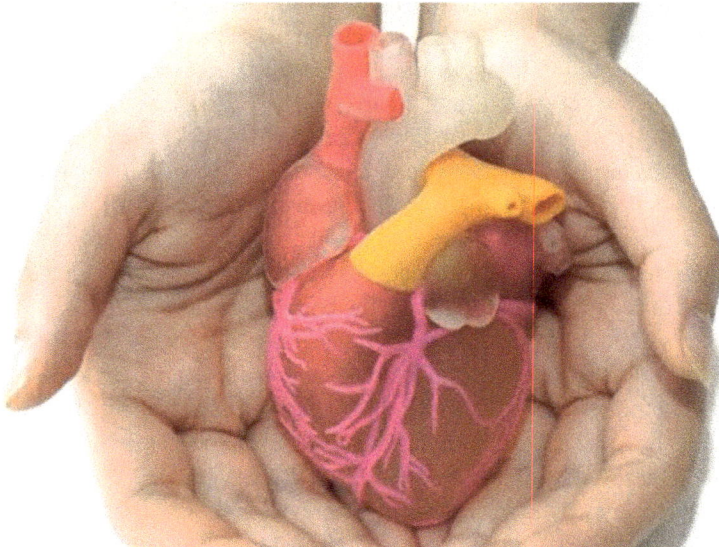

Figure 7.15: 3D bioprinted artificial heart [97].

Besides the VAD, there are still cardiovascular-related AOs which are implanted in cases where the heart, its valves, or another part of the circulatory system is in disorder; artificial pacemakers represent another cardiovascular device that can be implanted to either intermittently augment (defibrillator mode), continuously augment, or completely bypass the natural living cardiac pacemaker as needed.

Artificial lung
Artificial lung device, which can fully substitute the lungs of living bodies, has not been developed and utilized yet; however, as there is a mechanical heart already, soon the world may see artificial lungs, although it would be much harder than an artificial heart because a very large surface area is needed to exchange enough oxygen. Definitely, there exists a growing demand for new technology that can take over the function of the human lung, from assisting an injured or recently transplanted lung to completely replacing the native organ [98–100]. Artificial lung is a prosthetic mechanical device with membranes made of synthetic material to transfer oxygen to the blood and eliminate carbon dioxide from the blood and is connected to blood vessels through tubes and cannulas of silicone. The blood passing through the device is oxygenated and cleared of carbon dioxide. There are basically two types of artificial lungs: a bubble-type artificial lung which mixes oxygen directly into blood to perform the gas exchange and a membrane-type in which gas exchange is achieved through the membrane made of synthetic material, well known as extracorporeal membrane oxygenation (ECMO). The most well-known term for artificial lung is ECMO. Because of this technology, it can also be applied for patients who do not have oxygenation problems, such as patients with hypercapnic respiratory failure, or to patients with severe pulmonary hypertension (PAH), and right ventricular failure, the most current term is extracorporeal lung support (ECLS) [101, 102].

There are many requirements for successful artificial lung: (1) An artificial lung must be able to sustain the gas exchange requirements of a normal functioning lung. The gas exchange requirements in an adult can vary. Light exercise can increase the average 240 mL/min of required oxygen to 800 mL/min. Because lungs process air, they are the only internal organs that are constantly exposed to the external environment. They breathe in between 2,100 and 2,400 gallons of air every day. Inhaled air is divided between two airways called bronchi that lead to either lung. Inside the lungs, the bronchi branches out into thousands of smaller and smaller tubes that connect to tiny sacs known as alveoli. An average person has enough alveoli in a single lung to cover an area the size of a tennis court [103]. Thus, even minor variations need to be accounted for when creating an artificial respiratory system. Artificial lungs use the right ventricle as a pump, as opposed to an external device. The pressure must, therefore, not drop significantly, as there is no external pump to help maintain the pressure of the system and ensure proper blood flow. (2) Another consideration in artificial lung design is the minimization of injury to the blood cells. Injured cells could potentially lead to clotting and immunologic response. The shear stress within the system must be kept within defined parameters. If the shear stress is too high, it can stimulate platelets and white cells, and it can also cause hemolysis. Conversely, if the shear stress is too low, it is suggested that the artificial lung will become a depository for thromboses [98]. (3) Additional problems include the attachment of phospholipids to the fiber surfaces, which can alter surface tension and promote plasma leakage so that fibers with surface coatings, such as silicon, could inhibit phospholipid adsorption, reduce activation of the coagulation cascade, and decrease the inflammatory response [98]. As the artificial lung relies on the right ventricle as its pump, the artificial lung itself must not cause undue stress on the heart. Right heart failure could result in and has been a major consideration of artificial lung design. The power needed to drive blood through the artificial lung and the native lungs is affected by impedance, which is dependent on the structure of the artificial lung and the way it is connected to circulation [99]. The artificial lung must have a similar impedance to that of the native lung; it must mimic the opposition to pulsing blood flow, which in turn affects cardiac load [100]. To create a matching impedance and to help decrease the workload of the right ventricle, a compliance chamber is often used with the artificial lung [98].

The main part of artificial lung system is a gadget is consisting of a blood cannula connection to the heart and pump-lung unit, as seen in Figure 7.16 [104], which is wearable in a patient's backpack. Figure 7.17 explains principles of artificial lung [105].

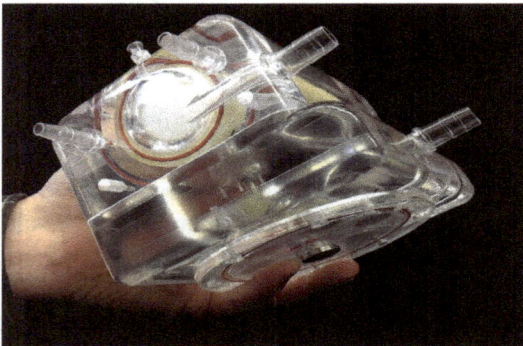

Figure 7.16: Main gadget of artificial lung [104].

In patients with severe lung disease caused by viral infections, physicians sometimes turn to ECMO – a life support machine that takes over the functions of the lungs, heart, or both when other support options appear to be failing. But initial reports of ECMO use in patients with COVID-19 described very high mortality [106, 107], and some physicians recommended against its use. However, it was reported that new

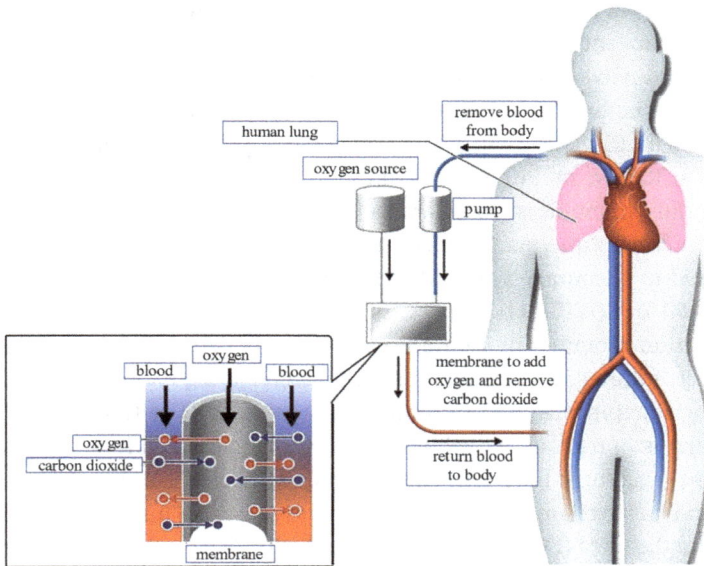

Figure 7.17: Main function of artificial lung system [105].

data from Columbia University and other ECMO centers throughout the world now show that more than 60% of severe COVID-19 patients who received ECMO have survived. Instead, the results support the use of ECMO in COVID patients with acute respiratory distress syndrome (ARDS) in experienced centers, although only a randomized clinical trial can provide a definitive answer [106]. Barbaro et al. [107] used data available from the Extracorporeal Life Support Organization (ELSO) registry to characterize the epidemiology, hospital course, and outcomes of patients aged 16 years or older with confirmed COVID-19 who had ECMO support initiated between January 16 and May 1, 2020, at 213 hospitals in 36 countries. The primary outcome was in-hospital death in a time-to-event analysis assessed at 90 days after ECMO initiation. A multivariable Cox model was applied to examine whether patient and hospital factors were associated with in-hospital mortality. It was concluded that (i) in patients with COVID-19 who received ECMO, both estimated mortality 90 days after ECMO and mortality in those with a final disposition of death or discharge were less than 40%, and (ii) these data from 213 hospitals worldwide provide a generalizable estimate of ECMO mortality in the setting of COVID-19 [107]. Badulak et al. [108] updated a guideline from the ELSO for the role of ECMO for patients with severe cardiopulmonary failure due to COVID-19. The great majority of COVID-19 patients (>90%) requiring ECMO have been supported using veno-venous (V–V) ECMO for ARDS. While COVID-19 ECMO-run duration may be longer than in non-COVID-19 ECMO patients, published mortality appears to be similar between the two groups. However, data collection is ongoing, and there is a signal that overall mortality may be increasing. Conventional selection criteria for COVID-19-

related ECMO should be used; however, when resources become more constrained during a pandemic, more stringent contraindications should be implemented. Formation of regional ECMO referral networks may facilitate communication, resource sharing, expedited patient referral, and mobile ECMO retrieval. There are no data to suggest deviation from conventional ECMO device or patient management when applying ECMO for COVID-19 patients. Rarely, children may require ECMO support for COVID-19-related ARDS, myocarditis, or multisystem inflammatory syndrome in children (MIS-C); conventional selection criteria and management practices should be the standard. It was strongly suggested to encourage participation in data submission to investigate the optimal use of ECMO for COVID-19 [108].

Lung transplantation remains the definitive curative treatment for end-stage lung disease. However, future applications of tissue bioengineering could overcome the donor organ shortage and the need for immunosuppression. The final goal of lung tissue engineering is to recreate the whole spectrum of specialized lung tissues and thereby provide physiologic functions through bioengineered conducting airways, vasculature, and gas exchange tissue. End-stage lung disease, namely chronic obstructive pulmonary diseases (COPD), represent the fourth leading cause of death worldwide. The increasing rates of tobacco smoking and exposure to air pollutants will further raise the number of COPD patients thus creating an urgent need for new therapeutic strategies [109]. ECMO and mechanical ventilation can be temporarily used in this scenario as a bridge to lung transplantation that remains the only definitive treatment, but the need for immunosuppression and the donor organ shortage are major limits for a larger clinical impact [110]. The final goal of lung tissue engineering is to recreate the whole spectrum of specialized lung tissues and thereby provide physiologic functions through bioengineered conducting airways, vasculature, and gas exchange tissue [110]. One of the most challenging tasks in lung bioengineering is reproduction of the extracellular matrix, whose proteins are necessary for host-derived defense and graft homeostasis. Synthetic scaffolds provide gas exchange but lack extracellular matrix proteins and hence do not offer all the elements required for successful replacement of pulmonary function. Another major problem to solve is the need to generate increasingly compact vascular flow networks capable of physiologic blood flow and gas exchange, thus recreating the lung's architectural hierarchy [110]. One way to overcome these limits is human donor lung decellularization, providing exactly the complex hierarchical structure of vascular and airway lung architecture. Unfortunately, even this option presents several drawbacks: in case of incomplete recellularization, extracellular matrix proteins will be exposed and may initiate pathological reparative responses in vivo; disruption of the extracellular matrix during decellularization may result in scaffold degradation; lastly, as each decellularization process requires preexisting native human lung, this approach does not solve the human donor shortage, thus making the use of xenogeneic scaffolds unavoidable [111]. Petrella et al. [109] concluded that (i) although experimental transplantation of bioartificial lung developed by perfusing decellularized or synthetic scaffolds has been shown to provide gas exchange in vivo over a prolonged period, it should be clearly acknowledged that its clinical application is still far from reality, and (ii) as an alternative to artificial lungs, the lung stem cell pathway and plasticity may be targeted by novel compounds to stimulate their contribution to lung regeneration.

It was announced that the thinnest ever in vitro model of the human lung was fabricated utilizing inkjet bioprinting, with potential applications for screening the safety and efficacy of new drug candidates and studying disease progression [112–114]. Figure 7.18 shows artificial lungs fabricated by 3D bioprinting technology [113].

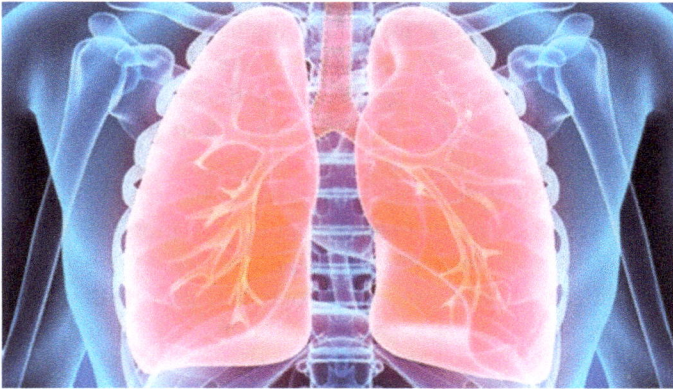

Figure 7.18: 3D-printed artificial lungs using bioinks [113].

The most important parameter for the 3D bioprinting is the bioink. Bioink is defined as bioprintable material used to produce engineered (artificial) live tissue in 3D bioprinting processes, where cells and other biologics are deposited in a spatially controlled pattern to fabricate living tissues and organs. It can be composed only of cells, but in most cases, an additional carrier material that envelops the cells is also added. This carrier material is usually a biopolymer gel, which acts as a 3D molecular scaffold. Cells attach to this gel, and this enables them to spread, grow, and proliferate. The first generation of bioinks include hydrogels, cell aggregates, microcarriers, and decellularized matrix components used in extrusion-, droplet-, and laser-based bioprinting processes. A detailed comparison of these bioink materials is conducted in terms of supporting bioprinting modalities and bioprintability, cell viability and proliferation, biomimicry, resolution, affordability, scalability, practicality, mechanical and structural integrity, bioprinting and post-bioprinting maturation times, tissue fusion and formation postimplantation, degradation characteristics, commercial availability, immune-compatibility, and application areas [115–118]. To increase the amounts of lungs available for transplantation, fabricating lungs in the lab by combining cells with a bioengineered scaffold could be one solution. It was reported that the researchers at Lund

University in Sweden first designed a new bioink made by combining two materials: a material derived from seaweed, alginate, and extracellular matrix derived from lung tissue [119].

Artificial thymus

The thymus is a specialized primary lymphoid organ of the immune system, where T lymphocytes mature. A dysfunctional thymus can lead to severe immunodeficiency and there is currently an unmet clinical need for suitable thymic therapeutic strategies [120]. The thymus is a primary lymphoid organ, essential for T cell maturation and selection. There has been long-standing interest in processes underpinning thymus generation and the potential to manipulate it clinically because alterations of thymus development or function can result in severe immunodeficiency and autoimmunity [121]. An implantable machine that performs the function of a thymus does not exist. However, researchers have been able to grow a thymus from reprogrammed fibroblasts. They expressed hope that the approach could one day replace or supplement neonatal thymus transplantation. The artificial thymus would play an important role in the immune system; it would use blood stem cells to produce more T cells, which would help the body fight infections; it would also grant the body the ability to eliminate cancer cells. Since when people become old, their thymus does not work well, an artificial thymus would be a good choice to replace the old, non-functioning thymus [122].

7.3.5 Digestive organs

Artificial liver

The complex function of the liver makes it challenging since it not only detoxifies toxic by-products but also participates in numerous other synthetic and metabolic functions of the body; its failure therefore constitutes a life-threatening condition [123, 124]. It is said that the liver, when it is likened to a biological chemical plant, performs more than 500 kinds of metabolic reactions (chemical reactions). It is extremely difficult to supplement the function of a liver overall with artificial equipment, and it has come to be thought that there is no other hand than using the hepatocyte of the living body. Liver failure can either occur without preceding liver disease, usually caused either by intoxication or as acute decompensation of chronic liver-related illness, and in both cases its symptoms include icterus, hepatic encephalopathy, and impairment of coagulation status and may result in multiorgan failure [125]. The only long-term therapy in most cases is orthotopic liver transplantation, unless the liver is able to regenerate. Many patients, especially those who are not listed for high-urgency transplantation, may not survive until a suitable donor organ is available, since donor organs are rare. In other cases, contraindications do not permit liver transplantation. For these indications, extracorporeal liver-assist devices have been developed in order to either bridge the patient to transplantation or temporarily support the failing organ until it is able to regenerate. In the course of liver failure, water-soluble toxins (e.g., ammonia, mercaptans) and albumin-bound toxins (e.g., bilirubin, bile acids, aromatic amino acids, fatty acids) may accumulate and cause encephalopathy and dysfunction of other organs. While the field of detoxification and partially also of regulation can be addressed by artificial devices similar to dialysis (artificial systems, detoxification devices), the synthetic function of the liver can only be provided by living cells. In order to temporarily apply these cells in a safe and convenient way, biohybrid artificial liver (BAL) support devices were developed [124]. With artificial systems, cell-free artificial systems make use of the processes of adsorption and filtration, assuming that removal of toxins from the patient's plasma will improve the clinical state of the patient. HD, being the common treatment for renal failure, is also used for treatment of patients with liver failure to remove water-soluble toxins. Since liver failure is often accompanied by renal failure, HD is part of the standard intensive care treatment [125]. Liver support systems are divided into an artificial liver assist device (ALD) and a BAL-assist device. ALDs include molecular adsorbent recirculating system (MARS), Prometheus, single-pass albumin dialysis, and selective plasma filtration therapy. These devices work as a blood purification system of the liver. On the other hand, bioartificial liver device (BLD) has hepatic cell lines incorporated in its equipment, which aims to function as a complex biological liver system providing support to its biochemical processes [123].

MARS is an extracorporeal HD system composed of three different circuits: blood, albumin, and low-flux dialysis [126]. MARS banks on the recycling of albumin solution via an anion exchanger and active charcoal. The patient's blood is led through the hollow fiber capillaries of a high-flux dialysis filter. Albumin solution, which is circulated in the extracorporeal circuit, passes the membrane counter directionally, allowing albumin-bound toxins in the blood to cross the membrane and bind to the albumin of the MARS circuit. The membrane is, however, impermeable to albumin. When passing through the adsorber and filter cartridges, the toxins are cleared by the filter and albumin is regenerated and able to accept new toxins when passing through the membrane again. Additionally, the albumin circuit itself is dialyzed in the method of continuous veno-venous HD or continuous veno-venous hemodiafiltration, resulting in diminishing the load of water-soluble toxins [125].

A BAL system is an artificial extracorporeal supportive device which represents an important therapeutic strategy for patients with acute liver failure [127]. Generally, a BAL system consists of functional liver cells supported by an artificial cell culture material. In particular, it incorporates hepatocytes into a bioreactor in which the cells are immobilized, cultured, and induced to perform the hepatic functions by processing the blood or plasma of liver failure patients. The BAL system acts as a bridge for the patients until a donor organ is available for transplantation or until liver regeneration [128]. Demetriou et al. [128], after clinical experiences with BAL system, indicated that BAL treatment is safe and beneficial and can be successfully used as a "bridge" to transplantation. Figure 7.19 shows a typical BAL system with human pluripotent stem cell-derived hepatic cells using double filtration plasmapheresis (DFPP). In a BAL system, patient plasma is first separated from whole blood by DFPP. Plasma then perfuses a bio-artificial device using hydrophilic hollow fibers. The human pluripotent stem cells (hPSCs)-derived hepatic cells are inoculated at the outside of the hollow fibers. The de-toxified patient plasma is filtered once more before returning to the patient's blood stream. The hollow fiber membranes and safety filter provide two layers of separation between the patient's blood stream and the hPSC-derived hepatic cells [129].

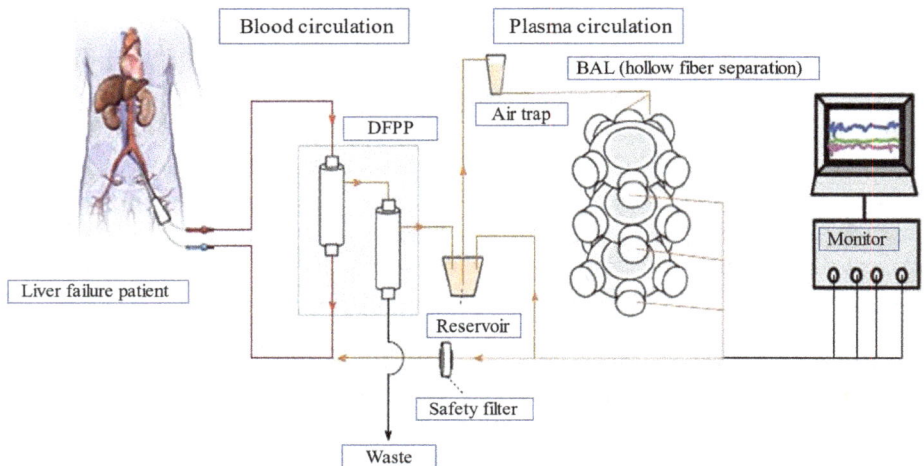

Figure 7.19: Bioartificial liver system for patient whose lung has been clinically failed [129].

Artificial pancreas

Diabetes or diabetes mellitus (DM) is a chronic disease that occurs when the pancreas is no longer able to make insulin (which acts like a key to let glucose from the food pass from blood stream into cells in the body to produce energy), or when the body cannot make good use of the insulin it produces. Not being able to produce insulin or use it effectively leads to raised glucose levels in the blood (known as hyperglycemia). Over the long term high glucose levels are associated with damage to the body and failure of various organs and tissues, resulting in diabetes. There are several types of diabetes. Type 1 diabetes can develop at any age but occurs most frequently in children and adolescents. When you have type 1 diabetes, your body produces very little or no insulin, which means that you need daily insulin injections to maintain blood glucose levels under control. Type 2 diabetes is more common in adults and accounts for around 90% of all diabetes cases. When you have type 2 diabetes, your body does not make good use of the insulin that it produces. The cornerstone of type 2 diabetes treatment is healthy lifestyle, including increased physical activity and healthy diet. However, over time most people with type 2 diabetes will require oral drugs and/or insulin to keep their blood glucose levels under control [130]. Recently, the term "type 3 diabetes" for Alzheimer's disease was proposed because of the shared molecular and cellular features among type 1 diabetes, type 2 diabetes, and insulin resistance associated with memory deficits and cognitive decline in elderly individuals [131]. In addition to these, there is the gestational diabetes (GDM), which is a type of diabetes that consists of high blood glucose during pregnancy and is associated with complications to both mother and child. GDM usually disappears after pregnancy, but women are affected and their children are at increased risk of developing type 2 diabetes later in life [130].

In order to prevent the crisis and progress of chronic vascular complications of DM, it is necessary to strictly control blood glucose levels for a lifetime. Therefore, artificial pancreas system which replaces the lost pancreatic endocrine ability (pancreatic β cells secreting insulin and pancreatic α cells secreting glucagon) of diabetic patients by machine has been developed. In healthy persons, blood glucose level rises when eating, insulin is secreted from pancreatic β cells, and glucose is used by working on muscle, adipose tissue, and liver, resulting in lowering the blood sugar levels. However, since the insulin secretion β pancreatic cells is lowered in DM patients, glucose in each tissue is not available, and blood sugar level rises. Insulin injection therapy for present DM treatment is only self-injecting the insufficient insulin amount according to the doctor's instruction, and this glycemic control cannot control the condition effectively and sufficiently. Accordingly, a proper usage of the artificial pancreas enables the physiological glycemic control like a healthy person's [131].

There are several types of artificial pancreas systems. The system started with a large-scale bed-side artificial pancreas unit recognized as the first generation, following the small-scale bed-side system as the second generation. Then a portable-type closed-loop insulin delivery system has been developed as the third generation [132]. The fourth and current generation of system is known as an embed-type system. An ultra-small light sensor is attached to the body (like an earring or denture) to measure blood sugar and send the results by radio waves to the implanted artificial pancreas embedded under the skin of the abdomen. Then, the artificial pancreas calculates the required amount of insulin based on the blood sugar level sent to it and automatically injects insulin into the body. As a result, it can contribute to improving HRQoL and extending life expectancy in DM patients. Using stem cells as main bioink material, 3D printed functional pancreas has been fabricated [133, 134], as seen in Figure 7.20. The 3D printed artificial pancreas model consists in superposing layers of cells taken from a type 1 diabetes patient with hydrogel until a 3D tissue is created.

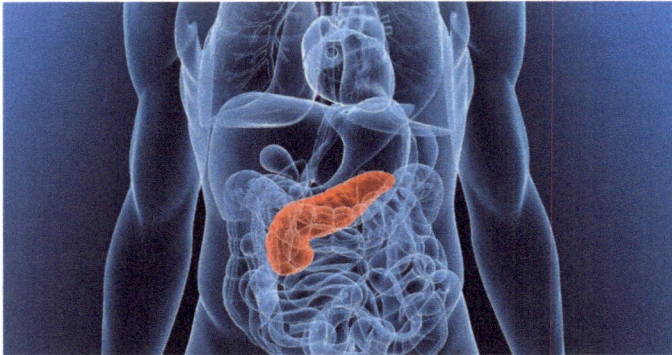

Figure 7.20: 3D bioprinted artificial pancreas [134].

Artificial stomach

Chemical digestion, the decomposition of macromolecules by the action of enzymes, begins in the mouth and stomach but occurs primarily in the small intestine. The digestive tract is a single long tube that functions to move nutrients, water, and electrolytes from the external to the internal environment of an animal. Chewing in the mouth and peristaltic churning in the stomach involves a mechanical digestion of food into much smaller particles. Bile, secreted by the liver, coats globs of fats and mechanically reduces the fats into small globules and tiny droplets. The breakdown of large bulky foods into microscopic particles increases the surface area available for chemical digestion. Decomposition of macromolecules by enzymes is the principal means of chemical digestion. The stomach is an expanded portion of the digestive tract. Although there are variations, essentially a stomach is defined as an organ that produces gastric juices, particularly pepsinogen (pepsin in inactive form) and hydrochloric acid. Since pepsin is a protease, the stomach begins protein digestion. Amino acids and monosaccharides are absorbed through the small intestine and enter the capillaries associated there. These digested nutrients are transported by the hepatic portal system to the liver, processed, and eventually released again into the blood supply. Fats are not absorbed by this route, however. Once fats have been digested into monoglycerides and fatty acids, they are absorbed into the epithelial cells that line the small intestine. Inside these cells, the lipid components are reconverted into triglycerides and combined with cholesterol. The reconstituted fats are transported into the lymph vessels closely associated with the intestinal wall and eventually absorbed into the vascular (venous) system [134, 135]. Bile salts are reabsorbed through the intestine and returned to the liver for recycling [135], as illustrated in Figure 7.21 [136].

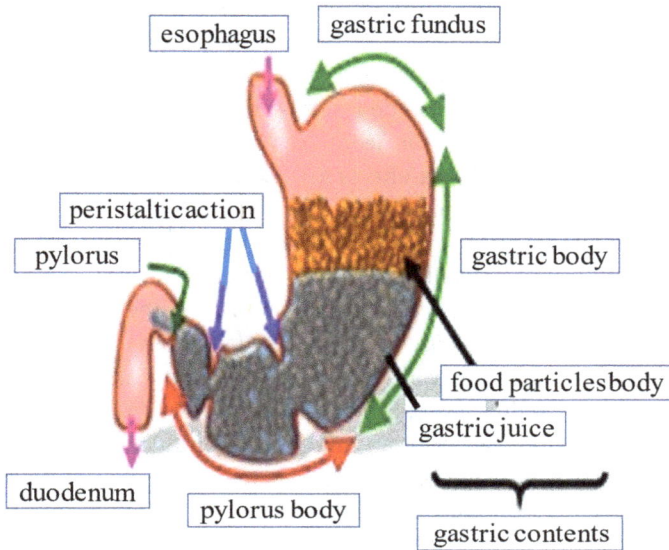

Figure 7.21: Schematic view of food digestive action inside a human stomach [136].

Besides these chemical digestive reactions, there are two major mechanical issues involved in the digestive actions: fluid dynamics of gastric juice for food digestion [137] and peristaltic movement [136, 138, 139]. In efforts to fight obesity and enhance drug absorption, scientists have extensively studied how gastric juices in the stomach break down ingested food and other substances. The relevant parts of the stomach are the corpus, where food is stored; the antrum, where food is ground; and the pylorus or pyloric sphincter, the tissue valve that connects to the small intestine. Slow-wave muscle contractions begin in the corpus, with wave speed and amplitude increasing to form the antral contraction waves (ACWs) as they propagate toward the pylorus [136, 138].

In 2006, British scientists built a physiologically correct artificial stomach that simulates human digestion, complete with food, stomach acid (gastric juice), digestive enzymes, hormones, and the ability to vomit [140].

7.3.6 Urinary organs

Artificial kidney

The kidneys are important organs that remove waste products and excess fluids in the body and regulate blood so that it does not lean acidic. If for some reason(s) the kidneys no longer function, humans cannot survive. When the function of the kidneys decreases, the blood has to be cleaned with a device called an artificial kidney.

Artificial kidney devices are available in many forms, but HD is the most widely used. This is a device that removes moisture and waste products from the blood through a special membrane called a semipermeable membrane and regulates the composition of the electrolyte so that the blood does not become acidic.

Artificial kidneys have been on dialysis for nearly 30 years, and although they function in their own way, there are still side effects associated with current methods and materials. Hence, there are advanced approaches HD unit which is available at home, portable artificial kidney, or implantable artificial kidney. Curley [141] reported that scientists at the University of California, San Francisco, are developing an implantable artificial kidney that can replicate the work of the real organs and potentially eliminate the need for dialysis. UCSF (Roy S)-Vanderbilt (Fissell WH) jointly have made significant advancements with the technology but are still identifying methods to prevent the blood clotting associated with their machine [142, 143].

Figure 7.22 illustrates schematic design for an implantable artificial kidney device, using iliac vessels for arterial blood inflow and venous return, with ultrafiltrate draining into the bladder. The detailed view of the main filtration unit (left side) is designed to accommodate up to a liter of blood per minute, filtering it through an array of silicon membranes. The filtered fluid contains toxins, water, electrolytes, and sugars. The fluid then undergoes a second stage of processing in a bioreactor of lab-grown cells of the type normally lining the tubules of the kidney. These cells reabsorb most of the sugars, salts, and water back into the bloodstream. The remainder becomes urine that is directed to the bladder and out of the body [144]. New research at the University of Arkansas has given hope for artificial kidneys. They created a mechanical device that enables to behave with the blood the same way as kidneys do. The kidney works in two steps: (1) first, clusters of blood vessels called glomeruli separate blood and proteins, from waste and water, followed by sending back the blood with essential constituents, and (2) it transfers the waste to the nephron network, which further goes under the filtration process – the ion transport [145], as seen on right side of Figure 7.22.

7.3.7 Others

There is, furthermore, an artificial urinary bladder which possesses two main methods for replacing bladder function involving either redirecting urine flow or replacing the bladder. Methods to grow bladders using stem cells had been attempted in clinical research but this procedure was not part of medicine [146, 147].

When considering future human organs, the development of small- and high-power actuator acting as an artificial muscle becomes important. Research on artificial muscles is studied using micromachining technology to conduct electromagnetic, thermal, and mechanical actuators.

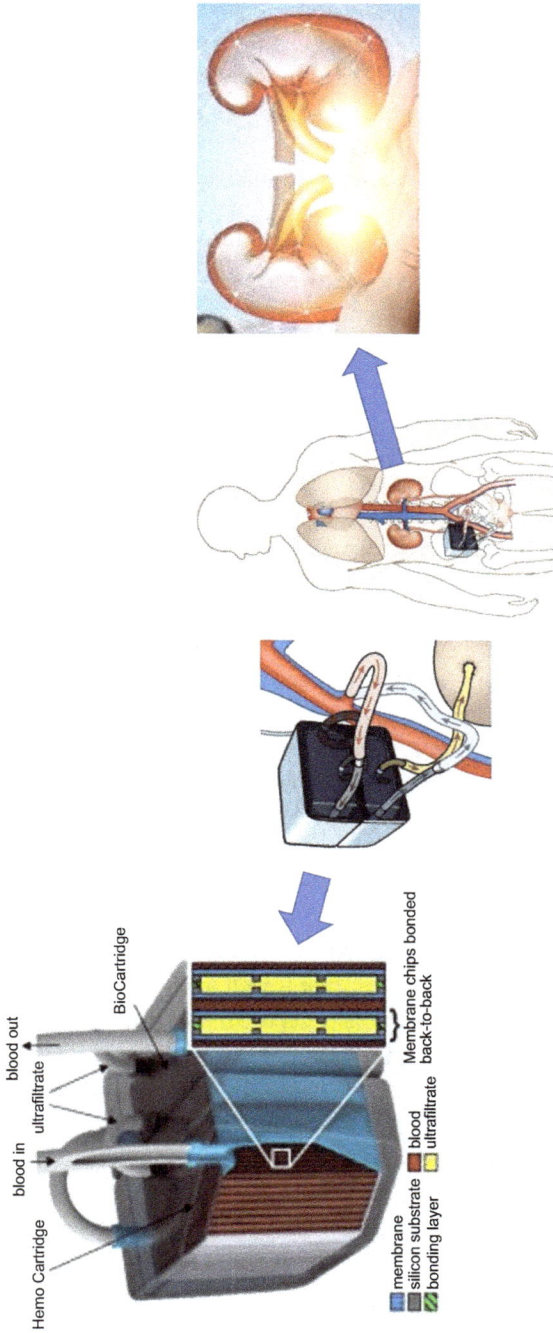

Figure 7.22: Schematic design of an implantable artificial kidney device [143–145].

Artificial skin belongs to a regenerative medicine and is a wound-covering material consisting of biomaterials used to prevent the invasion and proliferation of bacteria from the outside and the leakage of body fluids in skin defects caused by burns. For example, it is known that artificial skin in which the patient's fibroblasts are cultured in collagen and then planted the patient's epidermal cells shows a hierarchical structure similar to biological tissues and does not cause rejection. Artificial skins formed from cow collagen to sheet-like sponges and laminated with silicone thin films have also been developed.

Although there are still artificial bones and joints, since they are responsible to bearing load, they will be discussed in the next section.

7.3.8 Future of artificial organs

As we have been reviewing, due to diseases and wounds, AOs have been utilized to temporarily maintain life and restore organ function or to act on behalf of the function semipermanently, and many AOs are currently used in clinical practice. However, there are still problematic issues, including (i) incomplete biocompatibility, (ii) relatively large scale, heavy weight and less flexibility, (iii) no long-term durability, (iv) only single functionality, (v) incomplete control system, and (vi) relatively complicated operation and less energy efficiency. Accordingly, it becomes very difficult for human body to recover from homeostasis. In future, AOs should possess the following characteristics, including (a) multifunctionality, (b) auto-diagnosis and auto-repair functionalities, (c) functionally distributed AOs, (d) flexible and soft device, (e) tissue-engineered AOs, and (f) super organs which can transcend biological organ function. To realize these future characteristics, surrounding technology should also further develop to assist them. Such required technologies can include (1) new materials development, (2) advanced biomechanics, (3) sensor technology, (4) simulation engineering, (5) micromachining, and (6) man–machine interface AI technology [148, 149].

Among these supporting technologies, 3D bioprinting is well advanced [150–152]. Klak et al. [150] mentioned that the most important aspects of 3D bioprinting include (i) preparation of a bioink for biological printing (e.g., with appropriate printability – viscosity, consistency, shear rate) that has a composition mimicking the native cell environment, (ii) optimization of the conditions of the bioprinting process, for example, printing speed, pressure applied to the bioinks in the extrusion method, inner diameter of the needle, (iii) assessment of the final bioconstruct, for example, mechanical strength, biodegradation, diffusion, and (iv) monitoring the condition of cells subjected to the bioprinting process, for example, viability, proliferation, and functionality.

An organ-on-a-chip (OoC) or organ-on-chips (OoCs) is a multichannel 3D microfluidic cell culture chip that simulates the activities, mechanics, and physiological response of entire organs and organ systems, a type of AO. OoC is in the list of top 10 emerging technologies and refers to a physiological organ biomimetic system built on a microfluidic chip. Through a combination of cell biology, engineering, and biomaterial technology, the microenvironment of the chip simulates that of the organ in terms of tissue interfaces and mechanical stimulation [153, 154]. This reflects the structural and functional characteristics of human tissue and can predict response to an array of stimuli including drug responses and environmental effects. OoC has broad applications in precision medicine and biological defense strategies [154]. The 3D in vitro models, such as OoC platforms, are an emerging and effective technology that allows the replication of the function of tissues and organs, bridging the gap amid the conventional models based on planar cell cultures or animals and the complex human system [155, 156]. Bioprinting technology is the most advanced technology for producing biomimetic cellular constructs. This technology can produce computer-designed 3D structures with multiple types of cells, biomaterials, and biomolecules. Recently, bioprinting has been actively applied in the development of OoC. The study has usually focused on developing biomimetic tissue models with living cells for the liver, microvascular network, skin, heart, and so on, and studies have shown promising results. A recent study has also shown that bioprinting can be applied to cell-laden-microfluidic systems and biosensors, which are the main components of OoC. This technical flexibility should enhance the completeness of OoC, and automated process with bioprinting should improve the system's repeatability [157, 158].

It is believed that both 3D bioprinting and OoCs will have a significant impact on health care around the world. Developments in the field of 3D bioprinting and chip systems provide novel platforms to follow trends toward nonanimal testing and fulfill unmet medical needs such as regeneration medicine and transplantation. These technologies are of great potential to eliminate testing on animals and provide patient-specific drug testing. Other benefits include enhancing experimentation capabilities and savings in funding. The 3D cell culture obtained in the bioprinting process mimics the spatial organization of cells in a living organism, while OoCs allow monitoring intercellular interactions; thus, testing the activity of drug candidates is more predictive and valuable in these assay systems. Therefore, 3D bioprinting and OoC shall become an important tool in preclinical studies as well as research activities. The bioprinting of organs that resemble human nature can limit the animal usage in research and pharma studies. The utilization of 3D bioprinting and OoC in drug candidate screening allows for the better selection of potential therapeutics for further development in clinical studies, hence reducing the failure rate and providing an alternative for animal models [154–156].

7.3.9 Apheresis

Apheresis (ἀφαίρεσις: a taking away or a separation) is a medical technology in which the blood of a person is passed through an apparatus that separates out one particular constituent and returns the remainder to the circulation. It is thus an extracorporeal therapy [159]. A substitute organ is an organ used as a substitute for a damaged organ. Most AOs, such as artificial kidneys and artificial hearts, are substitute organs that act as substitute organs for replacing original function of the organs. However, apheresis is not a substitute organ, it is rather a real AO to perform new functions. Today, extra-body circulation separates plasma and cell components from the blood, as well as the liquid factors that cause disease from isolated plasma components. Some examples of apheresis (also referred to as hemapheresis or pheresis) include (i) plasmapheresis (removal of plasma), (ii) leukapheresis (removal of white blood cells), (iii) granulocytapheresis (removal of granulocytes: neutrophils, eosinophils, and basophils), (iv) lymphocytapheresis (removal of lymphocytes), (v) lymphoplasmapheresis (removal of lymphocytes and plasma), and (vi) plateletpheresis or thrombocytapheresis (removal of platelets) [160].

7.4 Load-bearing orthopedic implants

7.4.1 In general

Load-bearing implants are considered more likely structural implants, as opposed to those which were discussed in the previous section and can be considered as functional implants. Once orthopedic implant was placed and started to healing process through an osseointegration, such a placed implant will be subjected to complicated loading situation. In dental implant treatments, there are basically three protocols in terms of loading, which should include (i) immediate loading, (ii) early loading, and (iii) delayed loading. On the other hand, with orthopedic implants such as TKR and THR, a loading situation will be more complicated. The loading situation can include static vs. dynamic [161], cycling loading or fatigue loading [162–164], or wear [165] which might cause the secondary adverse effect of the wear debris toxicity.

There is still another important issue involved in orthopedic implants, that is, a biomechanical situation. When bone and implant are loaded, this stiffness mismatch results in stress shielding and as a consequence, degradation of surrounding bony structure can lead to disassociation of the implant [166–168]. To establish a biomechanical compatibility of placed implant to surrounding hard tissues, there are several approaches in material developments. Kesteven et al. [169] studied low elastic modulus Ti-Ta alloy, and Brar et al. [170] evaluated and Oshida [171] discussed biodegradable magnesium materials. Composite materials include composites with zirconium dioxide [172], with bioactive glass fiber [173], or others [174]. HA-based biomaterials were also investigated [175–178]. Calcium-silicate coating would be another answer [179]. It is known that a control of porosity is a promising method to lower the modulus of elasticity, resulting in a good biomechanical compatibility [180–183]. Evans et al. [184] developed porous polyether-etherketone (PEEK) for load-bearing orthopedic implants. Porous pure Ti [185] and Ti-6Al-4V alloy [186] were also fabricated.

7.4.2 Artificial limbs

In medicine, a prosthesis (or prosthetic implant) is recognized as an artificial device that replaces a missing body part, which may be lost through disease (e.g., diabetes mellitus, peripheral vascular disease, infection, tumors), traumatic, congenital disorder, or a car accident. Prostheses are intended to restore the normal functions of the missing body part [187]. There are studies on relationship between amputation, prosthesis installation, and HRQoL [187–189]. The HRQoL is related to adjustments to amputation and artificial limb, and QoL, and to analyze the influence of sociodemographic, medical, and amputation-related factors on this relationship [188]. Horne at el. [189] reported that (i) the type and quality of the prosthesis affect the patient's physical and mental ability of adaptation, and (ii) rehabilitation practitioners and researchers need measures that can distinguish between levels of disability, predict prognosis, assist in patient care, and map changes in functional status as the result of interventions.

As to indications of prosthetics, in general, young traumatic or neoplastic amputee, and motivated perceived vulnerability to disease (PVD)/neuropathy amputee with cardiac reserves would be indications. On the other hand, contraindications should include (i) no ambulation potential, (ii) severe cardiac disease, (iii) poor vision, (iv) poor motivation or compliance, (v) poor stump – infected, ulceration, poor skin, and (vi) fixed flexion deformities (FFD) knee or hip [190, 191].

Artificial hand/arm
There are upper-arm prosthetic arms and forearm prosthetic arms, depending on the amputation location. And moreover, depending on function and operational manner, there are decorative prosthetics, active prosthetics, and myoelectric prosthetics [192]. For the upper-arm prosthetic arm, there is an elbow fitting/connector between the hand portion and socket portion [192, 193], as seen in Figure 7.23.

Artificial leg
Depending on the amputation locations, there are three types of prosthesis: lower leg prosthetic leg, thigh prosthetic leg, and crotch prosthetic leg, as shown in Figure 7.24 [192].

7.4.3 Joints

We have joints connecting upper and lower portions of body parts, as illustrated in Figure 7.25 [194]. When pressure is applied to the knee(s) with OA, the nerves may become pinched and that can cause many other problems in the knee such as pain, tingling, and functionality issues. OA is a common form of arthritis that affects

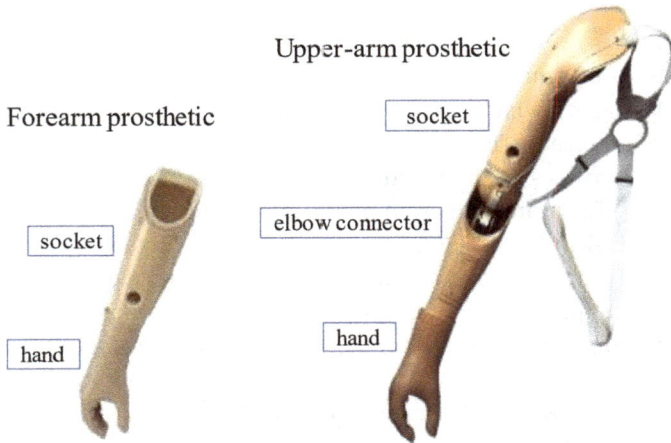

Figure 7.23: Two types of prosthetic hand/arms [192].

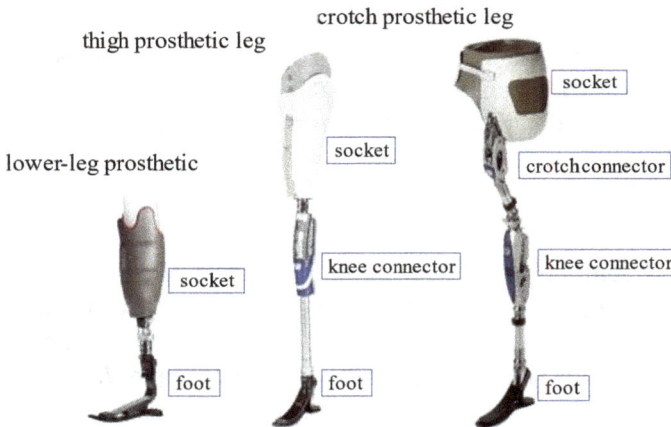

Figure 7.24: Three type of leg prosthesis [192].

several joints in the body including the knees. It causes the cartilage in the joints to break down and the bones then begin to rub together as a result. OA in the knees is commonly caused by being overweight, old age, and knee injuries [194]. OA is the most common form of arthritis, affecting millions of people worldwide. It occurs when the protective cartilage that cushions the ends of the bones wears down over time. Although OA can damage any joint, the disorder most commonly affects joints in hands, knees, hips, and spine [195]. OA symptoms can usually be managed, although the damage to joints cannot be reversed. OA symptoms often develop slowly and worsen over time. Signs and symptoms of OA include: pain (affected joints might hurt during or after movement), stiffness (joint stiffness might be most noticeable upon awakening or after being inactive), tenderness (joint might feel tender when you apply light pressure to or near it), loss of flexibility (unable to move your joint through its full ROM), grating sensation (feel a grating sensation when you use the joint, and you might hear popping or crackling), bone spurs (these extra bits of bone, which feel like hard lumps, can form around the affected joint), and swelling (caused by soft tissue inflammation around the joint) [195].

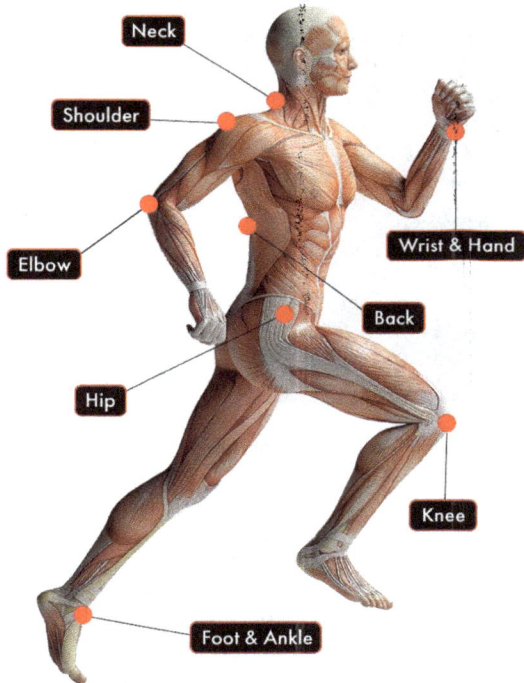

Figure 7.25: Joints in our body [194].

Pain which can be sensed and recognized might differ among the different joint portions. There is neck pain (cervical joint degeneration, cervical arthritis), back pain (facet syndrome, degenerative disc disease, lumbar arthritis, OA), shoulder pain (rotator cuff tears, shoulder bursitis), elbow pain (lateral epicondylitis, golfer's elbow, tennis elbow, distal biceps tendon tear), wrist and hand pain (carpal tunnel syndrome, wrist arthritis), hip pain (labrum tear, hip OA, hip bursitis), knee pain (meniscus tear, knee degeneration, anterior cruciate ligament (ACL) tear, medical collateral ligament (MCL) tear, chondromalacia), and foot and ankle pain (planter fasciitis, ankle sprain, rolled ankle) [195].

In next two sections, we will be discussing three major orthopedic implants (total shoulder replacement (TSR), THR, and TKR), as indicated in Figure 7.26 [196].

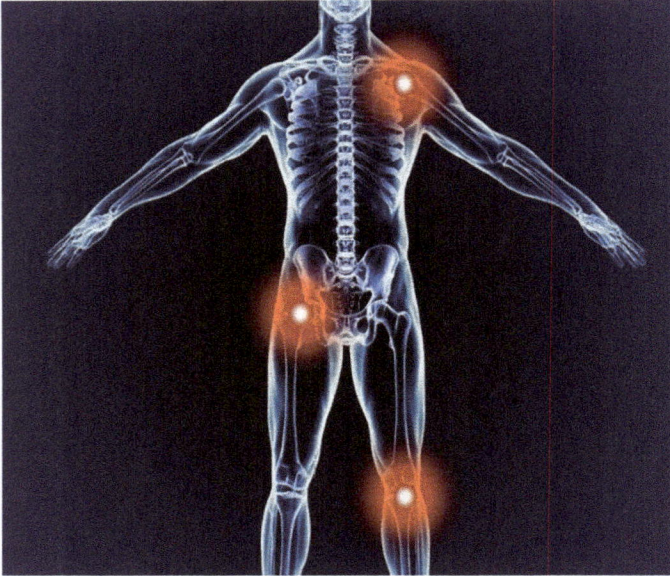

Figure 7.26: Three major joints – shoulder joint, hip joint, and knee joint [196].

7.4.4 Total shoulder replacement (TSR) and reverse TSR (rTSR)

TSR, which is also known as TSA, involved a removal of portions of the shoulder joint, which is replaced with artificial implants to reduce pain and restore range of rotation and mobility. It is very successful for treating the severe pain and stiffness caused by end-stage arthritis. Although shoulder joint replacement is less common than knee or hip replacement, it is just as successful in relieving joint pain. It is reported that about 53,000 people in the United States have shoulder replacement surgery each year, according to the Agency for Healthcare Research and Quality. This compares to more than 900,000 Americans a year who have knee and hip replacement surgery [197]. Referring to Figurer 7.27 [198], a healthy shoulder is made up of three bones: humerus (upper arm bone), scapula (shoulder blade), and clavicle (collarbone). The shoulder is a ball-and-socket joint: The ball, or head, of upper arm bone fits into a shallow socket (or glenoid) in the shoulder blade. With a healthy shoulder, center of the humerus head should be at the same level of that of a glenoid fossa, as shown in Figure 7.27 [240].

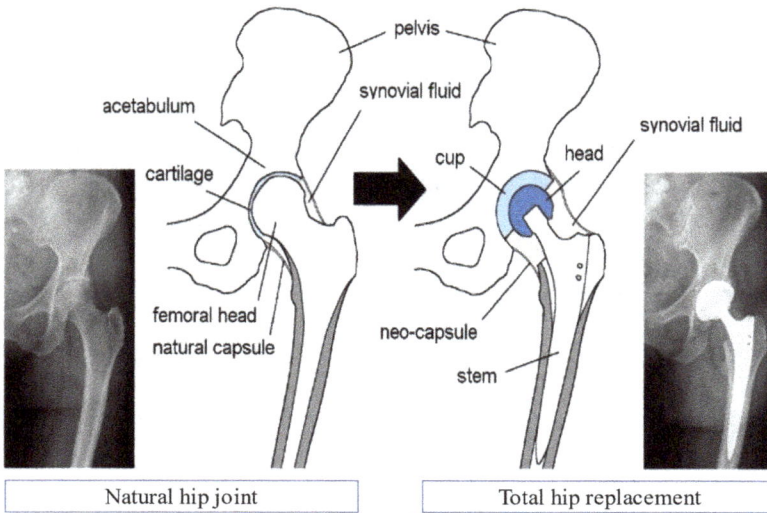

Figure 7.27: Anatomy of healthy shoulder [198].

In shoulder replacement surgery, the damaged parts of the shoulder are removed and replaced with artificial components, called a prosthesis. The treatment options are either replacement of just the head of the humerus bone (ball), or replacement of both the ball and the socket (glenoid). There are several diseases required for TSA treatments. They include (1) OA or degenerative joint disease with worn cartilage and bone spurs, which is an age-related "wear and tear" type of arthritis. The cartilage that cushions the bones of the shoulder softens and wears away. The bones then rub against one another. Over time, the shoulder joint slowly becomes stiff and painful. (2) RA is a disease in which the synovial membrane that surrounds the joint becomes inflamed and thickened. This chronic inflammation can damage the cartilage and eventually cause cartilage loss, pain, and stiffness. RA is the most common form of a group of disorders termed "inflammatory arthritis." (3) Post-traumatic arthritis can follow a serious shoulder injury. Fractures of the bones that make up the shoulder or tears of the shoulder tendons or ligaments may damage the articular cartilage over time. This causes shoulder pain and limits shoulder function. (4) Rotator cuff tear arthropathy is when a patient with a very large, long-standing rotator cuff tear may develop cuff tear arthropathy. In this condition, the changes in the shoulder joint due to the rotator cuff tear may lead to arthritis and destruction of the joint cartilage. (5) Avascular necrosis (or osteonecrosis (ON)) is a painful condition that occurs when the blood supply to the bone is disrupted. Because bone cells die without a blood supply, ON can ultimately cause destruction of the shoulder joint and lead to arthritis. Chronic steroid use, deep sea diving, severe fracture of the shoulder, sickle cell disease, and heavy alcohol use are risk factors for avascular necrosis. (6) Severe fractures of the shoulder is another common reason why people have shoulder replacements. When the head of the upper arm bone is shattered, it may be very difficult for a doctor to put the pieces of bone back in place. In addition, the blood supply to the bone pieces can be interrupted. In this case, a surgeon may recommend a shoulder replacement. Older patients with osteoporosis are most at risk for severe shoulder fractures. In addition to the above, although uncommon, some shoulder replacements fail, most often because of implant loosening, wear, infection, and dislocation. When this occurs, a second joint replacement surgery (which is called a revision surgery) may be necessary [197].

Orthopedic surgeons use the Walch classification to assess the glenoid morphology. The Walch classification of glenoid morphology is the most commonly used system for describing glenohumeral pathology in primary OA [199]. The Walch classification is used to stratify the outcomes of shoulder arthroplasty for varying pathologic glenoid types as well as assisting in preoperative planning to recognize morphologies that may pose intraoperative difficulties [200]. There are basically four major types (A, B, C, and D). Type A is characterized by centered humeral head, concentric wear, and no subluxation of the humeral head. There are subtype A1 (with minor central erosion) and A2 (with major central erosion, humeral head protruding into the glenoid cavity). Type B is recognized as humeral head subluxated posteriorly, biconcave glenoid with asymmetric wear and further B1 (with narrowing of the posterior joint space, subchondral sclerosis, osteophytes), B2 (with biconcave aspect of the glenoid with posterior rim erosion and retroverted glenoid), and B3 (with monoconcave and posterior wear with > 15° retroversion or >70% posterior humeral head subluxation, or both). Type C has two subtypes: C1 is characterized by dysplastic glenoid with >25° retroversion regardless of the erosion, and C2 is characterized by biconcave, posterior bone loss, posterior translation of the humeral head. Type D is characterized by any level of glenoid anteversion or with humeral head subluxation of less than 40% [200–202]. Figure 7.28 shows schematic illustration of Walch classifications [203].

There are different types of shoulder replacements [197, 204].

Anatomic (or standard) total shoulder replacement (TSR)
The typical TSR involves replacing the arthritic joint surfaces with a highly polished metal ball (CoCrMo alloy, 316 L stainless steel or Ti-6Al-4V alloy) attached to a stem, and a plastic (e.g., polyethylene) socket (see Figure 7.29). The implants resemble the natural shape of the bones. These components come in various sizes. They may be either cemented or "press fit" into the bone. If the bone is of good quality, your surgeon may choose to use a non-cemented (press-fit) humeral component. If the bone is soft, the humeral component may be implanted with bone cement. In most cases, an all-plastic glenoid (socket) component is implanted with bone cement.

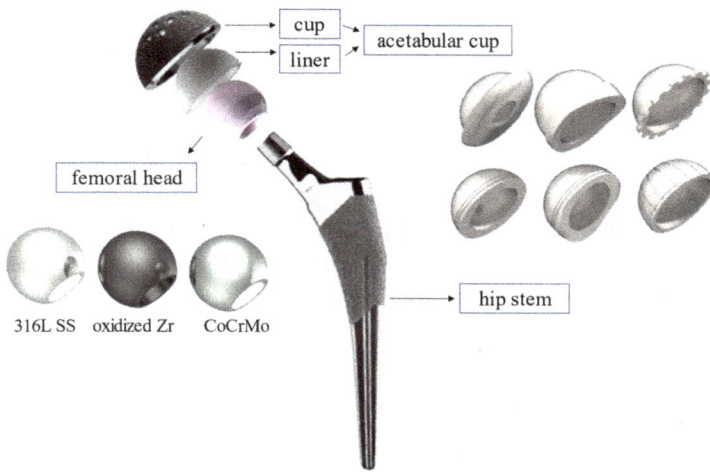

Figure 7.28: Schematic illustration of Walch classification of glenoid morphology [203].

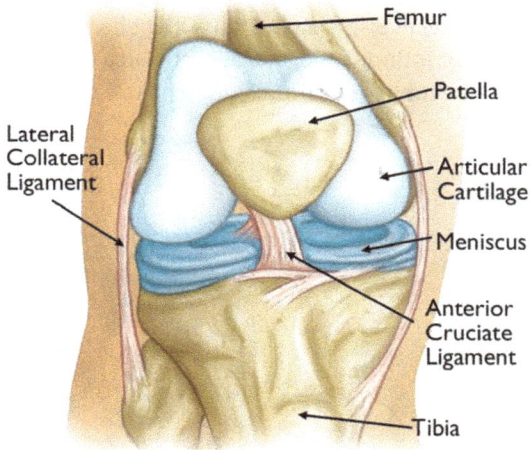

Figure 7.29: Anatomic (standard) total shoulder replacement [205, 206].

Partial shoulder replacement

Only the head (ball) of the joint is replaced. It may be recommended when only the ball side of the joint is damaged.

Hemiarthroplasty

In this procedure only the ball and stem are replaced. The stem is connected to the ball and articulated with your natural socket. There are stemmed and stemless procedures. In a traditional hemiarthroplasty, the head of the humerus is replaced with a metal ball and stem, similar to the component used in a TSR. This is called a stemmed hemiarthroplasty. The stemless TSA is a bone-preserving version of the TSA where the metallic ball is attached to the upper arm without a stem.

Resurfacing hemiarthroplasty

Resurfacing hemiarthroplasty involves replacing just the joint surface of the humeral head with a cap-like prosthesis without a stem. With its bone preserving advantage, it offers those with arthritis of the shoulder an alternative to the standard stemmed shoulder replacement.

Reverse total shoulder replacement (rTSR)

In the rTSR, the joint is literally reversed, namely the metal ball is placed where the glenoid socket was placed and a plastic cup is attached to the stem and is moved to the upper arm bone (humerus) (see Figure 7.30). This allows the patient to use the deltoid muscle instead of the torn rotator cuff to lift the arm. This option typically is preferred if the rotator cuff is severely damaged. The rTSR is indicated for patients who have (i) completely torn rotator cuffs with severe arm weakness, (ii) the effects of severe arthritis and rotator cuff tearing (cuff tear arthropathy), and /or (iii) had a previous shoulder replacement that failed. There are controversial pros and cons of the rTSR treatments [207–211].

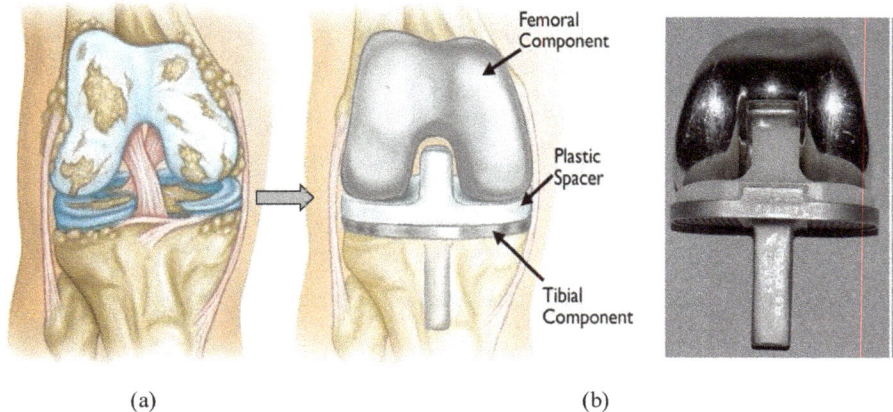

(a) (b)

Figure 7.30: Reverse TSR [205, 206].

7.4.5 Total hip replacement (THR)

THR (aka total hip arthroplasty (THA)) is one of the most cost-effective and consistently successful surgeries performed in orthopedics [212–214]. THR provides reliable outcomes for patients' suffering from end-stage degenerative hip OA, specifically pain relief, functional restoration, and overall improved QoL. The OA affects millions of Americans, and with an incidence of 88 symptomatic cases per 100,000 patients per year, translating to hip OA claiming the top underlying diagnosis leading to THR. Unlike the knee where the kinematics is highly driven by the soft tissues, it naturally represents a ball-in-socket configuration with three rotational degrees of freedom [161]. Bergmann et al. [215] reported that (i) the average patient loaded his hip joint with 238% of body weight (PBW) when walking at about 4 km/h and with a slightly less percentage when standing on one leg; (ii) when climbing upstairs the joint contact force is 251% BW which is less than 260% BW when going downstairs; (iii) inwards torsion of the implant is probably critical for the stem fixation; (iv) on average it is 23% larger when going upstairs than during normal level walking; and (v) the inter- and intra-individual variations during stair climbing are large, and the highest torque values are 83% larger than during normal walking. Because the hip joint loading during all other common activities of most hip patients is comparably small (except during stumbling), implants should mainly be tested with loading conditions that mimic walking and stair climbing [215]. Anatomically, the hip is a ball-and-socket type diarthrodial joint. Hip joint stability is achieved via a dynamic interplay from osseous and soft tissue anatomic components. Osseous components include the proximal femur (head, neck, trochanters) and the acetabulum, which is formed from three separate ossification centers (the ilium, ischium, and pubic bones) [216]. Loading conditions are now affecting the artificial replaced hip joint. It is obvious to see the significant difference in biotribological environment from the natural hip joint to THR, as seen in Figure 7.31 [217].

Contemporary THA techniques have evolved into press-fit femoral and acetabular components. Options for bearing surfaces include: metal-on-polyethylene (MoP), which has the longest track record of all bearing surfaces at the lowest cost; ceramic-on-polyethylene (CoP), which is becoming an increasingly popular option , ceramic-on-ceramic (CoC), which and has the best wear properties of all THR bearing surfaces, and metal-on-metal (MoM), which is although falling out of favor, MoM has historically demonstrated better wear properties than its MoP counterpart. MoM has lower linear-wear rates and decreased volume of particles generated. However, the potential for pseudotumor development as well as metallosis-based reactions (type-IV delayed hypersensitivity reactions) has resulted in a decline in the use of MoM. MoM is also contraindicated in pregnant women, patients with renal disease, and patients at risk of metal hypersensitivity [218, 219].

Summarizing, in the United States, there are currently four types of THR devices available with different bearing surfaces. These are [220]: (1) MoP: The ball is made of

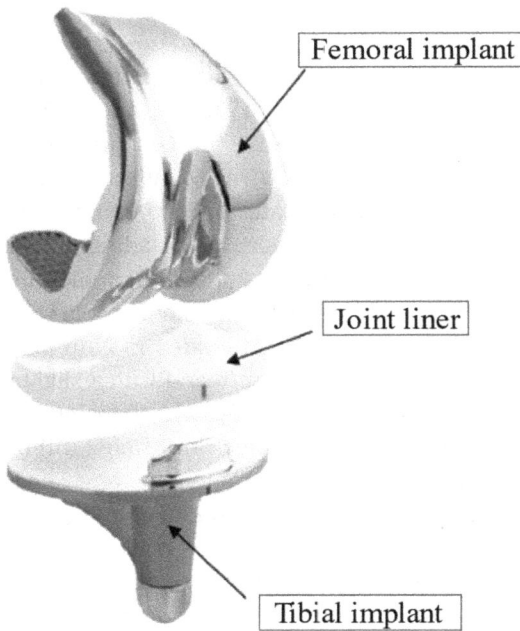

Figure 7.31: Comparison between natural hip joint and total hip replacement [217].

metal and the socket is made of plastic (polyethylene) or has a plastic lining, (2) CoP: The ball is made of ceramic and the socket is made of plastic (polyethylene) or has a plastic lining, (3) CoC: The ball is made of ceramic and the socket has a ceramic lining, and (4) CoM: The ball is made of ceramic and the socket has a metal lining, as shown in Figure 7.32 [221, 222]. An orthopedic surgeon should determine which hip implant will offer the most benefit and least risk for each patient. As of May 16, 2016, the effective date of the final order requiring premarket approval applications for these devices, there are no FDA-approved metal-on-metal THR devices marketed for use in the US. However, there are some patients who received a metal-on-metal THR prior to May 16, 2016. The FDA's Metal-on-Metal Total Hip Replacement Implant webpage provides specific information on metal-on-metal total hip replacements [223].

Under these complicated combinations of biomaterials which are exposed to biotribological (friction, wear, and lubricant actions in biological environment), it can be anticipated that an artificial hip joint is faced to risk of the failure or fracture due to wear debris (sometimes further causing wear debris toxicity) and high frictional torque in the case of poor lubrication that may cause loosening of the implant. Table 7.2 compares overall wear rates in various combinations of cup and head materials (table was slightly rearranged from the original [223]).

Besides materials in listed in Table 7.2, current materials used for the THR systems are more varied. For polymers, which are the first choice for low-friction hip

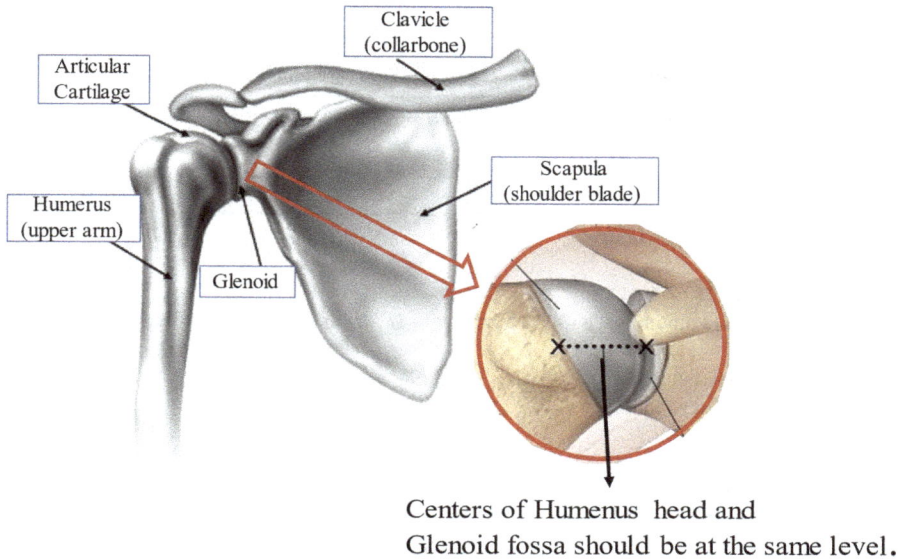

Centers of Humenus head and
Glenoid fossa should be at the same level.

Figure 7.32: Total hip replacement system [221] and acetabular cups and femoral heads [222].

Table 7.2: Socket and ball combinations and their overall wear rate.

Type of combination	Materials	Wear rate (mm^3/Mc)
Soft bearing couples		
MoP	CoCr – XLPE	6.71 ± 1.03
	CoCrMo – XLPE	4.09 ± 0.64
CoP	Alumina composite – XLPE	2.0 ± 0.3
	Alumina – XLPE	3.35 ± 0.29
	Alumina – PE	34
	ZTA – PE	80
Hard bearing couples		
CoM	Alumina composite – CoCrMo	0.02 ~ 0.87
CoC	Alumina – alumina	0.03 ~ 0.74
	ATZ – ATZ	0.024 ~ 0.06
	ATZ – ZTA	0.18
	ZTA – ZTA	0.14 ± 0.10
	ATZ – Alumina	0.20

Table 7.2 (continued)

Type of combination	Materials	Wear rate (mm³/Mc)
	Alumina composite – alumina composite	0.10
MoM	CoCrMo – CoCrMo	0.11 ~ 0.60

XLPE: cross-linked polyethylene
CoCr: typically, Co-33Cr alloy
CoCrMo: typically, Co-28 Cr-6Mo alloy
Alumina composite: Aluminum oxide (80%) matrix with 17% zirconia and 3% strontium oxide
PE: polyethylene
ZTA: Zirconia-toughened alumina
ATZ: Alumina-toughened zirconia

replacements, highly stable polymeric systems such as PTFE, ultra-high molecular weight polyethylene (UHMWPE), or PEEK have been investigated due to their excellent mechanical properties and their high wear resistance. Metallic materials have wide applications in the medical and bioengineering fields and are widespread as orthopedic implant components. The most common traditional metals used for THA are stainless steels, titanium alloys (Ti-6Al-4V), and mainly CoCrMo alloys. The latter have good corrosion resistance compared to other metals, and high toughness, high wear resistance, and higher hardness (HV = 350) than other metals and polymers. MoM articulation is typically produced from CoCrMo alloys (composed of 58.9–69.5% Co, 27.0–30% Cr, 5.0–7.0% Mo, and small amounts of other elements such as Mn, Si, Ni, Fe, and C). Metallic materials have high module of elasticity, which limits stress distribution from implant to bone. Therefore, new metallic components have been developed with lower elastic modulus and higher corrosion and wear resistance. Ti-6Al-4V alloys improve biocompatibility and mechanical resistance; this Ti-6Al-4V (or Ti64) alloy was replaced with iron (Fe) or niobium (Nb), realizing the improved alloys Ti-5Al-2.5Fe and Ti-6Al-7Nb (or Ti67). Oxinium (oxidized zirconium) can be added to this category. Alumina (Al_2O_3) has a long history since being introduced in THR implants. Recently zirconia (ZrO_2) with high toughness and good mechanical properties and zirconia-toughened alumina ceramic can be included [222].

7.4.6 Total knee replacement (TKR)

TKR, also called TKA, is a surgical procedure to resurface a knee damaged by arthritis or severe knee injury. Various types of arthritis may affect the knee joint. OA, a degenerative joint disease that affects mostly middle-aged and older adults, may cause the breakdown of joint cartilage and adjacent bone in the knees. Damage to the cartilage and bones limits movement and may cause pain. People with severe degenerative joint disease may be unable to do normal activities that involve bending at the knee, such as walking or climbing stairs, because they are painful. RA, which causes inflammation of the synovial membrane and results in excessive synovial fluid, can lead to pain and stiffness. Traumatic arthritis, arthritis due to injury, may cause damage to the cartilage of the knee [224]. Anatomically, the knee is the largest joint in the body and having healthy knees is required to perform most everyday activities. The knee is made up of the lower end of the thighbone (femur), the upper end of the shinbone (tibia), and the kneecap (patella). The ends of these three bones are covered with articular cartilage, a smooth substance that protects the bones and enables them to move easily within the joint. In a healthy knee, these structures work together to ensure smooth, natural function and movement. Referring to Figure 7.33 [225], joints are the areas where two or more bones meet. Most joints are mobile, allowing the bones to move. Basically, the knee is two long leg bones held together by muscles, ligaments, and tendons. Each bone end is covered with a layer of cartilage that absorbs shock and protects the knee. There are two groups of muscles involved in the knee, including the quadriceps muscles (located on the front of the thighs), which straighten the legs, and the hamstring muscles (located on the back of the thighs), which bend the leg at the knee. Tendons are tough cords of connective tissue that connect muscles to bones. Ligaments are elastic bands of tissue that connect bone to bone. Some ligaments of the knee provide stability and protection of the joints, while other ligaments limit forward and backward movement of the tibia (shin bone) [224].

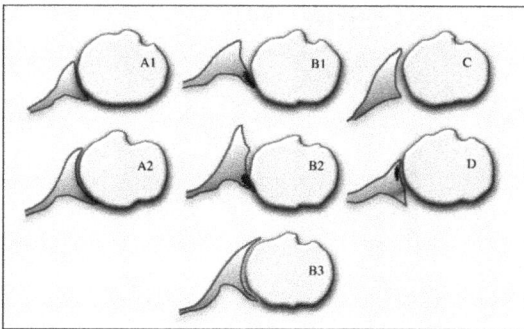

Figure 7.33: Normal knee anatomy [225].

Summarizing, the knee consists of tibia (the shin bone or larger bone of the lower leg), femur (the thighbone or upper leg bone), patella (the kneecap), cartilage (a type of tissue that covers the surface of a bone at a joint and helps reduce the friction of movement within a joint), synovial membrane (a tissue that lines the joint, sealing it into a joint capsule, and secretes synovial fluid (a clear, sticky fluid) around the joint to lubricate it), ligament (a type of tough, elastic connective tissue that surrounds the joint to give support and limits the joint's movement), tendon (a type of tough connective tissue that connects muscles to bones and helps to control movement of the joint), and meniscus (a curved part of cartilage in the knees and other joints that acts as a shock absorber, increases contact area, and deepens the knee joint) [224, 226].

The TKA might be more accurately termed as knee "resurfacing" because only the surface of the bones is replaced. Basically, there are four steps involved in TKA: (i) preparing the bone; (ii) positioning the metal implants. The removed cartilage and bone are replaced with metal components that recreate the surface of the joint. These metal parts may be cemented or "press-fit" into the bone; (iii) resurfacing the patella. The undersurface of the patella (kneecap) is cut and resurfaced with a plastic button. Some surgeons do not resurface the patella, depending upon the case; and (iv) insertion of a spacer. A medical-grade plastic spacer is inserted between the metal components to create a smooth gliding surface [224, 227]. Figure 7.34 shows typical TRA treatment [224], in which the arthritic cartilage and underlying bone are removed and resurfaced with metal implants on the femur and tibia. A plastic spacer is placed in-between the implants.

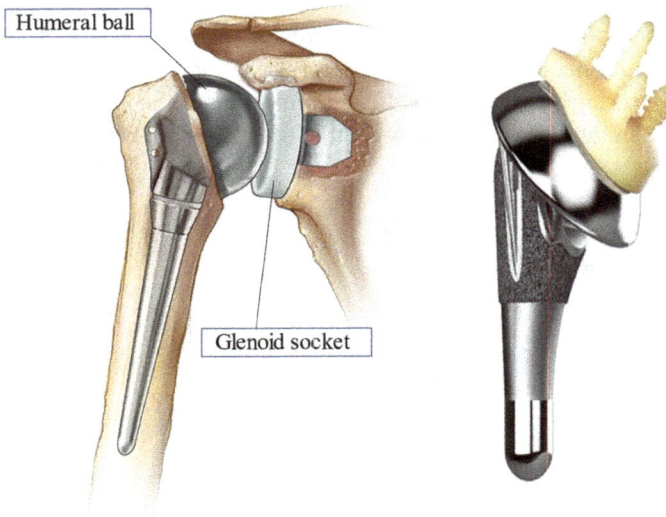

Figure 7.34: Severe osteoarthritis (a) and TKA treatment (b) [224].

Depending on selected material type, there are basically four different types of total knee implants [228]: (1) Metal-on-plastic (MoP) type. This is the most common type of implant. It features a metal femoral component that rides on a polyethylene plastic spacer attached to the tibial component. The metals commonly used include cobalt-chromium, titanium, zirconium, and nickel. Metal-on-plastic is the least expensive type of implant and has the longest track record for safety and implant life span.

However, one problem that can happen with plastic implants is an immune reaction triggered by tiny particles that wear away from the spacer. This can cause bone to break down, leading to loosening and failure of the implant. Advances in manufacturing have greatly reduced the rate of wear in the plastic. (2) Ceramic-on-plastic (CoP) type. This type uses a ceramic femoral component instead of metal (or a metal component with a ceramic coating). It also rides on a plastic spacer. People who are sensitive to the nickel used in metal implants might get the ceramic type. Or there are hypoallergenic coating options available. Plastic particles from this type of implant also can lead to an immune reaction. (3) Ceramic-on-ceramic (CoC) type. The femoral and tibial components are both made of ceramic; ceramic parts are the least likely to react with the body. However, ceramic joint prostheses can make a squeaking noise when you walk. In rare cases, they can shatter under heavy pressure into pieces that must be removed by surgery. (4) Metal-on-metal (MoM) type. The femoral and tibial components are both made of metal. Metal-on-metal implants have been used much less often in recent years because of concerns over traces of metal leaking into the bloodstream. The metal comes from the chemical breakdown of the implant hardware. All metal implants were originally developed to provide longer-lasting joint replacements for younger people. But the traces of metal can cause inflammation, pain, and possibly organ damage. Metal-on-metal implants may be considered only for young, active men, because they may last longer than other materials [228–230]. Metals include typical biometallic materials including 316 L type stainless steel, Ti-based alloys (Ti-6Al-4V or Ti-6Al-7Nb), and CoCrMo alloys, same as those used for THR implants. As for polymer, cross-linked polyethylene (XLPE) is the most frequently used. Alumina and zirconia are typical materials choice as ceramic materials, and these are often reinforced with some additives.

When these different material combinations are applied to the real total knee implants, as seen in Figure 7.35 [231], the major tribological contact surface should be femoral implant and joint liner surfaces. The tibial implant is composed of a tibial tray and a liner material and these are connected firmly, so that there should not be any risk for biotribological deterioration therebetween.

There are still variations in implant materials. Bahraninasab et al. [232] mentioned that an aseptic loosening of femoral components is a significant problem affecting the life of current TKRs. Hence, to prevent the problem of aseptic loosening, a new metal-ceramic porous functionally graded biomaterial has been designed to replace the existing metal alloy material normally used. It was indicated that the use of the new functionally graded biomaterial improves the performance of knee prostheses by solving or modifying three currently leading causes of failure: (1) stress shielding of the bone by the implant, (2) wear of the articular surfaces, and (3) the development of soft tissue at the bone/prosthesis interface as a result of relative implant motion.

As mentioned before, to a certain type of patients who developed the cutaneous hypersensitivity rate against metallic materials, the surface of metallic implants (which particularly contains nickel element) is coated with the hypoallergenic

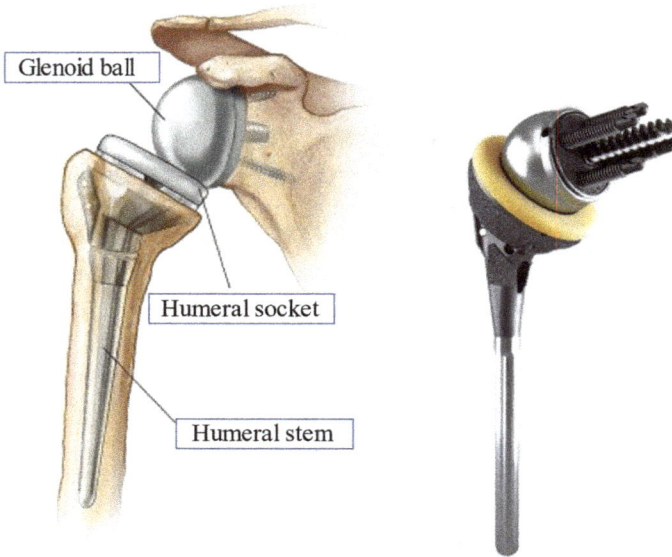

Figure 7.35: Three major components in the total knee implant system [231].

material. Bader et al. [233] mentioned that on apparent allergy against metallic implant components different alternative solutions to standard endoprostheses should be taken into account for primary implantation or revision of TKR, for example, the application of implant components without metallic elements (e.g., ceramics), the use of nonallergic metallic implants, such as titanium or ZrNb alloys, or potential allergy-inducing metallic materials after masking the implant surface using a suitable coating. In the case of primary or revision surgery, most patients with metal allergy are treated with a Ti(Nb)N-coated knee implant made of CoCrMo or Ti-based alloys [233].

Recently, new bearing materials, new methods in in vitro wear simulation, specific cell culture, animal models to evaluate the response to particulate debris, and dedicated retrieval analysis programs to learn more about material degradation in vivo have been developed in the field of biotribology. Improvements in knee arthroplasty design, materials, sterilization techniques, oxidation resistance, and articulating surface treatments have led to superior performance of total knee prostheses by reducing the prevalence of disastrous wear, delamination, and structural fatigue and are expected to show substantial benefits in decreasing wear and osteolysis in the future [234]. When the combined adverse effects of wear and corrosion due to the biological environments are considered, biotribocorrosion takes place. In this scenario, the biotribology and the biotribocorrosion properties of modern implantable biomaterials are extremely important, especially for the situations where there is relative movement between the implanted biomaterials or between the implanted biomaterial and the natural tissues under physiological environment [235]. This is the case, for example, of the TKR or THR which represents one of the most investigated biotribological system due to their growing diffusion in all the world [236, 237].

Lubricant function is one of three tribological elements and is important in knee joint replacements [238]. A knee joint simulator was designed and equipped with optical module based on fluorescent optical method for film thickness observation, and the contact between the femoral knee metal implant and real-shaped polymer insert mimicking actual contact nature was observed [239]. Simple solutions of albumin and γ-globulin proteins as well as its mixture were used while the film thickness was studied as a function of time considering simplified flexion/extension motion with variable load over the cycle. A clear importance of the interaction of proteins was observed since the mixtures showed different results compared to simple solutions. It was then concluded that (i) especially considering albumin protein, its behavior was substantially affected by adding γ-globulin, and (ii) a satisfactory compliance with previous findings related to hip joint lubrication in terms of the behavior of both proteins was found.

It is reported that approximately 10% of TKAs require revision surgery and loosening and wear account for approximately 21% of all revisions [240, 241]. Loosening is related to stresses at the bone fixation site, whereas wear is mainly due to a lack of congruency during implant motion [242]. Fixed-bearing and mobile-bearing are two kinds of bearing designs for TKA. A fixed-bearing knee design has round femoral components that articulate with a relatively flat tibial articular surface. Although this configuration allows for some axial rotation, it results in high contact stress between the femoral and tibial surface [243]. Because of these circumstances, the concept of a mobile-bearing knee design was developed. Due to its motion at the tibia-insert interface, greater tibiofemoral congruency can be achieved to reduce wear on the implants and reproducing more natural kinematics of the knee, and these processes are not accompanied by an increase in the stress at the bone-implant interface [244], resulting in increased durability and knee function. To compare the benefits and harms of fixed-bearing versus mobile-bearing TKAs, numerous comparative studies have been conducted in the past two decades, and most studies concluded that there is no difference between fixed- and mobile-bearing designs with regard to pain, ROM, or function [243].

Aglietti et al. [245] compared the postoperative recovery and early results of two groups of patients undergoing TKA: 107 patients received an established fixed-bearing posterior-stabilized prosthesis, and 103 patients the meniscal-bearing prosthesis. At an average follow-up of 36 months, knee, function, and patellar scores were comparable in both groups. It was reported that (i) the fixed-bearing group showed a significantly higher maximum flexion than the meniscal-bearing group (112° vs 108°), and (ii) using a fixed-bearing or a mobile-bearing design did not seem to influence the short-term recovery and early results after knee arthroplasty.

Poirier et al. [246] compared the outcomes of fixed and mobile bearings in the same type of TKA model after a longer follow-up. It was concluded that (i) there are no significant differences in the clinical outcomes between fixed- and mobile-bearing inserts of the same TKA model; (ii) although the mobile-bearing knees had a better radiographic appearance, this did not translate to better clinical outcomes; (iii) in practice, the superiority of mobile bearings is solely theoretical. Çatma et al. [247] investigated the impact of fixed- or mobile-bearing tibial inserts on patellofemoral arthrosis and evaluated which one is to be preferred for patients with patellofemoral arthrosis. Patellofemoral joints of patients were evaluated according to the scoring system defined by Fulkerson-Shea. Unicondylar knee arthroplasty (UKA) with 22 fixed-bearing tibial inserts (66.6%) (male: 3, female: 19) and UKA with 11 mobile-bearing tibial inserts (33.9%) (male: 2, female: 9) were implanted. It was found that average knee flexion was found to be 116.5 (100–135) degrees in 22 patients with mobile-bearing tibial inserts, and 114.5 (95–135) in 11 patients with fixed-bearing tibial inserts. It was concluded that (i) patellofemoral arthrosis is an important factor for UKA prognosis and one of the determinants of patient satisfaction, and (ii) significantly less patellofemoral complaints were seen with UKA with fixed-bearing tibial insert compared to mobile bearing tibial insert. A comparison was made between tibial component migration measured with radiostereometric analysis and clinical outcome of otherwise similarly designed cemented fixed- and mobile-bearing single-radius TKAs [248]. It was found that (i) both groups showed comparable migration, with a mean migration at six-year follow-up of 0.90 mm for the fixed-bearing group compared with 1.22 mm for the mobile-bearing group, and (ii) clinical outcomes were similar between groups. It was hence concluded that (iii) fixed- and mobile-bearing single-radius TKAs showed similar migration, and (iv) the latter may, however, expose patients to more complex surgical techniques and risks such as insert dislocations inherent to this rotating-platform design [248].

When reliability between THA and TKA is compared, it is reported that THA, in general, provides even more reliable and consistent positive results compared to its counterpart procedure (i.e., TKA) [249, 250].

References

[1] Gebelein CG. The basics of artificial organs. Polym Mater Artif Organs. 1984, 1–11, https://pubs.acs.org/doi/abs/10.1021/bk-1984-0256.ch001.
[2] What are artificial organs? Frost & Sullivan Japan. https://www.sbbit.jp/article/cont1/34868.
[3] Papaioannou TG. Artificial organs. In: Golemati S, et al., ed. Cardiovascular Computing – Methodologies and Clinical Applications. Series in BioEngineering. Singapore: Springer, 2019, 247–57, doi: https://doi.org/10.1007/978-981-10-5092-3_12.
[4] Orthopedic trauma implants. GPC Medical Ltd. 2016; https://orthopedicimplantsindia.wordpress.com/2016/05/18/implants-in-orthopedic-trauma/.

[5]   Dutta RC, Dutta AK, Basu B. Engineering implants for fractured bones-metals to tissue constructs. J Mater Eng Appl. 2017, 1, 9–13.

[6]   Imachi K. Future artificial organs. BioMed Eng. 1998, 12, 8–13, https://www.jstage.jst.go.jp/article/jsmbe1987/12/1/12_1_8/_pdf.

[7]   Park J, Lakes RS. Biomaterials: An Introduction. New York NY USA: Springer, 2007, 564.

[8]   https://jadasingleton.weebly.com/organ-donation-overview.html.

[9]   https://www.healthdirect.gov.au/organ-transplants.

[10]   Christoperson LK. Quality of life. Organ transplantation and artificial organs. Int J Technol Assess Health Care. 1986, 2, 553–62.

[11]   Lu Z, MacDermid JC, Rosenbaum P. A narrative review and content analysis of functional and quality of life measures used to evaluate the outcome after TSA: An ICF linking application. BMC Musculoskelet Disord. 2020, 21, doi: https://doi.org/10.1186/s12891-020-03238-w.

[12]   Deshmukh AV, Koris M, Zurakowski D, Thornhill TS. Total shoulder arthroplasty: Long-term survivorship, functional outcome, and quality of life. J Shoulder Elb Surg. 2005, 14, 471–79.

[13]   Leite LMB, Lins-Kusterer L, Belangero PS, Patriota G, Ejinisman B. Quality of life in patients who have undergone reverse shoulder arthroplasty. Acta Orthop Bras. 2019, 27, doi: https://doi.org/10.1590/1413-785220192705222929.

[14]   Liu X-W, Zi Y, Xiang L-B, Wang Y. Total hip arthroplasty: A review of advances, advantages and limitations. Int J Clin Exp Med. 2015, 8, 27–36.

[15]   Knutsson S, BergbomEngberg I. An evaluation of patients' quality of life before, 6 weeks and 6 months after total hip replacement surgery. J Adv Nurs. 1999, 30, 1349–59.

[16]   Robinson AH, Palmer CR, Villar RN. Is revision as good as primary hip replacement? A comparison of quality of life. J Bone Joint Surg Br. 1999, 81B, 42–45.

[17]   Feeny D, Blanchard C, Mahon JL, Bourne R, Rorabeck C, Stitt L. Comparing community-preference-based and direct standard gambleutility scores: Evidence from elective total hip arthroplasty. Int J Technol Assess Health Care. 2003, 19, 362–72.

[18]   Räsänen P, Paavolainen P, Sintonen H, Koivisto AM, Blom M, Ryynänen OP. Effectiveness of hip or knee replacement surgery in terms of quality adjusted life years and costs. Acta Orthop. 2007, 78, 108–15.

[19]   Shan L, Shan B, Graham D, Saxena A. Total hip replacement: A systematic review and meta-analysis on mid-term quality of life. Osteoarthr Cartil. 2014, 22, 389–406.

[20]   Shi H-Y, Khan M, Culbertson R, Chang J-K, Wang J-W, Chiu H-C. Health-related quality of life after total hip replacement: A Taiwan study. Int Orthop. 2009, 33, 1217–22.

[21]   Gordon M, Paulsen A, Overgaard S, Garellick G, Pedersen AB, Rolfson O. Factors influencing health-related quality of life after total hip replacement-a comparison of data from the Swedish and Danish hip arthroplasty registers. BMC Musculoskelet Disord. 2013, 14, 316, doi: https://doi.org/10.1186/1471-2474-14-316.

[22]   Villalobos PA, Navarro-Espigares JL, Hernández-Torres E, Martínez-Montes JL, Villalobos M, Arroyo-Morales M. Body mass index as predictor of health-related quality-of-life changes after total hip arthroplasty: A cross-over study. J Arthroplast. 2013, 28, 666–70.

[23]   Poulsen NR, Mechlenburg I, Søballe K, Lange J. Patient-reported quality of life and hip function after 2-stage revision of chronic periprosthetic hip joint infection: A cross-sectional study. HIP Int. 2018, 28, 407–14.

[24]   Brown TS, Fehring KA, Ollivier M, Mabry TM, Hanssen AD, Abdel MP. Repeat two-stage exchange arthroplasty for prosthetic hip re-infection. Bone Joint J. 2018, 100-B, 1157–61.

[25]   Van Der K, Tirumala V, Box H, Oganesyan R, Klemt C, Kwon YM. One-stage revision is as effective as two-stage revision for chronic culture-negative periprosthetic joint infection after total hip and knee arthroplasty. Bone Joint J. 2021, 103-B, 515–21.

[26] Poulsen NR, Mechlenburg I, Søballe K, Troelsen A, Lange J. Improved patient-reported quality of life and hip function after cementless 1-stage revision of chronic periprosthetic hip joint infection. J Arthroplasty. 2019, 34, 2763–69.

[27] Abdelaziz H, Grüber H, Gehrke T, Salber J, Citak M. What are the factors associated with re-revision after one-stage revision for periprosthetic joint infection of the hip? A case-control study. Clin Orthop Relat Res. 2019, 477, 2258–63.

[28] Shan L, Shan SA, Nouh F, Saxena A. Intermediate and long-term quality of life after total knee replacement: A systematic review and meta-analysis. J Bone Jt Surg. 2015, 97, 156–68.

[29] Liebs TR, Herzberg W. Better quality of life after medial versus lateral unicondylar knee arthroplasty. Clin Orthop Relat Res. 2013, 471, 2629–40.

[30] Hell AK, Braunschweig L, Behrend J, Lorenz HM, Tsaknakis K, Von Deimling U, Mladenov K. Health-related quality of life in early-onset-scoliosis patients treated with growth-friendly implants is influenced by etiology, complication rate and ambulatory ability. BMC Musculoskelet Disord. 2019, 20, 588, doi: 10.1186/s12891-019-2969-2.

[31] Doany ME, Olgun ZD, Kinikli GI, Bekmez S, Kocyigit A, Demirkiran G, Karaagaoglu AE, Yazici M. Health-related quality of life in early-onset scoliosis patients treated surgically: EOSQ scores in traditional growing rod versus magnetically controlled growing rods. Spine. 2018, 43, 148–53.

[32] Bourghli A, with 11 co-authors. Lack of improvement in health-related quality of life (HRQOL) scores 6 months after surgery for adult spinal deformity (ASD) predicts high revision rate in the second postoperative year. Eur Spine J. 2017, 26, 2160–66.

[33] Núñez-Pereira S, Vila-Casademunt A, Domingo-Sàbat M, Kleinstück F, Pellisé F. Impact of early unanticipated revision surgery on health-related quality of life after adult spinal deformity surgery. Spine J. 2018, 18, 926–34.

[34] Irimia JC, with 13 co-authors. Spinal fusion achieves similar two-year improvement in HRQoL as total hip and total knee replacement. A prospective, multicentric and observational study. SCICOT-J. 2019, 5, 26, doi: https://doi.org/10.1051/sicotj/2019027.

[35] Avramovic M, Stefanovic V. Health-related quality of life in different stages of renal failure. Artif Organs. 2012, 36, 581–89.

[36] Turkmen K, Yazici R, Solak Y, Guney I, Altintepe L, Yeksan M, Tonbul HZ. Health-related quality of life, sleep quality, and depression in peritoneal dialysis and hemodialysis patients. Hemodial Int. 2012, 16, 198–206.

[37] Artificial Organs. https://ja.wikipedia.org/w/index.php?title=%E4%BA%BA%E5%B7%A5%E8%87%93%E5%99%A8&action=edit&section=3.

[38] Edwards CA, Kouzani A, Lee KH, Ross EK. Neurostimulation devices for the treatment of neurologic disorders. Mayo Clin Proc. 2017, 92, 1427–44.

[39] National institute of neurological disorders and stroke. IH. Brain stimulation therapies for epilepsy. https://www.ninds.nih.gov/About-NINDS/Impact/NINDS-Contributions-Approved-Therapies/Brain-stimulation-therapies-epilepsy.

[40] Tan H-Y, Cho H, Lee LP. Human mini-brain models. Nat Biomed Eng. 2021, 5, 11–25.

[41] Gamillo E. Mini brains grown from stem cells developed light-sensitive, eye-like features. Smart News, 2021; https://www.smithsonianmag.com/smart-news/mini-brains-grown-stem-cells-developed-eyes-can-sense-light-180978478/.

[42] Artificial nerve. https://ja.wikipedia.org/w/index.php?title=%E4%BA%BA%E5%B7%A5%E7%A5%9E%E7%B5%8C_(%E5%86%8D%E7%94%9F%E5%8C%BB%E7%99%82)&action=edit&section=2.

[43] Zhang A. One Step Closer to Cyborgs: The development of artificial nerves. 2018; https://sitn.hms.harvard.edu/flash/2018/artificial-nerves/.

[44] Service RF. New artificial nerves could transform prosthetics. 2018; https://www.science mag.org/news/2018/05/new-artificial-nerves-could-transform-prosthetics.

[45] Desarrollo IY. First prosthesis with direct connection to bone, nerves and muscles. Medical Express. 2016; https://medicalxpress.com/news/2016-03-prosthesis-bone-nerves-muscles.html.

[46] Cochlear Implants; https://www.csun.edu/ncod/cochlear-implants.

[47] Cochlear Implant Program; https://www.rwjbh.org/rwj-university-hospital-new-brunswick /treatment-care/speech-and-audiology/audiology/cochlear-implants/.

[48] Visual Prosthesis; https://en.wikipedia.org/wiki/Visual_prosthesis.

[49] Santos M, Fernandes JR, Piedade M. A microelectrode stimulation system for a cortical neuroprosthesis. 2006; https://www.inesc-id.pt/ficheiros/publicacoes/3630.pdf.

[50] Jackson GR, Owsley C, Curio CA. Photoreceptor degeneration and dysfunction in aging and age-related maculopathy. Ageing Res Rev. 2002, 1, 381–96.

[51] Provis JM, Van Driel D, Billson FA, Russell P. Human fetal optic nerve: Overproduction and elimination of retinal axons during development. J Comp Neurol. 1985, 238, 92–100.

[52] Hanlon M. The Bionic Eye approaches: the next generation of Retinal Implants. 2007; https://newatlas.com/the-bionic-eye-approaches-the-next-generation-of-retinal-implants /6855/.

[53] Intraocular lens; https://en.wikipedia.org/wiki/Intraocular_lens.

[54] Intraocular Lens (IOL) Options; https://oregoneyeconsultants.com/intraocular-lens-iol-options/.

[55] Ocular prosthesis. https://en.wikipedia.org/wiki/Ocular_prosthesis

[56] Base curve radius. https://en.wikipedia.org/wiki/Base_curve_radius.

[57] Piercy R. Postoperative Care for Enucleation or Evisceration Surgery. 2016; https://ocularpro. com/postoperative-care-for-enucleation-or-evisceration-surgery/.

[58] Eye prosthesis. https://www.medicalartresources.com/eye.

[59] The Fundamentals of Prosthetic Eye Movement. https://ocularpro.com/the-fundamentals-of-prosthetic-eye-movement/.

[60] Custom Made Prosthetic Eye; https://ocularpro.com/prosthetic-eye-services-los-angeles/cus tom-prosthetic-eye/.

[61] Nose Prosthesis. https://www.medicalartresources.com/nose.

[62] Life-like prosthetics. https://www.medicalartresources.com/prosthetic-options.

[63] Nasal – Nose Prosthetics. Medical Art Prosthetics. https://www.medicalartprosthetics.com/ prosthetics/nose/.

[64] Electrolarynx. https://en.wikipedia.org/wiki/Electrolarynx.

[65] How to Use & Easily Maintain an Electrolarynx. 2021; https://thancguide.org/2021/06/the-journey/how-to-use-easily-maintain-an-electrolarynx/.

[66] Electrolarynx. 2018; https://www.cancerresearchuk.org/about-cancer/laryngeal-cancer/liv ing-with/speaking-after-laryngectomy/electrolarynx.

[67] Tang CG, Sinclair CF. Voice restoration after total laryngectomy. Otolaryngol Clin North Am. 2015, 48, 687–702.

[68] Kaye R, Tang CG, Sinclair CF. The electrolarynx: Voice restoration after total laryngectomy. Med Dev (Auckl). 2017, 10, 133–40.

[69] Carr C. Artificial Trachea. https://openwetware.org/wiki/Artificial_Trachea,_by_Chris_Carr.

[70] Bae S-W, Lee K-W, Park J-H, Lee JH, Jung C-R, Yu JJ, Kim H-Y, Kim D-H. 3D bioprinted artificial trachea with epithelial cells and chondrogenic-differentiated bone marrow-derived mesenchymal stem cells. Int J Mol Sci. 2018, 19, 1624, doi: 10.3390/ijms19061624.

[71] Delaere P, Van Raemdonck D. Tracheal replacement. J Thoracic Dis. 2016, 8, S186–96.

[72]   Matsumoto K, Nagayasu T. Artificial trachea: Past, present, and future. In: Nakayama K, ed. Kenzan Method for Scaffold-Free Biofabrication. Cham: Springer, 2021, 91–108, doi: https://doi.org/10.1007/978-3-030-58688-1_7.

[73]   Osada H. Artificial trachea. J Bronchol. 2006, 13, 39–43.

[74]   Etienne H, with 9 co-authors. Tracheal replacement. Eur Resp J. 2018, 51, 1702211, doi: 10.1183/13993003.02211-2017.

[75]   Coghlan A. Man receives world's first synthetic windpipe. 2011. https://www.newscientist.com/article/dn20671-man-receives-worlds-first-synthetic-windpipe/.

[76]   Melnick M. Cancer Patient Gets World's First Artificial Trachea. 2011. https://healthland.time.com/2011/07/08/cancer-patient-gets-worlds-first-artificial-trachea/.

[77]   Park J-H, with 11 co-authors. Experimental tracheal replacement using 3-dimensional bioprinted artificial trachea with autologous epithelial cells and chondrocytes. Sci Rep. 2019, 9, 2103, doi: https://doi.org/10.1038/s41598-019-38565-z.

[78]   Chung E-J. Bioartificial esophagus: Where are we now? Adv Exp Med Biol. 2018, 1064, 313–32.

[79]   Taira Y, Shiraishi Y, Miura H, Yambe T, Homma D, Kamiya K, Miyata G. An implantable artificial esophagus to propel food simulating natural anatomical esophageal function. *10th Asian Control Conference (ASCC)*. 2015, 1–4; doi: 10.1109/ASCC.2015.7244863.

[80]   Medical Illustrations. https://www.medicinenet.com/image-collection/esophagus_picture/picture.htm.

[81]   Wang D, Xu Y, Li Q, Turng L-S. Artificial small-diameter blood vessels: Materials, fabrication, surface modification, mechanical properties, and bioactive functionalities. J Mater Chem B. 2020, 8, 1801–22.

[82]   Service R. Artificial Blood Vessels Prove Effective. 2009. https://www.sciencemag.org/news/2009/04/artificial-blood-vessels-prove-effective.

[83]   Syedain Z, Reimer J, Lahti M, Berry J, Johnson S, Bianco R, Tranquillo RT. Tissue engineering of acellular vascular grafts capable of somatic growth in young lambs. Nat Commun. 2016, 7, 12951, doi: https://doi.org/10.1038/ncomms12951.

[84]   Artificial blood vessels. Japanese Society for Artificial Organs. https://www.jsao.org/public/what/what06/.

[85]   Artificial blood vessels. https://www.encyclopedia.com/medicine/medical-journals/artificial-blood-vessels.

[86]   Gao G, with 12 co-authors. Tissue-engineering of vascular grafts containing endothelium and smooth-muscle using triple-coaxial cell printing. Appl Phys Rev. 2019, 6, 041402, doi: 10.1063/1.5099306.

[87]   Vorp D. Artificial Blood Vessel Research at the McGowan Institute. https://www.upmc.com/services/regenerative-medicine/research/artificial-organs/artificial-blood-vessels.

[88]   Bergmeister H, with 14 co-authors. Biodegradable, thermoplastic polyurethane grafts for small diameter vascular replacements. Acta Biomater. 2015, 11, 104–13.

[89]   Syedain ZH, Graham ML, Dunn TB, O'Brien T, Johnson SL, Schumacher RJ, Tranquillo RT. A completely biological "off-the-shelf" arteriovenous graft that recellularizes in baboons. Sci Transl Med. 2017, 9, eaan4209, https://stm.sciencemag.org/content/scitransmed/9/414/eaan4209.full.pdf.

[90]   Artificial blood vessels developed in lab can grow with recipient. Science News. 2016, https://www.tasnimnews.com/en/news/2016/09/28/1198429/artificial-blood-vessels-developed-in-lab-can-grow-with-recipient.

[91]   Cheng S, with 10 co-authors. Electronic blood vessel. Matter. 2020, 3, 1664–84.

[92]   Artificial heart. https://en.wikipedia.org/wiki/Artificial_heart

[93]   Ventricular assist device. https://www.jsao.org/public/what/what03/.

[94] Artificial heart. https://www.britannica.com/science/artificial-heart.

[95] Birla RK, Williams SK. 3D bioprinting and its potential impact on cardiac failure treatment: An industry perspective. APL Bioeng. 2020, 4, 010903, doi: 10.1063/1.5128371.

[96] Smith J. Researchers in Israel 3D Print a Personalized Heart from Donor's Tissue. 2019, https://www.labiotech.eu/more-news/tissue-3d-printing-heart/.

[97] Goodwin P. The future of 3D bioprinting in precision health. 2018, https://www.epmmaga zine.com/pharmaceutical-industry-insights/three-challenges-one-big-opportunity-the-future-of-3d-biopri/.

[98] Nolan H, Wang D, Zwischenberger JB. Artificial lung basics. Organogenesis. 2011, 7, 23–27.

[99] Boschetti F, Perlman CE, Cook KE, Mockros LF. Hemodynamic effects of attachment modes and device design of a thoracic artificial lung. ASAIO J. 2000, 46, 42–48.

[100] Ha RR, Wang D, Zwischenberger JB, Clark JW. Hemodynamic analysis and design of a paracorporeal artificial lung device. Cardiovasc Eng. 2006, 6, 11–30.

[101] Artificial Lungs is a Rapidly Growing Type of Artificial Organs, Global Market is Expected to Grow at a CAGR of 6.7% through 2031. 2021, https://www.globenewswire.com/en/news-release/2021/07/07/2259127/0/en/Artificial-Lungs-is-a-Rapidly-Growing-Type-of-Artificial-Organs-Global-Market-is-Expected-to-Grow-at-a-CAGR-of-6-7-through-2031.html.

[102] Cypel M, Keshavjee S. Regenerative Medicine Applications in Organ Transplantation. 2014, 683–89; https://doi.org/10.1016/B978-0-12-398523-1.00047-1.

[103] Fries JH. Within Breathing Distance. 2008. https://www.inlander.com/spokane/within-breathing-distance/Content?oid=2186636.

[104] Wilson C. Artificial lungs in a backpack may free people with lung failure. Technology. 2017, https://www.newscientist.com/article/2125422-artificial-lungs-in-a-backpack-may-free-people-with-lung-failure/.

[105] ECMO. https://www.pinterest.jp/pin/474496510744292956/.

[106] ECMO Support May Save Lives in COVID-19. 2020, https://www.cuimc.columbia.edu/news/ecmo-support-may-save-lives-covid-19.

[107] Barbaro RP, with 19 co-authors. Extracorporeal membrane oxygenation support in COVID-19: An international cohort study of the Extracorporeal Life Support Organization registry. The Lancet. 2020, 296, 1071–78.

[108] Badulak J, with 21 co-authors. Extracorporeal membrane oxygenation for COVID-19: Updated 2021 guidelines from the extracorporeal life support organization. ASAIO J. 2021, 67, 485–95.

[109] Petrella F, Spaggiari L. Artificial lung. J Thorac Dis. 2018, 10, S2329–32.

[110] Song JJ, Ott HC. Bioartificial lung engineering. Am J Transplant. 2012, 12, 283–88.

[111] Badylak SF. Xenogeneic extracellular matrix as a scaffold for tissue reconstruction. Transpl Immunol. 2004, 12, 367–77.

[112] 3D-printed artificial lung model holds promise for drug screening and studying viral response. 2021, https://www.3dmednet.com/3d-printed-artificial-lung-model-holds-promise-for-drug-screening-and-studying-viral-response/.

[113] Amelia H. POSTECH 3D Print First-of-its-Kind Lung Model. 2021, https://www.3dnatives.com/en/postech-3d-print-lung-model-300320215/.

[114] 3D-printed artificial lung model. 2021, https://www.eurekalert.org/news-releases/603037.

[115] Bio-ink. https://en.wikipedia.org/wiki/Bio-ink.

[116] Hospodiuk M, Dey M, Sosnoski D, Ozbolat IT. The bioink: A comprehensive review on bioprintable materials. Biotech Adv. 2017, 35, 217–39.

[117] Ramiah P, Du Toit LC, Choonara YE, Kondiah PPD, Pillay V. Hydrogel-based bioinks for 3D bioprinting in tissue regeneration. Front Mater. 2020, doi: https://doi.org/10.3389/fmats.2020.00076.

[118] Ashammakh N, Ahadian S, Xu C, Montazerian H, Ko H, Nasiri R, Barros N, Khademhosseini A. Bioinks and bioprinting technologies to make heterogeneous and biomimetic tissue constructs. Mater Today Bio. 2019, 1, 100008, https://www.sciencedirect.com/science/article/pii/S2590006419300146.

[119] De Santis MM, with 17 co-authors. Extracellular-matrix-reinforced bioinks for 3D bioprinting human tissue. Adv Mater. 2020, 33, 2005476, doi: 10.1002/adma.202005476.

[120] Building a functional artificial thymus. 2020, https://biotechscope.com/building-functional-artificial-thymus/.

[121] Campinoti S, with 21 co-authors. Reconstitution of a functional human thymus by postnatal stromal progenitor cells and natural whole-organ scaffolds. Nat Commun. 2020, 11, 6372, doi: 10.1038/s41467-020-20082-7.

[122] Vogt-James M. Artificial thymus developed at UCLA can produce cancer-fighting T cells from blood stem cells. 2017, https://newsroom.ucla.edu/releases/artificial-thymus-developed-at-ucla-can-produce-cancer-fighting-t-cells-from-blood-stem-cells.

[123] Mandal AK, Garlapati P, Tiongson B, Gayam V. Liver Assist Devices for Liver Failure, Liver Pathology. IntechOpen, 2020, https://www.intechopen.com/chapters/71584.

[124] Holt AW. Acute liver failure. Crit Care Resusc. 1999, 1, 25–38.

[125] Pless G. Artificial and bioartificial liver support. Organogenesis. 2007, 3, 20–24.

[126] Liver support system. https://en.wikipedia.org/wiki/Liver_support_system#Bioartificial_liver_devices.

[127] Morelli S. Biohybrid Artificial Liver (BAL) systems. Encyclopedia Membranes. 2016, https://link.springer.com/referenceworkentry/10.1007%2F978-3-662-44324-8_57.

[128] Demetriou AA, with 13 co-authors. Early clinical experience with a hybrid bioartificial liver. Scand J Gastroenterol. 1995, 208, 117–7.

[129] Sakiyama R, Blau BJ, Miki T. Clinical translation of bioartificial liver support systems with human pluripotent stem cell-derived hepatic cells. World J Gastroenterol. 2017, 23, 1974–79.

[130] What is diabetes? Diabetes Research Institute. 2020. https://www.diabetesresearch.org/what-is-diabetes.

[131] Kandimalla R, Thirumala V, Reddy PH. Is alzheimer's disease a type 3 diabetes? A critical appraisal. Biochim Biophys Acta. 2017, 1863, 1078–89.

[132] Bergnnstal RM, Garg S, Weinzimer SA, Buckingham BA, Bode BW, Tamborlane WV, Kaufman FR. Safety of a hybrid closed-loop insulin delivery system in patients with type 1 diabetes. JAMA. 2016, 316, 1407–08.

[133] 3D bioprinted artificial pancreas for type 1 diabetes on the horizon. 2017, https://www.diabetes.co.uk/news/2017/apr/3d-bioprinted-artificial-pancreas-for-type-1-diabetes-on-the-horizon-96776467.html.

[134] https://idarts.co.jp/3dp/3d-bioprinted-pancreas/.

[135] Culp M. How to construct an artificial stomach. Am Biol Teach. 2010, 72, 444–46.

[136] Kozu H, Kobayashi I, Nakajima M, Neves MA, Uemura K, Isoda H, Ichikawa S. Mixing characterization of liquid contents in human gastric digestion simulator equipped with gastric secretion and emptying. Biochem Eng J. 2017, 122, 85–90.

[137] American Institute of Physics. Prototype Artificial Stomach Reveals Fluid Dynamics of Food Digestion. 2021, https://scitechdaily.com/prototype-artificial-stomach-reveals-fluid-dynamics-of-food-digestion/.

[138] Dufour D, Tanner FX, Feigl KA, Windhab EJ. Investigation of the dispersing characteristics of antral contraction wave flow in a simplified model of the distal stomach. Phys Fluids. 2021, 33, doi: https://doi.org/10.1063/5.0053996.

[139] Wang Z, Kozu H, Uemura K, Kobayashi I, Ichikawa S. Effect of hydrogel particle mechanical properties on their disintegration behavior using a gastric digestion simulator. Food Hydrocoll. 2021, 110, 106166, doi: 10.1016/j.foodhyd.2020.106166.

[140] NBC News. Scientists build world's first artificial stomach. 2006, https://www.nbcnews.com/health/health-news/scientists-build-worlds-first-artificial-stomach-flna1c9475100.

[141] Curley B. Implantable Artificial Kidney Moves Closer to Reality. Healthline. 2018, https://www.healthline.com/health-news/implantable-artificial-kidney-moves-closer-to-reality.

[142] Artificial kidney development advances, thanks to collaboration by NIBIB Quantum grantees. National Institute of Biomedical Imaging and Bioengineering (BIBIB). 2018, https://www.nibib.nih.gov/news-events/newsroom/artificial-kidney-development-advances-thanks-collaboration-nibib-quantum.

[143] Fissell WF. Vanderbilt-UCSF entry among 15 dialysis innovations advancing in national accelerator competition. 2019, https://discover.vumc.org/2019/06/artificial-kidney-project-advances-in-kidneyx-competition/.

[144] Alpin A. Artificial kidney. 2017, https://www.troab.com/ucsf-bioartificial-kidney-project/.

[145] Hassan A. A New Technology Increases Hope For An Implantable Artificial Kidney. 2020, https://healthwriteups.com/2020/06/25/a-new-technology-increases-hope-for-an-implantable-artificial-kidney/.

[146] Adamowicz J, Pokrywczynska M, Van Breda SV, Kloskowski T, Drewa T. Concise review: Tissue engineering of urinary bladder; we still have a long way to go?. Stem Cells Transl Med. 2017, 6, 2033–43.

[147] Iannaccone PM, Galat V, Bury MI, Ma YC, Sharma AK. The utility of stem cells in pediatric urinary bladder regeneration. Pediatr Res. 2017, 83, 258–66.

[148] Imichi H. Artificial Organs of the Future – From Near-Future Artificial Organs to Super-Artificial Organs. 1987, https://www.jstage.jst.go.jp/article/jsmbe1987/12/1/12_1_8/_pdf.

[149] Gurland HJ, Mujais SK. The future of artificial organs. Int J Artif Org. 1995, 18, 64–68.

[150] Galeon D. Artificial Organs: We're entering an era where transplants are obsolete. 2017. https://futurism.com/neoscope/artificial-organs-entering-era-transplants-obsolete.

[151] Klak M, Bryniarski T, Kowalska P, Gomolka M, Tymicki G, Kosowska K, Cywoniuk P, Dobrzanski T, Turowski P, Wszola M. Novel strategies in artificial organ development: What is the future of medicine? Micromachines. 2020, 11, 646, doi: 10.3390/mi11070646.

[152] Handa N, Mochizuki S, Fujiwara Y, Shimokawa M, Wakao R, Arai H. Future development of artificial organs related with cutting edge emerging technology and their regulatory assessment: PMDA's perspective. J Artif Org. 2020, 23, 203–06.

[153] Organ-on-a-chip. https://en.wikipedia.org/wiki/Organ-on-a-chip.

[154] Wu Q, Liu J, Wang X, Feng L, Wu J, Zhu X, Wen W, Gong X. Organ-on-a-chip: Recent breakthroughs and future prospects. Biomed Eng Online. 2020, 9, doi: https://doi.org/10.1186/s12938-020-0752-0.

[155] Carvalho V, Gonçalves I, Lage T, Rodrigues RO, Minas G, Teixeira SFCF, Moita AS, Hori T, Kaji H, Lima RA. 3D printing techniques and their applications to organ-on-a-chip platforms: A systematic review. Sensors (Basel). 2021, 21, 3304, doi: 10.3390/s21093304.

[156] Low LA, Mummery C, Berridge BR, Austin CP, Tagle DA. Organs-on-chips: Into the next decade. Nat Rev Drug Discov. 2021, 20, 345–61.

[157] Park JY, Jang J, Kang H-W. 3D Bioprinting and its application to organ-on-a-chip. Microelectron Eng. 2018, 200, 1–11.

[158] Hwang DG, Choi C-S, Jang J. 3D bioprinting-based vascularized tissue models mimicking tissue-specific architecture and pathophysiology for in vitro studies. Front Bioeng Biotechnol. 2021, 9, doi: 10.3389/fbioe.2021.685507.

[159] Apheresis; https://en.wikipedia.org/wiki/Apheresis.

[160] What Is Apheresis? https://www.ahn.org/services/medicine/bloodless-medicine/faq/what-is-apheresis.
[161] Sonntag R, Reinders J, Rieger JS, Heitzmann DWW, Kretzer JP. Hard-on-hard lubrication in the artificial hip under dynamic loading conditions. PLoS ONE. 2013, 8, e71622, doi: https://doi.org/10.1371/journal.pone.0071622.
[162] Oldani CR, Dominguez AA. Simulation of the mechanical behavior of a HIP implant. Implant fixed to bone by cementation under arbitrary load. J Phys Conf Ser. 2007, 90, 012007, doi: 10.1088/1742-6596/90/1/012007.
[163] De Barros E Lima Bueno R, Dias AP, Ponce KJ, Wazen R, Brunski JB, Nanci A. Bone healing response in cyclically loaded implants: Comparing zero, one, and two loading sessions per day. J Mech Behav Biomed Mater. 2018, 85, 152–61.
[164] Zanetti EM, Aldieri A, Terzini M, Calì M, Franceschini G, Bignardi C. Additively manufactured custom load-bearing implantable devices: Grounds for caution. Aust Med J. 2017, 18, doi: 10.21767/AMJ.2017.3093.
[165] Kovochich M, Fung ES, Donovan E, Unice KM, Paustenbach DJ, Finley BL. Characterization of wear debris from metal-on-metal hip implants during normal wear versus edge-loading conditions. J Biomed Mater Res B Appl Biomater. 2018, 106, 986–96.
[166] Rahmanian R, Shayesteh Moghaddam N, Haberland C, Dean D, Miller M, Elahinia M. Load bearing and stiffness tailored NiTi implants produced by additive manufacturing: A simulation study. Proc SPIE. 2014, 9058, https://ui.adsabs.harvard.edu/abs/2014SPIE.9058E.14R/abstract.
[167] Stoppie N, Van Oosterwyck H, Jansen JA, Wolke JGC, Wevers M, Naert I. The influence of Young's modulus of loaded implants on bone remodeling: An experimental and numerical study in the goat knee. J Biomed Mater Res A. 2009, 90, 792–803.
[168] Mwangi JC, Admani AA. Management of bilateral fracture femur with implant failure: A case report. Afr J Online (AJOL). 2011, 5, 58–62.
[169] Kesteven J, Kannan MB, Walter R, Khakbaz H, Choe H-C. Low elastic modulus Ti-Ta alloys for load-bearing permanent implants: Enhancing the biodegradation resistance by electrochemical surface engineering. Mater Sci Eng C Mater Biol Appl. 2015, 46, 226–31.
[170] Brar HS, Platt MO, Sarntinoranont M, Matin PI, Manuel MV. Magnesium as a biodegradable and bioabsorbable material for medical implants. JOM. 2009, 61, 31–34.
[171] Oshida Y. Magnesium Materials. De Gruyter Pub., 2021.
[172] Meischel M, with 9 co-authors. Adhesive strength of bone-implant interfaces and in-vivo degradation of PHB composites for load-bearing applications. L Mech Behav Biomed Mater. 2016, 53, 104–18.
[173] Vallittu PK, Närhi TO, Hupa L. Fiber glass-bioactive glass composite for bone replacing and bone anchoring implants. Dent Mater. 2015, 31, 371–81.
[174] Evans SL, Gregson PJ. Composite technology in load-bearing orthopaedic implants. Biomaterials. 1998, 19, 1329–42.
[175] Zhao G-H, Aune RE, Mao H, Espallargas N. Degradation of Zr-based bulk metallic glasses used in load-bearing implants: A tribocorrosion appraisal. I Mech Behav Biomed Mater. 2016, 60, 56–67.
[176] McNamara SL, Rnjak-Kovacina J, Schmidt DF, Lo TJ, Kaplan DL. Silk as a biocohesive sacrificial binder in the fabrication of hydroxyapatite load bearing scaffolds. Biomaterials. 2014, 35, 6941–53.
[177] Garai S, Sinha A. Biomimetic nanocomposites of carboxymethyl cellulose-hydroxyapatite: Novel three dimensional load bearing bone grafts. Colloids Surf B Biointerfaces. 2014, 115, 182–90.
[178] Oshida Y. Hydroxyapatite – Synthesis and Applications. Momentum Press, 2015.

[179] Xie Y, Li H, Zhang C, Gu X, Zheng X, Huang L. Graphene-reinforced calcium silicate coatings for load-bearing implants. Biomed Mater. 2014, 9, 025009, doi: 10.1088/1748-6041/9/2/025009.

[180] Oshida Y. Bioscience and Bioengineering of Titanium Materials. Elsevier Pub., 2007.

[181] Wauthle R, Ahmadi SM, Yavari SA, Mulier M, Zadpoor AA, Weinans H, Humbeeck JV, Kruth J-P, Schrooten J. Revival of pure titanium for dynamically loaded porous implants using additive manufacturing. Mater Sci Eng. 2015, 54, 94–100.

[182] Lewallen EA, with 9 co-authors. Biological strategies for improved osseointegration and osteoinduction of porous metal orthopedic implants. Tissue Eng Part B Rev. 2015, 21, 218–30.

[183] Melancon D, Bagheri ZS, Johnston RB, Liu L, Tanzer M, Panini D. Mechanical characterization of structurally porous biomaterials built via additive manufacturing: Experiments, predictive models, and design maps for load-bearing bone replacement implants. Acta Biomater. 2017, 63, 350–68.

[184] Evans NT, with 12 co-authors. High strength, surface porous polyether-ether-ketone for load-bearing orthopaedic implants. Acta Biomater. 2015, 13, 159–67.

[185] Bandyopadhyay A, Shivaram A, Tarafder S, Shasrabudhe H, Banerjee D, Bose S. In vivo response of laser processed porous titanium implants for load-bearing implants. Ann Biomed Eng. 2017, 45, 249–60.

[186] Sallica-Leva E, Caram R, Jardini AL, Fogagnolo JB. Ductility improvement due to martensite α′ decomposition in porous Ti–6Al–4V parts produced by selective laser melting for orthopedic implants. J Mech Behav Biomed Mater. 2016, 54, 149–58.

[187] Desmond D, Gallagher P. Quality of life in people with lower-limb amputation. In: Preedy VR, et al., ed. Handbook of Disease Burdens and Quality of Life Measures. New York, NY: Springer, 2010, 3785–96, doi: https://doi.org/10.1007/978-0-387-78665-0_219.

[188] Sinha R, Van Den Heuvel WJA, Arokiasamy P, Van Dijk JP. Influence of adjustments to amputation and artificial limb on quality of life in patients following lower limb amputation. Int J Rehabil Res. 2014, 37, 74–79.

[189] Horne CE, Beil JA. Quality of life in patients with prosthetic legs: A comparison study. J Prosthet Orthot. 2009, 21, 154–59.

[190] Prosthetics. http://52.62.202.235/book/export/html/1219.

[191] Maryse B. Guided growth: Novel applications in the hip, knee, and ankle. J Pediatr Orthop. 2017, 37, S32–6.

[192] Prosthetic hand and arm. https://www.ottobock.co.jp/prosthetic_ue/info/structure/.

[193] Elbow fitting/connector. https://www.ottobock.co.jp/prosthetic_ue/movo/elbow/12k19_12k5/.

[194] Are you living with a chronic joint pain? https://www.jointsrestoration.com/ca/san-ramon/.

[195] Osteoarthritis. https://www.mayoclinic.org/diseases-conditions/osteoarthritis/symptoms-causes/syc-20351925.

[196] Joint Replacement for the Knee, Hip, and Shoulder. https://www.orthopaedicsurgerynyc.com/nycorthopaedicsurgeon/joint-replacement-for-the-knee-hip-shoulder.htm.

[197] Shoulder Joint Replacement, https://orthoinfo.aaos.org/en/treatment/shoulder-joint-replacement/.

[198] Basic Anatomy of the Shoulder. 2017, https://www.acropt.com/blog/2017/7/29/basic-anatomy-of-the-shoulder.

[199] Shoulder replacement surgery. https://www.mayoclinic.org/tests-procedures/shoulder-eplacement/about/pac-20519121.

[200] Weerakkody Y. Walch classification of glenoid morphology. 2020, https://radiopaedia.org/articles/walch-classification-of-glenoid-morphology-1#:~:text=The%20Walch%20classifica

tion%20is%20used%20to%20stratify%20the,to%20recognize%20morphologies%20that%20may%20pose%20intraoperative%20difficulties.

[201] Walch G, Badet R, Boulahia A, Khoury A. Morphologic study of the glenoid in primary glenohumeral osteoarthritis. J Arthroplast. 1999, 14, 756–60.

[202] Bercik MJ, Kruse K, Yalizis M, Gauci M-O, Chaoui J, Walch G. A modification to the Walch classification of the glenoid in primary glenohumeral osteoarthritis using three-dimensional imaging. J Shoulder Elb Surg. 2016, 25, 1601–06.

[203] Samim M, Virk M, Mai D, Munawar K, Zuckerman J, Gyftopoulos S. Multilevel glenoid morphology and retroversion assessment in Walch B2 and B3 types. Skeletal Radiol. 2019, 48, 907–14.

[204] Shoulder Replacement. https://my.clevelandclinic.org/health/treatments/8290-shoulder-replacement.

[205] Groh G. A Patient's Guide to Anatomic Total Shoulder Replacement. https://www.drgordongroh.com/orthopaedic-injuries-treatment/shoulder/standard-total-shoulder-replacement/.

[206] Gombera M. What to Expect After Your Shoulder Replacement (Arthroplasty). 2019, https://www.gomberamd.com/blog/what-to-expect-after-your-shoulder-replacement-arthroplasty-12906.html.

[207] McClintock K. Pros and Cons of Revere Shoulder Replacement Surgery. 2021, https://drmcclintock.com/pros-and-cons-of-reverse-shoulder-replacement-surgery/.

[208] Pros And Cons Of Reverse Shoulder Replacement. https://whatt.org/pros-and-cons/reverse-shoulder-replacement/.

[209] Plancher K. Understanding the Pros and Cons of Reverse Total Shoulder Replacement. 2017, https://plancherortho.com/understanding-the-pros-and-cons-of-reverse-total-shoulder-replacement/.

[210] Pros And Cons Of Reverse Shoulder Replacement. https://prosancons.com/human/pros-and-cons-of-reverse-shoulder-replacement/.

[211] Churchill JL, Garrigues GE. Current controversies in reverse total shoulder arthroplasty. JBJS Rev. 2016, 14, e4, doi: 10.2106/JBJS.RVW.15.00070.

[212] Varacallo M, Luo TD, Johanson NA. Total Hip Arthroplasty Techniques. StatPearls Publishing, 2021.

[213] Varacallo MA, Herzog L, Toossi N, Johanson NA. Ten-year trends and independent risk factors for unplanned readmission following elective total joint arthroplasty at a large urban academic hospital. J Arthroplasty. 2017, 32, 1739–46.

[214] Varacallo M, Chakravarty R, Denehy K, Star A. Joint perception and patient perceived satisfaction after total hip and knee arthroplasty in the American population. J Orthop. 2018, 15, 495–99.

[215] Bergmann G, Deuretzbacher G, Heller M, Graichen F, Rohlmann A, Strauss J, Duda N. Hip contact forces and gait patterns from routine activities. J Biomech. 2001, 34, 859–71.

[216] Gold M, Munjal A, Varacallo M. Anatomy, bony pelvis and lower limb, hip joint. StatPearls. 2021; https://www.ncbi.nlm.nih.gov/books/NBK470555/.

[217] https://journals.plos.org/plosone/article/figure?id=10.1371/journal.pone.0071622.g001.

[218] Lau YJ, Sarmah S, Witt JD. 3rd generation ceramic-on-ceramic cementless total hip arthroplasty: A minimum 10-year follow-up study. Hip Int. 2017, doi: 10.5301/hipint.5000541.

[219] Peters RM, Van Steenbergen LN, Stevens M, Rijk PC, Bulstra SK, Zijlstra WP. The effect of bearing type on the outcome of total hip arthroplasty. Acta Orthop. 2018, 89, 163–69.

[220] FDA. General Information about Hip Implants. https://www.fda.gov/medical-devices/metal-metal-hip-implants/general-information-about-hip-implants.

[221] Understanding Total Hip Replacement. https://www.microportortho.com/patients/hip-replacement-solutions/understanding-total-hip-replacement.

[222] Merola M, Affatato S. Materials for hip prostheses: A review of wear and loading considerations. Materials (Basel). 2019, 12, 495, doi: 10.3390/ma12030495.

[223] FDA. Metal-on-Metal Hip Implant Systems. https://www.fda.gov/medical-devices/metal-metal-hip-implants/metal-metal-hip-implant-systems.

[224] Knee Replacement Surgery Procedure. https://www.hopkinsmedicine.org/health/treatment-tests-and-therapies/knee-replacement-surgery-procedure.

[225] Total Knee Replacement. https://orthoinfo.aaos.org/en/treatment/total-knee-replacement.

[226] Varacallo M, Luo TD, Johanson NA. Total knee arthroplasty techniques. StatPearls. 2021. https://www.ncbi.nlm.nih.gov/books/NBK499896/.

[227] Matanky B. Total Knee Replacement. https://www.drmatanky.com/total-knee-replacement-orthopaedic-sports-surgeon-case-grande-chandler.html.

[228] 4 Types of Knee Implants. Harvard Health Publishing, 2019, https://www.health.harvard.edu/pain/4-types-of-knee-implants.

[229] Jamali A. What are the best materials for joint replacement? Joint Preservation Institute. 2021, https://www.jointpreservationinstitute.com/blog/what-are-the-best-materials-for-joint-replacement-24337.html.

[230] McClure G, North T. Different Types of Knee Replacement Implants. 2016, https://peerwell.co/blog/different-types-of-knee-replacement-implants/.

[231] What is a prosthetic joint? Total Knee Arthroplasty (TKA). https://www.amplitude-ortho.com/en/total-knee-arthroplasty-tka.

[232] Bahraminasab M, Sahari BB, Edwards KL, Farahmand F, Hong TA, Naghibi H. Material tailoring of the femoral component in a total knee replacement to reduce the problem of aseptic loosening. Mater Des. 2013, 52, 441–51.

[233] Bader R, Bergschmidt P, Fritsche A, Ansorge S, Thomas P, Mittelmeier W. Alternative materials and solutions in total knee arthroplasty for patients with metal allergy. Orthopade. 2008, 37, 136–42.

[234] Grupp TM, Ultzschenider S, Wimmer MA. Biotribology in knee arthroplasty. Biomed Res Int. 2015, 618974, doi: https://doi.org/10.1155/2015/618974.

[235] Ruggiero A, Zhang H. Editorial: Biotribology and biotribocorrosion properties of implantable biomaterials. Front Mech Eng. 2020, doi: https://doi.org/10.3389/fmech.2020.00017.

[236] Kurtz S, Ong K, Lau E, Mowat F, Halpern M. Projections of primary and revision hip and knee arthroplasty in the United States from 2005 to 2030. J Bone Jt Surg. 2007, 89, 780–85.

[237] Learmonth ID, Young C, Rorabeck C. The operation of the century: Total hip replacement. Lancet. 2007, 370, 1508–19.

[238] Spencer ND, Crockett R. Biotribology in natural and artificial joints. In: Mang T, ed. Encyclopedia of Lubricants and Lubrication. Berlin, Heidelberg: Springer, doi: https://doi.org/10.1007/978-3-642-22647-2_289.

[239] Nečasa D, Sadeckáa K, Vrbkaa M, Gallob J, Galandákovác A, Křupkaa I, Hartla M. Observation of lubrication mechanisms in knee replacement: A pilot study. Biotribology. 2019, 17, 1–7.

[240] Bozic KJ, Kurtz SM, Lau E, Ong K, Chiu V, Vail TP, Rubash HE, Berry DJ. The epidemiology of revision total knee arthroplasty in the United States. Clin Orthop Relat Res. 2010, 468, 45–51.

[241] Chang MJ, Lim H, Lee NR, Moon YW. Diagnosis causes and treatments of instability following total knee arthroplasty. Knee Surg Relat Res. 2014, 26, 61–67.

[242] Goodfellow J, O'Connor J. The mechanics of the knee and prosthesis design. J Bone Joint Surg (Br). 1978, 60-B, 358–69.

[243] Hao D, Wang J. Fixed-bearing vs mobile-bearing prostheses for total knee arthroplasty after approximately 10 years of follow-up: A meta-analysis. J Orthop Surg Res. 2021, 16, 437, doi: https://doi.org/10.1186/s13018-021-02560-w.

[244] Callaghan JJ, Insall JN, Greenwald AS, Dennis DA, Komistek RD, Murray DW, Bourne RB, Rorabeck CH, Dorr LD. Mobile-bearing knee replacement: Concepts and results. Instr Course Lect. 2001, 50, 431–49.

[245] Aglietti P, Baldini A, Bussi R, Lup D, De Luca L. Comparison of mobile-bearing and fixed-bearing total knee arthroplasty: A prospective randomized study. J Arthroplasty. 2005, 20, 145–53.

[246] Poirier N, Graf P, Dubrana F. Mobile-bearing versus fixed-bearing total knee implants. Results of a series of 100 randomised cases after 9 years follow-up. Orthop Traumatol Surg Res. 2015, 101, S187–92.

[247] Çatma MF, Aksekili MAE, Kılınçarslan K, Işık Ç, Anaforoğlu B, Altay M. Impact of fixed-bearing and mobile-bearing tibial insert in unicondylar knee arthroplasty. Acta Medica Anatolia. 2016, doi: 10.5505/ACTAMEDICA.2016.96729.

[248] Van Hamersveld KT, Marang-Van De Mheen PJ, Van Der Heide HJL, Van Der Linden-Van Der Zwaag HMJ, Valstar ER, Nelissen RGHH. Migration and clinical outcome of mobile-bearing versus fixed-bearing single-radius total knee arthroplasty: A randomized controlled trial. Acta Orthop. 2018, 89, doi: 10.1080/17453674.2018.1429108.

[249] Varacallo M, Luo TD, Johanson NA. Total knee arthroplasty techniques. StatPearls. 2020, https://pubmed.ncbi.nlm.nih.gov/29763071/.

[250] Learmonth ID, Young C, Rorabeck C. The operation of the century: Total hip replacement. Lancet. 2007, 370, 1508–19.

Chapter 8
Biomechanics and implant failure

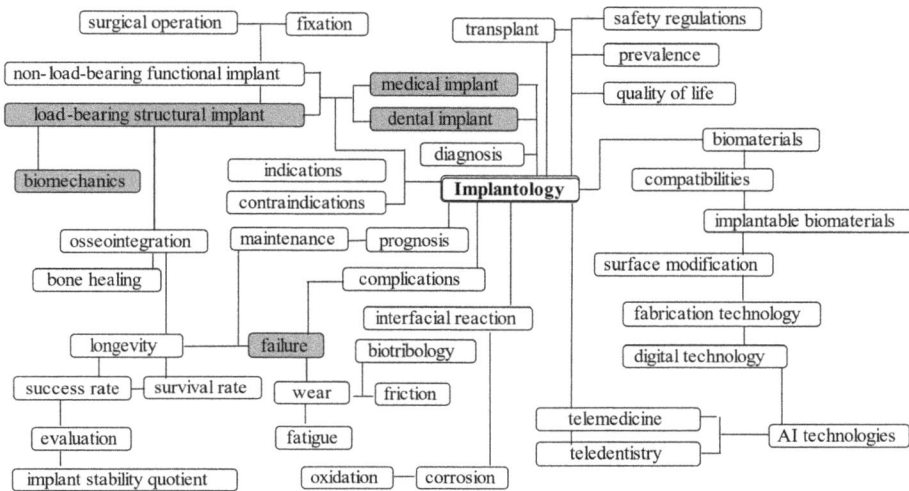

8.1 In general

In our body, there are six types of synovial joints [1–4], as illustrated in Figure 8.1 [1]. The shoulder joint, hip joint, and knee joint are all categorized in a ball/socket joint, which is affected by biotribological action and its resultant wear debris toxicity, and biotribocorrosion.

Orthopedic and dental implant treatments have allowed to enhance the QoL of many patients. The most important implantable orthopedic devices are total joint replacements, including total shoulder replacement (TSR), total hip replacement (THR), and total knee replacement (TKR). Nowadays, total hip and knee replacement surgeries are considered routine procedures with generally excellent outcomes. Given the increasing life expectancy of the world population, however, many patients will require revision or removal of the artificial joint during their lifetime. The most common cause of failure of hip and knee replacements is mechanical instability, secondary to wear of the articulating components. Thus, tribological and biomechanical aspects of joint arthroplasty are of specific interest in addressing the needs of younger, more active patients. The most significant improvements in the longevity of artificial joints have been achieved through the introduction of more wear resistant bearing surfaces. These innovations, however, brought about new tribocorrosion phenomena, such as fretting corrosion at the modular junctions of hip implants. Stiffness mismatch between the prosthesis components, nonphysiological stress transfer, and uneven

https://doi.org/10.1515/9783110740134-009

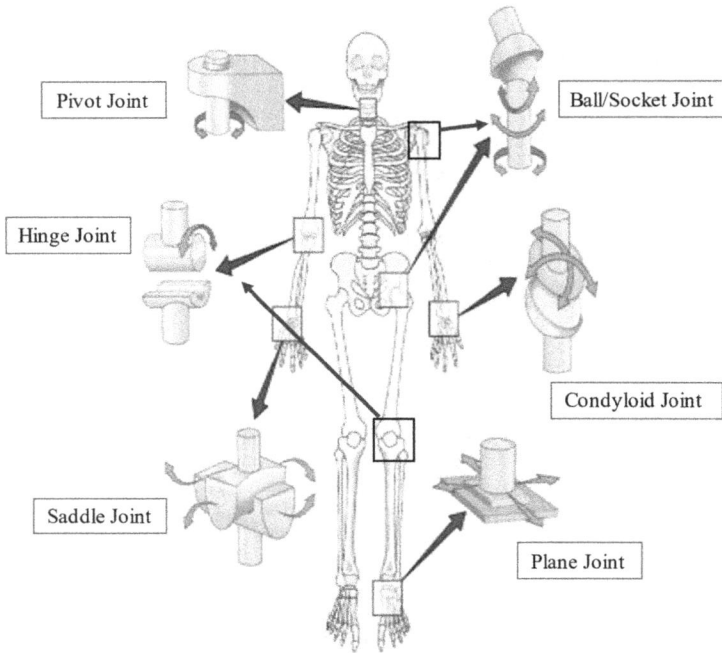

Figure 8.1: Six types of synovial joints [1].

implant–bone stress distribution are all involved in premature failure of hip arthroplasty. The development of more durable hip and knee prostheses requires a comprehensive understanding of biomechanics and tribocorrosion of implant materials. Some of these insights can also be applied to the design and development of dental implants [5]. The extent of these deteriorative actions is influenced by the mechanical environment. If the stress is unevenly distributed over these couplings or if the loading stress upon these placed implants is not properly transferred to surround hard tissue, placed implants face either deosseointegration or failure, and hard tissue will be subject to a detrimental bone resorption.

8.2 Biomechanics in orthopedic implant systems

Orthopedic biomechanics is the study of mechanical systems in the body to further the prevention and treatment of musculoskeletal disorders and is a typical example of multidisciplinary in nature, which can integrate information and knowledge from bioengineering, orthopedics, bone–implant interface (BII) characterization, and more [6]. Consequently, the strategy for conducting cutting-edge experimental research in orthopedic biomechanics in hospitals, universities, and industry includes a combination of orthopedic surgery, mechanical testing, and medical imaging, as illustrated in Figure 8.2 [7].

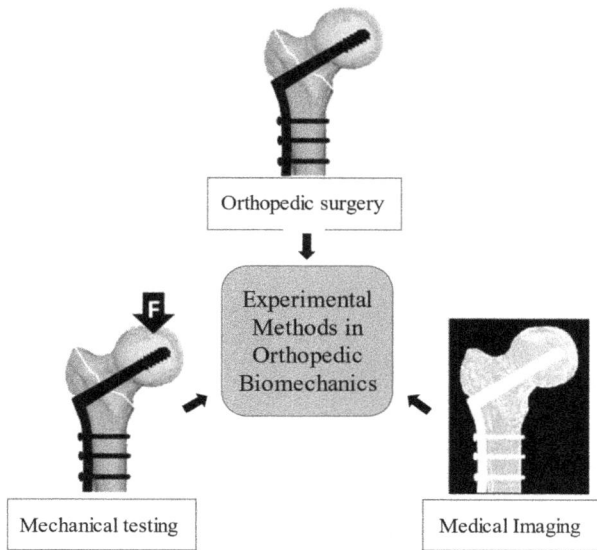

Figure 8.2: Experimental research activities in orthopedic biomechanics [7].

In general, the movement of a body is composed of two types: rotation in which a defined point in the body rotates about a defined axis and translation in which motion occurs along a line. Almost without exception, human joints have more than one axis of rotation [8]. The joints of the fingers, while they might superficially be viewed as hinge joints, allow small out-of-plane rotations and translations. There are a variety of types of joint movements in a body [9, 10]. Flexion is bending a joint and occurs when the angle of a joint decreases; for example, the elbow flexes when performing a biceps curl and the knee flexes in preparation for kicking a ball. Extension is straightening a joint and occurs when the angle of a joint increases, for example, the elbow when throwing a shot put. Abduction is a movement away from the midline of the body and occurs at the hip and shoulder joints during a jumping jack movement. Adduction is a movement toward the midline of the body and occurs at the hip and shoulder, returning the arms and legs back to their original position from a jumping jack movement or when swimming breaststroke. Circumduction is where the limb moves in a circle and occurs at the shoulder joint during an overarm tennis serve or a cricket bowl. Rotation is where the limb turns round its long axis, like using a screwdriver and occurs in the hip joint in golf while performing a drive shot or the shoulder joint when playing a topspin forehand in tennis. It has internal rotation and external rotation. Translational movements are linear movements or, simply, movements in a straight line. Gliding movements of the joint are translational in character. The term slide has also been used in referring to translational movements between joint surfaces. It directs to anterior, posterior, medial, and lateral direction.

Table 8.1 summarizes joint types and locations and their associated involved bones and movements [9, 10].

Table 8.1: Body joint locations and types of joint movements [9].

Joint	Joint type	Bones involved	Joint movements
Knee	Hinge	Femur, tibia	Flexion, extension
Elbow	Hinge	Humerus, ulna, radius	Flexion, extension
Hip	Ball and socket	Femur, pelvis	Flexion, extension, abduction, adduction, rotation, circumduction
Shoulder	Ball and socket	Humerus, scapula	Flexion, extension, abduction, adduction, rotation, circumduction

The knee hinge joint is a complex system of bones, cartilage, tendons, and other soft tissue, that, in a healthy state, work together to provide mobility and support to the body. However, in severe states of osteoarthritis, total replacement of the knee joint is essential not only to relieve the pain, but most importantly to restore the mobility lost due to the degeneration of the cartilage and also to protect the surfaces of the articulating bones from further wear. In TKR systems available today, a major cause of implant failure is wear of the articulating materials. It is therefore essential to study the modes of wear in order to understand better how to resolve this critical problem. Because the femur and tibia have the freedom to move somewhat independently of one another, there is not a simple solution for contact between the two surfaces [11]. Figure 8.3 shows three possible orientations of the femoral surface on the tibial surface and compares shear forces for each position [12]. Recent advances in interventional MRI have allowed for imaging of the internal anatomy of the knee joint and tibio-femoral contact regions. The posterior femoral condyles have been shown to have close to circular surfaces in the sagittal plane, and these points have been used as reference for femoral condylar position relative to the tibia [12]. Johal et al. [12] study the tibio-femoral movement during flexion in the living knee. It was reported that (i) femoral external rotation (or tibial internal rotation) occurs with knee flexion under loaded and unloaded conditions, but the magnitude of rotation is greater and occurs earlier on weight bearing, (ii) with flexion plus tibial internal rotation, the pattern of movement follows that in neutral, and (iii) with flexion in tibial external rotation, the lateral femoral condyle adopts a more anterior position relative to the tibia and, particularly in the nonweight-bearing knee, much of the femoral external rotation that occurs with flexion is reversed.

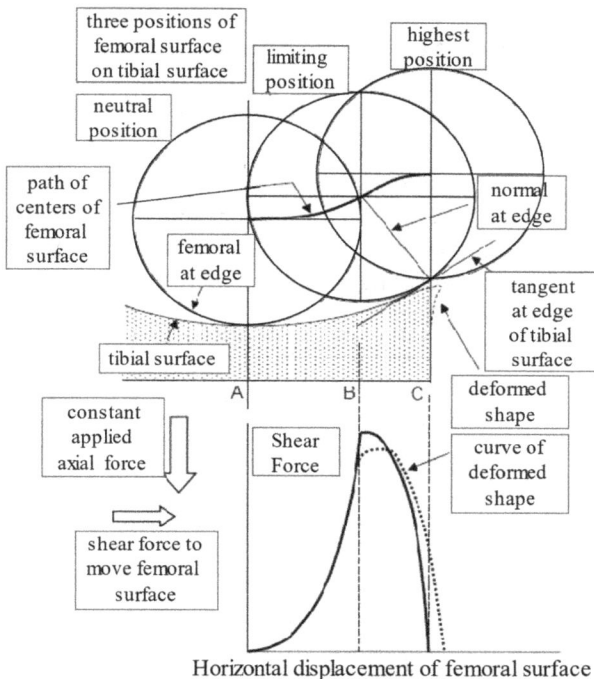

Figure 8.3: Tibio-femoral surface contact [12].

The movement of joint surfaces is known as the arthrokinematics, which differs from osteokinematics (in general, the osteokinematics means bone movement). Three fundamental motions occur at joint surfaces – roll, slide, and spin [13]. Roll: multiple points along one rotating articular surface contact multiple points on another articular surface (such as a tire rotating over the road). Slide: a single point on one articular surface contacts multiple points on another articular surface (such as a tire that is not rotating skids on the icy road). Spin: a single point on one articular surface rotates on a single point on another articular surface (such like a rotating top spinning on one spot). The angular movement of bones in the human body occurs as a result of a combination of rolls, spins, and slides.

Bagheri et al. [14] mentioned that experimental stress analysis of whole bones, implants, and whole bone–implant interfacial region is an important experimental approach in orthopedic biomechanics. High stresses in whole bones and implants may cause mechanical failure, but low stresses in whole bone may cause the stress shielding, which leads to bone atrophy, and bone resorption, resulting in possibly an implant loosening. There are several experimental tools for strain (stress) analysis, including thermographic stress analysis [14], a finite element analysis (FEM) [15], or CT-image based FEM [16–18]. In order to assess the post-operative safety of the implant, the equivalent stress distribution of three models (intact bone model, resurfacing hip arthroplasty (RHA) model, and total hip arthroplasty (THA) model) was investigated [18]. Drucker–Prager yield criterion was chosen as bone is a brittle material and the compressive strength is higher than the tensile strength. As seen in Figure 8.4 [18], it was reported that (i) the equivalent stress for RHA shows almost identical distribution as the intact bone model, (ii) the THA model equivalent stress distribution reveal that the proximal lateral section experiences high stress, while the distal lateral section experiences low stress, (iii) however, the stress increases around the implant tip, and (iv) proximal medial region, around the porous coating of the implant, also experiences high stress, although it appears to decrease distally.

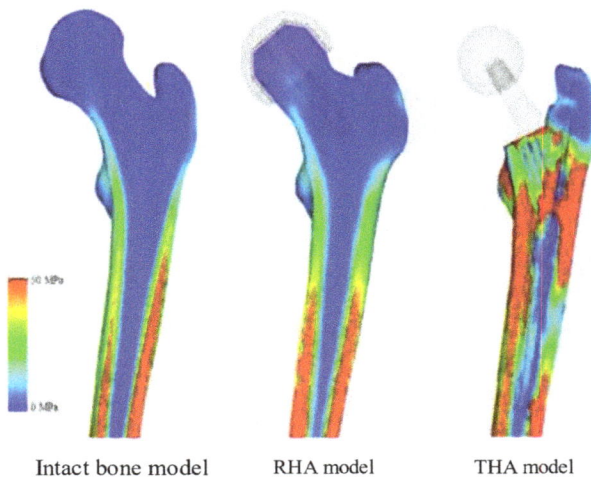

Intact bone model   RHA model   THA model

Figure 8.4: Drucker–Prager equivalent stress distribution in three models [18].

For a better understanding the stress shielding which causes bone resorption, Jung et al. [19] investigated the stress distribution at the BII of implant models custom-fitted to Asian individuals, using a finite-element method. Based on the standard geometry of Asian femurs, four different custom-fitted implant stems and applied boundary conditions were designed, including a stationary loading of 1,750 N. It was reported, referring to Figure 8.5 [19], that (i) stress distributed along the length of the femur generally increased toward the distal region, (ii) no significant difference was observed among the four types at both the distal-lateral and the distal-medial regions; however, the prostheses with supplementary structures showed particular difference in stress distribution, (iii) the stress distributed in the type 3 implant increased at the proximal-medial region (circle b) when compared to the custom-designed prosthesis (the type 1 implant; circle a); however, stress in the type 4 implant was reduced at this region (circle c) [19].

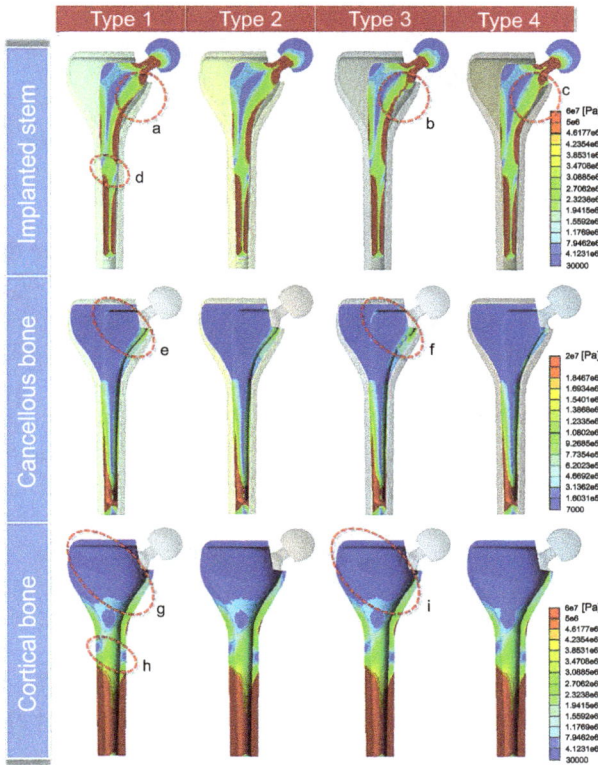

Figure 8.5: Equivalent stress (von Mises) distribution at the cross section along the length of the implanted stem, cancellous bone, and cortical bone [19].

To maintain and understand the adequate initial fixation which is prerequisite for osseointegration and the secondary stability of cementless cups, besides the stress distribution, the load transfer is another important factor. Physiologic force transmission between the cup and acetabulum guarantees the best long-term fixation in the THA. To study load transfer within the natural hip joint and in the BII of two different hemispherical noncemented press-fit cups, Widmer et al. [20] investigated 10 hips were in an experimental setup simulating single-leg stance. It was reported that (i) main load transfer occurs in the cranial region of the acetabulum, where it is buttressed by the iliac bone; the second location is at the posterior-inferior region at the ischial facet, and the third location is at the anterior region, where support is provided by the pubic bone, (ii) peak local forces were found at the ischial and iliac facets and local forces can be grouped into an iliac, an ischial, and a pubic group contributing 55%, 25%, and 20% to the total hip joint force, (iii) pole contact was not present in the natural hip and with the biradial press-fit cup with flattened pole area but was observed with the pure hemispherical cup; accordingly, stable fixation of an acetabular cup is achieved best by a three-point-like bony support at the iliac, ischial, and pubic bone, and (iv) the acetabular fovea does not provide functional support of the femoral head or endoprosthetic socket [20].

8.3 Stress shielding in orthopedic implant systems

The chief complication of THR with cementless fixations is a bone resorption at peri-implant, which results in failure of the bond at the BII due to a foreign body reaction between the vital bone and implant. Stress shielding, a mechanical phenomenon that refers to the reduction of load transferred to the surrounding bone, is one of the possible factors to cause bone resorption [19, 21–24]. Stress shielding (or stress protection) takes place when metallic implants are used to repair fractures or in joint replacement surgery and this phenomenon refers to the reduction in bone density (osteopenia) as a result of removal of typical stress from the bone by an implant (for instance, the femoral component of a hip prosthesis) [25, 26]. According to Wolff's law [27], bone in a healthy person or animal will remodel in response to the loads it is placed under. Therefore, if a loading on a bone decreases, the bone will become less dense and weaker because there is no stimulus for continued remodeling that is required to maintain bone mass.

Short- and long-term stabilities of cementless implants are strongly determined by the interfacial load transfer between implants and bone tissue. Stress-shielding effects arise from shear stresses due to the difference of material properties between bone and the implant. Raffa et al. [28] investigated the dependence of the stress field in peri-implant bone tissue in terms of the surface roughness, material properties of bone and the implant, and the bone–implant contact ratio. It was reported that (i) the isostatic pressure is not affected by the presence at BII while shear stresses arise due to the combined effects of a geometrical singularity (for low surface roughness) and of shear stresses at the BII (for high surface roughness), and (ii) stress-shielding effects are likely to be more important when the bone–implant contact ratio value is low, which corresponds to a case of relatively low implant stability.

As shown in Figure 8.6 [23], when the transfer of a weight load line is concerned, the prosthesis area exhibits a typical example of a stress-shielding situation.

Figure 8.7 explains an occurrence of the stress shielding in more detail. Referring to Figure 8.7 [29–31], when stress (weight load) is applied to the top of the femur, in general the stress is transmitted through the trabeculae of the cancellous bone and further transferred through the cortical bone. However, there is a mismatching in mechanical property (particularly, modulus of elasticity (MOE); stiffness) and the applied stress is transmitted down the implant body, resulting in that less stress is carried by the bone in this region, so that the bone resorption is likely to take place.

There are several ideas proposed to reduce or minimize the extent of the adverse stress-shielding effect. Since a main cause for the stress shielding is a mismatching of

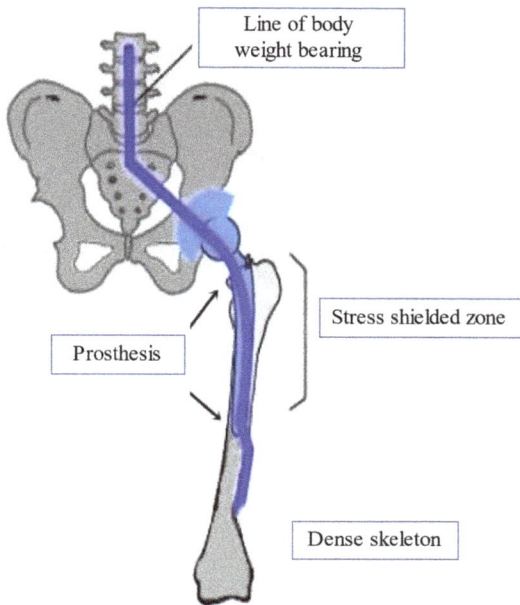

Figure 8.6: Unevenness of weight load transfer at the area where THA is installed [23].

stiffness (or MOE) between placing implant material and receiving vital hard tissue (or bone), main measures proposed to minimize stress shielding is the development of new biocompatible material(s) having reasonable level of MOE. Bobyn et al. [32] mentioned that (i) in experimental canine model studies of stiff versus flexible, fully porous-coated, metallic femoral stems (differing by three- to fivefold in stiffness characteristics) revealed markedly different resorptive bone remodeling patterns, (ii) the flexible stem resulted in about 30% more cortical bone retention adjacent to the implant at one-year post-implantation and larger differences in dogs, (iii) the stiffness characteristics of the human femur were established as a function of canal size and compared with those of noncemented hip prostheses, (iv) increased mechanical compatibility was found for stems made of titanium alloy and with design features that reduce cross-sectional area and moment of inertia, and (v) clinical data suggest that to reduce the likelihood of pronounced bone resorption, it would be beneficial for the implant to possess a bending stiffness of about one half to one third that of the human femur. Accordingly, new implant alloys have been developed. Yamako et al. [33] developed Ti–33.6Nb–4Sn alloy, which was subjected to the local heat treatment to manipulate the MOE in ranging from 82.1 GPa (proximal end) to 51.0 GPa (distal end) in a modulus gradient manner. Niiomi et al. [34] developed β-type Ti–29Nb–13Ta–4.6Zr, having a MOE of around 60 GPa. Antonialli et al. [35] developed a similar Ti–Nb–Ta–Zr alloy with 47 GPa of stiffness. Moussa et al. [36] used functionally graded Ti–6Al–4 V

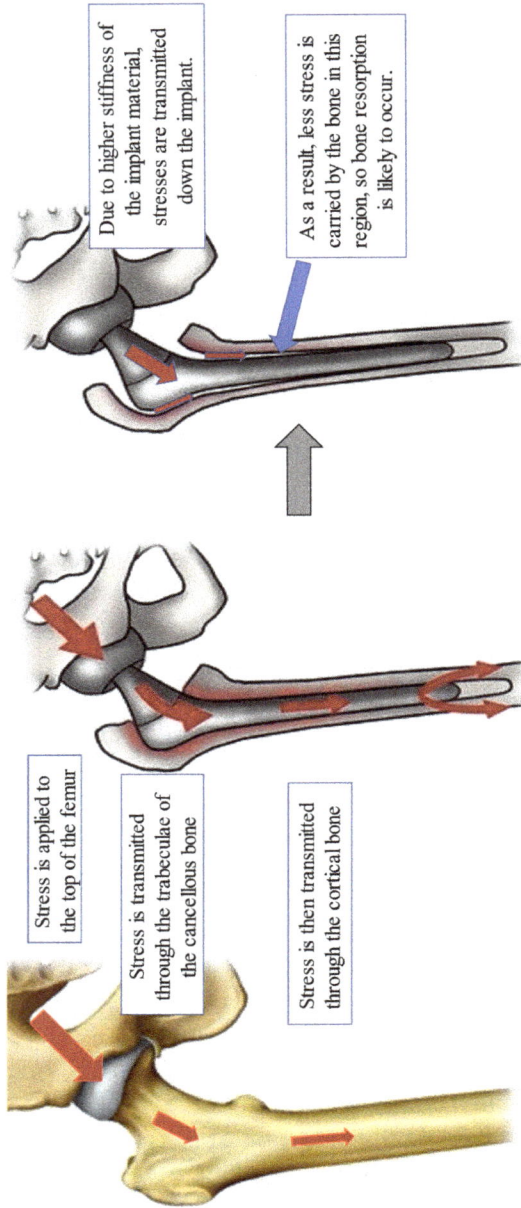

Due to higher stiffness of the implant material, stresses are transmitted down the implant.

As a result, less stress is carried by the bone in this region, so bone resorption is likely to occur.

Stress is applied to the top of the femur

Stress is transmitted through the trabeculae of the cancellous bone

Stress is then transmitted through the cortical bone

Figure 8.7: Situation for development of the stress-shielding phenomenon.

with varying MOE from 110 GPa to 55 GPa. Hannon [37] developed a more complicated Ti-Al-V-Fe-Nb-Mo-Sn alloy system, with stiffness ranging from 70 GPa to 50 GPa.

Stiffness mismatching can be easily understood when various biomaterials are compared in terms of strength and stiffness, as seen in Figure 8.8 [38].

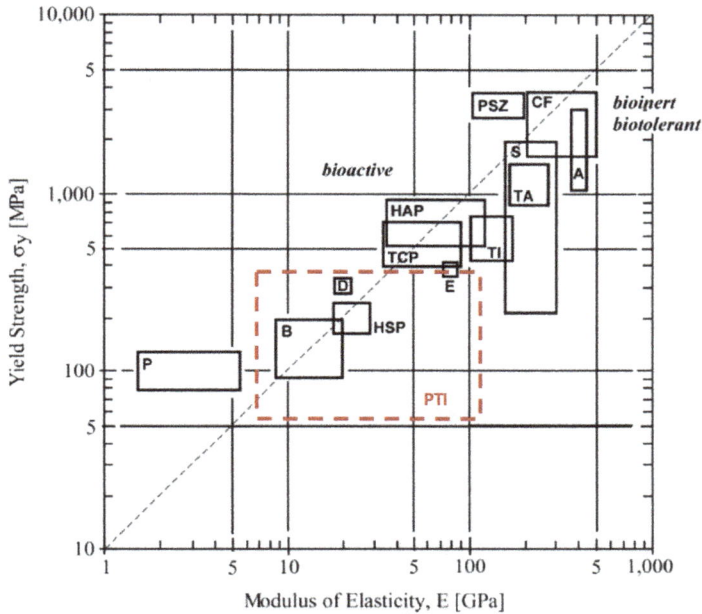

Figure 8.8: Relationship between strength and stiffness of various biomaterials in log–log plot. P: polymeric materials, B: bone, HSP: high strength polymers (Kelvar, Kapton, PEEK, etc.), D: dentin, TCP: tricalcium phosphate, HAP: hydroxyapatite, E: enamel, TI: commercially pure titanium (all unalloyed grades), TA: titanium alloys (e.g., Ti–6Al–4V), S: steels (e.g., 304-series stainless steel), A: alumina, PSZ: partially stabilized zirconia, CF: carbon fiber.

Adding an appropriate porosity nature on solid biomaterial(s), it is easy to speculate that mechanical properties can be altered depending on the degree of the porosity. Red broken area in Figure 8.8 represents porous Ti (PTI) with porosity ranging from 35% to 70% [39–41], indicating that the porous Ti material exhibits biomechanical compatibility to surround hard tissue. Based on the information, Oshida proposed a conceptual implant design as illustrated in Figure 8.9 [42].

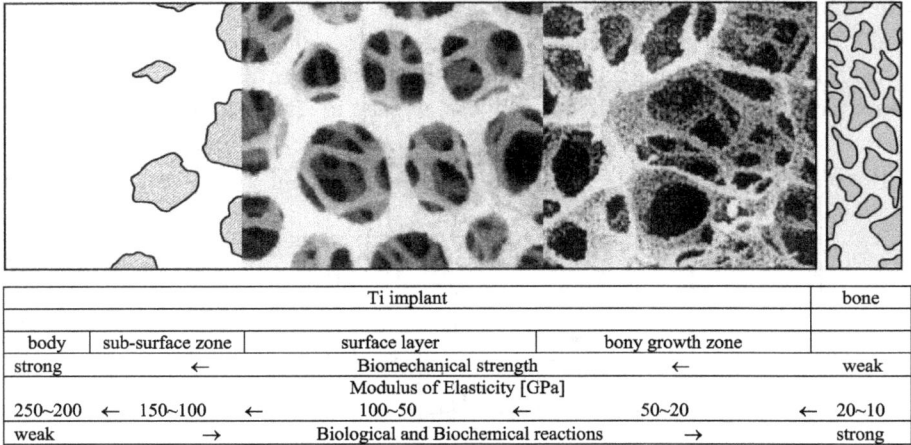

Ti implant				bone
body	sub-surface zone	surface layer	bony growth zone	
strong	←	Biomechanical strength	←	weak
		Modulus of Elasticity [GPa]		
250~200 ← 150~100 ←		100~50 ←	50~20	← 20~10
weak →		Biological and Biochemical reactions →		strong

Figure 8.9: Schematic and conceptual implant design with graded functionality [42].

8.4 Biomechanics in dental implant systems

Biomechanics involved in implantology should include at least (1) the nature of the biting forces on the implants, (2) transferring of the biting forces to the interfacial tissues, and (3) the interfacial tissues reaction, biologically, to stress transfer conditions. Interfacial stress transfer and interfacial biology represent more difficult, interrelated problems. While many engineering studies have shown that variables such as implant shape, elastic modulus, extent of bonding between implant and bone, and so on, can affect the stress-transfer conditions, the unresolved question is whether there is any biological significance to such differences. The successful clinical results achieved with osseointegrated dental implants underscore the fact that such implants easily withstand considerable masticatory loads. In fact, it was shown that bite forces in patients with these implants were comparable to those in patients with natural dentitions [43]. A critical aspect affecting the success or failure of an implant is the manner in which mechanical stresses are transferred from the implant to bone at the BII. It is essential that neither implant nor bone be stressed beyond the long-term fatigue capacity. It is also necessary to avoid any relative motion that can produce abrasion of the bone or progressive loosening of the implants. An osseointegrated implant provides a direct and relatively rigid connection of the implant to the bone. This is an advantage because it provides a durable interface without any substantial change in form or duration. There is a mismatch of the mechanical properties and mechanical impedance at the interface of Ti and bone that would be evident at ultrasonic frequencies. It is interesting to observe that from a mechanical standpoint, the shock-absorbing action would be the same if the soft layer were between the metal implant and the bone. In the natural tooth, the periodontum, which forms a shock-absorbing layer, is in this position between the tooth and a jawbone [43]. Natural teeth and implants have different force transmission characteristics to bone. Compressive strains were induced around natural teeth and implants as a result of static axial loading, whereas combinations of compressive and tensile strains were observed during lateral dynamic loading. Strains around the natural tooth were significantly lower than the opposing implant and occluding implants in the contralateral side for most regions under all loading conditions. There was a general tendency for increased strains around the implant opposing natural tooth under higher loads and particularly under lateral dynamic loads [44].

It was reported that normal bite force in human is ranged from 250 to 1,200 N in posteriors, while from 80 to 400 N in anterior region, and the lateral component of the bite force is about 20 N. Moreover, the maximum contact stress on teeth is around 20 MPa [45]. During a normal masticatory action, occlusal forces act along a cusp tip or flat surface such as a fossa or marginal ridge directs the force through the along axial of the tooth (as seen in Figure 8.10 [46]). There are unfortunately abnormalities during the tooth contacting action. Occlusal malfunctional (or parafunctional) forces on placed implants generate complicated stress–strain environment in the peri-implant area. The most common cause of implant failure after successful surgical fixation or early loss of rigid fixation during the first year of implant loading is the result of parafunction. Such complications occur with greater frequency in the maxilla because of a decrease in bone density and an increase in the moment of force [47, 48]. Parafunction in dentistry refers to those activities of the stomatagnathic system that would be considered to fall outside of functional activities. Lip and cheek chewing, fingernail biting, and teeth clenching are examples of parafunctional activity. Much focus in dentistry has been given to teeth grinding, clenching, or bruxism, as a parafunctional activity [49].

Masticatory forces acting on dental implants can result in undesirable stress in adjacent bone, which in turn can cause defects and the eventual failure of implants. It was reported that the maximum stress areas were located around the implant neck. The decrease in stress was the greatest (31.5%) for implants with a diameter ranging from 3.6 mm to 4.2 mm. An increase in the implant length also led to a decrease in the maximum von Mises equivalent stress values; the influence of implant length, however, was not as pronounced as that of implant diameter. It was further mentioned that an increase in the implant diameter decreased the maximum von Mises equivalent stress around the implant neck more than an increase in the implant length, as a result of a more favorable distribution of the simulated masticatory forces applied [50]. At present, load-bearing implants are designed on a deterministic basis in which the structural strength and applied loading are given fixed values, and global safety factors are applied to (1) cover any uncertainties in these quantities, and (2) to design against failure of the component [51]. This approach will become increasingly inappropriate as younger and more active patients demand more exactness, and as devices become more complex. Browne et al. [51] described a preliminary investigation in

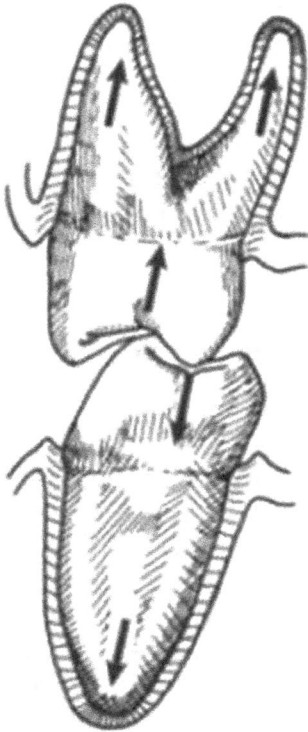

Figure 8.10: Occlusal force distribution during a normal masticatory action [46].

which a scientific and probabilistic technique is applied to assess the structural integrity of the knee tibial tray. It was envisaged that by applying such a technique to other load-bearing biomedical devices, reliability theory may aid in future lifting procedures and materials/design optimization [51].

When an implant is placed into alveolar bone, the situation is slightly different. As illustrated in Figure 8.11 [52], there are two major areas which show clear differences between the natural tooth and placed implant; (1) collagen fibers are attached to the tooth root surface in the natural tooth, while the implant surface is not covered with collagen fibers; (2) in the natural tooth, there is a periodontal ligament (PDL), but no PDL with the dental implant. The PDL is a group of specialized connective tissue fibers that essentially attach a tooth to the alveolar bone and it inserts into root cementum on one side and onto alveolar bone on the other. McCormack at al. [53] mentioned that (i) the principal function of the PDL is to connect the tooth to the jaw, which it must do in such a way that the tooth will withstand the considerable forces of mastication, and (ii) the PDL fibers have a very important role in load transfer between the teeth and alveolar bone, and (iii) orthodontic tooth movement occurs as a result of resorption and formation of the alveolar bone due to an applied load, so that the case with placed implant, orthodontic mechanotherapy cannot take place.

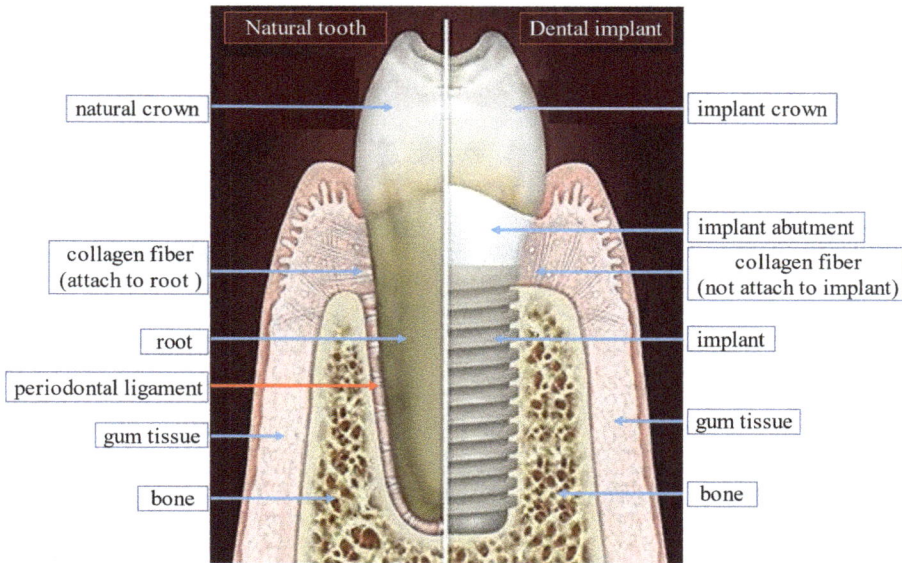

Figure 8.11: Comparison between natural tooth and dental implant [52].

Dental implants are subjected to occlusal loads when placed in function. Such loads may vary dramatically in magnitude, frequency, and duration, depending on the patient's parafunctional habits. Passive mechanical loads also may be applied to dental implants during the healing stage because of mandibular flexure, contact with the first-stage cover screw, and second-stage permucosal extension [54, 55]. A force applied to a dental implant rarely is directed longitudinally along a single axis. In fact, three dominant clinical loading axes exist in implant dentistry: (1) mesiodistal, (2) faciolingual, and (3) occlusoapical, as seen in Figure 8.12 [55]. A single occlusal contact most commonly results in a three-dimensional (3D) occlusal force. Importantly, this 3D force may be described in terms of its component parts (fractions) of the total force that are directed along the other axes. For example, if an occlusal scheme on an implant restoration that results in a large magnitude of force component directed along the faciolingual axis (lateral loading) is used, then the implant is at extreme risk for fatigue failure (described later in this chapter). The process by which 3D forces are broken down into their component parts is referred to as vector resolution and may be used routinely in clinical practice for enhanced implant longevity [55].

Figure 8.13 demonstrates a significant influence of a presence of the PDL in force-deformation curves under compression loading (simulating normal occlusal force) [56]. While both dental implants show an almost linear relationship until implant failure (if the osseointegration is established), the natural tooth behaves like a normal ductile metal, exhibiting elastic deformation, followed by plastic deformation.

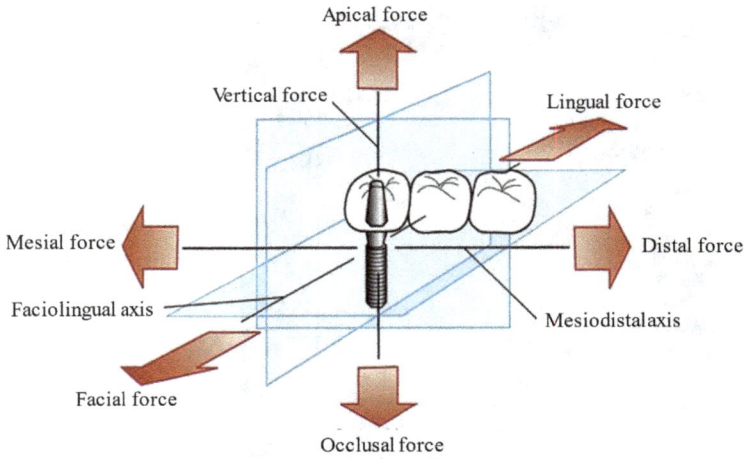

Figure 8.12: Three-dimensional illustration of force distribution [55].

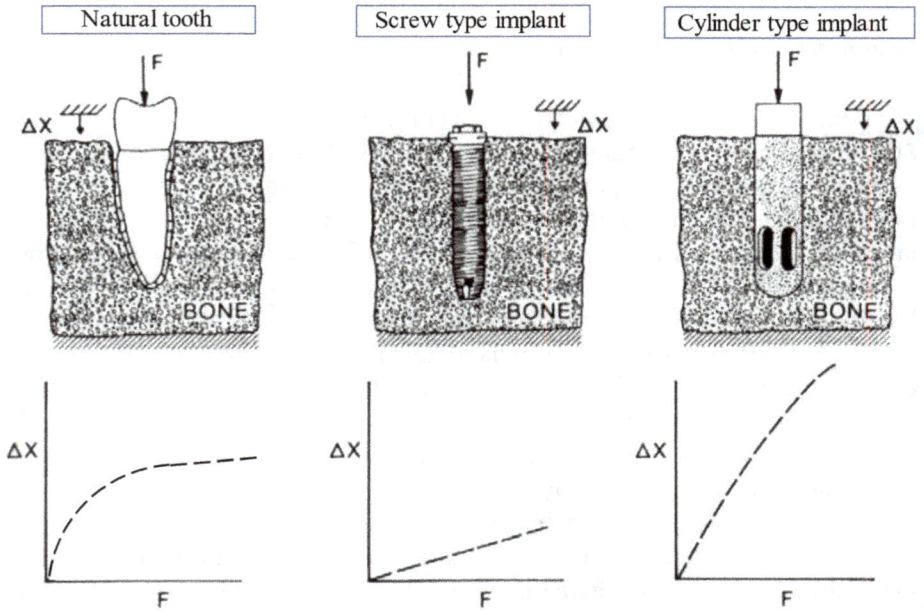

Figure 8.13: Comparison of force–displacement under compression between natural tooth and dental implants [56].

By means of finite element method (FEM) analysis, stress distribution in bone around implants was calculated with and without a stress-absorbing element [57]. A freestanding implant and an implant connected with a natural tooth were simulated. For the freestanding implant, it was concluded that the variation in the MOE of the stress-absorbing element had no effect on the stresses in bone. Changing the shape of the stress-absorbing element had little effect on the stresses in cortical bone. For the implant connected with a natural tooth, it was concluded that a more uniform stress was obtained around the implant with a low MOE of the stress-absorbing element. It was also concluded that the bone surrounding the natural tooth showed a decrease in the height of the peak stresses [57]. Mechanical in vitro tests of the Brånemark implant disclose that the screw joint which attaches the prosthetic gold cylinder and the transmucosal abutment to the fixture forms a flexible system. This inherent flexibility seems to match well with the vertical mobility of a supporting tooth connected to the implant. Calculations of vertical load distribution based on measured flexibility data demonstrate that the forces are shared almost equally between the tooth and the implant, even without taking the flexibility of the surrounding bone or the prosthesis into account. The therapy of a single Brånemark implant connected to a natural tooth should be considered without any additional element of a flexible nature. Mechanical tests and theoretical considerations, however, indicate that the transverse mobility of the connected tooth should be limited, and that the attachment of the prosthesis to the tooth should be a rigid design to avoid gold-screw loosening [58]. The stress distribution pattern clearly demonstrated a transfer of preload force from the screw to the implant during tightening. A preload of 75% of the yield strength of the abutment screw was not established using the recommended tightening torques. Using FEM, a torque of 32 Ncm applied to the abutment screws in the implant assemblies was studied in the presence of a coefficient of friction of 0.26 and resulted in a lower than optimum preload for the abutment screws. It was then mentioned that in order to reach the desired preload of 75% of the yield strength, using the 32 Ncm torque applied to the abutment screws in the implant assemblies studied, the coefficient of friction between the implant components should be 0.12 [59].

Bahrami et al. [60] investigated, by means of FEM analysis, the effect of surface roughness treatments on the distribution of stresses at the BII in immediately loaded mandibular implants. Finite element models were created with one of the four roughness treatments on the implant fixture surface. Of these, three were surface treated to create a uniform coating determined by the coefficient of friction (μ); these were either (1) plasma sprayed or porous-beaded ($\mu = 1.0$), (2) sandblasted ($\mu = 0.68$), or (3) polished ($\mu = 0.4$). The fourth implant had a novel two-part surface roughness consisting of a coronal polished component ($\mu = 0.4$) interfacing with the cortical bone, and a body plasma treated surface component ($\mu = 1$) interfacing with the trabecular bone. Finite element stress analysis was carried out under vertical and lateral forces. It was found that (i) the type of surface treatment on the implant fixture affects the stress at the BII of an immediately loaded implant complex, and (ii) von Mises stress data showed that the two-part surface treatment created the better stress distribution at the BII, suggesting that (iii) the two-part surface treatment for implants creates lower stresses than single uniform treatments at the BII, which might decrease peri-implant bone loss. Pérez-Pevida et al. [61] evaluated how the elastic properties of the fabrication material of dental implants influence peri-implant bone load transfer in terms of the magnitude and distribution of stress and deformation. A 3D FEM analysis was performed; the model used was a section of mandibular bone with a single implant containing a cemented ceramic-metal crown on a titanium abutment. The following three alloys were compared: rigid (Y-TZP), conventional (Ti–6Al–4V), and hyperelastic (Ti–Nb–Zr). A 150-N static load was tested on the central fossa at 6° relative to the axial axis of the implant. It was reported that there were no differences in the distribution of stress and deformation of the bone for any of the three types of alloys studied, mainly being concentrated at the peri-implant cortical layer; however, there were differences found in the magnitude of the stress transferred to the supporting bone, with the most rigid alloy (Y-TZP: yttrium-stabilized tetragonal zirconia) transferring the least stress and deformation to cortical bone; concluding that there is an effect of the fabrication material of dental implants on the magnitude of the stress and deformation transferred to peri-implant bone [61].

Impact of the implant shape on the biomechanical performance of all-on-four treatment of dental implant is still unclear. Hence, Wu et al. [62] evaluated the all-on-four treatment with four osseointegrated implants in terms of the biomechanical effects of implant design and loading position on the implant and surrounding bone by using both in vitro strain gauge tests and 3D FEM analyses. Both in vitro and 3D FEM models were constructed with placing NobelSpeedy and NobelActive implants as well as a titanium framework in an edentulous jawbone based on the concept of all-on-four treatment. Three types of loads were applied at the central incisor area (loading position 1) and at the molar regions with (loading position 2) and without (loading position 3) the denture cantilever. For the in vitro tests, the principal bone strains were recorded by rosette strain gauges. The 3D FE simulations analyzed the peak von Mises stresses in the implant and surrounding cortical bone. It was found that (i) the peak stress and strain in the surrounding bone were typically 36–62% (3D FEM analysis) and 47–57% (in vitro test) higher for loading position 3 than for loading positions 1 and 2, and (ii) between those two implant designs, the bone strains and bone stresses did not differ significantly, concluding that for all-on-four treatment with four osseointegrated dental implants, altering the implant design does not appear to affect the biomechanical performance of the entire treatment, especially in terms of the stresses and strains in the surrounding bone [62].

Chang et al. [63] investigated the stress distributions in an implant, abutment, and crown restoration with different implant systems, in various bone qualities, and with different loading protocols using a 3D FEM model. Four types of dental implants were embedded in two different bone qualities (types II and IV) under 100-N axial and 30° oblique loading forces were applied to analyze the stress distribution in the crown restoration, abutment, abutment screw, implant, and supporting bone. It was obtained that (i) the highest maximum von Mises stress was noted in the abutment of a tissue-level implant with the Straumann system (1203.04 MPa) under a 30° oblique loading force, (ii) with axial load application, stresses in the screw and abutment of the NobelBiocare system were greater in the tissue-level implant (MK III) than in the bone-level implant (active), and (iii) the von Mises stresses in the cortical bone were mostly greater in the tissue-level implant (MK III) than in the bone-level implant (Active) of the NobelBiocare system; however, von Mises stresses in cancellous bone were mostly greater in the bone-level implant (active) than in the tissue-level implant (MK III) of the NobelBiocare system, concluding that (iv) the Straumann system produced greater stresses than the NobelBiocare system in type IV cortical bone, but they were almost equal in type II bone, and (v) by contrast, the NobelBiocare system produced greater stresses than the Straumann system in cancellous bone, regardless of the type of loading angle or bone quality. Figure 8.14 compares stress distribution among three implant systems; A: Straumann tissue-leveled implant, B: NobelBiocare MK III tissue-leveled implant, and C: Straumann bone-leveled implant.

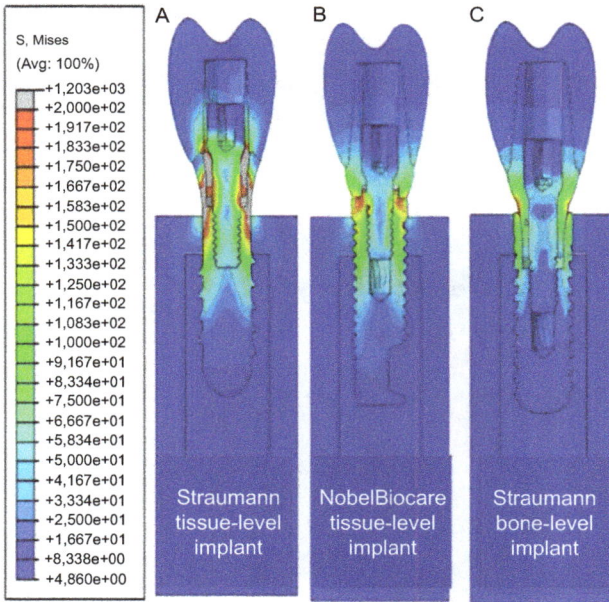

Figure 8.14: The stress concentrated in abutment and implant connection for three different implant systems [63].

Kumar et al. [64] evaluated the stress formed around an implant and a natural tooth under occlusal forces, on different tooth implant-supported fixed prosthesis (TIFP) designs in order to suggest a design, which transmits less stress to the bone. A distal extension situation was utilized in this study to evaluate stress distribution around a natural tooth and an implant in TIFP models with three connection designs: (1) rigidly connected to an abutment tooth, (2) connected to an abutment tooth with a nonrigid connector (NRC), and (3) connected to an abutment implant with an NRC. The stress values of the three models loaded with vertical forces (300 N) were analyzed using 3D FEM analysis. It was found that (i) the highest level of stress around the implant and natural tooth was noted on the TIFP models with the RC, (ii) on the other hand, NRCs incorporated into the prostheses reduced the stress in the bone around the implant and natural tooth, and (iii) in model 3 (where the second premolar and the implant are connected rigidly), the highest stress values recorded were located in the cortical bone region of the implant along lines 3 (zones 7, 8, 9) and 4 (zones 10, 11, 12) with values recorded as 23.19 and 20.46 MPa, respectively, and the maximum stress values generated around the natural tooth were 9.20 and 6.57 MPa in zones 1 and 6, respectively. Figure 8.15 shows the von Mises stress contours for the rigid connection configuration [64]. Based on these findings, it was concluded that (iii) the use of NRCs on the implant abutment-supported site is recommended, if the tooth and implant abutment are to be used together as fixed prosthesis supports, and (iv) the NRC placed on the implant abutment site reduces the stress around the implant and natural tooth in a fixed prosthesis supported by tooth and implant increasing the life span of both.

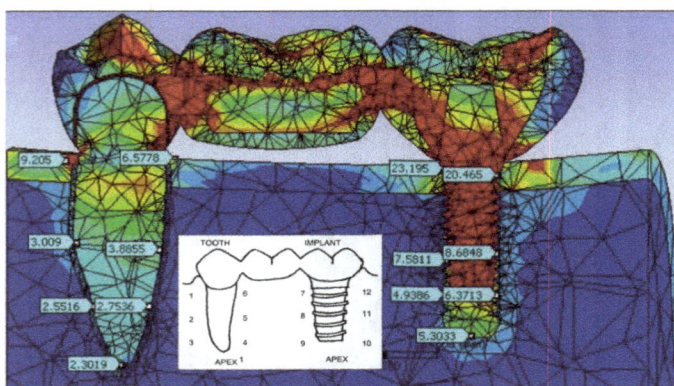

Figure 8.15: Shows the von Mises stress contours for the rigid connection configuration [64].

Shirazi et al. [65] mentioned that, in a dental implant system, the value of stress and its distribution play a pivotal role on the strength, durability, and life of the implant-bone system. A typical implant consists of a titanium core and a thin layer of biocompatible material such as the hydroxyapatite (HA). This coating has a wide range of clinical applications in orthopedics and dentistry due to its biocompatibility and bioactivity characteristics. Low bonding strength and sudden variation of mechanical properties between the coating and the metallic layers are the main disadvantages of such common implants. To overcome these problems, a radial distributed functionally graded biomaterial was proposed. Oshida [38, 66] clearly indicated that an application of HA to titanium substrate has two-fold purposes; (1) there should be an acceptable level of biocompatibility between HA and bone, due to similarity of chemistry of HA and receiving hard tissue and (2) there should be an excellent biomechanical compatibility, since there is similar levels of MOE of both materials. As a result, a stiffness mismatching issue can be overcome to minimize the stress shielding.

8.5 Biomechanics in orthodontics

Normally, the orthodontic mechanotherapy exhibits several characteristic tooth movements (see Figure 8.16 [67]). Translation (or bodily movement) under a controlled tipping is to move the teeth horizontally to eliminate interdental space and is the most common movement in orthodontic treatment. Tipping and torqueing are uncontrolled tipping. During the tipping movement, the root of the tooth does not move so much, and the crown (head of the tooth) moves. It is effective when the next tooth falls down by keeping the missing tooth space. By torqueing, the crown portion does not move much, while the root of the tooth mainly moves in labio-

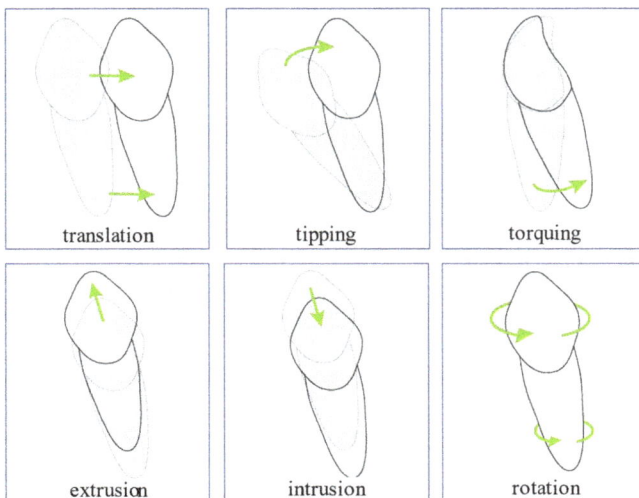

Figure 8.16: Typical orthodontic tooth movements [67].

lingual direction. Extrusion and intrusion are axial movements, and extrusion is to move tooth toward the coronal part of the tooth while the intrusion is to move along the axis toward the apex of the root. By rotational movement, the teeth rotate along the long axis of the tooth (the axis connecting the tip of the tooth to the root of the tooth) to twist tooth to correct the direction of the teeth [68].

Under such a variation of movement, there are only two essential components which are responsible to tooth movement: tension component and compression component. Biological actions are directly affected by these force components, as shown in Figure 8.17 [69]. Orthodontic mechanotherapy for a tooth movement is relied on localized tissue resorption and formation in the affecting bone structure and PDL, which attaches the tooth to the adjacent bone and are dense fibrous connective tissue structures, consisting of collagenous fiber bundles, cells, neural and vascular components and tissue fluids. On average, the PDL occupies a space about 0.2 mm wide. Depending on its location along the root, PDL width can range from 0.15 to 0.38 mm, with its thinnest part located in the middle third of the root. PDL space also decreases progressively with age [70]. Orthodontic force causes local hypoxia and fluid flow, initiating an aseptic inflammatory cascade culminating in osteoclast resorption in areas of compression and osteoblast deposition in areas of tension [71, 72]. Tension and compression components in the orthodontic force are associated with particular signaling factors, establishing local gradients to regulate remodeling of the bone and PDL for tooth displacement. In Figure 8.17, major reactions are compared between tension side PDL and compression side PDL [73, 74].

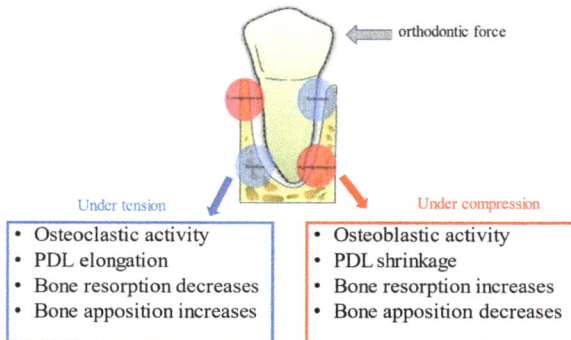

Figure 8.17: Tension and compression components caused by an orthodontic force [69].

References

[1] 6 types of joints in the body. https://quizlet.com/ca/538927699/6-types-of-joints-in-the-body-diagram/.

[2] Martin RB, Burr DB, Sharkey NA, Fyhrie DP. Synovial joint mechanics. In: Skeletal Tissue Mechanics. New York, NY: Springer, 2015, 227–73, doi: https://doi.org/10.1007/978-1-4939-3002-9_5.

[3] Anatomy of a joint. https://www.stanfordchildrens.org/en/topic/default?id=anatomy-of-a-joint-85-P00044.

[4] https://www.vedantu.com/question-answer/explain-any-five-movable-joints-with-examples-class-11-biology-cbse-5f6252edeee2a36606a391ef.

[5] Gotman I. Biomechanical and tribological aspects of orthopaedic implants. In: Ostermeyer GP, et al., ed. Multiscale Biomechanics and Tribology of Inorganic and Organic Systems. Springer Tracts in Mechanical Engineering. Cham.: Springer, 25–44, doi: https://doi.org/10.1007/978-3-030-60124-9_2.

[6] McMahon M. What Is Orthopedic Biomechanics? https://www.wise-geek.com/what-is-orthopedic-biomechanics.htm.

[7] Zdero R. What is orthopaedic biomechanics?. In: Experimental Methods in Orthopaedic Biomechanics. 2017, xxi–xxvi, doi: https://doi.org/10.1016/B978-0-12-803802-4.02001-1.

[8] Johnson GR. Biomechanics of joints. In: Joint Replacement Technology. Woodhead Publishing Series in Biomaterials. 2008, 3–30, doi: https://doi.org/10.1533/9781845694807.1.3Get rights and content.

[9] Joint Anatomy and Basic Biomechanics. https://cnx.org/resources/470adb6ee6172760e b50e49bfb920808/11-Reading%20-%20Harmony.pdf.

[10] Skeletal system. https://www.bbc.co.uk/bitesize/guides/zxc34j6/revision/1.

[11] Silva M, Snow K. Biotribology considerations of total knee replacement. Corpus. 2010, 53533299, http://www1.coe.neu.edu/~smuftu/docs/2009/ME5656_TermProject_%20Tribol ogy%20of%20Total%20Knee%20Replacements_(Silva,%20Snow).

[12] Johal P, Williams A, Wragg P, Hunt D, Gedroyc W. Tibio-femoral movement in the living knee. A study of weight bearing and non-weight bearing knee kinematics using 'interventional' MRI. J Biomech. 2005, 38, 269–76.

[13] Joint motion (arthrokinematics). Podia Paedia, https://podiapaedia.org/wiki/biomechanics/biomechanics-principles/joint-motion-arthrokinematics/.

[14] Bagheri ZB, Bougherara H, Zdero R. Thermographic stress analysis of whole bones and implants. Exp Methods Orthop Biomech. 2017, 49–64, doi: https://doi.org/10.1016/B978-0-12-803802-4.00004-4.

[15] Eberle S, Gerber C, von Oldenburg G, Högel F, Augat P. A biomechanical evaluation of orthopaedic implants for hip fractures by finite element analysis and in-vitro tests. Proc Inst Mech Eng H. 2010, 224, 1141–52.

[16] Bessho M, Ohnishi I, Matsuyama J, Matsumoto T, Imai K, Nakamura K. Prediction of strength and strain of the proximal femur by a CT-based finite element method. J Biomech. 2007, 40, 1745–53.

[17] Wakao N, Harada A, Matsui Y, Takemura M, Shimokata H, Mizuno M, Ito M, Matsuyama Y, Ishiguro N. The effect of impact direction on the fracture load of osteoporotic proximal femurs. Med Eng Phys. 2009, 31, 1134–39.

[18] Todo M, Fukuoka K. Biomechanical analysis of femur with THA and RHA implants using CT-Image based finite element method. Orthop J Sports Med. 2020, 2, 89–107.

[19] Jung JM, Kim CS. Analysis of stress distribution around total hip stems custom-designed for the standardized Asian femur configuration. Biotechnol Biotechnol Equip. 2014, 4, 525–32.

[20] Widmer K-H, Zurfluh B, Morscher EW. Load transfer and fixation mode of press-fit acetabular sockets. J Arthroplasty. 2002, 17, 926–35.

[21] Luo C, Wu X-D, Wan Y, Liao J, Cheng Q, Tian M, Bai Z, Huang W. Femoral stress changes after total hip arthroplasty with the ribbed prosthesis: A finite element analysis. BioMed Res Int. 2020, 6783936, doi: https://doi.org/10.1155/2020/6783936.

[22] Huiskes R, Weinans H, van Rietbergen B. The relationship between stress shielding and bone resorption around total hip stems and the effects of flexible materials. Clin Orthop Relat Res. 1992, 274, 124–34.

[23] Ridzwan MIZ, Shuib S, Hassan AY, Shokri AA, Mohamad Ibrahim MN. Problem of stress shielding and improvement to the hip implant designs: A review. J Med Sci. 2007, 7, 460–67.

[24] Cheruvu B, Venkatarayappa I, Goswami T. Stress shielding in cemented hip implants assessed from computed tomography. Biomed J Sci Tech Res. 2019, 18, 13637–41.

[25] Ibrahim H, Esfahani SN, Poorganji B, Dean D, Elahinia M. Resorbable bone fixation alloys, forming, and post-fabrication treatments. Mater Sci and Eng: C. 2017, 70, 870–88.

[26] Stress shielding. https://en.wikipedia.org/wiki/Stress_shielding.

[27] Frost HM. Wolff's Law and bone's structural adaptations to mechanical usage: An overview for clinicians. Angle Orthod. 1994, 64, 175–88.

[28] Raffa ML, Nguyen V-H, Hernigou P, Flouzat-Lachaniette C-H, Haiat G. Stress shielding at the bone-implant interface: Influence of surface roughness and of the bone-implant contact ratio. J Orthop Res. 2021, 39, 1174–83.

[29] Arifin A, Sulong AB, Muhamad N, Syarif J, Ramli MI. Material processing of hydroxyapatite and titanium alloy (HA/Ti)composite as implant materials using powder metallurgy: A review. Mater Des. 2014, 55, 165–75.

[30] Ødegaard KS, Torgersen J, Elverum CW. Structural and biomedical properties of common additively manufactured biomaterials: A concise review. Metals. 2020, 10, 1677, doi: https://doi.org/10.3390/met10121677.

[31] Total hip replacement images. https://www.shutterstock.com/search/total+hip+replacement.

[32] Bobyn JD, Mortimer ES, Glassman AH, Engh CA, Miller JE, Brooks CE. Producing and avoiding stress shielding. Laboratory and clinical observations of noncemented total hip arthroplasty. Clin Orthop Relat Res. 1992, 274, 79–96.

[33] Yamako G, Janssen D, Hanada S, Anijs T, Ochiai K, Totoribe K, Chosa E, Verdonschot N. Improving stress shielding following total hip arthroplasty by using a femoral stem made of β type Ti-33.6Nb-4Sn with a Young's modulus gradation. J Biomech. 2017, 63, 135–43.

[34] Niiomi M, Nakai M. Titanium-based biomaterials for preventing stress shielding between implant devices and bone. Intl J Biomater. 2011, 836587, doi: https://doi.org/10.1155/2011/836587.

[35] Antonialli AÍS, Bolfarini C. A numerical evaluation of reduction of stress shielding in laser coated hip prostheses. Mater Res. 2011, 14, 331–34.

[36] Moussa AA, Fischer J, Yadav R, Khandaker M. Minimizing stress shielding and cement damage in cemented femoral component of a hip prosthesis through computational design optimization. Adv Orthop. 2017, 8437956, doi: https://doi.org/10.1155/2017/8437956.

[37] Hannon P. A brief review of current orthopedic implant device issues: Biomechanics and biocompatibility. Open Access Text. doi: 10.15761/BEM.1000102.

[38] Oshida Y. Bioscience and Bioengineering of Titanium Materials. Elsevier, 2007.

[39] Wang X-H, Li J-S, Hu R, Kou H-C, Zhou L. Mechanical properties of porous titanium with different distributions of pore size. Trans Nonferrous Metals Soc China. 2013, 23, 2317–22.

[40] Oh I-H, Nomura N, Hanada S. Microstructures and mechanical properties of porous titanium compacts prepared by powder metallurgy. Mater Trans. 2002, 43, 443–46.

[41] Zou C, Zhang E, Li M. Preparation, microstructure and mechanical properties of porous titanium sintered by Ti fibres. J Mater Sci Mater Med. 2008, 19, 401–05.
[42] Oshida Y, Tominaga T. Nickel-Titanium Materials. De Gruyter Pub, 2020.
[43] Skalak R. Biomechanical considerations in osseointegrated prostheses. J Prosthet Dent. 1983, 49, 843–48.
[44] Hekimoglu C, Anil N, Cehreli MC. Analysis of strain around endosseous implants opposing natural teeth or implants. J Prosthet Dent. 2004, 92, 441–46.
[45] Tomkun J, Nguyen B. Biomechanics of dental implants. https://www.ece.mcmaster.ca/~ibruce/courses/EE3BA3_2010/EE3BA3_2008_presentation08.pdf.
[46] Optimal occlusion and muscles of mastication (2) /certified fixed orthodontic courses by Indian dental academy; https://www.slideshare.net/indiandentalacademy/optimal-occlusion-and-muscles-of-mastication-2-certified-fixed-orthodontic-courses-by-indian-dental-academy.
[47] Elsayed MD. Biomechanical factors that influence the bone-implant-interface. Oral Maxillofac Surg. 2019, 3, doi: 10.23937/2643-3907/1710023.
[48] Jaffin R, Berman C. The excessive loss of Brånemark fixtures in type IV bone: A 5 year analysis. J Periodontol. 1991, 62, 2–4.
[49] Bender SD. Occlusion, Function, and Parafunction: Understanding the Dynamics of a Healthy Stomatagnathic System; http://www.chairsidesplint.com/docs/Occlusion-Function-Parafunction-BENDER.pdf.
[50] Himmlová L, Dostálová T, Kácovský A, Konvičková S. Influence of implant length and diameter on stress distribution: A finite element analysis. J Prosthet Dent. 2004, 91, 20–25.
[51] Browne M, Langley RS, Gregson PI. Reliability theory for load bearing biomedical implants. Biomaterials. 1999, 20, 1285–92.
[52] The Benefits of Saving Your Natural Tooth. 2018; https://yorkhillendodontics.com/benefits-saving-natural-tooth/.
[53] McCormack SW, Witzel U, Watson PJ, Fagan MJ, Gröning F. The biomechanical function of periodontal ligament fibres in orthodontic tooth movement. PLoS ONE. 2014, 9, e102387, doi: https://doi.org/10.1371/journal.pone.0102387.
[54] Manea A, with 16 co-authors. Principles of biomechanics in oral implantology. Med Pharm Rep. 2019, 92, S14–9.
[55] Bidea MW, Misch CE. Clinical Biomechanics in Implant Dentistry. https://dentistrykey.com/library/clinical-biomechanics-in-implant-dentistry/
[56] Oshida Y. Class Note for G911. Indiana University School of Dentistry, Graduate School.
[57] van Rossen IP, Braak LH, de Putter C, de Groot K. Stress-absorbing elements in dental implants. J Prosthet Dent. 1990, 64, 198–205.
[58] Rangert B, Gunne J, Sullivan DY. Mechanical aspects of a Brånemark implant connected to a natural tooth: An in vitro study. Int J Oral Maxillofac Implants. 1991, 6, 177–86.
[59] Lang LA, Kang B, Wang R-F, Lang BR. Finite element analysis to determine implant.
[60] Bahrami B, Shahrbaf S, Mirzakouchaki B, Ghalichi F, Ashtiani M, Martin N. Effect of surface treatment on stress distribution in immediately loaded dental implants – a 3D finite element analysis. Dent Mater. 2014, 30, e89–97.
[61] Pérez-Pevida E, Brizuela-Velasco A, Chávarri-Prado D, Jiménez-Garrudo A, Sánchez-Lasheras F, Solaberrieta-Méndez E, Diéguez-Pereira M, Fernández-González FJ, Dehesa-Ibarra B, Monticelli F. Biomechanical consequences of the elastic properties of dental implant alloys on the supporting bone: Finite element analysis. Biomater Dental Appl. 2016, 1850401, doi: https://doi.org/10.1155/2016/1850401.
[62] Wu Y-J, Hsu J-T, Fuh L-J, Huang H-L. Biomechanical effect of implant design on four implants supporting mandibular full-arch fixed dentures: In vitro test and finite element analysis. J Formos Med Assoc. 2020, 119, 1514–23.

[63] Chang H-S, Chen Y-C, Hsieh Y-D, Hsu M-L. Stress distribution of two commercial dental implant systems: A three-dimensional finite element analysis. J Dental Sci. 2013, 8, 261–71.

[64] Kumar GA, Kovoor LC, Oommen VM. Three-dimensional finite element analysis of the stress distribution around the implant and tooth in tooth implant-supported fixed prosthesis designs. J Dental Implants. 2011, 1, 75–79.

[65] Shirazi HA, Ayatollahi MR, Asnafi A. To reduce the maximum stress and the stress shielding effect around a dental implant-bone interface using radial functionally graded biomaterials. Comput Methods Biomech Biomed Eng. 2017, 20, 750–59.

[66] Oshida Y. Hydroxyapatite – Synthesis and Applications. Momentum Press, 2015.

[67] https://www.kuroda-kyousei.com/o-idou.html.

[68] Watted N, Proof P, Péter B, Muhamad A-H. Medication and tooth movement. Am J Pharm Pharmacol. 2014, 2014, https://www.researchgate.net/publication/292901913_Medication_ and_tooth_movement.

[69] d'Apuzzo F, with 9 co-authors. Review article biomarkers of periodontal tissue remodeling during orthodontic tooth movement in mice and men: Overview and clinical relevance academic editors. Sci World J. 2013, 105873, doi: http://dx.doi.org/10.1155/2013/105873.

[70] Nanci A, Bosshardt DD. Structure of periodontal tissues in health and disease. Periodontol. 2000, 40, 11–28.

[71] Li Y, Jacox LA, Little SH, Ko -C-C. Orthodontic tooth movement: The biology and clinical implications. Kaohsiung J Med Sci. 2018, 34, 207–14.

[72] Hussain A. How Does Invisalign Move Teeth? 2019, https://www.pureorthodontics.ca/blog/ how-invisalign-works-part-two.

[73] Jeon HH, Teixeira H, Tsai A. Mechanistic insight into orthodontic tooth movement based on animal studies: A critical review. J Clin Med. 2021, 10, 1733, doi: https://doi.org/10.3390/ jcm10081733.

[74] Li Y, Zhan Q, Bao M, Yi J, Li Y. Biomechanical and biological responses of periodontium in orthodontic tooth movement: Up-date in a new decade. Int J Oral Sci. 2021, 13, 20, doi: https://doi.org/10.1038/s41368-021-00125-5.

Chapter 9
Dental implants

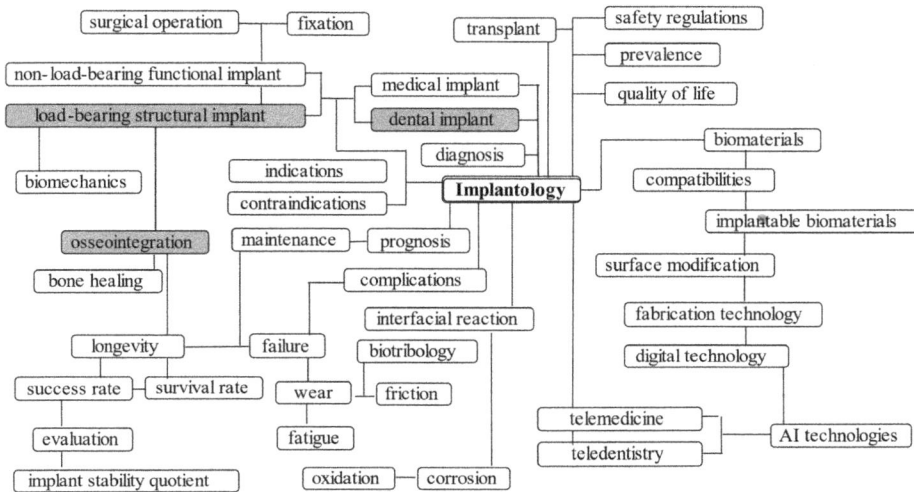

9.1 Introduction

Although both orthopedic implants and dental implants share many important aspects such as biomechanics and biofunctionality as well as required compatibilities (biological, mechanical, and morphological) [1], there are several significant differences between these implant systems. One of them is obviously aesthetics. Besides, there is unique tooth numbering systems. Teeth notation is used by dentists to identify each tooth. This helps in recording relevant data under a standard system as well as clear communication. There are different tooth notation systems, each with their own set of symbols. The most commonly used dental notation systems are Palmer notation method, Universal Numbering System, and International Organization for Standardization system, or Fédération Dentaire Internationale system [2] (see Figure 9.1). A choice out of these three systems depends on the (i) nationality of dentist(s) and place of practicing, (ii) journal or book publisher's policy, and (iii) academic society's policy, to make the clinical as well as academic communications much smooth and easy without further explanation.

Every patient has a reason to visit a dental clinic with chief complaint(s). The chief complaint is a concise statement describing the symptom, problem, condition, and/or diagnosis. Dental chief complaints can be classified into (1) maintenance (for cleaning, follow-up check for previous and ongoing treatments), (2) tooth conditions

https://doi.org/10.1515/9783110740134-010

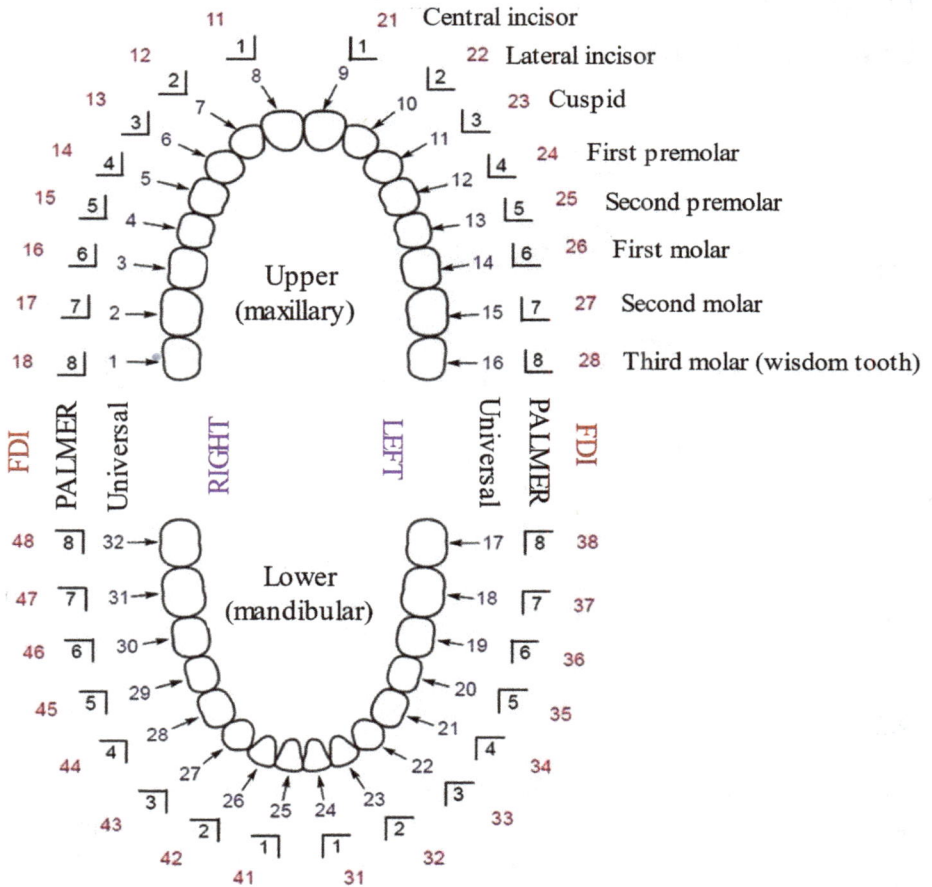

Figure 9.1: Three different systems for tooth numbering [2].

(pain – acute or chronic, toothache, mobile tooth, sensitivity, dental caries, missing tooth, chipping tooth, whitening, rearranging arch-form, etc.), (3) gum condition (pain, bleeding, swelling), and (4) intraoral condition (bruxism, other malocclusions, bad breath, ill-fit of denture, etc.) [3–6]. Besides these clinical chief complaints directly related to treatments, there is still another reason to visit a dental clinic, namely, it is for obtaining or counseling for the second opinion. For all these reasons, it would be much better to use the tooth numbering to communicate between a dentist and a patient. Although these complaints can be heard at the general practice dental clinic, there should be more if the patient specified his/her chief complaint(s) at specified disciplines such as orthodontics, periodontics, endodontics, prosthodontics, or cosmetic dentistry.

According to Grand View Research [7], the valuation of the global dental implant market was USD 3,563.80 million in 2016. It is likely to reach USD 6.82 billion by 2024, progressing at a compound annual growth rate of 7.7% during the forecast period. The results of a survey conducted by the American Association of Oral and Maxillofacial Surgeons in the United States were as follows: (i) over 30 million people do not have teeth in one or both jaws, (ii) 26.0% people aged above 75 years do not have teeth at all, (iii) 69.0% of people aged between 35 and 45 years have one tooth missing, (iv) more than 15 million people have crowns or bridges, (v) approximately 500,000 people undergo dental implant procedure each year, and (vi) dental implants have a success rate of about 95.0% [8]. Dental implants are long-term replacements preserving adjacent teeth. Implanting a tooth is equivalent to receiving a new tooth (teeth). In addition, it is considered as the only restorative technique that preserves and stimulates the natural bone. They also give steady support to dentures (prosthetics). Furthermore, dental implants provide convenience, comfort, and improve patient's appearance (unlike removable dentures). Similarly oral implants have different functions, depending on the tooth number (out of total 8 teeth × 4 quadrants). Roughly, more aesthetic appearance is emphasized in an anterior tooth zone, while biomechanical concerns become more important in a posterior tooth area due to occlusal force.

Several important differences between natural tooth and dental implant were described [90], as illustrated in Figure 8.11. In addition, Froum et al. [10] pointed out that (i) physiologic probing depths are different and natural teeth are shallow (1–3 mm), with strong junctional epithelial attachment at the base, and the biological width is between 2.04 and 2.91 mm, while dental implants possess deep (4–5 mm) and weaker sulcular attachment and the average biological width can be 3.08 mm in average based on abutment length and restoration margin; (ii) with regard to a connective tissue adhesion and mechanical resistance to probing, natural teeth on enamel have with the connective tissue fiber running perpendicularly to tooth surfaces, while dental implants on titanium have the connective tissue fiber parallel and circularly; there is no attachment to the implant or bone.

Here is a typical sequential procedure of the implant treatment:

1. A patient visits with various reasons, including (i) asking for an immediate treatment of chief complaints, looking for a consultation, and/or asking for a second opinion
2. Intraoral examination and systemic evaluation (especially the internal medicine examination)
3. Explanation of detailed implant treatment planning
4. Exchanging the informed consent form

5. Initial treatment
   - treatment decayed tooth
   - extracting unpreservable tooth
   - perform periodontal treatment
   - occlusion treatment and adjustment of occlusion position and height, if necessary
6. Pretreatment medicine administration
7. Main surgical procedure (e.g., two-stage implantation)
   - local anesthesia
   - tooth extraction
   - site preparation
8. Additional surgical procedure
   - bone augmentation, if necessary
   - implant placement
   - suturing
   - checking the keratinized gingiva and adjusting its form
9. Opening for abutment installation
10. Impression taking
11. Superstructure prosthesis
    - fabrication
    - installation
12. Prognosis checking
13. Maintenance and routine follow-up

In the following portion of this chapter, several important actions that are directly related to bioscience and bioengineering of dental implantology will be discussed.

9.2 OHRQoL evaluations for dental implants

Fifty years after the first oral treatment with titanium dental implants, the parameters conditioning osseointegration, such as oral hygiene, occlusal force, and type of implant appear to be well controlled [11, 12]. Much of the research to assess the improved quality of life (QoL) with dental implants uses a construct, the oral health-related quality of life (OHRQoL) survey. The OHRQoL measures oral and dental health, functional well-being, emotional and social health, and expectations and satisfaction with care. OHRQoL is an integral part of general health and well-being. Rates of success and surgical procedures have been described extensively. Furthermore, many studies have explored the efficacy of implant treatment using objective parameters (retention, stability, and mastication parameters): implants have been shown to improve denture stability and retention, consequently improving oral comfort and OHRQoL for patients [13, 14]. Despite its relatively recent emergence over the past few decades, OHRQoL has important implications for the clinical practice of dentistry and dental research [15]. OHRQoL is an integral part of general health and well-being and is recognized by the WHO as an important segment of the Global Oral Health Program [16]. Common dimensions in OHRQoL instruments are given in Figure 9.2 [17], along with specific examples of items associated with each dimension. Patient-oriented outcomes like OHRQoL will enhance our understanding of the relationship between oral health and general health and demonstrate to clinical researchers and practitioners that improving the quality of a patient's well-being goes beyond simply treating dental maladies. OHRQoL is associated with functional factors, psychological factors, social factors, and experience of pain or discomfort [18].

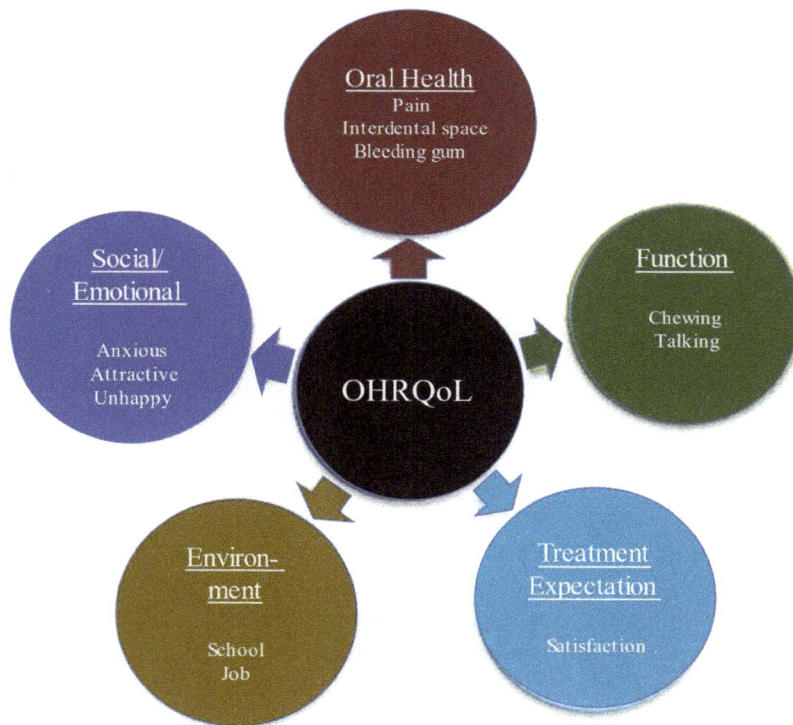

Figure 9.2: Five major factors related to oral health-related quality of life [17].

Fillion et al. [11] analyzed the improvement of OHRQoL of patients who underwent dental implant treatment using the "functional," "psychosocial," and "pain and discomfort" categories of the Geriatric Oral Health Assessment Index (GOHAI). Within a prospective cohort of patients rehabilitated with Straumann dental implants, the OHRQoL of 176 patients (104 women and 72 men) was assessed using the GOHAI questionnaire, at two different times, before and after implant placement. The degree of oral treatment was categorized into three classes: "single tooth implant" ($n = 77$), "fixed partial denture" ($n = 75$), "fixed or retained full prostheses" ($n = 24$). The participants' characteristics (gender, age, tobacco habits, periodontal treatment, and time between both evaluations) were assessed. It was obtained that (i) before treatment, the GOHAI score was lower for participants with fewer teeth, while after treatment, no difference was observed between participants; significant improvements were observed in the GOHAI scores obtained for each of the GOHAI fields studied (functional, psychosocial, and pain and discomfort), regardless of the degree of treatment, (iii) the best improvement was observed in patients who needed complete treatment, (iv) the presence of preliminary periodontal treatment, tobacco habits, age, and gender of the participants did not have a significant impact on OHRQoL, and (v) changing the time between the two evaluations (before and after treatment) had no impact on the changes in the GOHAI score. Based on these findings, it was concluded that implants enhanced the OHRQoL of participants that needed oral treatment.

DeBaz et al. [19] compared the QoL in partially edentulous osteoporotic women who have missing teeth restored with dental implant retained restorations with those who do not and, secondarily, to report the rate of osteonecrosis in this sample. A total of 237 participants completed the Utian QoL survey, a 23-question document measuring across psychosocial domains of well-being including occupational, health, emotional, and sexual domains which together contribute to an overall score. The subset of participants having dental implant-supported prosthesis (64) was compared with the subset having nonimplant-supported fixed restorations (47), the subset having nonimplant-supported removable restorations (60), and the subset having no restoration of missing teeth (66). A significant difference in all QoL domains between the four subsets was found. Although 134 reported oral bisphosphonate and 51 reported IV bisphosphonate use, no signs of the osteonecrosis of the jaw were identified in any participants. It was, therefore, concluded that implant retained oral rehabilitation has a statistically significant impact over nonimplant and traditional fixed restorations, removable restorations, and no restoration of missing teeth in far-reaching areas including occupational, health, emotional, sexual, and overall QoL.

Using various data sources, Nelson et al. [20] included the randomized controlled trials (RCTs), non-RCTs, and cohort studies measuring pretreatment to posttreatment change in OHRQoL score using validated measures. It was concluded that (i) the tooth-supported fixed dental prosthesis (TFDP) and implant-supported fixed dental prosthesis (IFDP) had short- and long-term positive effects on OHRQoL; (ii) removable partial dentures (RPDs) positively affected OHRQoL in the short term, and (iii) IFDP showed greater short-term improvement in OHRQoL than RPD and TFDP.

Yoshida et al. [21] evaluated the changes in OHRQoL during implant treatment for partially edentulous patients and to evaluate the influence of the type of partially edentulous arch. Twenty patients with a small number of lost teeth (fewer than four teeth) who underwent implant treatment were selected. Chronological QoL change during implant treatment was measured. The subjects completed the shortened Japanese version of the Oral Health Impact Profile (OHIP-J14) before the surgery (T0), 1 week after the surgery (T1), 1 week after interim prosthesis placement (T2), and 1 week after definitive prosthesis placement (T3). It was obtained that (i) the total OHIP-J14 score was significantly reduced only at T3, (ii) "physical pain" and "physical disability" scores significantly decreased at T3, and "psychological discomfort" scores also significantly dropped at T2; however, "functional limitation" scores significantly increased at T1, (iii) "psychological disability," "social disability," and "handicap" scores remained the same; on the other hand, in the comparison depending on the type of partially edentulous arch, the total OHIP-J14 score significantly decreased at T3 in the unilateral free-end edentulous space, whereas no significant difference was observed in the bounded edentulous space. Based on these findings, it was concluded that although there is a temporary functional limitation after implant placement in overall OHRQoL improvement was observed after the definitive prosthesis placement, and implant treatment was more effective in the unilateral free-end edentulous space. Peri-implant tissue health is a requisite for success of dental implant therapy. Plaque accumulation leads to initiation of gingivitis around natural teeth and peri-implantitis around dental implants. Peri-implantitis around dental implants may result in implant placement failure. Hence, for obtaining long-term success, timely assessment of dental implant site is mandatory. Alzarea et al. [22] assessed and evaluated OHRQoL of individuals with dental implants using the OHIP-14. A total of 92 patients were evaluated for assessment of the health of peri-implant tissues by recording, plaque index (PI), probing pocket depth (PD), bleeding on probing (BOP) and probing attachment level (PAL) as compared to contralateral natural teeth (control). In the same patients QoL Assessment was done by utilizing OHIP-14. It was concluded that similar inflammatory conditions are present around both natural teeth and implant prostheses as suggested by the result of mean PI, mean BOP, mean PD, and mean PAL, hence, reinforcing the periodontal health maintenance both prior to and after incorporation of dental implants. Influence of implant prostheses on patient's OHRQoL (as depicted by OHIP-14) and patients' perceptions and expectations may guide the clinician in providing the best implant services. Elsyad et al. [23] evaluated patient satisfaction and OHRQoL of conventional denture, fixed prosthesis, and milled bar overdenture for all-on-4 implant rehabilitation. It was concluded that (i) "all-on-4" implant rehabilitation of edentulous mandible with FP and MB achieves high patient satisfaction and OHRQoL compared to complete dentures (CD), (ii) no significant difference in OHRQoL between FP and MB was observed. Regarding VAS, FP rated greater satisfaction with retention, stability, and chewing compared to MB;

however, MB rated greater satisfaction with ease of cleaning and handling compared to FP [23].

Because nutritional status can be related to chewing ability, it is an area in which oral health has a significant impact on QoL. Unfortunately, studies analyzing the influence of OHRQoL on nutritional status have been plagued by significant differences among indicators and assessment tools employed. Using noninvasive tests that eliminated personal preference of foods, Lee et al. [24] investigated interrelationship among chewing ability, nutritional status, and general QoL. It was found that (i) a regression analysis showed that people with better chewing ability had lower scores on the OHIP-14T, meaning they had a better OHRQoL and were less likely to be underweight, and (ii) better chewing ability also correlated with better physical and mental health scores on the SF-36, suggesting that chewing ability is the most significant OHRQoL indicator for both nutritional status and general QoL among the elderly, and steps that restore, maintain, or improve a patient's ability to chew should be taken to promote physical and mental health in this population. Sivakumar et al. [25] assessed the impact of CD therapy on overall OHRQoL in elderly edentulous patients and evaluated the possible role of the patient's initial expectation toward OHRQoL. It was concluded that (i) elderly edentulous patients had an improved overall OHRQoL after CD therapy, (ii) female patients had appreciably better OHRQoL than their male counterparts, and (iii) a patient's initial expectation did not have significant influence on overall OHRQoL. Aarabi et al. [26] described for patients treated with fixed dental prostheses, removable dental prostheses, and CD how they perceived their oral health over a period of 2 years using the concept OHRQoL. It was indicated that (i) patients reported better OHRQoL after conventional prosthodontic rehabilitation, with the positive effect continuing at 2 years, and (ii) while many factors influence a patient's OHRQoL, prosthodontic treatment appears to be a substantial factor in patients' perceptions of their oral health.

Diseases and disorders that damage the mouth and face can disturb well-being and patient's self-esteem. OHRQoL is a relatively new, but rapidly growing phenomenon [27]. Bennadi et al. [27] mention that, since single measures of clinical disease cannot document the full impact of oral disorders [28], the use of sociodental indicators in oral epidemiology has been widely advocated. Several methods have been developed to minimize the complexity and social and cultural relative aspects of QoL as well as to provide indexes capable to capture data beyond the biological and pathological disease process. In general, HRQoL can be determined by two approaches: The first includes an interpretative and qualitative explanatory method and the second, which is the most common approach usually based on the questionnaires that emphasize the subject's perception on physical and psychological health and functional capacity [27, 29]. As we have reviewed earlier, the results obtained by using these instruments are usually reported as a score system, which indicates the severity of the outcome measures or oral diseases [30]. Bennadi et al. [27] concluded with important messages that the OHRQoL can provide the basis for any oral healthcare program and the perception of QoL has a subjective component and therefore varies from one culture to another. Therefore, research at the conceptual level is needed in countries where the OHRQoL has not been described. This is a necessary step because adapting models developed and validated in other cultures could lead to inaccurate measurement of OHRQoL and may not address the important issues pertaining to other culture [27].

A dental implant can replace one or more permanent teeth that have been lost to an injury, gum disease, tooth decay, or infection. The basic implant treatment procedures can be divided into three stages: preoperation, intraoperation, and postoperation.

During a preoperation period, an implant dentist and/or oral surgeon conduct extensive and comprehensive examination on the condition of jawbone (quality and quantity) and the best dental implant procedure (operation strategy for either one-stage or two-stage, implant type and design selection, site preparation, and others). Other prosthetic options such as dentures and bridges should be explained. Contraindications should also be explained. This initial evaluation includes X-rays, taking impressions, and matching the color of your teeth to make your implant look as natural as possible. The implantologist may determine how many teeth you want to replace with implants, which could take some additional planning with other dental specialists, like periodontists, depending on your oral health condition. All medical history, medical current condition, and medications should be discussed. After all discussions and explanations between an implant dentist and a patient, finally the patient can be entitled as an implant patient candidate.

During an intraoperation period, after applying a local anesthetic of novocaine (or lidocaine), if a patient still has a remaining tooth that needs replacing, it should be extracted prior to the dental work. In most of the cases, the implant placement can be done right after the tooth extraction. Special caution should be mentioned about the dry socket creation and plain development. During the implant placement, bone grafting might be done if necessary. Then, abutment placement and final prosthesis installation are followed.

During a postoperation period, regular postsurgery cares for recovery stage should be carefully taken on feeling a groggy from anesthetic and some discomfort experiences, which are natural after the surgery, such as bruising, swelling, bleeding, and pain. The most important is a long-term care. Dental implants typically require the same dental hygiene as natural teeth. To keep them healthy, brush teeth twice a day, floss, and see a dentist for regular follow-up appointments. Dental implants do not get tooth decay, but they can be impacted by periodontal diseases (like peri-implant mucositis or peri-implantitis), so it is crucial to practice good dental care.

Among important items listed earlier, some of them are discussed in different chapters of this book and these are "indication and contraindications" in Chapter 10, "complications" in Chapter 11, and "osseointegration and loading protocol" in Chapter 12. Hence, in this chapter, mainly intraoperation issues will be discussed.

9.3 Three different strategies

In general, depending on how many teeth is missing and how the superstructure is designed and installed to placed implants, there are three main strategies. (1) Replacing a single tooth, using a single dental implant. If you have one missing tooth or multiple that are not adjacent to each other, then a single tooth dental implant may be your best option. However, if you have multiple missing teeth adjacent to each other, then this may not be the best option. (2) Replacing several teeth, using an implant-supported bridge. When you have multiple missing teeth adjacent to each other, you may find that the best option is an implant-supported bridge. Typically, a bridge consists of two crowns on either side of your missing tooth gap with an artificial tooth held by those crowns in between. Now, instead of having the crowns attach to teeth, an implant-supported bridge has crowns that connect to dental implants. The benefits of an implant-supported bridge are you can securely replace multiple missing teeth in a row – without the cost of replacing each tooth. The downside is that not all teeth will receive an implant, and therefore you will lose some bone mass. (3) Replacing all your teeth, using an implant-retain denture. If you are missing a majority or all of your teeth in either upper or lower (aka edentulous), then an implant-retained denture may be the best option. A denture is an artificial arch of teeth. It rests on your gum line and gives you the appearance of a full set of teeth. The problem with traditional dentures is that they are removable, which means they can slip, slide, click, fall out, and make daily tasks uncomfortable like eating and talking. To fix this problem, you can permanently secure denture with dental implants. If you have a lot of missing teeth, this may also be an excellent option to restore your smile and confidence. But the key thing to all these types of dental implants is whether or not you are a good candidate for dental implant surgery [31–33]. In this class, there are still two methods: fixed (screw-in type) implant-supported denture and removable (snap-on-type) implant-supported denture [34, 35]. Figure 9.3 depicts compare and demonstrates these three different usages of implant and subgroups of fixed and removable implant-supported denture plans.

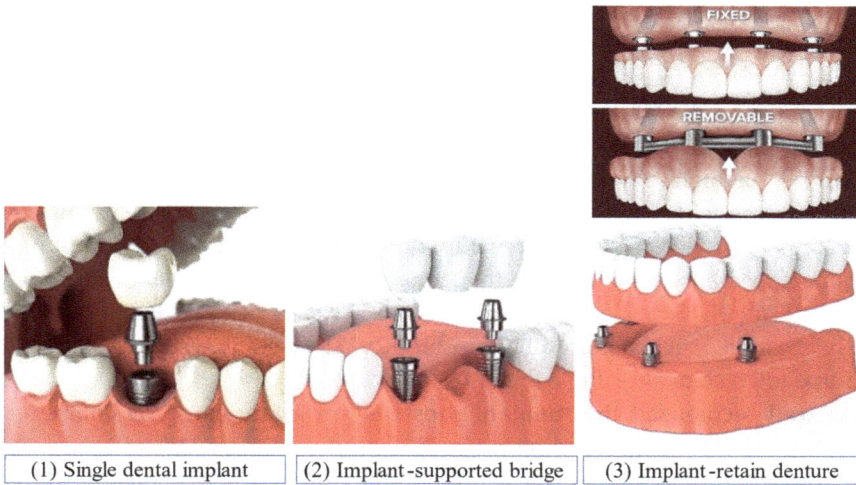

| (1) Single dental implant | (2) Implant-supported bridge | (3) Implant-retain denture |

Figure 9.3: Three main implant strategies [33, 34].

9.4 Materials, types, and designs of dental implant system

9.4.1 Superstructure

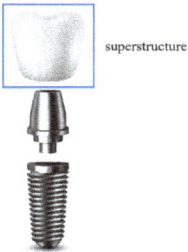

A dental implant is composed of three components: superstructure, abutment, and fixture as shown in an attached figure [36]. Implants can basically be used in the same manner as natural teeth for supporting all types of dental restorations (single crowns, bridges, fixed-removable restorations, and overdentures). A respective restoration is called a superstructure. It can be designed as removable, fixed/removable, or only fixed. Superstructure is one of important factors of successful osseointegration along with bone quality and quantity, implant material and surface, implantation planning, surgery, occlusion, and aftercare.

Various metallic materials and ceramics are available for fabricating superstructures, and selection depends on surface hardness, fracture toughness, formability, and aesthetics. Currently, metallic materials (Ti-based alloys and Co–Cr alloy), ceramics (zirconia, Y-TZP – yttrium tetragonal zirconia polycrystal, alumina, and lithium disilicate), and resin materials (PMMA and composite resin). Although all metallic superstructures can be fabricated by the classic lost-wax casting process because of aesthetic appearance, these superstructures are installed in posterior area. Hence, tooth-colored ceramics and resin superstructures are preferentially employed in the anterior area where the aesthetics become more important [37]. Particularly, zirconia superstructure has been prevailingly utilized due to its good endurance, acceptable appearance, and not ease to plaque and tartar accumulation. Moreover, lithium disilicate ($Li_2Si_2O_5$) is a glass ceramic and is widely used as dental ceramics due to its strength, machinability, and translucency. The color is aesthetically closest to a natural tooth color [38–40].

It is well recognized that implant prostheses are fabricated by computer-assisted design and computer-assisted manufacturing (CAD/CAM) with conventionally fabricated implant prostheses when assessing aesthetics, complications (biologic and mechanical), patient satisfaction, and economic factors [41–45]. All bioceramics mentioned earlier can be fabricated by CAD/CAM technology. Moreover, it was pointed out that (i) CAD/CAM manufacturing helps clinicians and laboratory technicians create implant restorations ranging from single-unit abutments and crowns to multiunit, full-arch complex frameworks in predictable fashion, (ii) milling with CAD/CAM techniques eliminates the distortion and porosity induced by the classic lost-wax casting process, and (iii) the larger the frameworks and the more implants involved, the more obvious CAD/CAM advantages become [46]. The use of materials with elastic properties for the fabrication of dental implant superstructures seems to be a promising way to reduce the functional occlusal forces on implants. A hybrid ceramic material for CAD/CAM technology is available in a special form that can be relatively easily combined with titanium (Ti) base connectors for the fabrication of abutment crowns and mesostructures [47–50].

9.4.2 Abutment

abutment

According a source [37], three components of dental implant system can be divided into the tertiary element (exostructure) for the superstructure and the secondary element (mesostructure) for the abutment, and the primary element (endostructure) for the fixture body. The abutment is the part of a one- or two-phase implant system which is connected to the implant or fixed to it. It is the build-up that protrudes into the oral cavity, which is either directly included into the superstructure or which serves as a connection element between the implant and the superstructure. If a patient has a dental bridge, crowns will be placed on two abutments, connected by other replacement teeth called pontics that rest on top of the gums [51].

A wide variety of abutment materials are available on the dental implant market. A major challenge for clinicians today is understanding the biologic response to each material, as well as the best indication for using each of the different types. The mucosal seal surrounding a dental implant abutment is an essential factor in preventing bacterial penetration into the crestal bone and around the implant neck. A mucosal seal surrounding dental implants is also essential to avoid the peri-implantitis. The biologic width surrounding dental implants also contains a junctional epithelium, followed apically by a connective tissue layer. Unlike the natural dentition, in implant abutments the apical connective tissue fibers do not have the same quality of attachments. The natural dentition has dentogingival fibers running perpendicular to the tooth from the bone to the cementum. The connective tissue layer surrounding a dental implant abutment has fibers running in a parallel fashion. Due to the weakened connective tissue support around implant abutments, the junctional epithelium is believed to be more susceptible to apical migration. In other words, a dental implant is more susceptible to peri-implantitis than a natural tooth is to periodontitis. Hence, it is important to note that this biologic width or the peri-implant seal protects the implant against peri-implantitis and provides an aesthetic result [52, 53].

Abutments are usually made in a dental lab and are most commonly made from titanium, gold, stainless steel, zirconia, resin, or polyether ether ketone (PEEK). Titanium is either machined, polished, or Laser-Lok, and the long-term studies supported favorable soft tissue maintenance with machined or polished titanium. Strongest peri-implant seal permitting improved long-term soft tissue maintenance (comparable mucosal seal to the natural dentition) [37]. All four grades (grades I, II, III, and IV) of unalloyed commercially pure titanium (cpTi) are used as abutment materials. For hardening and coloring purposes, surface layer of cpTi is modified by nitridation [54, 55]. By controlling the partial pressure of treating nitrogen gas, cpTi can be nitride with either golden color TiN or silver color Ti_2N, as shown in Figure 9.4. Ti-based alloys such as Ti–6Al–4V or Ti–6Al–7Nb (which is significantly stronger than cpTi and offers better tensile strength and fracture resistance) can be utilized to fabricate abutments.

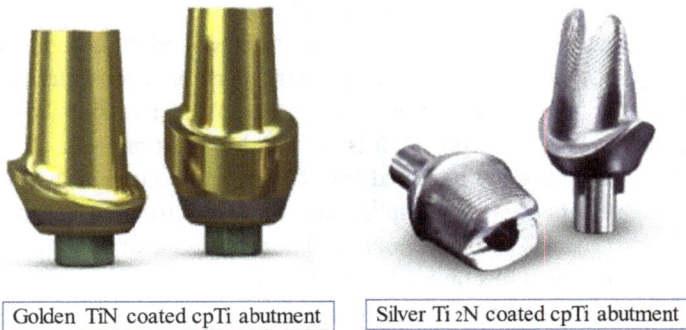

Golden TiN coated cpTi abutment | Silver Ti₂N coated cpTi abutment

Figure 9.4: Two types of nitrided cpTi abutments [37].

Surgical-grade stainless steel can also be employed as an abutment material. Surgical stainless steel is a grade of stainless steel used in biomedical applications, exhibiting an excellent corrosion resistance in biological environment. The most common surgical steels are austenitic SAE 316 (18Cr–10Ni–2Mo) stainless and martensitic SAE 440 (19Cr–1Mo), SAE 420 (12Cr–1Mo), and 17-4 (17Cr–4Ni) stainless steels [56]. SAE 316 and 17-4 stainless steels are known to show as investment castability. Besides these materials, gold metal, zirconia ceramics, and PEEK resin are also used. Traditional cast gold abutments can be used for screw- and cement-retained single crowns and bridges and are available for implants placed at soft tissue level (TL) or bone level (BL). Their advantages consist in the facilitation of the screw retention with bridges. Disadvantages, however, are that gold abutments are technique-sensitive, require more time, and generate higher manufacturing costs. An in vivo histological study in dogs has demonstrated that gold alloys also have disadvantages in terms of soft tissue integration. Histologically, an apical shift of the barrier epithelium and the marginal bone around gold alloy abutments has been shown [57]. Zirconia (ZrO_2) have comparable ability to form a peri-implant seal to that of machined or polished titanium, and most hygienic abutment on the market allows improved long-term maintenance of the peri-implant seal. With the background of the available clinical evidence and systematic reviews, no differences were found between zirconium dioxide and metal abutments in clinical performance based upon aesthetic, technical, or biological outcomes [58–60]. The in vitro studies have shown statistically significant greater wear of zirconium dioxide than that of titanium abutments inside the titanium implant [61]. PEEK is also recognized to be comparable soft tissue and hygienic properties to titanium.

Implant-based prostheses can be cement-retained, screw-retained, or a combination of both. By definition, an abutment is a component that is intermediate between the implant and the restoration and it is usually screw-retained to the implant. The abutment provides the retention, support, stability, and optimal position necessary for the definitive restoration. There are other available options (locking taper, one-piece (1P) implants) where the abutment is part of the implant itself; however, they are less popular as these concepts are less flexible regarding prosthetic restoration options [62]. Besides the straight abutment, there are angulated abutments, which are used when dental implants are not placed parallel to the adjacent teeth or contiguous implants to achieve proper restorative contours. These are normally used for cement-retained single crowns or bridges. The abutments are available with angles of 15°, 25°, 35°, and 45°. The angulated abutments facilitate paralleling nonaligned implants, therefore, making prosthesis fabrication easier. These abutments can also aid the clinician in avoiding anatomical structures when placing the implants. In addition, the use of angled abutments can reduce treatment time, fees, and the need to perform guided bone regeneration (GBR) procedures [63].

Using CAD/CAM technology, abutments can be fabricated with titanium and ceramics (zirconia or alumina) [46, 62]. When compared to conventional abutments, CAD/CAM abutments demonstrate good survival and success rates, provide superior soft tissue reaction (i.e., less recession) [64], and reduce the incidence of screw loosening [65]. Long et al. [66] reported that, overall, CAD/CAM abutments provide comparable, if not better, clinical outcomes than conventional abutments.

The selection of the implant abutment for each individual patient case is an important part of the implant-prosthetic treatment phase [67]. Long-term clinical studies on fixed implant-supported reconstructions show low technical complication rates regarding the abutment itself [68]. Implant abutments are located in a transition zone where they are in contact with the implant and the surrounding peri-implant tissues. Therefore, the choice of abutment is of major importance, especially in a sensitive region like the aesthetic zone. For single-unit reconstructions, zirconia abutments are indicated, which can be either standard or customized depending on the prosthetic position of the implant. For multiunit reconstructions, zirconia abutments are recommended for cement-retained bridges, and gold titanium abutments for screw-retained bridges [67]. It was also mentioned that (i) the clinical indication of each implant abutment type depends primarily on the prosthetic position of the implant and whether single or multiple units need to be replaced; (ii) angulated abutments, individualized CAD/CAM abutments made of titanium, or cast abutments in gold are indicated in cases where the implant is not placed in an ideal prosthetic position; and (iii) in multiunit reconstructions, standard titanium or individualized gold abutments are recommended.

9.4.3 Fixture

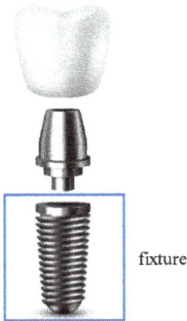

fixture

Macroscopically dental implant (fixture portion) has various body designs, including cylinder type, thread type, plateau type, perforated type, solid type, and hollow type. Furthermore, based on surface conditions and modifications, dental implants can be classified into smooth surface, machined surface, textured surface, and coated surface. Most dental implants can be found in various thread shapes developed for effective inserting and force transmission [69]. Thread shapes available for screw-retained implants include square shape, V-shape, buttress, and reverse buttress threads, which are defined by the thread thickness and face angle [70]. Once an implant is inserted, bone undergoes constant remodeling against external stress, called bone homeostasis (toward osseointegration). While an implant receives an appropriate level of stress and its distribution, the surrounding bone will be remodeling and produces woven bone. However, under extreme adverse stresses, microfractures occur in the alveolar bone inducing the osteoclast genesis [71], leading to the unwanted implant failure. Since it would be not easy to achieve the optimal stress distribution and too little or much stress can induce bone resorption [71, 72], fixture threads should be fabricated to increase the surface contact area and favorable forces while reducing adverse stimuli [69]. Thread design should maximize implant surface area and create a better spreading of stress and primary stability [73]. Like thread shape, pitch is another important geometric factor to determine the bone-to-implant contact (BIC) biomechanical load distribution [69]. In addition to the thread shape and pitch, its depth and width are important design parameters that affect the stress distribution around endosteal implants [69]. While the fixture length is not significant at the crestal bone interface, the surface area of each implant is related directly to the width of the implant. Wider root form implants have a greater area of bone contact than narrow implants (of similar design) resulting from their increased circumferential bone contact areas. Each 0.25 mm increase in implant diameter may increase the overall surface area 5–10% in a cylinder implant body [69]. Kong et al. [74, 75] employed a nonlinear finite element method to examine the effects of implant diameter and length on the maximum von Mises stresses in the jaw, and to evaluate the maximum displacement of the implant–abutment complex in immediate loading models. The implant diameter (D) ranged from 3.0 to 5.0 mm and implant length (L) ranged from 6.0 to 16.0 mm. It was found that (i) the maximum von Mises stress in the cortical bone decreased by 65.8% under a buccolingual load with an increase in D; (ii) in cancellous bone, it was decreased by 71.5% under an axial load with an increase in L; (iii) the maximum displacement in the implant–abutment complex decreased by 64.8% under a buccolingual load with an increase in D; and (iv) the implant was found to be more sensitive to L than to D under axial loads, while D played a more important role in enhancing its stability under buccolingual loads. Based on these results, it was concluded that when D exceeded 4.0 mm and L exceeded 11.0 mm, both minimum stress and displacement were obtained; accordingly, these dimensions were the optimal biomechanical selections for immediate loading implants in type B/2 bone, and the implant diameter is more important than length in reducing bone stress.

Despite the high success rates of endosseous oral implants, restrictions have been advocated to their placement with regard to the bone available in height and volume. The use of short or nonstandard diameter implants could be one way to overcome this limitation. For the purpose of exploring the relationship between implant survival rates and their length and diameter, Renouard et al. [76] conducted a literature review covering the period 1990–2005. Papers were included, which reported (1) relevant data on implant length and diameter; (2) implant survival rates, either clearly indicated or calculable from data in the paper; (3) clearly defined criteria for implant failure; and in which (4) implants were placed in healed sites; and (5) studies were in human subjects (i.e., a total of 53 human studies fulfilled the inclusion criteria). It was found that (i) concerning implant length, a relatively high number of published studies (12) indicated an increased failure rate with short implants which was associated with operators' learning curves, a routine surgical preparation (independent of the bone density), the use of machined-surfaced implants, and the placement in sites with poor bone density; (ii) recent publications (22) reporting an adapted surgical preparation and the use of textured-surfaced implants have indicated survival rates of short implants comparable with those obtained with longer ones; (iii) considering implant diameter, a few publications on wide-diameter implants have reported an increased failure rate, which was mainly associated with the operators' learning curves, poor bone density, implant design and site preparation, and the use of a wide implant when primary stability had not been achieved with a standard diameter implant; and (iv) more recent publications with an adapted surgical preparation, new implant designs, and adequate indications have demonstrated that implant survival rate and diameter have no relationship. Based on these findings, it was concluded that (1) when surgical preparation is related to bone density, texture-surfaced implants are employed, operators' surgical skills are developed, and indications for implant treatment duly considered, the survival rates for short and for wide-diameter implants have been found to be comparable with those obtained with longer implants and those of a standard diameter; (2) the use of a short or wide implant may be considered in sites thought unfavorable for implant success, such as those associated with bone resorption or previous injury and trauma; and (3) while in these situations implant failure rates may be increased, outcomes should be compared with those associated with advanced surgical procedure such as bone grafting, sinus lifting, and the transposition of the alveolar nerve [76].

For specific purposes, there are still mini-implants and orthodontic implants. Mini-dental implants have the same structure as regular implants, but they are smaller. Unlike conventional implants, they are comprised of a 1P screw that is <3 mm in diameter and includes a ball-shaped end that protrudes from the jawbone. While full-sized dental implants are between 3.4 and 5.8 mm wide, mini-implants are only 1.8–3.0 mm in diameter and include a ball-shaped end that protrudes from the jawbone. Since mini-dental implants are smaller, they do not need as much bone density to work. Mini-dental implants are placed through less-invasive techniques than conventional dental implants. These toothpick-sized implants are put over the gum surface when placed into the bone, while conventional implants are placed under the gums [77]. The use of mini-dental implants is implied in cases where conventional implants cannot be placed. When a patient does not have (i) the ability to undergo invasive surgery, (ii) enough time to go to repeat dental visits, and (iii) enough bone in the jaw for a full-sized implant, the mini-implant treatment is indicated [78].

When an orthodontist puts braces on a patient's teeth, they are usually bonding tooth colored or metal attachments on each tooth (called as brackets), then tying the brackets together with thin wires (mostly, superelastic NiTi or stainless steel wire) of various sizes and strengths. Gentle slow pushing, pulling, or rotational forces are then applied to the teeth to move them in a desired direction in a slowly controlled manner. This is a general orthodontic mechanotherapy. There are some orthodontic tooth movements that are more difficult to accomplish than others. Sometimes patients may have missing teeth that would otherwise be used for stable anchorage during treatment. Occasionally, a patient's bone may be less dense in one area than in another. In adults, periodontal bone loss can affect the stability of a tooth needed for anchorage. These and other situations can create circumstances where there is less than ideal anchorage present for the type of tooth movement planned. In these cases, a small skeletal orthodontic anchorage attachment is recommended to help in the efficient orthodontic teeth movements [79]. Orthodontic implants have become a reliable method in orthodontic practice for providing temporary additional anchorage and are useful to control skeletal anchorage in less compliant patients or in cases where absolute anchorage is necessary. A wide range of devices may be implanted in and around the jaws therapeutically, accidentally, or for social reasons, many of which are not endosseous dental implants. The orthodontic implants are used for specific time periods and do not always have osseointegration [80]. Other terms such as miniscrews, miniscrew implants, and temporary anchorage devices have been used [81, 82]. A typical orthodontic anchorage device is shown in Figure 9.5 [83]. Removal of the orthodontic implant is done without anesthesia and twisting in the opposite direction to that used in the insertion. Histological analyses showed that when the implant loading is made immediately after insertion, osseointegration in the implant–bone interface is smaller than when loading is delayed [80]. On the other hand, early loading did not compromise the stability of the orthodontic implants during clinical treatment [82]. Some authors think this phenomenon is favorable because it facilitates the removal after treatment [84, 85].

Figure 9.5: Typical orthodontic anchorage device [83].

While traditional implants are placed in the jawbone, surgeons utilize the zygomatic implant's length to securely place them in the patient's cheekbone (zygoma), which is a very dense bone and provides excellent support for the lifetime of the patient. Severe maxillary atrophy poses a difficult challenge when considering implant-supported dental rehabilitation. For cases of severe maxillary atrophy, zygomatic implants will allow for effective and predictable implant-supported dental rehabilitation without the need for bone augmentation, such as sinus augmentation or GBR [86]. The 12-year survival rate of these implants has been ~95% [87].

Zygomatic implants can be placed in an edentulous arch or at the time of extraction. Zygomatic implants are also useful in complex reconstructions after pathology and trauma. Immediate provisionalization of the implants is preferred because it provides cross-arch stabilization of the zygomatic implants. If immediate loading is not an option, zygomatic implants can be "buried" for 4–6 months to allow for osseointegration before fabrication of the final prosthesis [86]. The traditional placement of zygomatic implants has been intrasinus, with a palatal emergence of the implant head. However, the palatal emergence is less desirable prosthetically because it can increase the palatal prosthesis bulk, negatively affecting the patient's speech and comfort. However, this placement is necessary to keep the implant body within the sinus. Placing extrasinus implants can improve the intraoral emergence toward a more crestal location and also reduces the risk of the development of chronic sinusitis [88]. The most common complication associated with zygomatic implants is sinusitis. Appropriate presurgical diagnostics and evaluation of the sinus as well as using the extra-sinus surgical approach and immediate loading of the implants seem to reduce or even eliminate this complication. Other complications reported during and after the insertion of zygoma implants include infraorbital nerve paresthesia, orosinusal fistula, and perforation of the orbit [80]. Figure 9.6 shows a typical zygomatic implant [89].

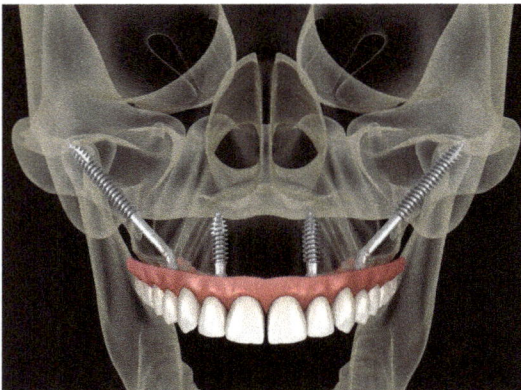

Figure 9.6: Typical zygomatic implant [89].

Biomaterials can be divided, depending on the biological responses, into biotolerant materials (such as Co–Cr alloys, stainless steel, Nb, Ta, PEEK, or PMMA), bioinert materials (such as cpTi, Ti-based alloys, alumina, and zirconia), and bioactive materials (such as HA, TCP, or bioglass) [90]. To select an appropriate material, several important parameters should be taken into consideration. They can include properties that can meet requirements for biocompatibility including biocorrosion resistance, biotribocorrosion resistance, or biotribology-debris nontoxicity. The biocorrosion is not limited to general corrosion, but localized corrosion (such as pitting corrosion, crevice corrosion, and galvanic corrosion) should be added, since the latter type of corrosion is more corrosive due to localized increased corrosion current. Bulk mechanical properties (such as modulus of elasticity (MOE), tensile/compressive/shear strength, yield strength, ductility, fatigue strength and limit, hardness, and fracture toughness) as well as surface properties (such as surface roughness, surface energy, and tension) should be included in the material selection process. Among variety of material candidates for fixture component, currently commonly utilized materials are not many. They include cpTi (grades I, II, III, and IV), Ti–6Al–4V, Ti–6Al–7Nb, alumina (Al_2O_3), and zirconia (ZrO_2), and very recently a new alloy of Ti and Zr (Ti with 13–17% zirconium: TiZr1317) has been developed, claiming for better mechanical properties such as increased elongation and fatigue strength [91]. Najeeb et al. [92] found a potential applicability of PEEK for clinical uses in implant dentistry because of their physical and mechanical properties being close to that of bone (which can achieve the requirement for biomechanical compatibility [93]), along with improved biocompatibility and antimicrobial properties. In addition, Ti and ZrO_2 composite [94–96] has been developed for implant material application, claiming that the composite can possess both properties of metallic Ti and ceramic zirconia [97].

Furthermore, surface layer of metallic dental implants is subjected to modifications to alter the physical and morphological properties accommodate and promote the early osseointegration. Surface engineering can include (1) blasting with various media. Blasting media is mainly alumina powder, but titania or HA powder can be applied. (2) Acid etching is performed on the sand-blasted surface to create a multiscale roughness (i.e., macroscale roughness by blasting and microscale roughness by acid etching). (3) Oxidation treatment can be applied to furtherly create a nanoscale roughness [98]. (4) Mechanical polishing to create relatively smooth surface. (5) Coating substrate layer with mainly HA powder to promote biointegration.

Recently, two subjects have been extensively investigated and discussed: debate on metallic Ti implant versus ceramic ZrO_2 and gradual functionality added implant surface. Table 9.1 compares advantages and disadvantages of cpTi implant and zirconia implant [99–107].

Nanotechnology (aka molecular engineering) has created a revolution in implant dentistry. Various dental products, materials and processes have greatly improved since the introduction of nanotechnology and dental implants are undergoing a similar transformation [108, 109]. As Oshida mentioned [110], there are several important factors controlling surface physiochemistry, and they are surface roughness (macroscale, microscale, and nanoscale), surface material's chemistry, surface topographic configuration, and surface energy, which are all modified and enhanced by the appropriate surface engineering. Nanotechnology for roughening of zirconia dental implant surfaces is still under progress. A significant rise in osseointegration levels compared to machined-smooth zirconia related with enhanced differentiation of osteoblasts was obtained by roughening yttria-stabilized tetragonal zirconia polycrystal (Y-TZP) with mesoscale grooves, microscale valleys, and nanoscale nodules [111], by HA coating [112], or by introducing selective infiltration etching technique (such as coating with infiltration glass, thermal heating, and washing of glass residues) [113].

A concept of the function gradation has been applied to titanium dental implant surfaces. Functionally graded material (FGM) is a heterogeneous composite

Table 9.1: Comparison between metallic cpTi implant and ceramic ZrO$_2$ implant.

	Titanium	Zirconium dioxide	Ref.
Appearance			[99, 100]
Microstructure (note: magnification is not comparable)			[101, 102]
Pros	– Titanium dental implants have been in use for over 65 years – High success rate of 97% for 10-year prognosis – Titanium is a strong material with durability, making implant fractures rare	– As a white material, the aesthetic properties of ceramics are self-evident, especially in patients with a thin or delicate soft tissue biotype or in cases of soft tissue recession – Less mucosal discoloration than titanium	[100, 102, 103, 105, 107]

			[100, 102, 103, 105, 107]
	– Titanium dental implants can last up to 25 years with proper dental care and maintenance – Titanium implants have excellent biocompatibility with bone and gum tissues. It is highly resistant to corrosion and has low electrical conductivity – Titanium implants can come in two-piece varieties, which is helpful if angled implants are needed to correct your implant positioning	– Soft tissue attachment, low inflammatory responses, and osseointegration are similar as Ti materials – Low affinity for attracting and retaining plaque has also been demonstrated as well as less bacterial adhesion than titanium – Hypoallergenic – Highly aesthetic and can be colored to match the natural tooth color – Low bacterial attraction, which can limit the risk of postoperative peri-implant diseases – Biocompatible and resistant to corrosion and wear – High strength and decent fracture resistance – Originally, they were only available as a one-piece implant, but the introduction of two-piece and screw-retained zirconia implants now allows for abutments to be fully customized, creating the best outcomes	
Cons	– Allergy or hypersensitivity can cause discomfort and inflammation. However, it is important to note that allergic reactions appear to be very rare. In one study, just 0.6% of patients displayed reactions to titanium in an allergy test. – For patients with certain autoimmune conditions (such as rheumatoid arthritis, Crohn's disease, or diabetes), metal ions released from the implant can cause local inflammation and irritation	– Low-temperature degradation as the product ages, resulting in the mechanical properties of the material becoming degraded, reducing the strength, density, and toughness of the material – Zirconia implants may have higher failure rates compared to titanium. Most of the failures recorded for two-piece dental implants were due to aseptic loosening	

(continued)

Table 9.1 (continued)

	Titanium	Zirconium dioxide	Ref.
	– For those patients with soft tissues or a receding gum line, the darker metal part of a titanium implant can be visible or show through the gums – Because the titanium screw and abutment are two separate pieces, there is potential for moisture and bacteria to get in between the two pieces of the implant	– If a patient needs any adjustment of fitting of the implant, they should avoid zirconia, as any grinding on the surface of the implant can weaken its fracture resistance	
Advantages	– High success rate (ca. 95%) – Long-lasting (30 years or more) – Strong and durable yet lightweight (high specific strength) – These implants are made up of two pieces and require a much simpler process to place	– Aesthetics due to their tooth-like color, zirconia implants do not have any dark color showing through the gums – No hypoallergic reaction – Lower plaque accumulation around implants – Zirconia is excellent at resisting corrosion – It is a poor electrical and thermal conductor so that there is no concern of a galvanic or battery effects with zirconia implants	[103, 104, 106]

| Disadvantages | – Interference with autoimmune disease
– Galvanic toxicity (metal taste, sensation of electrical charge when near other metals, chronic insomnia)
– Allergic reaction to titanium | – Limited variety of components and designs – zirconia implants are still relatively early in their development cycle. Some critical design improvements such as two-piece screw-retained abutments have only become available in the US market in 2019
– Long-term success – the long-term performance and success of zirconia implants have not been proven.
– Low fracture strength and resistance
– Zirconia implants with a small diameter are prone to fracture
– Not practical for complex oral rehabilitations or implant-supported dentures. Zirconia implants are not ideal when treating patients who are missing all of their teeth or need all of their teeth replaced with implant dentures | [103, 106] |

material including a number of constituents that exhibit a compositional gradient from one surface of the material to the other subsequently, resulting in a material with continuously varying properties in the thickness direction. FGMs are gaining attention for biomedical applications, especially for implants, owing to their reported superior composition. Dental implants can be functionally graded to create an optimized mechanical behavior and achieve the intended biocompatibility and osseointegration improvement [110, 114].

Here we have some biomechanistic issue. Having discussed on mechanistic environment around the osseointegrated implant and vital bone, it would be helpful to compare visually basic mechanical properties of these materials to understand the importance of mechanical mismatching, buffering effect, early and long-term stability of osseointegration by referring to Figure 4.4, which shows yield strength versus MOE of various vital or nonvital materials which we can see intraorally; all are shown in a log–log scale diagram.

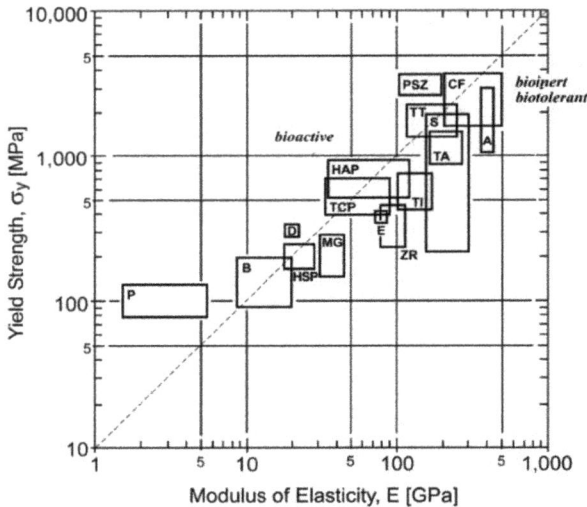

(Figure 4.4 in Chapter 4)

First of all, bone (covering all types of 206 different bones in our body, occupying 12% of entire body weight) can be found about 100–200 MPa on yield strength and 10–20 GPa in MOE value. cpTi and Ti–6Al–4V or Ti–6Al–7Nb alloy (TA) are found in stronger and more rigid zone, indicating that about one decade (in other words, 10 times higher) in both strength and rigidity. There can be found a big gap between bone and titanium materials (TI and TA). Even if, with this big gap, the initial osseointegration (or fusing vital bone tissue into surface zone of the placed implant) were established, the interface at the surface zone should respond to the loading transmitting function. The placed implant and receiving tissues establish a unique [stress–strain] relationship. Between them, there should be an interfacial layer. During the loading with implant/bone couple at BII, the strain-field continuity should be held (if not, it should indicate that implant is not fused to the vital bone), although the stress field is obviously in a discrete manner due to different values of MOE between the host tissue and the foreign implant material; namely, stress at bone $\sigma_B = E_B\varepsilon_B$ and stress at implant $\sigma_I = E_I\varepsilon_I$. Under the continuous strain field, $\varepsilon_B = \varepsilon_I$. However, $E_B \neq E_I$ due to dissimilar material couple condition. If the magnitude of the difference in MOE is large, then the interfacial stress, accordingly, could become so large that the placed implant system will face a risky failure or detachment situation. In other words, if the interfacial stress due to stress difference $\Delta\sigma = (\sigma_I - \sigma_B)$ is larger than the osseointegrated fused implant retention strength, the placed implant will be failed [115].

Accordingly, there have been several ideas proposed and practiced. Hydroxyapatite (HA) coating onto titanium implant has been widely adopted since both HA and receiving vital bone possess similar chemical compositions; hence, early adaptation can be highly expected [116]. At the same time, E_{HA} is positioned in between the values of E_B and E_I; as a result, HA coating will have a second function for mechanical compatibility to make the stress a smooth transfer (or to minimize the interfacial stress). This is one of the typical hindsight, because HA coating is originally and still now performing due to its similarity of its chemical composition to the receiving bone – biointegration [116].

HA deposition is not only a method to minimize the gap between bone and implant. Creating a foam structure at the surface zone of the implant material is an effective alternative technique [117–119]. Since the foam structure, depending on the extent of porosity and pore size thereof, exhibits its mechanical strength reduce down to 10–50% of those of solid structure [120, 121], indicating that the original mechanical gap between the bone and the implant can be remarkably reduced.

Recently, two new implantable materials have been introduced: TiZr alloy [122–124] and zirconia ceramic [125, 126]. The typical chemical composition of TiZr implantable alloy is Ti–15Zr–4Nb–4Ta [122] and exhibits mechanical property range is shown in the above figure as marked with TZ. Although mechanical strength of TZ appears to be similar to that of TA, TZ shows higher MOE value than TA, possibly resulting in creating higher level of interfacial stress than the case of bone and TA as discussed previously. Another new implantable material of zirconia is also seen in the figure, marked with PSZ. As seen clearly, PSZ possesses higher strength as well as rigidity, indicating that the risk of stress discrete situation between the bone and zirconia should be the highest among any combinations foreseen from the figure. In addition, ceramic material is brittle so that surface modification is not easily accomplished such as HA coating or foam-structure texturing.

9.5 Treatment planning

If a patient has not been practicing good oral hygiene care and has been a heavy smoker, or develops diabetics, which are all at greater risk for a variant of gum disease that affects implants called peri-implantitis and increases the chance of long-term failures. Long-term steroid use, osteoporosis, and other diseases that affect the bones can increase the risk of early failure of placed implants. After passing all these patient evaluation, if the patient is considered as a dental implant candidate, the subsequent treatment planning for implants should focus on the general health condition of the patient, the local health condition of the mucous membranes and the jaws, and the shape, size, and position of the bones of the jaws, adjacent and opposing teeth. The long-term success of implants is determined, in part, by the forces they have to support. As implants have no periodontal ligament (as shown in Figure 8.11), there is no sensation of pressure when biting so the forces created are higher. To overcome this, the location of implants must distribute forces evenly across the prosthetics they support. If stress distribution is not nicely controlled, localized stress concentration can then result in fracture of the bridgework, implant components, or loss of bone adjacent to the implant. The ultimate location of implants is based on both biologic (bone type, vital structures, and health) and mechanical factors. Implants placed in thicker, stronger bone like that found in the front part of the bottom jaw have lower failure rates than implants placed in lower density bone, such as the back part of the upper jaw. A patient who has a malocclusion habit can also increase the force on implants and increase the likelihood of failures.

9.5.1 Bone quantity and quality

A final decision for choosing appropriate type of dental implant (including various parameters such as macroscopic design, materials, surface conditions, and dimensions) should be relied on the bone quality and quantity of placement site(s), since osseointegration process is affected by these factors and primary stability and site preparation as well [127–130].

During a masticatory action, mandible and maxilla are moving sharing the temporomandibular joint (TMJ) as a pivot point under corresponding muscle movements as shown in Figure 9.7 [131]. The TMJ, or jaw joint, is a synovial joint that allows the complex movements necessary for breathing, eating, and speech. It is the joint between condylar head of the mandible and the mandibular fossa of the temporal bone. The TJM is defined as a ginglymoarthrodial joint because it has a rotational movement in the sagittal plane and a translation movement on its own axis – this translation movement generates more movement [132].

In general, the mandible plays an important role as a force absorption unit, if the force is applied on anterior area, tensile stress is created at the superior border,

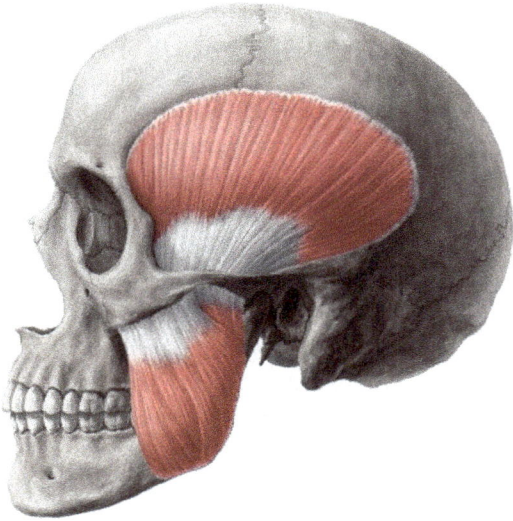

Figure 9.7: Muscles contributed to masticatory movements [131].

and compressive forces will be generated on the inferior zone, while if the load is applied on a posterior area, compressive forces are created at the superior zone and tensile forces are formed at the inferior area [133]. On the other hand, the maxilla possesses a biomechanical function of fore distribution as shown in Figure 9.8 [134].

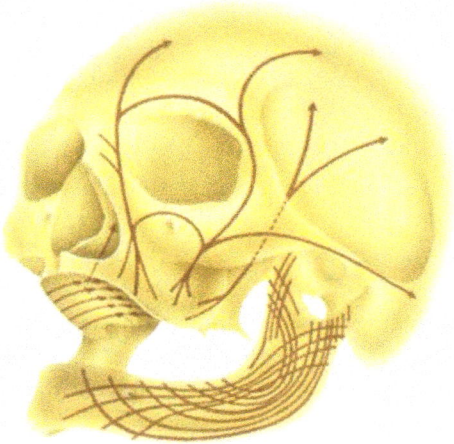

Figure 9.8: Stress distribution contour lines of upper portion of mouth under mastication [134].

The bone density testing has normally several purposes, including (1) to identify decreases in bone density (aka osteopenia) before breaking a bone or a vertebral deformity, (2) to determine your risk of broken bones (fractures), (3) to confirm
a diagnosis of osteoporosis, and (4) to monitor osteoporosis treatment [135, 136]. If
a patient recently had a barium examination or had contrast material (iodine-based
or barium-sulfate compounds) for a CT (computed tomography), nuclear medicine
test, or X-ray test, these contrast materials might interfere with the bone density
test results [135–137]. There are several procedures that can be used to measure
bone density [127, 135–138], including (i) dual-energy X-ray absorptiometry, (ii) sin-
gle-energy X-ray absorptiometry, (iii) ultrasound, or (iv) micro-CT. The test results
will be in the form of two scores [137]: T score (showing the amount of bone, compared with a young adult of the same gender with peak bone mass) and Z score (indicating the amount of bone, compared with other people in your age group and of
the same size and gender). Oshida et al. [139] investigated the surface roughness and
fractal dimension to correlate between these parameters to provide a useful indicator
for dental implant selection task. Two mandibular alveolar bones obtained from
human cadavers (one is from a dentulous case and the other is from an edentulous
case) were section-cut and subjected to roughness measurements and the fractal dimension analysis by the box counting method. It was concluded that (i) cross-
sectioned surface of the dentulous mandible exhibited the fractal dimension (D_F) of
1.81 while the edentulous mandible showed D_F of 1.55, indicating that the former has
more complicated surface texture, and (ii) surface roughness appears to be linearly
related to the fractal dimension, suggesting that D_F can serve an important parameter
for selecting appropriate type of dental implant(s).

There are two classifications for measured bone density: Lekholm and Zarb
(L&B) classification and Misch classification. Table 9.2 complies data sources regarding the L&Z bone classification [140–146]. Table 9.3 summarizes information
about the Misch bone classification [69, 147–149].

There are few reports of L&B bone classification. Seriwatanachai et al. [150] review common techniques and reference used in dental bone classification as well
as the recent reports from histomorphometric analysis and molecular components.
It is well acknowledged that clinical awareness of evaluating the amounts of bone
surrounding the implant site by appropriate method is critical for a successful outcome. It was concluded that (i) accurate and thorough measurement of the jawbone
density is a crucial information to support clinician decision regarding patient selection, implant shape/structure, and surgical technique used; and (ii) although
many techniques commonly used to determine alveolar bone quality, the correlation between radiographic techniques and bone type (L&Z) is the most reliable for
evaluating alveolar bone type. Oliveira et al. [151] evaluated the bone quality of the
maxilla and mandible by using the classification proposed by L&Z and histomorphometry. Sixty edentulous areas were evaluated. The classification by L&Z was obtained through the evaluation of periapical and panoramic radiographs associated

Table 9.2: Lekholm and Zarb bone classification, according to the structure of alveolar bone.

	Type I	Type II	Type III	Type IV
Appearance				
Definition	Implant placement as part of the same surgical procedure and immediately following tooth extraction	Complete soft tissue coverage of the socket	Substantial bone fill of the socket	Healed site
Terminology	Immediate implant placement	Early implant placement	Delayed implant placement	Late implant placement
Time after extraction	Immediately	4–8 weeks	3–4 months	>4–6 months
Clinical findings	Freshly extracted socket	Healed soft tissue	Healed soft tissue and substantial bone healing	Completely healed bone
Characteristics	Almost the entire bone is composed of homogenous compact bone and its resistance fled during drilling is comparable to oak wood	Thick layer of compact bone surrounds a core of dense trabecular bone and the resistance is comparable to pine wood. Considered to be best bone for osseointegration	Thin layer of cortical bone surrounds a core of dense trabecular bone and the resistance is comparable to balsa wood	A thin layer of cortical bone surrounds a core of low-density trabecular bone, and resistance is comparable to styrofoam
Advantages	– Reduced number of surgical procedures – Reduced overall treatment time – Optimal availability of the existing bone	– Increased soft tissue and volume facilitates the soft tissue flap management – Allows assessment of resolution of local pathology	– Substantial bone fill of the socket facilitates implant placement – Mature soft tissues facilitate flap management	– Clinically healed ridge – Mature soft tissues facilitate flap management

Table 9.2 (continued)

	Type I	Type II	Type III	Type IV
Disadvantages	– Site morphology may complicate optimal placement and anchorage – Thin tissue phenotype may compromise optimal outcome – Potential lack of keratinized mucosa for flap adaptation – Adjunctive surgical procedures may be required – Technique-sensitive procedure	– Site morphology may complicate optimal placement and anchorage – Increased treatment time – Varying amounts of resorption of the socket walls – Adjunctive surgical procedures may be required – Technique-sensitive procedure	– Increased treatment time – Adjunctive surgical procedures may be required – Varying amounts of resorption of the socket walls	– Increased treatment time – Adjunctive surgical procedures may be required – Large variation in the available bone volume

CB, cortical bone; TB, trabecular bone.

Table 9.3: Misch bone classification.

	D 1	D 2	D 3	D 4
Appearance				
Description	Dense cortical bone	Porous cortical bone and dense trabecular bone	Thin and porous cortical bone and thin trabecular bone	Thin and fine trabecular bone

Table 9.3 (continued)

	D 1	D 2	D 3	D 4
Bone density (HU)	>1,250	850–1,250	350–850	150–350
Localization	Anterior mandible	Anterior and posterior mandible Anterior maxilla	Anterior and posterior maxilla Mandible	Posterior maxilla
Tactile analogue	Oak/maple	White pine/spruce	Balsa wood	Styrofoam

HU, Hounsfield unit is a relative quantitative measurement of radio density used by radiologists in the interpretation of computed tomography (CT) images.

with the surgeon's tactile perception during milling and implant installation. Before implant installation, bone biopsies of standardized sizes were performed for histological evaluation. It was obtained that (i) type III bone quality was more frequent in the posterior (73.33%) and anterior (73.33%) maxilla, whereas type II bone quality was more frequent in the posterior (53.33%) and anterior (60.00%) mandible; (ii) through histometry, statistical difference was observed for the amount of bone tissue of the posterior region of the maxilla in relation to the anterior and posterior regions of the mandible; (iii) however, there was no difference in osteocyte counts between alveolar regions; and (iv) in the female gender, age showed a low positive correlation with the L&Z classification and in the male gender, a moderate negative correlation was observed. It was hence concluded that both methods detected differences in the bone quality of the alveolar regions of the maxilla/mandible, and the classification by L&Z is a reliable method, since it was consistent with histomorphometry, considering the gold standard method for the evaluation of bone quality, and greater bone density was observed in older men.

Several factors such as implant geometry, preparation technique, and quality and quantity of local bone influence primary stability, and primary implant stability is one of the main factors influencing implant survival rates [134, 152]. Turkyilmaz et al. [153] determined the local bone density in dental implant recipient sites using CT and investigated the influence of local bone density on implant stability parameters and implant success, using a total of 300 implants which were placed in 111 patients between 2003 and 2005. The bone density in each implant recipient site was determined using CT. It was concluded that (i) CT is a useful tool to determine the bone density in the implant recipient sites, and (ii) the local bone density has a prevailing influence on primary implant stability, which is an important determinant for implant success. Jemt et al. [154] reported early implant failures as great as 35%, especially in poor bone quality, after successful surgical survival of implants. Schnitman et al. reported 22% failure in the soft bone of the posterior maxilla and a 0% failure in the good bone of the anterior mandible during a 3-year period, ranging from 100% in the anterior mandible to 78% in the posterior maxilla, as shown in Figure 9.9 [155].

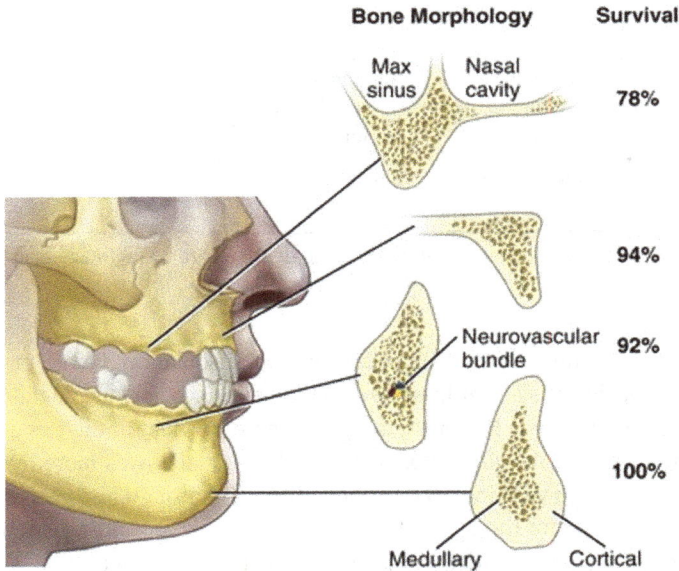

Figure 9.9: The implant survival rate over a 3-year period [155].

Juodzbalys et al. [156] reviewed the classifications for assessment of the jawbone anatomy and evaluated the diagnostic possibilities of mandibular canal identification and the risk of inferior alveolar nerve injury, and aesthetic considerations in the aesthetic zone. It was indicated that the classification system has been proposed, based on anatomical and radiological jawbone quantity and quality evaluation, and the proposed system should be a helpful tool for planning of treatment strategy and collaboration among specialists, as illustrated in Figure 9.10 [156].

9.5.2 Decision for one-stage or two-stage implantation

For proper conducting dental implant treatments, there are various areas where an implant dentist should take into consideration (sometimes through discussion with a patient), and a choice for a dental system is one of them, such as 1P implant or two-piece (2P) implant, or the full mouth reconstruction.

A general comparison between 1P and 2P implant is shown in Figure 9.11 [157, 158]. As clearly shown in the figure, an abutment and fixture body portions consist of one unit to make 1P implant, while a fixture body of a 2P implant is inserted

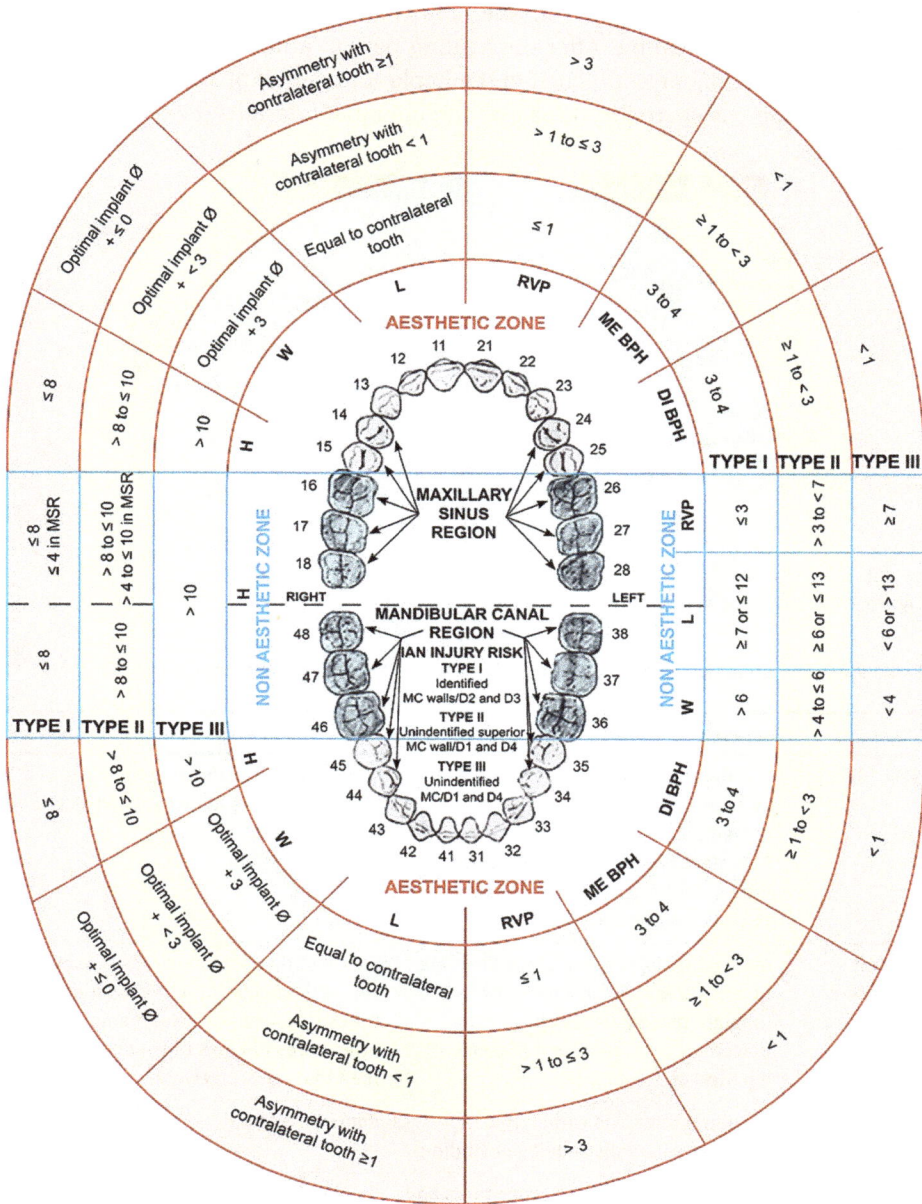

Figure 9.10: Classification system of the jawbone anatomy in endosseous dental implant [156]. H, height; W, width; L, length; RVP, alveolar ridge vertical position; ME BPH, mesial interdental bone peak height; DI BPH, distal interdental bone peak height; MC, mandibular canal; IAN, inferior alveolar nerve; MSR, maxillary sinus region (all linear measurements are in mm).

into the bone after raising a soft tissue flap, which is then repositioned to cover the implant during healing. After the healing period, a new flap is raised, and a transmucosal abutment is attached to the implant to permit placement of a super-structure. Table 9.4 compares these two dental implant designs [159–161].

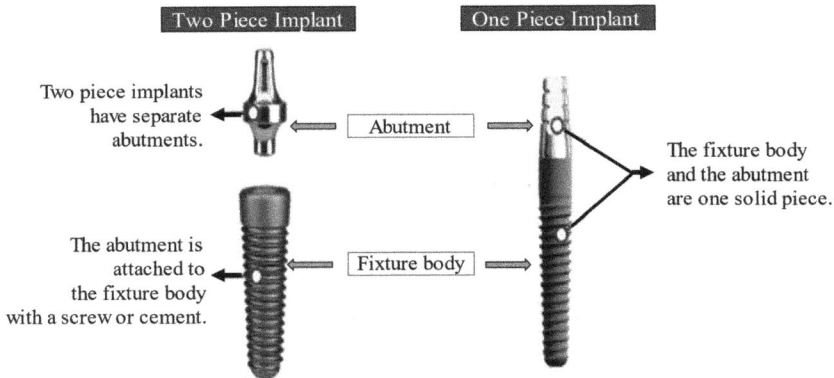

Figure 9.11: Two-piece dental implant versus one-piece dental implant.

Table 9.4: Comparison between two-piece dental implant and one-piece dental implant.

	Two-piece implant	One-piece implant
Basic design	The implant and the abutment are separate. The abutment is either cemented or cold welded. If the abutment is secured with a screw onto the implant, then it can be considered to be three-piece implant	The implant and the abutment are fused – they are manufactured as one-piece implant
Implant placement	More complex surgical procedures are necessary for two or three sittings (implant placement, healing screw placement, and abutment placement) in a period of 3–6 months	Single sitting surgical procedure and very often flapless (no open surgical procedures are necessary). Implant procedures are less time-consuming than that for bridgework
Loading	Loading protocol is varied from immediate loading to delayed loading	Immediate loading
Prosthodontic procedures	Requires more complex procedures and is more time-consuming	Conventional impressions of the implants can be made just as is the case with conventional bridgework. Less time-consuming

Table 9.4 (continued)

	Two-piece implant	One-piece implant
Sizes and designs	Limited sizes and designs are available, thereby limiting their application	A wide range of sizes and designs are available suiting various bone types and measurements. The designs even help avoid bone augmentation and sinus lifts
Pros	If failure of osseointegration can be found early and can recommend a bone graft or another course of action	The one-step dental implant allows to get a smile looking naturally attractive in a short period of time
Cons	Relative long bone healing time would add more time and appointments to the treatment process, including impression taking and superstructure fabrication	One-step implants are already fully in place when the osseointegration happens, so a dentist will need to remove the entire piece in order to potentially fix the underlying jawbone
Advantages	– Applicable in most cases such as those who have thin jaw bones, seeking aesthetics, or with systemic diseases – Depending on the state of the mouth and the course of treatment, abutment can be selected and connected from several types, and the form of the artificial tooth can be changed as necessary – If the implant part is subjected to a strong impact, a structure (failsafe) is provided that can avoid damage to the bond between the implant and the bone by breaking the screw of the abutment	More mechanical durability One-piece implant minimizes bone damage due to no microgap between implant and abutment so that the lack of alveolar bone around the implants is reduced because it is unable to have bacteria
Disadvantages	– If two surgeries are required, the burden on the patient on the body will be applied – Since there are more parts than one piece, the cost is high – Since the process until artificial teeth enter is complicated, it takes more time to treat than one piece – After installing the abutment, the screw connected to the implant may loosen	– Abutments cannot be customized – Implants are so thin that more implants might need to be placed – As crowns are made manually, we cannot use CAD/CAM software; therefore, a precision in these implants cannot be achieved

Table 9.4 (continued)

	Two-piece implant	One-piece implant
Long-term maintenance	Loosening and micromovement can occur between the root portion and the abutment portion. This might be due to the fact that two-piece implants experience higher mechanical stress under oblique loading. Maintenance of these implants are more complex and very often screws (when used) are to be tightened at periodic intervals	Being a single piece, the strength provided by the implant is excellent and there is no separate root portion and abutment portion. Maintenance is very simple and is just the same as that of conventional bridgework

Endosseous, root-form dental implants distribute occlusal stresses into the supporting bone as a function of their overall design and the amount of bone-to-implant interface achieved [162]. Rieger et al. [163] suggested that both high and low stresses can lead to marginal bone resorption. Zamani [162] compared the level of stresses generated by 1P and 2P implant designs in the simulated homogenous bone to determine if load distributions were significantly different. It was concluded that 1P implants create similar stresses to 2P implants in the same length and diameter. Wu et al. [164] analyzed the stresses and strains in both the implant and the surrounding bone when using 1P (NobelDirect) and 2P (NobelReplace) small-diameter implants using both 3D FEM simulations and in vitro experimental tests, with the aim of understanding the underlying biomechanical mechanisms. It was reported that (i) the usage of a 1P small-diameter G-NP implant might increase the stress and/or strain in peri-implant bones and increases the risk of overloading-induced bone loss, and (ii) the mechanical stress in the implant itself is higher in a 2P small-diameter implant.

Abdelwahed et al. [165] evaluated and compared the bacteriological effect of 2P implants and 1P implants in complete overdenture cases on supporting structures. Ten male completely edentulous patients were selected and randomly divided into two equal groups according to the implant design and surgical technique for this study. Group 1: patients were rehabilitated with complete mandibular overdenture supported by 2P implants on each side of the lower arch following two-stage surgical technique; and group 2: patients were rehabilitated with complete mandibular overdenture supported by 1P implants on each side. Evaluation was made at the time of insertion, 6, 12, and 18 months after overdenture insertion, by measuring bacteriological changes around implant abutments. It was found that complete overdenture supported by 1P implants showed better effect on the bacteriological changes as compared to that supported by 2P implants, concluding that complete overdenture supported by 1P implants on each side of the lower arch showed better effect on the bacteriological changes than using the same prosthesis supported by

2P implants. Duda et al. [166] compared the time-dependent outcome of immediately loaded 1P implants with delayed loaded 1P and 2P implants. A cohort of 33 patients is divided into 3 groups: group A, 13 patients, 49 immediately placed and loaded 1P implants; group B, 11 patients, immediately placed, and delayed loaded 1P implants; and group C, 10 patients, 39 2P implants delayed placed, and loaded in a two-stage procedure. Marginal bone loss (MBL) was analyzed using X-ray radiography every 6 months, 1 year, and 3 years. It was obtained that (i) a statistically significant mean MBL was observed between baseline, 6 months, 1 year, and 3 years in all groups; and (ii) there was no statistically significant difference in MBL between immediate and delayed loaded 1P implants. MBL around mandibular implants was lower compared with maxillary implants. It was therefore concluded that 2P implants showed less MBL compared with 1P implants in both maxilla and mandible, and there was no statistical difference in MBL between immediate and delayed loaded 1P implants. Immediate loaded implants show more MBL in maxilla. Systematic review and meta-analysis were conducted to compare the use of 1P versus 2P implants in terms of MBL and implant survival rate, and it was concluded that (i) the meta-analysis did not reveal a significant difference in relation to the implant survival rate, as well as to MBL and (ii) 1P and 2P implants demonstrated effectiveness in the rehabilitation of patients requiring dental implants [167].

9.5.3 Full mouth reconstruction

Dentistry seeks to increase the life span of the functioning dentition. As medicine increases the life span of the functioning individual, the responsibility of dentistry increases proportionately. For this reason, full mouth rehabilitation (FMR: or full mouth reconstruction) of the neglected adult mouth is regarded as of increasing importance, for it is only through this procedure that adult patients with dentitions in varying stages of degeneration can be restored to dental functioning and dental health [168, 169]. A full mouth reconstruction is a highly personalized treatment plan designed to restore complete dental function and oral health, using multidisciplinary various restorative procedures, implant designs, and other issues. To address all areas of decay, damage, or tooth loss, we can combine any number of restorative procedures. As no two patients are the same, each full mouth reconstruction will be unique. Patients who are suffering from multiple oral health issues are typically excellent candidates for a full mouth reconstruction. This includes patients who (i) have missing teeth, (ii) suffer from periodontal disease, (iii) have tooth decay, (iv) experience bite issues, (v) have TMJ disorder, and/or (vi) are looking for better aesthetic appearance. During an initial consultation, a dentist can perform a full oral assessment to determine the patient's eligibility, using 3D CT scans or other advanced images, which provide detailed pictures of patient's teeth, gums, and jawbone. Restorative treatments are designed to replace or repair teeth, so patients can speak clearly, eat comfortably, and smile with confidence. Procedures commonly incorporated into a full mouth reconstruction include onlays, veneers, dental implants, dental bridges, crowns, fillings, and/or dentures. In some patients, an orthodontic specialist may join the treating team to facilitate the best possible outcome [169–177]. Figure 9.12 provides some good idea of the full mouth reconstruction [169].

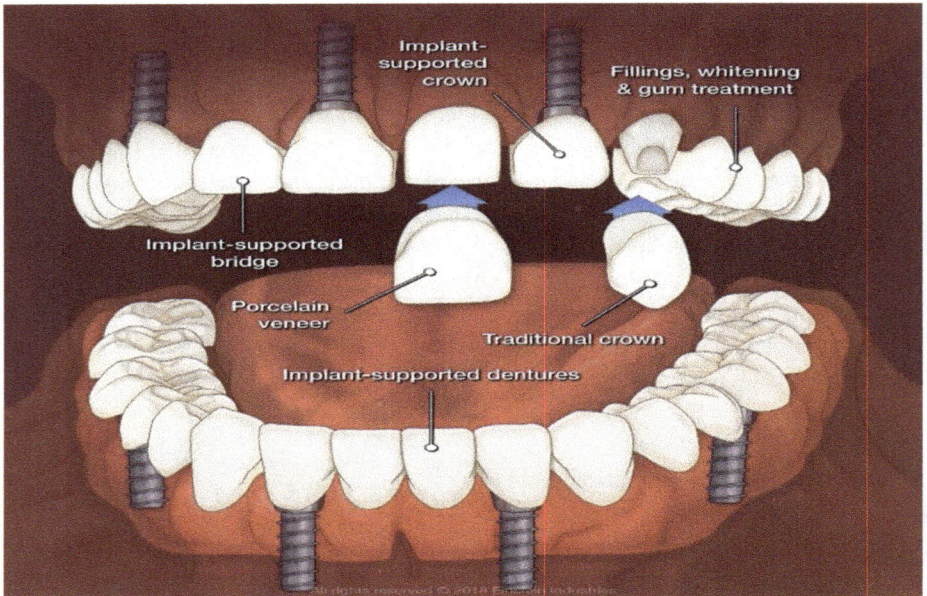

Figure 9.12: A typical intraoral view of case receiving a full mouth reconstruction [169].

Full mouth reconstruction provides numerous advantages to patients who require a significant amount of restorative work including (1) affordability and efficiency: Because all procedures are part of the same treatment plan, a full mouth reconstruction is both convenient and budget-friendly; (2) improved oral health: restorative treatments can eliminate tooth decay and help you manage the symptoms of gum disease. Once your full mouth reconstruction is complete, you can enjoy optimal oral health, and (3) enhanced oral function: comprehensive treatment can help you chew, eat, and speak with ease [169, 170, 177].

Brayman [178] further discussed the medical conditions for FMR indications and the following points can be good reasons for a patient to be evaluated as an FMR candidate. (1) Ectodermal dysplasia: this involves the abnormal development of skin, hair, or teeth. Some patients with ectodermal dysplasia may have missing teeth or defective tooth enamel. Full mouth reconstruction can help people address some of the symptoms of ectodermal dysplasia. (2) Amelogenesis: this is a rare inherited disorder that causes tooth decay and defective tooth enamel. The main symptom is teeth that do not have working enamel, or are missing enamel altogether. Beyond this, patients with amelogenesis may have cracked teeth, accelerated tooth decay, and gum disease. Full mouth reconstruction is one of the best ways to address the complications associated with amelogenesis. (3) Dentinogenesis imperfecta: this is a disorder that can cause teeth to look discolored, often turning yellow-brown or blue-gray. Dentinogenesis imperfecta can also cause teeth to appear translucent. The disorder can both affect infant teeth and adult teeth. Full mouth reconstruction can help address discoloration and replace unhealthy teeth if necessary. (4) Malocclusion: this is a misalignment of the jaws which can result in pain. Full mouth reconstruction can use implants and grafting to realign the jaws and reduce associated pain. Moreover, (5) oral cancer: this is the cancer of the mouth where cells uncontrollably grow. Cancer cell growth can spread to other areas of the body and potentially be fatal if not treated. Oral cancer can present as sores or bumps in the mouth. While oral cancer must be treated by an oncologist, once the cancer has been removed, full mouth reconstruction can help restore the mouth to health. For example, gum grafting can help fix parts of the gums removed during an operation [178].

Some case reports after FMR treatments will be introduced here. Oral rehabilitation for a patient with severe loss of alveolar bone and soft tissue resulting from severe periodontitis presents a challenge to clinicians. Bencharit et al. [179] treated such a patient by FMR with implant-supported fixed prostheses. They demonstrated that early placement of implants (3 weeks after extractions) with minimal bone grafting may be an alternative to conventional bone grafting followed by implant placement, and primary stability during implant placement may contribute to this outcome. In addition, composite resin gingival material may be indicated in cases of large fixed implant prostheses as an alternative to pink porcelain. Song et al. [180] reported on a 77-year-old female, who had the loss of anterior guidance, the severe wear of dentition, and the reduction of the vertical dimension. Occlusal overlay splint was used after the decision of increasing vertical dimension by anatomical landmark, facial and physiologic measurement. Once the compatibility of the new vertical dimension had been confirmed, interim fixed restoration and the permanent reconstruction was initiated. This case reports that a satisfactory clinical result was achieved by restoring the vertical dimension with an improvement in aesthetics and function. Shimizu et al. [181] reported for the case of a patient with bilateral mandibular condylar neck fractures that resulted from a traffic accident, and that the conservative treatment was performed before fixed implant-supported prostheses were placed in the maxilla for occlusal reconstruction. The report included follow-up observations at 5 years and 4 months after the placement of the final implant superstructure and at 6 years and 8 months after the bilateral mandibular condylar fractures were incurred. Zeighami et al. [182] reported the FMR of a 36-year-old bruxer with severely worn dentition and other dental problems such as unfavorable restorations. A diagnostic work-up was performed and provisional restorations were made; then, they were clinically evaluated and adjusted based on the criteria dictating aesthetics, phonetics, and vertical dimension. After endodontic therapy, clinical crown lengthening was performed. Two short implants were inserted in the posterior mandible. Custom-cast dowel cores and metal-ceramic restorations were fabricated, and a full occlusal splint was used to protect the restorations. Stable contacts on all teeth were ensured with equal intensity in centric relation and anterior guidance in accordance with functional jaw movements. Moreira et al. [183] described the management of a 70-year-old man with a history of bruxism and excessive wear, loss of the occlusal vertical dimension, limited space for restoration, aesthetic complaints and compromised dental function due to reduced tooth structure. A multidisciplinary approach was applied with tooth and implant-supported full-ceramic restorations. The patient used two full arch provisional bridges during the osseointegration of the dental implants. Maxillary and mandibular teeth and implants were restored with monolithic zirconia crowns with feldspathic veneers. An occlusion mouth guard was given to protect the restorations. After 36 months of function, no major complications were registered. The restoration of worn dentition in cases of bruxism requires proper planning and a multidisciplinary approach in order to ensure the prognosis and the success of prosthetic treatment. Partially veneered monolithic zirconia appears to be a reliable treatment option with satisfactory clinical results and minimal technical complications [183].

All-on-4 dental implants are a full set of implants designed to replace entire maxilla or mandible set of teeth. They act as a more permanent and natural-looking alternative to dentures. The implants offer several benefits that allow patients to enjoy the smile of a lifetime, including (i) entire set of missing teeth can be replaced in just one day with only one surgery, (ii) the implants permanently secure to a mouth, so they would not loosen or shift like dentures, (iii) only four implants are needed to restore the entire set of teeth, (iv) all-on-4 implants cost less than replacing each tooth individually, (v) all-on-4 dental implants are a bone graftless, cost-effective solution that provides patients with a fixed full-arch prosthesis all in the same day, (vi) restore the full chewing function, and (vii) aesthetically pleasing. On the other hand, some potential drawbacks which a patient should consider before choosing all-on-4 implants are as follows: (i) a patient will need to have a high level of bone available to hold the implants in place, (ii) it might be difficult to test how the teeth feel or look before your procedure, and (iii) the implants cannot be placed in the mouth's molar area, where you have the greatest bite force [184–189].

Figure 9.13 shows all-on-4 for maxilla case which is looked from inside the mouth [189] and all-on-4 for mandible case is viewed from outside the mouth [190].

Hodges [192] pointed out five important factors that are concerned as a regular basis for implant treatment. (1) Speech issues: speech issues are a major concern for patients. Consequently, this can have an impact on speech sounds such as "D," "T," and "N," as the tongue contacts the hard palate lingual to the central incisors

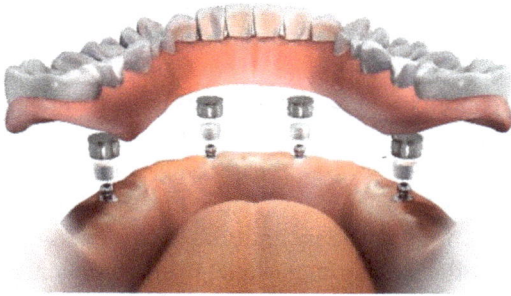

All-On-Four for mandible,
viewed from intraoral direction

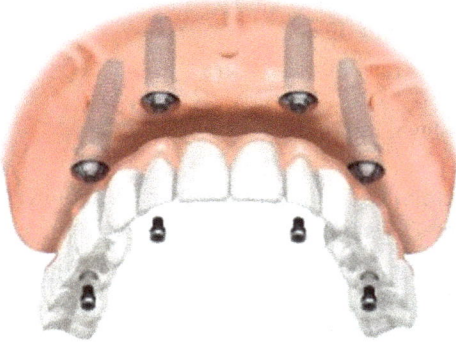

All-On-Four for maxilla,
viewed from extraoral direction

Figure 9.13: Two examples of all-on-4 cases [190, 191].

to form the sound. Likewise, horizontal bulk in the posterior can affect the "S" sound, where the lateral borders of the tongue flare upon making the sound. All-on-4 patients need to be made aware of this compromise. (2) Difficulty adapting to bridge thickness: It can be very difficult for patients to adapt psychologically to the different feel of an all-on-4 bridge. They are used to feel the transition from their soft tissue to their teeth. With the all-on-4 bridge, patients feel real soft tissue and then fake soft tissue (the tissue portion of the bridge) prior to transitioning to the teeth. This can be a big problem for some patients, and sadly they often figure out that it is a problem after surgery when the ridge has been reduced. At that point, nothing can be done to get them back to that natural-feeling transition. (3) Proprioception: this is a crucial issue. A dental implant does not have a periodontal ligament, which might make teeth sensitivity to low forces (<1 to 4 N), and it takes approximately 10 times more force to register the same proprioception as a tooth. The proprioception associated with a dental implant is similar to that of a tooth affected by local anesthesia. As a result, patients will have difficulty recognizing premature or excessive occlusal contacts. They may generate excessive biting forces due to the lack of feedback. Patients with dual-arch all-on-4 restorations are more likely to bite excessively than patients with the remaining teeth. This can result in fractures of the restoration or bone loss. (4) Parafunction: parafunctional habits (such as bruxism, clenching, and irregular chewing cycles) may impact your treatment decision. Since natural teeth can detect forces much more readily than implants, it might be better to include a few saved natural teeth in a treatment plan so that the patient may enable to sense the parafunction better, which will increase the chances of them modifying the negative behavior. Finally, (5) high caries index: it is also important to notify a patient about the high risk of developing caries on the remaining teeth since an evidence that a patient needs an implant treatment obviously implies that the patient might be prone to develop caries soon or later on the remaining teeth [192].

9.5.4 Tissue-level implant versus bone-level implant

Another important variable in dental implant selection is whether we should use a TL or a BL implant. Figure 9.14 compares TL implants and BL implants [193].

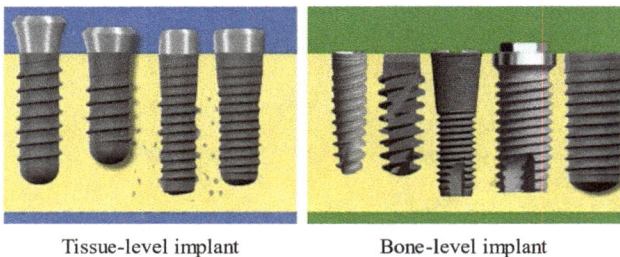

Tissue-level implant   Bone-level implant

Figure 9.14: Tissue-level implants and bone-level implants [193].

The TL implants (represented by the original ITI/Straumann implants) are not placed at the crest of the bone but a few mm above and in some cases where aesthetics is not critical will be even at the level of the mucosa. These implants are referred to as 1P implants. A TL implant is easier to clean and maintain for the patient because the connection between the implant screw and the tooth restoration sits at the level of the gum line which is easy to access with a normal toothbrush. The only disadvantage of a TL dental implant is that the silver metal collar of the implant "may" be visible around the gum line which would provide an aesthetic concern. BL implants are implants that are placed flush with the crest of the alveolar bone or even slightly below and this was a classical surgical feature of the traditional Bråne-mark implants, countersinking was necessary then to sink the slightly wider im-plant head within the bone and have the cover-screw placed without protruding through the mucosa after flap closure. The literature usually refers to these im-plants as 2P implants as a transmucosal part (abutment) will then be connected to the implant to protrude through the mucosa to carry the restoration. A BL dental im-plant is advantageous because it sits approximately 3–6 mm below the visible gum line which means that the connection between the implant screw and the restoration sits well below the visible gum margin. For this reason, these implants provide a more predictable and consistent aesthetic outcome for patients. The downside to this feature is that the connection between the implant and the restoration is harder to clean and maintain as it is below the gum margin and therefore harder to access with a normal toothbrush. Despite this, with thorough daily oral hygiene using flossing and brushes, good cleaning can be maintained [193–197].

There are comparative studies between TL implants and BL implants. Kumar et al. [198] compared the amount of MBL in a BL and a soft TL implant system, both of which have similar intra-bony shape and surface composition. It was concluded that (i) BL implants had statistically significant lesser MBL as compared to TL in time periods above 12 months; (ii) although the difference is statistically significant, the difference may not be clinically significant; and (iii) the initial depth of implant placement had an influence on the amount of MBL, with deeper placed implants and screw structure of the implant placed below the bone, having more MBL in the period of study. After the literature review, Parke et al. [199] found that (i) current literature reveals 50% less MBL surrounding BL implants as compared to TL im-plants; (ii) microbiological and immunological studies revealed differences between the environment surrounding healthy implants and teeth; (iii) of significance, nu-merous species common to subgingival plaque are found in the sulcus adjacent to otherwise healthy implants; (iv) there is evidence indicating increased levels of proinflammatory mediators, as well as anaerobic bacterial populations around im-plant sites as compared to natural tooth sites; (v) furthermore, the literature reveals lower levels of MBL and lower incidences of peri-implantitis in patients that ad-hered to 6-month implant hygiene protocols; and (vi) in a randomized control trial, TL possessed 1.1 mm more keratinized tissue than BL on average. It was then con-cluded that BL implants preserve more marginal bone compared to TL implants. The implant–abutment junction for BL implants is closer to the crest of alveolar bone and consequently favors pathogenic anaerobic bacteria that cause the upregu-lation of host inflammatory cells following the final restoration; however, TL im-plants present with transgingival access for periodontal pathogens to elicit release of host's proinflammatory cytokines that disrupt osseointegration during implant healing periods. TLs have more keratinized tissues than BL surrounding the im-plant, and the presence of keratinized tissue has been associated with improved peri-implant hygiene. Hadzik et al. [200] evaluated implantation effectiveness for

BL and TL short implants provided in lateral aspects of partially edentulous mandible and limited alveolar ridge height. It was reported that (i) the MBL of BL implants was significantly lower compared to TL implants, (ii) BL implants had greater primary and secondary stability in comparison with TL implants (primary: 77.8 ISQ versus 66.5 ISQ; secondary: 78.9 ISQ versus 73.9 ISQ, respectively), (iii) since short BL implants showed a significantly decreased MBL 12 and 36 weeks after implantation as well as better results for the primary stability compared to TL implants, they should preferentially be used for this mentioned indication. Canullo et al. [201] assessed the outcomes of crowns designed as per the biologically oriented preparation technique cemented on conical titanium abutments on TL and BL implants. It was obtained that (i) a statistically significant difference was found in the mean ± standard deviation bone resorption between TL implants (0.38 ± 0.46 mm) and BL implants (0.83 ± 0.58 mm), and (ii) higher values for both PES and WES were obtained in the TL implant group, concluding that (iii) TL implants with a conical transmucosal portion seem to provide a suitable alternative to BL implants in the anterior area.

Based on what has been reported in individual manuscripts and case reports in the above, a generalized rule can be found. In most cases, we find that TL implants are ideal to replace teeth at posterior area where access for daily cleaning with flossing is generally more difficult than the anterior area of the mouth. A possible silver collar around the gum line at the back of the mouth is not a disaster as it cannot be seen in normal function. On the other hand, the BL implants are used in most cases at the anterior zone because of the better aesthetics and also the fact that the front of the mouth is easier to clean more thoroughly due to the ease of access. Miyazaki sets forth a treatment policy in such a way that for both maxilla and mandible aesthetic zones (central incisor, lateral incisor, and cuspid) BL implants are placed, for first premolar (which is considered as semi-aesthetic zone), either TL or BL implant can be used, while for posterior zones (second molar to molar), basically TL is employed. For overdenture cases, TL implants are selected since aesthetic issue is not considered to be essential [202].

9.6 Main surgical procedures

Dentists and oral surgeons perform tooth extractions for many reasons, including for painful wisdom tooth, a tooth that has been badly damaged by decay, a tooth to make space for dental prosthetics or orthodontic mechanotherapy, or implant treatment. For, especially, implant treatment, detailed examination in terms of remaining bone quality and quantity at the postextracted socket area will be crucial to determine subsequent success of osseointegration and resultant survival rate.

9.6.1 Tooth extraction

Selection of right type of tooth extraction methods (such as a simple tooth extraction, impacted tooth extraction, or others) depends on the tooth's shape, size, position, and location in the mouth. Following necessary preparations including administration of antibiotics if necessary, anesthesia (normally local one by nitrous oxide, also known as laughing gas, an oral sedative medication, intravenous, or IV, sedation, or general anesthetic), a surgical extraction operation can proceed. Postoperational cares include appropriate controlling pain and swelling [203].

One complication of tooth extraction is dry socket. This is not an infection. After a dentist removes a tooth, a blood clot usually forms where the tooth was. The clot protects the underlying bone, tissues, and nerves as the site heals. In some cases, however, the blood clot does not form or becomes dislodged, leaving the bone and nerves exposed. This is known as dry socket, or alveolar osteitis [204]. The symptoms of dry socket can vary but may include (i) severe pain at the site of the extraction, (ii) a missing blood clot at the extraction site, (iii) visible bone at the extraction site, (iv) a foul smell coming from the mouth, (v) a bad taste in the mouth, and (vi) pain radiating from the tooth socket to the ear, eye, temple, or neck on the same side [204]. Infection is another complication, and it can occur when bacteria infect the gumline in and around the socket within 1–2 days after surgery.

Although more than 98% of the time the implant treatments are very successful and work for the life of the patient, there can be complications with dental implants. Implants are made mostly of metallic substance but under extreme biting forces even a metal can break. The implant itself can break, get infected, or get loose, and therefore, may need to be removed by a qualified surgeon. In this case, placed implant is needed to remove and usually place another larger sized implant at the same time if the area is evaluated to be in perfect condition and there must be suitable healthy bone; being similar to having a tooth removed and an implant placed at the same time [205].

The best timing for the postextraction implant placement depends on how much bone and tissue structures are present. In general, there are three distinct timings for implant placements [205, 206]. (1) Immediate placement. If there are sufficient bone amount present around the extraction site to stabilize the dental implant, placing an implant can be done immediately after extracting the tooth. It will take normally 3–6 months to establish the osseointegration, followed by installing prosthesis. If the bone is healthy, the immediate placement strategy can work for the full mouth reconstruction. For example, the all-on-4 system is selected, and four implants are immediately placed. (2) Early placement. Early implant placement refers to placing the implant 2–3 months after extraction, during which the bone heals, and the site is better prepared. Early implants must be placed within this timeframe since two-thirds of bone resorption occurs within the first 3 months after extraction. It might need to wait for the implant to heal and bond with the bone before placing the prosthetic tooth, which usually requires 3–6 additional months. (3) Delayed placement. If there is an insufficient amount of bone at the extraction site, the bone augmentation is required before placing the implant. After the procedure, it requires 3–6 months to heal, and then the area will be ready for the implant. An additional 3–6 months should be waited before a permanent prosthetic crown is installed.

9.6.2 Site preparation

Implant sites should be prepared using gentle, atraumatic surgical techniques with a constant reminder to avoid overheating the bone. Conventional implant site preparation by rotary drilling of the bone without proper cooling might generate debris, increased temperature, bleeding, and hematoma formation that could injure the bone and increase the risk of implant failure. Sakuma et al. [207] evaluated histomorphometrically early healing at implants placed in sites prepared with either a sonic device or conventional drills. Sixteen volunteer patients were recruited. Two titanium mini-implants were placed in the distal segments of the maxilla in recipient sites prepared with either a sonic device or conventional drills. Biopsy specimens containing the mini-implants were retrieved after 2 weeks in eight patients, and after 6 weeks in the other eight patients. Histomorphometric analyses were performed. It was obtained that (i) histologic slides were available from seven patients for both 2-week and 6-week periods; (ii) after 2 weeks of healing, small amounts of new bone were found in contact with the implant surface, with 5.5% ± 7.3% and 3.8% ± 10.0% at the sonic and drill groups, respectively; (iii) after 6 weeks of healing, new bone was 46.9% ± 15.5% at the sonic group, and 46.4% ± 14.9% at the drill group, and (iv) none of the differences was statistically significant, concluding that (v) the percentage of new bone in contact with the implant surface was similar in the sonic and drill groups. An alternative for site preparation is piezosurgery, an ultrasonic vibrating cutting technology, which achieves accurate cuts while avoiding the disadvantages of conventional drilling [208–210]. Vercellotti et al. [211] reported on the multicenter study, indicating that piezoelectric preparation of more than 3,579 implants has proven successful in a wide range of indications. All implants were placed into 1,885 subjects and it was found that (i) no surgical complications related to the implant site preparation were reported for any of the implant sites, (ii) 78 implants (59 maxillary, 19 mandibular) failed within 5 months of insertion, for an overall osseointegration percentage of 97.82% (97.14% maxilla, 98.75% mandible), and (iii) three maxillary implants failed after 3 years of loading, with an overall implant survival rate of 97.74% (96.99% maxilla, 98.75% mandible). Stacchi et al. [212] investigated the stability changes of implants inserted using traditional rotary instruments or piezoelectric inserts, and to follow their variations during the first 90 days of healing and concluded that (i) ultrasonic implant site preparation results in a limited decrease of ISQ values and in an earlier shifting from a decreasing to an increasing stability pattern, when compared with the traditional drilling technique and (ii) from a clinical point of view, a favorable effect of piezoelectric preparation has been identified on osseointegration, which results in an earlier transition from primary to secondary implant stability. Recently, Miyazaki and his research groups [98, 213, 214] reported that (i) UV surface alteration and enough blood supply by the piezosurgery preparation exhibited synergistic effects on improvement of ISQ scales, indicating that these dual techniques appear applicable to implant treatments, (ii) it is important to keep thin bone on buccal side, and (iii) it is also important to keep continuity of flatness and attaching position of periodontal membrane on the remaining tooth.

Besides these site preparation techniques, there are still newly developed methods. The recently introduced technique of osseodensification (OD technique) for dental implant involves the use of special drills (Densah) run in a counterclockwise direction at the osteotomy site. It is claimed that this causes expansion of the osteotomy site and increases density of the bone in immediate vicinity of the osteotomy [215, 216]. Article review was conducted on the primary stability attained using this drilling technique. As a secondary finding, the BIC and the bone area fraction occupancy (BAF) were also compared between the conventional drilling protocol and the OD protocol, among these articles. A Systematic search was performed in PubMed-Medline, Embase, and Google Scholar for clinical/animal studies up to November 2018 [215]. A total of 12 articles, from a database of 132 articles, consisting of 8 animal histologic studies, 2 human based clinical studies, 1 case series, and 1 case report were assessed. About 10/12 articles measured the insertion torque values, 7/12 articles measured the BIC, and 6/12 articles estimated the BAF between the two techniques. Quality assessment of 8 studies performed using AR-RIVE (Animal Research: Reporting of In Vivo Experiment) guidelines showed that 6/8 studies had a high score. It was noted that an average increase in the insertion torque, BIC and BAF, was noted in the OD group as compared to the conventional drilling group. Since most of these studies are nonclinical, it can be inferred that OD is an efficient way to enhance primary stability of implants in low-density bone in an animal model. However, extrapolation to long-term clinical success cannot be ascertained until further evidence becomes available [215]. Trisi et al. [217] evaluated the OD surgical technique for implant site preparation that could allow to enhance bone density, ridge width, and implant secondary stability. The edges of the iliac crests of two sheep were exposed and ten 3.8 × 10 mm Dynamix implants (Cortex) were inserted into the left sides using the conventional drilling method (control group). Ten 5 × 10 mm Dynamix implants (Cortex) were inserted into the right sides (test group) using the OD procedure (Versah). After 2 months of healing, the sheep were killed, and biomechanical and histological examinations were performed. It was found that (i) no implant failures were observed after 2 months of healing, (ii) a significant increase of ridge width and bone volume percentage (approximately 30% higher) was detected in the test group, and (iii) significantly better removal torque values and micromotion under lateral forces (value of actual micromotion) were recorded for the test group in respect with the control group. It was then concluded that OD technique used in the in vivo study was demonstrated to be able to increase the bone volume percentage around dental implants inserted in low-density bone with respect to conventional implant drilling techniques, which may play a role in enhancing implant stability and reduce micromotion.

Proper angulations and bone preservation while preparing the osteotomy site for implant placement is mandatory for the successful and long-term success of implants [218]. Drills should be used in such a way that excellent primary stability is achieved [219]. Rai et al. [219] evaluated the effectiveness of trephine drills in preparation of osteotomy for dental implant recipient site. Trephine drills are used very frequently in implant dentistry nowadays. Forty patients were divided into 2 groups (I and II) of 20 each and group I patients treated with conventional drills and group II with trephine drills. It was reported that, in order to preserve the bone during osteotomy, (i) the trephine drills provide better primary stability and good long-term results, as compared to conventional drills. Originally, the trephination is the surgical procedure in which a hole is created in the skull by the removal of circular piece of bone. Stajic et al. [220] reviewed five different explantation techniques for the removal of failing implants among the bur forceps, neo bur elevator forceps, trephine drill, high-torque wrench, and scalpel forceps techniques and reported that the high-torque wrench technique appeared to be the most elegant technique with the highest predictability for insertion of another implant. Suriyan et al. [221] evaluated the accuracy of implant placement using the novel guided trephine drill protocol with and without a surgical sleeve and concluded that (i) the trephination-based, guided implant surgery protocol produces accurate surgical guides that permit guided surgery in limited vertical access and with the same guided surgery protocol for multiple implant systems, and (ii) guided sleeves, although not always necessary, improve depth control and reduce surgical time in implant placement. Deeb et al. [222] compared the time and accuracy of implant-site preparation and implant placement using a trephine drill versus a conventional drilling technique under dynamic navigation and mentioned that (i) there was no significant difference in time or accuracy between the trephine and conventional drilling techniques, and (ii) implant-site preparation with a single trephine drill using dynamic navigation was as accurate under in vitro experimental conditions as a conventional drilling sequence. Bhargava et al. [223] compared and evaluated the osteotomy sites created using standard drill, bone trephine, and alveolar expanders for dental implant surgery, suing ten goat hemimandibles. Three osteotomy sites were prepared at the inferior border of the mandible using standard drill, trephine, and alveolar expander in each hemimandibles, and the sites were subjected to cone-beam CT (CBCT). It was concluded that (i) the CBCT images showed minimum bone loss with the use of alveolar expander which may be due to the lateral bone condensation rather the removal of the marrow, and (ii) trephine showed less marrow removal in comparison to the standard drill used for dental implant surgery.

Advent of osseointegration has rapidly led to use of dental implants over recent years. Implant complications are often inadvertent sequelae of improper diagnosis, treatment planning, surgical method, and placement. This can be overcome by using surgical guides for implant positioning. Although conventionally made surgical guide are used, the clinical outcome is often unpredictable, and even if the implants are well placed, the location and deviation of the implants may not meet the optimal prosthodontic requirements. High accuracy in planning and execution of surgical procedures is important in securing a high success rate without causing iatrogenic damage. This can be achieved by CT, 3D implant planning software, image-guided template production techniques, and computer-aided surgery. This chapter evaluates about the various systems of conventionally made surgical guide using radiograph and also the newer computer-generated surgical guide in detail. The success of dental implants in the treatment of patients is directly related to patient evaluation and good treatment planning. Earlier dentists were intended to place implants where the greatest amount of bone was present, with less regard to placement of final definitive restoration. In most of the times, the placement of implant is not as accurate as intended. Failures arise as a result of lack of consideration of the superstructure during presurgical planning. Accurate placement is required to achieve best functional and aesthetic result. Since the oral cavity is a relatively restricted space, a high degree of accuracy in placement of implant is very important for success of the prostheses [224–226]. Guided surgery, in which the osteotomy is created through a digitally designed and printed surgical guide, has the potential to afford the highest level of precision and control, and can be invaluable depending on the complexity of the case and the anatomy of the patient [227].

Freehand surgery

Freehand surgery, in which a flap is reflected and the implant is placed according to the available diagnostic information, is a cost-effective approach that is advantageous in many cases. In freehand surgery cases, periapical and panoramic radiographs are used to assess the bone available for implant placement as well as the surrounding anatomy [228]. Traditionally, periodontal probes, gauges, or calipers have been used during the intraoral examination for bone sounding, which offers a reasonable idea of the height and thickness of the ridge. Further, the surrounding teeth can be used as guides for determining the correct positioning of the implant, noting, for example, that the implant should be placed at least 1.5 mm from any neighboring dentition and 2 mm apical to their cementoenamel junction. One of the advantages associated with the freehand surgery is the capability to manage the keratinized soft tissue, which is often less available after a tooth is lost and the residual ridge has resorbed. One method of reestablishing the gingival contours is to position the surgical flap apically and buccally during freehand surgery [229]. This can help establish a situation in which the neck of the implant is surrounded by the 2–4 mm of keratinized tissue needed for healthy and aesthetic gingival contours, margins, and interdental papillae [228].

Guided surgery

In guided surgery cases, CBCT scanning and digital intraoral impressions are used to generate a virtual representation of the patient's jaw and oral anatomy. This is utilized to develop a digital treatment plan in which the exact position of the implant is determined in advance of treatment. A surgical guide is fabricated that controls the osteotomy in precise accordance with the preplanned implant position. The depth, angulation, and mesial–distal and buccal–lingual location of the implant osteotomy are precisely controlled by titanium sleeves situated within the surgical guide [228]. A dental implant is positioned properly 2 mm apical to the cementoenamel junction of and 1.5 mm from the adjacent teeth. This can be achieved by means of a surgical guide which provides adequate information regarding implant placement and at the time of surgery it fits on to the existing dentition or on to the edentulous span [230, 231]. The surgical template (as a guide used to assist in proper surgical placement and angulation of dental implants) enables a predictable and a safe minimally invasive surgery and possesses a function to direct the implant drilling system and provide an accurate placement of the implant according to the surgical treatment plan [224]. A typical type of a surgical guide is the union of two components of the guiding cylinders and the contact surface. The contact surface fits either on an element of a patient's gums or on the patient's jaw (i.e., the bone and the teeth). Cylinders within the drill guides help in transferring the plan by guiding the drill in the exact location and orientation [232]. The implant must be placed such that firstly the bottom and sides are covered fully by the bone or bone-replacement material. Secondly, care should be taken for not damaging any neighboring anatomic structures. These are in particular the mandibular nerve in case of mandible and the Schneiderian membrane of the maxillary sinus in maxilla and also the roots of adjacent teeth. Thirdly, position of the implant has to be compatible with the intended final prosthodontic restoration [224, 233].

There are three types of surgical guide [224, 234]. They include (1) bone supported (or bone-borne guide). A bone-borne surgical guide is generally used for full-arch edentulous implant cases. This type of guide derives its support from load-bearing areas and/or the natural divergence of the alveolar ridge seen in both maxillary and mandibular arches in a corono-apical direction. Bone-borne guides are difficult to fabricate and accurately place intraorally due to the extent of flap reflection needed for access, as well as inaccuracies that can result from the digital workflow. Although the difficulties associated with bone-borne guides are well reported [235], they may be mitigated with proper planning, accurate imaging, and operator experience. (2) Mucosa supported (or mucosal-borne guide). As the name suggests, a mucosal-borne surgical guide uses intraoral soft tissues for support and stability. This guide is commonly fabricated based on an existing removable prosthesis. Possible concerns with mucosal support involve the lack of retention during surgery and variation in tissue thickness and quality – which is a common complaint also seen with CDs [236, 237]. A mucosal-borne guide has the potential to undergo movement in varying directions that could increase the inaccuracy of implant placement. This, along with flap reflection, may introduce additional complexity during implant placement. (3) Tooth supported (or tooth-borne guide). Tooth-borne surgical guides are common and easily adaptable to various clinical situations. One benefit is that consistently reproducible landmarks (namely, teeth) can be used for support, stability and retention. The improved accuracy and precision of implants placed through tooth-borne guides have been evaluated in both benchtop and clinical models [235, 238]. The most common utilization of tooth-borne guides is for single or multiple implants for fixed restorations. Although these guides have been shown to result in predictable restorative outcomes, it is important for the practitioner to be cognizant of available space and additional guided drill length, especially when placing posterior implants [239].

Three techniques commonly used for preparing the guide holes and fabricating the radiographic and surgical implant guide are conventional freehand, milling, and CAD/CAM technology [240]. Various techniques have been proposed for the fabrication of surgical guide templates in implant dentistry. Surgical guide template fabrication involves a diagnostic tooth arrangement through one of the following ways: (1) a diagnostic waxing, (2) a trial denture teeth arrangement, or (3) the duplication of a preexisting dentition/restoration [241]. The fabrication of the surgical guide templates is on one of the following design concepts [242, 243]: (1) nonlimiting design, (2) partially limiting design, and (3) completely limiting design. These design concepts are classified based on the amount of surgical restriction offered by the surgical guide templates [242].

Recently, 3D-guided implant surgery has been developed [244]. 3D-guided implant surgery is a revolutionary surgical technique that takes digital treatment planning one step further. 3D-guided implant surgery is a combination of CT scanning technology and CAD/CAM technology. There are multiple benefits associated with 3D-guided implant surgery. These benefits include: (i) greater precision, as well as greater accuracy, in the placement of your dental implants and your dental restorations, and (ii) increased surgical predictability. Guided surgery allows us to visualize every step of your surgical procedures, greatly increasing the predictability of the process. This helps to make treatment safer, reduce the risks of complications, both during surgery as well as during the recovery period; customized surgical guides help provide more exact implant and restoration placement; guided surgery is typically faster than freehand surgery; and reduced downtime after oral surgery.

It is important to ensure to drill from a drill with a small diameter to a large drill in order. Using drills with a large diameter from the beginning has greater frictional resistance between the bone and the drill surface, and there is a risk of heat generation and thermal damage to the bone [245–247]. Möhlhenrich et al. [247] reviewed articles related to an increase in temperature during preparation of the site and reported that the highest temperature measured was 64.4 °C and the lowest was 28.4 °C. The influence of bone density, drill diameter, drilling speed, and irrigation on temperature rise was investigated and reported that both external irrigation with higher drilling speeds and no irrigation with lower speeds were effective methods to avoid excessive heat generation during implant osteotomies [248]. Although many implant systems recommend drilling at speeds of 800–1,200 rpm because within this range of drill speed there would not be vibration along the drilling axis [249], Miyazaki recommended that it should be in a range between 700 and 800 rpm for patients having relatively soft bone [202].

Parallelism

Parallel placement of dental implants is an accepted surgical and prosthodontic norm when multiple implants are used. Use of paralleling pins supplied with the implant kit is an arbitrary method. An instrument is designed that helps in maintaining the parallelism between implants intraorally. The instrument consists of a horizontal plate with millimeter graduation. A slot is created in the horizontal plate to house a vertical rod. The vertical rod can move mesiodistally as well as occluso-gingivally. It must be ensured that the vertical rod is exactly parallel to the long axis of the drill in all directions. The horizontal plate is then securely attached to the handpiece. After the first osteotomy is made, the vertical rod is placed in the site. The position of vertical rod will guide the second drill parallel to the osteotomy site. This instrument will guide us to correctly determine the inter-implant distance, and also depth of the preparation [250]. Transmission of masticatory load through the implant to the supporting bone within physiologic parameters dictates the success of implant treatment. Unfavorable concentration of stress results in resorption of bone and eventually the failure of implant. Vertical loading applied on an angled abutment is a classic example of untoward stress concentration. Stress-induced damage to the bone attains undesirable proportions when more than one implant is splinted, especially when the implants are not placed in a mutually parallel position. More than the predictable control of stresses, parallel placement of implants favors aesthetics [251]. After placement of multiple dental implants, a radiograph is usually taken to demonstrate the implants' relative positions. The implants can be relatively parallel or not parallel. There may be a certain aesthetic satisfaction in parallel placement. There may be prosthetic fabrication issues if the implants are extremely off from parallel [252]. Nonetheless, from a biomedical engineering perspective, nonparallel placement may better resist occlusal loads [253]. Nunes et al. [254]

and Dario [255] pointed out that (i) the true parallel placement may be impossible when implants are placed in the curved edentulous arch, especially when there is a facial concavity, and (ii) the importance of parallel placement may be overrated since there are techniques to correct difficulties in prosthetic fabrication. Nonparallel placement of multiple implants may increase the stability of a prosthesis when the implants are splinted in the prosthesis. Under load, this placement distributes stresses through more of the containing bone than parallel implants. In-line implants may induce screw loosening under functional load, especially when there is a cantilever [253]. Parallelism may be desirable for ease of prosthetic fabrication, but parallelism is not mandatory for a successful treatment outcome. Thus, if a dental implant surgeon is criticized for nonparallel implant placement, the response should be that of an adherent of the engineering school of thought [252].

9.7 Additional surgical procedures

The postextraction available bone is very crucial at the bone–implant contact zone. With a greater surface area of implant–bone contact, less stress is transmitted to the bone, improving the implant prognosis. Elsayed [69] introduced the Misch's classification for maxillary and mandibular atrophy followed by tooth extraction into the following four classes. (1) Division A (abundant bone): Division A abundant bone forms soon after the tooth is extracted. The abundant bone volume remains for a few years, although the interseptal bone height is reduced and the original crestal width is usually reduced by more than 30% within 2 years. Division A bone corresponds to the abundant available bone in all dimensions. (2) Division B: As the bone resorbs, the width of the available bone first decreases at the expense of the facial cortical plate because the cortical bone is thicker on the lingual aspect of the alveolar bone, especially in the maxilla. A 25% decrease in bone width occurs the first year, and 40% decrease in bone width occurs within the first 1–3 years after tooth extraction. The resulting narrower ridge is often inadequate for many 4-mm diameter root-form implants. Slight-to-moderate atrophy is often used to describe this clinical condition. Once the bone reaches this division B bone volume, it may remain at this level for more than 15 years in the anterior mandible. (3) Division C (compromised bone): division C available bone is deficient in one or more dimensions (height, length, width, angulation, or crown height/ bone height ratio) Therefore, the width may be <2.5 mm, the crown height >15 mm, and the bone angulation >30°. (4) Edentulous division D (deficient bone): Long-term bone resorption may result in the complete loss of the alveolar process, accompanied by basal bone atrophy. Severe atrophy describes the clinical condition of division D ridges site.

The main problem in peri-implantitis is often the combination of severe peri-implant bone loss with a contaminated implant surface and an insufficient soft tissue situation [256]. This may be caused by ill-osseointegration which needs a healthy quantity of bone. In order for a placed implant to exhibit a long-term survival, it needs to have a thick healthy soft tissue (gingiva) envelope around it. It is common for either the bone or soft tissue to be so deficient that the surgeon needs to reconstruct it either before or during implant placement [256–258]. Accordingly, besides the main surgical procedures, we have additional surgical procedures for (1) hard tissue reconstruction and (2) soft tissue reconstruction.

9.7.1 Hard tissue reconstruction – sinus floor elevation

If bone width is not adequate, it can be regrown using either artificial or cadaveric bone pieces to act as a scaffold for natural bone to grow around. When a greater amount of bone is needed, it can be taken from another site (commonly the back of the bottom jaw) and transplanted to the implant site. The maxillary sinus can limit the amount of bone height in the back of the upper jaw. With a sinus lift technique, bone can be grafted under the sinus membrane increasing the height of bone. Hence, bone reconstruction can be achieved by either the sinus lift technique or grafting method.

Implant placement is often challenging in the posterior maxilla due to various problems. They should include as follows. (1) Normal sinus anatomy. It's possible that the extent of the person's maxillary sinus is just naturally comparatively large. As such, its floor may dip down to a level that approximates the area of the roots of the patient's upper molars and/or premolars. (2) Post-extraction bone resorption. This process refers to the loss of bone that naturally occurs over time following a tooth extraction. If enough time passes, the thickness of the bone in the region of the sinuses may be too thin to accommodate the length of an implant. (3) Pneumatization of maxillary sinuses. This is a naturally occurring process where following a surgical procedure in the area of a sinus, its size increases due to the pressure it exerts (possibly related to the pressures of respiration) on the healing tissues. Continued enlargement may continue after bone healing has occurred. (4) Other causes. Additional reasons include the removal of bone associated with previous surgical procedures (like a difficult tooth extraction), or the loss of bone caused by periodontal (gum) disease, or simply poor quality and quantity of the alveolar bone. Bone resorption frequently causes severe loss of bone both vertically and horizontally which may limit the use of dental implants. When the distance between the sinus floor and the alveolar crest is less than 8–10 mm, sinus floor elevation may be indicated. Different augmentation procedures have been suggested for the placement of dental implants in resorbed posterior maxilla [127, 259, 260].

There are two distinct approaches, including lateral approach (or sinus lift, external sinus floor elevation) and crestal approach (or socket lift, internal sinus floor elevation). Table 9.5 compares some characteristics between these two approaches [259–266].

Table 9.5: Comparison between the lateral approach and crestal approach for sinus floor elevation.

	Procedure	Indications	Merits	Demerits
Sinus lift: lateral approach, external elevation	Perforate laterally the maxillary sinus side wall	– Remaining bone at sinus bottom is <3–5 mm – Many teeth are missing	– Massive bone regeneration can be achieved – Surgery can be performed while visually checking – Long-term success	– High surgical difficulty – Implant cannot be placed until the grafted bone heals – High risk of infection and prolonged treatment period
Socket lift: crestal approach, internal elevation	Lift the alveolar crest upwardly	– Remaining bone at sinus bottom is >3–5 mm	– Minimally invasive surgery is possible – Surgery is relatively easy – Implant can be placed right after the bone grafting – Minimizes postsurgical complications	– Less amount of regenerated bone – Operation field cannot be seen due to the groping surgery – High risk to overlook the membrane perforation

Tomruk et al. [259] assessed the prevalence of sinus-lifting procedures and survival rates of implants placed in the posterior maxilla, to which implants were placed later on. The study included 302 patients at a mean age of 5.2 years, who received a total of 609 dental implants. A total of 380 (62.3%) implants were inserted in native areas, 203 (33.3%) in external sinus-lifted areas (by lateral approach), and 26 (4.4%) in internal lifted areas (by crestal approach). It was reported that (i) the survival rate in native or internal lifted areas were 100% and 95.6% in external sinus lifted implants (10 implant failures/203 implants), and (ii) almost half of the implants were examined radiologically with a mean duration of 30 months and the mean MBL was 0.64 ± 1.2 mm, indicating that the survival rates of native bone and the internal sinus lifting were slightly higher than that of external sinus lifting, and implants placed with sinus augmentation exhibited more MBL than implants in the native bone.

The Loma Linda pouch technique (by lateral approach) was developed and was introduced for repairing the perforated maxillary sinus membrane during sinus grafting procedures [267], by which (i) a collagen membrane is placed against the perforated site and subsequently covers the internal surface of the maxillary sinus and (ii) the collagen membrane is then folded along the lateral access window to form a pouch that surrounds and isolates the graft material. This method has been employed for cases when the Schneiderian membrane is perforated. Moreover, using a characteristic of piezosurgery of cutting only hard tissue effectively, the piezolift method has been developed, in which tissue is cut till the membrane by the piezosurgery and the membrane was peeled off slightly by water pressure provided from the tip of surgical device, and the bone grafting material is inserted [268, 269].

9.7.2 Hard tissue reconstruction – bone augmentation

Bone grafting is necessary when there is a lack of bone. Also, it helps to stabilize the implant by increasing survival of the implant and decreasing marginal BL loss. The application of barrier membranes to promote bone regeneration was first described by Hurley et al. [270]. While there are always new implant types, such as short implants, and techniques to allow compromise, a general treatment goal is to have a minimum of 10 mm in bone height, and 6 mm in width. Alternatively, bone defects are graded from A to D (A = 10+ mm of bone, B = 7–9 mm, C = 4–6 mm, and D = 0–3 mm) where the extent of osseointegration is related to the grade of bone [271, 272]. Basically, there are two approaches: the lateral alveolar augmentation to enhance the width of the implantation site, and the vertical alveolar augmentation to improve the height of the implantation site. The final decision about which bone grafting technique that is best is based on an assessment of the degree of vertical and horizontal bone loss that exists, each of which is classified into mild (2–3 mm loss), moderate (4–6 mm loss), or severe (>6 mm loss) [273]. Orthodontic extrusion or orthodontic implant site development can be used in selected cases for vertical/ horizontal alveolar augmentation [274].

Postextraction crestal bone resorption is common and unavoidable, which can lead to significant ridge dimensional changes. To regenerate enough bone for successful implant placement, GBR is often required. Since the promising results in periodontology, GBR has been developed to regenerate bone defects in the alveolar process, in which a defect is filled with either natural (harvested or autograft) bone or allograft (donor bone or synthetic bone substitute), covered with a semipermeable membrane and allowed to heal.

GBR-assisted treatment should start with the evaluation of the bony defect [275, 276]. Referring to Figure 9.15 [275], the use of CBCT imaging, along with interactive CBCT treatment planning (e.g., nerve drawing, bone density measurements, defining bone graft requirements) is imperative to develop a definitive surgical and prosthetic treatment plan. A careful review of the topography of the recipient graft site should include BLs around adjacent teeth, associated bony protuberances and concavities, the dimensions of the defect, the number of remaining bony walls if present, and the condition of the surrounding soft tissue. Bony defects may include (i) small depression, (ii) concavity defect, (iii) convex/straight bone deficiency, (iv) vertical (height) defects, and (v) buccal-lingual defects [275]. It continues to the flap design and incision which are crucial steps in obtaining a

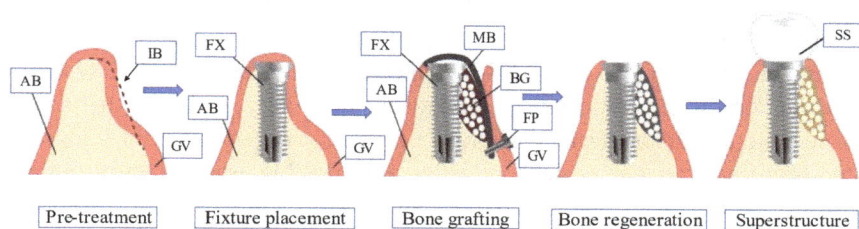

Figure 9.15: Typical GBR procedures [275]. AB, alveolar bone; IM, insufficient bone area; GV, gingiva; FX, fixture; MB, membrane; BG, bone grafting; FP, fixing pin; and SS, superstructure.

predictable regenerative result. Ideally, complete access to the surgical site without compromising the integrity of the surrounding tissue should be obtained. It is indispensable that a continuous full-thickness incision be made on bone through the tissue and the periosteum [276]. Then, the tissue is reflected and the flap has complete release of tension to prevent incision line opening. It is followed by the preparation of the site, including the soft tissue removal and decortication. The main operation is the next step of placing an implant fixture body, applying bone graft materials and artificial membrane. The last but not least important is the final closure of the bone graft site. Most importantly, a tension-free flap adaptation is the key to predictable and consistent results. If incision line opening occurs, the morbidity of the procedure will increase. Therefore, meticulous principles should be adhered to with respect to a tension-free flap design, ideal suture technique, and close postoperative evaluation of the surgical site. The suture selected should be made from a material that has high tensile strength, ideally polyglycolic acid (absorbable), or PTFE (nonabsorbable) [277]. To complete the implant treatment, after the completion of the bone regeneration, which may take 4–6 months, a superstructure as a final restorative prosthesis is installed.

For successful GBR treatment, there are two crucial material selections: bone grafting materials and artificial membrane materials.

Bone graft materials

Bone regeneration can be accomplished through three different mechanisms: osteogenesis, osteoinduction, and osteoconduction. Osteogenesis is the formation and development of bone, even in the absence of local undifferentiated mesenchymal stem cells. Osteoinduction is the transformation of undifferentiated mesenchymal stem cells into osteoblasts or chondroblasts through growth factors that exist only in the living bone. Osteoconduction is the process that provides a bioinert scaffold, or physical matrix, suitable for the deposition of new bone from the surrounding bone or encourage differentiated mesenchymal cells to grow along the graft surface

[278–280]. The primary types of bone graft material are autogenous bone, allografts, xenografts, and alloplasts. All grafting materials have one or more of these three mechanisms of action. The mechanisms by which the grafts act are normally determined by their origin and composition. Autogenous bone harvested from the patient forms new bone by osteogenesis, osteoinduction, and osteoconduction. Allografts harvested from cadavers have osteoconductive and possibly osteoinductive properties, but they are not osteogenic. Xenografts/alloplasts are typically only osteoconductive [278]. In addition to these graft materials, there are various types of synthetic bone graft materials, all of which are derived from man-made materials. Synthetic bone graft materials were specifically designed to provide an alternative to autograft. They are manufactured from biocompatible ceramics, glasses, polymers, and collagen (animal-derived) and are intended to function as a scaffold for bone formation. Synthetic bone grafts are also resorbable and will eventually be replaced by the patient's own bone following the healing response. Synthetic bone graft products are found in granule and block form. Granules may be further processed into putties and sheet-based products when combined with a carrier [277].

There are still different types of bone graft products available. Growth factors are a type of bone graft product typically consisting of specific proteins, such as bone morphogenetic protein. These proteins have been shown to positively influence the bone formation response. In bone grafting procedures, a porous scaffold is needed to support the bone formation response. Therefore, growth factor products consist of the recombinantly manufactured growth factor component, which is rehydrated into a solution, and a porous material that absorbs the fluid and functions as a scaffold to support bone formation [277]. Platelet-rich plasma (PRP, aka autologous platelet gel) is essentially an increased concentration of autologous platelets suspended in a small amount of plasma after centrifugation [280]. Platelets play a fundamental role in hemostasis and are a natural source of growth factors. Basically, patient's blood is collected and centrifuged at varying speeds until it separates into three layers: platelet poor plasma (PPP), PRP, and red blood cells. Two spins are normally used. The first spin ("hard spin") separates the PPP from the red fraction and PRP. The second spin ("soft spin") separates the red fraction from the PRP. The material with the highest specific gravity (PRP) will be deposited at the bottom of the tube. Immediately prior to application, a platelet activator/agonist (topical bovine thrombin and 10% calcium chloride) is added to activate the clotting cascade, producing a platelet gel. The whole process takes approximately 12 min and produces a platelet concentration of 3–5× that of native plasma [281–283]. No evidence of benefit associated with PRP technique was found on the use in sinus lift during the implant placement [282] or bone healing [284].

Membrane materials

The membrane used for GBR is an essential component of the treatment. The desirable characteristics of the membrane utilized for GBR therapy include biocompatibility, cell occlusion properties, integration by the host tissues, clinical manageability, space-making ability, and adequate mechanical and physical properties [279]. There are several properties required as an ideal membrane material and structure. (1) Mechanical properties (stiffness and plasticity). The amount of regenerated bone in the bone defect would be reduced if the membranes collapse into the defect space. Therefore, the ideal GBR membrane should be sufficiently rigid to withstand the compression of the overlying soft tissue. It should also possess a degree of plasticity in order to be easily contoured and mold to the shape of the defect. A balance between these mechanical properties is required to achieve an adequate space-making capacity. (2) Porosity is an important property of the GBR membrane. Studies have addressed the role of this property in the biological response in vivo using nonresorbable and resorbable membranes. The pore size of the membrane influences the degree of bone regeneration in the underlying secluded space [279].

There are two types of membranes: resorbable type and nonresorbable type. Since nonresorbable membranes need a second surgical intervention for membrane removal, subsequently, a second generation of membranes made of resorbable materials was developed and became widely used in different clinical situations. Currently, there are two kinds of resorbable membranes: polymeric and collagen derived from different animal sources [278, 284]. Polymeric membranes are valuable in preserving alveolar bone in extraction sockets and preventing alveolar ridge defects, as well as ridge augmentation around exposed implants. Polymeric membranes are made up of synthetic polyesters, polyglycolides, polylactides, or copolymers [285]. Most of the commercially available collagen membranes are developed from type I collagen or a combination of type I and type III collagen. The source of collagen comes from tendon, dermis, skin, or pericardium of bovine, porcine, or human origin [286].

As nonabsorbable membrane materials, there are polymers, metals, and ceramics. The first reported synthetic polymer used for GBR was e-PTFE (expanded form of PTFE) and it is considered to be one of the most inert, stable polymers in the biological system. It resists breakdown by host tissues and does not elicit immunological reactions [287]. As natural polymers, collagen-based membranes are the most commonly used naturally derived membranes for GBR [286]. These membranes have received major attention by virtue of collagen being the principal component of connective tissues and having important roles with respect to structural support and being an important component in cell-matrix communication [286]. Chitosan is another natural-derived polymer used for preparation of GBR membranes [288]. Alginate membranes have also been introduced for GBR. Alginate is a biocompatible, anionic polymer that can be obtained from brown seaweed and achieves a similar structure to extracellular matrices when cross-linked to hydrogels [289].

The use of titanium mesh which can maintain the space can be a predictable and reliable treatment modality for regenerating and reconstructing a severely deficient alveolar ridge [290–294]. The main advantages of the titanium mesh are that it maintains and preserves the space to be regenerated without collapsing and it is flexible and can be bent. It can be shaped and adapted so it can assist bone regeneration in nonspace-maintaining defects. Due to the presence of holes within the mesh, it does not interfere with the blood supply directly from the periosteum to the underlying tissues and bone-grafting material. It is also completely biocompatible to oral tissues [294, 295]. Titanium mesh can be used before placing dental implants (staged approach) to gain bone volume or in conjunction with dental implant placement (nonstaged approach) [293]. Jovanovic et al. [296] developed titanium-reinforced e-PTFE membrane and reported that the reinforcement of e-PTFE membrane with titanium were able to maintain a large, protected space for blood clot stabilization without the addition of bone grafts and provided superior preservation of the original form of the regenerated ridge during the healing period.

Furthermore, Co–Cr–Mo alloy has also been suggested for GBR [297]. Although this alloy is known to be less biocompatible than titanium and titanium alloy, it has superior mechanical properties (e.g., stiffness and toughness). The potential use of CoCrMo alloy for GBR has been evaluated in a recent animal study but it has not yet been documented in any clinical report. It was mentioned that placement of CoCrMo membrane on a rabbit tibial defect provides sufficient space and promotes bone regeneration [297]. Ceramic materials such as calcium sulfate [298, 299], HA [300–302], and beta-tricalcium phosphate [303, 304] have been incorporated in resorbable membranes.

9.7.3 Soft tissue reconstruction

An adequate understanding of the relationship between periodontal tissues and restorative dentistry is paramount to ensure adequate form, function, aesthetics, and comfort of the dentition [305, 306]. Maintenance of gingival health constitutes one of the keys for tooth and dental restoration longevity [307]. The biologic width is the distance which is established by the junctional epithelium and the connective tissue attachment to the tooth's root surface. This may also be used to describe the height between the deepest point of the gingival sulcus and the alveolar bone crest [308]. This distance is a vital consideration in the creation of dental restorations as they must respect the natural architecture of the gingival attachment in order to avoid harmful consequences. The biologic width is specific to each patient and ranges from about 0.75–4.3 mm [309], and the mean biologic width was 2.04 mm, of which 1.07 mm is comprised of connective tissue attachment and another 0.97 mm is occupied by the junctional epithelium, as shown in Figure 9.16 [310].

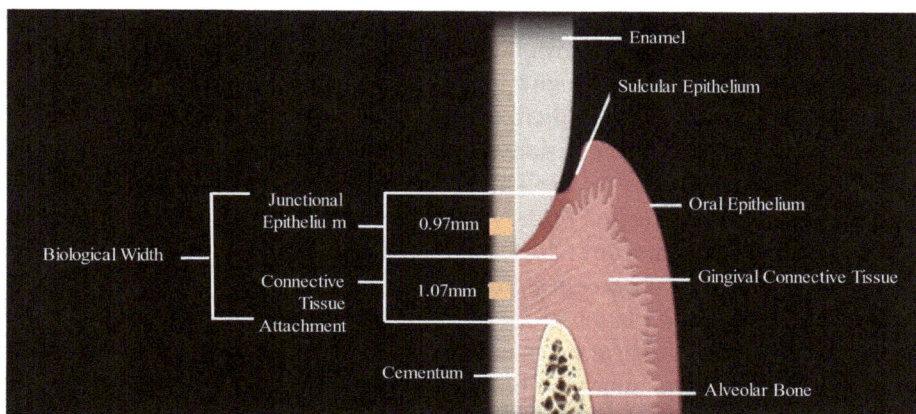

Figure 9.16: Biological width and surrounding tissues [310].

The gingiva surrounding a tooth has a 2–3 mm band of bright pink, very strong attached mucosa, then a darker, larger area of unattached mucosa that folds into the cheeks. Upon placing a dental implant, a band of strong, attached gingiva is needed to keep the implant healthy in the long-term [311]. This is especially important with implants because the blood supply is more precarious in the gingiva surrounding an implant, and is theoretically more susceptible to injury because of a longer attachment to the implant than on a tooth (a longer biologic width) [312].

Fuchigami et al. [313] examined peri-implant mucosal thickness at different sites of peri-implant crevice around 70 implants placed in 35 patients. It was found that (i) the overall mean peri-implant mucosal thickness was 3.6 ± 1.4 mm, wherein maxillary anterior implants, maxillary posterior implants and mandibular posterior implants had significantly different dimensions of median thickness of 4.25, 3.75, and 3.0 mm, respectively, and (ii) the mesial and distal sites of those positioned implants measured unevenness in the thickness especially in the maxillary posterior region with statistical significance. It was then mentioned that (i) the peri-implant mucosal thickness was measured with a big variation from overall 3.6 mm with a big variation from 1.6 to 7.0 mm in healthy volunteers; (ii) significant difference was found in the depth among the three regions and, statistically, dispersion of individual peri-implant mucosal thickness resulted in lack of consistency; and (iii) although dental implants have been well developed, predictable, and prevailing prosthetics, onset of peri-implantitis might be inevitable in some cases, concluding that establishment of a standardized dimensional diagnosis of peri-implant tissues followed by pathologic ascertainment could be taken into account for the prevention or curing of peri-implantitis. Because it is nearly impossible to perfectly restore a tooth to the exact coronal edge of the junctional epithelium, dentists often opt to remove enough bone needed to maintain 3 mm between the restorative margin and the crest of alveolar bone. When the restoration does not account for these considerations and violate the biologic width of the tooth, it can result in the below issues such as chronic pain, chronic gingiva inflammation, or loss of alveolar bone [314].

Implant dentistry has been definitively established as a predictable treatment modality for replacing missing or nonrestorable teeth which yields excellent clinical success rates. The aesthetics of implant restorations is dictated by two fundamental components: the reproduction of the natural tooth characteristics on the implant crown and the establishment of a soft tissue housing that will intimately embrace the crown. Therefore, the success of implant rehabilitation in the aesthetic zone relies heavily on the preservation or the augmentation of peri-implant soft tissue by means of periodontal surgical procedures [315]. For the soft tissue (gingiva) reconstruction, there are two major methods available: free gingival grafts (FGG) and subepithelial connective tissue grafts (SCTG) to reconstruct the buccal dimensions of the site improving the tissue thickness. In addition, they create the illusion of root prominence and increase the width of the crestal peri-implant mucosa in order to provide an emergence profile for the restoration and enable the constructed site to closely resemble a natural tooth [315].

Free gingival graft

Gingival grafting (aka gum grafting) is a procedure used to thicken gingiva and, in some cases, to cover recession under teeth. This procedure is used for a variety of reasons including reinforcing gums and teeth and for patients who have sensitivity due to recession. There are several procedures for the gingival grafting, including free gingival grafting, connective tissue grafting, alloderm (no palate) grafting, or pedicle grafting [316]. Among them, the free gingival grafting has been employed extensively. FGGs are considered a reliable and efficacious approach for augmenting peri-implant soft tissue defects and are most often utilized to increase the amount of keratinized tissue around an implant. The most common donor site of a FGG is the highly keratinized hard palate so that the color and shade of the augmented recipient site do not often blend naturally with the adjacent soft tissues, causing a nonaesthetic result, contradicting the initial purpose of the procedure. FGG can be used for patients with low smile lines, when extensive soft tissue augmentation is needed, or where the color of an FGG will not compromise the aesthetic appearance of the implant site. It is also used to augment the tissue and to try to prevent further bone loss and recession.

Subepithelial connective tissue graft

SCTG procedures have been used successfully throughout the years for the management of recession and soft tissue defects around natural teeth and for augmenting alveolar ridge contours [317, 318]. Some may argue that the traditional approaches for connective tissue grafting do not fare well when one attempts to graft and achieve cover of a nonvital implant surface since the soft tissues around the implant do not respond in the same manner as a vital tooth. Nonetheless, many of these procedures can be translated directly to peri-implant soft tissue modification and aesthetic optimization. When indicated and properly utilized, these surgical procedures can provide stable and significant gains in soft tissue volume and contour that can contribute to the successful aesthetic management of implant sites [315].

The main problem in peri-implantitis is often the combination of severe peri-implant bone loss with a contaminated implant surface and an insufficient soft tissue situation. Classic surgical concepts with crestal access to the bony defect and debridement of the surface most often lead to partial defect regeneration and a soft tissue recession. An incision directly above the pathologic bony lesion is contrary to general surgical treatment rules [319]. Based on this background, Noelken et al. [319] introduced an innovative regenerative treatment approach for severe periimplantitis defects. After diagnosis and non-surgical pre-treatment of a severe peri-implantitis lesion, the following treatment protocol was applied: horizontal mucosal incision 5 mm apical to marginal mucosa, supraperiosteal preparation in apical direction, cutting through periosteum at the level of the implant apex, subperiosteal coronal flap elevation, exploration and cleaning of the peri-implant defect, thorough debridement of the implant surface with the Er:YAG laser, subperiosteal grafting with connective tissue, grafting of the bony defect with autogenous bone chips from the mandibular ramus, and bilayered suturing of periosteum and mucosa. Implant survival, marginal BLs, peri-implant probing depths, recession, and facial mucosa thickness were evaluated in a pilot case at 1-year follow-up examination. It was obtained that (i) inter-proximal, oral, and buccal marginal BLs increased significantly to the level of the implant shoulder from preoperative to 1-year follow-up examination; (ii) no signs of suppuration or peri-implant infection were present; and (iii) probing depths and recession decreased significantly, while the facial mucosa thickness improved from preoperative to final examination. It was therefore concluded that (i) marginal BLs and soft tissue improvement suggest feasibility for the regeneration of severe peri-implant hard and soft tissue deficiencies by this new treatment approach, and (ii) the simultaneous implant surface cleansing and improvement of hard and soft tissues seem to be possible and unfavorable postoperative exposition of titanium surface might be prevented.

Survival of dental implants depends on several factors; soft tissue management around dental implants is one of the foremost. Kadkhodazadeh et al. [320] reviewed published articles related to the timing of soft tissue management around dental implants. Out of articles published from January 1995 to July 2015, only in vivo studies and clinical trials in relation to the terms soft tissue management, management timing, keratinized mucosa, FGG, connective tissue graft, soft tissue, augmentation, and dental implant were included. A total of 492 articles were reviewed, and eventually 42 articles were thoroughly evaluated. Those with treatment protocols in terms of the timing of soft tissue grafting were selected and classified. It was found that (i) the soft tissue management around dental implants may be done prior to the surgical phase, after the surgical phase, before loading, or even after loading; (ii) a thick gingival biotype is more suitable for implant placement, providing more favorable aesthetic results; (iii) a treatment plan should be based on individual patient needs as well as the knowledge and experience of the clinician; and (iv) the width and thickness of keratinized tissues, the need for bone management, and local risk factors that influence aesthetic results determine the appropriate time for the soft tissue augmentation procedures.

9.8 Impression taking

From braces and retainers to veneers, mouthguards and restorative superstructure as a final stage of dental implant treatment, dental impressions play an important role in fitting them correctly to a certain location and purposes. One prevalent reason for getting dental impressions is to have a crown or bridge (as superstructure for implant treatment) placed in a mouth. Traditional impression-taking potentially involves multiple materials and occasionally more steps. Impression materials include elastic impression materials such as aqueous materials (agar or alginate) and nonaqueous elastomers (polysulfides, silicones, or polyethers). Impression materials are also classified as reversible (compounds and hydrocolloids) or irreversible (silicones, polyethers, and alginates) [321]. Because this is a highly delicate and skilled process, it is easier to introduce error throughout any of the numerous steps involved, either from the human element, or material defects such as voids, air bubbles, or improper setting or distortions. A traditional impression is captured in the negative, making it more difficult to identify mistakes. If mistakes are identified, the dentist will need to take another impression, which means patients must undergo the procedure again, resulting in greater inconvenience and a longer appointment, as well as time lost, and added cost and material use for the dentist. One of the most common areas of concern when creating a crown, bridge, or veneer involves the tiny space between the tooth and the gum tissue that surrounds the tooth, known as the subgingival margin. Without an accurate imprint of this area, the final restoration may not fit the tooth appropriately, resulting in a myriad of problems down the road [322].

On the other hand, a digital scan enables dentists to see the "positive" image and magnify and evaluate it carefully. Errors can be corrected immediately before submitting the impression to the laboratory. With digital impression taking, when the final impression is scanned and the bite registration obtained, a virtually articulated model of the preparation is visible on the monitor [323]. Digital impressions represent cutting-edge technology that allows dentists to create a virtual, computer-generated replica of the hard and soft tissues in the mouth using lasers and other optical scanning devices. The digital technology captures clear and highly accurate impression data in mere minutes, without the need for traditional impression materials that some patients find inconvenient and messy. There are two types of digital impression technology currently available for dentists to use. One type captures the images as digital photographs, providing dentists and dental laboratories with a series of images; the other type captures images as digital video. The images can be captured using lasers or digital scanning. Laser scanning uses concentrated light that is safe and highly precise. It captures all details of the teeth and gums while eliminating the patient's need to hold unpleasant, distasteful material in his or her mouth. Digital optical scanners are also safe and highly accurate but require teeth to be powder-coated with a special spray before scanning to ensure all parts of the impression are recorded properly [323]. As to benefits of digital impressions [322–324], the digital optical impressions significantly increase efficiency, productivity, and accuracy, and make it possible for dentists to e-mail the virtual impression to the laboratory, rather than send a traditional impression or stone model via regular mail. Also, digital impressions can be used to make same day dentistry restorations, thereby speeding up patient treatment and reducing the need for multiple office visits. Other benefits of digital impressions include: (i) improved image/impression quality for better fitting restorations; (ii) no need for distasteful impression materials that cause some patients to gagging, or claustrophobic; (iii) reduced possibility of impression-taking errors and elimination of material inaccuracies for fewer restoration mistakes; (iv) the scan of the teeth being restored, as well as the opposing teeth and bite, can be completed in just 3–5 min; and (v) the digital impression can be stored electronically indefinitely, which saves space, contributes to efficient recordkeeping, and supports a paper-free environment. These benefits are recognized and reported by various comparative studies between the traditional impression taking and a digital impression taking [46, 325–327]. None of advanced technology is perfect and still needs improvements and the digital technology application to the impression taking task is not an exception. Chi [328] pointed out that concerning issues with the digital impression taking should include (i) missing scan information, (ii) distortion caused by moisture, and (iii) misalignment in the bite scan.

At the very last step of long procedures of the implant treatment, the digital impression is transmitted electronically to the dental laboratory or in-office dental CAD/CAM system for use in restoration fabrication, which could be done by full usage of CAD/CAM system to complete a digital dentistry. A final superstructure will be installed to make a patient back to the normal occlusal function and/or recovery of aesthetics.

References

[1]   Kumar M, Kumar R, Kumar S, Prakash C. Biomechanical properties of orthopedic and dental implants: A comprehensive review. In: Handbook of Research on Green Engineering Techniques for Modern Manufacturing. IGI Global Pub., 2019, doi: 10.4018/978-1-7998-8050-9.ch026.

[2]   Tooth numbering systems in dentistry. 2015, https://dentagama.com/news/dental-numbering-systems.

[3]   Masiga MA. Presenting chief complaints and clinical characteristics among patients attending the Department of Paediatric Dentistry Clinic at the University of Nairobi Dental Hospital. East Afr Med J. 2005, 82, 652–55.

[4]   Al-Johani K, Lamfon HA, Beyari MM. Common chief complaints of dental patients at Umm Al-Qura University, Makkah City, Saudi Arabia. Medicine. 2017, https://www.semanticscholar.org/paper/Common-Chief-Complaints-of-Dental-Patients-at-Umm-Al-Johani-Lamfon/255581f10fbc59789258656c724d4fba99025b60

[5]   Dhaimade PA, Banga KS. Evaluation of chief complaints of patients and prevalence of self-medication for dental problems: An institutional study. Intl J Community Med Public Health. 2018, 5, doi: http://dx.doi.org/10.18203/2394-6040.ijcmph20180249.

[6]   What kind of dental treatment is such a symptom and a chief complaint? https://osaka-dental-clinic.com/complaint/.

[7]   Grand View Research. https://www.grandviewresearch.com/?utm_source=referral&utm_medium=getreferralmd&utm_campaign=guestpost_11october18_dental-implants-market.

[8]   Agarwal V. Factors Impacting Dental Implant Market Growth. 2018, https://getreferralmd.com/2018/11/factors-impacting-dental-implants-market-growth/.

[9]   Dental-implant-vs-natural-tooth. https://laserandholisticdental.com/dental-implant-vs-natural-tooth.

[10] Froum S, Kurtzman GM. Top 5 anatomical differences between dental implants and teeth that influence treatment outcomes. 2017, https://www.perioimplantadvisory.com/clinical-tips/periodontal-complications/article/16412223/top-5-anatomical-differences-between-dental-implants-and-teeth-that-influence-treatment-outcomes.

[11] Fillion M, Aubazac D, Bessadet M, Allègre M, Nicolas E. The impact of implant treatment on oral health related quality of life in a private dental practice: A prospective cohort study. Health Qual Life Outcomes, 2013, 11, 197, doi: https://doi.org/10.1186/1477-7525-11-197

[12] Pavel K, Seydlová M, Dostalova T, Zdenek V, Chleborad K, Zvárová J, Feberova J, Hippmann R. Dental implants and improvement of oral health-related quality of life. Community Dent Oral Epidemiol. 2012, 40, 65–70.

[13] Feine JS, with 20 co-authors. The McGill consensus statement on overdentures. Int J Prosthodont. 2002, 15, 413–14.

[14] Feine JS, with 21 co-authors. The McGill consensus statement on overdentures. Mandibular two-implant overdentures as first choice standard of care for edentulous patients. Gerodontology. 2002, 19, 3–4.

[15] Sischo L, Broder HL. Oral health-related quality of life: What, why, how, and future implications. J Dent Res. 2011, 90, 1264–70.

[16] Petersen PE. The World Oral Health Report 2003: Continuous improvement of oral health in the 21st century-the approach of the WHO Global Oral Health Programme. Community Dent Oral Epidemiol. 2003, 31, 2–34.

[17] Mehta D, Deshpande N, Dandekar S. Evaluation of Oral Health Related Quality of Life of Tuberculosis Patients in Vadodara – A Questionnaire Study. 2017, https://www.research

gate.net/publication/319628984_EVALUATION_OF_ORAL_HEALTH_RELATED_QUALITY_OF_LIFE_OF_TUBERCULOSIS_PATIENTS_IN_VADODARA-A_QUESTIONNAIRE_STUDY.

[18] Inglehart MR, Bagramian RA, Inglehart MR, Bagramian RA. Oral Health Related Quality of Life. Illinois: Quintessence Publishing Co. Inc., 2002.

[19] DeBaz C, Hahn J, Lang L, Palomo L. Dental implant supported restorations improve quality of life in osteoporotic women. Intl J Dentistry. 2015, 451923, doi: https://doi.org/10.1155/2015/451923.

[20] Nelson J, Holland N, Moore C, McKenna G. Implant-supported fixed prostheses give greatest OHRQoL improvement. Evid Based Dent. 2019, 20, 119–20.

[21] Yoshida T, Masaki C, Komai H, Misumi S, Mukaido T, Kondo Y, Nakamoto T, Hosokawa R. Changes in oral health-related quality of life during implant treatment in partially edentulous patients: A prospective study. J Prosthodont Res. 2016, 60, 258–64.

[22] Alzarea BK. Assessment and evaluation of quality of life (OHRQoL) of patients with dental implants using the oral health impact profile (OHIP-14) – A clinical study. J Clin Diagn Res. 2016, 10, ZC57–60.

[23] Elsyad MA, Elgamal M, Askar OM, Al-Tonbary GY. Patient satisfaction and oral health-related quality of life (OHRQoL) of conventional denture, fixed prosthesis and milled bar overdenture for All-on-4 implant rehabilitation – A crossover study. Clin Oral Implants Res. 2019, 30, 1107–17.

[24] Lee I-C, Yang Y-H, Ho P-S, Lee I-C. Chewing ability, nutritional status and quality of life. J Oral Rehabil. 2014, 41, 79–86.

[25] Sivakumar I, Sajjan S, Ramaraju AV, Rao B. Changes in oral health-related quality of life in elderly edentulous patients after complete denture therapy and possible role of their initial expectation: A follow-up study. J Prosthodont. 2015, 24, 452–56.

[26] Aarabi G, John MT, Schierz O, Heydecke G, Reissmann D. The course of prosthodontic patients' oral health-related quality of life over a period of 2 years. J Dent. 2015, 43, 261–68.

[27] Bennadi D, Reddy CVK. Oral health related quality of life. J Int Soc Prev Community Dent. 2013, 3, 1–6.

[28] Locker D, Miller Y. Evaluation of subjective oral health status indicators. J Public Health Dent. 1994, 54, 167–76.

[29] McGrath C, Broder H, Wilson-Genderson M. Assessing the impact of oral health on the life quality of children: Implications for research and practice. Community Dent Oral Epidemiol. 2004, 32, 81–85.

[30] Slade GD, Spencer AJ. Development and evaluation of the Oral Health Impact Profile. Community Dent Health. 1994, 11, 3–11.

[31] Raval N, Kohnen S, Lee E. 3 types of dental implants (which one is best for you?). Restorative Dent. 2021, https://www.implantandperiodonticspecialists.com/blog/dental-implant-types/

[32] Lopez JR. Implant Supported Dentures vs Bridges. 2017, https://www.lawtoncosmeticdentistry.com/blog/2017/4/18/implant-supported-dentures-vs-bridges/.

[33] Sahwil H. Implant retained versus implant supported dentures. Dental Technology. https://blog.ddslab.com/implant-retained-versus-implant-supported-dentures

[34] Fixed or Removable: Deciding Which Implant-Supported Bridge Is Best For You. 2017, https://www.cjdentalarts.com/blog/48315-fixed-or-removable-deciding-which-implant-supported-bridge-is-best-for-you.

[35] Fixed vs. Removable Dental Implant Options. 2020, https://www.susanhodds.com/fixed-vs-removable-dental-implant-options/#:~:text=The%20difference%20between%20the%20two%20options%20is%20that,and%20can%20only%20be%20removed%20by%20a%20dentist.

[36]   Cost of Dental Implants, https://www.memphisdenturesandimplants.com/cost-of-dental-implants-memphis.
[37]   Competence in Implant Esthetics. Manual. Implant Superstructures for Crown and Bridge Restorations, https://docplayer.net/23543919-Competence-in-implant-esthetics-manual-implant-superstructures-for-crown-and-bridge-restorations.html.
[38]   Succaria F, Morgano SM. Prescribing a dental ceramic material: Zirconia vs lithium-disilicate. Saudi Dent J. 2011, 23, 165–66.
[39]   Zarone F, Di Mauro MI, Ausiello P, Ruggiero G, Sorrentino R. Current status on lithium disilicate and zirconia: A narrative review. BMC Oral Health, 2019, 19, 134, doi: https://doi.org/10.1186/s12903-019-0838-x
[40]   Fasbinder DJ, Joseph B, Dennison JB, Donald Heys D, Gisele Neiva G. A clinical evaluation of chairside lithium disilicate CAD/CAM crowns. JADA. 2010, 141, 10S–14S.
[41]   Kapos T, Evans C. CAD/CAM technology for implant abutments, crowns, and superstructures. Int J Oral Maxillofac Implants. 2014, 29, 117–36.
[42]   Bachhav VC, Aras MA. Zirconia-based fixed partial dentures: A clinical review. Quintessence Int. 2011, 42, 173–82.
[43]   Fuster-Torres MA, Albalat S, Luis Alcañiz Raya M, Penarrocha-Diago M. CAD / CAM dental systems in implant dentistry: Update. Med Oral Patol Oral y Cirugía Bucal. 2009, 14, E141–5.
[44]   Kanazawa M, Inokoshi M, Minakuchi S, Ohbayashi N. Trial of a CAD/CAM system for fabricating complete dentures. Dent Mater J. 2011, 30, 93–96.
[45]   Miyazali T, Hotta Y. CAD/CAM systems available for the fabrication of crown and bridge restorations. Aust Dent J. 2011, 56, 97–106.
[46]   Al Farawati F. Digital Fabrication of Dental Implant Prostheses. 2018, https://decisionsindentistry.com/article/digital-fabrication-of-dental-implant-prostheses/.
[47]   Kurbad A. Final restoration of implants with a hybrid ceramic superstructure. Int J Comput Dent. 2016, 19, 257–79.
[48]   Rosentritt M, Hahnel S, Engelhardt F, Behr M, Preis V. In vitro performance and fracture resistance of CAD/CAM-fabricated implant supported molar crowns. Clin Oral Investig. 2017, 21, 1213–19.
[49]   Weyhrauch M, Igiel C, Scheller H, Weibrich G, Lehmann KM. Fracture strength of monolithic all-ceramic crowns on titanium implant abutments. Int J Oral Maxillofac Implants. 2016, 31, 304–09.
[50]   Conejo J, Kobayashi T, Anadioti E, Blatz MB. Performance of CAD/CAM monolithic ceramic Implant-supported restorations bonded to titanium inserts: A systematic review. Eur J Oral Implantol. 2017, 10, 139–46.
[51]   The Role An Abutment Plays In A Dental Implant. https://www.colgate.com/en-us/oral-health/bridges-and-crowns/the-role-an-abutment-plays-in-a-dental-implant.
[52]   Shafie HR, White BA. Clinical and laboratory manual of dental implant abutments. In: Shafie HR, ed. Clinical and Laboratory Manual of Dental Implant Abutments. John Wiley & Sons, Inc., 2014, 1–16.
[53]   Shafie HR, White BA. Implant Abutment Materials, https://pocketdentistry.com/1-implant-abutment-materials/.
[54]   Oshida Y, Hashem A. Titanium-porcelain system Part I: Oxidation kinetics of nitrided pure titanium, simulated to porcelain firing process. Biomed Mater Eng. 1993, 3, 185–98.
[55]   Oshida Y, Fung LW, Isikbay SC. Titanium-porcelain system. Part II. Bond strength of fired porcelain on nitrided pure titanium. Biomed Mater Eng. 1997, 7, 13–34.
[56]   Surgical Steel vs Stainless steel: What are the differences? https://www.rocheindustry.com/surgical-steel-vs-stainless-steel/.

[57] Welander M, Abrahamsson I, Berglundh T. The mucosal barrier at implant abutments of different materials. Clin Oral Implants Res. 2008, 19, 635–41.

[58] Kapos T, Ashy LM, Gallucci GO, Weber HP, Wismeijer D. Computer-aided design and computer-assisted manufacturing in prosthetic implant dentistry. Int J Oral Maxillofac Implants. 2009, 24, 110–17.

[59] Sailer I, Philipp A, Zembic A, Pjetursson BE, Hämmerle CH, Zwahlen M. A systematic review of the performance of ceramic and metal implant abutments supporting fixed implant reconstructions. Clin Oral Implants Res. 2009, 20, 4–31.

[60] Zembic A, Bösch A, Jung RE, Hämmerle CH, Sailer I. Five-year results of a randomized controlled clinical trial comparing zirconia and titanium abutments supporting single-implant crowns in canine and posterior regions. Clin Oral Implants Res. 2013, 24, 384–90.

[61] Wismeijer D, Brägger U, Evans C, Kapos T, Kelly JR, Millen C, Wittneben JG, Zembic A, Taylor TD. Consensus statements and recommended clinical procedures regarding restorative materials and techniques for implant dentistry. Int J Oral Maxillofac Implants. 2014, 29, 137–40.

[62] Abutment materials. https://www.for.org/en/treat/treatment-guidelines/edentulous/treat ment-options/implant-prosthetics-fixed/cement-retained-restorations/abutment-materials.

[63] Angled Abutments. https://dsisrael.com/index.php/15-deg-abutment.

[64] Lops D, Bressan E, Parpaiola A, Sbricoli L, Cecchinato D, Romeo E. Soft tissues stability of CAD/CAM and stock abutments in anterior regions: 2-year prospective multicentric cohort study. Clin Oral Implants Res. 2015, 26, 1436–42.

[65] Korsch M, Walther W. Prefabricated versus customized abutments: A retrospective analysis of loosening of cement-retained fixed implant-supported reconstructions. Int J Prosthodont. 2015, 28, 522–26.

[66] Long L, Alqarni H, Masri R. Influence of implant abutment fabrication method on clinical outcomes: A systematic review. Eur J Oral Implantol. 2017, 10, 67–77.

[67] Wittneben J-G. Abutment selection and long-term success. https://www.straumann.com/ ca/en/shared/news/prosthetic-excellence/wittneben-article-abutment-selection-long-term-success.html.

[68] Wittneben JG, Buser D, Salvi GE, Bürgin W, Hicklin S, Brägger U. Complication and failure rates with implant-supported fixed dental prostheses and single crowns: A 10-year retrospective study. Clin Implant Dent Relat Res. 2014, 16, 356–64.

[69] Elsayed MD. Biomechanical factors that influence the bone-implant-interface. Res Rep Oral Maxillofac Surg. 2019, 3, 023, http://www.org/10.23937/iaoms-2017/1710023.

[70] Boggan RS, Strong JT, Misch CE, Bidez MW. Influence of hex geometry and prosthetic table width on static and fatigue strength of dental implants. J Prosthet Dent. 1999, 82, 436–40.

[71] Hansson S, Werke M. The implant thread as a retention element in cortical bone: The effect of thread size and thread profile: A finite element study. J Biomech. 2003, 36, 1247–58.

[72] Frost HM. Skeletal structural adaptations to mechanical usage (SATMU): 1. Redefining Wolff's law: The bone modeling problem. Anat Rec. 1990, 226, 403–13.

[73] Ivanoff CJ, Gröndahl K, Sennerby L, Bergström C, Lekholm U. Influence of variations in implant diameters: A 3-5 year retrospective clinical report. Int J Oral Maxillofac Implants. 1999, 14, 173–80.

[74] Kong L, Sun Y, Hu K, Li D, Hou R, Yang J, Liu B. Bivariate evaluation of cylinder implant diameter and length: A three-dimensional finite element analysis. J Prosthodont. 2008, 17, 286–93.

[75] Kong L, Gu Z, Li T, Wu J, Hu K, Liu Y, Zhou H, Liu B. Biomechanical optimization of implant diameter and length for immediate loading: A nonlinear finite element analysis. Int J Prosthodont. 2009, 22, 607–15.

[76] Renouard F, Nisand D. Impact of implant length and diameter on survival rates. Clin Oral Implants Res. 2006, 17, 35–51.

[77] What are Mini Dental Implants? https://www.northpointefamilydental.com/what-are-mini-dental-implants.php.

[78] Dental Implants Vs Mini Implants. https://bitelock.com/p/Dental-Implants-Vs-Mini-Implants-p30904.asp.

[79] Skeletal Orthodontic Anchorage. https://oralsurgeryny.com/p/dental-implants-Scarsdale-NY-Skeletal-Orthodontic-Anchorage-p48695.asp.

[80] Ellias CN, de Oliveira Ruellas AC, Fernandes DJ. Orthodontic implants: Concepts for the orthodontic practitioner. Intl J Dent. 2012, 549761, doi: 10.1155/2012/549761.

[81] Papadopoulos MA, Tarawneh F. The use of miniscrew implants for temporary skeletal anchorage in orthodontics: A comprehensive review. Oral Surg Oral Med Oral Pathol Oral Radiol Endod. 2007, 103, e6–15.

[82] Morais LS, Serra GG, Muller CA, Muller CA, Andrade LR, Palermo EFA, Elias CN, Meyers M. Titanium alloy mini-implants for orthodontic anchorage: Immediate loading and metal ion release. Acta Biomater. 2007, 3, 331–39.

[83] https://www.medicalexpo.com/medical-manufacturer/orthodontic-anchor-dental-mini-implant-51611.html.

[84] Favero L, Brollo P, Bressan E. Orthodontic anchorage with specific fixtures: Related study analysis. Am J Orthod Dentofacial Orthop. 2002, 122, 84–94.

[85] Park YC, Lee SY, Kim DH, Jee SH. Intrusion of posterior teeth using mini-screw implants. Am J Orthod Dentofacial Orthop. 2003, 123, 690–94.

[86] Weyh AM, Nocella R, Salman SO. Commentary – Step-by-step: Zygomatic implants. J Oral & Maxillofac Surg. 2020, 78, E6–9.

[87] Chrcanovic B, Albrektsson T, Wennerberg A. Survival and complications of zygomatic implants: An updated systematic review. J Oral Maxillofac Surg. 2016, 74, 1949–64.

[88] Davó R, David L. Quad zygoma: Technique and realities. Oral Maxillofac Surg Clin North Am. 2019, 31, 285–97.

[89] Atcha I. Everything You Need to Know About Zygomatic Implants. 2021, https://newteethchicagodentalimplants.com/everything-you-need-to-know-about-zygomatic-implants/.

[90] Saini M, Singh Y, Arora P, Arora V, Jain K. Implant biomaterials: A comprehensive review. World J Clin Cases. 2015, 16, 52–57.

[91] Chiapasco M, Casentini P, Zaniboni M, Corsi E, Anello T. Titanium-zirconium alloy narrow-diameter implants (Straumann Roxolid(®)) for the rehabilitation of horizontally deficient edentulous ridges: Prospective study on 18 consecutive patients. Clin Oral Implants Res. 2012, 23, 1136–41.

[92] Najeeb S, Mali M, Yaqin Syed AU, Zafar MS, Khurshid Z, Alwadaani A, Matinlinna JP. Dental implants materials and surface treatments. Advanced Dent Biomater. 2019, 581–98.

[93] Oshida Y. Bioscience and Bioengineering of Titanium Materials. Elsevier, 2007.

[94] Teng LD, Wang FM, Li WC. Thermodynamics and microstructure of Ti–ZrO2 metal–ceramic functionally graded materials. Mater Sci Eng A. 2000, 293, 130–36.

[95] Lin KL, Lin CC. Effected of annealing temperature on microstructural development at the interface between zirconia and titanium. J Am Ceram Soc. 2007, 90, 893–99.

[96] Lada P, Miazga A, Bazarnik P, Konopka K, Szafran M. Zirconia–titanium interface in ceramic based composite. Powder Metall Metal Ceram. 2018, 57, 458–64.

[97] What Materials Are Dental Implants Made from? 2020, https://sikesoms.com/what-materials-are-dental-implants-made-from/.

[98] Miyazaki T, Yutani T, Murai N, Kawata A, Shimizu H, Uejima N, Miyazaki Y, Oshida Y. Early osseointegration attained by UV-photo treated implant into Piezosurgery-prepared site.

Report III. Influence of surface treatment by hydrogen peroxide solution and determination of early loading timing. Int J Dent Oral Health. 2021, 7, http://www.dx.doi.org/10.16966/2378-7090.381

[99] Sahwil H. Zirconia vs Titanium Dental Implants. https://blog.ddslab.com/zirconia-vs-titanium-dental-implants.

[100] Daneshmand N. Titanium vs. Ceramic Dental Implants, What You Need to Know in 2021. https://mdperio.com/blog/ceramic-vs-metal-dental-implants/.

[101] Microstructure of titanium. 2012, https://www.physicsforums.com/threads/microstructure-of-titanium.609116/.

[102] Ceramic vs. titanium implants: When to choose which? 2019, https://www.nobelbiocare.com/blog/tips-and-techniques/ceramic-vs-titanium-implants-when-to-choose-which/.

[103] Huang J. Dental Implants: Titanium or Zirconia, Which Is Better? 2018, https://allin1dental.com/dental-implants-titanium-or-zirconia-which-is-better/.

[104] Titanium Versus Ceramic Dental Implants. https://cornerstonedentistrync.com/titanium-vs-ceramic-dental-implant/.

[105] Atcha I. Dental Implants: Zirconia or Titanium? 2021, https://newteethchicagodentalimplants.com/dental-implants-zirconia-or-titanium/.

[106] Nejad M, Stabley K. Zirconia Dental Implants Vs. Titanium Implants. https://www.beverlyhillsladentist.com/blog/are-zirconia-implants-better-than-titanium/.

[107] Zirconia vs. Titanium Implants: Which One Is Right for You? https://www.colgate.com/en-us/oral-health/implants/zirconia-vs-titanium-implants-which-one-is-right-for-you.

[108] Gupta S, Noumbissi S, Kunrath MF. Nano modified zirconia dental implants: Advances and the frontiers for rapid osseointegration. Medical Devices & Sensors. 2020, doi: https://doi.org/10.1002/mds3.10076.

[109] AlKahtani RN. The implications and applications of nanotechnology in dentistry: A review. Saudi Dent J. 2018, 30, 107–16.

[110] Oshida Y. Surface Engineering and Technology for Biomedical Implants. Momentum Press, 2014.

[111] Rezaei NM, with 10 co-authors. Biological and osseointegration capabilities of hierarchically (meso-/micro-/nano-scale) roughened zirconia. Int J Nanomedicine. 2018, 13, 3381–95.

[112] Schünemann FH, Galárraga-Vinueza ME, Magini R, Fredel M, Silva F, Souza JCM, Zhang Y, Henriques B. Zirconia surface modifications for implant dentistry. Mater Sci Eng: C. 2019, 98, 1294–305.

[113] Aboushelib MN, Feilzer AJ, Kleverlaan CJ. Bonding to zirconia using a new surface treatment. J Prosthodont. 2010, 19, 340–46.

[114] Mehrali M, Shirazi FS, Mehrali M, Metselaar HSC, Kadri NAB, Osman NAA. Dental implants from functionally graded materials. J Biomed Mater Res Part A. 2013, 101A, 3046–57.

[115] Oshida Y, Miyazaki T, Tominaga T. Some biomechanistic concerns on newly developed implantable materials. J Oral Dent Health. 2018, 4, 0117, https://scientonline.org/open-access/some-biomechanistic-concerns-on-newly-developed-implantable-materials.pdf

[116] Oshida Y. Hydroxyapatite – Synthesis and Applications. Momentum Press, 2015.

[117] Banhart J, Baumeister J. Deformation characteristics of metal foams. J Mater Sci. 1998, 33, 1431–40.

[118] Banhart J. Manufacture, characterization and application of cellular metals and metal foams. Prog Mater Sci. 2001, 46, 559–632.

[119] Lefebvre LP, Banhart J, Dunand DC. Porous metals and metallic foams: Current status and recent developments. Adv Eng Mater. 2008, 10, 775–87.

[120] Gibson LJ. Mechanical behavior of metallic foams. Ann Rev Mater Sci. 2000, 30, 191–227.

[121] Papadopoulos DP, Konstantinidis ICh, Papanastasiou N, Skolianos S, Lefakis H, Tsipas DN. Mechanical properties of Al metal foams. Mat Lett. 2004, 58, 2574–78.

[122] Ikarashi Y, Toyoda K, Kobayashi E, Doi H, Yoneyama T, Hamanaka H, Tsuchiya T. Improved biocompatibility of titanium–zirconium (Ti–Zr) alloy: Tissue reaction and sensitization to Ti–Zr alloy compared with pure Ti and Zr in rat implantation study. Mater Trans. 2005, 46, 2260–67.

[123] Grandin HM, Berner S, Dard M. A review of titanium zirconium (TiZr) alloys for use in endosseous dental implants. Materials. 2012, 5, 1348–60.

[124] Cordeiro JM, Faveranic Leonardo P, Grandini Carlos R. Characterization of chemically treated Ti-Zr system alloys for dental implant application. Mat Sci Eng C. 2018, 92, 849–61.

[125] Karapataki S. From titanium to zirconia implants. Ceramic Implants. 2017, 1, 6–12.

[126] Donaca R, Rausch P. Shifting of dental implants through ISO standards. Ceramic Implants. 2017, 1, 34–39.

[127] Rues S, Schmitter M, Kappel S, Sonntag R, Ketzer JP, Ndorf J. Effect of bone quality and quantity on the primary stability of dental implants in a simulated bicortical placement. Clin Oral Investig. 2021, 25, 1265–72.

[128] Sundell G, Dahlin C, Andersson M, Thuvander M. The bone-implant interface of dental implants in humans on the atomic scale. Acta Biomater. 2017, 48, 445–50.

[129] O'Sullivan D, Sennerby L, Jagger D, Meredith N. A comparison of two methods of enhancing implant primary stability. Clin Implant Dent Relat Res. 2004, 6, 48–57.

[130] Chrcanovic BR, Albrektsson T, Wennerberg A. Bone quality and quantity and dental implant failure: A systematic review and meta-analysis. Int J Prosthodont. 2017, 30, 219–37.

[131] Ocran E. Muscles of mastication. 2021, https://www.kenhub.com/en/library/anatomy/the-muscles-of-mastication.

[132] Maini K, Dua A. StatPearls [Internet]. Treasure Island (FL): StatPearls Publishing, 2021, https://pubmed.ncbi.nlm.nih.gov/31869076/

[133] Bohluli B, Mohammad E, Oskui I, Moharamnejad N. Treatment of mandibular angle fracture: Revision of the basic principles. Chin J Traumatol. 2019, 22, 117–19.

[134] Misch CE. Progressive bone loading. Pocket Dentistry. https://pocketdentistry.com/32-progressive-bone-loading/.

[135] Bone density test. https://www.mayoclinic.org/tests-procedures/bone-density-test/about/pac-20385273.

[136] Bone density testing. https://www.betterhealth.vic.gov.au/health/conditionsandtreatments/bone-density-testing.

[137] Bone Densitometry (DEXA, DXA). https://www.radiologyinfo.org/en/info/dexa.

[138] Bone Density Exam/Testing. https://www.nof.org/patients/diagnosis-information/bone-density-examtesting/.

[139] Oshida Y, Hashem A, Nishihara T, Yapchulay MV. Fractal dimension analysis of mandibular bones: Toward a morphological compatibility of implants. BioMed Mater Eng. 1994, 4, 397–407.

[140] Lekholm U, Zarb GA. Patient Selection and Preparation. Tissue Integrated Prostheses: Osseointegration in Clinical Dentistry. Chicago, USA: Quintessence Publishing Company, 1985, 199–209.

[141] Al-Sabbagh M, Kutkut A. Immediate implant placement: Surgical techniques for prevention and management of complications. Dent Clin North Am. 2015, 59, 73–95.

[142] Infected Sites. Pocket Dentistry. https://pocketdentistry.com/4-infected-sites/.

[143] Alghamdi H. Methods to improve osseointegration of dental implants in low quality (Type-IV) bone: An overview. J Funct Biomater, 2018, 9, 7, doi: 10.3390/jfb9010007

[144] Vidyasagar L, Apse P. Biological response to dental implant loading / overloading. Implant overloading: Empiricism or science? Stomatologija. 2003, https://www.researchgate.net/publication/229055692_Biological_response_to_dental_implant_loadingoverloading_Implant_overloading_Empiricism_or_science.

[145] Hämmerle CH, Chen ST, Wilson TG. Consensus statements and recommended clinical procedures regarding the placement of implants in extraction sockets. Int J Oral Maxillofac Implants. 2004, 19, 26–28.

[146] Infected sites. Pocket Dentistry. https://pocketdentistry.com/4-infected-sites/.

[147] Goncalves SB, Correia J, Costa AC. Evaluation of dental implants using Computed Tomography. 2013 IEEE 3rd Portuguese Meeting in Bioengineering (ENBENG). https://ieeexplore.ieee.org/document/6518387.

[148] Bone density for dental implants. Pocket Dentistry. https://pocketdentistry.com/bone-density-for-dental-implants/.

[149] Does Bone Density Impact the Success of Dental Implants? 2017, https://www.dooleydental.com/blog/dental-implants//.

[150] Seriwatanachai D, Kiattavorncharoen S, Suriyan N, Boonsiriseth K, Wongsirichat N. Reference and techniques used in alveolar bone classification. J Interdiscipl Med Dent Sci, 2015, 3, 2, doi: http://dx.doi.org/10.4172/2376-032X.1000172

[151] Oliveira MR, Gonçalves A, Gabrielli MAC, de Andrade CR, Vieira EH, Pereira-Filho VA. Evaluation of alveolar bone quality: Correlation between histomorphometric analysis and Lekholm and Zarb classification. J Craniofac Surg. 2021, 32, doi: 10.1097/SCS.0000000000007405.

[152] Meredith N. Assessment of implant stability as a prognostic determinant. Int J Prosthodont. 1998, 11, 491–501.

[153] Turkyilmaz I, McGlumphy EA. Influence of bone density on implant stability parameters and implant success: A retrospective clinical study. BMC Oral Health, 2008, 8, 32, doi: 10.1186/1472-6831-8-32

[154] Jemt T, Lekholm U. Oral implant treatment in posterior partially edentulous jaws: A 5-year follow-up report. Intl J Oral Maxillofac Implants. 1993, 8, 635–40.

[155] Schnitman PA, Wohrle PS, Rubenstein JE. Immediate fixed interim prostheses supported by two-stage threaded implants: Methodology and results. J Oral Implantol. 1990, 16, 96–105.

[156] Juodzbalys G, Kubilius M. Clinical and radiological classification of the jawbone anatomy in endosseous dental implant treatment. J Oral Maxillofac Res, 2013, 4, e2, doi: 10.5037/jomr.2013.4202

[157] Mini Dental Implants: Uses, Surgery & Risks. https://sharedentalcare.com/mini-dental-implants/.

[158] Types of Zirconia Implants. https://naturaldentistrycenter.com/zirconiaimplants/zirconia-implant-types/.

[159] Comparison of Two/Three Piece & Single Piece Dental Implants. https://www.dentalimplantskerala.com/options/comparison-of-two-piece-and-single-piece-dental-implants/.

[160] Dona PA. Pros and Cons of One-Piece Dental Implants. 2016, http://cincinnatinewvoices.com/2016/12/06/pros-and-cons-of-one-piece-dental-implants/.

[161] Batra S. Dental Single piece Implants: Its Advantages And Disadvantages. 2018, https://dentistinnagpurblog.wordpress.com/2018/01/17/dental-single-piece-implants-its-advantages-and-disadvantages/.

[162] Zamani S. One-piece and two-piece implants demonstrate comparable stress levels in bone: Preliminary results of an FEA study. 2008, http://www.zimmerbiomet.co.il/images/lib_art ZOPZOPA991%2018.pdf.

[163] Rieger MR, Mayberry M, Brose MO. Finite element analysis of six endosseous implants. J Prosthet Dent. 1990, 63, 671–76.
[164] Wu AY-J, Hsu J-T, Chee W, Lin Y-T, Fuh L-J, Huang H-L. Biomechanical evaluation of one-piece and two-piece small-diameter dental implants: In-vitro experimental and three-dimensional finite element analyses. J Formos Med Assoc. 2016, 115, 794–800.
[165] Abdelwahed A, Mahrous AI, Abadallah MF, Asfour H, Aldawash HA, Alagha EI. Bacteriological evaluation for one-and two-piece implant design supporting mandibular overdenture. Nigerian Med J. 2015, 56, 400–03.
[166] Duda M, Matalon S, Lewinstein I, Harel N, Block J, Ormianer Z. One piece immediately loaded implants versus 1 piece or 2 pieces delayed – 3 years outcome. Implant Dent. 2016, 25, 109–13.
[167] de Oliveira Limírio JPJ, Lemos CAAL, de Luna Gomes JM, Minatel L, Rezende MCRA, Pellizzer EP. A clinical comparison of 1-piece versus 2-piece implants: A systematic review and meta-analysis. J Prosthet Dent. 2019, 124, 439–45.
[168] Goldman I. The goal of full mouth rehabilitation. J Prosthet Dent. 1952, 2, 246–51.
[169] Restore Your Oral Health and Dental Function. https://www.elitedentalcare.com/procedures/full-mouth-reconstruction.
[170] Glerum K. Full Mouth Reconstruction. https://www.smilesbyglerum.com/full-mouth-reconstruction/.
[171] Quiec C. What procedures are involved in a full mouth reconstruction. https://drqdental.net/what-procedures-are-involved-in-a-full-mouth-reconstruction/.
[172] Rutherford R. Full mouth reconstruction. Cosmetic Town J. 2017, https://www.cosmetictown.com/journal/articles/Teeth/Full-Mouth-Reconstruction.
[173] What are the Steps of a Full Mouth Reconstruction? https://advanceddentalofnewwindsor.com/blog/what-are-the-steps-of-a-full-mouth-reconstruction/.
[174] Full Mouth Reconstruction: Benefits, Risks, Cost, & Steps. 2021, https://www.rejuvdentist.com/cosmetic-dentistry/full-mouth-reconstruction/.
[175] Full Mouth Reconstruction Defined. https://www.iowadentalgroup.com/full-mouth-reconstruction-defined/.
[176] Full Mouth Reconstruction. American College of Prosthodontists. https://www.gotoapro.org/full-mouth-reconstruction/.
[177] Hedge TK. Full Mouth Reconstruction. https://www.yourdentistryguide.com/fmr/.
[178] Brayman K. A Definitive Guide to Full Mouth Reconstruction. 2019, https://www.katebraymandds.com/cosmetic-dentistry/full-mouth-reconstruction/.
[179] Bencharit S, Schardt-Sacco D, Border MB, Barbaro CP. Full mouth rehabilitation with implant-supported prostheses for severe periodontitis: A case report. Open Dent J. 2010, 4, 165–71.
[180] Song M-Y, Park J-M, Park E-J. Full mouth rehabilitation of the patient with severely worn dentition: A case report. J Adv Prosthodont. 2010, 2, 106–10.
[181] Shimizu T, Mizushiri M, Fukunishi K, Taniike N, Takenobu T, Nishimura M, Murata H. Full mouth reconstruction with dental implants in the conservative treatment of bilateral condylar fractures: A clinical letter. J Oral Implantol. 2015, 41, 89–92.
[182] Zeighami S, Siadat H, Nikzad S. Full mouth reconstruction of a bruxer with severely worn dentition: A clinical report. Case Report in Dentistry. 2015, 531618, doi: https://doi.org/10.1155/2015/531618.
[183] Moreira A, Freitas F, Nabais J, Caramâs J. Full mouth rehabilitation of a patient with bruxism using implant and tooth-supported monolithic zirconia with feldspathic veneers. J Clin Diagn Res. 2018, 12, ZD07–11.
[184] Guide To All-On-4 Dental Implants. 2018, https://hiossen.com/news/guide-to-all-on-4-dental-implants/.

[185] All-On-Four Full Mouth Reconstruction. https://thecdgofhouston.com/all-on-four-full-mouth-reconstruction/.

[186] Full Mouth Dental Implants. https://www.clearchoice.com/all-on-four-implants/.

[187] All-on-4, 5, 6 or 8 Full Mouth Dental Implants. https://www.handcraftedsmilesvp.com/cosmetic-dentist/full-mouth-dental-implants-all-on-four.

[188] What Is All On 4? https://bexleydental.com.au/dental-implants/what-is-all-on-4/.

[189] What To Expect During All-on-4 Dental Implant Recovery. https://www.elitedentalanddenture.com/blog/what-to-expect-during-all-on-4-dental-implant-recovery.

[190] Traditional Options Vs. All-on-4® or Teeth-in-a-day® Dental Implants. https://www.shorelineperio.com/traditional-options-vs-all-on-4-or-teeth-in-a-day-dental-implants/.

[191] Dental Implant Techniques. https://www.indovinaoralsurgery.com/procedures/dental-implants.

[192] Hodges JA. When is All-on-4 the best option? 5 factors to consider when creating your dental treatment plan. 2017, https://www.dentistryiq.com/dentistry/implantology/article/16365915/when-is-allon4-the-best-option-5-factors-to-consider-when-creating-your-dental-treatment-plan.

[193] Bone level or Tissue Level? 2020, https://mattheos.net/2020/12/18/.

[194] Patel M. How We Choose the Best Type of Implant for Each Individual Case. 2019, https://infinitydentalclinic.co.uk/blog/choose-best-type-implant-individual-case/.

[195] Bone Level or Tissue Level? 2020, https://mattheos.net/bone-level-or-tissue-level/.

[196] Price M. Bone Level vs Tissue Level. 2012; https://www.campbellacademy.co.uk/blog/blog/bone-level-tissue-level-2.

[197] Buser D, Schmid B, Belder UC, Cochran DL. The new bone level implants – Clinical rationale for the development and current indications for daily practice. Intl Dent SA. 2010, 12, http://www.moderndentistrymedia.com/may_june2010/buser.pdf

[198] Kumar VV, Sagheb K, Kämmerer PW, Al-Nawas B, Wagner W. Retrospective clinical study of marginal bone level changes with two different screw-implant types: Comparison between tissue level (TE) and bone level (BL) implant. J Maxillofac Oral Surg. 2014, 13, 259–66.

[199] Parke T, Sidiura T, Khandelwal N. A Comparison of Bone-level and Tissue-level Implants. Division: AADR/CADR Annual Meeting. 2018, https://iadr.abstractarchives.com/abstract/47am-2848409/a-comparison-of-bone-level-and-tissue-level-implants.

[200] Hadzik J, Botzenhart U, Krawiec M, Gedrange T, Heinemann F, Vegh A, Dominiak M. Comparative evaluation of the effectiveness of the implantation in the lateral part of the mandible between short tissue level (TE) and bone level (BL) implant systems. Ann Anat. 2017, 213, 78–82.

[201] Canullo L, Menini M, Bagnasco F, Yullio N, Pesce P. Tissue-level versus bone-level single implants in the anterior area rehabilitated with feather-edge crowns on conical implant abutments: An up to 5-year retrospective study. J Prosthet Dent. 2021, doi: https://doi.org/10.1016/j.prosdent.2021.01.031.

[202] Miyazaki T. unpublished data, 2021.

[203] What to know about tooth extraction. https://www.medicalnewstoday.com/articles/327170.

[204] Everything you need to know about dry socket. https://www.smilesofvalparaiso.com/blog/3-types-of-tooth-extractions-from-your-dentist-in-valparaiso/.

[205] Norkiewicz D. Can a Dental Implant be Placed at the Same Time as the Tooth Removal? 2015, https://rockvilledentalarts.com/can-a-dental-implant-be-placed-at-the-same-time-as-the-tooth-removal/.

[206] How Long Will I Need to Wait for a Dental Implant After an Extraction? 2019, https://maycontedds.com/how-long-will-i-need-to-wait-for-a-dental-implant-after-an-extraction/.

[207] Sakuma S, Piattelli A, Baldi N, Ferri M, Iezzi G, Botticelli D. Bone healing at implants placed in sites prepared either with a sonic device or drills: A split-mouth histomorphometric randomized controlled trial. Int J Oral Maxillofac Implants. 2020, 35, 187–95.

[208] Venkatakrishnan CJ, Bhuminathan S, Chandran CR. Dental implant site preparation – A review. Biomed Pharmacol J. 2017, 10(3), http://biomedpharmajournal.org/?p=16420

[209] Baker JA, Vora S, Bairam L, Kim H-I, Davis EL, Adreana S. Piezoelectric vs. conventional implant site preparation: Ex vivo implant primary stability. Clin Oral Implants Res. 2012, 23, 433–37.

[210] Atieh MA, Alsabeeha NHM, Tawse-Smith A, Duncan WJ. Piezoelectric versus conventional implant site preparation: A systematic review and meta-analysis. Clin Implant Dent Relat Res. 2018, 20, 261–70.

[211] Vercellotti T, Stacchi C, Russo C, Rebaudi A, Vincenzi G, Pratella U, Baldi D, Mozzati M, Monagheddu C, Sentineri R, Cuneo T, Di Alberti L, Carossa S, Schierano G. Ultrasonic implant site preparation using piezosurgery: A multicenter case series study analyzing 3,579 implants with a 1- to 3-year follow-up. Intl J Period Restor Dent. 2014, 34, 11–18.

[212] Stacchi C, Vercellotti T, Torelli L, Furlan F, Di Lenarda R. Changes in implant stability using different site preparation techniques: Twist drills versus piezosurgery. A single-blinded, randomized, controlled clinical trial. Clin Implant Dent Relat Res. 2013, 15, 188–97.

[213] Miyazaki T. Early osseointegration attained by UV-photo treated implant into piezosurgery-prepared site. Report I. Retrospective study on clinical feasibility. Int J Dent Oral Health. 2020, 6, http://www.dx.doi.org/10.16966/2378-7090.344

[214] Miyazaki T, Yutani T, Murai N, Kawata A, Shimizu H, Uejima N, Miyazaki Y. Early osseointegration attained by UV-photo treated implant into piezosurgery-prepared site Report II. Influences of age and gender. Int J Dent Oral Health. 2021, 7, http://www.dx.doi.org/10.16966/2378-7090.351

[215] Osseodensification – A systematic review and qualitative analysis of published literature. J Oral Biol Craniofac Res. 2020, 10, 375–80.

[216] Brown IS. Densah Bur Technology: A New Way to Place Dental Implants. https://www.theperiogroup.com/dental-implants/dansah-bur-teachnology-a-new-way-to-place-dental-implants/.

[217] Trisi P, Berardini M, Falco A, Vulpiani MP. New osseodensification implant site preparation method to increase bone density in low-density bone: In vivo evaluation in sheep. Implant Dent. 2016, 25, 1–8.

[218] Gapski R, Wang H-L, Mascarenhas P, Lang NP. Immediate implant loading. Clin Oral Implant Res. 2003, 14, 515–27.

[219] Rai A, Agrawal S, Dakarkar A. Utility of trephine drills in implant dentistry. J Maxillofac Oral Surg. 2015, 14, 506–08.

[220] Stajcic Z, Stajćić LS, Kalanovic M, Krasavcevic AD, Divekar N, Rodić M. Removal of dental implants: Review of five different techniques. Int J Oral Maxillofac Surg. 2015, 45, doi: 10.1016/j.ijom.2015.11.003.

[221] Suriyan N, Sarinnaphakorn L, Deeb GR, Bencharit S. Trephination-based, guided surgical implant placement: A clinical study. J Prosthetic Dent. 2018, 121, doi: 10.1016/j.prosdent.2018.06.004.

[222] Deeb JG, Frantar A, Deeb GR, Carrico CK, Rener-Sitar K. In vitro comparison of time and accuracy of implant placement using trephine and conventional drilling techniques under dynamic navigation. J Oral Implantol. 2021, 47, 199–204.

[223] Bhargava D, Thomas S, Pandey A, Deshpande A, Mishra SK. Comparative study to evaluate bone loss during osteotomy using standard drill, bone trephine, and alveolar expanders for implant placement. J Indian Prosthodont Soc. 2018, 18, 226–30.

[224] Ramasamy M, Giri RR, Subramonian K, Narendrakumar R. Implant surgical guides: From the past to the present. J Pharm Bioall Sci. 2013, 5, 98–102.

[225] Vercruyssen M, Fortin T, Widmann G, Jacobs R, Quirynen M. Different techniques of static/dynamic guided implant surgery: Modalities and indications. Periodontol. 2014, 66, 214–27.

[226] The Benefits of Dental Implant Surgical Guides. 2019, https://www.cedarwalkdentistry.com/blog/the-benefits-of-dental-implant-surgical-guides.

[227] Noharet R, Pettersson A, Bourgeois D. Accuracy of implant placement in the posterior maxilla as related to 2 types of surgical guides: A pilot study in the human cadaver. J Prosthet Dent. 2014, 112, 526–32.

[228] Abai S. Freehand vs. Guided Surgery: Clinical Considerations and Case Examples. 2016, https://glidewelldental.com/education/inclusive-dental-implant-magazine/volume-6-issue-4/freehand-vs-guided-surgery-clinical-considerations-and-case-examples/.

[229] Bruschi GB, Crespi R, Capparé P, Gherlone E. Clinical study of flap design to increase the keratinized gingiva around implants: 4-year follow-up. J Oral Implantol. 2014, 40, 459–64.

[230] Marchack CB, Moy PK. The use of a custom template for immediate loading with the definitive prosthesis: A clinical report. J Calif Dent Assoc. 2003, 31, 925–29.

[231] Chiu WK, Luk WK, Cheung LK. Three-dimensional accuracy of implant placement in a computer-assisted navigation system. Int J Oral Maxillofac Implants. 2006, 21, 465–70.

[232] Drill guides for every case scenario: SurgiGuide Cookbook. https://www.duluthdentalimplants.com/files/2016/04/Cookbook-credits-to-materialise.compressed.pdf.

[233] Brief J, Edinger D, Hassfeld S, Eggers G. Accuracy of image-guided implantology. Clin Oral Implants Res. 2005, 16, 495–501.

[234] Trobough KP, Garrett PW. Surgical guide techniques for dental implant placement. Decision Dent 2018, 4, 11–13.

[235] Vermeulen J. The accuracy of implant placement by experienced surgeons: Guided vs freehand approach in a simulated plastic model. Int J Oral Maxillofac Implants. 2017, 32, 617–24.

[236] Cassetta M, Giansanti M, Di Mambro A, Stefanelli LV. Accuracy of positioning of implants inserted using a mucosa-supported stereolithographic surgical guide in the edentulous maxilla and mandible. Int J Oral Maxillofac Implants. 2014, 29, 1071–78.

[237] Ochi M, Kanazawa M, Sato D, Kasugai S, Hirano S, Minakuchi S. Factors affecting accuracy of implant placement with mucosa-supported stereolithographic surgical guides in edentulous mandibles. Comput Biol Med. 2013, 43, 1653–60.

[238] Nickenig HJ, Wichmann M, Hamel J, Schlegel KA, Eitner S. Evaluation of the difference in accuracy between implant placement by virtual planning data and surgical guide templates versus the conventional free-hand method – A combined in vivo – In vitro technique using cone-beam CT (Part II). J Craniomaxillofac Surg. 2010, 38, 488–93.

[239] Raico Gallardo YN, da Silva-Olivio IR, Mukai E, Morimoto S, Sesma N, Cordaro L. Accuracy comparison of guided surgery for dental implants according to the tissue of support: A systematic review and meta-analysis. Clin Oral Implants Res. 2017, 28, 602–12.

[240] Arfai NK, Kiat-Amnuay S. Radiographic and surgical guide for placement of multiple implants. J Prosthet Dent. 2007, 97, 310–12.

[241] Misch CE, Dietsh-Misch F Diagnostic casts, preimplant prosthodontics, treatment prostheses, and surgical templates. In: Misch CE ed., Contemporary Implant Dentistry. 2nd ed. St Louis, Mo, Mosby; 1999, 135–50.

[242] D'Souza KM, Ara MA. Types of implant surgical guides in dentistry: A review. J Oral Implantol. 2012, 38, 643–52.

[243] Stumpel LJ. Cast-based guided implant placement: A novel technique. J Prosthet Dent. 2008, 100, 61–69.

[244] 3D Guided Dental Implants. https://chendental.com/our-services/3d-guided-dental-implants

[245] Cordioli G, Majzoub Z. Heat generation during implant site preparation: An in vitro study. Int J Oral Maxillofac Implants. 1997, 12, 186–93.

[246] Abouzgia MB, James DF. Temperature rise during drilling through bone. Int J Oral Maxillofac Implants. 1997, 12, 342–53.

[247] Möhlhenrich SC, Modabber A, Steiner T, Mitchell DA, Hölzle F. Heat generation and drill wear during dental implant site preparation: Systematic review. Br J Oral Maxillofac Surg. 2015, 53, 679–89.

[248] Salomó-Coll O, Auriol-Muerza B, Lozano-Carrascal N, Hernández-Alfaro F, Wang H-L, Gargallo-Albiol J. Influence of bone density, drill diameter, drilling speed, and irrigation on temperature changes during implant osteotomies: An in vitro study. Clin Oral Investig. 2021, 25, 1047–53.

[249] Kawana H. Fundamental techniques in dental implant treatment. J Japanese Soc Oral Surgery. 2008, 56, 144–49, https://www.jstage.jst.go.jp/article/jjoms/56/3/56_144/_pdf/-char/ja

[250] Pal H, Nair KC, Hedge D. Custom-made tool for parallel placement of implants. Interdiscip Dent. 2018, 8, 23–26.

[251] Brosh T, Pilo R, Sudai D. The influence of abutment angulation on strains and stresses along the implant/bone interface: Comparison between two experimental techniques. J Prosthet Dent. 1998, 79, 328–34.

[252] Flanagan D. Parallelism of dental implants. J Oral Implantol, 2019, 45, 87, doi: https://doi.org/10.1563/aaid-joi-D-18-00291

[253] Renuard F, Rangert B. Biomechanical risk factors. In: Risk Factors In Implant Dentistry. UK: Quintessence, 1999, 44–45.

[254] Nunes DB, da Silva P, Pereira-Cenci T, Garbin CA, Schuh C, Boscato N. Correction of nonparallel implants for an implant-retained overdenture. Gen Dent. 2010, 58, e168–71.

[255] Dario LJ. A maxillary implant overdenture that utilizes angle-correcting abutments. J Prosthodont. 2002, 11, 41–45.

[256] Noelken R, Al-Nawas B. A modified surgical approach for hard and soft tissue reconstruction of severe periimplantitis defects: Laser-assisted periimplant defect regeneration (LAPIDER). Int J Implant Dent, 2020, 10, 22, doi: 10.1186/s40729-020-00218-6

[257] Lindhe J, Lang NP, Karring T. Clinical Periodontology and Implant Dentistry, 5th ed., in English. Oxford, UK: Blackwell Munksgaard, 2008.

[258] Harris CM, Laughlin R. Reconstruction of hard and soft tissue maxillofacial defects. Atlas Oral Maxillofac Surg Clin North Am. 2013, 21, 127–38.

[259] Tomruk CO, Sençift MK, Capar GD. Prevalence of sinus floor elevation procedures and survival rates of implants placed in the posterior maxilla. Med Biotechnol. 2015, 134–39.

[260] Krasny K, Krasny M, Kaminski A. Two-stage closed sinus lift: A new surgical technique for maxillary sinus floor augmentation. Cell Tissue Bank. 2015, 16, 579–85.

[261] Woo I, Le BT. Maxillary sinus floor elevation: Review of anatomy and two techniques. Implant Dent. 2004, 13, 28–32.

[262] Uckan S, Deniz K, Dayangac E, Araz K, Ozdemi BH. Early implant survival in posterior maxilla with or without beta-tricalcium phosphate sinus floor graft. J Oral Maxillofac Surg. 2010, 68, 1642–45.

[263] Summers RB. A new concept in maxillary implant surgery: The osteotome technique. Compendium. 1994, 15, 152–58.

[264] Galindo-Moreno P, Avila G, Fernandez-Barbero JE, Aguilar M, Sánchez-Fernández E, Cutando A, Wang H-L. Evaluation of sinus floor elevation using a composite bone graft mixture. Clinical Oral Implants Res. 2007, 18, 376–82.

[265] Pjetursson BE, Tan WC, Zwahlen M, Lang NP. A systematic review of the success of sinus floor elevation and survival of implants inserted in combination with sinus floor elevation. J Clin Periodontol. 2008, 35, 216–40.

[266] Schleier P, Øyri H, Törpel J. The internal sinus floor elevation procedure is comparable to the conventional sinus floor elevation procedure in highly atrophic alveolar ridges: Results four years after loading in a randomized, controlled, blind pilot study. Dent Health Curr Res, 2018, 7, 3, doi: 10.4172/2470-0886.1000139

[267] Proissaefs P, Lozada J. The "Loma Linda pouch": A technique for repairing the perforated sinus membrane. Int J Periodontics Restorative Dent. 2003, 23, 593–97.

[268] Kirste M. Internal sinus lift – Patient-friendly with piezo surgery. EDI. 2019, 1, 108–10, https://www.wh.com/en_global/dental-newsroom/reportsandstudies/new-article/10631

[269] Toscano NJ, Holtzclaw D, Rosen PS. The effect of piezoelectric use on open sinus lift perforation: A retrospective evaluation of 56 consecutively treated cases from private practices. J Periodontol. 2010, 81, 167–71.

[270] Hurley LA, Stinchfield FE, Bassett AL, Lyon WH. The role of soft tissues in osteogenesis. An experimental study of canine spine fusions. J Bone Joint Surg. American Volume, 1959, 41-A, 1243–54.

[271] Hermann JS, Buser D. Guided bone regeneration for dental implants. Curr Opin Periodontol. 1996, 3, 168–77.

[272] Buser D, Schenk RK. Guided Bone Regeneration in Implant Dentistry. Hong Kong: Quintessence Books, 1994, ISBN 978-0867152494.

[273] Laskin D. Decision Making in Oral and Maxillofacial Surgery. Chicago, IL: Quintessence Pub. Co, 2007.

[274] Borzabadi-Farahani A, Zadeh HH. Orthodontic therapy in implant dentistry: Orthodontic implant site development. In: Tolstunov L, ed.. Vertical Alveolar Ridge Augmentation in Implant Dentistry: A Surgical Manual. John Wiley & Sons, 2016, 30–37.

[275] https://www.implant.ac/knowledge/article/217/.

[276] Resnik RR. Guided Bone Regeneration: 8 Steps to Successful Ridge Augmentation (1 CEU). 2020, https://glidewelldental.com/education/chairside-dental-magazine/volume-15-issue-2-implant-edition/guided-bone-regeneration-8-steps.

[277] 4 Common Types of Bone Grafts. 2021, https://biogennix.com/bone-grafting/4-common-types-of-bone-grafts-bx/.

[278] Liu J, Kerns DG. Mechanisms of guided bone regeneration: A review. Open Dent J. 2014, 8, 56–65.

[279] Misch CE, Dietsh F. Bone-grafting materials in implant dentistry. Implant Dent. 1993, 2, 158–67.

[280] Wang H-L, Avila G. Platelet rich plasma: Myth or reality?. Eur J Dent. 2007, 1, 192–94.

[281] Marx RE, Carlson ER, Eichstaedt RM, Schimmele SR, Strauss JE, Georgeff KR. Platelet-rich plasma: Growth factor enhancement for bone grafts. Oral Surg Oral Med Oral Pathol Oral Radiol Endod. 1998, 85, 638–46.

[282] Esposito M, Grusovin MG, Rees J, Karasoulos D, Felice P, Alissa R, Worthington H, Coulthard P. Effectiveness of sinus lift procedures for dental implant rehabilitation: A Cochrane systematic review. Eur J Oral Implantol. 2010, 3, 7–26.

[283] Pocaterra A, Caruso S, Bernardi S, Scagnoli L, Continenza MA, Gatto R. Effectiveness of platelet-rich plasma as an adjunctive material to bone graft: A systematic review and meta-analysis of randomized controlled clinical trials. Int J Oral Maxillofac Sur. 2016, 45, 1027–34.

[284] Hammerle CH, Jung RE. Bone augmentation by means of barrier membranes. Periodont 2000. 2003, 33, 36–53.

[285] Hutmacher D, Hurzeler MB, Schliephake H. A review of material properties of biodegradable and bioresorbable polymers and devices for GTR and GBR applications. Int J Oral Maxillofac Implants. 1996, 11, 667–78.

[286] Bunyaratavej P, Wang HL. Collagen membranes: A review. J Periodontol. 2001, 72, 215–29.

[287] Sheikh Z, Abdallah M, Hamdan N, Javaid M, Khurshid Z, Matilinna K. Barrier membranes for tissue regeneration and bone augmentation techniques in dentistry. In: Matinlinna J, ed.. Handbook of Oral Biomaterials. Singapore: Pan Stanford Publishing, 2014, 605–36.

[288] Pillai C, Paul W, Sharma CP. Chitin and chitosan polymers: Chemistry, solubility and fiber formation. Prog Polym Sci. 2009, 34, 641–78.

[289] Lee KY, Mooney DJ. Alginate: Properties and biomedical applications. Prog Polym Sci. 2012, 37, 106–26.

[290] Sumi Y, Miyaishi O, Tohnai I, Ueda M. Alveolar ridge augmentation with titanium mesh and autogenous bone. Oral Surg Oral Med Oral Pathol Oral Radiol Endod. 2000, 89, 268–70.

[291] Malchiodi L, Scarano A, Quaranta Metal. Rigid fixation by means of titanium mesh in edentulous ridge expansion for horizontal ridge augmentation in the maxilla. Int J Oral Maxillofac Implants. 1998, 13, 701–05.

[292] Brunette DM. Titanium in Medicine: Material Science, Surface Science, Engineering, Biological Responses and Medical Applications. Berlin: Springer, 2001.

[293] Elgali I, Oamr O, Dahlin C, Thomsen P. Guided bone regeneration: Materials and biological mechanisms revisited. Eur J Oral Sci. 2017, 125, 315–37.

[294] Malchiodi L, Scarano A, Quaranta M. Rigid fixation by means of titanium mesh in edentulous ridge expansion for horizontal ridge augmentation in the maxilla. Int J Oral Maxillofac Implants. 1998, 13, 701–05.

[295] Steflik DE, Corpe RS, Young TR, Buttle K. In vivo evaluation of the biocompatibility of implanted biomaterials morphology of the implant-tissue interactions. Implant Dent. 1998, 7, 338–50.

[296] Jovanovic SA, Schenk RK, Orsini Metal. Supracrestal bone formation around dental implants an experimental dog study. Int J Oral Maxillofac Implants. 1995, 10, 23–31.

[297] Decco O, Cura A, Beltran V, Lezcano M, Engelke W. Bone augmentation in rabbit tibia using microfixed cobalt-chromium membranes with whole blood, tricalcium phosphate and bone marrow cells. Int J Clin Exp Med. 2015, 8, 135–44.

[298] Harris RJ. Clinical evaluation of a composite bone graft with a calcium sulfate barrier. J Periodontol. 2004, 75, 685–92.

[299] Melo LG, Nagata MJ, Bosco AF, Ribeiro LL, Leite CM. Bone healing in surgically created defects treated with either bioactive glass particles, a calcium sulfate barrier, or a combination of both materials. A histological and histometric study in rat tibias. Clin Oral Implants Res. 2005, 16, 683–91.

[300] Anderud J, Jimbo R, Abrahamsson P, Isaksson SG, Adolfsson E, Malmstrom J, Kozai Y, Hallmer F, Wennerberg A. Guided bone augmentation using a ceramic space-maintaining device. Oral Surg Oral Med Oral Pathol Oral Radiol. 2014, 118, 532–38.

[301] Basile MA, d'Ayala GG, Malinconico M, Laurienzo P, Coudane J, Nottelet B, Ragione FD, Oliva A. Functionalized PCL/HA nanocomposites as microporous membranes for bone regeneration. Mater Sci Eng C Mater Biol Appl. 2015, 48, 457–68.

[302] Ribeiro N, Sousa SR, van Blitterswijk CA, Moroni L, Monteiro FJ. A biocomposite of collagen nanofibers and nanohydroxyapatite for bone regeneration. Biofabrication. 2014, 6, doi: 10.1088/1758-5082/6/3/035015.

[303] Shim JH, Huh JB, Park JY, Jeon YC, Kang SS, Kim JY, Rhie JW, Cho DW. Fabrication of blended polycaprolactone/poly(lactic-co-glycolic acid)/beta-tricalcium phosphate thin membrane

using solid freeform fabrication technology for guided bone regeneration. Tissue Eng Part A. 2013, 19, 317–28.

[304] Mota J, Yu N, Caridade SG, Luz GM, Gomes ME, Reis RL, Jansen JA, Walboomers XF, Mano JF. Chitosan/bioactive glass nanoparticle composite membranes for periodontal regeneration. Acta Biomater. 2012, 8, 4173–80.

[305] Nugala B, Kumar BBS, Sahitya S, Krishna PM. Biologic width and its importance in periodontal and restorative dentistry. J Conserv Dent. 2012, 15, 12–17.

[306] Khuller N, Sharma N. Biologic width: Evaluation and correction of its violation. J Oral Health Comm Dent. 2009, 3, 20–25.

[307] Felippe LA, Monteiro Júnior S, Vieira LC, Araujo E. Reestablishing biologic width with forced eruption. Quintessence. 2003, 34, 733–38.

[308] Rajendran M, Rao G, Logarani A, Sudagaran M, Badgujar S. Biologic width – Critical zone for a healthy restoration. IOSR J Dent Med Sci. 2014, doi: 10.9790/0853-13249398.

[309] Biological Width. https://www.stonerperiospecialists.com/periodontology-anatomy-biologic -width/.

[310] What is the Biological Width around a Tooth? 2017, https://www.toorongadentistry.com.au/ biologic-width-around-tooth/.

[311] Newman MG, Takei H, Klokkevold PR, Carranza FA. Newman and Carranza's Clinical Periodontology, 13th ed. Elsevier Saunders, 2019, ISBN: 9780323523004.

[312] Spear F. A Comprehensive Guide To Biologic Width. 2021, https://www.speareducation.com/ spear-review/2014/12/biologic-width-part.

[313] Fuchigami K, Munakata M, Kitazume T, Tachikawa N, Kasugai S, Kuroda S. A diversity of peri-implant mucosal thickness by site. Clin Oral Implants Res. 2017, 28, 171–76.

[314] Dental Implant. https://en.wikipedia.org/wiki/Dental_implant#cite_note-Sclar2003-36.

[315] Ioannou A, Kotsakis GA, McHale MG, Lareau DE, Hinrichs JE, Romanos GE. Soft tissue surgical procedures for optimizing anterior implant esthetics. Int J Dent. 2015, doi: https://doi.org/ 10.1155/2015/740764.

[316] Different Types of Gingival Grafting. https://www.holzingerperio.com/different-types-of-gingival-grafting/.

[317] Nemcovsky CE, Artzi Z, Tal H, Kozlovsky A, Moses O. A multicenter comparative study of two root coverage procedures: Coronally advanced flap with addition of enamel matrix proteins and subpedicle connective tissue graft. J Periodont. 2004, 75, 600–07.

[318] Happe A, Stimmelmayr M, Schlee M, Rothamel D. Surgical management of peri-implant soft tissue color mismatch caused by shine-through effects of restorative materials: One-year follow-up. Int J Periodont Restorative Dent. 2013, 33, 81–88.

[319] Noelken R, Al-Nawas B. A modified surgical approach for hard and soft tissue reconstruction of severe periimplantitis defects: Laser-assisted periimplant defect regeneration (LAPIDER). Int J Implant Dent, 2020, 6, 22, doi: https://doi.org/10.1186/s40729-020-00218-6

[320] Kadkhodazadeh M, Amid R, Kermani E, Mitakhori M, Hosseinpour S. Timing of soft tissue management around dental implants: A suggested protocol. General Dent. 2017, https:// www.agd.org/docs/default-source/self-instruction-(gendent)/gendent_mj17_amid.pdf? sfvrsn=757c7ab1_0

[321] Eichmiller FC. Dental materials. Contemporary Esthetic Dentistry. 2012, 33–50, https://www. sciencedirect.com/science/article/pii/B9780323068956000025.

[322] Watson S, Khoo E. An alternative to traditional dental impressions. Dental Health Procedures & Treatments. 2021, https://www.verywellhealth.com/digital-versions-of-dental-impressions-1059374.

[323] Digital Impressions: Virtually Perfect. https://www.yourdentistryguide.com/digital-impressions/.

[324] Warner B. The Benefits of Digital Dental Impressions. https://blog.ddslab.com/the-benefits-of-digital-dental-impressions.

[325] Papaspyridakos P, Gallucci GO, Chen CJ, Hanssen S, Naert I, Vandenberghe B. Digital versus conventional implant impressions for edentulous patients: Accuracy outcomes. Clin Oral Implants Res. 2016, 27, 465–72.

[326] Rutkūnas V, Gečiauskaitė A, Jegelevičius D, Vaitiekūnas M. Accuracy of digital impressions with intraoral scanners. A systematic review. Eur J Oral Implantol. 2017, 10, 101–20.

[327] Joda T, Brägger U. Patient-centered outcomes comparing digital and conventional implant impression procedures: A randomized crossover trial. Clin Oral Implants Res. 2016, 27, e185–9.

[328] Chi J. Common Problems with Digital Impressions – And How to Avoid Them. 2020, https://glidewelldental.com/education/chairside-dental-magazine/volume-15-issue-1/common-problems-digital-impressions.

Chapter 10
Success rate, fixation, osseointegration, and longevity

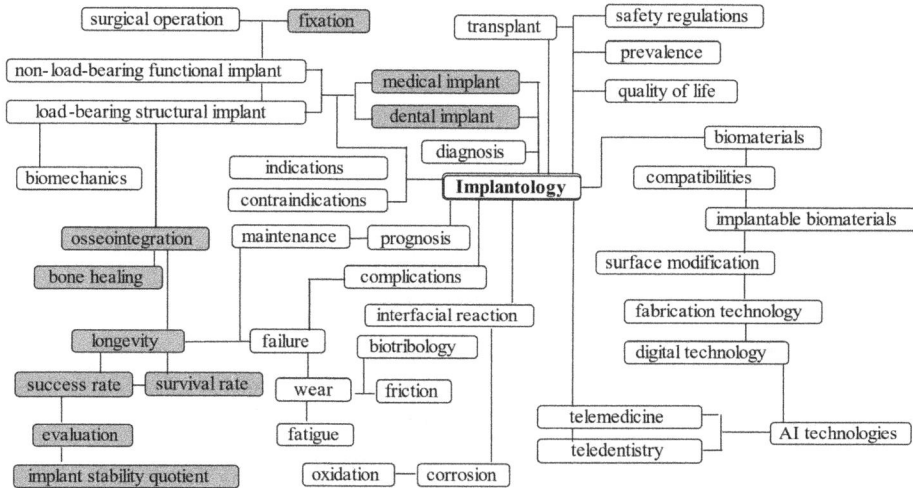

Implant fixation (or retention), bone healing promotion, osseointegration, loading timing, post-treatment hygiene management and others are crucial parameters to control and affect the overall implant success rate, survival rate, and longevity. Orthopedics and dental implants share some of these, while others do not since orthopedics emphasizes on biomechanics and biofunctionality but dental implantology, in addition to these, emphasizes the esthetics too.

10.1 Success rate and survival rate

Implants are often evaluated in terms of either success rate or survival rate. There is a definitive difference between these two terms [1]. The term success is denoted if a particular implant meets the success criteria it is being evaluated with, while the term survival simply implies that the implant exists in the body or mouth. The latter appears not to include evaluated biofunction. There are four major factors influencing the success rates of placed implants. They include (1) correct indication and favorable anatomic conditions (bone and mucosa), (2) good operative technique, (3) patient cooperation (oral hygiene), and (4) adequate superstructure design and fabrication.

https://doi.org/10.1515/9783110740134-011

10.1.1 Orthopedic implants

Total shoulder arthroplasty

Since it is reported that the total shoulder arthroplasty (TSA) would lead to improvement in a functional outcome and the implant survival would decline between 5 years and 10 years postoperatively, Denard et al. [2] examined the mid- to long-term functional outcome and implant survival of TSA in adults aged 55 years or younger with primary glenohumeral arthritis. It was found that (i) after TSA adjusted constant scores (which represent the gender- and age-matched function) 80.0 in patients free of revision of the glenoid compared with 43.6 in the group requiring revision of the glenoid and (ii) survivorship of the glenoid component with revision surgery for glenoid loosening as the endpoint was 98% at 5 years and 62.5% at 10 years. Several studies have reported revision rates following TSA, a common effective surgery for the treatment of refractory symptomatic shoulder disease [2, 3]. The revision rates for 43-month follow-up were 6.5% [4] and 6% for 59-month post-operative examination [5]. On the other hand, a wide range has been noted for 5- and 10-year implant revision rates, ranging 2–20% and 3–27%, respectively [6–8]. Singh et al. [9] reported that (i) 2,207 patients underwent 2,588 TSAs, with 63% of patients with underlying diagnosis and (ii) 212 TSAs were revised during the follow-up. At 5-, 10- and 20-years, implant survival rates were 94.2%, 90.2% and 81.4%, respectively.

Total hip arthroplasty

According to the United Nations, the world's population is aging rapidly with the number of people older than 60 years of age projected to double from 11% to 22% (2 billion) by 2050 [10]. This will fuel an increasing incidence of osteoarthritis (OA; which is a serious public health issue with symptomatic disease prevalent in 9% of men and 11% of women [11]) and demand for THR [12]. THR achieves excellent technical outcomes with 10-year survival exceeding 95%, 25-year implant survival greater than 80%, and significant benefits for pain, mobility, and physical function [13, 14].

In the UK, the National Institute of Health and Care Excellence set a benchmark in 2014, that individual components making up a total hip replacement (THR) are only recommended for people with end-stage arthritis, if they have 10-year revision rates of 5% or lower [15], since in an immortal cohort, all hip replacements will eventually fail because of processes such as infection, fracture, or a combination of normal tribological and biological processes, such as loosening and wear [16]. Evans et al. [16] identified 140 eligible articles reporting 150 series and included 44 of these series (13,212 total hip placements); while the national joint replacement registries from Australia and Finland provided data for 92 series (215,676 THRs) and reported that the 25-year pooled survival of hip replacements from case series was 77.6% and from joint replacement registries was 57.9%. Nearly 100,000 people underwent THR in the UK in 2018, and most can expect it to last at least 25 years.

However, some THRs fail and require revision surgery [17]. Evans et al. [17] analyzed a national, mandatory, prospective, cohort study (National Joint Registry for England, Wales, Northern Ireland and the Isle of Man (NJR)) of all THRs performed in England and Wales, including 664,761 patients with records in the NJR who have received a stemmed primary THR between 1 April 2003 and 31 December 2017 in one of the 461 hospitals, with OA as the only indication. It was reported that (i) the crude analyses including all THRs demonstrated better implant survival at the exemplar unit with an all-cause construct failure of 1.7% compared with 2.9% in the rest of the country after 13.9 years, (ii) the same was seen in analyses adjusted for age, sex, and American Society of Anesthesiology score, indicating that adjusted analyses restricted to the same implants as the exemplar unit show no demonstrable difference in restricted mean survival time between groups after 13.9 years. Hauer et al. [18] compared primary total hip arthroplasty (THA) implantations after fractured neck of femur between different countries in terms of THA number per inhabitant. All clinical studies on THA and HA for femoral neck fractures between 1999 and 2019 were reviewed and evaluated with a special interest on revision rate. Revision rate was calculated as the revision per 100 component years. THA registers were compared between different countries with respect to the number of primary implantations per inhabitant. It was reported that (i) THA studies showed a mean revision rate of 11.8% after 10 years, which was lower than a 24.6% 10-year revision rate for HA and (ii) eight arthroplasty registers were identified, revealing an annual average incidence of THA for fractured neck of femur of 9.7 per 100,000 inhabitants.

Evans et al. [16] mentioned that the typical patient who had a hip replacement in the UK in 2016 was 69.8 years old if female or 67.6 years old if male and had a BMI of 28 · 8. 90% of hip replacements were done for OA and 60% of recipients were female. Despite successful outcomes, THA revision rates have grown steadily in recent years [19]. Increased life expectancy in a globally aging population is associated with the increased use of THA, resulting in increased revision rates. Common causes of revision THA are wear, loosening, dislocation, or instability and infection. It should be stressed that a changing pattern of modes of failure has been observed; failures are separated into either early or late, and factors related to failure modes have been identified, as seen in Figure 10.1 [20]. With a more thorough understanding of reasons for failure, of revision timing and identifiable risk factors, surgeons are better placed to improve their THA outcomes.

Total knee arthroplasty
Hip and knee replacements have been routinely done for the treatment of end-stage arthritis over the past 40 years [21, 22]. Total of 76,000 THRs and 82,000 total knee replacements were done in 2014 in the UK alone [23]. Bayliss et al. [24] conducted the implant survival analysis on all patients within the Clinical Practice Research Datalink who had undergone THR or total knee replacement. These data were adjusted

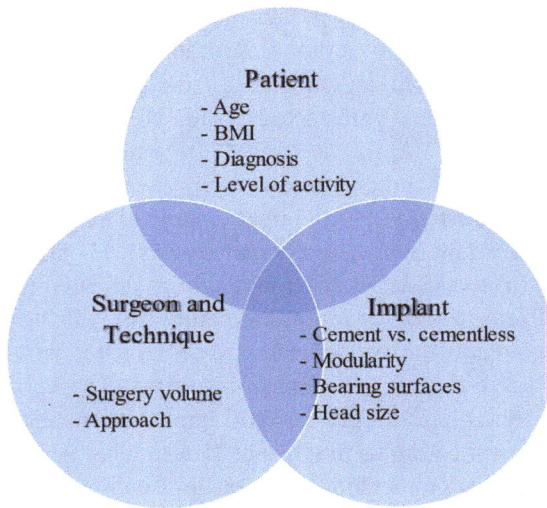

Figure 10.1: Factors related to total hip arthroplasty failure modes [20].

for all-cause mortality with data from the Office for National Statistics and used to generate lifetime risks of revision surgery based on increasing age at the time of primary surgery. All together 63,158 patients were identified who had undergone THR and 54,276 who had total knee replacement between 1 January 1991, and 10 August 2011, and followed up these patients to a maximum of 20 years. It was reported that (i) for THR, 10-year implant survival rate was 95.6% and 20-year rate was 85.0%, and (ii) for total knee replacement, 10-year implant survival rate was 96.1%, and 20-year implant survival rate was 89.7%. Argenson et al. [25] conducted a retrospective study to analyze 846 TKA patients at a minimum 10 years' follow-up. It was mentioned that the diagnosis was principally OA ($n = 752$ (89%)), and most TKAs were cemented ($n = 704$ (83%)), replacing the patella ($n = 668$ (79%)) and sacrificed the posterior cruciate ligament (PCL) ($n = 707$ [84%]), 65% being posterior-stabilized and 35% ultracongruent, with fixed (39%) or mobile bearing (61%). It was reported that (i) 63 (7.5%) revision surgeries were required, mainly for loosening ($n = 18$ (2%)) or infection ($n = 18$ (1.8%)) and (ii) overall 10-year survivorship was 92%. Between January 1988 and December 2006, a total of 3,014 primary total knee arthroplasties (TKAs) in 2,042 patients were performed, and survivorship analysis was performed [26]. Survivorship analysis showed a 10-year survival of 93.8% and a 20-year survival of 70.9%. Bae et al. [27] performed, from September 1990 to June 2009, 224 revision TKAs in 194 patients and reported that (i) the 5-, 8-, and 10-year survival rates were 97.2%, 91.6%, and 86.1%, respectively, (ii) rerevision TKAs were performed in 20 knees because of infection (seven knees), loosening (six knees), polyethylene wear (six knees), and periprosthetic fractures (one knee), and (iii) the long-term survival rate of revision TKA was satisfactory, but careful attention is necessary to detect the late failure.

Since there are conflicting reports of early and mid-term results of the high-flexion TKAs, Kim et al. [28] determined the long-term (minimum 20 years) clinical and radiographic and CT scan results, and the survival rates of high-flexion versus standard TKAs, on 95 patients (190 knees) with the mean follow-up 20.3 years. It was found that (i) revision of the TKA was performed in five knees (5.2%) with high-flexion TKA and in three knees (3.2%) with standard TKA and (ii) the rate of survival at 20 years was 94.8% in the high-flexion TKA group and 96.8% in the standard TKA group with reoperation for any reason. Although TKA has contributed to improvement of patient's quality of life, it has been associated with some complications including osteolysis, component loosening, and polyethylene wear. To prevent these postoperative complications, various fixation techniques have been employed [29, 30]. Currently, hybrid fixation has become the most preferred alternative to cementless fixation that has been receiving conflicting reviews. Survivorship of hybrid TKA is high according to many short-term follow-up studies [31, 32], but it has not been fully investigated in long-term studies [33, 34]. Based on this background, Choi et al. [35] evaluated the survivorship and clinical and radiographic results of hybrid TKA during ≥10 years of follow-up. It was reported that the survival rate was 93.8% at 12 years for all knees, and 96.5% when septic loosening was excluded.

10.1.2 Dental implants

As the living standard of the population improves, dental restoration has become the definitive therapy for most dental defects. Implants have been recognized as the "third set of teeth," since they are beautiful, comfortable, and have good chewing efficiency, making them feel like natural teeth [36]. Large-scale studies have reported that the long-term survival rates of implants are between 93.3% and 98% [37, 38], indicating that dental implants are an effective treatment for edentulousness. Busenlechner et al. [37] placed, from 2004 to 2012, a total of 13,147 implants in 4,316 patients at the Academy for Oral Implantology in Vienna and computed the survival rates after 8 years of follow-up and the impact of patient- and implant-related risk factors was assessed. It was found that (i) overall implant survival was 97%, independent of various factors of implant length, implant diameter, jaw location, implant position, local bone quality, previous bone augmentation surgery, or patient-related factors including osteoporosis, age, or diabetes mellitus. Krebs et al. [38] placed, between April of 1991 and May of 2011, 12,737 implants in 4,206 patients for a variety of clinical indications and reported that the Kaplan–Meier cumulative survival rate was 93.3% after 204 months. Artzi et al. [39] differentiated between the survival and success definitions of functional hydroxyapatite (HA)-coated implant prosthesis by evaluating a total of 248 implants (62 patients), 5–10 years in function. It was reported that (i) the accumulative survival rate after 5 and 10 years was 94.4% and 92.8%, respectively, (ii) accumulative success rates were 89.9%

and 54%, respectively, (iii) implants that were 13 and 15 mm in length (97.9% and 96.4%, respectively) had the highest survival rate, which was higher over implants of 8 and 10 mm in length (75% and 88.2%, respectively), and (iv) the survival rate of 4 mm diameter implants compared with 3.25 mm was 96.5% and 90.3%, respectively, concluding that (v) a distinguishable observation between survival and success rate was noted particularly in long-term observations, and (vi) implant length and diameter have an influence on the survival rate. Clinical parameter scores expressed an influence on the defined implant status.

Moraschini et al. [40] evaluated the survival and success rates of osseointegrated implants determined in longitudinal studies that conducted a follow-up of at least 10 years and reported that survival rates were 94.6%, indicating that osseointegrated implants are safe and present high survival rates and minimal marginal bone resorption in the long term. Thirty-five placed implants in 19 patients were subjected to assess the success and survival rates and it was reported that (i) there was a success rate of 74% after definitive prosthetic rehabilitation, while six implants showed bone loss of between 2 and 4 mm, being classified as satisfactory survival, which was 100%, (ii) there was no relationship between the success and/or survival rate and any of the parameters evaluated, (iii) four implants presented with peri-implant mucositis, while peri-implantitis was observed in two implants, and (iv) regarding the definitive restorations, 17 prostheses were classified as successful, while there were complications in eight prostheses [41]. Evaluating on 590 patients with 990 implants, Cochran et al. [42] reported that (i) the majority of implants were 10 and 12 mm long (78.7%) and were placed in type II and III bone (87%) and 73% of the implants were placed in the mandible, and 27% were placed in the maxilla, (ii) the cumulative survival rate was 99.56% at 3 years and 99.26% at 5 years, and (iv) the overall success rate was 99.12% at 3 years and 97.38% after 5 years. Studying on 185 patients with 271 implants, it was reported that (i) three implant failures were recorded, resulting in a cumulative survival rate of 98.6% after 5 years post-loading, (ii) at 5-year follow-up, the mean crestal bone loss was -0.28 ± 0.60 mm, and (iii) over 99% of patients reported satisfaction with the restoration as excellent or good [43]. After studying 1,078 cases (601 males and 477 females) with 2,053 implants, it was found that after implantation, 1,974 implants were retained, and the early survival rate was 96.15% [36].

10.2 Fixation

In general, implants of either orthopedic or dental implants can be mechanically fixed with screws, pins, wires, mechanical interference, glue, or cement. Biological fixation is also possible with surface treatments and improvements with biomaterials and chemical modifications. Augmenting of surface roughness and area with microspheres, meshes, and filaments is also of great importance when biological attachment through a tissue ongrowing or ingrowing is expected [44].

10.2.1 Fixation of orthopedic implants

The fixation of orthopedic implants (which includes major TSAs, THAs, TKAs, and bone plate, nails, bone screws, and other various connecting devices) has been one of the most difficult and challenging problems. The fixation can be achieved via: (a) direct mechanical fixation using screws, pins, wires, and so on; (b) passive or interference mechanical fixation where the implants are allowed to move or merely positioned onto the tissue surfaces; (c) bone cement fixation which is actually a grouting material; (d) biological fixation by allowing tissues to grow into the interstices of pores or textured surfaces of implants; (e) direct chemical bonding between implant and tissues; or (f) any combination of the above techniques [45].

For orthopedic fixations of TSAs, THAs, and TKAs implants, there are mainly cementation and cementless retention, and there has been debate between two methods. Table 10.1 compares both fixation methods [46–61].

Table 10.1: Comparison between cemented implants and cementless implants [46–61].

	Cementation	Cementless	Ref
Characteristics	A polymethylmethacrylate used to produce an interlocking fit between cancellous bone and prosthesis.	Biological fixation of bone to a surface coating on the prosthesis. Initial fixation is achieved by inserting a prosthesis slightly larger than the prepared bone-bed, generating compression hoop stresses, and obtaining the press-fit.	[46]
Features	Taper-slip stems or force-closed types are collarless with polished surface finish and achieve stability through micromotion at the prosthesis-cement interface promoting slight subsidence of the stem within the cement mantle, the generation of radial stresses, and ultimately compression at the bone-cement and prosthesis-cement interfaces. Composite beam stems or shape-closed types achieve stability through a solid bond between stem, cement and bone, maintaining the position of the stem within the mantle.	Uncemented stems exhibit a large range of designs, employing wedged, tapered, cylindrical, modular and anatomic shapes, and with the addition of proximal fins and ribs for added stability, and splines, flutes, and slots to reduce modulus of elasticity. More recently, shorter stem designs have been introduced with the aim of creating a more "physiological" loading of the proximal femur and reducing the problems of stress shielding.	[47–49]

Table 10.1 (continued)

	Cementation	Cementless	Ref
Advantages	– Allows a surgeon to affix prosthetic joint components to a slightly porous bone from osteoporosis. – Longer longevity, lasting 10–20 years. – A better option for patients with poor-quality bone due osteoporosis and have less chance of healthy bone regrowth. – Cement dries within 10 min of application, so soon as the surgery is complete, the implant is securely in place. – A small amount of antibiotic material can be added to the bone cement, helping to decrease the risk of post-surgical infection.	– Cementless total knee replacement, aka press-fit, may have a benefit over cemented implants, specifically for younger joint replacement patients and larger patients. – Cementless components offer a better long-term bond between the prostheses and bones. – Cementless eliminates the concern for a breakdown in cement. – Cementless implants use the patient's natural bone to hold the implants in place; they will last longer and form a more permanent bond with the patient's bones than cemented implants.	[50–51]
Disadvantages	– A breakdown of the cement can cause the artificial joint to come loose, which may prompt the need for revision surgery. – The cement debris can irritate the surrounding soft tissue and cause inflammation. – While rare, the cement can enter the bloodstream and end up in the lungs, a condition that can be life-threatening. This risk is greatest for people who undergo spinal surgeries.	– Uncemented fixation, compared with cemented fixation, was associated with a statistically significantly higher risk of aseptic revision. – Press-fit prostheses require healthy bones. Patients with low bone density due to osteoporosis may not be eligible for these components. – It can take up to 3 months for bone material to grow into a new joint component.	[51–53]

Table 10.1 (continued)

	Cementation	Cementless	Ref
Overall	Cement prosthesis enhances early fixation and prevents the periprosthetic bone resorption. Cemented implants (TKA) have been the gold standard in TKA with improved outcome and implant survivorship as long as 20 years; however, toxic effect may be a major concern and it is more difficult for revision surgery.	Cementless fixation in implants (TKA) has become more popular because it is associated with a long term of survival, particularly in younger patients. Use of cementless fixation could achieve a physiological bond between bone and implant which results in a prolonged survival from aseptic loosening; however, evidence of osteolysis has also been shown with cementless implants; thus, it has not been widely accepted in the field of joint surgery.	[54–58]
	The debate between cementless and cemented total knees is important as more than 1 million total knees are performed worldwide each year. Survivorship is one outcome but may not provide valuable information regarding age related factors which may suggest the use of one technique over another.		[54, 59–61]

Total hip arthroplasties

Harrison et al. [62] evaluated a new surface architecture for cementless orthopedic implants (OsteoAnchor), which incorporates a multitude of tiny anchor features for enhancing primary fixation, in an ovine hemiarthroplasty pilot study. It was reported that (i) intraoperative surgeon feedback indicated that superior primary fixation was achieved for the implant stems and rapid return to normal gait and load bearing was observed post-operation, and (ii) following a 16-week recovery time, histological evaluation of the excised femurs revealed in-growth of healthy bone into the porous structure of the implant stems, indicating that the potential for the OsteoAnchor surface architecture to enhance both the initial stability and long term lifetime of cementless orthopedic implants.

While a standard hybrid THA is recognized as a cementless hemi-spherical acetabular component combined with cemented femoral stem, the reverse hybrid total hip arthroplasty (rTHA) is composed of cemented socket and cementless stem. THA is the treatment of choice for primary and secondary arthritis of the hip joint. Uncemented total hip is the implant of choice in young adults with good bone stock in the acetabulum and femur. Correspondingly, cemented implants are preferred in patients with poor bone quality and in elderly. Uncemented prosthesis are relatively expensive when compared to cemented but have longer viability [63]. Nawfal et al. [63] evaluated the rTHAs and reported that (i) the mean functional activity score during preoperative stage was 5.29 ± 2.47 and during post-operative stage was 12.0 ± 1.41, (ii) there was a statistically significant difference between the pre- and post-operative scores in the study group, and (iii) the mean total score during preoperative stage was 33.43 ± 12.33 and during post-operative stage was 89.43 ± 8.73, indicating that (iv) rTHA is a good alternative to uncemented THA.

Lindalen et al. [64] analyzed the results from the Norwegian Arthroplasty Register, with up to 10 years of follow-up, on 3,963 operations in 3,630 patients. The rTHAs were used mainly since 2000. It was obtained that (i) an equal survival of cemented THR at 5 years (cemented: 97.0%); reverse hybrid: 96.7% and at 7 years (cemented: 96.0%); reverse hybrid: 95.6%, and (ii) with a follow-up of up to 10 years, rTHRs performed well, and similarly to all-cemented THRs from the same time period, suggesting that (iii) the reverse hybrid method might therefore be an alternative to all-cemented THR. Data extracted on the rTHAs, from January 2000 until December 2013, of 38,415 cases were analyzed and compared with cemented THAs [65]. It was reported that (i) a higher rate of revision for rTHAs than for cemented THAs, with an adjusted relative risk of revision (RR) of 1.4, (ii) at 10 years, the survival rate was 94% for cemented THAs and 92% for rTHA hybrids, (iii) a higher rate of early revision due to periprosthetic femoral fracture for reverse hybrids than for cemented THAs in patients aged 55 years or more (RR = 3.1), indicating that (iv) the rTHAs had a slightly higher rate of revision than cemented THAs in patients aged 55 or more; the difference in survival was mainly caused by a higher incidence of early revision due to periprosthetic femoral fracture in the reversed hybrid THAs [65]. Jain et al. [66] evaluated an implant survival with all cause revision and revision for aseptic loosening of either component as endpoints. Data was collected prospectively on 1,088 operation cases of rTHAs. It was reported that (i) 10-year implant survival (122 hips at risk) was 97.2% for all cause revision, 100% for aseptic acetabular loosening, and 99.6% for aseptic stem loosening, (ii) there was no difference in implant survival by age, gender, head size, or surgeon grade for all cause revision, and (iii) there was no difference in survival by gender, head size, or surgeon grade, concluding that (iv) the rTHA offers highly successful outcomes, irrespective of age, gender, head size and surgeon grade.

The average life expectancy of many people undergoing THRs exceeds 25 years and the demand for implants that increase the load-bearing capability of the bone without affecting the short- or long-term stability of the prosthesis is high [67]. Between the two alternatives of THRs, the cemented fixation method was mostly adopted owing to offering the immediate stability from cement-stem and cement-bone bonding interfaces after implant surgery [68]. Mechanical failure owing to cement damage and stress shielding of the bone are the main factors affecting the long-term survival of cemented hip prostheses and implant design must realistically adjust to balance between these two conflicting effects. Hence, the long-term survival of the cemented prosthesis is contingent on achieving a balance between stress shielding and cement damage [67]. Moussa et al. [67] had introduced a novel evolutionary technique to optimizing stem designs in a cemented hip prosthesis with the objective of minimizing stress shielding and cement damage. The self-regulated technique combines realistic and solid modeling of the implant and embedding medium, finite element to assess the levels of stress shielding, and cement damage in addition to a fast global optimization using orthogonal arrays and probabilistic restarts. The concept of functionally graded materials was included in this technique, which exhibit progressive change in composition, structure, and properties as a function of position within the material. Several processes have been reported to allow fabrication of such composites including plasma spraying, powder metallurgy, and physical vapor deposition [68–71]. The methodology can also be extended to other orthopedic joint implants, such as TKAs and TSAs, as well as dental implants [71–75].

Total knee arthroplasties

TKA implants may be cemented or cementless, depending on the type of fixation used to hold the implant in place. The majority of knee replacements are generally cemented into place. There are also implants designed to attach directly to the bone without the use of cement. These cementless designs rely on bone growth into the surface of the implant for fixation. Most implant surfaces are textured or coated so that the new bone actually grows into the surface of the implant to establish the osseointegration [76]. There is another type of a fixation, called as the hybrid TKA, which is accomplished with cemented tibial and cementless femoral/patellar components. Many studies of hybrid fixation have shown positive results [34, 77–82].

The extensive clinical experience with cement fixation and the long-term studies reported in literature justify its worldwide use. Cementation allows an easier surgical technique, ensures greater primary stability as demonstrated by radiostereometry analyses (RSAs) studies, may be useful for the delivery of local antibiotics (given the diffusion of antibiotic-loaded cements), and, finally, may produce a barrier able to prevent the diffusion of wear particles over the periprosthetic bone tissue, known to be the most frequent cause of aseptic failure of knee implants [83]. The majority of knee implants are cemented implants. On average, a cemented implant will last 10 to 20 years or more before it needs to be replaced. The cement dries very quickly, so the implant is securely in place when the surgery is complete. Cemented implants may be a better option for patients who have poor-quality bone due to conditions like osteoporosis; for these patients, bone growth may not be sufficient enough to hold the implant in place. Cemented implants are often recommended for patients who are older, overweight, and less active, as well [84].

Although it was mentioned that cemented TKA has been employed in the vast majority (89%) of the operations and it still deserves the status of gold standard for TKA in working-age patients [85], there are still several concerns associated with the cementation method. Implant–bone inducible micromotions and permanent migrations were noticed due to localized fatigue damage which could occur in a 6-degrees-of-freedom motion, leading to loosening the placed implants [86]. It was mentioned that cemented knee replacements in younger patients may not last as long as those in older patients, simply due to the increased activity and demand they place on the implants and cement which may lead to premature loosening. As the bone cement breaks down, it can also leave debris behind, which can irritate the tissues surrounding the joint. This can trigger an inflammatory response within the body as it tries to remove the debris. Eventually, it can lead to a condition called osteolysis, in which the body begins to remove small bits of bone around the implant as well. Osteolysis weakens the bone, causing further loosening of the knee implant [84]. Several studies have addressed the risks of extra-articular impingement of the cement mantle on the tibial insert, and third body wear induction by the release of particles in the articular space [85]. It was mentioned that, in the study of 162 revision TKAs, significant abrasive wear was found in 35% of retrieved cemented components (vs. 25% of retrieved cementless ones); however, the risk of residual particles after cementation may be prevented by thorough washing and cement removal before closing [87]. The thermal necrosis, which may be induced during the polymerization of the cement, carries a specific risk of tissue damage. The use of a cement mantle (generally 2 mm thick) introduces an additional surface (cement/bone plus cement/component) in the implant, increasing the risk of mobilization or of wear production [88]. It was measured that the exothermal temperature raise during the polymerization of polymethyl methacrylate (PMMA) was up to the most 65 degrees in C, which would cause a permanent and fatal damage on surrounding tissues [89]. There is still a debate on which of the two interfaces is the source of failure: the cement/bone interface is generally suggested to be the critical zone. An important study of the complications of THA demonstrated a high risk of fat embolism during the pressurization of the cement in the femoral canal [90]. Clarke et al. [91] described a significantly increased risk of deep venous thromboembolism in cemented TKAs with respect to cementless implants. The revisions of failed cemented TKAs are technically more demanding with respect to cementless implants, particularly on account of the frequent bone loss after removal of the components [83].

A cementless TKAs has been shown to provide reliable results and is a good alternative treatment option for patients with gonarthrosis [92]. A cementless TKA was first used in 1970s and has gradually become a preferred procedure over time. In comparison, cementless fixation has potential benefits of preserving the bone stock, shorter operating time and none of the cement-related complications [93]. Tibial component fixation can be performed with or without screws for a cementless TKR. Although using screws for a tibial component fixation enhances the early stability of the implant [94, 95], it has been reported to be related to increasing osteolysis around the screws [96–98]. Synovial fluid and polyethylene debris reaching the cancellous bone through the screw holes are thought to be the potential causes of this failure [93]. Cementless knee implants, also known as press-fit knee implants, have a rough, porous surface that encourages new bone growth. The new bone grows into the spaces in the implant, holding it in place without the need for cement. The bones within the knee are shaped with special tools so that they fit snugly with the implant. In some cases, screws or pegs may be used to hold the prosthetics in place while the bone grows. Currently, cementless implants are only available for total knee replacements [84]. The cementless fixation possesses several problems, including (i) it may require a longer healing time, because it can take time for new bone growth that is sufficient enough to hold the implant in place, (ii) cementless implants are not suitable for patients who have poor bone quality due to a condition like osteoporosis; strong, healthy bone is needed to hold the knee implants in place, (iii) because the knees take on quite a bit of stress from daily activity, microscopic debris can be created from wear on the implant, possibly triggering an inflammatory response that leads to osteolysis [83, 84].

Press-fit fixation is now feasible in an increasing number of knee replacement cases. As in the hip, the press-fit prosthesis gives us the potential of a permanent fixation: it becomes part of the patient's bone because it is directly attached; there is no interface holding it in place that could break down or become loose over time. With these newer metals, we have a better likelihood of long-term ingrowth on the tibial component. Among the benefits of the press-fit technology is that it can be applied using either conventional or robotic-assisted surgical techniques. Physicians and surgeons are seeing younger patients presenting with arthritis, whose active lifestyles makes them candidates for total knee replacement at a much earlier age. For these patients, press-fit knee replacement offers a more permanent method of fixation of their components, which we believe will demonstrate a lower incidence of loosening or failure requiring revision surgery [99]. Press-fit fixation is appropriate for most patients who are active, healthy, and without multiple comorbidities, and who have healthy bone in the knee. It may not be ideal for patients with osteoporosis or low bone-mineral density, because when we press-fit the prosthesis, there is a risk of fracture, or subsiding into the bone. As people maintain more active lifestyles later into life, bone density improvements may allow this technology to be performed successfully in much older patients [99].

One of the main indications for using a cementless TKA is good bone quality with high metabolic activity, in order to promote biological fixation. Indeed, a younger age (under 65 years old) and an adequate bone stock are the most typical indications [83]. Although cementless implants are up to three times more expensive than cemented ones due to the high technology required to produce bioactive surfaces, it reduces the pneumatic ischemia time (there is no need for complete exposure of the trabecular bone ready to receive the cement), and finally it allows an easier bone-sparing revision in the event of failure [83]. Over the decades, in-vitro studies have demonstrated that the use of rotating platforms in cementless TKAs is associated with a better tribologic performance and survival of the implant, related to the reduction of stresses at the bone–metal interface. Several studies in

the clinical setting have also shown long-term survivorship of press fit TKAs with rotating platforms, ranging from 83% to as high as 99.4% [100–102]. Ersan et al. [103]
evaluated the long-term clinical and radiological results of cementless TKAs. A total
of 51 knees of 49 patients (mean age: 61.6 years) who underwent TKA surgery with a
posterior stabilized HA coated knee implant were included. All of the tibial components were fixed with screws. The hospital for special surgery (HSS) scores were examined preoperatively and at the final follow-up. Radiological assessment was
performed with Knee Society (KS) evaluating and scoring system. Kaplan–Meier
survival analysis was performed to rule out the survival of the tibial component. It
was reported that (i) the mean HSS scores were 45.8 and 88.1, preoperatively and
at the final follow-up, respectively, (ii) lucent lines at the tibial component were
observed in four patients; one of these patients underwent a revision surgery due
to the loosening of the tibial component, and (iii) the 10-year survival rate of a tibial
component was 98%, concluding that (iv) cementless TKA has satisfactory long-term
clinical results, and (v) primary fixation of the tibial component with screws provides
adequate stability even in elderly patients with good bone quality.

Hybrid fixation, which combines a cemented component (generally the tibial
plate) with a cementless one (usually the femoral component), has been proposed on
the strength of the high osteoconductive properties of the modern component coatings [83]. In a randomized controlled study, Gao et al. [79], using RSA, found results
in terms of migration, clinical outcomes, and survival rates of 41 TKAs in young patients (<60 years) undergoing knee replacement: 22 with fully cemented implants and
19 with hybrid fixated implants. Yang et al. [104], following up 235 TKAs, performed
with a hybrid fixation technique and, using five different knee systems, reported a
survival rate of 95% at 10 years, and then of 92% at 15 years. Cementation of the patellar component is crucial: it is now clear that cementless patella are associated with
a high risk of failure due to early loosening of the component. The decision to press
fit can be easily and quickly made intraoperatively. If the bone is healthy, the press-
fitting the prosthesis should be selected, and for a case of poorer bone quality, cementation should be performed. The key, as always, is proper patient selection and
patient education. It is vital to talk to patients before the procedure, so they understand the potential long-term benefit of press fit biologic fixation, as well as the reliability of a traditional cemented knee arthroplasty [99].

There are, as reviewed in the above, four different TKAs, including (1) cement
fixation, (2) cementless (either with or without screw retention), (3) hybrid fixation,
and (4) reverse THAs. The question of whether to use cemented or cementless fixation for a TKA is still debated. Discouraging preliminary results of cementless TKAs
have determined the worldwide use of cemented implants. However, with the development of biotechnologies and new biomaterials with high osteoconductive
properties, biological fixation is now becoming an attractive option for improving
the longevity of TKAs, especially in young patients [83, 105–107].

Gililland et al. [108] tested their hypothesis that hybrid TKA yields similar clinical, radiographic, and survivorship results compared to fully cemented TKA. The clinical and radiographic outcomes of 304 cruciate retaining TKAs with minimum 2-year follow-up, including 193 hybrid (mean follow-up of 4.1 years) and fully cemented TKAs (mean follow-up of 3.2 years) were evaluated. It was reported that (i) KS scores were similar between the two groups, (ii) the total number of femoral radiolucency was also similar between the two groups, while a greater number of femoral zone 4 radiolucency was seen in the cemented group (9% versus 1.6%,), (iii) the hybrid group demonstrated a 99.2% survival rate of the femoral component out to 7 years for aseptic loosening, and (iv) no significant difference in survivorship was seen between the groups for all cause or aseptic failure at 7 years; concluding that hybrid fixation leads to similar intermediate-term outcomes as fully cemented components and remains a viable option in TKA [108]. A total of 755 knees (356 knees underwent cemented fixation and 399 underwent cementless fixation) were subjected to meta-analysis and it was reported that no significant difference was found in revision rates and knee function in cemented versus cementless TKR at up to 16.6-year follow-up [109]. A register-based study to assess the survivorship of cemented, uncemented, hybrid, and inverse hybrid TKAs in patients aged < 65 years was conducted on 115,177 unconstrained TKAs performed for patients aged < 65 years with primary knee OA over 2000–2016 [85]. It was reported that (i) the 10-year survivorship of cemented TKAs was 93.6%, uncemented 91.2%, hybrid 93.0%, and inverse hybrid 96.0% and was concluded that (i) cemented TKA still deserves the status of gold standard in TKA irrespective of the patients' age, (ii) in addition to age, the optimal fixation method in younger patients may also be influenced by patients' other characteristics such as level of activity, anatomy, or bone quality, (iii) even though hybrid/inverse hybrid versions of the well-performing contemporary TKA designs provided younger patients with a good mid-term outcome in our study, these results do not support systematic use of these more expensive components in TKA for younger patients. Nakama et al. [110] compared effects of cemented, cementless, or hybrid fixation of total knee replacement (arthroplasty) for OA and other nontraumatic diseases, on 216 participants. It was found that (i) the risk of future aseptic loosening with uncemented fixation is approximately half that of cemented fixation in people with knee OA and other nontraumatic diseases, (ii) 16 fewer people out of 100 had a future prediction of arthroplasty instability with uncemented fixation (16% fewer, ranging from 27% fewer to 5% fewer), (iii) 13 people out of 100 had a future prediction of arthroplasty instability with uncemented fixation, and (iv) 29 people out of 100 had a future prediction of arthroplasty instability with cemented fixation.

10.2.2 Retention of dental implants

A dental implant system is made up of three components: fixture (dental implant), abutment, and implant prosthesis, as illustrated in Figure 10.2 [111]. Three possible connection sites are added to the figure. A connecting the fixture (dental implant main body, or artificial tooth root) with the alveolar bone is basically controlled by the osseointegration which can be shared with dental implants and orthopedic implants. Hence, the bone-fusion issue will be discussed later in this chapter. In addition, it suffices to mention that (i) there are three connecting methods between abutment and fixture, and (ii) they are all screw connection and include an external joint design, an internal joint design, and a taper joint design. Accordingly, we will be reviewing the connection of implant prosthesis (or superstructure of crown or others) to abutment head portion.

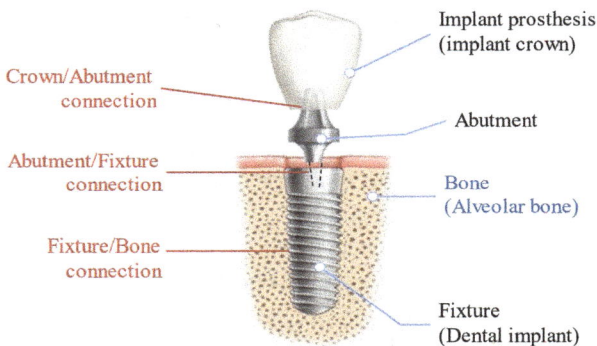

Figure 10.2: Typical dental implant system with three components and three connecting sites [111].

There is a choice for implant-retained prosthesis and implant-supported prosthesis, in which the former refers that dentures are anchored directly to the jawbone through abutment and fixture, while for the latter patients must first undergo dental implant placement [112]. When placing dental implants (aka superstructures), choosing the right method of component retention can be a complicated decision, depending on retrievability, stress, esthetics, accessibility, and/or suitability for optimal implant placement. Normally, the superstructure denture and abutment connection can be divided into two classes: (1) removable implant-supported denture and (2) fixed implant-supported denture [113, 114].

Removable implant-supported superstructure

For various reasons (such as implant inspection, servicing, hygiene, and maintenance), the installed superstructure is needed to be easily retrieved. Hence, attachment mechanism should accommodate this need, and basically, there are several technologies developed and clinically employed.

(1) Friction retention

Friction-retained implant systems (where the abutment and/or prosthesis are retained by the force of friction between components) can provide a high level of stability and allow for a tension-free fit. It must be removed on a daily basis to avoid prosthesis cold welding onto the abutments, so it is not a good option with noncompliant patients, and because of exceptional friction retention, removal requires the patient to have good dexterity. Changing tissue position, overgrowth, or compression greatly may influence the retention of a friction-fit system [115, 116]. Zhang et al. [117] investigated 18 Ankylos SynCone conical crowns with 4-degree angle and 18 SynCone conical crowns with 6-degree angle were tested in vitro for a total of 5,000 insertion–separation cycles to evaluate their retentive characteristics and found that (i) under 20 N insertion force, the retentive force of Ankylos SynCone conical crown system was between 5 and 10 N and (ii) the retentive force kept almost constant during the entire testing cycles.

The snap-on-type retention technique is similar to the friction method. If jawbone has already been lost, snap-on dentures are a better option than fixed because they require less support than a permanent restoration. Another distinct advantage for many patients is that removable implant denture options are much more affordable than fixed implant dentures. In addition, implant supported removable dentures are easier to keep clean since it can be taken out every day for hygiene [113, 118].

(2) Stud method

Stud overdentures provide retention, allow for independent servicing, and do not provide as much stability. Stud precision attachments are primarily used on roots and implants for retaining removable partial dentures or overdentures. All stud attachments must be parallel to each other to provide ease of insertion and removal and reduce wear potential. Do not engage labial soft tissue undercuts with the denture base flange, as this will alter the path of insertion and cause excessive wear and servicing requirements. Stud attachments are low in profile to reduce leverage upon the retaining abutments, are easy for patient hygiene maintenance, allow physiologic independent movement of abutments, and are easy to independently service [119, 120].

(3) Bar and clip technique

Bar and clip technique provides outstanding stability but takes more space and requires better patient hygiene [119]. Petropoulos et al. [121] and Naert et al. [122–124] showed that the bar and clip design are the most retentive overdentures. The metal clips are adjustable by the clinician, so retention of metal clips might be variable in different studies. In this study, the distance between the parallel sleeves was measured to be 2 mm, and they were adjusted after every eight consecutive pulls.

(4) Magnet systems

Magnets have become very popular as retainers for removable prosthesis overdentures for both root and implant abutments. The advantages of magnetic retention over either bars or stud attachments are: (i) no specific path of insertion is required (good for limited dexterity patients), (ii) no abutment parallelism is required, (iii) soft tissue undercuts may be engaged, (iv) potentially pathologic lateral or rotating forces are basically eliminated providing maximum abutment protection, (iv) no path of insertion for patients with limited dexterity, and (v) prosthesis may engage undercut for increased stability [119, 125, 126].

The degree of retention for overdenture attachments depends on design, location, and alignment of supporting dental implants and the type of attachments. Retention decreased over the course of consecutive pulls for all attachments in both directions. It was reported that a decrease in the retention forces after 5 years of follow-up was for magnet (70%), ball attachments (33%), and bar attachments (44%) [122, 125].

Fixed implant-supported superstructure

There are basically two techniques: cement retention and screw retention.

(1) Cement retention

During the cementation of superstructure, an abutment had been previously screw retained to the fixture, as seen in Figure 10.3. Then the superstructure is cemented to the abutment.

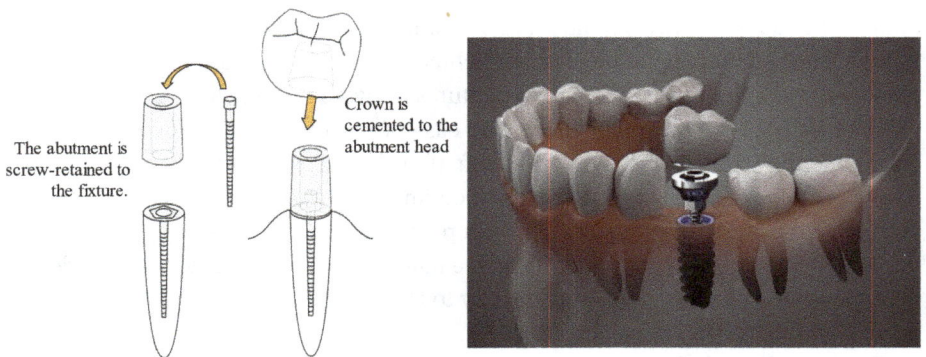

The abutment is screw-retained to the fixture.

Crown is cemented to the abutment head

Figure 10.3: Typical cement retention [127, 128].

(2) Screw retention

In screw retention, a screw is applied through a central hole of the superstructure to the fixture to connect these components, and later on the opening hole will be filled with resin, as demonstrated in Figure 10.4 [127, 128].

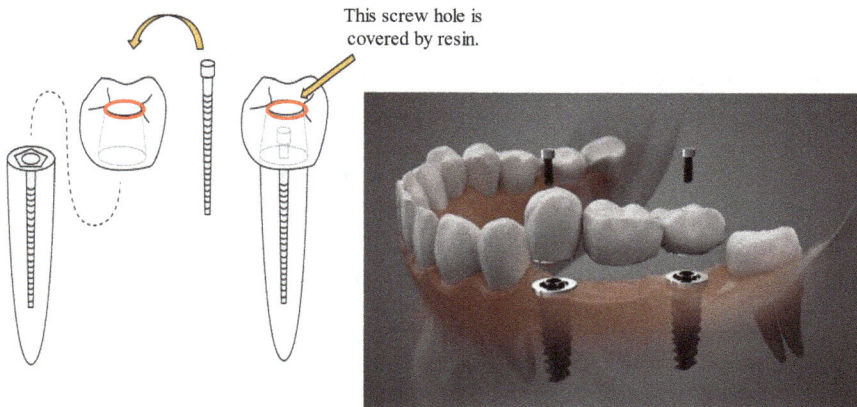

Figure 10.4: Typical screw retention [127, 128].

There are advantages and disadvantages to using a screw-retained versus a cement-retained crown. Table 10.2 lists these pros and cons of these two techniques as well as their characteristics [115, 128–134].

Table 10.2: Comparison among three implant retention methods [115, 128–134].

	Advantages and characteristics	Disadvantages and characteristics
Cement retention	Esthetics: Cemented implants provide a superior appearance versus screw-retained crowns. This is because there is no need for an access hole when cementing the prosthesis directly to the implant abutment. The resulting appearance is more like that of a natural tooth. In anterior, an angulated abutment is used, which eliminates the ridge lap and replicates a more natural emergence profile; while in posterior, the cementable crown obviously has no entrance cavity; allowing the forces of occlusion to be distributed along the axial inclination, congruent with the long axis of the tooth, is easier. Resistance: Cement-retained implants are known for their increased resistance against porcelain fracture, so theoretically the restoration could retain its esthetics and structure for longer.	Excess cement: This can easily harbor bacteria, which could compromise osseointegration (bone loss and peri-implantitis) and lead to an increased risk of dental implant cement failure. Poor removal of excess cement has also been linked to increased risk of peri-implantitis and peri-implant mucositis. Debonding: This can lead to the repeated loosening of the prosthesis; using a permanent cement may reduce debonding episodes but makes recovery more difficult or impossible. Retrievability: This is a challenge with cemented prosthetics. Because the restoration is cemented to a screw-retained abutment, there's no way to remove the restoration it if the screw becomes loose. Generally, this results in total destruction of the restoration to access the screw.

Table 10.2 (continued)

	Advantages and characteristics	Disadvantages and characteristics
		Esthetics: The cemented implants are more esthetic, if the material chips or anything goes wrong in the future, they are much more difficult to undo.
Screw retention	Accessibility: Screw-retained prosthetics are far easier to access, if they ever become damaged or worn, by removing the screw for a repair, replacement, or cleaning.	Accessibility: Restoring a screw-retained restoration in a patient with a limited opening and/or in the posterior of the mouth can be challenging. The implant abutment connection must line up with the interproximal contacts to allow seating of the one-piece restoration. Screw-retained restorations present mainly a butment fracture.
	Retrievability: One huge advantage of screw-retained restorations is retrievability, which allows for simplified removal for implant repair, cleaning, direct visualization of the implant, soft tissue exam, or retightening the connection.	Achieving a passive fit: This is significantly more difficult to attain with a screw-retained restoration due to stress being introduced into the restoration through the tightening of screws.
	Lack of cement: Because screw-retained implants do not use cement to connect the implant with the abutment and the abutment with the crown, they are exempt from the potential complications due to improper cement cleanup.	Complications: Since the material around the screw access hole is left unsupported, the surrounding porcelain is more likely to fracture. The screws might also loosen from day-to-day cyclical loading due to food chewing.
		Esthetics: Many patients are concerned with the cosmetic problems posed by the screw access hole – particularly in the esthetic zone. In anterior screw-retained crowns, the implant is placed lingually to allow screw emergence through the cingulum area. The restoration is cantilevered facially from the implant body, which results in offset loading of the implant. Lingual implant placement also results in a porcelain ridge lap, which compromises hygiene. In posterior screw-retained restorations, the access hole will exit through the central fossa of the prosthetic tooth. This is not only a cosmetic compromise but an occlusal one.

Campbell et al. [133] mentioned that many dental professionals would conclude that cement-retained crowns are finer for esthetics and occlusion; similarly, many would conclude that screw-retained crowns are a necessity for multiple units requiring retrievability [135, 136].

10.3 Union bone healing

Bone healing can be roughly divided into two groups depending on fracture pattern: union bone healing and nonunion (traumatized) bone healing. Implant bone healing is typical traumatized bone healing, in particular, implant site surgical preparation is crucial during the dental implant treatment, so that the subsequent bone healing control the following stability of the placed implant(s).

Although this type of bone healing is not directly relevant to the implant bone healing, the basic bone healing biological mechanism can be shared between union and nonunion bone healing.

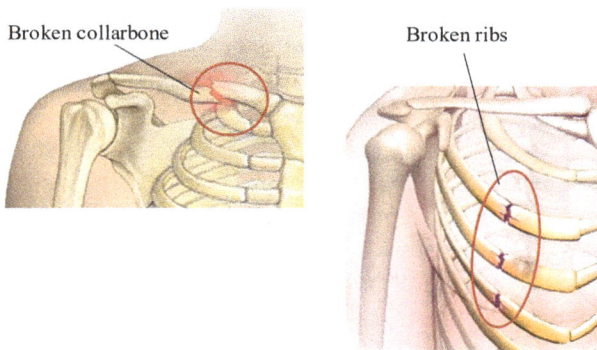

Figure 10.5: Typical images of broken collarbone and ribs [137, 138].

As seen in Figure 10.5, typical broken collarbone (aka clavicle bone) and ribs represent union bone fracture. The common causes of these simple bone fracture include falls, vehicle trauma, or impact during contact sports. It can be added a birth collarbone injury from passing. In most cases, broken collarbone or ribs usually heal on their own in 1 or 2 months, so that there is no specific treatment for these fractures, but various supportive measures such as a sling or bandage can be taken to immobilized fractured components [137–140].

A similar situation (like bone-to-bone fracture mode at the healing site) can be created intentionally for medical and dental purposes. It is called the corticotomy. In bone surgery, a corticotomy is a cutting of the bone that may or may not split it into two pieces (bone fracture) but involves cortex only, leaving intact the medullary vessels and periosteum. Corticotomy is particularly important in distraction osteogenesis or surgically facilitated orthodontic therapy. A correctly performed corticotomy is essential for the success of distraction osteogenesis and prepares for deformity correction, limb lengthening, elimination of bone and soft tissue defects, reshaping of bones, and treatment of cavitary osteomyelitis [141, 142]. The complications of corticotomy include damage to the osteogenic elements through rough surgical technique, displacement of the fragments after corticotomy, and incomplete corticotomy [142]. Corticotomy was introduced as a surgical procedure to shorten orthodontic treatment time. Corticotomy removes the cortical bone that strongly resists orthodontic force in the jaw and keeps the marrow bone to maintain blood circulation and continuity of bone tissues to reduce risk of necrosis and facilitate tooth movement [143, 144].

Bone regeneration is a complex process that involves several interacting biologic mechanisms and is a critical component of many aspects of musculoskeletal care, including fracture healing, spinal fusion, and osseointegration of implants [145, 146]. Taking a bone healing of broken femur as an example, Einhorn et al. [145] demonstrated the biological events and activities, as well as the cells involved in typical bone fracture healing at different phases (i.e., bleeding, inflammation, proliferation, and remodeling [145–149]) are illustrated in Figure 10.6, in which bone morphogenetic protein (BMP), bone morphogenetic protein receptor, Dickkopf-related protein 1, LDL-receptor-related protein, mesenchymal stem cell (MSC), polymorphonuclear leukocyte (PMN), parathyroid hormone, parathyroid-hormone-related protein, and receptor activator of nuclear factor κB ligand. The time scale of healing is equivalent to a mouse's closed femur fracture fixed with an intramedullary rod [145]. During the proliferation phase, the endochondral ossification involves the replacement of hyaline cartilage with bony tissue. Most of the bones of the skeleton are formed in this manner. These bones are called endochondral bones. In this process, the future bones are first formed as hyaline cartilage models.

10.4 Nonunion bone healing

A major difference of nonunion bone healing from the above union bone healing is that the former is directly related to the implant bone healing involves traumatized bone healing adjacent to the foreign implant material(s) [150]. For most patients, the placement of dental implants (most of which are made of commercially pure titanium grades I through IV) involves two surgical procedures. First (surgical stage), implants are placed within the jawbone. For the first 3 to 6 months following surgery, the implants integrate with the jawbone – osseointegration (due to successful growing of osteoblasts on and into the rough surfaces of the implant, forming a structural and functional connection between the hard tissue and placed implant). After some months, the implant is uncovered and a healing abutment and temporary crown is placed onto the implant. This encourages the gum to grow in the right scalloped shape to approximate a natural tooth's gums and allows assessment of the final aesthetics of the restored tooth. After the implant has integrated/ bonded to the jawbone, the second (prosthetic) phase begins. The placed implants are uncovered and small posts are attached which will act as anchors for the artificial teeth. These posts protrude through the gums. When the artificial teeth (or permanent crowns) are placed, these posts will not be seen. The entire procedure

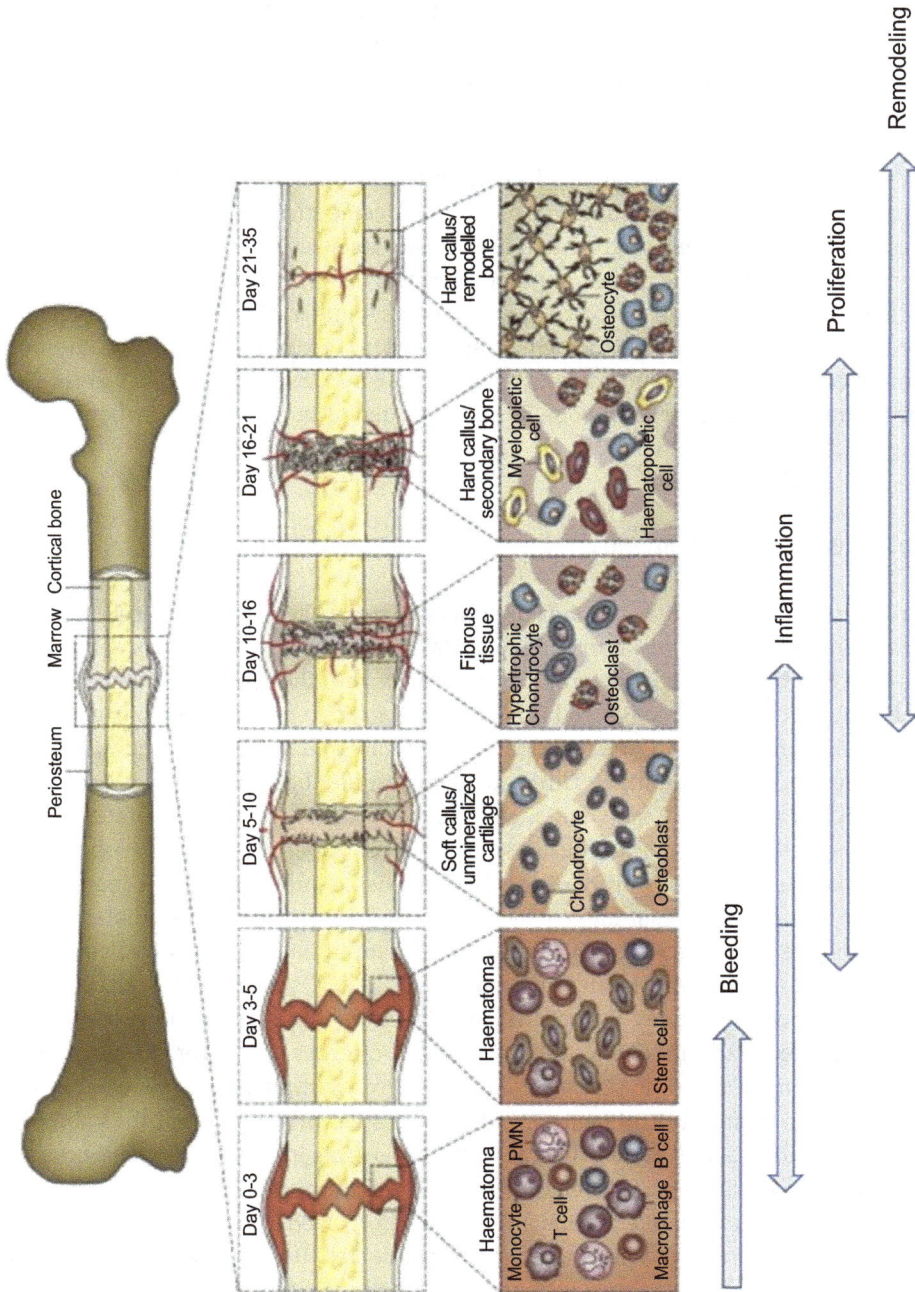

Figure 10.6: Illustration of a typical fracture healing process, biological events, and cellular activities at four different phases [145].

usually takes 4 to 8 months. On the other hand, orthopedic implants are used to replace damaged or troubled joints. Each implant procedure involves removal of the damaged joint and an artificial prosthesis replacement, including hip, knee, finger, elbow, shoulder, ankle, and trauma products. Orthopedic implants are mainly constructed of metallic alloys (Co–Cr–Mo alloy, Fe–Cr–Ni–Mo stainless steel, or Ti–6Al–4V alloy) for strength and lined with plastic to act as artificial cartilage. Some are cemented into place and others (i.e., cementless) are pressed to fit and allow the patient's bone to grow into the implant for strength. Once cytocompatibility is established on the placed implant's surface, the attached cells should grow to achieve the initial stage of osseointegration. Osseointegration is a time-dependent process and success of the osseointegration is strongly related to the quality and condition of receiving hard tissue. In some clinical cases, patients need to receive bone grafting to the alveolar bone structure in dental implantology, or bone grafting for orthopedic implants.

10.4.1 Bone–implant interface

Eriksson et al. [151] pointed out the importance of surface topography on cellular reactions. Surface plays a crucial role in biological interactions for four reasons. First, the surface of a biomaterial is the only part in contact with the bioenvironment. Second, the surface region of a biomaterial is almost always different in morphology and composition from the bulk. Differences arise from molecular rearrangement, surface reaction, and contamination. Third, for biomaterials that do not release nor leak biologically active or toxic substances, the characteristics of the surface govern the biological response. And fourth, some surface properties, such as topography, affect the mechanical stability of the implant–tissue interface [152]. Biomaterials used in a living organism may come into contact with cells in the related tissue for a long period of time. For this reason, they should naturally be harmless to the organism, and their mechanical properties should be suited to the purpose and compatible with receiving tissue surfaces. Furthermore, they should possess a biological effect capable of providing favorable circumstances for the properties and functions of the cells at the implant site. For example, materials used in the construction of an artificial heart or heart valve must provide for antithrombogenesis, which for a dental or bone implant must be suitable for cell attachment, because both the connective and epithelial cells (with which these materials mainly come into contact) are anchorage dependent, and therefore need a cell attachment scaffold for cell division and cell differentiation to be conducted. Broadly speaking, two types of anchorage mechanisms have been described: biomechanical and biochemical. Biomechanical binding is when bone in-growth occurs into micrometer-sized surface irregularities. The term osseointegration is probably, realistically, this biomechanical phenomenon. Biochemical bonding may occur with certain bioactive materials where there is primarily a chemical bonding, with possible supplemental biomechanical interlocking. The distinct advantage with the biochemical bonding is that the anchorage is accomplished within a relatively short period of time, while biomechanical anchorage takes weeks to develop. This would clinically translate into the possibility of earlier restorative loading of implants. Most commercially available implants depend on biomechanical interlocking for anchorage. All implants must exhibit biomechanical as well as morphological compatibilities [153, 154]. It is therefore important to investigate attachability of the cells to the materials, which is one of the parameters in the evaluation of biomaterials.

At the bone–implant interface (BII), there are four important biological activities taking place, including the cell adhesion, adsorption, spreading, and proliferation. When a biological system encounters an implant, reactions are induced at the implant–tissue interface. Such interfacial reactions show that (i) the body wants to keep the implant isolated; (ii) a protective layer of macrophages, monocytes, and giant cells is formed; and (iii) a wall of collagen fibers (capsule) is established, and the type of material may influence the fibrous capsule thickness. Surface and interface properties of interest are chemical composition, contamination and cleanliness, microarchitecture and structure, and so on [155]. Fibroblasts lay down dense fibrous tissue (collagen, elastin, reticular) that is gradually replaced by less dense, normal connective tissue, but the body cannot build normal tissue all the way up to an implant. Host response to biomaterials should be considered first as an inflammatory reaction, as mentioned before. Any biomaterial implanted into the body will be perceived as a threat. The host defenses will attempt to eliminate it. This will not happen if the biomaterial is inert and cannot be degraded. Eventually inert biomaterials will be integrated into the tissue. Biodegradable biomaterials are slowly resorbed over time and are eventually eliminated. Some biomaterials have chemical reactivity and will continue to stimulate over long periods of time. The inflammatory processes should include four main stages: initial events (redness, swelling, pain), cellular invasion (white cells invade tissues), tissue remodeling, repair (being orchestrated by macrophages; occurs differently in different tissues; bone may completely remodel; usually complete within 3 to 4 weeks), and resolution (extrusion, resorption, integration, and encapsulation). Furthermore, cellular invasion has neutrophil action (main function is phagocytes; die at tissue site; release further inflammatory medicators; prostaglandins, leukotrinets) and macrophages (phagocytes; removal of dead cells; number and activity depend on pressure of particulate material). Moreover, resolution is an attempt to return to the original condition: extrusion and resorption of implant material for reestablishment of homeostasis, and integration and encapsulation with a layer of fibrous tissue to establish a steady state. The response, however, resembles normal wound healing regarding cell recruitment and persistence [156, 157]. During the surgical procedure of implantation, the biomaterial most likely will encounter blood. Almost instantly following contact with blood, the implant surface will be covered with plasma proteins that become adsorbed to the surface [158, 159].

Ti in contact with serum or plasma is known to adsorb high molecular weight kininogen, Factor XII, fibrinogen, IgG, prekallikrein, and Clq [160, 161]. After 5 s, platelets are found on Ti surfaces exposed to whole blood, while it takes around 10 min before polymorphonuclear granulocytes adhere [162]. Protein–platelet [163], leukocyte–platelet [164], and protein–leukocyte [165] are possible interactions at the surface. Furthermore, release of inflammatory mediators at the surface may influence both recruitment and activation of cells. The polymorphonuclear granulocytes are the first leukocytes to be recruited to a Ti surface exposed to blood [162]. Adhesion and activation are two processes polymorphonuclear granulocytes may go through upon material contact [165, 166].

Most of the articles on BII are presented and discussed based on interfacial reaction between metal surface and bone. However, this is not right, since the immediate contacting species of the implant surface is not metal, but its oxide. If the implant material is one of titanium materials, titanium oxide should be considered as an extreme surface, contacting the living hard/soft tissue. Titanium reacts immediately with oxygen when exposed to air, forming a very thin surface oxide film. This film, which increases during prolonged exposure to oxygen, consists primarily of titanium dioxide (TiO_2), but Ti_2O_3 and other oxides are also present [167, 168]. TiO_2 has physical/chemical characteristics that differ from metallic Ti, characteristics which are more closely related to ceramics than to metals. As Ti dioxide particles have been used in chromatography, it is well established that titanium oxide surfaces bind cations, particularly polyvalent cations [169]. TiO_2 dioxide surfaces have a net negative charge at the pH values encountered in animal tissues (around 4.0). This binding of cations is based on electrostatic interactions between titanium-linked O^- on the implant surface and cations. The oxide is highly polar and attracts water and water-soluble molecules in general [170]. The immediate reactions on the implant surface after exposure in the tissues may influence later events and are conceivably important for the clinical success of the implantation. An understanding of these immediate biological reactions may therefore be of significance. The chemical properties of the titanium oxide surface suggest that calcium ions may be attached to the oxide-covered surface by electrostatic interaction with O^-, as just discussed. Calcium deposits have been observed in direct contact with the titanium oxide [171]. According to the same model, calcium-binding macromolecules may adsorb selectively to the implant surface in vivo as the next sequence of events. Calcium-binding molecules are often acidic with surface-exposed carboxyl, phosphate, or sulfate groups. Proteoglycans, which contain carboxyl and sulfate groups, might bind to the TiO_2 surface by this mechanism. HA, the major mineral component of bone, also exhibits a surface dominated by negatively charged oxygen (P-bound) that can attract cations and subsequently anion calcium-binding macromolecules [172, 173]. Ellingsen [174] treated CpTi surfaces with (1) 100 m mol/L $CaCl_2$ for 5 min, water-washed and air-dried, and (2) 100 m mol/L $CaCl_2$ for 5 min, followed by treating for 24 h with 0.2 mol/L EDTA, water-washed and air-dried, with untreated CpTi as a control. The reaction of human serum proteins with TiO_2 was examined and compared with the reaction with HA. It was found that (i) the oxide-covered Ti surfaces reacted with calcium when exposed to $CaCl_2$ and calcium was identified to a depth of 17 nm into the oxide layer, (ii) surface adsorbed serum proteins were dissolved by EDTA and surfaces of TiO_2 and HA appeared to take up the same selectively from human serum, and (iii) albumin, prealbumin, and IgG were identified by immunoelectrophoresis. It was, therefore, suggested that calcium binding may be one mechanism by which proteins adsorb to TiO_2. It is well established that this is the case with HA [174].

The modern range of medical devices presents contrasting requirements for adhesion in biological environments. For artificial blood vessels, the minimum adhesion of blood is mandatory, whereas the maximum blood cell adhesion is required at placed implant surfaces. Strong bioadhesion is desired in many circumstances to assure device retention and immobility. Minimal adhesion is absolutely essential in others, where thrombosis or bacteria adhesion would destroy the utility of the implants. In every case, primary attention must be given to the qualities of the first interfacial conditioning films of biomacromolecules deposited from the living systems. For instance, fibrinogen deposits from blood may assume different configurations on surfaces of different initial energies, and thus trigger different physiological events [175, 176]. Sunny et al. [177] showed that the Ti oxide film on Ti affects the adsorption rate of albumin/fibrinogen significantly. Multinucleated giant cells have been observed at interfaces between bone marrow and Ti implants in mouse femurs, suggesting that macrophage-derived factors might perturb local lymphopoiesis, possibly even predisposing to neoplasia in the B lymphocyte lineage. It has been found that (i) an implant–marrow interface with associated giant cells persists for at least 15 years, (ii) precursor B cells show early increase in number and proliferative activity; however, (iii) at later intervals they do not differ significantly from controls. Rahal mentioned, in mice study, that following initial marrow regeneration and fluctuating precursor B cell activity, and despite the presence of giant cells, Ti implants apparently become well-tolerated by directly apposed bone marrow cells in a lasting state of the so-called myelointegration [178].

Ossification mechanisms that occur following the placement of the implant are very important for understanding the biologic response to endosseous implants [179]. Osborn [180] categorized this bioresponse into the following three groups: (1) biotolerant type, characterized by distance osteogenesis, the implant is not rejected from the tissue, but it is surrounded by a fibrous connective tissue, (2) bioinert type, characterized by contact osteogenesis, the osteogenic cells migrate directly to the surface where they will establish de novo bone formation, and (3) bioreactive type, the implant allows new bone formation around itself, thereby exchanging ions to create a chemical bond with the bone. Upon insertion, various implant materials exhibit different biologic responses. While biotolerant materials, such as gold, cobalt–chromium alloys, stainless steel, polyethylene, and polymethylmethacrylate, exhibit distance osteogenesis, titanium and titanium alloys are accepted to be bioinert according to their surface oxides [181]. Besides, Zhao et al. [182] mentioned that the rutile-type oxide, which is formed on titanium as a titanium dioxide, is described as a stable crystalline form similar to ceramics in its bioreactive behavior. Although titanium has superior characteristics compared to other implant metals, the osteoconductivity of titanium is lower than calcium phosphate (CaP)-based bioceramics [183]. Therefore, CaP-based ceramics are referred to be bone-bonding materials, whereas titanium is a nonbonding material to bone [184]. Approaches have mainly focused on enhancing the bioactivity of titanium and providing a higher osteoconductivity to the bulk material by altering the surface properties. The character of the host tissue also plays an important role on the ossification mechanism following implantation. Understanding the different peri-implant healing cascades of the cortical and trabecular bone is crucial for better orientating the osseointegration in poor quality bone [185]. Following surgical trauma, the vascular injury of the cortex results in death of the peri-implant cortical bone, and followed by a slow proceeding osteoclastic remodeling. The removal of the injured tissue by osteoclasts and the subsequent formation of the new bone is a long lasting process. Therefore, the healing around the implant in cortical bone results in distance osteogenesis. Although this slow remodeling phase provides early stability in cortical bone leading to low rate of implant failure [186], especially in the parasymphyseal mandible, it is a handicap for the surface science approaches for enhancing the osseointegration histologically. On the other hand, the trabecular bone enables the migration of osteogenic cells due to its marrow component. The colonization of differentiating progenitor cells on the implant surface and de novo bone formation provides the evidence that peri-implant healing in trabecular bone occurs via contact osteogenesis. Actually, as Marco et al. [187] pointed out, the presence of osteoprogenitor populations in the spongious bone, which is characterized to be of poor quality in implant dentistry (Lekholm and Zarb type III and IV bone; see Table 9.2 in Chapter 9), favors the migration and bone forming activity of these cells directly on the surface when the implant is considered. In the recent decades, the development of novel osteoconductive titanium surfaces, that increased the local quantity and quality of osseous tissue at the interface, thereby improved the success of implants, especially in regions of the jaw such as the edentulous posterior maxillae where the cortical thickness is frequently insufficient for the primary stability.

The surgical placement of the implant results in injury of the host bone. If the implant is considered to be bioinert, the body responds to this injury with physiological mechanisms similar to the bone fracture healing. Following implant placement, the implant surface first gets in contact with the blood originating from the injured vessels facing the implant cavity. After several seconds, the surface is completely covered with a thin layer of serum proteins. This protein modification of the surface occurs for all implant materials in the same way. However, the type and surface characteristics of the material have a major influence on the structure and conformation of this protein layer [188]. Shortly after protein adsorption, the surface becomes associated with thrombocytes. As a result of thrombocyte aggregation and degranulation on the surface, coagulation mechanisms take place and cytokines (e.g., transforming growth factor-β (TGF-β) and platelet-derived growth factor (PDGF)) and several vasoactive factors (e.g., serotonin and histamine) are released from cytoplasmic granules of thrombocytes. These chemoattractants stimulate proliferation and migration of various cells, thereby orientating the peri-implant healing mechanisms [189]. For example, Heldin et al. [190] mentioned that PDGF has important mitogenic and migrative effects on several cell types, such as inflammatory leukocytes, osteoblasts, smooth muscle cells, and fibroblasts.

Polymorphonuclear neutrophils (PMNs) are also first group of cells that play an important role in the inflammatory response. PMNs dominate the BII at the first and second days. The number of PMNs tends to decrease when bacteria and endotoxins are not present at the interface. At the second day of healing, monocyte migration and macrophage accumulation starts to take place [191]. PMNs and macrophages remove dead cells, extracellular matrix (ECM) residues and bacteria. Besides their role on the initial inflammatory phase, another mission of macrophages is the expression of cytokines, such as fibroblast growth factor (FGF), PDGF, and vascular endothelial growth factor (VEGF). Thus, they provide important signals in order to stimulate the recruitment of osteogenic and endothelial progenitors for the next proliferative phase. The release of vasoactive amines, thrombocyte and leukocyte infiltration, the establishment of the coagulum and fibrin network, and macrophage actions are important events that occur at the inflammatory phase. This first phase, which can sometimes extend to 5 days, is followed by the removal of the coagulum by PMNs and subsequently by monocytes, at the same time angiogenesis starts also to take place [192]. The growth of new capillaries into the fibrin network is mostly stimulated by the growth factors (primarily FGF and VEGF) expressed by macrophages and endothelial cells as a response to hypoxic and acidic nature of the BII [193]. In this way, the proliferation, maturation, and organization of endothelial cells to new capillary tubes take place, thereby providing oxygen and nutrients to the newly formed tissue at the interface. The behavior of blood cells inside the fibrin-based structural matrix has a major impact on the healing mechanisms at the BII. Besides, the quality of bone healing around an implant is also affected by the capacity of osteogenic cells to proliferate and migrate. Meyer et al. [194] have demonstrated that the osteoprogenitor cells started to attach the implant surface after one day following insertion. This was a similar finding, as stated by Davies [185], showing that early recruitment and colonization of MSCs cells occur on an implant surface in a short time through modulation of white blood cells, fibrin network and thrombocytes [195]. The three-dimensional structure of fibrin matrix and the migrating effects of growth factors expressed by the first arriving cells play an important role in the establishment of an osteoprogenitor reservoir at the interface. Therefore, the chemistry of the implant material and its surface characteristics are of special interest in implantology, since they initially influence the binding capacity of fibrin and the release of growth factors, thereby affecting the migration of mesenchymal cells directly [196].

Titanium implant materials possess ideal fibrin retention on their surface. Through this fibrin matrix, osteogenic cells having the migration ability arrive at the implant surface and start to produce bone directly on the surface. Davies [191] termed this phenomenon as de novo bone formation through contact osteogenesis. Upon arrival to the surface, the differentiated osteogenic cells secrete the collagen-free matrix (cement lines/lamina limitans) for the mineralization through calcium and phosphate precipitation. This layer, where the initial mineralization occurs, consists of noncollagenous proteins (mostly osteopontin and bone sialoprotein) and proteoglycans [197]. Following calcium phosphate precipitation, the formation and mineralization of collagen fibers take place. Thus, a noncollagenous tissue is established between the implant surface and the calcified collagen compartment through contact osteogenesis. This intermediary tissue is very important for the understanding the bonding mechanism between bone and a bioinert titanium implant. Following the establishment of the calcified matrix on the implant surface, woven bone formation and organization of the bone trabeculae start to take place for the reconstitution of the damaged bone at the peri-implant area [187]. Since the woven bone mostly consists of irregularly shaped and loosely packed collagen fibers, it does not provide sufficient mechanical stability compared to the organized the lamellar bone. However, most of woven bone usually remodels in three months and replaced by the lamellar bone. At 3 months of healing, the implant is mostly surrounded by a mixture of woven and lamellar bone [198]. The formation and remodeling of the new lamellar bone around the implant occur more rapidly in the regions where there is denser marrow component present. Therefore, the biologic fixation of the implant is achieved faster in the trabecular bone, while a better primary stability is obtained in the cortical bone following implantation.

An implant surface is considered to be clean following fabrication processes. If not stored under special conditions, contaminations (e.g., hydrocarbon, sulfur dioxide and nitric oxide) occur from the atmosphere [199]. In order to decrease and eliminate such risk of contamination, commercial implant surfaces are usually subjected to passivation treatments and stored carefully in optimal packages until usage. If such an implant is placed into the bone, its surface first get in contact with the blood, which is mostly composed of water molecules. Differently from the liquid water, the water molecules bind to the surface and form water mono- or bilayer [200]. The organization of water molecules differs according to the wettability characteristics of the surface [201]. While on hydrophilic surfaces, the interaction with water molecules results in the dissociation of molecules and in the formation hydroxyl groups, the water-binding capacity of hydrophobic surfaces is very low. Following the establishment of water overlayer, the ions (e.g., Cl^- and Na^+) enter the layer and become hydrated. The characteristics of an implant surface have a major implant on this arrangement of ions and their water shells. After the establishment of an intermediate layer composed of ions and water molecules, the biomolecules arrive at the surface in milliseconds. Proteins absorb first onto the surface, then change their conformation, denaturize and desorb from surface leaving their place to other proteins that have more affinity to the surface. Thus, a biologic layer having a different arrangement and conformation surrounds the surface. Surface characteristics have an important effect on the adsorption of biomolecules by changing the arrangement of water molecules and ions. While on hydrophobic surfaces proteins bind with their hydrophobic regions, on hydrophilic surfaces the connection is established with the help of hydrophilic regions [200]. This protein overlayer is never considered to be static. It is subjected to structural and conformational changes in time. Normally the protein, which is found in higher concentration in the biological fluid, reaches and adsorbs to the surface first. Usually, this protein is afterward replaced with another one that has a more affinity to the surface, although its concentration is low in the biological fluid. As a result of these adsorption and desorption mechanisms, a diverse layer which is composed of different protein is formed and maintained at the surface. The major role of this protein layer is the attachment of functionary cells of the healing process. If a bone implant is planning to be developed, the establishment of a surface, that generates an optimal protein composition and conformation for the attachment of osteogenic cells on itself, is one most important strategies of the production. Several proteins (e.g., fibronectin, vitronectin, laminin, serum albumin, and collagen) facilitate the attachment of osteogenic cells on titanium surfaces [202–204]. Therefore, the protein binding capacity of an implant surface is considered to be an important factor for successful osseointegration, since surface properties, such as micro- and nanotopography [205], physicochemical composition [203], and surface free energy [206], have an influence on the extend of protein adsorption. It has been documented that osteogenic cells preferably attach to the specific protein sequences, such as the arginine-glycine-aspartic acid (RGD) motif. This motif is found in various ECM proteins, including fibronectin, vitronectin, laminin, and osteopontin [207]. Osteogenic cells attach to these binding motifs using their membrane receptors, termed as integrins. Integrin-mediated cell attachment is crucial for physiological and pathological mechanisms, such as the embryonic development, maintenance of tissue integrity, circulation, migration and phagocytic activity of leukocytes, wound healing, and angiogenesis. Integrins are obligate heterodimers composed of two distinct glycoprotein subunits; α and β subunits [208]. Integrin subunits cross the plasma membrane with a long extracellular ligand, while generally a very short domain remains in the cytoplasm. For the integrin family, 18 α and 8 β subunits have been characterized in mammals until now. Through the combination of these different α and β subunits, 24 distinct integrins can be assembled. A cell can modulate more than one integrin receptor and change their location, thereby modifying its capacity to bind to different protein sequences [188]. As mentioned before, adhesion-promoting proteins in blood (e.g., fibronectin, vitronectin, and various collagen types) bind to integrins through an RGD-dependent pathway [209]. But, there are also different domains within these proteins that have the ability to bind to integrins and provoke integrin-mediated cellular signaling cascades. Briefly, integrin-mediated cell attachment to ECM initiate several intracellular events, including protein kinase C and Na^+/H^+ antiporter, phosphoinositide hydrolysis, tyrosine phosphorylation of membrane, and intracellular proteins [210]. These mechanisms result in mitogen-stimulated protein kinase activation by altering the cellular pH and calcium concentration. Thus, intracellular communication is established and the extracellular signal is transmitted to the nucleus. The cell responds to this integrin-mediated signal through migration, proliferation, and differentiation. The response of osteogenic cells to the initial protein layer on the implant surface is very important for the activation of osteoblastic pathways through integrin-mediated signaling, thereby for optimal osseointegration. Therefore, the development of an implant surface, that favors an osteogenic protein conformation on itself, has been one of the major areas of implant surface science. In the recent decades, various approaches have focused on understanding the effect of the surface characteristics on the protein dependent mechanisms of cell adhesion, proliferation, differentiation, and bone matrix deposition, aiming at the development of novel implant surfaces [211].

The success of total joint arthroplasty is dependent on the interface between the metal and bone. Although bone cement is still frequently used today for femoral stems, the cementless femoral stem has become the standard for primary THA in young, active patients. Unlike cemented implants that rely on a cement mantle to fix metal to bone, cementless implants rely on bone healing to secure the implant to the host bone to establish an osseointegration. The biological process behind bone bonding occurs within the normal human bone and the bones of all vertebrates, as a part of physiological bone remodeling. In bone remodeling, osteoclasts remove the old bone and in so doing provide a surface on the old bone with the appropriate topography, with which the newly formed bone may bond. This bonding occurs as the cement line interdigitates with the surface of the old bone [212]. The cement line is the first matrix developed during de novo bone formation. Evidence suggests that this process may occur at an implant surface, and thus result in bone bonding, if the implant's surface topography is three-dimensionally complex, with pores and undercuts [213]. Peri-implant bone healing is not limited to bone bonding and certainly stable implant fixation occurs in the absence of bone bonding, although bone bonding may be regarded as the ideal situation. Contact and distance osteogenesis have been used to explain peri-implant bone healing. In contact osteogenesis, de novo bone forms on the implant surface, while in distance osteogenesis, the bone grows from the old bone surface toward the implant surface in an appositional manner. Contact osteogenesis may lead to bone bonding if the surface of the implant displays the appropriate surface topography. The early stage of peri-implant bone healing is very important and involves the body's initial response to a foreign material: protein adsorption, platelet activation, coagulation, and inflammation. This results in the formation of a stable fibrin clot that is a depot for growth factors and allows for osteoconduction. Osteoconduction is the migration and differentiation of osteogenic cells, such as pericytes, into osteoblasts. Osteoconduction allows for contact osteogenesis to occur at the implant surface. The late stage of healing involves the remodeling of this woven bone. In many respects, this process is similar to the bone healing occurring at a fracture site [212]. Although contact and distance osteogenesis have been defined and described, studies have not clearly elucidated these bone formation mechanisms around implants inserted into bone [214]. Therefore, it remains unclear whether contact and distance osteogenesis act independently or interactively to achieve peri-implant bone healing. However, some limited evidence suggests that they act independently. First, while both types of osteogenesis occur around topographically microroughened implant surfaces, this is not true for interfaces with smooth or polished implants; with these implants, bone only forms in the same direction as the old host bone. This suggests that distance osteogenesis can occur in the absence of contact osteogenesis [185, 215]. Second, a study showed that when a microroughened implant is placed, there is an initial increase in the adsorption of proteins from the blood on the surface of the implant, and that these proteins direct bone formation from the implant surface toward the host bone [216–218]. This suggests that, to some extent, contact bone formation may act independently of distance osteogenesis in peri-implant bone healing [219]. Lavenus et al. [220] mentioned that implants are commonly used in dental surgery for restoring teeth. One of the challenges in implantology is to achieve and maintain the osseointegration as well as the epithelial junction of the gingival with implants. An intimate junction of the gingival tissue with the neck of dental implants may prevent the bacteria colonization leading to peri-implantitis while direct bone bonding may ensure a biomechanical anchoring of the artificial dental root, as seen in Figure 10.7 [220].

10.4.2 Effects of surface condition on biological events

The biological events occurring at the BII are influenced by the topography, chemistry, and wettability of the implant surface, as seen in the above. A goal of biomaterials research has been, and continues to be, the development of implant materials which are predictable, controlled, guided, with rapid healing of the interfacial tissues, both hard and soft. The performance of biomaterials can be classified in terms of (1) the response of the host to the implant, and (2) the behavior of the material in the host. This is actually related to which side we are looking at the host (vital tissue)/foreign materials (implant) interface. The event that occurs almost immediately upon implantation of metals, as with other biomaterials, is adsorption of proteins. These proteins come first from blood and tissue fluids at the wound site, and later from cellular activity in the interfacial region. Once on the surface,

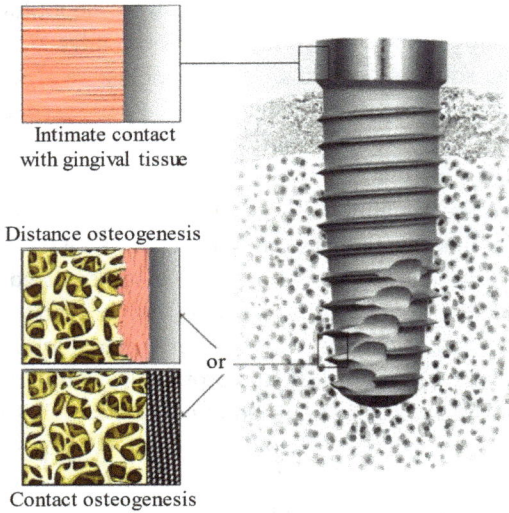

Figure 10.7: Tissue integration of dental implant.
Note the intimate contact with gingival tissue in the upper part and the desired contact osteogenesis in the tapered lower part rather than distance osteogenesis [220].

proteins can desorb (undenatured or denatured, intact or fragment), remain, or mediate tissue–implant interaction [221]. The host response to implants placed in bone involves a series of cell and matrix events, ideally culminating in tissue healing that is as normal as possible, and that ultimately leads to intimate apposition of bone to the biomaterials, that is, the operative definition of osseointegration. For this intimate contact to occur, gaps that initially exist between the bone and implant at surgery must be filled initially by a blood clot, and bone damaged during preparation of the implant site must be repaired. During this time, unfavorable conditions, for example, micromotion as a biomechanical factor, will disrupt the newly forming tissue, leading to formation a fibrous capsule [222]. The criteria for clinical success of osseointegration are based on functionality and compatibility, which depend on the control of several factors including: (1) biocompatible implant material using commercially pure titanium, (2) design of the fixture; a threaded design is advocated creating a larger surface per unit volume as well as evenly distribution loading forces, (3) the provision of optimal prosthodontic design and implant maintenance to achieve ongoing osseointegration, (4) specific aseptic surgical techniques and a subsequent healing protocol which are reconcilable with the principles of bone physiology; this would incorporate a low heat/low trauma regimen, a precise fit, and the two-stage surgery program, (5) a favorable status of host-implant site from a health and morphologic standpoint, (6) nonloading of the implant during healing is a basic tenet of osseointegration, and (7) the defined macro-microscopic surface of the implant [186].

The implantation of any foreign material in soft tissue initiates an inflammatory response. The cellular intensity and duration of the response is controlled by a variety of mediators and determined by the size and nature of the implanted material, site of implantation, and reactive capacity of the host. Dental implants vary markedly in the topography of the surfaces that contact cells. Four principles of cell behavior first observed in cell culture explain to some extent the interactions of cells and implants; (1) contact guidance aligns cells and collagen fibers with fine grooves, such as those produced by machining, (2) rugophilia describes the tendency of macrophages to prefer rough surfaces, (3) the two-center effect can explain the orientation of soft connective tissue cells and fibers attached to porous surfaces, and (4) haptotaxis may be involved in the formation of capsules around implants with low-energy surfaces [223].

Surface roughness is an influencing parameter for cell response. Bigerele et al. [224] compared the effect of roughness organization of Ti–6Al–4V or CpTi on human osteoblast response (proliferation and adhesion). Surface roughness is extensively analyzed at scales above the cell size (macroroughness) or below the cell size (microroughness) by calculation of relevant classic amplitude parameters and original frequency parameters. It was found that (i) the human osteoblast response on electro-erosion Ti–6Al–4V surfaces or CpTi surface was largely increased when compared to polished or machine-tooled surfaces after 21 days or culture, and (ii) the polygonal morphology of human osteoblast on these electro-erosion surfaces was very close to the aspects of human osteoblast in vivo on human bone trabeculae. It was concluded that electro-erosion (creating a rough surface) is a promising method for preparation of bone implant surfaces, as it could be applied to the preparation of most biomaterials with complex geometries.

The role of surface roughness on the interaction of cells with titanium model surfaces of well-defined topography was investigated using human bone-derived cells (MG63 cells). The early phase of interactions was studied using a kinetic morphological analysis of adhesion, spreading, and proliferation of the cells. SEM and double immunofluorescent labeling of vinculin and actin revealed that the cells responded to nanoscale roughness with a higher cell thickness and a delayed apparition of the focal contacts. A singular behavior was observed on nanoporous oxide surfaces, where the cells were more spread and displayed longer and more numerous filopods. On electrochemically microstructured surfaces, the MG63 cells were able to go inside, adhere, and proliferate in cavities of 30 or 100 µm in diameter, whereas they did not recognize the 10 µm diameter cavities. Cells adopted a 3D shape when attaching inside the 30 µm diameter cavities. It was concluded that nanotopography on surfaces with 30 µm diameter cavities had little effect on cell morphology compared to flat surfaces with the same nanostructure, but cell proliferation exhibited a marked synergistic effect of microscale and nanoscale topography [225]. Rat bone marrow stromal cells were cultured on either a smooth surface (roughness: 0.14 µm) or a rough surface (5.8 µm) of Ti–6Al–4V discs for 24 or 48 h.

Cells on the smooth surface showed typical fibroblastic morphology, whereas cells on the rough surface were in clusters with a more epithelial appearance. RNA was extracted from the cells at both time points, and gene expression was analyzed by using a rat gene microarray. At 24 and 48 h, a similar number of genes were both up- and down-regulated at least twofold on the rough surface compared to the smooth surface. Roughness did not appear to be a specific stimulator of osteogenesis because genes of both the bone and cartilage lineage were upregulated on the tough surface. It was shown that surface roughness alters the expression of a large number of genes in marrow stromal cells, which are related to multiple pathways of mesenchymal cell differentiation [226]. A highly porous, strong Ti–6Al–4V was fabricated by a polymeric sponge replication method [227]. A polymeric sponge, impregnated with a Ti–6Al–4V slurry prepared from Ti–6Al–4V powders and binders, was subjected to drying and pyrolyzing to remove the polymeric sponge and binders. The porous Ti–6Al–4V made by this method had a three-dimensional trabecular porous structure with interconnected pores mainly ranging from 400 to 700 μm and a total porosity of about 90%. The compressive strength was 10.3 ± 3.3 MPa and the elastic constant 0.8 ± 0.3 GPa. It was also reported that MC3T3-E1 cells attached and spread well in the inner surface of pores [227].

Webster et al. [228] claimed that $CaTiO_3$ is a strong candidate to form at the interface between HA and titanium implants during many coating procedures. The ability of bone-forming cells (osteoblasts) to adhere on titanium coated with HA, resulting in the formation of $CaTiO_3$, was investigated. For formation of $CaTiO_3$, titanium was coated on HA discs and annealed either under air or a $N_2 + H_2$ environment. Results from cytocompatibility tests revealed increased osteoblast adhesion on materials that contained $CaTiO_3$ compared to both pure HA and uncoated titanium. The greatest osteoblast adhesion was observed on titanium-coated HA annealed under air conditions. Because adhesion is a crucial prerequisite to the subsequent functions of osteoblasts (such as the deposition of calcium containing minerals), it was suggested that orthopedic coatings that form $CaTiO_3$ could increase osseointegration with juxtaposed bone needed for increased implant efficacy [228]. Improving the biological performance of engineered implants apposing interfacing tissues is a critical issue in implantology and its related science and engineering. Micromotion at the soft tissue–implant interface has been shown to sustain an inflammatory response. To eliminate micromotion, it is desirable to promote cellular and ECM adhesion to the implant surface. Surfaces are modified topographically or chemically to effect cellular adhesion and to influence cellular interactions and function. An in-vitro study was conducted to compare the independent effects of surface chemistry and topography on fibroblast-test specimen proximity [229]. Titanium was sputter coated in stepwise manner, increasing thicknesses (20–350 nm) onto a series of either smooth or microtextured polyethylene terephthalate (PET) (resulting in a stepwise change from a PET surface to Ti surface – as gradually functioning material. It was found that (i) fibroblast proximity to the coverslip surface increases as the Ti thickness increases on either smooth or textured test specimens, and (ii) fibroblasts were firmly attached to the ridge tops on the coated textured test specimens. Therefore, fibroblast apposition is strongly enhanced by microtextured surfaces and Ti rather than smooth surfaces and PET [229]. Osteoblast adhesion on the implant material surface is essential for the success of any implant in which osseointegration is required. Surface properties of implant material have a critical role in the cell adhesion progress. The microarc oxidizing (MAO) and hydrothermally synthesizing methods were used to modify the TiO_2 layer on the titanium surface, on which the mouse osteoblastic cell line (MC3T3-E1) was seeded to evaluate their effect on cell behavior [230]. The surface structure of MAO samples exhibited micropores with a diameter of 1–3 µm, whereas the MAO/hydrothermal synthesized samples showed additional multiple crystalline microparticles on the microporous surface. Both treated surfaces possessed higher energy than that of untreated titanium. It was concluded that the MAO and MAO/hydrothermal synthesization methods change the surface energy of the TiO_2 layer on the titanium surface. This may have an influence on the initial cell attachment [230]. Bioactivation of the titanium surface by sodium plasma immersion ion implantation and deposition was compared to that of the untreated, Na beam-line implanted and NaOH-treated titanium samples [231]. It was found that (i) from a morphological point of view, cell adherence on the NaOH-treated titanium is the best, (ii) on the other hand, the cell activity and protein production were higher on the nonbioactive surfaces, and (iii) the active surfaces support an osteogenic differentiation of the bone marrow cells at the expense of lower proliferation, due to the high alkaline phosphatase (ALP) activity per cell [231].

Living bone cells are responsive to mechanical loading. Consequently, numerous in vitro models have been developed to examine the application of loading to cells. However, not all systems are suitable for the fibrous and porous three-dimensional materials, which are preferable for tissue repair purposes, or for the production of tissue engineering scaffolds. For three-dimensional applications, mechanical loading of cells with either fluid flow systems or hydrodynamic pressure systems has to be considered. The response of osteoblast-like cells to hydrodynamic compression, while growing in a three-dimensional titanium fiber mesh scaffolding material, was evaluated by Walboomers et al. [232]. Bone marrow cells were obtained from the femora of young (12 days old) or old (1 years old) rats, and precultured in the presence of dexamethasone and β-glycerophosphate to achieve an osteoblast-like phenotype. Subsequently, cells were seeded onto the titanium mesh scaffolds, and subjected to hydrodynamic pressure, alternating from 0.3 to 5.0 MPa at 1 Hz, at 15-min intervals for a total of 60 min per day for up to 3 days. After pressurization, cell viability was checked. Afterward, DNA levels, ALP activity, and extracellular calcium content were measured. Finally, all specimens were observed with scanning electron microscopy. Cell viability studies showed that (i) the applied pressure was not harmful to the cells, and (ii) the cells were able to detect the compression forces [232].

10.4.3 Bone grafting

Both bone cement and graft are currently used to treat nonunions and bone defects [233]. Bone cements have been used to retain artificial joints in TSAs, THAs, or TKAs. The bone cement fills the free space between the prosthesis and the bone and plays the important role of an elastic zone, exhibiting a shock-absorbing function, because the human hip is acted on by approximately 10–12 times the body weight and therefore the bone cement must absorb the forces acting on the hips to ensure that the artificial implant remains in place over the long term. Bone cement chemically is nothing more than PMMA. As mentioned previously, during the free-radical polymerization process, the exothermal reaction causes temperature raise up to 82–86 °C in the body [234, 235]. The thus localized temperature raise is higher than the critical level for protein denaturation in the body. Besides this exothermic reaction problem, the bone cement implantation syndrome has been described [236]. It was believed that the incompletely converted monomer released from bone cement was the cause of circulation reactions and embolism. However, it is now known that this monomer (residual unpolymerized monomer) is metabolized by the respiratory chain and split into carbon dioxide and water and excreted. In dentistry, polymer-based cements are also used as fillers of cavities. They are generally cured photochemically using UV radiation in contrast to bone cements [237]. More recently bone cement has been used in the spine kyphoplasty procedures. The composition of these types of cement is mostly based on calcium phosphate and more recently magnesium phosphate. A novel biodegradable, nonexothermic, self-setting orthopedic cement composition based on amorphous magnesium phosphate was developed [238]. The occurrence of undesirable exothermic reactions was avoided through using the amorphous magnesium phosphate as the solid precursor [239]. New bone cement formulations require characterization according to ASTM F451 [239]. This standard describes the test methods to assess cure rate, residual monomer, mechanical strength, benzoyl peroxide concentration, and heat evolution during cure. Bone cement has proven particularly useful because specific active substances, for example, antibiotics, can be added to the powder component. The active substances are released locally after implant placement of the new joint, that is, in the immediate vicinity of the new prosthesis and have been confirmed to reduce the danger of infection. The antibiotics act against bacteria precisely at the site where they are required in the open wound without subjecting the body in general to unnecessarily high antibiotic levels. This makes bone cement a modern drug delivery system that delivers the required drugs directly to the surgical site [234].

Bone grafting is a surgical procedure that replaces missing bone in order to repair bone fractures that are extremely complex, pose a significant health risk to the patient, or fail to heal properly. Some small or acute fractures can be cured without bone grafting, but the risk is greater for large fractures like compound fractures [240]. Bone generally has the ability to regenerate completely but requires a very small fracture space or some sort of scaffold to do so. Bone grafts may be autologous (bone harvested from the patient's own body, often from the iliac crest), allograft (cadaveric bone usually obtained from a bone bank), or synthetic (often made of HA or other naturally occurring and biocompatible substances) with similar mechanical properties to bone [241]. Most bone grafts are expected to be reabsorbed and replaced as the natural bone heals over a few months' time. It is estimated that around half of the number of patients seeking implant treatment have bone deficiencies, or bone that is frequently insufficient to accommodate the correct size of the implant(s). The choice of bone grafting material will depend on the patient's wishes and the complexity of the bone deficiency [242]. The rationales involved in successful bone grafts include osteoconduction (guiding the reparative growth of the natural bone), osteoinduction (encouraging undifferentiated cells to become active osteoblasts), and osteogenesis (living bone cells in the graft material contribute to bone remodeling) [243, 244]. Osteoconduction is the ability to provide an environment capable of hosting the indigenous MSCs, osteoblasts, and osteoclasts, it is essential for the function of bone graft. Osteoconduction is the process by which a graft acts as a scaffold, passively hosting the necessary cells [245]. Microscopically, the porous osteoconductive lattice of bone graft resembles the structure of cancellous bone [246]. All bone grafts provide some degree of osteoconductive scaffold. The bioceramic bone graft substitutes, such as calcium sulfate and calcium phosphate, behave exclusively as osteoconductive scaffolds [247]. Osteoinduction has been defined as the process of recruitment, proliferation, and differentiation of host MSCs into chondroblasts and osteoblasts. Extensive research has identified BMPs (specifically BMP $-2, -4, -6, -7, -9$, and -14), FGF, PDGF, and VEGF as common growth factors involved in the osteoinductive process of new bone formation [245, 247, 248]. In order for a bone graft to possess the property of osteogenesis, it must contain viable MSCs, osteoblasts, and osteocytes [245]. Osteogenic bone grafts have all the cellular elements, growth factors and scaffolding required to form new bone. The most widely used osteogenic bone graft is autogenous bone, which is commonly harvested from the iliac crest. Additionally, the decortication performed during a spinal fusion is considered to be an osteogenic process as it exposes cancellous bone that is rich in osteogenic cells. Bone marrow aspirate in combination with allograft has also been employed to provide osteogenesis while limiting the morbidity of iliac crest bone graft [245, 249].

Table 10.3 compares classification and basic properties among four bone graft material groups, compiling data from references [250–252].

There are several studies in debating the bone repair efficiency and efficacy between bone cement and bone grafts [253, 254].

Table 10.3: Bone graft materials classification and properties [250–252].

Source		Advantages	Disadvantages	Osteoconductive	Osteoinductive	Osteogenic
Autograft	Patient	– True osteogenic – Living cells – Growth factors – No disease transmission – Good with cortical bone	– Pain – Infection – Complex surgery – Limited supply	+	+	+
Allograft	Other human	– Osteoinductive – Osteoconductive – Effective as shells	– Risk of disease transmission	+	+/–	–
Xenograft	Other species (mostly bovine)	– Similar to human and volume stability (HA) – Accelerates bone formation (collagen)	– Osteoconductive only	+	–	–
Alloplast	Synthetic (HA, TCP, bioglass)	– No risk of disease transmission	– Osteoconductive only	+	–	–

10.5 Osseointegration and stability

The phenomena "osseointegration" can be shared between orthopedic implants and dental implants. Although a majority portion of implant is submerged into the bone structure, the actual area(s) which is responsible to the osseointegration is limited. For example, in Figure 10.8(a) of THR, acetabular cup outer liner hemisphere and only shoulder portion of long femur stem are required to integrated to surrounding bone structure; whilst, in a dental implant (as seen in Figure 10.8(b)), only fixture portion is osseointegrated.

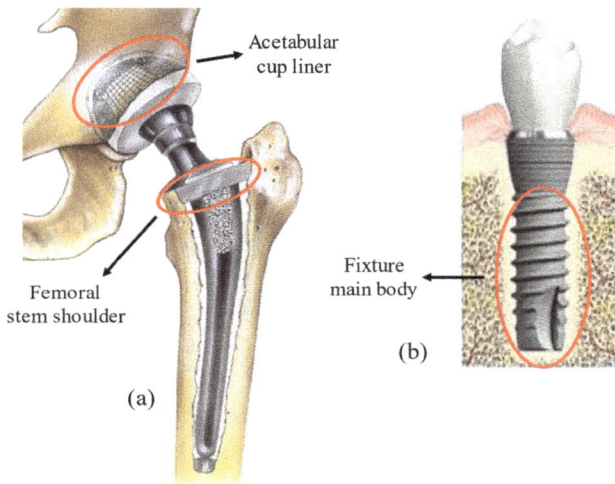

Figure 10.8: Locations where good osseointegration is required in orthopedic and dental implants, marked with red circles [255, 256].

10.5.1 Osseointegration

The surface of an implant ultimately determines its ability to integrate into the surrounding tissue. The composite effect of surface energy, composition, roughness, and topography plays a major role during the initial phases of the biological response to the implant, such as protein adsorption and cellular adherence, as well as during the later and more chronic phases of the response. For bone, the successful incorporation (and hence rigid fixation) of an alloplastic material within the surrounding bony bed is called osseointegration. The exact surface characteristics that lead to optimal osseointegration, however, remain to be elucidated. This review will focus on how surface characteristics, such as composition and roughness, affect cellular responses to implant materials. Data from two different culture systems suggest that surface characteristics play a significant role in the recruitment and maturation of cells along relevant differentiation pathways. In the case of osseointegration, if the implant surface is inappropriate or less than optimal, cells will be unable to produce the appropriate complement of autocrine and paracrine factors required to adequately stimulate osteogenesis at the implant site. In contrast, if the surface is appropriate, cells at the implant surface will stimulate interactions between cells at the surface and those in distal tissues. This, in turn, will initiate a timely sequence of events which include cell proliferation, differentiation, matrix synthesis, and local factor production, thereby resulting in the successful incorporation of the implant into the surrounding bony tissue [257]. The characteristics of high-quality osseointegration are an accelerated healing process, high stability, and the durability of the dental implant. A major factor that determines the success of a dental implant is the stable anchorage of the implant in living bone [258, 259]. In most cases, osseointegration is incomplete. An analysis of retrieved titanium implants showed that bone-to-implant attachment averaged 70%–80%, with a minimum of 60%, even for successful implants that had lasted for up to 17 years [260]. This leaves much room for improvement. Bettering the surface quality of titanium dental implants may help bridge the gap. The surface energy of a biomaterial is determined by the material's surface-charge density and the net polarity of the charge. Compared to an electrically neutral surface, a surface with net positive or negative charge may be more hydrophilic [261]. The surface charge of a dental implant is known to be a key factor for bone cell adhesion and early-stage bone mineralization at the BII. Thus, surface-charge modification seems to be a promising new direction for improving the osseointegration of titanium dental implants. Although surface-charge modification is a relatively new methodology, it has been rapidly gaining research attention in recent years. The main challenge, however, lies in effective modification of the surface charge of the dental implant material. The main objective, hence, is to develop effective and practical techniques that create a long-lasting electric field on the implant's surface, in order to promote the implant's osseointegration without incurring the drawbacks of existing surface-treatment methods. Implant stability and osseointegration are important factors in the success of treatment. Commercially pure titanium and its alloys were coated with HA and the healing response of various nanocoating surface implants by ion-beam-assisted deposition placed in surgically created circumferential gaps 12 weeks post-placement [262]. It was reported that calcium phosphate coatings improved the bone response; they may therefore be suitable for implant designs with complex surface geometries [262]. In order to improve fixation in bone, HA has been extensively coated on titanium implants [263–266] and Co–Cr implants [267]. Similar to the previous study, Ca–P compound was coated on implant surfaces to promote fixation [268, 269]. Schiegnitz et al. [269] coated Ca–P compound onto sand-blasted/acid-etched (SLA) implants in a rabbit model and used the percentage of linear bone fill as well as bone-to-implant contact (BIC) as osseointegration assessment. It was concluded that (i) the calcium phosphate coated surfaces on supracrestal-inserted implants have vertical osteoconductive characteristics and increase the BIC at the implant shoulder significantly in a rabbit model and (ii) in clinical long-term settings, these implants may contribute to a better vertical bone height [269].

Compounds that encourage osseointegration can be created by oxidizing titanium surfaces [270]. The β titanium was subjected to MAO, in order to create anatase-rich (A-TiO$_2$) and rutile-rich (R-TiO$_2$) titanium dioxide samples. It was found that osteoblasts adhered more tenaciously and grew more lamellipodia on R-TiO$_2$, than on either A-TiO$_2$ or raw β-Ti. The number of cells that adhered and proliferated on the R-TiO$_2$ surface was visually greater than on the others. These results showed two things that osteoblasts can grow into the porous structure of TiO$_2$ surfaces, and that β-Ti alloys with rutile-rich surfaces may be ideal for orthopedic and dental implants [270]. Other studies have also reported similar beneficial results from titanium surface oxidation [271]; both TiO$_2$ [272] and ZrO$_2$ surfaces exhibited a similar effect on osseointegration [273]. Plecko et al., [274] evaluated Co–Cr alloy, Ti-coated Co–Cr, Ti/Zr-coated Co–Cr, CpTi, and steel in terms of their osseointegration and biocompatibility in a pelvic implantation model in sheep. It was reported that the Co–Cr screws showed significantly lower removal torque values than CpTi screws and generally tended toward lower torque values than the other materials, with the exception of steel screws, which showed no significant difference. Histomorphometrically, there were no significant differences of bone area between the groups. Based on these findings, it was concluded that Co–Cr and steel show less osseointegration than the other tested materials. However, the osseointegration of Co–Cr was improved by zirconium- and/or titanium-based coatings; their osseointegrative behavior is similar to that of pure titanium [274]. A TiN coating also affects bone formation, as it was reported that the healing observed around TiN-coated implants was similar to that of uncoated surfaces, and that TiN coating demonstrated good biocompatibility [275]. It is important to acknowledge that not all coatings encourage osseointegration. Freeman et al. [276] evaluated the effect of titanium aluminum nitride (TiAlN) coating on osseointegration and on the biological response and reported that TiAlN coatings may result in reduced osseointegration between bone and implant. Surface modification of Ti-based metals is an important issue in improving the bone cell responses and bone–implant integration [277]. Hacking et al. investigated whether the superimposition of a microtexture on a porous-coated Ti substrate affected the extent of bone ingrowth [278]. Cylindrical titanium intramedullary implants were coated with sintered titanium beads to form a porous finish using commercial sintering techniques. A control group of implants was left in the as-sintered condition. The test group was etched in a boiling acidic solution to create an irregular surface over the entire porous coating. The results of showed that bone formation was positively affected by microtextured implant surface [278]. Rough implant surfaces have been shown to improve osseointegration rates [279, 280]. This phenomenon is commonly referred to as the rugophilia. In the majority of dental implants, microrough surfaces are obtained by sandblasting and/or acid etching (SAL treatment). The surfaces with increasing roughness show more osteoblastic adhered cells and the combination of the grit-blasted and acid-etched accelerated lightly bone regeneration at the different periods of implantation in comparison with the grit-blasted implants [279]. SAL technique offers good morphological and biological management on implant surfaces [281–283]. Alkaline etching shows an effect similar to acid etching via SAL technique. An et al. [284] treated Ti substrates with alkali- and heat-treatments and reported that osteoblastic cells showed better compatibility on the pre-etched surface compared to the pure Ti surface or alkali- and heat-treated surface and (ii) the pre-etched surface showed better pull-off tensile adhesion strength against the deposited apatite. [284]. A study by Stadlinger et al. [285] tested whether ossification could be improved by coating titanium implants with ECM components. They coated the implants with collagen alone, collagen and chondriotin sulfate, or chondroitin sulfate and growth factor BMP4. Peri-implant bone formation was measured within a defined recess along the long axis of the implant. BIC and bone volume density (BVD) were determined using histomorphometry and synchrotron radiation microcomputed tomography. It was concluded that collagen and collagen/chondroitin sulfate implant coatings impart an advantage to peri-implant bone formation [285].

Titanium surface modifications that simultaneously prevent bacterial adhesion and promote bone-cell functions could be highly beneficial to improving implant osseointegration. Alcheikh et al. [286] studied the effect of sulfonate groups on titanium surfaces with respect to both *Staphylococcus aureus* adhesion and osteoblast functions pertinent to new bone formation. Oxidized CpTi was covalently bonded with poly(sodium styrene sulfonate) groups via radical polymerization. It was reported that grafting poly(sodium styrene sulfonate) groups to CpTi simultaneously inhibited bacteria adhesion and promoted osteoblast functions pertinent to new bone formation. Such modified titanium surfaces offer a promising strategy for preventing biofilm-related infections and enhancing osseointegration of implants in orthopedic and dental applications [286]. Titanium and its alloys represent the gold standard for orthopedic and dental prosthetic devices because of their good mechanical properties and biocompatibility. Recent research has focused on surface treatments designed to promote the rapid osseointegration of titanium implants, even when bone quality of the implant site is poor. Ferraris et al. [287] investigated a new surface treatment designed to improve the tissue integration of titanium-based implants. The surface treatment was able to induce bioactive behavior (without the introduction of a coating) and preserved the mechanical properties of Ti–6Al–4V substrates (e.g., fatigue resistance). The technique created a complex surface topography that was characterized by a combination of microroughness and nanotexture. Further surface texture was created by coupling this technique with conventional macroroughness blasting. The modified metallic surfaces were rich in hydroxyls groups – this feature is extremely important for inorganic bioactivity (in vitro and in vivo apatite precipitation) and for further functionalization procedures, such as the grafting of biomolecules. Modified Ti–6Al–4V induced HA precipitation after soaking for 15 days in simulated body fluid. It was reported that the process was optimized in order to not induce cracks or damages on the surface and the surface oxide layer presents high scratch resistance [287].

Pham et al. [288] investigated the effect of fluoride on the surface chemistry of polycrystalline ceramic titanium dioxide (TiO_2) and metallic titanium (Ti), as well as its effect on the proliferation and differentiation of primary human osteoblasts. The osteoblasts were exposed to fluoride-modified and unmodified samples for 1, 3, 7, 14, and 21 days. It was reported that cell differentiation, with regard to gene expression, showed no significant difference between fluoride-modified and unmodified samples and less effect on protein release for all groups. Primary human osteoblast cells did not differ on fluoride-treated Ti and TiO_2 surfaces, indicating that fluoride is not the unique factor for the bioactivity of Ti and TiO_2 surfaces [288]. Osseointegration was examined using titanium implants that had hydrophilic phosphate ion-incorporated oxide surfaces [289]. The implants were grit blasted and assessed in rabbit cancellous bone. Bone healing was compared between these samples and commercially available MAO, phosphate-incorporated clinical implants (TiUnite, TU implant). The hydrophilic phosphate-incorporated Ti surface (P implant) was produced by hydrothermal treatment on grit-blasted, moderately rough-surfaced clinical implants. The TU implant was used as a control. Thirty-two threaded implants (16 P implants and 16 TU implants), 10 mm in length x 3.3 mm in diameter, were placed in the femoral condyles of 16 New Zealand white rabbits. Histomorphometric analysis, removal torque tests, and surface analysis of the torque-tested implants were performed 4 weeks after implantation. It was found that the P and TU implants displayed microrough surface features. Torque-tested P and TU implants had a considerable amount of bone attached to their surfaces, yet P implants exhibited significantly higher BIC percentages, both in the thread region and the total lateral length of the implants, as compared to the TU implants. However, no statistical difference was found for the removal torque values; suggesting that the phosphate-incorporated Ti oxide surface obtained by hydrothermal treatment achieves rapid osseointegration in cancellous bone by increasing the degree of BIC [289].

As have been reviewed in above, surfaces properties exhibit an important role in biological interactions. In particular, the nanometer-sized roughness and the chemistry have a determinate role in the interactions of surfaces with proteins and cells. These early interactions should show influences on subsequent tissue integration. As a result, nanotechnology on surface modification will enhance the peri-implant bone healing [220, 290–292]. Nanotechnology was originally defined by the National Nanotechnology Initiative as the study and controlled manipulation of individual atoms and molecules of size between 1 and 100 nm; however, the definition has since evolved to include a broader spectrum of research endeavors and applications [293]. Nanotechnology exists as a collaboration among multiple scientific disciplines including surface science, molecular biology, microelectronics, and tissue engineering [294–297]. When conventional macromaterials are engineered into much smaller nanosized particles, they may possess completely different physical and chemical properties in certain instances. Specifically, as the size of particulate matter decreases to 100 nm or smaller, phenomena such as the quantum size effect become more prominent [298]. This principle is observed when the electrical properties of a material change as a result of significant reductions in particle size. For example, materials that are insulators at the macroscale may possess conductive properties when reduced to the nanoscale. In addition to alterations in electrical properties, changes in mechanical properties may take place as well due to an increased surface area to volume ratio as particle size is reduced. This bears significance as nanophase materials are able to maintain relatively large surface area to volume ratios allowing for more favorable interactions with surrounding structures.

The application of nanotechnology to medicine, known as "nanomedicine," has been utilized in numerous novel therapies in the field of orthopedics. A few clinical applications include targeted drug delivery, implantable materials, vertebral disk regeneration, and diagnostic modalities [299]. Figure 10.9 summarizes the potential utilities of nanotechnology among different orthopedic subspecialties [296].

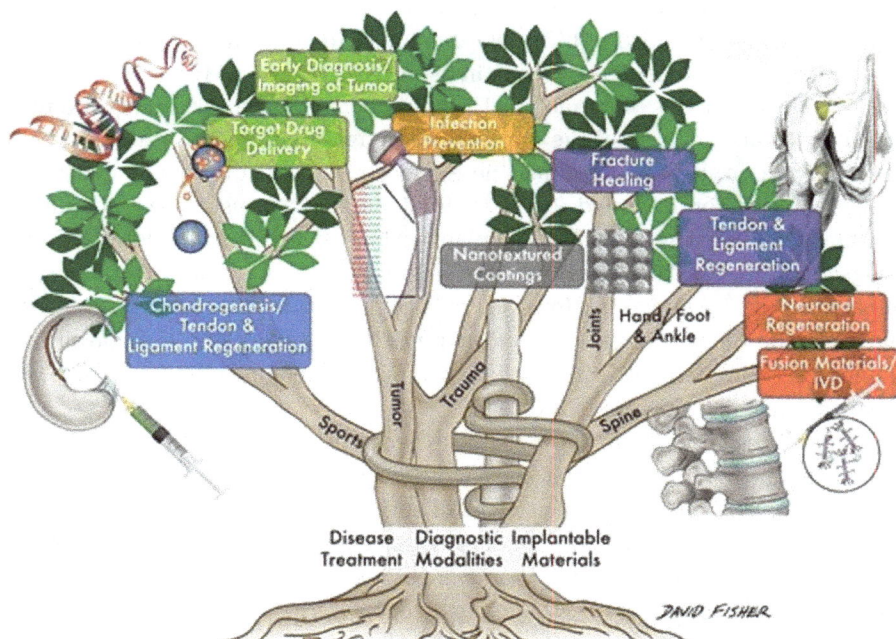

Nanotechnology in Orthopedics

Figure 10.9: A versatility of nanotechnology applications in orthopedics [296].

In the example of orthopedic implants, this allows for a greater degree of interaction between an implant and native bone, leading to more effective osseointegration

[300, 301]. Much of nanotechnologies' potential benefit with regards to medicine rests in the fact that nanotechnology may allow for more precise therapeutic applications at the subcellular level [302]. Given that many molecules involved in cellular processes exist and interact fundamentally at the nanometer scale, nanoengineered materials have the theoretical ability to target and modify these processes [303]. Applying this principle to orthopedics, bone, when broken down to the nanoscale, is naturally a nanostructure composite of collagen and HA [304]. The practical application of these principles and appreciation of these relationships have allowed for improvements in functionality and performance of a wide variety of products both in and outside of the medical field [296].

One of the prime aims of nanotechnology in the dental implant field is to increase osseointegration behavior [297, 305]. Surface modification techniques (such as acid etching, alkali surface treatment, sol–gel, and chemical vapor deposition) offer opportunities to implement better dental implants, especially in the micro- and nanoscales through design and interfacial engineering [306]. Nanoscale osseointegration modulation could be led to the following incidents [297]: (1) adhesion to osteoblast and reduced fibroblast adhesion, (2) the regulation by anisotropy and dimensional nanostructures of cell activity (adhesive proliferation and differentiation), (3) rapid cell distinction in the lineage of the osteoblast, (4) improved activity and mineralization of ALPs, (5) a drop in nanostructured ZnO or TiO_2 bacterial colonization, and (6) regulation and immunity response to protein adsorption. Osseointegration is considered as an acceptable contact between the dental implant and the surrounding tissue and is an important factor in implant design [300]. Moreover, the wettability and surface roughness are based on accelerating and enhancing osseointegration in the latest generation of dental implants [306, 307]. Sul [308] fabricated TiO_2 nanotubes on TiO_2 grit-blasted, screw-shaped rough titanium (CpTi grade IV) implants (3.75×7 mm) using potentiostatic anodization at 20 V in 1 M H_3PO_4 + 0.4 wt% HF. The resulting nanotunes were vertically aligned and approximately 700 nm in length, and highly ordered; the structures were spaced ≈ 40 nm apart and had a wall thickness of ≈ 15 nm. It was found that the fluorinated chemistry of the nanotubes of $F-TiO_2$, $TiOF_2$, and $F-Ti-O$ with F ion incorporation of ≈ 5 at%, and their amorphous structure was the same regardless of reaction time, while the average roughness gradually decreased and the developed surface area slightly increased with reaction time. It was shown that, despite their low roughness values, after 6 weeks, the fluorinated TiO_2 nanotube implants in rabbit femurs demonstrate significantly increased osseointegration strengths (41 vs. 29 Ncm) and new bone formation (57.5% vs. 65.5%). It was indicated that the method provided potential applications of the TiO_2 nanotubes in the field of bone implants and bone tissue engineering [308]. Hou et al. [309] investigated the surface characteristic, biomechanical behavior, hemocompatibility, bone tissue response, and osseointegration of the optimal MAO surface-treated titanium (MST-Ti) dental implant. It was found that (i) the hybrid volcano-like micro/nanoporous structure was formed on the surface of the MST-Ti dental implant, (ii) the hybrid volcano-like micro/nanoporous surface played an important role to improve the stress transfer between fixture, cortical bone, and cancellous bone for the MST-Ti dental implant, (iii) the MST-Ti implant was considered to have the outstanding hemocompatibility, (iv) the in vivo testing results showed that the BIC ratio significantly altered as the implant with micro/nanoporous surface, (v) after 12 weeks of implantation, the MST-Ti dental implant group exhibited significantly higher BIC ratio than the untreated dental implant group, and (vi) the MST-Ti dental implant group also presented an enhancing osseointegration, particularly in the early stages of bone healing. Based on these findings, it was concluded that (vi) the MAO approach induced the formation of micro/nanoporous surface and is a promising and reliable alternative surface modification for Ti dental implant applications.

Bone healing around implants involves a cascade of cellular and extracellular biological events that take place at the BII until the implant surface appears finally covered with a newly formed bone. These biological events include the activation of osteogenic processes similar to those of the bone healing process, at least in terms of initial host response [310, 311]. Factors enhancing osseointegration include biological and biomechanical factors, biological factors such as the status of the host bone bed (such as bone density, available remaining bone at the post-extrusion stage, and parafunctional habits) and its intrinsic healing potential, the biomechanical factors, including post-implantation loading conditions [312]. Osseointegration is also affected by implant-related factors as follows [312]: (1) Implant macrodesign (implant body, length and diameter, threads shape, pitch, lead, depth and width, and crest module), and implant design is mainly responsible for (i) increase the surface area of the implant, (ii) decrease the stress, and (iii) distributing the forces on the bone and convert the stresses into favorable compressive stresses. Li et al. [313] mentioned that the suitable depth, the steeper slope of the upper flanks, and flat roots of healing chambers can improve the bone ingrowth and osseointegration. (2) Chemical composition and biomaterial of the implant and its relation to biocompatibility, enhancing healing, and modulus of elasticity. (3) Implant surface treatment and coatings (surface topography), responsible for increase the surface area of the BII, decrease the stresses and enhance adhesion qualities to the BII at initial healing. Other factors (such as implant tilting, prosthetic passive fit, cantilever, crown high and occlusal table and loading time) should be added. It was pointed out [314] that occlusal forces affect an oral implant and the surrounding bone. Bones carrying mechanical loads adapt their strength to the load applied on it by bone modeling/remodeling. This also applies to the peri-implant bone structure. The response to an increased mechanical stress below a certain threshold will be a strengthening of the bone by increasing the bone density or apposition of bone. On the other hand, fatigue microdamage resulting in bone resorption may be the result of mechanical stress beyond this threshold. Hence, loading initiation timing at the post-operation phase and duration become a crucial factor to control the success and survival rate of osseointegrated implants, which will be discussed in the following section.

10.5.2 Stability

Population aging and the occurrence of road traffic, sports, and work accidents are the main reasons for the increasing interest of the research community in studying the osteoarticular system [315]. Implanting biomaterials within bone tissue to restore the functionality of the treated organ has become a common technique in orthopedic and dental surgery [316]. Implants and articular prostheses have led to important progress in the repair of joint degeneration (hip, knee, shoulder, etc.) and in maxillofacial surgery (to restore missing teeth or support craniofacial reconstructions). Modern orthopedic and dental implant treatments aim at a rapid, strong, and long-lasting attachment between implant and bone tissue. The surgical success of implant surgeries depends on the evolution of the biomechanical properties of the BII. To date, cemented and cementless implants are the two main types of implants used in orthopedic surgery, while bone cement is not used for the anchorage of oral implants. Although bone cement, acting as a bonding medium, can provide initial fixation, cementless implants are now often preferred because of the risks of cemented implant failures related to an accumulation of microcracks in and around the cemented area [317]. Moreover, systemic risks such as cement implantation syndrome during and after the cementation procedure have been noticed [318]. The biomechanical properties of the BII are the determinant for implant stability. A good quality of bone healing leads to: (i) direct contact between mineralized bone tissue and the implant and (ii) an important proportion of the implant surface in intimate contact with bone tissue [315], leading to successful osseointegration.

The implant stability is a prerequisite characteristic of the osseointegration [319]. Continuous monitoring in quantitative and objective manners is important to determine the implant stability [320, 321]. Osseointegration is also a measure of implant stability which can occur in two stages: primary and secondary [322, 323]. Implant stability can be defined as an absence of clinical implant mobility and consists of primary and secondary implant stability [323] and these two stabilities are competing phenomena, as illustrated in Figure 10.10 [324].

Referring to Figure 10.10, the primary implant stability (remodeling of existing bone) mostly occurs from a mechanical engagement with the cortical bone. A key factor for the implant primary stability is the BIC [325], so that the primary stability is affected by bone quality and quantity, the surgical technique, and implant geometry (length, diameter, surface characteristics) [326, 327]. Primary stability (remodeling of exiting bone) includes the mechanical attachment of an implant in the surrounding bone at the insertion, whereas secondary implant stability is the tissue response to the implant and subsequent bone remodeling processes [328]. Primary implant stability is known to be a crucial factor for successful osseointegration of dental implants [329, 330]. There is sufficient evidence to accept a positive correlation between primary implant stability and implant success, as the success relies on the sustainable integration of the implants into hard and soft tissues [330–333].

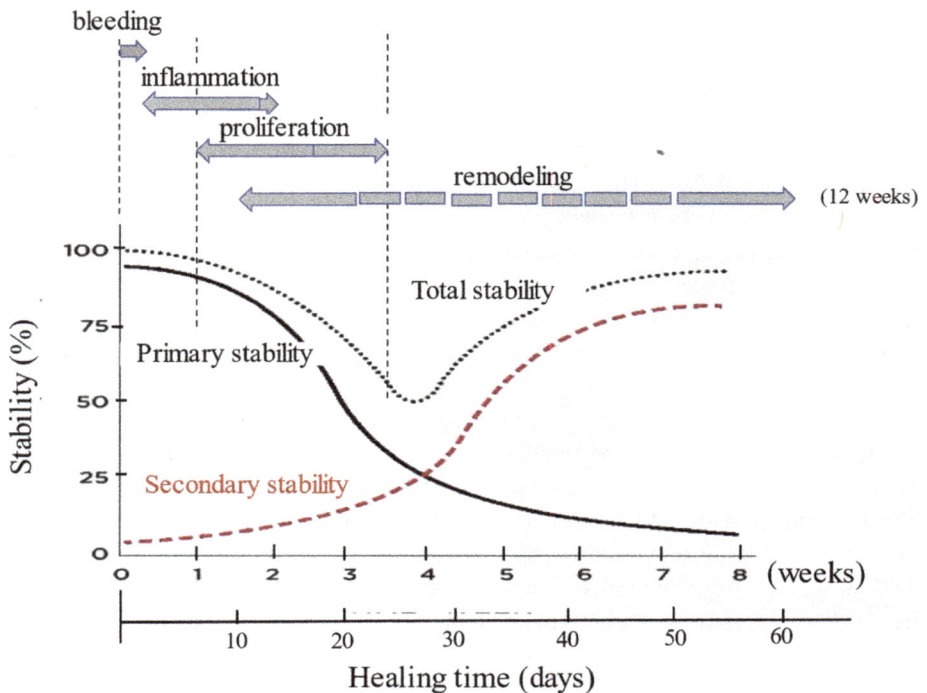

Figure 10.10: Conceptual diagram of placed implant stabilities [324].

The secondary stability (de novo bone formation) offers a biological stability through the bone regeneration and remodeling processes [334–337]. The secondary stability is affected by the primary stability [336, 338]. During the transition period from the primary to the secondary stability, placed implants may face the risk of micromotion; possibly leading to an implant failure. Secondary stability depends on primary stability and has been reported to increase 4 weeks after implant placement [327, 339]. Thus, in the first 2–3 weeks after implant placement, a stability gap with the lowest implant stability is expected [340].

In Figure 10.10, detailed processes involved in the primary stability and onset of the secondary stability [145] are inserted. During the bleeding phase, which lasts for a few hours, it is normal that the more vascular the tissues, the longer they will bleed. The followings are major characteristics associated with the inflammatory phase: (i) essential for tissue repair, (ii) rapid onset (few hours) and increases in magnitude for 2 to 3 days before gradually resolving over a few weeks, (iii) complex, chemically mediated amplification cascades should be responsible for the initiation and control of the inflammatory reaction, and (iv) vascular and cellular cascades are the two essential elements. The proliferation phase can be characterized by (i) involvement of repair material generation, (ii) rapid onset for 24 to 48 h, (iii) peak activity reached in 2 to 3 weeks, (iv) decrease over several months, (v) two fundamental processes: fibroplasia and angiogenesis, and (vi) chemical mediators, that is, macrophage-derived growth factors, PDGFs, lactic acid, FGF. During the remodeling stage, there are several important characteristics which (i) primarily involves collagen and the ECM, (ii) with maturity, collagen becomes more oriented in line with local stress, and (iii) the type III collagen, which is fine, weak, and highly cellular (it is the collagen of granulation tissue and produced by young fibroblasts) is converted to type I collagen, which is more cross linked with greater tensile strength and, therefore, more stable [145]. After roughly 2 to 3 weeks in post-implantation, the osteoclastic activity decreases the initial mechanical stability of the placed implant(s) but not enough new bone has been produced to provide an equivalent or greater amount of compensatory biological stability [341, 342]. This is related to the biologic reaction of the bone to a surgical trauma during the initial bone remodeling phase; bone and necrotic materials resorbed by osteoclastic activity is reflected by a reduction in implant stability quotient (ISQ) value. This process is followed by new bone apposition initiated by osteoblastic activity, therefore leading to adaptive peri-implant bone remodeling [343]. The total stability, as determined by the addition of primary stability and secondary stability, normally shows a merging gap, which is called the stability dip. The stability dip is considered unavoidable in current dental implants because the rate of losing primary stability is faster than the development of secondary stability. However, Suzuki et al. [344] reported that (i) the use of photofunctionalization eliminated the stability dip (in other words, no clear transient from primary to secondary stability) because of faster (when the primary stability is high) and even faster (when the primary stability is low) development of secondary stability and (ii) the dip can shift horizontally with respect to the healing time axis.

10.5.3 Evaluation

Albrektsson et al. [345] pointed out that there are several crucial parameters to establish successful osseointegration, including implant design, material, surface condition, receiving peri-implant bone condition, implant site preparation, or loading conditions. Historically, the gold standard method used to evaluate the degree of osseointegration was microscopic or histologic analysis [346]. And biological responses study can include cell morphology and cell activity (cell adhesion, differentiation, and proliferation) [347–350]. It is well documented that the biochemistry and topography of biomaterials' surfaces play a key role on success or failure upon placement in a biological environment [351]. Wettability, texture, chemical composition, and surface topography are properties of the biomaterials that directly influence their interaction with cells [352, 353]. Since cell interactions with ECM directly affect the cellular processes of adhesion, proliferation, and differentiation [354], the surface properties of biomaterials are essential to the response of cells at biomaterial interface, affecting the growth and quality of newly formed bone tissue [355]. In general, the common finding of the aforementioned references can be found as a fact that the cell activity is strongly related to implant surface morphology and topology as related to the process of osseointegration [356]. This surface characteristic is one of the three major requirements for placed implant to exhibit the subsequent retaining in the bone (namely, osseointegration), aka morphological compatibility [154, 306]. Two other requirements are biological compatibility and biomechanical compatibility [306, 357]. To manipulate surface structure as being morphologically compatible to receiving hard tissue surface configuration, there have been various methods and techniques proposed, including as-machined, blasted surface, acid or alkaline etching, chemical treating on blasted surface, HA coating, recently the biomimetic calcium phosphate coating [306]. Accordingly, when the obtained data is analyzed, all these contributing parameters should be considered for assessment of the osseointegration.

Implant stability is a requisite characteristic of osseointegration. Without it, long-term success cannot be achieved. Continuous monitoring in a quantitative and objective manner is important to determine the status of implant stability [346]. Osseointegration is also a measure of implant stability which can occur in two stages (primary stability and secondary stability), as described above. The intrasurgical assessment of the implant stability can be valuable and indicative for best timing for loading, while the post-surgical assessment is important not only for loading time decision but also for a longevity of the placed implant(s). Swami et al. [349] grouped various methods to assess implant stability into two groups: invasive/destructive methods and noninvasive/nondestructive methods.

Invasive/destructive methods
These methods can include biological and mechanical tests in both vitro and vivo. Histologic and/or histomorphologic analysis is obtained by calculating the peri-implant bone quantity and BIC from a dyed specimen of the implant and peri-implant bone. Accurate measurement is an advantage, but due to the invasive and destructive procedure, it is not appropriate for long-term studies. It is used in the nonclinical studies and experiments [349]. It is assessed at pre-, intra-, and post-surgical time points [358].

It is generally believed that outcomes on the initial biological behavior of implantable materials obtained in vitro cannot be fully correlated to in vivo performance. Cell cultures cannot reproduce the dynamic environment that involves the in vivo bone/implant interaction, and their results can only be confirmed in animal models and subsequently in clinical trials [359, 360]. Irrespective of the different animal models or surgical sites, valuable information can be retrieved from properly designed animal studies. Static and dynamic histomorphometric parameters plus biomechanical testing are recommended as measurable indicators of the host/implant response where different surface designs are compared. BIC, which is the most often evaluated parameter in in-vivo studies, together with bone density and amount and type of cellular content, are examples of static parameters [360]. In most majority of publications, canine [361, 362], sheep/goat [363–365], pig [366], rabbit [350, 367–370], rat, and mice [371–374] are popular kinds of animals as non-human primates which are mostly sacrificed for research purpose [375, 376].

Although animal tests appear to be well-established techniques to provide valuable and applicable information to human, there are still several concerns. One of the main problems associated with the in vivo tests using animal models is the test duration and its validity for application. Dziuba et al. [369], using the rabbit model, investigated the biological behavior of Mg implantable alloy for the longest of 12 months. Amerstorfer et al. [373] had conducted the in vivo with Sprague Dawley rats for 12 months to investigate the biodegradability for Mg alloy as a potential implant material. Furthermore, Akens et al. [363], using sheep for 18 months at the longest test duration, studied the efficacy of photooxidized bovine osteochondral transplant. Although the conclusive remarks common over these tests indicate that the result are promising, but still a longer period of testing time is required before applying the materials to human subjects.

The relevancy of results obtained from animal models has been subject to a debate with great extent. The use of animal models in the study of dental implants has contributed greatly to understand many different devices in use. Animal testing plays a major role in assessing the safety and efficacy of dental implants. To date, animal testing has shown the nature of soft tissue attachment to implants and the types of interfacial tissues within bone sites. There have been increased studies correlating animal tests with in vitro analysis and human studies. There is an important aspect to the question of the relevance of animal research to humans, namely the way the observations and results are evaluated and adopted [377, 378]. Since some results from in vitro studies can be difficult to extrapolate to the in vivo situation, the use of animal models is often an essential step in the testing of orthopedic and dental implants prior to clinical use in humans [376]. As mentioned before, variety of animals such as the dog, sheep, goat, pig, and rabbit models has been used for the evaluation of bone-implant interactions. There are differences in bone composition between the various species and humans. While no species fulfils all of the requirements of an ideal model, an understanding of the differences in bone architecture and remodeling between the species is likely to assist in the selection of a suitable species for a defined research question [376].

Besides the abovementioned histological analyses, there are the biomechanical tests (tensile test, torque, push-out, pullout, removal torque analysis) to measure the amount of force that a torque needs to fail the BII surrounding different implant surfaces [340, 379–381]. Considering the several factors that influence the osseointegration, the evaluation of the largest possible number of host/implant response parameters is desirable for better understanding the bone healing around different implant surface, clarifying their indications of use and supporting their immediate/early loading. Mechanical tests include tensile test [382], push-out/pull-out test [383], and removal torque tests [384–386].

Noninvasive/nondestructive methods

Assessing the implant stability is mainly performed for aiming the safest and earliest timing for loading, so that an implant patient will be able to recover back to his/her normal masticatory action. Hence, any devices to evaluate the placed (hopefully osseointegrated) implants should be nondestructive and noninvasive method. Most of nondestructive stability assessment tests can be performed at the chairside, so that a clinician will enable to make a decision for good timing of the loading initiation.

Besides torque tests (including cutting torque resistance analysis [387, 388], insertion torque measurement [389, 390], reserve torque test [380, 391], or seating torque test [392], there are several techniques involved in the vibration analysis and they include the percussion test [380, 390], pulsed oscillation waveform method [393, 394], the Periotest [395, 396], and a resonance frequency analysis (RFA) [397–400].

The ISQ which is obtained using RFA is the value on a scale that indicates the level of stability and osseointegration in dental implants. The scale ranges from 1 to 100, with higher values indicating greater stability [401, 402]. There can be found scattered data and clinical evidences regarding ISQ scale and its related implant mobility.

1) In general, higher values are generally found in the mandible than the maxilla [403].
2) High initial stability (ISQ values of 70 and above) tends to not increase with time, even if the high mechanical stability will decrease to be replaced by a developed biological stability. Lower initial stability will normally increase with time due to the lower mechanical stability being enforced by the bone remodeling process (osseointegration) [401, 404].
3) ISQ has a nonlinear correlation to micro mobility. Micromobility decreases >50% from 60 to 70 ISQ [405, 406].
4) It is suggested that implants with high ISQ values show lower micromotions and withstand higher forces in the mouth [407, 408].
5) Moreover, several studies have confirmed that RFA measurements are able to successfully predict implant failure [409–412].
6) Any observed decrease in stability is thought to be related to bone healing following surgical injury. Understanding the healing process and its effect on implant stability can assist in clinical decision-making related to the timing of implant restoration. In this regard, RFA can provide an objective assessment of implant stability, bone healing and osseointegration [413].

Figure 10.11 summarizes clinical activities related to a range of ISQ, based on numerous sources [328, 413–415].

Miyazaki et al. [319, 416, 417] investigated 127 cases (54 male and 73 female): age ranging with 69 cases (younger than 65 years old: 41 male + 28 female) and 58 cases (older than 65 years old: 13 male + 45 female). All received SLAed (sandblasted with large alumina grits, followed by acid etching) cpTi Grade IV implants. ISQ scale measured at right after implant placement was designated as ISQ-I, while

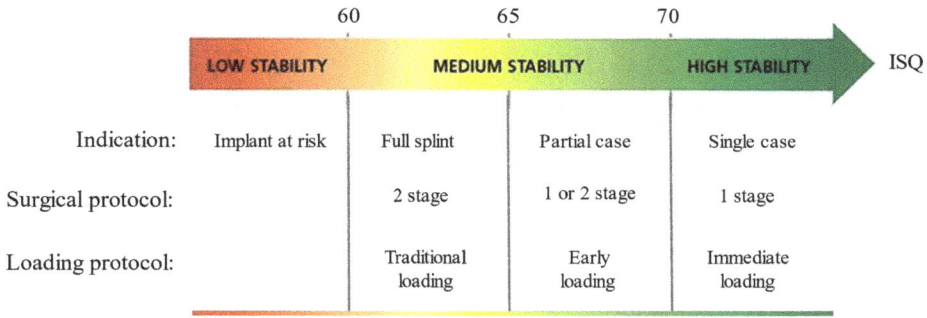

Figure 10.11: Surgical and loading protocols, depending on obtained ISQ scale range.

ISQ scale obtained at the loading was indicated as ISQ-L. Moreover, [(ISQ-L − ISQ-I)/ ISQ-I]x100% is defined as the change rate of ISQ during the bone healing process. Figure 10.12 shows a relationship between ISQ measured at implant placement (ISQ-I) and change rate of ISQ during the bone healing and Figure 10.13 depicts a relationship between ISQ-I and ISQ measured at initial loading (ISQ-L).

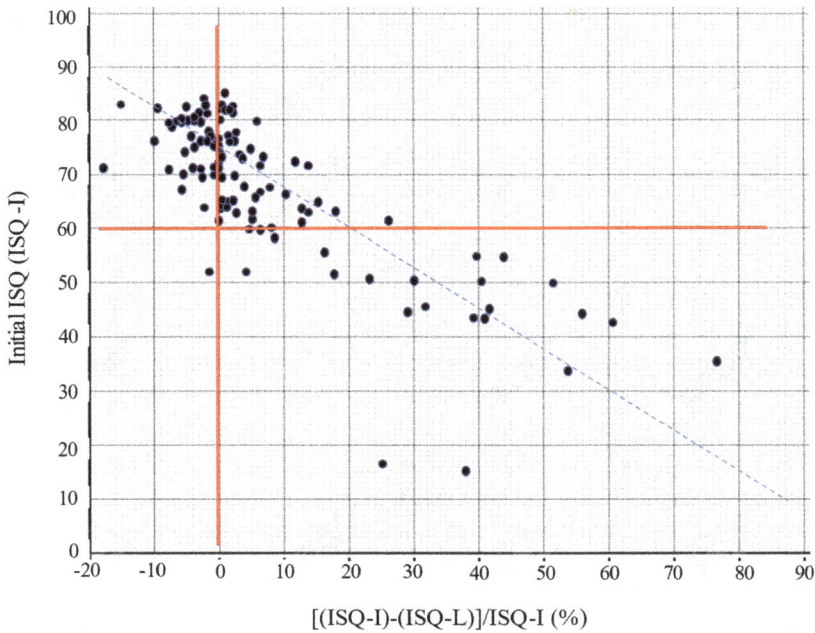

Figure 10.12: ISQ-I versus ΔISQ.

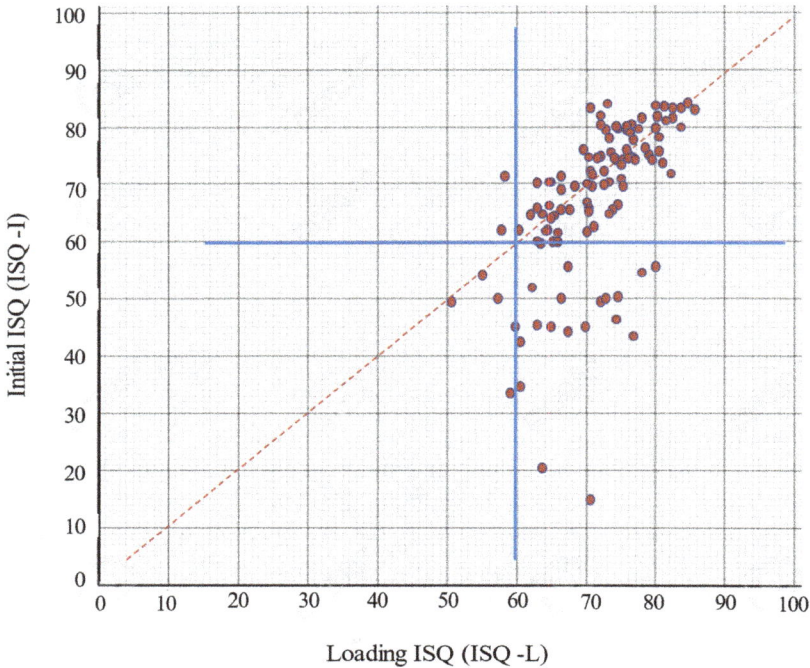

Figure 10.13: ISQ-I versus ISQ-L.

After roughly 2 to 3 weeks in post-implantation, the osteoclastic activity decreases the initial mechanical stability of the placed implant(s) but not enough new bone has been produced to provide an equivalent or greater amount of compensatory biological stability [418, 419]. This is related to the biologic reaction of the bone to a surgical trauma during the initial bone remodeling phase; bone and necrotic materials resorbed by osteoclastic activity is reflected by a reduction in ISQ value. This process is followed by new bone apposition initiated by osteoblastic activity, therefore leading to adaptive bone remodeling around the implant [343]. Figure 10.12 confirms this statement, indicating that the higher the initial ISQ value is, the lower the progressive decrease rate of ISQ value during the bone healing process. It was also found that the majority data belong to a zone where initial ISQ-I at implantation is over 60. Figure 10.13 depicts a semi-linear relationship between initial ISQ (ISQ-I) at the implantation and ISQ-L at loading and this finding confirms one of conclusions reported by Suzuki et al. [420]. It is true that implant stability is one of the most important factors for the success of implant treatments. Although most studies showed a correlation between the bone density and the implant stability, some studies suggest the opposite; due to the differences in the methods used. Recent studies suggest that the implant stability during the healing process only increases for implants with low initial stabilities; meanwhile, a loss of stability during the healing stage can be observed in implants with high initial stabilities [421].

10.5.4 Loading protocol

Immediate loading protocols are of interest for modern implant therapy since the technique of implant placements immediately following tooth extraction was introduced in the 1970s [422]. During the past few decades, clinicians have used two-piece implants as a one-piece system for immediate loading to accomplish early recovery for function and esthetics [423]. It was reported that the statistical analysis by the Institute of Clinical Materials demonstrates high survival rates of $92.3 \pm 8.3\%$ [424]. Several concerns with traditional two-stage approaches have been stated, such as alveolar bone loss, increased time of edentulism and surgery, need for a second surgical procedure, and psychological factors [425]. In contrast, the implants' immediate placement in the fresh extraction socket as a one-step procedure reduces the length of surgery as well as the number of interventions, and, consequently, has psychological benefits for patients [425–427]. In a recent systematic review conducted by Cosyn et al., immediate placement of a single tooth was associated with a higher risk for early implant loss compared to delayed implant placement [428]. All implant failures were early failures resulting from a lack of osseointegration. As the success of immediate loading approaches directly relies on primary stability, prognostic factors are of interest for preoperative risk-stratification models. Further, innovative surgical techniques, such as ultrasonic site preparation, can be included with the methodology provided to assess other predicting factors in the future [429]. Several recent studies focused on the assessment of implant stability based on radiofrequency analysis [430–432]. However, these studies focused on the differences between implant systems and characteristics without considering confounding factors to predict implant stability for future risk stratification models [328].

After placing implant(s), there are normally three protocols for a load timing: (1) immediate loading (within 48 h), (2) early loading (2 days to 3 months), and (3) conventional delayed loading (3 to 6 months). The success of the implant treatment depends upon its osseointegration with the surrounding bone; namely, there should not be any progressive relative movement between the placed implant and the bone with which it has direct contact in biomechanistic environment [319]. Immediate loading is proposed on an assumption that an initial primary stability can be achieved in the old bone and will change over time into the secondary stability in new and osseointegrated bone during the bone remodeling process [324]. Skvirsky [433] mentioned that there are five questions that dentists should ask themselves to determine whether to recommend delayed or immediate implant loading protocol to their patients and answers to each of question should be reflected to indication or contraindication for the immediate loading. They include (1) How important are the esthetics of the implanted teeth? (2) What is the condition of the patient's oral hygiene? (3) What is the occlusion and wear pattern of the patient's teeth? (4) How much bone grafting is required? (5) What implant surface treatment are you using? Questions 1 through 4 are related

to patient's conditions, while the question 5 concerns the implant surface characteristics, which directly affect the early osseointegration of placed implants. Implant
surfaces are modified to increase the rate of the osseointegration as well as the onset
of primary and secondary stabilities. Surface modifications can include various surface engineering technologies such as sandblasting, shot peening, acid- or alkaline
etching, metal powder or ceramic powder deposition, or a sequential combination of
the above [306].

Miyazaki et al. [319, 416, 417] complied ISQ data and superimposed them to the
conceptual two-stage stability diagram (as shown previously in Figure 10.10). Presented data in Figure 10.14 are selected only those are limited for healing time from 20
to 60 days. Although the time axis is arranged the same as that shown in Figure 10.10,
the right-side vertical axis position for ISQ change rate is chosen arbitrarily. It was observed that (i) a majority of cases stays around 0 ± 10 at the ISQ change rate, (ii) however, 40 days and beyond, there seems to be an increasing trend although it is not
clear, at this moment, what parameter is responsible to this trend, and (iii) this remarkable increasing trend starts at the onset of secondary stability phenomenon. To
try to find the appropriate timing for loading, same data are added to Figure 10.10,
resulting in Figure 10.15. Again, the location of vertical ISQ-L axis was arbitrarily determined. It was found that, between healing times from 20 to 60 days, all data (except

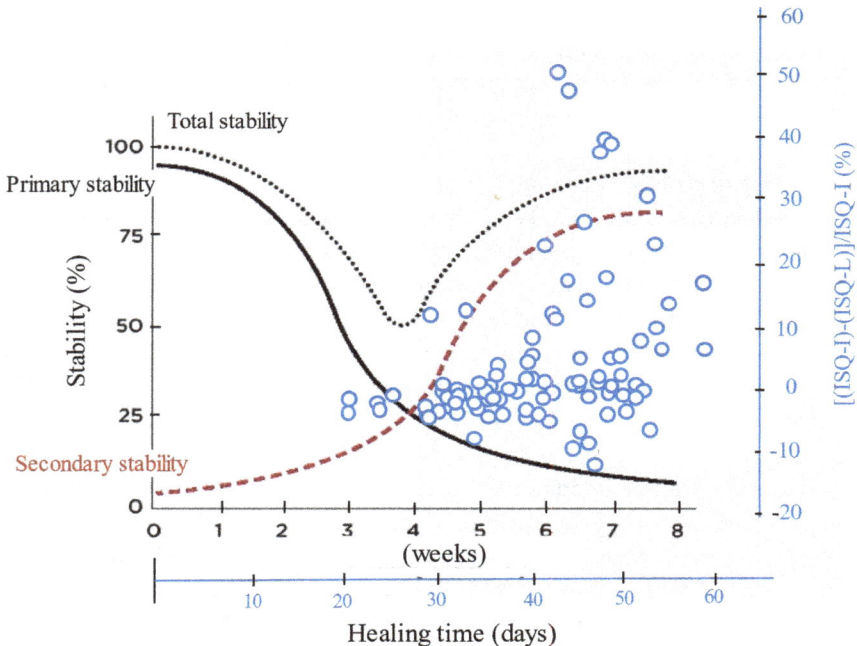

Figure 10.14: Superimposed change rate of ISQ data on two-stage stability diagram [319].

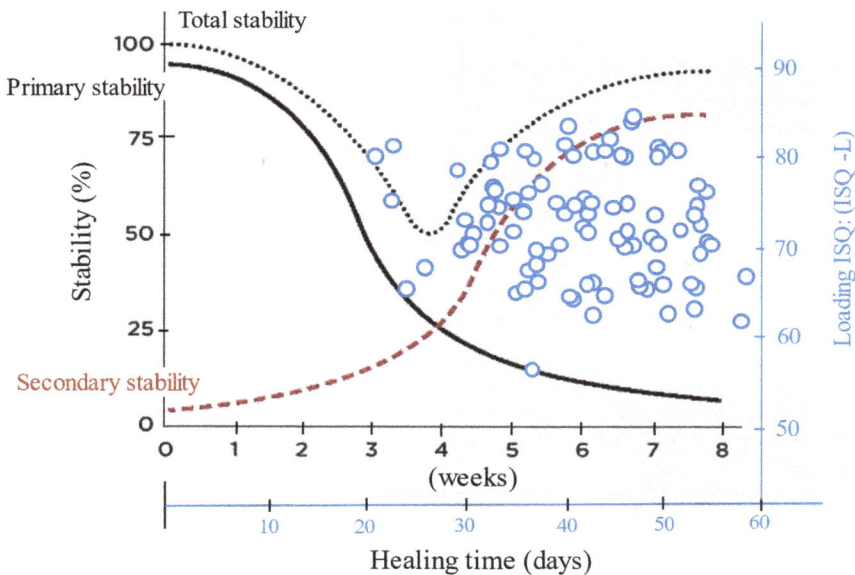

Figure 10.15: Superimposed ISQ at loading time on two-stage stability diagram [319].

one case with 54 of ISQ-L) show 60 or above of ISQ measured at the first loading. These findings lead to conclusions and indications for the safest and earliest loading protocol as follows. The proper timing for semiearly loading protocol can be determined and standardized as follows; (1) if ISQ-I > 60, as shown in the area of left upper square in Figure 10.12, loading can be conducted on or after 30 days, per suggestions from Figure 10.15, (2) if 60 > ISQ-I > 40, as shown in the area of right down square in Figure 10.13, from indications from Figures 10.13 and 10.14, 40-day post-implantation can be long enough for the loading, and (3) if ISQ-I < 40, follow-up checking ISQ is recommended until "40 days/60-ISQ criterion" is established.

References

[1]   Negm SAM. Implant success versus implant survival. Dentistry. 2016, 6, 1, doi: 10.4172/2161-1122.1000359.
[2]   Denard PJ, Raiss P, Sowa B, Walch G. Mid- to long-term follow-up of total shoulder arthroplasty using a keeled glenoid in young adults with primary glenohumeral arthritis. J Shoulder Elb Surg. 2013, 22, 894–900.
[3]   Singh JA, Sperling J, Buchbinder R, McMaken K. Surgery for shoulder osteoarthritis. Cochrane Database Syst Rev. 2010, 10, CD008089, doi: 10.1002/14651858.CD008089.pub2.
[4]   Radnay CS, Setter KJ, Chambers L, Levine WN, Bigliani LU, Ahmad CS. Total shoulder replacement compared with humeral head replacement for the treatment of primary glenohumeral osteoarthritis: A systematic review. J Shoulder Elbow Surg. 2007, 16, 396–402.

[5] van de Sande MA, Brand R, Rozing PM. Indications, complications, and results of shoulder arthroplasty. Scand J Rheumatol. 2006, 35, 426–34.

[6] Fevang BT, Lie SA, Havelin LI, Skredderstuen A, Furnes O. Risk factors for revision after shoulder arthroplasty: 1,825 shoulder arthroplasties from the Norwegian Arthroplasty Register. Acta Orthop. 2009, 80, 83–91.

[7] Martin SD, Zurakowski D, Thornhill TS. Uncemented glenoid component in total shoulder arthroplasty Survivorship and outcomes. J Bone Joint Surg Am. 2005, 87, 1284–92.

[8] Tammachote N, Sperling JW, Vathana T, Cofield RH, Harmsen WS, Schleck CD. Long-term results of cemented metal-backed glenoid components for osteoarthritis of the shoulder. J Bone Joint Surg Am. 2009, 91, 160–66.

[9] Singh JA, Sperling JW, Cofield RH. Revision surgery following Total Shoulder Arthroplasty: Analysis of 2,588 shoulders over 3 decades (1976–2008). J Bone Joint Surg Br. 2011, 93, 1513–17.

[10] United Nations. World population prospects: The 2008 revision. Department of Economic and Social Affairs/Population Division, New York; https://www.un.org/en/development/desa/population/publications/trends/population-prospects.asp#:~:text=World%20Population%20Prospects%2C%20the%202008%20Revision.%20World%20population,the%20official%20United%20Nations%20population%20estimates%20and%20projections.

[11] Zhang Y, Jordan JM. Epidemiology of osteoarthritis. Rheum Dis Clin North Am. 2008, 34, 515–29.

[12] Birrell F, Johnell O, Silman A. Projecting the need for hip replacement over the next three decades: Influence of changing demography and threshold for surgery. Ann Rheum Dis. 1999, 58, 569–72.

[13] Prime MS, Palmer J, Khan WS. The national joint registry of England and Wales. Orthopedics. 2011, 34, 107–10.

[14] Smith GH, Johnson S, Ballantyne JA, Dunstan E, Brenkel IJ. Predictors of excellent early outcome after total hip arthroplasty. J Orthop Surg Res. 2012, 7, 13–15.

[15] National Institute for Health and Care Excellence. Total hip replacement and resurfacing arthroplasty for end stage arthritis of the hip. NICE Technol Appraisal Guid. 2014, 304, https://www.nice.org.uk/guidance/ta304/resources/total-hip-replacement-and-resurfacing-arthroplasty-for-endstage-arthritis-of-the-hip-pdf-82602365977285.

[16] Evans JT, Evans JP, Walker RW, Blom AW, Whitehouse MR, Sayers A. How long does a hip replacement last? A systematic review and meta-analysis of case series and national registry reports with more than 15 years of follow-up. Lancet. 2019, 393, 647–54.

[17] Evans JT, Blom AW, Timperley AJ, Dieppe P, Wilson MJ, Sayers A, Whitehouse MR. Factors associated with implant survival following total hip replacement surgery: A registry study of data from the National Joint Registry of England, Wales, Northern Ireland and the Isle of Man. PLoS Med. 2020, 17, e1003291, doi: https://doi.org/10.1371/journal.pmed.1003291.

[18] Hauer G, Heri A, Klim S, Puchwein P, Leithner A, Sadoghi P. Survival rate and application number of total hip arthroplasty in patients with femoral neck fracture: An analysis of clinical studies and national arthroplasty registers. J Arthroplasty. 2020, 35, 1014–22.

[19] Kurtz S, Ong K, Lau E, Mowat F, Halpern M. Projections of primary and revision hip and knee arthroplasty in the United States from 2005 to 2030. J Bone Joint Surg (Am). 2007, 89-A, 780–85.

[20] Karachalios T, Komnos G, Koutalos A. Total hip arthroplasty: Survival and modes of failure. EFORT Open Rev. 2018, 3, 232–39, doi: 10.1302/2058-5241.3.170068.

[21] Carr AJ, Robertsson O, Graves S, Price AJ, Arden NK, Judge A, Beard DJ. Knee replacement. Lancet. 2012, 379, 1331–40.

[22] Learmonth ID, Young C, Rorabeck C. The operation of the century: Total hip replacement. Lancet. 2007, 370, 1508–19.

[23] NJR (National Joint Registry for England, Wales and Northern Ireland). 12th Annual Report. 2015, 41; http://www.njrcentre.org.uk/njrcentre/Portals/0/Documents/England/Reports/12th%20annual%20report/NJR%20Online%20Annual%20Report%202015.pdf.

[24] Bayliss LB, with 10 co-authors. The effect of patient age at intervention on risk of implant revision after total replacement of the hip or knee: A population-based cohort study. Lancet. 2017, 389, 1424–30.

[25] Argenson J-N, with 11 co-authors. Survival analysis of total knee arthroplasty at a minimum 10 years' follow-up: A multicenter French nationwide study including 846 cases. Orthop Traumatol Surg Res. 2013, 99, 385–90.

[26] Bae DK, Song SJ, Park MJ, Eoh JH, Song JH, Park CH. Twenty-year survival analysis in total knee arthroplasty by a single surgeon. J Arthroplasty. 2012, 27, 1297–304.

[27] Bae DK, Song SJ, Heo DB, Lee SH, Song WJ. Long-term survival rate of implants and modes of failure after revision total knee arthroplasty by a single surgeon. J Arthroplasty. 2013, 28, 1130–34.

[28] Kim Y-H, Park J-W, Jang Y-S. 20-year minimum outcomes and survival rate of high-flexion versus standard total knee arthroplasty. J Arthroplasty. 2021, 36, 560–65.

[29] Bauer GC. What price progress? Failed innovations of the knee prosthesis. Acta Orthop Scand. 1992, 63, 245–46.

[30] Goodfellow J. Knee prostheses: One step forward, two steps back. J Bone Joint Surg Br. 1992, 74, 1–2.

[31] Kraay MJ, Meyers SA, Goldberg VM, Figgie HE, Conroy PA. "Hybrid" total knee arthroplasty with the Miller-Galante prosthesis. A prospective clinical and roentgenographic evaluation. Clin Orthop Relat Res. 1991, 273, 32–41.

[32] Rorabeck CH, Bourne RB, Lewis PL, Nott L. The Miller-Galante knee prosthesis for the treatment of osteoarthrosis. A comparison of the results of partial fixation with cement and fixation without any cement. J Bone Joint Surg Am. 1993, 75, 402–08.

[33] Kim YH, Oh JH, Oh SH. Osteolysis around cementless porous-coated anatomic knee prostheses. J Bone Joint Surg Br. 1995, 77, 236–41.

[34] Faris PM, Keating EM, Farris A, Meding JB, Ritter MA. Hybrid total knee arthroplasty: 13-year survivorship of AGC total knee systems with average 7 years followup. Clin Orthop Relat Res. 2008, 466, 1204–09.

[35] Choi YJ, Lee KW, Kim CH, Ahn HS, Hwang JK, Knag JH, Han HD, Cho WJ, Park JS. Long-term results of hybrid total knee arthroplasty: Minimum 10-years follow-up. Knee Surg Relat Res. 2012, 24, 79084.

[36] Yang Y, Hu H, Zeng M, Chu H, Gan Z, Duan J, Rong M. The survival rates and risk factors of implants in the early stage: A retrospective study. BMC Oral Health. 2021, 21, 293, doi: https://doi.org/10.1186/s12903-021-01651-8.

[37] Busenlechner D, Fürhauser R, Haas R, et al. Long-term implant success at the Academy for Oral Implantology: 8-year follow-up and risk factor analysis. J Periodontal Implant Sci. 2014, 44, 102–08.

[38] Krebs M, Schmenger K, Neumann K, et al. Long-term evaluation of ANKYLOS® dental implants, Part I: 20-year life table analysis of a longitudinal study of more than 12,500 implants. Clin Implant Dent Relat Res. 2015, 17, e275–86.

[39] Artzi Z, Carmeli G, Kozlovsky A. A distinguishable observation between survival and success rate outcome of hydroxyapatite-coated implants in 5–10 years in function. Clin Oral Implants Res. 2006, 17, 85–93.

[40] Moraschini V, da C. Poubel LA, Ferreira VF, dos S.P.Barboza E. Evaluation of survival and success rates of dental implants reported in longitudinal studies with a follow-up period of at least 10 years: A systematic review. Int J Oral Maxillofac Surg. 2015, 44, 377–88.

[41] Bandeira de Almeida A, Prado Maia L, Demoner Ramos U, Luís Scombatti de Souza S, Bazan Palioto D. Success, survival and failure rates of dental implants: A cross-sectional study. J Oral Health Rehabil. 2017, https://www.dtscience.com/success-survival-and-failure-rates-of-dental-implants-a-cross-sectional-study.

[42] Cochran D, Oates T, Morton D, Jones A, Buser D, Peters F. Clinical field trial examining an implant with a sand-blasted, acid-etched surface. J Periodontol. 2007, 78, 974–82.

[43] Beschnidt SM, with 10 co-authors. Implant success and survival rates in daily dental practice: 5-year results of a non-interventional study using CAMLOG SCREW-LINE implants with or without platform-switching abutments. Int J Implant Dent. 2018, 4, 33, doi: https://doi.org/10.1186/s40729-018-0145-3.

[44] Bartolo P, Kruth J-P, da Silva JVL, Levy G, Malshe A, Rajurkar K, Mitsuishi M, Ciurana J, Leu MC. Biomedical production of implants by additive electro-chemical and physical processes. CIRP Ann Manuf Technol. 2012, 61, 635–55.

[45] Park JB. Orthopedic prosthesis fixation. Ann Biomed Eng. 1992, 20, 583–94.

[46] Maggs J, Wilson M. The relative merits of cemented and uncemented prostheses in total hip arthroplasty. Indian J Orthop. 2017, 51, 377–85.

[47] Davies N, Jackson W, Price A, Rees J, Lavy C. FRCS Trauma and Orthopaedics Viva. Oxford: Oxford University Press, 2012.

[48] Khanuja HS, Vakil JJ, Goddard MS, Mont MA. Cementless femoral fixation in total hip arthroplasty. J Bone Joint Surg Am. 2011, 93, 500–09.

[49] Castelli CC, Rizzi L. Short stems in total hip replacement: Current status and future. Hip Int. 2014, 24, S25–8.

[50] Understanding your Surgery: Cemented vs. Cementless Total Knee Replacement. 2021; https://www.drpauljacob.com/blog/understanding-your-surgery-cemented-vs-cementless-total-knee-replacement-25733.html.

[51] Sood V. Cemented vs. Cementless Alternatives in Joint Replacement. 2014; https://www.arthritis-health.com/surgery/shoulder-surgery/cemented-vs-cementless-alternatives-joint-replacement.

[52] Okike K, Chan PH, Prentice HA, Paxton EW, Burri R. Association between uncemented vs cemented hemiarthroplasty and revision surgery among patients with hip fracture. JAMA. 2020, 323, 1077–84.

[53] NIH. Hip Replacement Surgery. 2013, https://www.niams.nih.gov/health-topics/hip-replacement-surgery#5.

[54] Chen C, Li R. Cementless versus cemented total knee arthroplasty in young patients: A meta-analysis of randomized controlled trials. J Orthop Surg Res. 2019, 14, 262, doi: https://doi.org/10.1186/s13018-019-1293-8.

[55] Blake SM, Cox PJ. Obtaining an optimal bone–cement interface in total knee arthroplasty. Ann R Coll Surg Engl. 2006, 88, 317–19.

[56] Labutti RS, Bayers-Thering M, Krackow KA. Enhancing femoral cement fixation in total knee arthroplasty. J Arthroplasty. 2003, 18, 979–83.

[57] Badawy M, Espehaug B, Indrekvam K, Engesaeter LB, Havelin LI, Furnes O. Influence of hospital volume on revision rate after total knee arthroplasty with cement. J Bone Joint Surg Am. 2013, 95, e131, doi: 10.2106/JBJS.L.00943.

[58] Namba RS, Chen Y, Paxton EW, Slipchenko T, Fithian DC. Outcomes of routine use of antibiotic-loaded cement in primary total knee arthroplasty. J Arthroplasty. 2009, 24, 44–47.

[59] Prudhon J-L, Verdier R. Cemented or cementless total knee arthroplasty? Comparative results of 200 cases at a minimum follow-up of 11 years. SICOT-J. 2017, 3, 70, doi: https://doi.org/10.1051/sicotj/2017046.

[60] Matassi F, Carulli C, Civinini R, Innocenti M. Cemented versus cementless fixation in total knee arthroplasty. Joints. 2013, 1, 121–25.

[61] Romeo AA. Surgeons debate cemented vs uncemented TKA. 2014; https://www.healio.com/news/orthopedics/20140811/surgeons-debate-cemented-vs-uncemented-tka.

[62] Harrison N, Field JR, Quondamatteo F, Curtin W, McHugh PE, McDonnel P. Preclinical trial of a novel surface architecture for improved primary fixation of cementless orthopaedic implants. Clin Biomech (Bristol, Avon). 2014, 29, 861–68.

[63] Nawfal N, Karthik MN, Satish KC. Reverse hybrid total hip arthroplasty, a good alternative to uncemented total hip arthroplasty. Intl J Orthop Sci. 2020, 6, 455–59.

[64] Lindalen E, Havelin LI, Nordsletten L, Dybvik E, Fenstad AM, Hallan G, Furnes O, Høvik O, Stephan M, Rohr SM. Is reverse hybrid hip replacement the solution?. Acta Orthop. 2011, 82, 639–45.

[65] Wangen H, with 11 co-authors. Reverse hybrid total hip arthroplasty. Acta Orthop. 2017, 88, 248–54.

[66] Jain S, Magra M, Dube B, Veysi VT, Whitwell GS, Aderinto JB, Emerton ME, Stone MH, Pandit HG. Reverse hybrid total hip arthroplasty: A survival analysis of 1088 consecutive cases with five – to 10-year follow-up. Orthop Proc. 2018, 100-B, https://online.boneandjoint.org.uk/doi/abs/10.1302/1358-992X.2018.9.036.

[67] Moussa AA, Fischer J, Yadav R, Khandaker M. Minimizing stress shielding and cement damage in cemented femoral component of a hip prosthesis through computational design optimization. Adv Orthop. 2017, 8437956, doi: https://doi.org/10.1155/2017/8437956.

[68] Learmonth ID, Young C, Rorabeck C. The operation of the century: Total hip replacement. Lancet. 2007, 370, 1508–19.

[69] Shanmugavel P, Bhaskar GB, Chandrasekaran M, Mani PS, Srinivasan SP. An overview of fracture analysis in Functionally Graded Materials. Eur J Sci Res. 2012, 68, 412–39.

[70] Atai AA, Nikranjbar A, Kasiri R. Buckling and postbuckling behaviour of semicircular functionally graded material arches: A theoretical study. Proc Inst Mech Eng C J Mech Eng Sci. 2012, 226, 607–14.

[71] Oshida Y, Tuna EB. Science and technology integrated titanium dental implant systems. In: Basu B, et al., ed. Advance Biomaterials. Wiley, 2009, 143–77.

[72] Lin D, Li Q, Li W, Zhou S, Swain MV. Design optimization of functionally graded dental implant for bone remodeling. Compos Part B Eng. 2009, 40, 668–75.

[73] Zou J-P, Ruan J-M, Huang B-Y, Zhou Z-C, Liu Y-L. Preparation and microstructure of HA-316L stainless steel fiber asymmetrical functionally gradient biomaterial. J Inorg Mater. 2005, 20, 1181–88.

[74] Chu C, Xue X, Zhu J, Yin Z. In vivo study on biocompatibility and bonding strength of Ti/Ti-20 vol.% HA/Ti-40 vol.% HA functionally graded biomaterial with bone tissues in the rabbit. Mater Sci Eng A. 2006, 429, 18–24.

[75] Hedia HS, Shabara MAN, El-Midany TT, Fouda N. A method of material optimization of cementless stem through functionally graded material. Int J Mech Mater Des. 2005, 1, 329–46.

[76] Knee Replacement Implant Materials. https://bonesmart.org/knee/knee-replacement-implant-materials/.

[77] Pelt CE, Gililland JM, Doble J, Stronach BM, Peters CL. Hybrid total knee arthroplasty revisited: Midterm followup of hybrid versus cemented fixation in total knee arthroplasty. BioMed Res Int. 2013, doi: https://doi.org/10.1155/2013/854871.

[78] Illgen R, Tueting J, Enright T, Schreibman K, McBeath A, Heiner J. Hybrid total knee arthroplasty: A retrospective analysis of clinical and radiographic outcomes at average 10 years follow-up. J Arthroplasty. 2004, 19, 95–100.

[79] Gao F, Henricson A, Nilsson KG. Cemented versus uncemented fixation of the femoral component of the NexGen CR total knee replacement in patients younger than 60 years. A prospective randomised controlled RSA study. Knee. 2009, 16, 200–06.

[80] König A, Kirschner S, Walther M, Eisert M, Eulert J. Hybrid total knee arthroplasty. Arch Orthop Trauma Surg. 1998, 118, 66–69.

[81] Demey G, Servien E, Lustig S, Si Selmi TA, Neyret P. Cemented versus uncemented femoral components in total knee arthroplasty. Knee Surg Sports Traumatol Arthrosc. 2011, 19, 1053–59.

[82] Huddleston JI, Wiley JW, Scott RD. Zone 4 femoral radiolucent lines in hybrid versus cemented total knee arthroplasties: Are they clinically significant?. Clin Orthop Relat Res. 2005, 441, 334–39.

[83] Matassi F, Carulli C, Civinini R, Innocenti M. Cemented versus cementless fixation in total knee arthroplasty. Joints. 2013, 1, 121–25.

[84] Meneghini RM. Cemented vs. Cementless Knee Replacement. 2021; https://www.meneghi nimd.com/specialties/cemented-vs-cementless-knee-replacement.

[85] Niemeläinen MJ, with 9 co-authors. The effect of fixation type on the survivorship of contemporary total knee arthroplasty in patients younger than 65 years of age: A register-based study of 115,177 knees in the Nordic Arthroplasty Register Association (NARA) 2000–2016. Acta Orthop. 2020, 91, 184–90.

[86] Cristofolini L, Affatato S, Erani P, Leardini W, Tigani D, Viceconti M. Long-term implant-bone fixation of the femoral component in total knee replacement. Proc Inst Mech Eng H. 2008, 222, 319–31.

[87] Noble PC, Conditt MA, Thompson MT, Stein JA, Kreuzer S, Parsley BS, Mathis KB. Extraarticular abrasive wear in cemented and cementless total knee arthroplasty. Clin Orthop Relat Res. 2003, 416, 120–28.

[88] Bert JM, McShane M. Is it necessary to cement the tibial stem in cemented total knee arthroplasty?. Clin Orthop Relat Res. 1998, 356, 73–78.

[89] Panyayong W, Oshida Y, Andres CJ, Barco TM, Brown DT, Hovijitra S. Reinforcement of acrylic resins for provisional fixed restorations Part III: Effects of addition of titania and zirconia mixtures on some mechanical and physical properties. Bio-Med Mat Eng. 2002, 12, 353–66.

[90] McCaskie AW, Barnes MR, Lin E, Harper WM, Gregg PJ. Cement pressurisation during hip replacement. J Bone Joint Surg Br. 1997, 79, 379–84.

[91] Clarke MT, Green JS, Harper WM, Gregg PJ. Cement as a risk factor for deep-vein thrombosis. Comparison of cemented TKR, uncemented TKR and cemented THR. J Bone Joint Surg Br. 1998, 80, 611–13.

[92] Cooke C, Walter WK, Zicat B. Tibial fixation without screws in cementless total knee arthroplasty. J Arthroplast. 2006, 21, 237–41.

[93] Fricka KB, Sritulanondha S, McAsey C. To cement or not? Two-year results of a prospective, randomized study comparing cemented vs. cementless total knee arthroplasty (TKA). J Arthroplast. 2015, 30, 55–58.

[94] Lee RW, Volz RG, Sheridan DC. The role of fixation and bone quality on the mechanical stability of tibial knee components. Clin Orthop Relat Res. 2991, 273, 177–83.

[95] Summer DR, Turner TM, Dawson D, Rosenberg AG, Urban RM, Galante JO. Effect of pegs and screws on bone ingrowth in cementless total knee arthroplasty. Clin Orthop Relat Res. 1994, 309, 150–55.

[96]   Berger RA, Lyon JH, Jacobs JJ, Baren RM, Berkson EM, Sheinkop MB, Rosenberg AG, Galante JO. Problems with cementless total knee arthroplasty at 11 years followup. Clin Orthop Relat Res. 2001, 392, 196–207.

[97]   Goldberg VM, Kraay M. The outcome of the cementless tibial component. A minimum 14-year clinical evaluation. Clin Orthop Relat Res. 2004, 428, 214–20.

[98]   Surace MF, Berzzins A, Urban RM, Jacobs JJ, Berger RA, Natarajan RN, Andriacchi TP, Galante JO. Coventry Award paper. Backsurface wear and deformation in polyethylene tibial inserts retrieved postmortem. Clin Orthop Relat Res. 2002, 404, 14–23.

[99]   Grabill S. Press-fit fixation for total knee replacements: Hr magazine. Sport Medicine and Orthopaedic Center; https://smoc-pt.com/press-fit-fixation-for-total-knee-replacements-hr-magazine/.

[100] Sorrells RB, Voorhorst PE, Murphy JA, Bauschka MP, Greenwald AS. Uncemented rotating-platform total knee replacement: A five to twelve-year follow-up study. J Bone Joint Surg Am. 2004, 86-A, 2156–62.

[101] Ali MS, Mangaleshkar SR. Uncemented rotating-platform total knee arthroplasty: A 4-year to 12-year follow-up. J Arthroplasty. 2006, 21, 80–84.

[102] Sharma S, Nicol F, Hullin MG, McCreath SW. Long-term results of the uncemented low contact stress total knee replacement in patients with rheumatoid arthritis. J Bone Joint Surg Br. 2005, 87, 1077–80.

[103] Ersan Ö, Öztürk A, Çatma MF, Ünlü S, Akdoğan M, Ateş Y. Total knee replacement – cementless tibial fixation with screws: 10-year results. Acta Orthop Traumatol Turc. 2017, 51, 433–36.

[104] Yang JH, Yoon JR, Oh CH, Kim TS. Hybrid component fixation in total knee arthroplasty: Minimum of 10-year follow-up study. J Arthroplasty. 2012, 27, 1111–18.

[105] Lombardi AV Jr, Berasi CC, Berend KR. Evolution of tibial fixation in total knee arthroplasty. J Arthroplasty. 2007, 22, 25–29.

[106] Naudie DD, Ammeen DJ, Engh GA, Rorabeck CH. Wear and osteolysis around total knee arthroplasty. J Am Acad Orthop Surg. 2007, 15, 53–64.

[107] O'Rourke MR, Callaghan JJ, Goetz DD, Sullivan PM, Johnston RC. Osteolysis associated with a cemented modular posterior-cruciate-substituting total knee design: Five to eight-year follow-up. J Bone Joint Surg Am. 2002, 84, 1362–71.

[108] Gililland JM, Doble J, Stronach BM, Peters CL. Hybrid total knee arthroplasty revisited: Midterm followup of hybrid versus cemented fixation in total knee arthroplasty. BioMed Res Int. 2013, doi: https://doi.org/10.1155/2013/854871.

[109] Prasa AK, Tan JHS, Bedair HS, Dawson-Bowling S, Hanna SA. Cemented vs. cementless fixation in primary total knee arthroplasty: A systematic review and meta-analysis. Effort Open Rev. 2020, 5, doi: https://doi.org/10.1302/2058-5241.5.200030.

[110] Nakama GY, Peccin MS, Almeida GJM, Lira Neto ODA, Queiroz AAB, Navarro RD. Fixation options of total knee replacement for osteoarthritis and other non-traumatic diseases. Cochrane. 2012; https://www.cochrane.org/CD006193/MUSKEL_fixation-options-of-total-knee-replacement-for-osteoarthritis-and-other-non-traumatic-diseases.

[111] https://puredentalhealth.com/decatur-dental-implants.

[112] Altomare J. Traditional Dentures vs. Implant-supported Dentures. 2019; https://www.jamesaltomaredds.com/blog/2019/07/19/traditional-dentures-vs-implant-supported-dentures-200651.

[113] Fixed vs. Removable Dental Implant Options. 2020; https://www.susanhodds.com/fixed-vs-removable-dental-implant-options/.

[114] Selim K, Ali S, Reda A. Implant supported fixed restorations versus implant supported removable overdentures: A systematic review. Open Access Maced J Med Sci. 2016, 15, 726–32.

[115] Sanders P. Implant retention: making the right choice. https://implantpracticeus.com/implant-retention-making-the-right-choice/

[116] Rensburg CJ. The truth about friction-retained hybrid prostheses. Inside Dent Technol. 2019, 10, https://www.aegisdentalnetwork.com/idt/2019/03/the-truth-about-friction-retained-hybrid-prostheses.

[117] Zhang R-G, Hannak WB, Roggensack M, Freesmeyer WB. Retentive characteristics of Ankylos SynCone conical crown system over long-term use in vitro. Eur J Prosthodont Restor Dent. 2008, 16, 61–66.

[118] What are Snap-On Overdentures? https://hybridgeimplants.com/snap-on-overdentures.

[119] Dental Implant Abutments & Attachments. https://preat.com/attachment_systems/dental-implant-abutments/.

[120] Overdenture Studs; https://preat.com/attachment_systems/overdenture-studs/.

[121] Petropoulos VC, Smith W, Kousvelari E. Comparison of retention and release periods for implant overdenture attachments. Int J Oral Maxillofac Implants. 1997, 12, 176–85.

[122] Naert I, Gizani S, Vuylsteke M, Van Steenberghe D. A 5-year prospective randomized clinical trial on the influence of splinted and unsplinted oral implants retaining a mandibular overdenture: Prosthetic aspects and patient satisfaction. J Oral Rehabil. 1999, 26, 195–202.

[123] Naert IE, Gizani S, Vuylsteke M, van Steenberghe D. A randomised clinical trial on the influence of splinted and unsplinted oral implants in mandibular overdenture therapy. Clin Oral Investig. 1997, 1, 81–88.

[124] Naert I, Gizani S, Vuyesteke M, van Steenberghe D. A 5-year randomized clinical trial on the influence of splinted and unsplinted oral implants in the mandibular overdenture therapy. Part I: Peri-implant outcome. Clin Oral Implants Res. 1998, 9, 170–77.

[125] Magnet Retained Overdenture Attachment Systems. https://preat.com/attachment_systems/magnet-attachment-systems/.

[126] Savabi O, Nejatidanesh F, Yordshashian F. Retention of implant-supported overdenture with bar/clip and stud attachment designs. J Oral Implantol. 2013, 39, 140–47.

[127] Screw-retention and cement-fixation; https://www.kubokura-dc.jp/implant.

[128] Dorr L. How to choose between cement-retained or screw-retained implants. 2020; https://www.dentalproductsreport.com/view/how-to-choose-between-cement-retained-or-screw-retained-implants.

[129] Goodacre CJ, Bernal GB, Rungcharassaeng K, Kan JY. Clinical complications with implants and implant prostheses. J Prosthet Dent. 2003, 90, 121–32.

[130] Weber HP, Kim DM, Ng MW, et al. Peri-implant soft-tissue health surrounding cement- and screw-retained implant restorations: A multi-center, 3-year prospective study. Clin Oral Implants Res. 2006, 17, 375–79.

[131] Sheets JL, Wilcox C, Wilwerding T. Cement selection for cement-retained crown technique with dental implants. J Prosthodont. 2008, 17, 92–96.

[132] Cement Vs. Screw-Retained Implant Crowns. 2019; https://hiossen.com/news/cement-vs-screw-retained-implant-crowns/.

[133] Campbell WF, Herman MW. Choosing Between Screw-Retained and Cement-Retained Implant Crowns. 2011; https://glidewelldental.com/education/inclusive-dental-implant-magazine/volume-2-issue-2/choosing-between-screw-retained-and-cement-retained-implant-crowns/.

[134] Freitas AC Jr, Bonfante EA, Rocha EP, Silva NR, Marotta L, Coelho PG. Effect of implant connection and restoration design (screwed vs. cemented) in reliability and failure modes of anterior crowns. Eur J Oral Sci. 2011, 119, 323–30.

[135] Lee A, Okayasu K, Wang HL. Screw- versus cement-retained implant restorations: Current concepts. Implant Dent. 2010, 19, 8–15.

[136] Chee W, Felton DA, Johnson PF, Sullivan DY. Cemented versus screw-retained implant prostheses: Which is better?. Int J Oral Maxillofac Implants. 1999, 14, 137–41.

[137] Broken collarbone; https://www.mayoclinic.org/diseases-conditions/broken-collarbone/symptoms-causes/syc-20370311.

[138] Broken ribs; https://www.mayoclinic.org/diseases-conditions/broken-ribs/symptoms-causes/syc-20350763.

[139] Clavicle Fractures; https://www.hopkinsmedicine.org/health/conditions-and-diseases/clavicle-fractures.

[140] Rib fracture; https://en.wikipedia.org/w/index.php?title=Rib_fracture&action=edit&section=5.

[141] Corticotomy; https://en.wikipedia.org/wiki/Corticotomy.

[142] Schwartsman V, Schwartsman R. Corticotomy. Clin Orthop Relat Res. 1992, 280, 37–47.

[143] What is corticotomy? 2021; https://dentagama.com/news/what-is-corticotomy.

[144] Lee W. Corticotomy for orthodontic tooth movement. J Korean Assoc Oral Maxillofac Surg. 2018, 44, 251–58.

[145] Einhorn TA, Gerstenfeld LC. Fracture healing: Mechanism and interventions. Nat Rev Rheumatol. 2015, 11, 45–54.

[146] Bone healing after implantation; http://dentalis-implants.com/resources/bone-maintenance/bone-healing-after-implantation/.

[147] Irandoust S, Müftü S. The interplay between bone healing and remodeling around dental implants. Sci Rep. 2020, 10, 4335, doi: https://doi.org/10.1038/s41598-020-60735-7.

[148] Villar CC, Huynh-Ba G, Mills MP, Cochran DL. Wound healing around dental implants. Endod Topics. 2013, doi: https://doi.org/10.1111/etp.12018.

[149] Hernandez CB. Bone Healing after a Dental Implant. 2015; https://sanidentalgroup.com/blog/bone-healing-after-a-dental-implant.

[150] Marsh D. Concepts of fracture union, delayed union, and nonunion. Clin Orthop Relat Res. 1998, S22–30.

[151] Eriksson C, Lausmaa J, Nygren H. Interactions between human whole blood and modified TiO_2-surfaces: Influence of surface topography and oxide thickness on leukocyte adhesion and activation. Biomaterials. 2001, 22, 1987–96.

[152] Wen X, Wang X, Zhang N. Microsurface of metallic biomaterials: A literature review. J Biomed Mater Eng. 1996, 6, 173–89.

[153] Albrektsson T. Direct bone anchorage of dental implants. J Prosthet Dent. 1983, 50, 255–61.

[154] Oshida Y, Hashem A, Nishihara T, Yapchulay MV. Fractal dimension analysis of mandibular bones: Toward a morphological compatibility of implants. J Biomed Mater Eng. 1994, 4, 397–407.

[155] Kasemo B, Lausmaa J. Biomatewrials and implant surfaces: A surface science approach. Int J Oral Maxillofac Implants. 1988, 3, 247–59.

[156] Masuda T, Salvi GE, Offenbacher S, Fleton D, Cooper LF. Cell and matrix reactions at titanium implants in surgically prepared rat tibiae. Int J Oral Maxillofac Implants. 1997, 1, 472–85.

[157] Rosengren A, Johansson BR, Danielsen N, Thomsen P, Ericson LE. Immuno-histochemical studies on the distribution of albumin, fibrigen, fibronectin, IgG and collagen around PTFE and titanium implants. Biomaterials. 1996, 17, 1779–86.

[158] Chehroudi B, McDonnel D, Brunette DM. The effects of micromachined surfaces on formation of bonelike tissue on subcutaneous implants as assessed by radiography and computer image processing. J Biomed Mater Res. 1997, 34, 279–90.

[159] Larsson C, Thomsen P, Lausmaa J, Rodahl M, Kasemo B, Ericson LE. Bone response to surface modified implants: Studies on electropolsihed implants with different oxide thickness and morphology. Biomaterials. 1994, 15, 1062–74.

[160] Wälivaara B, Askendal A, Lundström I, Tengvall P. Blood protein interaction with titanium surfaces. J Biomater Sci Polym Edn. 1996, 8, 41–48.

[161] Kanagaraja S, Lundström I, Nygren H, Tengvall P. Platelet binding and protein adsorption to titanium and gold after short time exposure to heparinized plasma and whole blood. Biomaterials. 1996, 17, 2225–32.

[162] Nygren H, Eriksson C, Lausmaa J. Adhesion and activation of platelets and polymer-phonuclear granulocyte cells at TiO_2 surfaces. J Lab Clin Med. 1997, 129, 35–46.

[163] Tan P, Luscinskas FW. Homer-Vanniasinkam. Cellular and molecular mechanisms of inflammation and thrombosis. Eur J Vasc Endovasc Surg. 1999, 17, 373–89.

[164] Brown KK, Henson PM, Maclouf J, Moyle M, Ely JA, Worthen GS. Neutrophil-platelet adhesion: Relative roles of platelet P-selection and neutrophil β2 (CD18) integrins. Am J Respir Cell Mol Biol. 1998, 18, 100–10.

[165] De La C, Haimovich B, Greco RS. Immobilized IgG and fibrinogen differentially affect the cytoskeletal organization and bactericidal function of adherent neutrophils. J Surg Res. 1998, 80, 28–34.

[166] Berton G, Yan SR, Fumagalli L, Lowell CA. Neutropil activation by adhesion: Mechanisms and pathophysiological implications. Int J Clin Lab Res. 1996, 26, 160–77.

[167] Kasemo B. Biocompatibility of titanium implants: Surface science aspects. J Prosthet Dent. 1983, 49, 832–37.

[168] McQueen D, Sundgren J-E, Ivarsson B, Lundstrøm I, af Ekenstam B, Svensson A, Bränemark P-I, Albreksson T. Auger Electroscopic Studies of Titanium Implants. Clinical Applications of Biomaterials. Lee AJC, Abreksson T, Bränemark P-I, editors. Chichester UK: John Wiley, 1982, 179–85.

[169] Abe M. Oxides and Hydrous Oxides of Multivalent Metals as Inorganic Ion Exchangers, Inorganic Ion Exchange Materials. Clearfield A, editor,. Boca Raton FL: CRC Press, 1982, 161–273.

[170] Parsegian VA. Molecular forces governing tight contact between cellular surfaces and substrates. J Prosthet Dent. 1983, 49, 838–42.

[171] Albreksson T, Hansson H-A. An ultrastructural characterization of the interface between bone and sputtered titanium or stainless steel surfaces. Biomaterials. 1986, 7, 201–05.

[172] Bernardi G, Kawasaki T. Chromatography of polypeptides and proteins on hydroxyapatite columns. Biochim Biophys Acta. 1968, 160, 301–10.

[173] Embery G, Rølla G. Interaction between sulphated macromolcules and hydroxyapatite studied by infrared spectroscopy. Acta Odontol Scand. 1980, 38, 105–08.

[174] Ellingsen JE. A study of the mechanisms of protein adsorption to TiO2. Biomaterials. 1991, 12, 593–96.

[175] Baier RE. Modification of surfaces to meet bioadhesive design goals: A review. J Adhes. 1986, 20, 171–86.

[176] Glantz P-O, Baier RE. Recent studies on nonspecific aspects of intraoral adhesion. J Adhes. 1986, 20, 227–44.

[177] Sunny MC, Sharma CP. Titanium-protein interaction: Change with oxide layer thickness. J Biomater Appl. 1991, 5, 89–98.

[178] Rahal MD, Delorme D, Brånemark P-I, Osmond DG. Myelointegration of titanium implants: B lymphopoiesis and hemopoietic cell proliferation in mouse bone marrow exposed to titanium implants. Int J Oral Maxillofac Implants. 2000, 15, 175–84.

[179] Ramazanoglu M, Oshida Y. Osseointegration and bioscience of implant surfaces – current concepts at bone-implant interface. In: Turkyilmaz I, ed. Implant Dentistry – A Rapidly Evolving Practice. 2011, http://www.intechopen.com/books/implant-dentistry-a-rapidly-evolving-practice/osseointegration-andbioscience-of-implant-surfaces-current-concepts-at-bone-implant-interface.

[180] Osborn JF. Biomaterials and their application to implantation. SSO Schweiz Monatsschr Zahnmed. 1979, 89, 1138–39.

[181] Kienapfel H, Sprey C, Wilke A, Griss P. Implant fixation by bone ingrowth. J Arthroplasty. 1999, 14, 355–68.

[182] Zhao X, Liu X, Ding C. Acid-induced bioactive titania surface. J Biomed Mater Res A. 2005, 75, 888–94.

[183] Schliephake H, Scharnweber D, Dard M, Röbetaler S, Sewing A, Hüttmann C. Biological performance of biomimetic calcium phosphate coating of titanium implants in the dog mandible. J Biomed Mater Res A. 2003, 64, 225–34.

[184] Hench LL, Wilson J. Surface-active biomaterials. Science. 1984, 226, 630–36.

[185] Davies JE. In vitro modeling of the bone/implant interface. Anat Rec. 1996, 245, 426–45.

[186] Adell R, Lekholm U, Rockler B, Branemark PI. A 15-year study of osseointegrated implants in the treatment of the edentulous jaw. Int J Oral Surg. 1981, 10, 387–416.

[187] Marco F, Milena F, Gianluca G, Vittoria O. Peri-implant osteogenesis in health and osteoporosis. Micron. 2005, 36, 630–44.

[188] Dee KC, Puleo DA, Bizios R. An Introduction To Tissue-Biomaterial Interactions. New Jersey, US: John Wiley & Sons, 2002.

[189] Dereka XE, Markopoulou CE, Vrotsos IA. Role of growth factors on periodontal repair. Growth Factors. 2006, 24, 260–67.

[190] Heldin CH, Westermark B. Mechanism of action and in vivo role of platelet-derived growth factor. Physiol Rev. 1999, 79, 1283–316.

[191] Davies JE. Understanding peri-implant endosseous healing. J Dent Educ. 2003, 67, 932–49.

[192] Stanford CM, Keller JC. The concept of osseointegration and bone matrix expression. Crit Rev Oral Biol Med. 1991, 2, 83–101.

[193] Schliephake H. Bone growth factors in maxillofacial skeletal reconstruction. Int J Oral Maxillofac Surg. 2002, 31, 469–84.

[194] Meyer U, Joos U, Mythili J, Stamm T, Hohoff A, Fillies T, Stratmann U, Wiesmann HP. Ultrastructural characterization of the implant/bone interface of immediately loaded dental implants. Biomaterials. 2004, 25, 1959–67.

[195] Park JY, Davies JE. Red blood cell and platelet interactions with titanium implant surfaces. Clin Oral Implants Res. 2000, 11, 530–39.

[196] Puleo DA, Nanci A. Understanding and controlling the bone-implant interface. Biomaterials. 1999, 20, 2311–21.

[197] Klinger MM, Rahemtulla F, Prince CW, Lucas LC, Lemons JE. Proteoglycans at the bone-implant interface. Crit Rev Oral Biol Med. 1998, 9, 449–63.

[198] Chappard D, Aguado E, Huré G, Grizon F, Basle MF. The early remodeling phases around titanium implants: A histomorphometric assessment of bone quality in a 3- and 6-month study in sheep. Int J Oral Maxillofac Implants. 1999, 14, 189–96.

[199] Kasemo B, Lausmaa J. Biomaterial and implant surfaces: On the role of cleanliness, contamination, and preparation procedures. J Biomed Mater Res. 1988, 22, 145–58.

[200] Kasemo B, Gold J. Implant surfaces and interface processes. Adv Dent Res. 1999, 13, 8–20.

[201] Lim YJ, Oshida Y, Barco T, Andres CJ. Surface characterization of variously treated titanium materials. Int J Oral Maxillofac Implants. 2001, 16, 333–42.

[202] Degasne I, Basle MF, Demais V, Hure G, Lesourd M, Grolleau B, Mercier L, Chappard D. Effects of roughness, fibronectin and vitronectin on attachment, spreading, and proliferation of human osteoblast-like cells (Saos-2) on titanium surfaces. Calcif Tissue Int. 1999, 64, 499–507.

[203] Park BS, Heo SJ, Kim CS, Oh JE, Kim JM, Lee G, Park WH, Chung CP, Min BM. Effects of adhesion molecules on the behavior of osteoblast-like cells and normal human fibroblasts on different titanium surfaces. J Biomed Mater Res A. 2005, 74, 640–51.

[204] Yang Y, Cavin R, Ong JL. Protein adsorption on titanium surfaces and their effect on osteoblast attachment. J Biomed Mater Res A. 2003, 67, 344–49.

[205] Lee MH, Oh N, Lee SW, Leesungbok R, Kim SE, Yun YP, Kang JH. Factors influencing osteoblast maturation on microgrooved titanium substrata. Biomaterials. 2010, 31, 3804–15.

[206] MacDonald DE, Deo N, Markovic B, Stranick M, Somasundaran P. Thermal and chemical modification of titanium-aluminum-vanadium implant materials: Effects on surface properties, glycoprotein adsorption, and MG63 cell attachment. Biomaterials. 2004, 25, 3135–46.

[207] Ruoslahti E. The RGD story: A personal account. Matrix Biol. 2003, 22, 459–65.

[208] Hynes RO. Integrins: Bidirectional, allosteric signaling machines. Cell. 2002, 110, 673–87.

[209] Ruoslahti E. Rgd and other recognition sequences for integrins. Annu Rev Cell Dev Biol. 1996, 12, 697–715.

[210] Plupper GE, McNamee HP, Dike LE, Bojanowski K, Ingber DE. Convergence of integrin and growth factor receptor signalling pathways within focal adhesion complex. Mol Biol Cell. 1995, 6, 1349–65.

[211] Sawyer AA, Hennessy KM, Bellis SL. Regulation of mesenchymal stem cell attachment and spreading on hydroxyapatite by RGD peptides and adsorbed serum proteins. Biomaterials. 2005, 26, 1467–75.

[212] Kuzyl PRT, Schemitsch EH. The basic science of peri-implant bone healing. Indian J Orthop. 2011, 45, 108–15.

[213] Davies JE. Bone bonding at natural and biomaterial surfaces. Biomaterials. 2007, 28, 5058–67.

[214] Osborn J, Newesely H. Dynamic aspects of the implant-bone interface. In: Heimke G, editor. Dental Implants. Verlag: Münche, 1980, 111–23.

[215] Berglundh T, Abrahamsson I, Lang NP, Lindhe J. De novo alveolar bone formation adjacent to endosseous implants. Clin Oral Implants Res. 2003, 14, 251–62.

[216] Moreo P, García-Aznar JM, Doblaré M. Bone ingrowth on the surface of endosseous implants. Part 1: Mathematical model. J Theor Biol. 2009, 260, 1–12.

[217] Sela MN, Badihi L, Rosen G, Steinberg D, Kohavi D. Adsorption of human plasma proteins to modified titanium surfaces. Clin Oral Implants Res. 2007, 18, 630–38.

[218] Terheyden H, Lang NP, Bierbaum S, Stadlinger B. Osseointegration–communication of cells. Clin Oral Implants Res. 2012, 23, 1127–35.

[219] Choi J-Y, Sim J-H, Yeo I-S L. Characteristics of contact and distance osteogenesis around modified implant surfaces in rabbit tibiae. J Periodontal Implant Sci. 2017, 47, 182–92.

[220] Lavenus S, Louarn G, Layrolle P. Nanotechnology and dental implants. Int J Biomater. 2010, doi: 10.1155/2010/915327.

[221] Bruck SD. Biostability of materials and implants. Long-Term Eff Med Imp. 1991, 1, 89–106.

[222] Bannon BP, Mild EE. Titanium alloys for biomaterial application: An overview. In: Luckey HA, Kubli F, editors. Titanium Alloys in Surgical Implants. ASTM STP, Vol. 796, 1983, 7–15.

[223] Brunette DM. The effects of implant surface topography on the behavior of cells. Int J Oral Maxillofac Implants. 1988, 3, 231–46.

[224] Bigerele M, Anselme K, Noël B, Ruderman I, Hardouin P, Iost A. Improvement in the morphology of Ti-based surfaces: A new process to increase in vitro human osteoblast response. Biomaterials. 2002, 23, 1563–77.

[225] Zinger O, Anselme K, Denzer A, Habersetzer P, Wieland M, Jeanfils J, Hardouin P, Landolt D. Time-dependent morphology and adhesion of osteoblastic cells on titanium model surfaces featuring scale-resolved topography. Biomaterials. 2004, 25, 2695–711.

[226] Leven RM, Virdi AS, Sumner DR. Patterns of gene expression in rat bone marrow stromal cells cultured on titanium alloy discs of different roughness. J Biomed Mater Res. 2004, 70A, 391–401.

[227] Li JP, Li SH, Van Blitterswijk CA, de Groot K. A novel porous Ti6Al4V: Characterization and cell attachment. J Biomed Mater Res. 2005, 73A, 223–33.

[228] Webster TJ, Ergun C, Doremus RH, Lanford WA. Increased osteoblast adhesion on titanium-coated hydroxylapatite that forms $CaTiO_3$. J Biomed Mater Res. 2003, 67A, 975–80.

[229] Jain R, Von Recum AF. Fibroblast attachment to smooth and microtextured PET and thin cp-Ti films. J Biomed Mater Res. 2004, 68A, 296–304.

[230] Zhang YM, Bataillon-Linez P, Huang P, Zhao YM, Han Y, Traisnel M, Xu KW, Hildebrand HF. Surface analyses of micro-arc oxidized and hydrothermally treated titanium and effect on osteoblast behavior. J Biomed Mater Res. 2004, 68A, 383–91.

[231] Maitz MF, Poon RWY, Liu XY, Pham M-T, Chu PK. Bioactivity of titanium following sodium plasma immersion ion implantation and deposition. Biomaterials. 2005, 26, 5465–73.

[232] Walboomers XF, Elder SE, Bumgardner JD, Jansen JA. Hydrodynamic compression of young and adult rat osteoblast-like cells on titanium fiber mesh. J Biomed Mater Res. 2006, 76A, 16–24.

[233] Perry CR. Bone repair techniques, bone graft, and bone graft substitutes. Clin Orthop Relat Res. 1999, 360, 71–86.

[234] Bone cement; https://en.wikipedia.org/wiki/Bone_cement.

[235] Havelin LI, Espehaug B, Vollset SE, Engesaeter LB. The effect of the type of cement on early revision of Charnley total hip prostheses. A review of eight thousand five hundred and seventy-nine primary arthroplasties from the Norwegian Arthroplasty Register. J Bone Jt Surg. 1995, 77, 1543–50.

[236] Donaldson AJ, Thomson HE, Harper NJ, Kenny NW. Bone cement implantation syndrome. Br J Anaesth. 2009, 102, 12–22.

[237] Vert M, Doi Y, Hellwich K-H, Hess M, Hodge P, Kubisa P, Rinaudo M, Schué F. Terminology for biorelated polymers and applications (IUPAC Recommendations 2012). PureAppl Chem. 2012, 84, 377–410.

[238] Babaie E, Lin B, Goel VK, Bhaduri SB. Evaluation of amorphous magnesium phosphate (AMP) based non-exothermic orthopedic cements, Biomed. Mater. 2016, 11, 055010, doi: https://dx.doi.org/10.1088/1748-6041/11/5/055010.

[239] ASTM F451: Standard Specification for Acrylic Bone Cement; https://global.ihs.com/doc_detail.cfm?document_name=ASTM%20F451&item_s_key=00020838.

[240] Bone grafting; https://en.wikipedia.org/wiki/Bone_grafting.

[241] Anil S, Al-Sulaimani AF, Beeran AE, Chalisserry EP, Varma HPR, Al Amri MD. Drug delivery systems in bone regeneration and implant dentistry, current concepts in dental implantology, Ilser Turkyilmaz, IntechOpen, 2015, https://www.intechopen.com/chapters/48155.

[242] Sahwil H. Bone Grafting Materials in Implant Dentistry; https://blog.ddslab.com/bone-grafting-materials-in-implant-dentistry.

[243] Khan WS, Rayan F, Dhinsa BS, Marsh D. An osteoconductive, osteoinductive, and osteogenic tissue-engineered product for trauma and orthopaedic surgery: How far are we?. Stem Cells Int. 2012, 236231, doi: http://dx.doi.org/10.1155/2012/236231.

[244] Fillingham Y, Jacobs J. Properties of bone grafts. Bone Joint J. 2016, 98–B, doi: https://doi.org/10.1302/0301-620X.98B.36350.

[245] Khan SN, Cammisa FPJ, Sandhu HS, Diwan AD, Girardi FP, Lane JM. The biology of bone grafting. J Am Acad Orthop Surg. 2005, 13, 77–86.

[246] McKee MD. Management of segmental bony defects: The role of osteoconductive orthobiologics. J Am Acad Orthop Surg. 2006, 14, S163–7.

[247] Roberts TT, Rosenbaum AJ. Bone grafts, bone substitutes and orthobiologics: The bridge between basic science and clinical advancements in fracture healing. Organogenesis. 2012, 8, 114–24.

[248] Urist MR. Bone: Formation by autoinduction. Science. 1965, 150, 893–99.

[249] Kwong FN, Harris MB. Recent developments in the biology of fracture repair. J Am Acad Orthop Surg. 2008, 16, 619–25.

[250] Seth AK, Xiang T. Patella as an osteoarticular autograft for reconstructing the articular surface of knee after osteosarcoma in the proximal end of tibia: A case report. J Den Med Sci. 2014, doi: 10.9790/0853-13911116.

[251] Resnik RR. Bone Substitutes in Oral Implantology. 2018; https://glidewelldental.com/educa tion/chairside-dental-magazine/volume-12-issue-3/bone-substitutes.

[252] Goldstep F. Bone Grafts For Implant Dentistry: The Basics. 2015; https://www.oral healthgroup.com/features/1003918360/.

[253] Wallace MT, Henshaw RM. Results of cement versus bone graft reconstruction after intralesional curettage of bone tumors in the skeletally immature patient. J Pediatr Orthop. 2014, 34, 92–100.

[254] Yousefi A-M. A review of calcium phosphate cements and acrylic bone cements as injectable materials for bone repair and implant fixation. J Appl Biomater Funct Mater. 2019, 17, 2280800019872594, doi: https://doi.org/10.1177/2280800019872594.

[255] Rubin A. History of the Hip Replacement. 2017; https://jamaicahospital.org/newsletter/his tory-of-the-hip-replacement/.

[256] Albrektsson T, Chrcanovic S, Jacobsson M, Wennerberg A. Osseointegration of implants – a biological and clinical overview. JSM Dent Surg. 2017, 2, 1022, https://www.jscimedcentral. com/DentalSurgery/dentalsurgery-2-1022.pdf.

[257] Kieswetter K, Schwartz Z, Dean DD, Boyan BD. The role of implant surface characteristics in the healing of bone. Crit Rev Oral Biol Med. 1996, 7, 329–45.

[258] Albrektsson T, Johansson C. Osteoinduction, osteoconduction and osseointegration. Europ Spine J. 2001, 10, S96–101.

[259] Kakuta S, Miyaoka K, Fujimori S, Lee WS, Miyazaki T, Nagumo M. Proliferation and differentiation of bone marrow cells on titanium plates treated with a wire-type electrical discharge machine. J Oral Implantol. 2000, 26, 156–62.

[260] Albrektsson T, Eriksson AR, Friberg B, Lekholm U, Lindahl L, Nevins M, Oikarinen V, Roos J, Sennerby L, Astrand P. Histologic investigations on 33 retrieved nobelpharma implants. Clin Mater. 1993, 12, 1–9.

[261] Boyan BD, Hummert TW, Dean DD, Schwartz Z. Role of material surfaces in regulating bone and cartilage cell response. Biomaterials. 1996, 17, 137–46.

[262] Chae G-J, Jung U-W, Jung S-M, Lee I-S, Cho K-S, Kim C-K, Choi S-H. Healing of surgically created circumferential gap around Nano-coating surface dental implants in dogs. Surf Interface Anal. 2008, 40, 184–87.

[263] Johansson CB, Cosentino F, Tundo S, Milella E, Ramires PA, Wennerberg A. Biological behavior of sol-gel coated dental implants. J Mater Sci Mater Med. 2003, 14, 539–45.

[264] Harimoto K, Yoshida Y, Yoshihara K, Nagaoka N, Matsumoto T, Tagawa Y. Osteoblast compatibility of materials depends on serum protein absorbability in osteogenesis. Dent Mater J. 2012, 31, 674–80.

[265] Zagury R, Harari ND, Conz MB, Soares G, Vidigal G. Histomorphometric analyses of bone interface with titanium-aluminum-vanadium and hydroxyapatite -coated implants by biomimetic process. Implant Dent. 2007, 16, 290–96.

[266] Zechner W, Tangl S, Fürst G, Tepper G, Thams U, Mailath G, Watzek G. Osseous healing characteristics of three different implant types – A histologic and histomorphometric study in mini-pigs. Clin Oral Implants Res. 2003, 14, 150–57.

[267] Grandfield K, Palmquist A, Gonçalves S, Taylor A, Taylor M, Emanuelsson L, Thomsen P, Engqvist HJ. Free form fabricated features on CoCr implants with and without hydroxyapatite coating in vivo: A comparative study of bone contact and bone growth induction. J Mater Sci Mater Med. 2011, 22, 899–906.

[268] Junker R, Manders PJD, Wolke J, Borisov Y, Jansen JA. Bone reaction adjacent to microplasma-sprayed CaP-coated oral implants subjected to occlusal load, an experimental study in the dog. Part I: Short-term results. Clin Oral Implants Res. 2010, 21, 1251–63.

[269] Schiegnitz E, Palarie V, Nacu V, Al-Nawas B, Kämmerer PW. Vertical osteoconductive characteristics of titanium implants with calcium-phosphate-coated surfaces – a pilot study in rabbits. Clin Implant Dent Related Res. 2012, doi: 10.1111/j.17-8-8208.2012.00469.x.

[270] Chen H-T, Chung C-J, Yang T-C, Tang C-H, He J-L. Microscopic observations of osteoblast growth on micro-arc oxidized β titanium. Appl Surf Sci. 2012, 266, 73–80.

[271] Chung C-J, Su R-T, Chu H-J, Chen H-T, Tsou H-K, He J-L. Plasma electrolytic oxidation of titanium and improvement in osseointegration. J Biomed Mat Res B. 2013, 101B, 1023–30.

[272] Ballo AM, Bjöörn D, Åstrand M, Palmquist A, Lausmaa J, Thomsen P. Bone response to physical-vapour-deposited titanium dioxide coatings on titanium implants. Clin Oral Implants Res. 2013, 24, 1009–17.

[273] Chung SH, Kim H-K, Shon W-J, Park Y-S. Peri-implant bone formations around (Ti,Zr)O2-coated zirconia implants with different surface roughness. J Clin Periodon. 2013, 40, 404–11.

[274] Plecko M, Sievert C, Andermatt D, Frigg R, Kronen P, Klein K, Stübinger S, Nuss K, Bürki A, Ferguson S, Stoeckle U, Von Rechenberg B. Osseointegration and biocompatibility of different metal implants – a comparative experimental investigation in sheep. BMC Musculoskelet Disord. 2012, 13, 32–44.

[275] Scarano A, Piattelli M, Vrespa G, Petrone G, Iezzi G, Piattelli A. Bone healing around titanium and titanium nitride-coated dental implants with three surfaces: An experimental study in rats. Clin Implant Dent Related Res. 2003, 5, 103–11.

[276] Freeman CO, Brook IM. Bone response to a titanium aluminium nitride coating on metallic implants. J Mater Sci Mater Med. 2006, 17, 465–70.

[277] Choi C-R, Yu H-S, Kim C-H, Lee J-H, Oh C-H, Kim H-W, Lee -H-H. Bone cell responses of titanium blasted with bioactive glass particles. J Biomater Appl. 2010, 25, 99–117.

[278] Hacking SA, Harvey EJ, Tanzer M, Krygier JJ, Bobyn JD. Acid-etched microtexture for enhancement of bone growth into porous-coated implants. J Bone Joint Surg. 2003, 85-B, 1182–89.

[279] Herrero-Climent M, Lázaro P, Rios JV, Lluch S, Marqués M, Guillem-Martí J, Gil FJ. Influence of acid-etching after grit-blasted on osseointegration of titanium dental implants: In vitro and in vivo studies. J Mater Sci Mater Med. 2013, 24, 2047–55.

[280] Oshida Y. Bioscience and Bioengineering of Titanium Materials. London UK: Elsevier, 2007, 236–37.

[281] Wall I, Donos N, Carlqvist K, Jones F, Brett P. Modified titanium surfaces promote accelerated osteogenic differentiation of mesenchymal stromal cells in vitro. J Bone. 2009, 45, 17–26.

[282] Jeong R, Marin C, Granato R, Suzuki M, Gil JN, Granjeiro JM, Coelho PG. Early bone healing around implant surfaces treated with variations in the resorbable blasting media method. A study in rabbits. Med Oral Patol Oral Cir Bucal. 2010, 15, 119–25.

[283] Park JW, Jang IS, Suh JY. Bone response to endosseous titanium implants surface- modified by blasting and chemical treatment: A histomorphometric study in the rabbit femur. J Biomed Mater Res B. 2008, 84, 400–07.

[284] An S-H, Matsumoto T, Miyajima H, Sasaki J, Narayanan M, Kim K-H. Surface characterization of alkali- and heat-treated Ti with or without prior acid etching. Appl Surf Sci. 2012, 258, 4377–82.

[285] Stadlinger B, Pilling E, Huhle M, Mai R, Bierbaum S, Bernhardt R, Scharnweber D, Kuhlisch E, Hempel U, Eckelt U. Influence of extracellular matrix coatings on implant stability and osseointegration: An animal study. J Biomed Mater Res B. 2007, 83B, 222–31.

[286] Alcheikh A, Pavon-Djavid G, Helary G, Petite H, Migonney V, Anagnostou F. PolyNaSS grafting on titanium surfaces enhances osteoblast differentiation and inhibits Staphylococcus aureus adhesion. J Mater Sci Mater Med. 2013, 24, 1745–54.

[287] Ferraris S, Spriano S, Pan G, Venturello A, Bianchi CL, Chiesa R, Faga MG, Maina G, Vernè E. Surface modification of Ti-6Al-4V alloy for biomineralization and specific biological response: Part I, inorganic modification. J Mater Sci Mater Med. 2011, 22, 533–45.

[288] Pham MH, Landin MA, Tiainen H, Reseland JE, Ellingsen JE, Haugen HJ. The effect of hydrofluoric acid treatment of titanium and titanium dioxide surface on primary human osteoblasts. Clin Oral Implants Res. 2013, doi: 10.1111/clr.12150.

[289] Park J-W. Osseointegration of two different phosphate ion-containing titanium oxide surfaces in rabbit cancellous bone. Clin Oral Implants Res. 2012, doi: 10.1111/j.1600-0501.2011.02406.x.

[290] Le Guéhennec L, Soueidan A, Layrolle P, Amouriq Y. Surface treatments of titanium dental implants for rapid osseointegration. Dental Mater. 2007, 23, 844–54.

[291] Geesink RGT, De Groot K, Klein CPAT. Chemical implant fixation using hydroxyl-apatite coatings. The development of a human total hip prosthesis for chemical fixation to bone using hydroxyl-apatite coatings on titanium substrates. Clin Orthop Relat Res. 1987, 170, 147–70.

[292] Gupta A, Singh G, Afreen S. Application of nanotechnology in dental implants. Corpus. 2017; https://www.semanticscholar.org/paper/Application-of-Nanotechnology-In-Dental-Implants-Gupta-Singh/12086c7c18aedfecf07827844ee4985044771611.

[293] Health Quality Ontario. Nanotechnology: An evidence-based analysis. Ont Health Technol Assess Ser. 2006, 6, 1–43.

[294] Tomsia AP, Lee JS, Wegst UGK, Saiz E. Nanotechnology for dental implants. Int J Oral Maxillofac Implants. 2013, 28, e535–46.

[295] Rasouli R, Barhoum A, Uludag H. A review of nanostructured surfaces and materials for dental implants: Surface coating, patterning and functionalization for improved performance. Biomeater Sci. 2018, 6, 1312–38.

[296] Smith WR, Hudson PW, Ponce BA, Manoharan SRR. Nanotechnology in orthopedics: A clinically oriented review. BMC Musculoskelet Disord. 2018, 19, 67, doi: https://doi.org/10.1186/s12891-018-1990-1.

[297] Kandavalli SR, Wang Q, Ebrahimi M, Gode C, Djavanroodi F, Attarilar S, Liu S, Brief A. Review on the evolution of metallic dental implants: History, design, and application. Front Mater. 2021, doi: https://doi.org/10.3389/fmats.2021.646383.

[298] Sichert JA, Tong Y, Mutz N, Vollmer M, Fischer S, Milowska KZ, García Cortadella R, Nickel B, Cardenas-Daw C, Stolarczyk JK, Urban AS, Feldmann J. Quantum size effect in organometal halide perovskite nanoplatelets. Nano Lett. 2015, 15, 6521–27.

[299] Pleshko N, Grande DA, Myers KR. Nanotechnology in orthopaedics. J Am Acad Orthop Surg. 2012, 20, 60–62.

[300] Gusić N, Ivković A, VaFaye J, Vukasović A, Ivković J, Hudetz D, Janković S. Nanobiotechnology and bone regeneration: A mini-review. Int Orthop. 2014, 38, 1877–84.

[301] Karazisis D, Ballo AM, Petronis S, Agheli H, Emanuelsson L, Thomsen P, Omar O. The role of well-defined nanotopography of titanium implants on osseointegration: Cellular and molecular events in vivo. Int J Nanomedicine. 2016, 11, 1367–82.

[302] Mattei TA, Rehman AA. Extremely minimally invasive: Recent advances in nano-technology research and future applications in neurosurgery. Neurosurg Rev. 2015, 38, 27–37.

[303] Wong KK, Liu XL. Nanomedicine: A primer for surgeons. Pediatr Surg Int. 2012, 28, 943–51.

[304] Korkusuz F. Editorial comment: Nanoscience in musculoskeletal medicine. Clin Orthop Relat Res. 2013, 471, 2530–31.

[305] Coelho PG, Granjeiro JM, Romanos GE, Suzuki M, Silva NRF, Cardaropoli G, Thompson VP, Lemons JE. Basic research methods and current trends of dental implant surfaces. J Biomed Mater Res B Appl Biomater. 2009, 88, 579–96.

[306] Oshida Y. Surface Engineering and Technology for Biomedical Implants. Momentum Press, 2014.

[307] Ballo AM, Omar O, Xia W, Palmquist A. Dental implant surfaces – physicochemical properties, biological performance, and trends. Implant Dent A Rapidly Evol Pract. 2001, 19–56, https://www.intechopen.com/chapters/18414.

[308] Sul Y-T. Elecgtrochemical growth behavior, surface properties, and enhanced in vivo bone response of TiO_2 nanotubes on microstructured surfaces of blasted, screw-shaped titanium implants. Int I Nanomed. 2010, 5, 87–100.

[309] Hou P-J, Ou K-L, Wang -C-C, Hunag C-F, Ruslin M, Sugiatno E, Yang T-S, Chou -H-H. Hybrid micro/nanostructural surface offering improved stress distribution and enhanced osseointegration properties of the biomedical titanium implant. J Mech Behav Biomed Mater. 2018, 79, 173–80.

[310] Fini M, Giavaresi G, Torricelli P, Borsari V, Giardino R, Nicolini A, Carpi A. Osteoporosis and biomaterial osseointegration. Biomed Pharmacother. 2004, 58, 487–93.

[311] Rigo ECS, Boschi AO, Yoshimoto M, Allegrini S, Konig B, Carbonari MJ. Evaluation in vitro and in vivo of biomimetic hydroxyapatite coated on titanium dental implants. Mater Sci Eng C. 2004, 24, 647–51.

[312] Elsayed MD. Biomechanical factors that influence the bone-implant-interface. Res Rep Oral Maxillofac Surg. 2019, 3, 023, doi: 10.23937/iaoms-2017/1710023.

[313] Li M-J, Kung P-C, Chang Y-W, Tsou N-T. Healing pattern analysis for dental implants using the mechano-regulatory tissue differentiation model. Int J Mol Sci. 2020, 21, 9205, doi: 10.3390/ijms21239205.

[314] Isidor F. Influence of forces on peri-implant bone. Clin Oral Implants Res. 2006, 2, 8–18.

[315] Gao X, Fraulob M, Haïat G. Biomechanical behaviours of the bone–implant interface: A review. J R Soc Interface. 2019, 16, doi: https://doi.org/10.1098/rsif.2019.0259.

[316] Williams DL, Isaacson BM. The 5 hallmarks of biomaterials success: An emphasis on orthopaedics. Adv Biosci Biotechnol. 2014, 5, 283–93.

[317] Dodd CA, Hungerford DS, Krackow KA. Total knee arthroplasty fixation. Comparison of the early results of paired cemented versus uncemented porous coated anatomic knee prostheses. Clin Orthop Relat Res. 1990, 260, 66–70.

[318] Donaldson AJ, Thomson HE, Harper NJ, Kenny NW. Bone cement implantation syndrome. Br J Anaesth. 2009, 102, 12–22.

[319] Miyazaki T, Yutani T, Murai N, Kawata A, Shimizu H, Uejima N, Miyazaki Y, Oshida Y. Early osseointegration attained by UV-photo treated implant into piezosurgery-prepared site. Report III. Influence of surface treatment by hydrogen peroxide solution and determination of early loading timing. Int J Dent Oral Health. 2021, 7, doi: 10.16966/2378-7090.381.

[320] Albrektsson T, Zarb G, Worthington P, Eriksson AR. The long-term efficacy of currently used dental implants: A review and proposed criteria of success. Int J Oral Maxillofac Implants. 1986, 1, 11–25.

[321] Swami V, Vijayaraghavan V, Swami V. Current trends to measure implant stability. J Indian Prosthodont Soc. 2016, 16, 124–30.

[322] Meredith N. Assessment of implant stability as a prognostic determinant. Int J Prosthodont. 1998, 11, 491–501.

[323] Sennerby L, Meredith N. Implant stability measurements using resonance frequency analysis: Biological and biomechanical aspects and clinical implications. Periodontol 2000. 2008, 47, 51–66.

[324] Raghavendra S, Wood MC, Taylor TD. Early wound healing adjacent to endosseous dental implants: A review of the literature. Int J Oral Maxillofac Implants. 2005, 20, 425–31.

[325] Barikani H, Rashtak S, Akbari S, Fard MK, Rokn A. The effect of shape, length and diameter of implants on primary stability based on resonance frequency analysis. Dent Res J (Isfahan). 2014, 11, 87–91.

[326] Melsen B, Costa A. Immediate loading of implants used for orthodontic anchorage. Clin Orthod Res. 2000, 3, 23–28.

[327] Baumgaertel S. Predrilling of the implant site: Is it necessary for orthodontic mini-implants?. Am J Orthod Dentofacial Orthop. 2010, 137, 825–29.

[328] Vollmer A, Saravi B, Lang G, Adolphs N, Hazard D, Giers V, Stoll P. Factors influencing primary and secondary implant stability – a retrospective cohort study with 582 implants in 272 patients. Appl Sci. 2020, 10, 8084, doi: 10.3390/app10228084.

[329] Esposito M, Hirsch JM, Lekholm U, Thomsen P. Biological factors contributing to failures of osseointegrated oral implants, (I). Success criteria and epidemiology. Eur J Oral Sci. 1998, 106, 527–51.

[330] Lioubavina-Hack N, Lang NP, Karring T. Significance of primary stability for osseointegration of dental implants. Clin Oral Implants Res. 2006, 17, 244–50.

[331] Javed F, Romanos GE. The role of primary stability for successful immediate loading of dental implants. A literature review. J Dent. 2010, 38, 612–20.

[332] Meredith N. Assessment of implant stability as a prognostic determinant. Int J Prosthodont. 1998, 11, 491–501.

[333] Saravi BE, Putz M, Patzelt S, Alkalak A, Uelkuemen S, Boeker M. Marginal bone loss around oral implants supporting fixed versus removable prostheses: A systematic review. Int J Implant Dent. 2020, 6, 20, doi: 10.1186/s40729-020-00217-7.

[334] Brunski JB. Biomechanical factors affecting the bone-dental implant interface. Clin Mater. 1992, 10, 153–201.

[335] Sennerby L, Roos J. Surgical determinants of clinical success of osseointegrated oral implants: A review of the literature. Int J Prosthodont. 1998, 11, 408–20.

[336] Cochran DL, Schenk RK, Lussi A, Higginbottom FL, Buser D. Bone response to unloaded and loaded titanium implants with a sandblasted and acid-etched surface: A histometric study in the canine mandible. J Biomed Mater Res. 1998, 40, 1–11.

[337] Mistry G, Shetty O, Shetty S, Singh RD. Measuring implant stability: A review of different methods. J Dent Implant. 2014, 4, 165–69.

[338] Patil R, Bharadwaj D. Is primary stability a predictable parameter for loading implant?. J Int Clin Dent Res Organ. 2016, 8, 84–88.

[339] Atsumi M, Park S-H, Wang H-L. Methods used to assess implant stability: Current status. Int J Oral Maxillofac Implants. 2007, 22, 743–54.

[340] Sachdeva A, Dhawan P, Sindwani S. Assessment of implant stability: Methods and recent advances. BJMMR. 2016, 12, 1–10.

[341] Barikani H, Rashtak S, Akbari S, Fard MK, Rokn A. The effect of shape, length and diameter of implants on primary stability based on resonance frequency analysis. Dent Res J (Isfahan). 2014, 11, 87–91.

[342] Norton M. Primary stability versus viable constraint a need to redefine. Int J Oral Maxillofac Implants. 2013, 28, 19–21.

[343] Dos Santos MV, Elias CN, Cavalcanti Lima JH. The effects of superficial roughness and design on the primary stability of dental implants. Clin Implant Dent Relat Res. 2011, 13, 215–23.

[344] Suzuki S, Kobayashi H, Ogawa T. Implant stability change and osseointegration speed of immediately loaded photofunctionalized implants. Implant Dent. 2013, 22, 481–90.

[345] Albrektsson T, Brånemark PI, Hansson HA, Lindström J. Osseointegrated titanium implants. Requirements for ensuring a long-lasting, direct bone-to-implant anchorage in man. Acta Orthop Scand. 1981, 52, 155–70.

[346] Swami V, Vijayaraghavan V, Swami V. Current trends to measure implant stability. J Indian Prosthodont Soc. 2016, 16, 124–30.

[347] Sandrini E, Giordano C, Busini V, Signorelli E, Cigada A. Apatite formation and cellular response of a novel bioactive titanium. J Mater Sci: Mater Med. 2007, 18, 1225–37.

[348] Guéhennec L, Soueidan A, Layrolle P, Amouriq Y. Surface treatments of titanium dental implants for rapid osseointegration. Dent Mater. 2007, 23, 844–54.

[349] Thalji G, Cooper LF. Molecular assessment of osseointegration in vitro: A review of current literature. Int J Oral Maxillofac Implants. 2014, 29, e171–99.

[350] Justin D, Jin SH, Frandsen C, Brammer K, Bjursten L, Oh SH, Pratt C. In vitro and in vivo evaluation of implant surfaces treated with titanium oxide (TiO2) nanotube arrays to enhance osseointegration between arthroplasty implants and surrounding bone. Orthop Proc. 2018, 98-B, https://online.boneandjoint.org.uk/doi/abs/10.1302/1358-992X.98BSUPP_8. ISTA2015-071.

[351] Kasemo B. Biological surface science. Surf Sci. 2002, 500, 656–77.

[352] Lamers E, Walboomers XF, Domanski M, Riet J, van Delft FC, Luttge R, et al. The influence of nanoscale grooved substrates on osteoblast behavior and extracellular matrix deposition. Biomaterials. 2010, 31, 3307–16.

[353] Mendonça G, Mendonça DB, Aragão FJ, Cooper LF. The combination of micron and nanotopography by H(2)SO(4)/H(2) O(2) treatment and its effects on osteoblast-specific gene expression of hMSCs. J Biomed Mater Res A. 2010, 94, 169–79.

[354] Lincks J, Boyan BD, Blanchard CR, Lohmann CH, Liu Y, Cochran DL, Dean DD, Schwartz Z. Response of MG63 osteoblast-like cells to titanium and titanium alloy is dependent on surface roughness and composition. Biomaterials. 1998, 19, 2219–32.

[355] Von der Mark K, Park J, Bauer S, Schmuki P. Nanoscale engineering of biomimetic surfaces: Cues from the extracellular matrix. Cell Tissue Res. 2010, 339, 131–53.

[356] Novaes AB, de Souza SLS, de Barros RRM, Pereira KKY, Iezzi G, Piattelli A. Influence of implant surfaces on osseointegration. Braz Dent J. 2010, 21, 471–81.

[357] Oshida Y, Miyazaki T, Tominaga T. Some biomechanistic concerns of newly developed implantable materials. J Dent Oral Health. 2018, 4, 0117, https://scientonline.org/open-access/some-biomechanistic-concerns-on-newly-developed-implantable-materials.pdf.

[358] Nkenke E, Hahn M, Weinzierl K, Radespiel-Tröger M, Neukam FW, Engelke K. Implant stability and histomorphometry: A correlation study in human cadavers using stepped cylinder implants. Clin Oral Implants Res. 2003, 14, 601–09.

[359] Lemons JE. Biomaterials, biomechanics, tissue healing, and immediate-function dental implants. J Oral Implantol. 2004, 30, 318–24.

[360] Babuska V, Moztarzadeh O, Kubikova T, Moztarzadeh A, Hrusak D, Tonar Z. Evaluating the osseointegration of nanostructured titanium implants in animal models: Current

experimental methods and perspectives (Review). Biointerphases. 2016, 11, 030801, doi: https://doi.org/10.1116/1.4958793.

[361] van Oirschot BAJA, Alghamdi HS, Närhi TO, Anil SA, Aldosari AAF, van Den Beucken JJJP, Jansen JA. In vivo evaluation of bioactive glass-based coatings on dental implants in a dog implantation model. Clin Oral Impl Res. 2014, 25, 21–28.

[362] Im JH, Kim SG, Oh JS, Lim SC. A comparative study of stability after the installation of 2 different surface types of implants in the maxillae of dogs. Implant Dent. 2015, 24, 586–91.

[363] Akens MK, Von Rechenberg B, Bittmann P, Nadler D, Zlinszky K, Auer JA. Long term in-vivo studies of a photo-oxidized bovine osteochondral transplant in sheep. BMC Musculoskelet Disord. 2001, 2, doi: https://doi.org/10.1186/1471-2474-2-9.

[364] Plecko M, with 11 co-authors. Osseointegration and biocompatibility of different metal implants – a comparative experimental investigation in sheep. BMC Musculoskelet Disord. 2012, 32, doi: https://doi.org/10.1186/1471-2474-13-32.

[365] Yoo D, Marin C, Freitas G, Tovar N, Bonfante E, Teixeira HS, Janal MN. Surface characterization and in vivo evaluation of dual acid-etched and grit-blasted/acid-etched implants in sheep. Implant Dent. 2015, 24, 256–62.

[366] Pettersson M, Pettersson J, Thorén MM, Johansson A. Release of titanium after insertion of dental implants with different surface characteristics – An ex vivo animal study. Aca Biomatre Odontol Scandina. 2017, 3, 63–73.

[367] Hermida JC, Bergula A, Dimaano F, Hawkins M, Colwell C, D'Lime DD. An in vivo evaluation of bone response to three implant surfaces using a rabbit intramedullary rod model. J Orthop Surg Res. 2010, 5, 57–64.

[368] Tsetsenekou E, Papadopoulos T, Kalyvas D, Papaioannou N, Tangl S, Watzek G. The influence of alendronate on osseointegration of nanotreated dental implants in New Zealand rabbits. Clin Oral Implants Res. 2012, 23, 659–66.

[369] Dziuba D, Meyer-Lindenberg A, Seitz JM, Waizy H, Angrisani N, Reifenrath J. Long-term in vivo degradation behaviour and biocompatibility of the magnesium alloy ZEK100 for use as a biodegradable bone implant. Acta Biomater. 2013, 9, 8548–60.

[370] Liu Y, Zhou Y, Jiang T, Liang YD, Zhang Z, Wang YN. Evaluation of the osseointegration of dental implants coated with calcium carbonate: An animal study. Intl J Oral Sci. 2017, 9, 133–38.

[371] Ikuta K, Urakawa H, Kozawa E, Hamada S, Ota T, Kato R, Ishiguro N, Nishida Y. In vivo heat-stimulus-triggered osteogenesis. Int J Hypertherm. 2015, 31, 58–66.

[372] Dang Y, Zhang L, Song W, Chang B, Han T, Zhang Y, Zhao L. In vivo osseointegration of Ti implants with a strontium-containing nanotubular coating. Int J Nanomedicine. 2016, 11, 1003–11.

[373] Amerstorfer F, with 14 co-authors. Long-term in vivo degradation behavior and near-implant distribution of resorbed elements for magnesium alloys WZ21 and ZX50. Acta Biomater. 2016, 42, 440–50.

[374] Ota T, Nishida Y, Ikuta K, Kato R, Kozawa E, Hamada S. Heat-stimuli-enhanced osteogenesis using clinically available biomaterials. PLoS ONE. 2017, 12, e0181404, doi: https://doi.org/10.1371/journal.pone.0181404.

[375] Turner AS. Animal models of osteoporosis – Necessity and limitations. Eur Cells Mater. 2001, 1, 66–81.

[376] Pearce AI, Richards RG, Milz S, Schneider E, Pearce SG. Animal models for implant biomaterial research in bone: A review. Eur Cells Mater. 2007, 2, 1–10.

[377] Natiella JR. The use of animal models in research on dental implants. J Dent Educ. 1988, 52, 792–97.

[378] Pound P, Bracken MB, Bliss SD. Is animal research sufficiently evidence based to be a cornerstone of biomedical research?. BMJ (Formerly the Brit Med J). 2014, 348, doi: https://doi.org/10.1136/bmj.g3387.

[379] Mistry G, Shetty O, Shetty S, Raghuwar D, Singh R. Measuring implant stability: A review of different methods. J Dent Implants. 2014, 4, 165–69.

[380] Atsumi M, Park SH, Wang HL. Methods used to assess implant stability: Current status. Int J Oral Maxillofac Implants. 2007, 22, 743–54.

[381] Tabassum A, Meijer GJ, Walboomers XF, Jansen JA. Evaluation of primary and secondary stability of titanium implants using different surgical techniques. Clin Oral Impl Res. 2014, 25, 487–92.

[382] Meenakshi S, Raghunath N, Raju SN, Srividya S, Indira PN. Implant stability a key determinant in implant integration. Trends Prosthodont Dent Implantol. 2013, 4, 28–48.

[383] Brunski JB, Puleo DA, Nanci A. Biomaterials and biomechanics of oral and maxillofacial implants: Current status and future developments. Int J Oral Maxillofac Implants. 2000, 15, 15–46.

[384] Sullivan DY, Sherwood RL, Collins TA, Krogh PH. The reverse-torque test: A clinical report. Int J Oral Maxillofac Implants. 1996, 11, 179–85.

[385] Ivanoff CJ, Sennerby L, Lekholm U. Reintegration of mobilized titanium implants. An experimental study in rabbit tibia. Int J Oral Maxillofac Surg. 1997, 26, 310–15.

[386] Kose OD, Karatasli B, Demircan S, Kose TE, Cene E, Aya SA, Ali Erdem M, Cankaya AB. In vitro evaluation of manual torque values applied to implant-abutment complex by different clinicians and abutment screw loosening. BioMed Res Int. 2017, 7376261, doi: 10.1155/2017/7376261.

[387] Friberg B, Sennerby L, Meredith N, Lekholm U. A comparison between cutting torque and resonance frequency measurements of maxillary implants. A 20-month clinical study. Int J Oral Maxillofac Surg. 1999, 28, 297–303.

[388] Friberg B, Sennerby L, Gröndahl K, Bergström C, Bäck T, Lekholm U. On cutting torque measurements during implant placement: A 3-year clinical prospective study. Clin Implant Dent Relat Res. 1999, 1, 75–83.

[389] Irinakis T, Wiebe C. Initial torque stability of a new bone condensing dental implant. A cohort study of 140 consecutively placed implants. J Oral Implantol. 2009, 35, 277–82.

[390] Bayarchimeg D, Namgoong H, Kim BK, Kim MD, Kim S, Kim TI, et al. Evaluation of the correlation between insertion torque and primary stability of dental implants using a block bone test. J Periodontal Implant Sci. 2013, 43, 30–36.

[391] Roberts WE, Simmons KE, Garetto LP, DeCastro RA. Bone physiology and metabolism in dental implantology: Risk factors for osteoporosis and other metabolic bone diseases. Implant Dent. 1992, 1, 11–21.

[392] O'sullivan D, Sennerby L, Jagger D, Meredith N. A comparison of two methods of enhancing implant primary stability. Clin Implant Dent Relat Res. 2004, 6, 48–57.

[393] Kaneko T. Pulsed oscillation technique for assessing the mechanical state of the dental implant-bone interface. Biomaterials. 1991, 12, 555–60.

[394] Mathieu V, Vayron R, Richard G, Lambert G, Naili S, Meningaud JP, Haiat G. Biomechanical determinants of the stability of dental implants: Influence of the bone-implant interface properties. J Biomech. 2014, 47, 3–13.

[395] Schulte W, Lukas D. Periotest to monitor osseointegration and to check the occlusion in oral implantology. J Oral Implantol. 1993, 19, 23–32.

[396] Schulte W, Lukas D. The Periotest method. Int Dent J. 1992, 42, 433–40.

[397] Aparicio C, Lang NP, Rangert B. Validity and clinical significance of biomechanical testing of implant/bone interface. Clin Oral Implants Res. 2006, 17, 2–7.

[398] Patil R, Bharadwaj D. Is primary stability a predictable parameter for loading implant?. J Int Clin Dent Res Organ. 2016, 8, 84, doi: 10.4103/2231-0754.176264.

[399] Shokri M, Daraeighadikolaei A. Measurement of primary and secondary stability of dental implants by resonance frequency analysis method in mandible. Int J Dent. 2013, 506968, doi: 10.1155/2013/506968.

[400] Kanth KL, Swamy DN, Mohan TK, Swarna C, Sanivarapu S, Pasupuleti M. Determination of implant stability by resonance frequency analysis device during early healing period. J Dr NTR Univ Health Serv. 2014, 3, 169–75.

[401] Sennerby L, Meredith N. Implant stability measurements using resonance frequency analysis: Biological and biomechanical aspects and clinical implications. Periodontol 2000. 2008, 47, 51–66.

[402] Meredith N. Assessment of implant stability as a prognostic determinant. Int J Prosthodont. 1998, 11, 491–501.

[403] Implant stability quotient. https://en.wikipedia.org/wiki/Implant_stability_quotient.

[404] Glauser R, Lundgren AK, Gottlow J, Sennerby L, Portmann N, Ruhstaller P, Hämmerle CH. Immediate occlusal loading of Brånemark TiUnite implants placed predominantly in soft bone: 1-year results of a prospective clinical study. Clin Implant Dent Relat Res. 2003, 5, 47–56.

[405] Glauser R, Sennerby L, Meredith N, Ree A, Lundgren A, Gottlow J, Hammerle CH. Resonance frequency analysis of implants subjected to immediate or early functional occlusal loading. Successful vs. failing implants. Clin Oral Implants Res. 2004, 15, 428–34.

[406] Valderrama P, Oates TW, Jones AA, Simpson J, Schoolfield JD, Cochran DL. Evaluation of two different resonance frequency devices to detect implant stability: A clinical trial. J Periodontol. 2007, 78, 262–72.

[407] Pagliani L, Sennerby L, Petersson A, Verrocchi D, Volpe S, Andersson P. The relationship between resonance frequency analysis (RFA) and lateral displacement of dental implants: An in vitro study. J Oral Rehabil. 2013, 40, 221–27.

[408] Trisi P, Carlesi T, Colagiovanni M, Perfetti G. Implant stability quotient (ISQ) vs. direct in [vitro measurement of primary stability (micromotion): Effect of bone density and insertion torque. J Osteol Biomat. 2010, 1, 141–49.

[409] Andersson P, Pagliani L, Verrocchi D, Volpe S, Sahlin H, Sennerby L. Factors influencing resonance frequency analysis (RFA) measurements and 5-year survival of neoss dental implants. Int J Dent. 2019, 2019, doi: 10.1155/2019/3209872.

[410] Rodrigo D, Aracil L, Martin C, Sanz M. Diagnosis of implant stability and its impact on implant survival: A prospective case series study. Clin Oral Implants Res. 2010, 21, 255–61.

[411] Sjöström M, Sennerby L, Nilson H, Lundgren S. Reconstruction of the atrophic edentulous maxilla with free iliac crest grafts and implants: A 3-year report of a prospective clinical study. Clin Implant Dent Relat Res. 2007, 9, 46–59.

[412] Turkyilmaz I, McGlumphy EA. Influence of bone density on implant stability parameters and implant success: A retrospective clinical study. BMC Oral Health. 2008, 8, 32, doi: 10.1186/1472-6831-8-32.

[413] Schallhorn RA. Resonance Frequency Analysis in Implant Dentistry. 2017, https://decisionsin dentistry.com/article/resonance-frequency-analysis-implant-dentistry/.

[414] Sachdeva A, Dhawan P, Sindwani S. Assessment of implant stability: Methods and recent advances. BJMMR. 2016, 12, 1–10.

[415] Sennerby L, Meredith N. Resonance frequency analysis: Measuring implant stability and osseointegration. Compend Contin Educ Dent. 1998, 19, 493–98.

[416] Miyazaki T. Early osseointegration attained by UV-photo treated implant into piezosurgery-prepared site. Report I. Retrospective study on clinical feasibility. Int J Dent Oral Health. 2020, 6, doi: 10.16966/2378-7090.344.

[417] Miyazaki T, Yutani T, Murai N, Kawata A, Shimizu H, Uejima N, Miyazaki Y. Early osseointegration attained by UV-photo treated implant into piezosurgery-prepared site Report II. Influences of age and gender. Int J Dent Oral Health. 2021, 7, doi: 10.16966/2378-7090.351.

[418] Barikani H, Rashtak S, Akbari S, Fard MK, Rokn A. The effect of shape, length and diameter of implants on primary stability based on resonance frequency analysis. Dent Res J (Isfahan). 2014, 11, 87–91.

[419] Norton M. Primary stability versus viable constraint a need to redefine. Int J Oral Maxillofac Implants. 2013, 28, 19–21.

[420] Suzuki S, Kobayashi H, Ogawa T. Implant stability change and osseointegration speed of immediately loaded photofunctionalized implants. Implant Dent. 2013, 22, 481–90.

[421] Bajaj G, Bathiya A, Gade J, Mahale Y, Ulemale M, Atulkar M. Primary versus secondary implant stability in immediate and early loaded implants. Int J Oral Health Med Res. 2017, 3, 49–54.

[422] Schulte W, Kleineikenscheidt H, Lindner K, Schareyka R. The Tübingen immediate implant in clinical studies. Dtsch Zahnarztl Z. 1978, 33, 348–59.

[423] Abboud M, Rugova S, Orentlicher G. Immediate loading: Are implant surface and thread design more important than osteotomy preparation?. Compendium. 2020, 14, 384–86.

[424] Kawahara H, Kawahara D, Hayakawa M, Tamai Y, Kuremoto T, Matsuda S. Osseointegration under immediate loading: Biomechanical stress–strain and bone formation–resorption. Implant Dent. 2003, 12, 61–68.

[425] Koh RU, Rudek I, Wang H-L. Immediate implant placement: Positives and negatives. Implant Dent. 2010, 19, 98–108.

[426] Lazzara RJ. Immediate implant placement into extraction sites: Surgical and restorative advantages. Int J Periodontics Restor Dent. 1989, 9, 332–43.

[427] Schwartz-Arad D, Chaushu G. The ways and wherefores of immediate placement of implants into fresh extraction sites: A literature review. J Periodontol. 1997, 68, 915–23.

[428] Cosyn J, De Lat L, Seyssens L, Doornewaard R, Deschepper E, Vervaeke S. The effectiveness of immediate implant placement for single tooth replacement compared to delayed implant placement: A systematic review and meta-analysis. J Clin Periodontol. 2019, 46, 224–41.

[429] Schierano G, Vercellotti T, Modica F, Corrias G, Russo C, Cavagnetto D, Baldi D, Romano F, Carossa S. A 4-year retrospective radiographic study of marginal bone loss of 156 titanium implants placed with ultrasonic site preparation. Int J Periodontics Restor Dent. 2019, 39, 115–21.

[430] Nienkemper M, Wilmes B, Pauls A, Drescher D. Impact of mini-implant length on stability at the initial healing period: A controlled clinical study. Head Face Med. 2013, 9, 30, doi: 10.1186/1746-160X-9-30.

[431] Khandelwal N, Oates TW, Vargas A, Alexander PP, Schoolfield JD, McMahan CA. Conventional SLA and chemically modified SLA implants in patients with poorly controlled type 2 Diabetes mellitus – A randomized controlled trial. Clin Oral Impl Res. 2013, 24, 13–19.

[432] Abtahi J, Tengvall P, Aspenberg P. A bisphosphonate-coating improves the fixation of metal implants in human bone. A randomized trial of dental implants. Bone. 2012, 50, 1148–51.

[433] Skvirsky Y. Five (5) questions to ask before choosing between delayed and immediate dental implant loading. Dentistry IQ. 2020; https://www.dentistryiq.com/dentistry/implantology/article/14168513/5-questions-to-ask-before-choosing-between-delayed-and-immediate-dental-implant-loading.

Chapter 11
Indications and contraindications

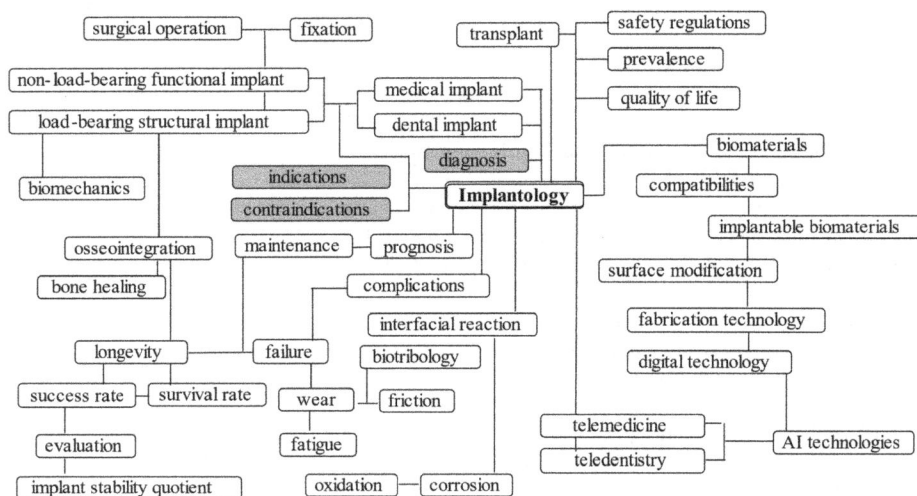

11.1 Absolute indication versus relative indication

Indication

In medicine, an indication is a valid reason to use a certain test, medication, procedure, or surgery. There can be multiple indications to use a procedure or medication. An indication can commonly be confused with the term "diagnosis," which is the assessment that a particular medical condition is present while an indication is a reason for use [1]. The opposite of an indication is a contraindication, a reason to withhold a certain medical treatment because the risks of treatment clearly outweigh the benefits.

It was stated that, to be able to decide and give consent for the procedure to be performed, it is necessary to distinguish between absolute indications and relevant indications for surgery [2].

A clear differentiation between absolute (absolutely necessary) indication and relative (could be done) indication should be done. These options should be explained to the patients, so that understanding the difference between an absolute and relative indication will assist the patient in the decision-making process. Otherwise, the patient would look for the second opinion. Absolute indication to do surgery is clearly evident. A patient with a condition presenting a risk to his/her life, to

https://doi.org/10.1515/9783110740134-012

preserving a limb of the patient, or preserving body function, can be classified as a mandatory or absolute indication for surgery. When it is evident that the risk will lead to irreversible damage should the operation not be performed, the surgeon should confirm that the indication is absolute and recommend surgery. Although the patient may refuse the surgery, the doctor's responsibility is to explain clearly the diagnosis and the risks of the disease. To explain precisely why the operation is necessary and to make sure that the patient also understands the consequences should the surgery not be performed. This is called an absolute indication for surgery [2].

On the other hand, if a condition exists where there is no immediate risk to life or limb, the surgeon may still recommend surgery; however, the operation does not need to be done. The medical condition may cause distress to the patient, however, by delaying the surgical option of treatment, or even omitting an operation entirely, no absolute risk to life or limb exists. After explaining the condition, with the risks attached to doing the surgery, the patient should understand that the operation is not absolutely necessary. The symptoms the patient complains about can be managed by alternative means, and surgery could be delayed for a period of time, even omitted as a treatment option. The decision to have the surgery is made by the patient. This decision is based on how the patient judges the effect of the disease on: (i) HRQoL – when life quality is diminished because of pain, stiffness, or loss of function, thereby preventing the patient from participating in activities they should typically be able to do (given their age and abilities), (ii) activities of daily living – when activities that are necessary for everyday functions, for example, washing, dressing, hair care, cooking, and cleaning, are becoming difficult because of the disease, and (iii) ability to work – when the ability to perform the functions in the typical working environment is threatened, whereby the income for sustenance is reduced. The expectation should be realistic, and the surgeon should not create impossible expectations for the suggested surgery. The patient decides on the surgery, as well as the timing of such an operation. This case is called a relative indication for surgery [2].

Contraindication

In medicine, a contraindication is a condition that serves as a reason not to take a certain medical treatment due to the harm that it would cause the patient [3]. Absolute contraindications are contraindications for which there are no reasonable circumstances for undertaking a course of action. Absolute contraindication means that event or substance could cause a life-threatening situation. A procedure or medicine that falls under this category must be avoided. For example, children and teenagers with viral infections should not be given aspirin because of the risk of Reye syndrome [4] and a person with an anaphylactic food allergy should never eat the food to which they are allergic. Similarly, a person with hemochromatosis should not be administered iron preparations [3]. Relative contraindications are con-traindications for circumstances in which the patient is at higher risk of complica-tions from treatment, but these risks may be outweighed by other considerations or mitigated by other measures. Relative contraindication means that caution should be used when two drugs or procedures are used together. For example, a pregnant woman should normally avoid getting X-rays, but the risk may be outweighed by the benefit of diagnosing (and then treating) a serious condition such as tuberculosis [3]. Taking another example of giving massage, the absolute contraindications can be considered as follows [5]. When something is considered an absolute contraindica-tion, it means that the client should not be given a massage in that particular area no matter what the circumstances. For example, if someone has deep vein thrombosis, under no circumstances should their lower calf muscles be massaged. Contrarily, when there is a level of caution or danger associated with a massage treatment, it is called a relative contraindication. With a relative contraindication, the massage may be performed, but simply modified to ensure the safety of the indicated area. This is typical where there has been a recent injury, or where there has been surgery. In these cases, a particular massage therapy treatment may be beneficial so long as it is performed by a knowledgeable and experienced massage therapist who knows how to avoid damaging the delicate tissues [5].

11.2 Indications and contraindications for orthopedic implants

Table 11.1 lists indications and contraindications for total shoulder arthroplasty, total hip arthroplasty, and total knee arthroplasty. Since these devices contain me-tallic materials (such as Fe-based stainless steel, Co–Cr–Mo alloy and Ti-based al-loys) with some percentage of the entire system, MRI contraindications are also included.

11.3 Indications and contraindications for dental implants

Table 11.2 lists indications and contraindications for dental implants.

Ghidrai [18] mentioned that some conditions or physiological changes, usually inside the mouth cavity, may temporarily prevent the placement of dental implants. Most of the times, these conditions can be remedied before the implants are in-serted in the jawbone. Such local contraindications include following concerns. There is insufficient bone to support the implants or bone structure is inadequate (due to some chronic infections or other conditions). To ensure a good prognosis, a dental implant must be surrounded by healthy bone tissue; (i) important anatomi-cal structures such as the maxillary sinus, the inferior alveolar nerve (located inside

Table 11.1: Indications and contraindications for TSA, THA and TKA.

Implant	Indications	Contraindications	Ref
TSA	The most common indications include osteoarthritis, inflammatory arthritis, complex proximal humerus fractures, irreparable tears of the rotator cuff, rotator cuff arthropathy, and avascular necrosis of the humeral head.		[6, 7]
	Indications include (i) pain (anterior to posterior), especially at night, and inability to perform activities for daily living, (ii) glenoid chondral wear to bone, (iii) preferred over hemiarthroplasty for osteoarthritis and inflammatory arthritis, and (iv) posterior humeral head subluxation.	Contraindications include (i) insufficient glenoid bone stock, (ii) rotator cuff arthropathy, (iii) deltoid dysfunction, (iv) irreparable rotator cuff (hemiarthroplasty or reverse total shoulder are preferable): risk of loosening of the glenoid prosthesis is high ("rocking horse" phenomenon), (v) active infection, and (vi) brachial plexus palsy.	[8]
		The principal contraindications include an infection in progress, Charcot's arthropathy, and severe neurological pathologies.	[9]
THA	The most common indication for THA includes (i) end-stage, symptomatic hip OA, including post-traumatic arthritis, rheumatoid arthritis, psoriatic arthritis and avascular necrosis, (ii) hip ON, congenital hip disorders including hip dysplasia, and inflammatory arthritic conditions.	THA is contraindicated in the following clinical scenarios: (i) local hip infection or sepsis, (ii) remote (i.e., extra-articular) active, ongoing infection or bacteremia, and (iii) severe cases of vascular dysfunction.	[10, 11]
TKA	TKA is indicated for (i) pain relief, (ii) functional impairment, (iii) limb alignment, (iv) failed conservative management (weight loss, activity modification, non-steroidal anti-inflammatories: NSAIDs, corticosteroid/viscosupplementation injections, or physical therapy).	TKA is contraindicated for (i) medical co-morbidities, (ii) active infection, (iii) non-functioning extensor mechanism, (iv) injury/incompetent extensor mechanism, (v) neurologic disease, (vi) chronic lower extremity ischemia, (vii) non-ambulatory, and (viii) skeletally immature.	[12]
		TKA is contraindicated in the following clinical scenarios: (i) local knee infection or sepsis, (ii) remote (extra-articular), active, ongoing infection or bacteremia, and (iii) severe cases of vascular dysfunction.	[13]

Table 11.1 (continued)

Implant	Indications	Contraindications	Ref
MRI		Potential contraindications for an MRI examination include, aneurysm clip(s), any metallic fragment or foreign body, coronary and peripheral artery stents, aortic stent graft, prosthetic heart valves and annuloplasty rings, cardiac occluder devices, Vena cava filters and embolization coils, hemodynamic monitoring and temporary pacing devices (such as Swan-Ganz catheter), hemodynamic support devices, cardiac pacemaker, implanted cardioverter defibrillator (ICD), retained transvenous pacemaker and defibrillator leads, electronic implant or device (e.g., insulin pump or other infusion pump), permanent contraceptive devices, diaphragm, or pessary, cochlear, otologic, or other ear implant, neurostimulation system, shunt (spinal or intraventricular), vascular access port and/or catheter, tissue expander (such as breast), joint replacement (e.g., hip, knee, shoulder), any type of prosthesis (e.g., eye, penile), tattoo or permanent makeup, known claustrophobia, body piercing jewelry, hearing aid, renal insufficiency, known/possible pregnancy or breast feeding.	[14, 15]
		Absolute contraindication should include (i) implantable pediatric sternum device, (ii) metallic foreign body in the eye, (iii) triggerfish contact lens, (iii) gastric reflux device, (iv) insulin pumps, and (v) temporary transvenous pacing leads.	[16]

Table 11.2: Indications and contraindications for dental implants.

Implant	Indications	Contraindications	Ref
Dental		Normal contraindicative factors include (i) endocrine disorders, such as uncontrolled diabetes mellitus, pituitary and adrenal insufficiency, and hypothyroidism can cause considerable healing problems, (ii) uncontrolled granulomatous diseases, such as tuberculosis and sarcoidosis may also lead to a poor healing response to surgical procedures, (iii) patients with cardiovascular diseases, taking blood-thinning drugs and patients with uncontrolled hematological disorders, such as generalized anemia, hemophilia (Factor VIII deficiency), Factor IX, X and XII deficiencies and any other acquired coagulation disorders are contraindicated to surgical procedures due to poor hemorrhage control, (iv) patients with bone diseases, such as histiocytosis X, Paget's disease and fibrous dysplasia may not be good candidates for implants, because there is a higher chance for the implant to fail due to poor osteointegration, (v) cigarette smoking, and (vi) patients receiving radio and chemotherapy should not have implants within the 6 months period of therapy,	[17]
	Following cases are indicated: (i) single unit toothless gap with healthy adjacent teeth, (ii) partial edentulism with the back (posterior) tooth missing, (iii) complete edentulism, and (iv) other situations when dental implants can be indicated such as when patients cannot tolerate a removable restoration (removable denture), or patients with high aesthetic and/or functional demands.	Absolute contraindications should include (i) heart diseases affecting the valves, recent infarcts, severe cardiac insufficiency, cardiomyopathy, (Ii) active cancer, certain bone diseases (osteomalacia, Paget's disease, brittle bones syndrome, etc.), (iii) certain immunological diseases, immunosuppressant treatments, clinical AIDS, awaiting an organ transplant, (iv) certain mental diseases, (v) strongly irradiated jaw bones (radiotherapy treatment), and (vi) treatments of osteoporosis or some cancers by bisphosphonates. Relative contraindications can include (i) diabetes (particularly insulin-dependent), (ii) angina pectoris (angina), (iii) significant consumption of tobacco, (iv) certain mental diseases, (v) certain auto-immunes diseases, (vi) drug and alcohol dependency, and (vii) pregnancy.	[18–20]

the mandible), have an abnormal position that can interfere with the dental implants. Adjunctive surgical procedures have to be performed before the placement of dental implants. These procedures aim to increase the amount of bone, so more bone is available to support the implants. It continues that, (ii) some local diseases of the oral mucosa or alveolar bone can temporarily prevent the placement of dental implants until the conditions are treated, (iii) hypersensitivity or other allergic reactions might rarely occur, (iv) poor oral hygiene, and (v) bruxism or involuntary grinding of the teeth.

References

[1]   Indication (medicine). https://en.wikipedia.org/wiki/Indication_(medicine).
[2]   Indications for surgery are Absolute or Relative. https://www.spinal-care.com/indications-absolute-vs-relative-surgery/.
[3]   Contraindication. https://en.wikipedia.org/wiki/Contraindication.
[4]   Reye syndrome. https://en.wikipedia.org/wiki/Reye_syndrome.
[5]   Tuchtan V. Indications and Contraindications for Massage: What You Need to Know. 2015; http://www.sagemassage.edu.au/blog/indications-and-contraindications-for-massage-what-you-need-to-know/.
[6]   Lin DJ, Wong TT, Kazam JK. Shoulder arthroplasty, from indications to complications: What the radiologist needs to know. Musculoskelet Imag. 2016, doi: https://doi.org/10.1148/rg.2016150055.
[7]   Buck FM, Jost B, Hodler J. Shoulder arthroplasty. Eur Radiol. 2008, 18, 2937–48.
[8]   Total Shoulder Arthroplasty; https://www.orthobullets.com/shoulder-and-elbow/3075/total-shoulder-arthroplasty.
[9]   Caniggia M, Fornara P, Franci M, Maniscalco P, Picinotti A. Shoulder arthroplasty. Indications, contraindications and complications. Panminerva Med. 1999, 41, 341–49.
[10]   Varacallo MA, Herzog L, Toossi N, Johanson NA. Ten-year trends and independent risk factors for unplanned readmission following elective total joint arthroplasty at a large urban academic hospital. J Arthroplasty. 2017, 32, 1739–46.
[11]   Varacallo M, Luo TD, Johanson NA. Total Hip Arthroplasty Techniques. StatPearls Publishing; 2021.
[12]   Pepper AM. Total Knee Arthroplasty: Indications, Contraindications, Post-operative Considerations. https://www.andrewsref.org/newsite/wp-content/uploads/2017/03/TKA-Indication-Contraindication-postop-AMP.pdf.
[13]   Varacallo M, Luo TD, Johanson NA. Total Knee Arthroplasty Techniques. StatPearls. 2021. https://www.ncbi.nlm.nih.gov/books/NBK499896/.
[14]   Shellock FG, Crues JV. MR procedures: Biologic effects, safety, and patient care. Radiology. 2004, 232, 635–52.
[15]   Ullrich P. Indications and contraindications for an MRI scan. Spine Health. 2009, https://www.spine-health.com/treatment/diagnostic-tests/indications-and-contraindications-mri-scan.
[16]   MRI Absolute Contraindications, https://radiology.ucsf.edu/patient-care/patient-safety/mri/absolute-contraindications.
[17]   Misch CE. Implant Dentistry. 2nd ed. St. Louis: Mosby, An Affiliate of Elsevier, 1999.

[18] Ghidrai G. Dental Implants. Indications, Contraindications. https://www.infodentis.com/den
tal-implants/indications-contraindications.php.

[19] Rafael Gómez-de Diego R, Mang-de la Rosa M, Romero-Pérez M-J, CutandoSoriano A, López-
Valverde-Centeno A. Indications and contraindications of dental implants in medically
compromised patients: Update. Med Oral Patol Oral Cir Bucal. 2014, 19, e483–9, https://
pdfs.semanticscholar.org/101d/5ae6326da8107e69fb9a3f11ccc8e177562f.pdf.

[20] Contraindications to dental implants. 2020, https://www.perfect-smile.pl/dental-implants/
contraindications-to-dental-implants/.

Chapter 12
Implant complications

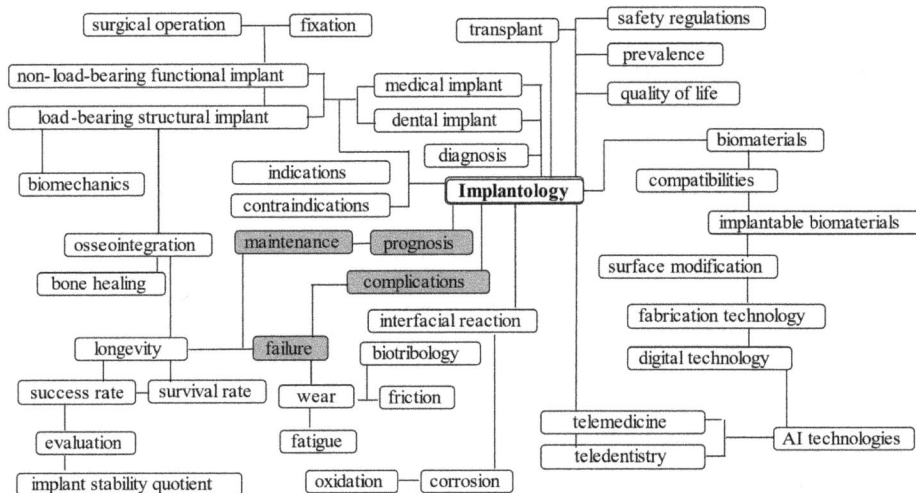

12.1 Introduction

A complication in medicine, or medical complication, is an unfavorable result of a disease, health condition, or treatment. Complications may adversely affect the prognosis, or outcome, of a disease. Complications generally involve a worsening in severity of disease or the development of new signs, symptoms, or pathological changes which may become widespread throughout the body and affect other organ systems. Thus, complications may lead to the development of new diseases resulting from a previously existing disease. Complications may also arise as a result of various treatments [1]. Some of possible complications can be anticipated in the preoperation and intraoperation periods but cannot be avoided in advance. Postsurgery complications include (i) shock (as a dangerous reduction of blood flow throughout the body, accompanied with reduced blood pressure), (ii) hemorrhage with rapid blood loss from the site of surgery, leading to shock, (iii) wound infection by which bacteria enter the site of surgery, causing an infection spreading to adjacent organs or tissue and delaying healing process, (iv) deep vein thrombosis (DVT), in which large blood clots can break free and clog an artery to the heart, leading to heart failure, (v) pulmonary complications due to lack of deep breathing within 48 h of surgery, maybe caused by inhaling food, water, or blood, or pneumonia, (vi) temporary urinary retention (or the inability to empty the bladder), caused by the anesthetic, and (vii) reaction to anesthesia, with symptoms ranging from lightheadedness to liver toxicity [2, 3].

https://doi.org/10.1515/9783110740134-013

Once an orthopedic or dental implant (a foreign material) is placed at a certain location (vital soft or hard tissue) and is initiated to perform or recover the biological function, there could be a variety of complications taking place. Such complications include (i) bone loss due to unfavorable surrounding biomechanical environment, (ii) implant loosening (which might be related to the previous bone necrosis), (iii) long-term material degradation due to biotribocorrosion, causing secondary wear debris toxicity. The occurrence of complications is a function of a postoperation time and the failure modes may differ in importance as time passes following the implant surgery [4]. Taking a total hip arthroplasty (THA) as an example, the infection is most likely soon after surgery, while loosening and implant fracture become progressively more important as time goes on. Failure modes also depend on the type of implant and its location and function in the body. For example, an artificial blood vessel is more likely to cause problems by inducing a clot or becoming clogged with thrombus than by breaking or tearing mechanically.

12.2 Implant loosening

One of the most important complications in implant treatments in both orthopedics and dentistry is an implant loosening. The risk of bacterial infections associated with open surgery and/or the implantation remains to be a major drawback in the implant treatments and these implant-associated infections cause the loosening of placed implants [5–8]. Besides the risk of infection, dental/medical implants can fail due to wear, fatigue, chemical degradation, and others which might lead to osteolysis and loosening implants [9]. Early implant instability has been proposed as a critical factor in the onset and progression of aseptic loosening and periprosthetic osteolysis in total joint arthroplasties.

Although it is known that macrophages stimulated with cyclic mechanical strain release inflammatory mediators, little is known about the response of these cells to mechanical strain with particles, which is often a component of the physical environment of the cell. Fujishiro et al. [10] studied the production of prostaglandin E2 (PGE2), an important mediator in aseptic loosening and periprosthetic osteolysis in total joint arthroplasties, for human macrophages treated with mechanical stretch alone, titanium particles alone, and mechanical stretch and particles combined. A combination of mechanical stretch and titanium particles resulted in a statistically synergistic elevation of levels of PGE2 compared with the levels found with either stretch or particles alone. It was hence suggested that, while mechanical strain may be one of the primary factors responsible for macrophage activation and periprosthetic osteolysis, mechanical strain with particles load may contribute significantly to the osteolytic potential of macrophages in vitro [10]. Corrosion of implant alloys releasing metal ions has the potential to cause adverse tissue reactions and implant failure. Lin et al. [11] hypothesized that macrophage cells and their released reactive chemical species affect the corrosion property of Ti–6Al–4V. It was reported that (i) there was no difference in the charge transfer in the presence and absence of cells, and (ii) the alloy had the lowest charge transfer and metal ion release with activated cells was as follows: Ti < 10 ppb, V < 2 ppb, attributing to an enhancement of the surface oxides by the reactive chemical species. It was concluded that macrophage cells and reactive chemical species reduced the corrosion of Ti–6Al–4V alloys [11].

Prosthetic and osteosynthetic implants from metal alloys will be indispensable in orthopedic surgery, as long as tissue engineering and biodegradable bone substitutes do not lead to products that will be applied in clinical routine for the repair of bone, cartilage, and joint defects. Therefore, the elucidation of the interactions between the periprosthetic tissues and the implant remains of clinical relevance, and several factors are known to affect the longevity of implants. Sommer et al. [12] studied the effects of metal particles and surface topography on the recruitment of osteoclasts in vitro in a coculture of osteoblasts and bone marrow cells. It was reported that (i) the cells were grown in the presence of particles of different sizes and chemical compositions, or on metal discs with polished or sand-blasted surfaces, respectively, (ii) at the end of the culture, newly formed osteoclasts were counted, (iii) osteoclastogenesis was reduced when particles were added directly to the coculture, and (iv) in cocultures grown on sand-blasted surfaces, osteoclasts developed at higher rates than they did in cultures on polished surfaces. It was therefore summarized that the wear particles and implant surfaces affect osteoclastogenesis, and thus may be involved in the induction of local bone resorption and the formation of osteolytic lesions, leading eventually to the loosening of orthopedic implants [12]. Nanomaterial safety and toxicity are of great importance for nanomaterial-based medical implants and a better understanding of the fate of nanomaterials after production and after implantation is clearly necessary. Yang et al. [13] pointed out that, in terms of implant degradation, nanoscale materials can be generated and released into peripheral host tissues regardless of their constituent grain sizes (or other characteristic features, such as particle size) and unfortunately, the biological responses to and toxicity of nanoscale implant materials have not been sufficiently investigated.

The load of the implant should preferably be in the longitudinal direction, so that it is important to avoid rotational or cantilever forces, after the implant has integrated. If forces are distributed in the longitudinal direction, even very high loads can be overcome by the implant during many years of function [14]. Loss of integration can be obtained by overload, torsional forces, or direct trauma to the implant. Overload might be the result of the long extensions of the bar construction or by misplaced implants. Minor torsion forces might induce microfractures in the bone that with time can rupture the delicate integration. Lack of adequate counterforces during tightening of the abutment might result in immediate implant loss. Tightening of individual magnets used as retentions without counter-holding might as well cause implant failures [14]. The quantity and quantity of bone might determine the force necessary to lose and implant. The most well-known cause for implant failure is the bone that has been irradiated as part of cancer treatment. Chemotherapy also affects implant survival to a similar degree as radiotherapy. Other factors that might affect implant survival are osteoporosis, steroid medication, and diabetes mellitus [15].

12.3 Orthopedic implant complications

A common factor in joint implants (including, wrist joint, shoulder joint, hip joint, and knee joint) is a ball–socket coupling situation. Karjalainen et al. [16] reported two cases of biotribology-related incidences of the total wrist arthroplasty which resulted in revision surgery. One was an adverse reaction to metal debris and blood cobalt and chrome levels were elevated and magnetic resonance imaging showed clear signs of a pseudotumor. The other case was with an adverse reaction to poly-etheretherketone (PEEK), which exhibited an extensive release of polyether ether ketone particles into the surrounding synovia due to adverse wear conditions in the cup, leading to the formation of a fluid-filled cyst sac with a black lining and diffuse lymphocyte-dominated inflammation in the synovia. Hansen et al. [17] measured the serum chrome and cobalt in 50 patients with trapeziometacarpal total joint replacement with MoM articulation and compared with serum chrome and cobalt values in 23 patients with trapeziometacarpal total joint replacement with MoP articulation. It was reported that (i) in 10 of 50 (20%) patients with MoM, slightly elevated serum chrome or cobalt values were found compared with only one in 23 (4%) patients with MoP articulation, (ii) all metal values were lower than accepted "normal values" for MoM hip arthroplasty and so considered not to be a general health risk; however, the mean disabilities of the arm, shoulder, and hand (DASH) score was 24 in patients with elevated serum chrome or cobalt compared with 10 in patients with normal metal values suggesting a local clinical effect of the elevated serum chrome or cobalt values, recommending that patients with trapeziometacarpal total joint replacement with MoM articulation are followed with DASH score and radiological examination every 3–5 years and serum chrome and cobalt should be analyzed in symptomatic cases to learn more about possible local complications leading to, or arising from, metal debris. PEEK is the typical polymeric material used in the MoP joint coupling. The biologic response to wear particles affects the longevity of total joint arthroplasties. Particles in the phagocytozable size range of 0.1–10 µm are considered the most biologically reactive, particularly particles with a mean size of <1 µm. Stratton-Powell et al. [18] conducted a systematic review aimed to identify the current evidence for the biologic response to PEEK-based wear debris from total joint arthroplasties. It was reported that (i) in the four studies that quantified PEEK-based particles produced using hip, knee, and spinal joint replacement simulators, the mean particle size was 0.23–2.0 µm (the absolute range reported was approximately 0.01–50 µm), (ii) qualitative histologic assessments showed immunologic cell infiltration to be similar for PEEK particles when compared with ultra-high-molecular-weight polyethylene (UHMWPE) particles in all six of the animal studies identified; however, increased inflammatory cytokine release (such as tumor necrosis factor-α) was identified in only one in vitro study, but without substantial suppression in macrophage viability, (iii) only one study tested the effects of particle size on cytotoxicity and found the largest unfilled PEEK particles (approximately 13 μm) to have a toxic effect; UHMWPE particles in the same size range showed a similar cytotoxic effect. Based on findings, it was concluded that wear particles produced by PEEK-based bearings were, in almost all cases, in the phagocytozable size range (0.1–10 μm); ad the studies that evaluated the biologic response to PEEK-based particles generally found cytotoxicity to be within acceptable limits relative to the UHMWPE control, but inconsistent when inflammatory cytokine release was considered [18].

Total shoulder arthroplasty

The potential risks and complications of shoulder joint replacement should be explained prior to the surgery, including those related to the surgery itself and those that can occur over time postsurgery period. Complications of shoulder arthroplasty depend on the prosthesis type used. Although when complications occur, most are successfully treatable, basically possible complications include the following issues [19–21]. (1) Infection is a complication of any surgery. In total shoulder arthroplasty (TSA), infection may occur in the wound or deep around the prosthesis. It may happen while in the hospital or after discharge. It may even occur years later. Minor infections in the wound area are generally treated with antibiotics. Major or deep infections may require more surgery and removal of the prosthesis. Any infection occurred in a body can spread to placed joint replacement. (2) Although prosthesis designs and materials, as well as surgical techniques, continue to advance, the prosthesis may wear down and the components may loosen. The components of a shoulder replacement may also dislocate. Excessive wear, loosening, or dislocation may require additional surgery (revision procedure). (3) In case of reverse total shoulder replacement, implant may dislocate which requires closed reduction or surgery to reduce or revise the implant. (4) Nerves in the vicinity of the joint replacement may be damaged during surgery, although this type of injury is infrequent. Over time, these nerve injuries often improve and may completely recover.

Wirth et al. [22] classified TSA complications into three groups from the most frequent complications. *Loosening* is the most frequent complication associated with the TSA. Loosening includes symptomatic loosening of the glenoid and humeral component represents one third of the complications from TSA. The most loosening (as glenoid loosening) is believed to be a result of aseptic loosening of the cement while new innovations of implants such as press fit, plasma sprayed, and tissue-ingrowth implants are promising alternatives that could prove to be more stable than traditional cemented implants. Although glenoid loosening accounts for the majority of loosening complications, radiolucent lines of 2 mm or more have been seen in the humeral component and most cases are in noncemented implants. *Glenohumeral instability* becomes the second cause of complications in the TSA and it includes anterior instability, superior instability, posterior instability, and inferior instability. *Rotator cuff tearing* is the third most frequent complication of TSA. There are still other important complications such as (i) periprosthetic, intraoperative, and postoperative fractures that can delay postoperative rehabilitation, and (ii) prosthetic complications, in which occasionally complications of the implant such as dissociation of the polyethylene glenoid insert from the metal backing or fracture of the keel insert can occur.

Buck et al. [21] mentioned that the (i) sonography is most commonly used postoperatively in order to demonstrate complications (hematoma and abscess formation) but may also be useful for the demonstration of rotator cuff tears occurring during follow-up, (ii) CT is useful for the demonstration of bone details both preoperatively and postoperatively, and (iii) MRI is mainly used preoperatively, for instance for demonstration of rotator cuff tears.

Total hip arthroplasty

There are similar complications associated with THA as follows [23–25].

(1) Aseptic loosening of a cemented femoral component after THA is a potential cause of pain and loss of function, resulting in the subsequent need for a revision. Several factors contributing to these adverse effects, which may eventually result in failure of the THA, include the selection of the patients and the materials and design of the implant. Probable underlying causes of aseptic loosening include excessive initial micromotion of the femoral component, which precludes bone ingrowth in the short term, and prosthetic materials and design that can result in adverse bone remodeling in the long term. THA aseptic loosening is the result of a confluence of steps involving particulate debris formation, prosthesis micromotion, and macrophage activated osteolysis. Treatment requires serial imaging and radiographs and/or CT imaging for preoperative planning. Persistent pain requires revision THA surgery [26].

(2) Dislocation-related instability is one of the most common complications after THA. Risk factors include neuromuscular conditions/disorders (e.g., Parkinson's disease) and patient noncompliance. Component malpositioning such as the excessive anteversion results in anterior dislocation and excessive retroversion results in posterior dislocation, previous hip surgery (most significant independent risk factor

for dislocation), or elderly age (older than 70 years old) [27]. Recent improvements in posterior soft tissue repair after primary THA have shown a reduced incidence of dislocation. When dislocation occurs, a thorough history, physical examination, and radiographic assessment help in choosing the proper intervention. Closed reduction usually is possible, and nonsurgical management frequently succeeds in preventing recurrence. When these measures fail, first-line revision options should target the underlying etiology. If instability persists, or if a primary THA repeatedly dislocates without a clear cause, a constrained cup or bipolar femoral prosthesis may be as effective as a salvage procedure [28–30]. Recurrent THA dislocations often result in revision THA surgery with component revision [24].

(3) THA prosthetic joint infection. Risk factors include patient specific lifestyle factors (morbid obesity, smoking, intravenous drug use and abuse, alcohol abuse, and poor oral hygiene). Other risk factors include patients with a past medical history consisting of uncontrolled diabetes, chronic renal and/or liver disease, malnutrition, and HIV [31]. The changing profile of antibiotic-resistant bacteria has made preventing and treating primary THA infections increasingly complex. Lindeque et al. reported that the pooled deep prosthetic joint infection rate was 0.9%. The pooled rate of methicillin-resistant *Staphylococcus aureus* infection was 0.5%. The pooled rate of intraoperative bacterial wound contamination was 16.9%. The postoperative risk of surgical site infection was significantly associated with intraoperative bacterial surgical wound contamination [32].

(4) Periprosthetic fractures are increasing in incidence with the overall increased incidence of procedures in younger patient populations [24]. Intraoperative fractures can occur and involve either the acetabulum and/or femur. Acetabular fractures occur in 0.4% of press-fit acetabular implant components, most often during component impaction. Risk factors include underreaming more than 2 mm, poor patient bone quality, and dysplastic conditions. Intraoperative femur fractures occur in up to 5% of primary THA cases as reported in some series. Risk factors include technical errors, press fit implants, poor patient bone quality, and revision surgery [33]. Treatment of fractures surrounding the femoral stem is reliably managed using the Vancouver classification system [24].

(5) The THA postoperative wound complication spectrum ranges from superficial surgical infections such as cellulitis, superficial dehiscence, and/or delayed wound healing, to deep infections resulting in full-thickness necrosis. Deep infections result in returns to the operating room for irrigation, debridement (incision and drainage) and depending on the timing of the infection, may require explant of THA components [24].

Other potential THA complications include sciatic nerve palsy, leg length discrepancy, iliopsoas impingement, heterotopic ossification, and vascular injury [24].

One in eight of all total hip replacements requires revision within 10 years, 60% because of wear-related biotribological complications. The bearing surfaces may be made of Co–Cr–Mo alloy, Fe–Cr–Ni–Mo stainless steel (typically, 316L SS), ceramic, or polyethylene. Friction between bearing surfaces and corrosion of nonmoving parts can result in increased local and systemic metal concentrations. Bradberry et al. [34] conducted systematic review and published reports of systemic toxicity attributed to metal released from hip implants. These searches identified 281 unique references, of which 23 contained original case data. Three further reports were identified from the bibliographies of these papers. As some cases were reported repeatedly, the 26 papers described only 18 individual cases. Ten of these eighteen patients had undergone revision from a ceramic containing bearing to one containing a metal component. The other eight had MoM prostheses. It was also reported that the systemic toxicity was first manifest months and often several years after placement of the metal-containing joint. The reported systemic features fell into three main categories: neuroocular toxicity (14 patients), cardiotoxicity (11 patients), and thyroid toxicity (9 patients). Neurotoxicity was manifest as peripheral neuropathy (8 cases), sensorineural hearing loss (7 cases), and cognitive decline (5 cases); ocular toxicity presented as visual impairment (6 cases). All these neurological features, except cognitive decline, have been associated with cobalt poisoning previously. Those patients reported to have systemic features who had received a metal-on-metal prosthesis ($n = 8$), had a median peak blood cobalt concentration of 34.5 (range, 13.6–398.6) μg/L; those with a metal-containing revision of a failed ceramic prosthesis ($n = 10$) had a median blood cobalt concentration of 506 (range, 353–6,521) μg/L. It was concluded that (i) rarely, patients exposed to high circulating concentrations of cobalt from failed hip replacements develop neurological damage, hypothyroidism, and/or cardiomyopathy, which may not resolve completely even after removal of the prosthesis, and (ii) the greatest risk of systemic cobalt toxicity seems to result from accelerated wear of a cobalt-containing revision of a failed ceramic prosthesis, rather than from primary failure of a MoM prosthesis.

Similarly, Matharu et al. [35] investigated whether blood metal ions could effectively identify patients with MoM hip implants with two common designs (Birmingham Hip Resurfacing [BHR] and Corail-Pinnacle) who were at risk of adverse reactions to metal debris. It was reported that (i) all ion parameters were significantly higher in the patients who had adverse reactions to metal debris compared with those who did not, and (ii) cobalt maximized the area under the curve for patients with the BHR implant (90.5%) and those with the Corail-Pinnacle implant (79.6%). For patients with the BHR implant, the area under the curve for cobalt was significantly greater than that for the cobalt–chromium ratio, but it was not significantly greater than that for chromium. For the patients with the Corail-Pinnacle implant, the area under the curve for cobalt was significantly greater than that for chromium, but it was similar to that for the cobalt–chromium ratio. It was concluded that (i) patients who underwent MoM hip arthroplasty performed with unilateral BHR or Corail-Pinnacle implants and who had blood metal ions below implant-specific thresholds were at low risk of adverse reactions to metal debris, (ii) these thresholds could be used to rationalize follow-up resources in asymptomatic patients, and (iii) implant-specific thresholds were more effective than currently recommended fixed authority thresholds for identifying patients at risk of adverse reactions to metal debris requiring further investigation.

Total knee arthroplasty

The rate of complications after total knee arthroplasty (TKA) is reported, in most publications, to range from 1.65% to 11.3% [36, 37]. Despite the importance of complications in evaluating patient outcomes after TKA, definitions of TKA complications are not standardized yet. Different investigators report different complications with different definitions when reporting outcomes of TKA. Accordingly, Healy et al. [38] developed a standardized list and definitions of complications and adverse events associated with TKA. It was obtained that the 22 TKA complications and adverse events include bleeding, wound complication, thromboembolic disease, neural deficit, vascular injury, medial collateral ligament injury, instability, malalignment, stiffness, deep joint infection, fracture, extensor mechanism disruption, patellofemoral dislocation, tibiofemoral dislocation, bearing surface wear, osteolysis, implant loosening, implant fracture/tibial insert dissociation, reoperation, revision, readmission, and death, concluding acceptance and utilization of these standardized TKA complications may improve evaluation and reporting of TKA outcomes. Koh et al. [39] determined whether the type of surgical strategy for bilateral TKA (staggered, staged, or simultaneous) influences the incidence of acute kidney injury and related complications and concluded that (i) the type of bilateral TKA strategy was an independent risk factor for the development of acute kidney injury, and (ii) the assessment of additional risk factors for the development of the acute kidney injury is essential before deciding on surgical strategy. Knee joint-preserving surgery includes osteotomy, ligament reconstruction, meniscus surgery, and cartilage repair procedures, often used in combination. Knee arthroplasty procedures consist of unicondylar, patellofemoral, and primary or revision total knee prosthesis. Pain after knee surgery is common and is influenced by the underlying pathology, the type of surgery, and the patient [40]. Lum et al. [41] reported that the mortality rate after two-stage total knee revision for infection is very high. Accordingly, although TKA remains a reliable and reproducibly successful surgery in patients suffering from debilitating advanced degenerative arthritic knees, reports still cite that up to one in five patients who have undergone primary TKA remain dissatisfied with the outcome [42].

Major types of complications are shared with previous TSA and THA complications.

(1) Aseptic loosening occurs secondary to a macrophage-induced inflammatory response resulting in eventual bone loss and TKA component loosening. Patients often present with pain that is increased during weight-bearing activity and/or recurrent effusions. Loosening (as a mechanical complication) is one of the most serious complications and pain is the most common one. Loosening can cause bone fractures, instability, and serious falls. Almost all serious complications require revision surgery. Patients may have minimal pain at rest or with range of motion. Serial imaging and infectious labs are required to appropriately work up these conditions which eventually are treated with revision surgery if symptom persists and the patient is considered a reasonable surgical candidate. The steps in aseptic loosening include particulate debris formation, macrophage-induced osteolysis, micromotion of the components, and dissemination of particulate debris. There are several causes for loosening. Infection is one of the most likely causes. Faulty design, product defects, or wear and tear on implant parts can also cause loosening [42, 43].

(2) Instability and dislocation are one of major complications. It is reported that implant instability is a chief reason for the revision surgery and it accounted for as many as one in five revision surgeries [43]. Instability can cause excessive wear on implant parts. It can also cause patients to fall, resulting in further injuries including fractures. Instability can cause the implant to dislocate for similar reasons as natural knees, including overextending the joint. Components in an artificial knee have to be precisely aligned to work. If they do not line up, they can wear out too soon. Misalignment of implant placement can fail without warning. Symptoms of misalignment and failure include instability, pain, and swelling. Other signs include reduced range of motions and warmth or heat around the joint. Patients almost always need revision surgery to fix misalignment problems. Excessive weight can cause components in a knee replacement to move out of their proper position and increase the risk of failure [43].

(3) Periprosthetic joint infection is a serious complication that threatens a patient's overall health. They are also a leading cause of implant failure. They can damage muscle or bone, weakening the implant [43]. The incidence of infection has been reported to be <1–2% with primary arthroplasty, 3–5% of revision knee arthroplasty and as high as 16% with hinged implants. Patients with rheumatoid arthritis, diabetes mellitus, poor nutrition, old age, and obesity are at higher risk of both superficial and deep infection [44, 45]. Surgical wound infections are often pain free with redness around the wound, discharging fluid, but with no joint effusion, joint stiffness, or restriction of movement. Risk factors include patient specific lifestyle factors (morbid obesity, smoking, intravenous drug use and abuse, alcohol abuse, and poor oral hygiene) and patients with a past medical history consisting of uncontrolled diabetes, chronic renal and/or liver disease, malnutrition, and HIV [42].

Early deep infections are often due to relatively virulent pathogens and present with an acute onset of symptoms, including joint pain, joint effusion, induration, erythema, wound oozing, and fever. Early orthopedic referral is paramount for the timely management of possible prosthetic joint infections, rather than immediately starting antibiotics. Prompt aspiration and tissue culture is essential to start appropriate antibiotics. The treatment for deep infection in a joint includes intravenous antibiotic therapy, debridement, polyethylene liner exchange, and revision surgery [44]. The most common offending bacterial organisms in the acute setting include *Staphylococcus aureus*, *Staphylococcus epidermidis*, and in chronic TKA periprosthetic joint infection cases, coagulase-negative *Staphylococcus* bacteria. More aggressive treatments, especially in the setting of presentation beyond the acute (3–4-week time point) include a one- or two-stage revision TKA procedure with interval antibiotic spacer placement. The surgeon must ensure and document evidence of infection eradication [42, 46].

(4) Periprosthetic bone fracture after TKA can happen in the thighbone, the patella, or the tibia. These are more likely to happen in older patients and those with low bone density issues such as osteoporosis. Loose components and malalignment or malposition of implants can cause bone fractures. This is a medical emergency and needs immediate treatment [43]. TKA periprosthetic fractures are further characterized by location and residual stability of the implants. Distal femur fractures occur at a 1–2% rate, and risk factors include compromised patient bone quality and increased constrained TKA components, while controversial, anterior femoral notching is a potential risk factor for postoperative fracture. Tibial fractures occur at a 0.5–1% rate, and risk factors include a prior tibial tubercle osteotomy, component malposition and/or loosening, as well as utilization of long-stemmed components. Patellar fractures occur less frequently in unresurfaced TKA cases, and incidence rates range from 0.2% up rates as high as 15% or 20%. Risk factors for fracture include osteonecrosis, technical errors in asymmetric or over-resection, and implant-related associations, including (i) central, single peg implants, (ii) uncemented fixation, and (iii) metal-backed components [42].

(5) Wound healing problems occur because of thin and soft tissue covering the knee, especially over the anterior aspect. Healing problems can be associated with hematoma, especially in the elderly, patients on steroids, and rheumatoid arthritis and psoriasis patients. Wound healing problems require urgent diagnosis and management to avoid more serious complications such as skin loss, infection, and possible loss of the prosthesis. An unhealthy looking wound should be referred back to the surgeon as soon as possible [44]. The TKA postoperative wound complication spectrum ranges from superficial surgical infections such as cellulitis, superficial dehiscence and/or delayed wound healing to deep infections resulting in full-

thickness necrosis resulting in returns to the operating room for irrigation, debridement (incision and drainage), and rotational flap coverage [42].

(6) Complications involving anesthesia. Like any major surgery involving general anesthesia, there is a low risk of strokes, heart attacks, pneumonia, and blood clots. Blood clots occurring in deep veins, or DVT, are a potential complication of knee replacement surgery. Left untreated, a blood clot can break free from the vein wall, a life-threatening condition known as pulmonary embolism. When caught in time, pulmonary embolism is treatable with anticlotting medication. Devices that can be wrapped around the affected leg to provide intermittent pneumatic compression can minimize the risk of DVT [46]. DVT potentially can be fatal, if the thrombosis embolizes to the lungs [47]. A DVT may be silent, presenting as a pulmonary embolism with shortness of breath, chest pain, and cyanosis, without limb symptoms. Alternatively, it may present with a painful calf or thigh usually 5–7 days postoperatively or earlier. A low threshold for lower limb ultrasound, chest X-rays, and spiral CT chest may help establish early diagnosis. Physical examination may reveal a unilateral swollen calf or thigh, erythema, tenderness, warmth, and a difference in calf diameters [44]. Prompt diagnosis and initiation of treatment can prevent further clot extension and pulmonary embolism. Anticoagulation therapy is indicated for patients with DVT and prompt referral to the orthopedic department or A&E should be organized for further investigations [44, 46].

(7) Nerve damage can happen during surgery, but it usually goes away within six months. Knee surgery may involve special tourniquets to restrict blood flow in the leg. Symptoms of nerve damage include radiating pain, "tingling" sensation in the leg, and numbness in the leg or foot. Some total knee replacement patients may experience nerve block complications. A nerve block is an anesthetic that surgeons inject close to the nerves around a surgical site to relieve pain following surgery [43, 44].

(8) Persistent pain and dissatisfaction. Although outcomes after total knee replacement are good, many patients continue to report pain and dissatisfaction. Clinically significant persistent pain and dissatisfaction has been reported in 20% of patients [48]. Night pain is quite common after knee replacement. Possible explanations for such pain include unrealistic expectations, technical flaws of the procedure, and pain from other sites. If the pain is persistent without a known cause then referral to the pain management team should be made [44]. Patients may present with postoperative limitation of motion that results in functional impairment of the joint. Many factors can be responsible including malrotation of the prosthesis and poor preoperative range of movement. Treatment may comprise physiotherapy or manipulation under anesthesia [44].

There are other potential complications [42, 43, 46]. (i) Legs may be of slightly different lengths after surgery. In some cases, a shoe insert can remedy this problem. (ii) The new knee may be stiff. Most people who have undergone knee replacement surgery can bend their knees at least 115 degrees. However, some people develop scar tissue that hinders flexibility. This limited flexibility is more common in people who had limited flexibility before surgery. (iii) Metal hypersensitivity or implant rejection. An allergic reaction to the prosthesis or bone cement can occur. In these cases, the bone cement and prosthesis must be removed. (iv) Heterotopic ossification should be included.

For the most part, complications can be treated. Complications tend to be higher in patients who use tobacco and who are older or obese. A surgery followed by complications may still be considered successful, if pain is alleviated and function improves over the long term. Complications (including anesthesia-related, exacerbation of comorbid medical issues, medication and allergic reactions) related to TKA, although uncommon, range from minor problems to devastating, life-threatening events. Efforts should be made to minimize the risk of complications with appropriate patient selection and optimization, meticulous surgical technique, and attentive postoperative management [49].

12.4 Dental implant complications

Dental implant failure can occur during the first stages after the procedure, or they can turn out to be a long-term failure. There are recommended items that patients should watch out for after getting dental implants and these checklists are directly or indirectly related to dental implant complications. These are patient's responsibilities for implant(s) placed into his/her own intraoral environment. They include as follows [50]. (1) Severe pain and discomfort. During the healing process, patient will experience the pain for the first few days. The pain is not as intense, and it can be controlled using some painkillers prescribed by the doctor. In case of a dental implant failure, patient will experience excruciating pain and discomfort that comes in the form of throbbing waves. This pain occurs long after the procedure. In this case, it's advisable to visit the dentist for a checkup before it is too late. (2) Gum recession around the implant. Two major reasons that cause gum recession around the implant include, poorly positioned implants and inadequate gum and bone tissue to hold the implant. The first sign will be an abnormally long implant crown. This is followed by painful inflammation around the implant. To prevent this, the dental surgeon should place the implant in the correct position. Angulation is also a crucial part of this procedure, which should be done using digital dental implant diagnostics and planning. Besides, it is important to ensure that healthy gums and bone tissue should be maintained with a healthy diet and oral hygiene. (3) Difficulty while chewing and biting. If a patient has a hard time chewing or biting food, this can be a sign of dental implant failure. Since implants are made to function and feel like natural teeth, any pain associated with a dental implant is not a good sign. It is just similar to experiencing pain while chewing with a tooth cavity. (4) Shifting and loose implant. If the implant is not properly seated on gums, a patient will feel it wobble while talking, eating, or touching it. This is the easiest sign to spot an implant failure. In case a patient discovers a shifting and loose dental implant, set up a consultation with a dentist immediately. If the loose implant is left unattended, it could interfere with the look of a smile or cause severe damage to the gums and jawbone. (5) Swollen gums. Minor swelling is normal after the procedure; however, this is expected to disappear within a few days. If it persists and even gets inflamed, itis a cause for concern. If gums appear extremely swollen and red, it is a sign of an infection. If untreated, this infection can spread to the rest of the mouth, and in severe cases, it can spread to the blood, resulting in a dangerous health condition. (6) Implant micromovements. Sometimes a dental implantologist can perform a tooth replacement immediately after implantation. This procedure requires less healing time as compared to doing the implantation and waiting until it integrates with jawbone before placing the tooth. If the jawbone is neither strong enough nor much enough, the procedure can put excess stress on it, which can cause implant failure. (7) Sudden allergic reactions. Dental implants are made of titanium alloy, which causes allergic reactions to some people. Some signs of allergic reactions include loss of taste, swelling around the gums, and a tingling sensation. Sudden allergic reactions are a sign of dental implant failure because they indicate that a patient's body is rejecting the implant. Allergic reaction should include the anaphylaxis. (8) Teeth grinding. This is a long-term dental implant failure. It happens due to several reasons, such as stress, missing teeth, or misaligned teeth. Some people also experience teeth grinding while sleeping. Teeth grinding makes it hard for the implant to integrate with the jawbone due to excessive pressure on teeth caused by the dental implant procedure. If you have episodes of teeth grinding, it may be a sign of dental implant failure so that it is recommended to see a dentist to check whether the implant is correctly aligned with the jawbone [50].

While dental implants can be beneficial and are long-term replacement for a missing tooth (or teeth), a dental implantologist and surgery can pose a risk. There are numerous potential complications that can occur during the intraoperation and postoperation of dental implant surgery. Although the most of complications are categorized into intraoperation or postoperation, there are several preoperation dental implant complications which can also be considered as contraindications and patient selection strategies. The following patients who have developed an anamnesis or present illness are, in the most of cases, contraindicated; terminal illness, long-term steroid therapy, radiation therapy to potential implant sites, recent heart surgery within the last 6 months, ASA IV (ASA: American Society of Anesthesiologists; a patient with severe systemic disease that is a constant threat to life) or ASA V (a moribund patient who is not expected to survive without the operation), uncontrolled systemic diseases, bleeding disorders, heavy smoking, patients of advanced age, patients still undergoing skeletal growth, postirradiation, psychological profiles, uncontrolled periodontal disease, and use of chronic bisphosphonate therapy [51–53].

In addition, the following should be checked as the preoperative examinations. For the anterior mandible, the location of the metal nerve should be identified. For the posterior mandible, the bone quality, inferior alveolar nerve (IAN), lingual artery, and lingual concavity are needed to be checked. For the anterior maxilla, nasal fossa, nasopalatine canal, bone quality, ridge width and shape, aesthetics, implant positioning, and smile line should be preexamined [53, 54].

Complications can be classified into intraoperation complications which can be further divided into common problems and less common problems, and postoperation complications which are characterized as the long-term issues [55–57].

12.4.1 Common intraoperation and short-term postoperation complications

Infection

At times, in spite of the better surgical methods, infections may occur and are considerable complications of dental implantation. The infection might occur as a result of the bacteria in the mouth getting into the area of surgery. In modern days, <1% of the surgeries result in an infection and most of those infections are minor. However, this is especially common after the surgery when the patient does not practice proper dental hygiene. Infection can occur in the affected area, presenting as an abscess, fistula, suppuration, inflammation, or radiolucency. It can also affect wider areas of the body to cause a systemic infection. Additionally, having thin gums, suffering from diabetes, and smoking can put you at more risk of contracting an infection. Practicing good dental hygiene has always been a rule of thumb in maintaining your teeth; that includes not smoking. However, if you have ailments or conditions, it is crucial that you discuss them with your dentist before you undergo any procedure. Although antibiotics are often given prior to the implantation procedure to help increase the chance of success of integration, they are unlikely to help reduce the risk of infection. An infection at the implant site during surgery or in the early days after surgery can increase the risk of implant failure. Preop antibiotics reduce the risk of implant failure but have little impact on the risk of infection. Proper oral hygiene after surgery is essential. A clean mouth will heal faster and the risk of infection is reduced if the mouth is not riddled with bacteria and food debris. Treatment for an infection depends on the severity and location of the infection. For example, a bacterial infection in the gum may require antibiotics or a soft tissue graft, while a bacterial infection in the bone may require removal of the infected bone tissue and possibly the implant, followed by a bone and soft tissue graft [58–61].

Nerve and tissue damage
When the implant is placed too close to the nerves, it will result in chronic pain, discomfort, and numbness in the area. The nerve damage might also be long-term or permanent [58, 61]. A nerve or tissue problem requires immediate attention. One of the most common and serious complications faced by the clinician and the patient following implant placement in the mandible is injury to the IAN in the lower jaw [61, 62]. It ranges from 0% to 40% of implant-related IAN injuries [63, 64]. These injuries may occur during preparation or placement of an implant. They may be directly related to the depth of preparation or implant length or width [65] and may result from local anesthetic application as well [66]. Some possible symptoms of an IAN injury include [60, 62]: (i) persistent numbness on the side of the implant, including the lower lip and chin, (ii) persistent pain or discomfort, and/or (iii) tingling, tickling, or burning sensations in the gums and skin. If the nerve damage is minor, patients will recover after a certain period of time.

Osseointegration failure and instability
In the first few weeks, the placed dental implant will be growing into and fusing with the jawbone [67]. This osseointegration, which is consisted of primary (mechanical) and secondary (biological) stability processes, is crucial to the long-term success of the implant. This is one of the most reported problems. If the implant fails to fuse with the bone, the implant should be removed and the patient may be able to reattempt the implant procedure once the area has healed [58, 60, 61]. Primary implant stability, which refers to the stability of a dental implant directly following the procedure, is a significant factor for the integration of the implant into the bone. Insufficient primary stability can lead to failure of the implant within the initial weeks. As the bone surrounding the implants begins to regrow and fuse with the dental implant, this supports the implant and provides secondary stability. This continues to strengthen as the fusion between the bone and implant increases, eventually leading to biological stability in the ideal scenario [59].

There are a variety of reasons as to why the implant may fail to properly fuse into the jawbone, including (i) wrong positioning of the implant, (ii) the absence of enough bone volume and density, (iii) the bone structure around the implant is damaged, (iv) general conditions such as uncontrolled diabetes, untreated osteoporosis, radiation exposure on the head and neck, and so on, (v) improper oral hygiene, (vi) heavy smokers or high alcohol consumption, and/or (vii) various accidents during surgery such as infection, puncture of a sinus cavity [58, 60]. Prophylactic antibiotics are usually given prior to the implantation procedure, as this can help to reduce the risk of implant failure by a third [59].

Others

Ghidrai [60] mentioned that some severe bleeding may occur especially if large blood vessels are injured intraoperation period. If such an accident should occur, treatment includes compression, vasoconstrictive medication, cautery, or ligation of arteries. Most of the times, bleeding is kept under control. Sinus problems occur when dental implants placed in the upper jaw protrude into one of the sinus cavities. Careful planning and precise surgery execution are essential to avoid this accident.

In some cases, a person may find that the gum tissue around the implant begins to recede. This can lead to inflammation and pain. Getting a prompt assessment from a dentist is essential to prevent the removal of the implant [61, 68].

Hanif et al. [69] listed complications in categorized areas such as mechanical complications and technical complications. Mechanical complications are usually a sequel to biomechanical overloading [70]. Factors contributing to the biomechanical overloading are poor implant position/angulation (cuspal inclination, implant inclination, horizontal offset of the implant, and apical offset of the implant) [71], insufficient posterior support (i.e., missing posterior teeth), and inadequate available bone or the presence of excessive forces due to the parafunctional habits, that is, bruxism [72]. These cause the screw loosening [73]. There are two major causes of implant fracture: biomechanical overloading and peri-implant vertical bone loss [74]. The risk of implant fracture increases multifold when the vertical bone loss is severe enough to concur with the apical limit of the screw [69]. Implant fractures are also attributable to flaws in the designs and manufacturing of implant itself [75]. It was also mentioned that the frequency of occurrences of technical complications is greater in implant-supported fixed partial dentures (FPDs) as compared to the implant-supported removable prosthesis [69, 76, 77].

12.4.2 Less common intraoperation and short-term postoperation complications

Sinusitis

Sinusitis is a common condition defined as inflammation of the paranasal sinuses. Sinus cavities produce the mucus that nasal passages need to work effectively. Although uncomfortable and painful, sinusitis often goes away without medical intervention. However, if symptoms are severe and persistent, a person should consult their doctor [78]. Sinusitis occurs when mucus builds up, and the sinuses become irritated and inflamed. It is often referred to sinusitis as rhinosinusitis [79] because inflammation of the sinuses nearly always occurs with rhinitis [80], which is an inflammation of the nose. An allergy occurs when the body's immune system overreacts to irritants and allergens. A sinus infection, sometimes called sinusitis, is usually bacterial or viral. The symptoms of allergies and sinus infections can be very similar. Both issues can cause sinus pain and pressure, a runny nose, congestion, and sneezing. Also, allergies can sometimes lead to sinus infections [81, 82]. Symptoms of the sinusitis vary depending on how long a condition lasts and how severe the symptoms are and they include, (i) nasal discharge, which may be green or yellow, (ii) a postnasal drip, where mucus runs down the back of the throat, (iii) facial pain or pressure, (iv) blocked or runny nose, (v) sore throat, (vi) cough, (vii) bad breath, (viii) fever, (ix) headaches, (x) a reduced sense of smell and taste, (xii) tenderness and swelling around the eyes, nose, cheeks, and forehead, and/or (xiii) toothache [61, 78]. The sinusitis can stem from various factors, but it always results from fluid becoming trapped in the sinuses, allowing germs to grow. The most common cause is a virus, but a bacterial infection can also lead to sinusitis. Triggers can include allergies and asthma, as well as pollutants in the air, such as chemicals or other irritants.

Sinusitis can be acute or chronic. Causes of sinus inflammation include viruses, bacteria, fungi, allergies, and an autoimmune reaction. Treatment options depend on how long the condition lasts. For acute and subacute sinusitis, if symptoms persist or are severe, a doctor may prescribe treatment. If a bacterial infection is present, a doctor may prescribe antibiotics. If symptoms remain after finishing the antibiotics, the person should return to the doctor. For chronic sinusitis, chronic sinusitis is not usually due to bacteria, so antibiotics are unlikely to help. Reducing exposure to triggers, such as dust mites, pollen, and other allergens, may relieve symptoms. Corticosteroid sprays or tablets may help manage inflammation, but these often need a prescription and medical supervision. Long-term use of these medications can lead to adverse effects [78].

Swallowing implant related instruments
Adverse outcomes resulting from aspiration or ingestion of instruments and materials can occur in any dental procedure. Accidental swallowing or aspiration of dental instruments and prostheses is a complication of dental procedures. Failure to manage these complications appropriately can lead to significant morbidity and possibly death [83, 84]. Clinical manifestation depends on the location, the obstructive potential of the foreign body, and the temporal factor since the accidental incident. Accidental inhalation of dental appliances can be an even more serious event than ingestion and must always be treated as an emergency situation [83]. Figure 12.1 depicts an X-ray image showing a swallowed implant screwdriver in the right main bronchus [83].

It was further reported that aspiration and ingestion of dental foreign objects are infrequent, but they can occur at large multidisciplinary dental procedures. These episodes have the potential to result in acute medical and life-threatening emergencies since the beginning of the event or at a late stage in proceeding in the underdiagnosed patient. Prevention of such incidents is, therefore, the best approach via the mandatory use of precautions during all dental procedures, and in case of suspicion with no retrievable material, patient must always be submitted to a radiographic study [83].

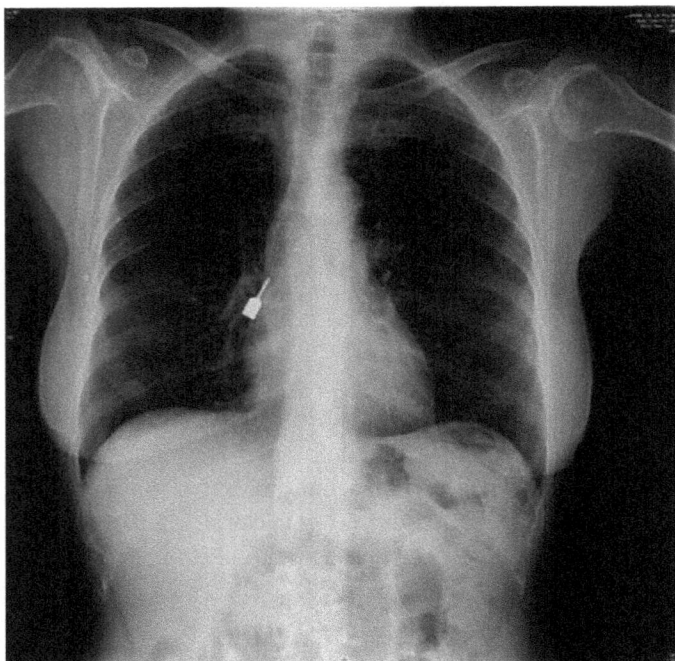

Figure 12.1: Implant screwdriver in right main bronchus [83].

12.4.3 Long-term postoperation complications

Although dental implants have been reported to have fairly high survival rates of 95.7% at 5 years and 92.8% at 10 years [85, 86], it is also known that progressive marginal bone loss and peri-implantitis remain a significant potential complication [86–88].

Bone loss

Excessive bone loss in the area of the dental implant can reduce the stability of the implant replacement and usually requires intervention. Bone loss between the implants and natural teeth can also lead to the appearance of black triangles between the teeth, which are not aesthetically pleasing and increase the difficulty of maintaining clean teeth [68]. There are numerous risk indicators associated with marginal bone loss, including (1) patient-related factors such as smoking, periodontal disease, diabetes, and plaque control/oral hygiene [89], (2) implant-related factors such as design of the implant–abutment complex, and implant shape [90–92], (3) surgical-related factors including (i) the use of bone grafting [93], (ii) immediate placement [94], (iii) site preparation and loading [95], (iv) the degree of separation between implants [96], (v) the presence of thin mucosal tissue [97], and (vi) soft tissue probing depth [98].

Peri-implantitis

Peri-implantitis is an infectious inflammation of the soft and hard tissues around a dental implant, and the long-term risks are significant [61, 99]. Approximately 1/3 of the patients and 1/5 of all implants experienced peri-implantitis. Ill-fitting/ill-designed fixed and cement-retained restorations, and history of periodontitis emerged as the principal risk factors for peri-implantitis [100]. According to Hanif et al. [69], biological failures include bacterial infections, microbial plaque buildup, progressive bone loss, and sensory disruptions and are subcategorized into early biological failures and late implant failures, where the early failures are attributed to the failure of placing the surgical implant under proper aseptic measures and the late complications are typically peri-implantitis and infections bred by bacterial plaque. It is very similar to gum disease. Severity can range from minor inflammation of the gums to severe degradation of the teeth and jaw. If left untreated, this often leads to losing their dental implants and developing other serious dental problems [99]. As a result, as seen in Figure 12.2, placed implant is loosened accompanied with bone loss.

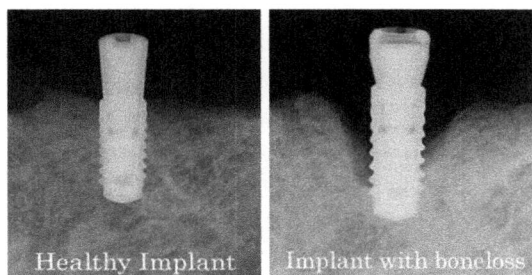

Figure 12.2: Comparison of images between healthy placed implant and implant accompanied with bone loss [99].

There are various factors that influence the susceptibility to peri-implantitis: prior disease (such as diabetes, another systemic disease, or heavy smokers), improper oral hygiene, and parafunctional habits such as bruxism causing an excessive mechanical load on the implant [60, 85, 99].

Symptoms associated with the peri-implantitis, ranging from minor to dangerous, include, (i) redness and inflammation of the surrounding gum tissue, (ii) deepening of the gum pockets around the implant, (iii) exposure or visibility of the implant threads, (iv) loosening of the implant, (v) pus discharging from the tissues around the implant, (vi) loss of supporting bone, (vii) bleeding upon being probed, and (viii) swollen lymph nodes around the neck [85, 99]. Peri-implantitis, if left un-treated can progress to severe stage and eventually lead to implant loss.

If peri-implantitis is diagnosed, treatment will depend on the amount of bone loss and the aesthetic impact of the implant in question. The therapeutic approach can range from local debridements around the implant fixture, antibiotics, antisep-tics, and ultrasonic and laser treatments to regenerative procedures using a bone graft [60]. The peri-implant disease treatment strategies have been explored and employed to prevent failure of the implant treatment, including nonsurgical me-chanical debridement, local antimicrobial delivery in periodontitis and peri-implan-titis, and surgical debridement with bone grafting [69, 101].

Implant fracture

One of the most important complications is the fracture of a dental implant that has undergone osseointegration by which the prosthesis is adversely affected by the loss of the supporting tissue [102, 103]. Although the success rate of this procedure is >90%, at times it was also reported with fractures at rare incidence [104]. It was also reported that a total 95.3% cumulative success rate of implants inserted in partial edentulous areas has been shown by previous studies after 3–7 years of loading [105]. It was reported fracture of eight implants out of 4,045 implants scoring up to 0.2% [106, 107]. Rangert et al. [108] found 39 patients with fractured implants after placing 10,000 implants, indicating that most of the fractures were in posterior partially eden-tulous segments, in which the generated occlusal forces can be greater, as opposed to anterior. In general, it has been reported that the incidence of implant fractures as 0.16–1.5% [109] or a fracture is an infrequent complication which affects two out of every 1,000 implants [110–114]. Implant fractures are a frustrating problem not only for patients, but also for clinicians, since they usually involve loss of both the implants and the prostheses [115]. Implant fractures constitutes clear implant failures and in most of the cases, they require implant removal [114].

According to analysis by Sanivarapu et al. [102], the cause of implant fracture may be broadly divided into (i) implant design and manufacturing defects (further including biomaterial and size effect), (ii) nonpassive fir of the prosthetic frame-work (including abutment screw design, abutment or screw loosening, bone loss), (iii) physiologic or biomechanical overload (including implant location, parafunc-tional habits, and prosthetic design), and (iv) others (such as localization of the im-plant, galvanic activity, iatrogenic implant placement, or manipulation). Other possible causes of fracture can also include failure in the production and design of dental implants, bruxism or large occlusal forces, superstructure design, implant localization, implant diameter, metal fatigue, and bone resorption around the im-plant [115]. The risk of fracture also increases with time [116].

It was suggested that, to reduce the rate of implant failure, (i) cantilever struc-tures should be avoided, (ii) as many large diameter implants as possible should be used in the molar and premolar areas, (iii) prostheses are required to decrease the buccolingual width of the crown, lower the inclination of cusp, and place the contact point during occlusion in the center of the implant fixture, and (iv) passive adjustment of the prosthesis through the proper path of insertion should be made [117].

Cement issue

Fixed partial dentures associated with dental implants can be either screw retained or cemented. Initial single tooth implant studies described problems of screw loosening associated with screw retention which was largely based on an implant design (external hex-top) which was not ideal for single tooth application [118]. Internal implant connections fared better in this regard with a significant reduction in screw loosening [119, 120]; however, the demand for prosthetic simplification leads to the more widespread use of cemented restorations [121]. According to Levine et al. [121], advantages of cement retention can include, (i) less demanding surgical placement, (ii) simpler laboratory techniques, (iii) passive fit, (iv) improved aesthetics, (v) improved control of the occlusion, (vi) elimination of screw loosening of the screwed retained crown, and (vii) lower initial cost of fabrication compared to screw retention; whilst the followings are considered as disadvantages of cement retention, including (i) inability in some cases to totally remove excess subgingival cement, (ii) lack of predictable retrievability depending upon the type of cement utilized, (iii) depending upon the design and dimensions of the abutment, resistance and retention can be unpredictable, and (iv) possibility of increased maintenance costs due to loss of retention.

There have been a number of documented cases in the dental literature indicating that excess (subgingival) cement or residual cement for retaining prostheses are strongly associated with peri-implant mucositis which should be a risk factor for increased probing depths crestal bone loss and peri-implantitis [121–128].

Wadhwani [129] pointed out that there are at least four potential causes of peri-implant diseases involving residual cement. (1) Microbiological factor. It was suggested that the disease process may be microbiological in nature based on the time it took for signs and symptoms to develop, which ranged from 4 months to 9.3 years after the cement-retained implant restoration was placed. It was also mentioned that excess dental cement was associated with signs of peri-implant disease in the majority (81%) of the cases. Clinical and endoscopic signs of peri-implant disease were absent in 74% of the test implants after the removal of excess cement [130]. (2) Host response as foreign body reaction. Ramer et al. [131] evaluated soft tissue removed from inflammatory sites adjacent to dental implants and have found foreign body reactions, some of which included giant cell formation. It is possible that in some cases, the tissue destruction is host-induced as a result of material incorporated within the tissues. It was also mentioned that a link has been established between peri-implant disease and excess cement extrusion in cement-retained implant restorations. The histologic findings of two patients with failed implants secondary to residual excess cement are reported here. If excess cement is detected early and adequately removed, resolution can occur in the majority of situations. Simple recommendations are proposed, with the intention of preventing further implant failures from residual excess cement [129]. (3) Allergic response. Nicholson et al. [132] reviewed the biological effects of resin-modified glass-ionomer cements used in clinical dentistry. The cement materials were resin-modified glass-ionomer and 2-hydroxyethyl methacrylate (HEMA). It was found that (i) HEMA is known to be released from these materials and has a variety of damaging biological properties, ranging from pulpal inflammation to allergic contact dermatitis, concluding that resin-modified glass-ionomers cannot be considered biocompatible to nearly the same extent as conventional glass-ionomers. Because subgingival restorative margins are frequently used with cement-retained implant restorations, complete barrier protection for the soft tissues against chemical insult from cements is rarely, if ever possible. It is quite conceivable that this material is leaching out of the cement prior to setting and producing an immune response [129]. Zinc oxide eugenol cements should be preferred to resin cements especially in patients with a history of periodontitis by inhibiting both planktonic and biofilm growth to the greatest degree compared with many other cement types [128, 129]. (4) Alterations in implant surfaces. Many cements developed for the natural dentition contain fluoride, which is added to prevent caries when used with a natural-tooth restoration; however, it should be noted that fluoride is a chemical known to etch titanium when used in conjunction with an acid [129]. Some cements clearly state in their instructions that they are not suitable for use with titanium structures, yet it appears many researchers overlook this [133, 134]. The omission must be considered a critical error. Tarica et al. [135] reported that 17% of US dental schools selected a polycarboxylate as the final cementing media for implant restorations. Durelon™, a popular polycarboxylate, contains fluoride, and a current investigation by the author has shown this material corrodes titanium, resulting in reactive oxidative species that are known to cause inflammation in surrounding tissues. There is valuable information with regard to avoiding and managing the implant complications [136–139].

References

[1]   Complication (medicine), https://en.wikipedia.org/wiki/Complication_(medicine).
[2]   General Surgery – Possible Complications, https://stanfordhealthcare.org/medical-treat
       ments/g/general-surgery/complications.html.
[3]   After Surgery: Discomforts and Complications. https://www.hopkinsmedicine.org/health/
       treatment-tests-and-therapies/after-surgery-discomforts-and-complications.
[4]   Van Humbeeck J. When does a material become a biomaterial? http://www.biotinet.eu/down
       loads/SummerSchool-Barcelona/SummerSchool-Humbeeck.pdf.

[5]   Cordero J, Munuera L, Folgueira MD. Influence of metal implants on infection. An experimental study in rabbits. J Bone Joint Surg Br. 1994, 76-B, 717–20.

[6]   Mombelli A. In vitro models of biological responses to implant microbiological models. Adv Dent Res. 1999, 13, 67–72.

[7]   Omri A, Anderson M, Mugabe C, Suntres Z, Mozafari MR, Azghani A. Artificial implants – new developments and associated problems. In: Mozafari MR, ed. Nanomaterials and Nanosystems for Biomedical Applications. Berlin: Springer-Verlag, Vol. 2007, 53–65.

[8]   Trebše R. Infected total joint arthroplasty. Biomater Artif Jt Replacement. 2012, doi: 10.1007/978-14471-2482-5_3.

[9]   Shtansky DV, Roy M. Surface Engineering for Enhanced Performance Against Wear. Springer, NY. 2013, 77–310.

[10]   Fujishiro T, Nishikawa T, Shibanuma N, Akisue T, Takikawa S, Yamamoto T, Yoshiya S, Kurosaka M. Effect of cyclic mechanical stretch and titanium particles on prostagland in E2 production by human macrophages in vitro. J Biomed Mater Res. 2004, 68A, 531–36.

[11]   Lin H, Bumgardner JD. In vitro biocorrosion of Ti-6Al-4V implant alloy by a mouse macrophage cell line. J Biomed Mater Res. 2004, 68A, 717–24.

[12]   Sommer B, Felix R, Sprecher C, Leunig M, Ganz R, Hofstetter W. Wear particles and surface topographies are modulators of osteoclastogenesis in vitro. J Biomed Mater Res. 2005, 72A, 67–76.

[13]   Yang L, Webster TJ. Biological responses to and toxicity of nanoscale implant materials. In: Eliaz N, ed. Degradation of Implant Materials. NY: Springer-Science, 2012, 481–508.

[14]   Tjellström A. Osseointegrated systems and their applications in the head and neck. Adv Otolaryyngol Head Neck Surg. 1989, 3, 39–70.

[15]   Granström G. Osseointegration in irradiated cancer patients. An analysis with respect to implant failures. J Oral Maxillofac Surg. 2005, 63, 579–85.

[16]   Karjalainen T, Pamilo K, Reito A. Implant failure after motec wrist joint prosthesis due to failure of ball and socket-type articulation-two patients with adverse reaction to metal debris and polyether ether ketone. J Hand Surg Am. 2018, 43, 1044e1–e4.

[17]   Hansen TB, Dremstrup L, Stilling MJ. Patients with metal-on-metal articulation in trapeziometacarpal total joint arthroplasty may have elevated serum chrome and cobalt. J Hand Surg Eur. 2013, 38, 860–65.

[18]   Stratton-Powell AA, Pasko KM, Brockett CL, Tipper JL. The biologic response to polyetheretherketone (PEEK) wear particles in total joint replacement: A systematic review. Clin Orthop Relat Res. 2016, 474, 2394–404.

[19]   Lin DJ, Wong TT, Kazam JK. Shoulder arthroplasty, from indications to complications: What the radiologist needs to know. Musculoskelet Radiol. 2016, doi: https://doi.org/10.1148/rg.2016150055.

[20]   Caniggia M, Fornara P, Franci M, Maniscalco P, Picinotti A. Shoulder arthroplasty. Indications, contraindications and complications. Panminerva Med. 1999, 41, 341–49.

[21]   Buck FM, Jost B, Hodler J. Shoulder arthroplasty. Eur Radiol. 2008, 18, 2937, doi: https://doi.org/10.1007/s00330-008-1093-8.

[22]   Wirth M, Rockwood C. Complications of total shoulder-replacement arthroplasty. J Bone Jt Surg. 1996, 78, 603–16.

[23]   Liu X-W, Zi Y, Xiang L-B, Wang Y. Total hip arthroplasty: A review of advances, advantages and limitations. Int J Clin Exp Med. 2015, 8, 27–36.

[24]   Varacallo M, Luo TD, Johanson NA. Total Hip Arthroplasty Techniques. StatPearls Publishing; 2021, https://www.ncbi.nlm.nih.gov/books/NBK507864/.

[25] Healy WL, Iorio R, Clair AJ, Pellegrini VD, Della Valle CJ, Berend KR. Complications of total hip arthroplasty: Standardized list, definitions, and stratification developed by The Hip Society. Clin Orthop Relat Res. 2016, 474, 357–64.

[26] Devane PA, Horne JG, Ashmore A, Mutimer J, Kim W, Stanley J. Highly cross-linked polyethylene reduces wear and revision rates in total hip arthroplasty: A 10-year double-blinded randomized controlled trial. J Bone Jt Surg Am. 2017, 99, 1703–14.

[27] Deak N, Varacallo M. Hip Precautions. Treasure Island (FL): StatPearls Publishing, May 12, 2021, Hip Precautions, https://pubmed.ncbi.nlm.nih.gov/30725716/.

[28] Opperer M, Lee YY, Nally F, Blanes Perez A, Goudarz-Mehdikhani K, Gonzalez Della Valle A. A critical analysis of radiographic factors in patients who develop dislocation after elective primary total hip arthroplasty. Int Orthop. 2016, 40, 703–08.

[29] Soong M, Rubash HE, Macaulay W. Dislocation after total hip arthroplasty. J Am Acad Orthop Surg. 2004, 12, 314–21.

[30] Shah SM, Walter WL, Tai SM, Lorimer MF, de Steiger RN. Late dislocations after total hip arthroplasty: Is the bearing a factor?. J Arthroplasty. 2017, 32, 2852–56.

[31] Senthi S, Munro JT, Pitto RP. Infection in total hip replacement: Meta-analysis. Int Orthop. 2011, 35, 253–60.

[32] Lindeque B, Hartman Z, Noshchenko A, Cruse M. Infection after primary total hip arthroplasty. Orthopedics. 2014, 37, 257–65.

[33] Gromov K, Bersang A, Nielsen CS, Kallemose T, Husted H, Troelsen A. Risk factors for post-operative periprosthetic fractures following primary total hip arthroplasty with a proximally coated double-tapered cementless femoral component. Bone Joint J. 2017, 99-B, 451–57.

[34] Bradberry SM, Wilkinson JM, Ferner RE. Systemic toxicity related to metal hip prostheses. Clin Toxicol (Phila). 2014, 52, 837–47.

[35] Matharu GS, Berryman F, Brash L, Pynsent PB, Treacy RBC, Dunlop DJ. The effectiveness of blood metal ions in identifying patients with unilateral Birmingham hip resurfacing and corail-pinnacle metal-on-metal hip implants at risk of adverse reactions to metal debris. J Bone Jt Surg Am. 2016, 98, 617–26.

[36] SooHoo NF, Lieberman JR, Ko CY, Zingmond DS. Factors predicting complication rates following total knee replacement. J Bone Jt Surg Am. 2006, 88, 480–85.

[37] Claus A, Asche G, Brade J, Bosing-Schwenkglenks M, Horchler H, Müller-Färber J, Schumm W, Weise K, Scharf H-P. Risk profiling of postoperative complications in 17, 644 total knee replacements. Unfallchirurg. 2006, 109, 5–12.

[38] Healy WL, Della Valle CJ, Iorio R, Berend KR, Cushner FD, Dalury DF, Lonner JH. Complications of total knee arthroplasty: Standardized list and definitions of the knee society. Clin Orthop Relat Res. 2013, 471, 215–20.

[39] Koh WU, Kim HJ, Park HS, Jang MJ, Ro YJ, Song JG. Staggered rather than staged or simultaneous surgical strategy may reduce the risk of acute kidney injury in patients undergoing bilateral TKA. J Bone Joint Surg Am. 2018, 100, 1597–604.

[40] van der Bruggen W, Hirschmann MT, Strobel K, Kampen WU, Kuwert T, Gnanasegaran G, Van den Wyngaert T, Paycha F. SPECT/CT in the postoperative painful knee. Semin Nucl Med. 2018, 48, 439–53.

[41] Lum ZC, Natsuhara KM, Shelton TJ, Giordani M, Pereira GC, Meehan JP. Mortality during total knee periprosthetic joint infection. J Arthroplasty. 2018, 33, 3783–88.

[42] Varacallo M, Luo TD, Johanson NA. Total knee arthroplasty techniques. StatPearls. 2021. https://www.ncbi.nlm.nih.gov/books/NBK499896/.

[43] Turner T. Knee Replacement Complications. https://www.drugwatch.com/knee-replacement/complications/.

[44] Waheed A, Dowd G. Complications after total knee replacement surgery. 2013, https://www.gponline.com/complications-total-knee-replacement-surgery/musculoskeletal-disorders/musculoskeletal-disorders/article/1209527.

[45] Blom AW, Brown J, Taylor AH, Pattison G, Whitehouse S, Bannister GC. Infection after total knee arthroplasty. J Bone Jt Surg Br. 2004, 86, 688–91.

[46] Bozic K. Total Knee Replacement Risks and Complications. 2018, https://www.arthritis-health.com/surgery/knee-surgery/total-knee-replacement-risks-and-complications.

[47] Januel JM, Chen G, Ruffieux C, Quan H, Douketis JD, Crowther MA, Colin C, Ghali WA, Burnand B. Symptomatic in-hospital deep vein thrombosis and pulmonary embolism following hip and knee arthroplasty among patients receiving recommended prophylaxis: A systematic review. JAMA. 2012, 307, 294–303.

[48] Beswick AD, Wylde V, Gooberman-Hill R, Blom A, Dieppe P. What proportion of patients report long-term pain after total hip or knee replacement for osteoarthritis? A systematic review of prospective studies in unselected patients. BMJ Open. 2012, 2, e000435, doi: 10.1136/bmjopen-2011-000435.

[49] Martin GM, Harris I. Complications of total knee arthroplasty. https://www.uptodate.com/contents/complications-of-total-knee-arthroplasty.

[50] What are the Signs of Dental Implant Failure? Smile Savers Dentistry. 2019; https://www.smilesaversdentistry.com/signs-of-dental-implant-failure.

[51] ASA (American Society Anaesthesiologist) Physical Status Classification System. 2020; https://www.asahq.org/standards-and-guidelines/asa-physical-status-classification-system.

[52] Ruggiero SL, Mehrotra B, Rosenberg TJ, Engroff SL. Osteonecrosis of the jaws associated with the use of bisphosphonates: A review of 63 cases. J Oral Maxillofac Surg. 2004, 62, 527–34.

[53] John V. Surgical Complications in Implant Dentistry and Solutions to Common Problems. Lecture Note. Indiana University School of Dentistry. 2021.

[54] Greenstein G, Cavallaro J, Romanos G, Tarnow D. Clinical recommendations for avoiding and managing surgical complications associated with implant dentistry: A review. J Periodontol. 2008, 79, 1317–29.

[55] Liaw K, Delfini RH, Abrahams JJ. Dental implant complications. Seminars in Ultrasound, CT and MRI. 2015, 36, 427–33.

[56] Oral Implantology: https://www.sciencedirect.com/topics/medicine-and-dentistry/oral-implantology.

[57] Annibali S, Ripari M, Ma Monaca G, Tonoli F, Cristalli MP. Local complications in dental implant surgery: Prevention and treatment. Oral Implantol (Rome). 2008, 1, 21–33.

[58] How to Avoid Potential Dental Implant Complications. 2021, https://www.absolutedental.com/blog/how-to-avoid-potential-dental-implant-complications/.

[59] Smith Y. Dental Implant Risks. https://www.news-medical.net/health/Dental-Implant-Risks.aspx.

[60] Ghidrai G. Dental Implants: Risks and Complications. 2019, https://www.infodentis.com/dental-implants/risks-and-complications.php.

[61] Archibald J. What problems can occur after dental implant surgery? Medical News Today. 2020; https://www.medicalnewstoday.com/articles/dental-implants-problems.

[62] Shavit I, Juodzbalys G. Inferior alveolar nerve injuries following implant placement – importance of early diagnosis and treatment: A systematic review. J Oral Maxillofac Implants. 2014, 5, e2, doi: 10.5037/jomr.2014.5402.

[63] Juodzbalys G, Wang HL, Sabalys G. Injury of the inferior alveolar nerve during implant placement: A literature review. J Oral Maxillofac Res. 2011, 2, e1, doi: 10.5037/jomr.2011.2101.

[64] Bartling R, Freeman K, Kraut RA. The incidence of altered sensation of the mental nerve after mandibular implant placement. J Oral Maxillofac Surg. 1999, 57, 1408–12.

[65] Khawaja N, Renton T. Case studies on implant removal influencing the resolution of inferior alveolar nerve injury. Br Dent J. 2009, 206, 365–70.

[66] Renton T, Janjua H, Gallagher JE, Dalgleish M, Yilmaz Z. UK dentists' experience of iatrogenic trigeminal nerve injuries in relation to routine dental procedures: Why, when and how often?. Br Dent J. 2013, 214, 633–42.

[67] Alghamdi HS. Methods to improve osseointegration of dental implants in low quality (type-IV) bone: An overview. J Funct Biomater. 2018, 9, 7, doi: 10.3390/jfb9010007.

[68] Everything You Need to Know about Receding Gums. 2020, https://baytowngentledental.com/everything-you-need-to-know-about-receding-gums/.

[69] Hanif A, Qureshi S, Sheikh Z, Rashid H. Complications in implant dentistry. Eur J Dent. 2017, 11, 135–40.

[70] Tolman DE, Laney WR. Tissue-integrated prosthesis complications. Int J Oral Maxillofac Implants. 1992, 7, 477–84.

[71] Abrahams JJ. Dental CT imaging: A look at the jaw. Radiology. 2001, 219, 334–45.

[72] Rieger MR, Mayberry M, Brose MO. Finite element analysis of six endosseous implants. J Prosthet Dent. 1990, 63, 671–76.

[73] Goodacre CJ, Bernal G, Rungcharassaeng K, Kan JY. Clinical complications with implants and implant prostheses. J Prosthet Dent. 2003, 90, 121–32.

[74] Gupta S, Gupta H, Tandan A. Technical complications of implant-causes and management: A comprehensive review. Natl J Maxillofac Surg. 2015, 6, 3–8.

[75] Sánchez-Pérez A, Moya-Villaescusa MJ, Jornet-Garcia A, Gomez S. Etiology, risk factors and management of implant fractures. Med Oral Patol Oral Cir Bucal. 2010, 15, e504–8.

[76] Jemt T. Failures and complications in 391 consecutively inserted fixed prostheses supported by Brånemark implants in edentulous jaws: A study of treatment from the time of prosthesis placement to the first annual checkup. Int J Oral Maxillofac Implants. 1991, 6, 270–76.

[77] Sahin S, Cehreli MC. The significance of passive framework fit in implant prosthodontics: Current status. Implant Dent. 2001, 10, 85–92.

[78] Everything you need to know about sinusitis. Medical News Today, https://www.medicalnewstoday.com/articles/149941.

[79] Osguthorpe JD. Adult rhinosinusitis: Diagnosis and management. Am Pam Phys. 2001, 63, 69–77.

[80] Fried MP. Rhinitis. 2020, https://www.merckmanuals.com/home/ear,-nose,-and-throat-disorders/nose-and-sinus-disorders/rhinitis.

[81] The Difference Between Sinusitis & Allergies. https://www.sinusandallergywellnesscenter.com/blog/the-difference-between-sinusitis-allergies.

[82] Dobson BC. What's the Difference Between Allergies and a Sinus Infection? https://entorlando.com/whats-the-difference-between-allergies-and-a-sinus-infection/.

[83] Martín LP, Soto MJM, Burgos RS, García MB. Bronchial impaction of an implant screwdriver after accidental aspiration: Report of a case and revision of the literature. Oral Maxillofac Surg. 2010, 14, 43–47.

[84] Abusamaan M, Giannobile WV, Jhawar P, Gunaratnam NT. Swallowed and aspirated dental prostheses and instruments in clinical dental practice: A report of five cases and a proposed management algorithm. J Am Dent Assoc. 2014, 145, 459–63.

[85] French D, Grandin HM, Ofec R. Retrospective cohort study of 4,591 dental implants: Analysis of risk indicators for bone loss and prevalence of peri-implant mucositis and peri-implantitis. J Periodontol. 2019, 90, 691–700.

[86] Pjetursson BE, Asgeirsson AG, Zwahlen M, Sailer I. Improvements in implant dentistry over the last decade: Comparison of survival and complication rates in older and newer publications. Int J Oral Maxillofac Implants. 2014, 29, 308–24.

[87] Lbrektsson T, Donos N. Implant survival and complications. The third EAO consensus conference 2012. Clin Oral Implants Res. 2012, 23, 63–65.

[88] Atieh MA, Alsabeeha NHM, Faggion CM, Duncan WJ. The frequency of peri-implant diseases: A systematic review and meta-analysis. J Periodontol. 2013, 84, 1586–98.

[89] Heitz-Mayfield LJA. Peri-implant diseases: Diagnosis and risk indicators. J Clin Periodontol. 2008, 35, 292–304.

[90] Albouy J-P, Abrahamsson I, Persson LG, Berglundh T. Spontaneous progression of peri-implantitis at different types of implants. An experimental study in dogs. I: Clinical and radiographic observations. Clin Oral Implants Res. 2008, 19, 997–1002.

[91] Ho DSW, Yeung SC, Zee KY, Curtis B, Hell P, Tumuluri V. Clinical and radiographic evaluation of NobelActive(TM) dental implants. Clin Oral Implants Res. 2013, 24, 297–304.

[92] Laurell L, Lundgren D. Marginal bone level changes at dental implants after 5 years in function: A meta-analysis. Clin Implant Dent Relat Res. 2011, 13, 19–28.

[93] Poli PP, Beretta M, Grossi GB, Maiorana C. Risk indicators related to peri-implant disease: An observational retrospective cohort study. J Periodontal Implant Sci. 2016, 46, 266–76.

[94] Rodrigo D, Martin C, Sanz M. Biological complications and peri-implant clinical and radiographic changes at immediately placed dental implants. A prospective 5-year cohort study. Clin Oral Implants Res. 2012, 23, 1224–31.

[95] Stavropoulos A, Cochran D, Obrecht M, Pippenger BE, Dard M. Effect of osteotomy preparation on osseointegration of immediately loaded, tapered dental implants. Adv Dent Res. 2016, 28, 34–41.

[96] Elian N, Bloom M, Dard M, Cho SC, Trushkowsky RD, Tarnow D. Radiological and micro-computed tomography analysis of the bone at dental implants inserted 2, 3 and 4 mm apart in a minipig model with platform switching incorporated. Clin Oral Implants Res. 2014, 25, e22–9.

[97] Linkevicius T, Apse P, Grybauskas S, Puisys A. Influence of thin mucosal tissues on crestal bone stability around implants with platform switching: A 1-year pilot study. J Oral Maxillofac Surg. 2010, 68, 2272–77.

[98] Serino G, Turri A, Lang NP. Probing at implants with peri-implantitis and its relation to clinical peri-implant bone loss. Clin Oral Implants Res. 2013, 24, 91–95.

[99] Implant Complications. https://www.mckinneyperioimplant.com/implant-complications/.

[100] Changi KK, Finkelstein J, Papapanou PN. Peri-implantitis prevalence, incidence rate, and risk factors: A study of electronic health records at a U.S. dental school. Clin Oral Implants Res. 2019, 30, 306–14.

[101] Rashid H, Sheikh Z, Vohra F, Hanif A, Glogauer M. Peri-implantitis: A re-view of the disease and report of a case treated with allograft to achieve bone regeneration. Dent Open J. 2015, 2, 87–97.

[102] Sanivarapu S, Moogla S, Kuntcham RS, Kolaparthy LK. Implant fractures: Rare but not exceptional. J Indian Soc Periodontol. 2016, 20, 6–11.

[103] Goiato MC, Haddad MF, Filho HG, Villa LMR, Dos Santos DM, Pesqueira AA. Dental implant fractures – aetiology, treatment and case report. J Clin Diagn Res. 2014, 8, 300–04.

[104] Albrektsson T, Zarb G, Worthington P, Eriksson AR. The long-term efficacy of currently used dental implants: A review and proposed criteria of success. Int J Oral Maxillofac Implants. 1986, 1, 11–25.

[105] Al Quran FA, Rashan BA, Al-Dwairi ZN. Management of dental implant fractures. A case history. J Oral Implantol. 2009, 35, 210–14.

[106] Balshi TJ, Hernandez RE, Pryszlak MC, Rangert B. A comparative study of one implant versus two replacing a single molar. Int J Oral Maxillofac Implants. 1996, 11, 372–78.

[107] Pommer B, Bucur L, Zauza K, Tepper G, Hof M, Watzek G. Meta-analysis of oral implant fracture incidence and related determinants. J Oral Implants. 2014, 7, doi: https://doi.org/10.1155/2014/263925.

[108] Rangert B, Krogh PH, Langer B, Van Roekel N. Bending overload and implant fracture: A retrospective clinical analysis. Int J Oral Maxillofac Implants. 1995, 10, 326–34.

[109] Berglundh T, Persson L, Klinge B. A systematic review of the incidence of biological and technical complications in implant dentistry reported in prospective longitudinal studies of at least 5 years. J Clin Periodontol. 2002, 29, 197–212.

[110] Eckert SE, Meraw SJ, Cal E, Ow RK. Analysis of incidence and associated factors with fractured implants: A retrospective study. Int J Oral Maxillofac Implants. 2000, 15, 662–67.

[111] Brägger U, Aeschlimann S, Bürgin W, Hämerle CH, Lang NP. Biological and technical complications and failures with fixed partial dentures (FPD) on implants and teeth after four to five years of function. Clin Oral Implants Res. 2001, 12, 26–34.

[112] Berglundh T, Persson L, Klinge B. A systematic review of the incidence of biological and technical complications in Implant Dentistry reported in prospective longitudinal studies of at least 5 years. J Clin Periodontol. 2002, 29, 197–212.

[113] Gargallo-Albiol J, Satorres-Nieto M, Puyuelo-Capablo JL, Sánchez Garcés MA, Pi Urgell J, Gay Escoda C. Endosseous dental implant fractures an analysis of 21 cases. Med Oral Patol Oral Cir Bucal. 2008, 1, E124–8.

[114] Sánchez-Pérez A, Moya-Villaescusa MJ, Jornet-Garcia A, Gomez S. Etiology, risk factors and management of implant fractures. S Med Oral Patol Oral Cir Bucal. 2010, 1, E504–8.

[115] Gealh W, Mazzo V, Barbi F, Camarini ET. Osseointegrated implants fracture: Causes and treatment. J Oral Implantol. 2010, 37, 499–503.

[116] Misch CE, Strong JT, Bidez MW. Scientific rationale for dental implant design. In: Misch CE, ed. Contemporary Implant Dentistry. 3rd ed. New Delhi: Elsevier, 2008, 220–29.

[117] Jin S-Y, Kim S-G, Oh J-S, You J-S, Jeong M-A. Incidence and management of fractured dental implants. Implant Dent. 2017, 27, 802–06.

[118] Avivi-Arber L, Zarb GA. Clinical effectiveness of implant-supported single-tooth replacement: The Toronto study. Int J Oral Maxillofac Implants. 1996, 11, 311–21.

[119] Levine RA, Ganeles J, Jaffin RA, Clem DS III, Beagle JR, Keller JW. Multicenter Retrospective analysis of wide-neck dental implants for single molar replacement. Int J Oral Maxillofac Implants. 2007, 22, 736–42.

[120] Jung RE, Pjetursson BE, Glauser R, Zembic A, Zwahlen M, Lang NP. A systematic review of the 5-year survival and complication rates of implant-supported single crowns. Clin Oral Impl Res. 2008, 19, 119–30.

[121] Levine RA, Present S, Wilson TG. Complications with excess cement & dental implants: Diagnosis, recommendations & treatment of 7 clinical cases. Implant Realities. 2014, 1, 51–59.

[122] Pauletto N, Lahiffe BJ, Walton JN. Complications associated with excess cement around crowns on osseointegrated implants: A clinical report. Int J Oral Maxillofac Implants. 1999, 14, 865–68.

[123] Wadhwani C, Rapoport D, La Rosa S, Hess T, Kretschmar S. Radiographic detection and characteristic patterns of residual excess cement associated with cement-retained implant restorations: A clinical report. J Prosthet Dent. 2012, 107, 151–57.

[124] Shapoff CA, Lahey BJ. Crestal bone loss and the consequences of retained excess cement around dental implants. Compend Contin Educ Dent. 2012, 33, 94–101.

[125] Ramer N, Wadhwani C, Kim A, Hershman D. Histologic findings within peri-implant soft tissue in failed implants secondary to excess cement: Report of two cases and review of literature. N Y State Dent J. 2014, 80, 43–46.

[126] Gapski R, Neugeboren N, Pomeranz AZ, Reissner MW. Endosseous implant failure influenced by crown cementation: A clinical case report. Int J Oral Maxillofac Implants. 2008, 23, 943–46.

[127] Quaranta A, Lim ZW, Tang J, Perrotti V, Leichter J. The impact of residual subgingival cement on biological complications around dental implants: A systematic review. Implant Dent. 2017, 26, 465–74.

[128] Pain and Bleeding Around Dental Implant From Residual Cement-Related Peri-implantitis; https://www.facialart.com/portfolio/dental-implants-stories/dental-implant-complications/pain-and-bleeding-around-dental-implant-from-residual-cement-related-peri-implantitis/.

[129] Wadhwani CPK. Peri-implant disease and cemented implant restorations: A multifactorial etiology. Compendium. 2013, 34, https://www.aegisdentalnetwork.com/cced/special-is sues/2013/10/periimplant-disease-cemented-implant-restorations.

[130] Wilson TG. The positive relationship between excess cement and peri-implant disease: A prospective clinical endoscopic study. J Periodontol. 2009, 80, 1388–92.

[131] Ramer N, Wadhwani C, Kim A, Hershman D. Histologic findings within peri-implant soft tissue in failed implants secondary to excess cement: Report of two cases and review of literature. N Y State Dent J. 2014, 80, 43–46.

[132] Nicholson JW, Czarnecka B. The biocompatability of resin-modified glass-ionomer cements for dentistry. Dent Mater. 2008, 24, 1702–08.

[133] Mansour A, Ercoli C, Graser G, Tallents R, Moss M. Comparative evaluation of casting retention using the ITI solid abutment with six cements. Clin Oral Implants Res. 2002, 13, 343–48.

[134] Mehl C, Harder S, Wolfart M, Kern M, Wolfart S. Retrievability of implant-retained crowns following cementation. Clin Oral Implants Res. 2008, 19, 1304–11.

[135] Tarica DY, Alvarado VM, Truong ST. Survey of United States dental schools on cementation protocols for implant crown restorations. J Prosthet Dent. 2010, 103, 68–79.

[136] Greenstein G, Cavallaro J, Romanos G, Tarnow D. Clinical recommendations for avoiding and managing surgical complications associated with implant dentistry: A review. J Periodontol. 2008, 79, 1317–29.

[137] Saltz AE. Avoiding and Managing Implant Complications. 2019, https://decisionsindentistry.com/article/avoiding-and-managing-implant-complications/.

[138] Common Dental Implant Problems and How to Treat Them, https://www.colgate.com/en-us/oral-health/implants/common-dental-implant-problems-and-treatment.

[139] Hempton T, Papathanasiou E. Managing implant complications. Decis Dent. 2017, 3, 44–47.

Chapter 13
Implantology in AI era

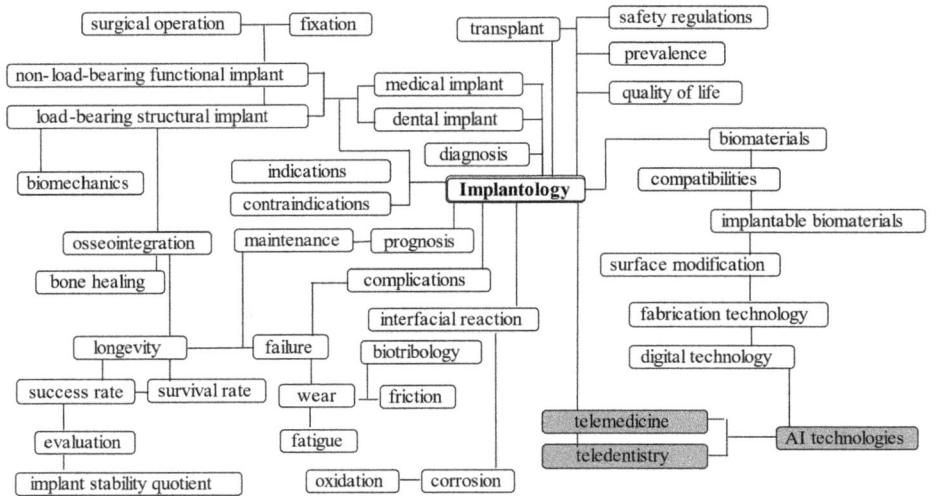

There are AI-powered juicers, AI-enabled wifi routers, AI-enhanced cameras, AI-assisted vacuum cleaners, AI-controlled vital sensors, and so on. It is true to say that no global company today survives without using the Internet, email, or mobile devices. Data, digital transformation, and machine intelligence will simply be table stakes for any organization that wants to stay competitive in an increasingly automated world [1]. In medicine and dentistry fields, there is no exceptions as AI-depended technologies have been developed and employed clinically. Implant treatments in both orthopedic clinics and dental treatments have been gradually relying on the AI technology.

13.1 AI and orthopedic implants

For certain reasons, accurate identification of metallic orthopedic implant systems (design, materials, manufacturer's name, and model number) is important for preoperative planning of revision arthroplasty [2]. Identifying implantable devices would be is normally cost, time-consuming, and inaccurate [3]. By some occasions, such surgical records of implant models are not available.

Revision arthroplasty is performed based on the fracture or failure detected, hence the detection accuracy becomes crucial to revision surgery planning. Ren et al. [4] developed and evaluated a two-stage deep convolutional neural network

https://doi.org/10.1515/9783110740134-014

(DCNN) system that mimics a radiologist's search pattern for detecting two small fractures: triquetral avulsion fractures and Segond fractures, which is an avulsion fracture of the knee that involves the lateral aspect of the tibial plateau and is very frequently (~75% of cases) associated with disruption of the anterior cruciate ligament. It was reported that (i) a two-stage pipeline increases accuracy in the detection of subtle fractures on radiographs compared with the one-stage classifier and generalized well to external test data; and (ii) focusing attention on specific image regions appears to improve detection of subtle findings that may otherwise be missed. A systematic review was performed [5] for searching answers for (i) the proportionality of correctly detected or classified fractures and the area under the receiving operating characteristic (AUC) curve of AI fracture detection and classification models and (ii) the performance of AI in this setting compared with the performance of human examiners. It was concluded that (i) preliminary experience with fracture detection and classification using AI shows promising performance, and (ii) AI may enhance processing and communicating probabilistic tasks in medicine, including orthopedic surgery, (iii) inadequate reference standard assignments to train and test AI is the biggest hurdle before integration into clinical workflow.

Yi et al. [6] developed and evaluated the performance of DCNN to detect and identify specific total shoulder arthroplasty (TSA) models. It was reported that (i) DCNNs can accurately identify the presence of and distinguish between TSA and reverse TSA, and classify five specific TSA models with high accuracy, and (ii) the proof of concept of these DCNNs may set the foundation for an automated arthroplasty atlas for rapid and comprehensive model identification. Employing deep learning models, Urban et al. [7] developed a novel way to automatically classify shoulder implants in X-ray images and compare their performance to alternative classifiers, such as random forests and gradient boosting. It was found that (i) DCNNs outperform other classifiers significantly if and only if out-of-domain data such as ImageNet is used to pretrain the models, and (ii) for a dataset containing X-ray images of shoulder implants from 4 manufacturers with 16 different models, DL is able to identify the correct manufacturer with an accuracy of approximately 80% in 10-fold cross-validation, while other classifiers achieve an accuracy of 56% or less. These results indicate that the approach will be a useful tool in clinical practice, and is likely applicable to other kinds of prostheses.

Using a predictive artificial neural network (ANN) model, Murphy et al. [8] developed a machine-learning algorithm using operative big data to identify an implant from a radiograph and compare algorithms that optimize accuracy in a timely fashion. It was concluded that ANNs offer a useful adjunct to the surgeon in preoperative identification of the prior implant. Kang et al. [9] developed a machine learning–based hip replacement implant recognition program to verify its accuracy. It was reported that (i) femoral stem identification in patients with total hip arthroplasty (THA) was very accurate, and (ii) the technology could be used to collect large-scale implant information. It was also mentioned that the system has the following clinical relevance: (i) the implants needed for revision surgery can be prepared by identifying the old types of implants, (ii) it can be used to diagnose peripheral osteolysis or periprosthetic fracture by further developing the ability to sensitize implant detection, and (iii) an automated implant detection system will help organize imaging data systematically and easily for arthroplasty registry construction. Karnuta et al. [10] trained, validated, and externally tested a deep learning system (DLS) to classify THA and hip resurfacing arthroplasty femoral implants as one of 18 different manufacturer models from 1972 retrospectively collected anteroposterior (AP) plain radiographs from four sites in one quaternary referral health system. It was concluded that a DLS using AP plain radiographs accurately differentiated between 18 hip arthroplasty models from four industry leading manufacturers.

Revisions and reoperations for patients who have undergone total knee arthroplasty (TKA), unicompartmental knee arthroplasty (UKA), and distal femoral replacement (DFR) necessitates accurate identification of implant manufacturer and model. Deep learning permits automated image processing to mitigate the challenges behind expeditious, cost-effective preoperative planning. Hence, Karnuta et al. [11] investigated whether a deep learning algorithm could accurately identify the manufacturer and model of arthroplasty implants of the knee from plain radiographs. It was found that (i) the training and validation datasets were comprised of 682 radiographs across 424 patients and included a wide range of TKAs from the four leading implant manufacturers, and (ii) after 1,000 training epochs by the deep learning algorithm, the model discriminated nine implant models with an AUC of 0.99, accuracy 99%, sensitivity of 95%, and specificity of 99% in the external-testing dataset of 74 radiographs. It was, therefore, concluded that (iii) a deep learning algorithm using plain radiographs differentiated between nine unique knee arthroplasty implants from four manufacturers with near-perfect accuracy, and (iv) the iterative capability of the algorithm allows for scalable expansion of implant discriminations and represents an opportunity to deliver cost-effective care for revision arthroplasty. Preoperative identification of knee arthroplasty is important for planning revision surgery. However, up to 10% of implants are not identified prior to surgery. Accordingly, Yi et al. [12] developed and examined the performance of a DLS for: (1) the automated radiographic identification of the presence or absence of a TKA, (2) classification of TKA vs. UKA, and (3) differentiation between two different primary TKA models. Total 237 AP knee radiographs were collected with equal proportions of native knees, TKA, and UKA, and 274 AP knee radiographs with equal proportions of two TKA models. Data augmentation was used to increase the number of images for DCNN training. A DLS based on DCNNs was trained on these images. Receiver operating characteristic (ROC) curves with AUC were generated. Heat maps were created using class activation mapping (CAM) to identify image features most important for DCNN decision-making. It was concluded that (i) DCNNs can accurately identify presence of TKA and distinguish between specific arthroplasty designs, and (ii) the proof-of-concept could be applied toward identifying other prosthesis models and prosthesis-related complications.

Patel et al. [13] had developed and evaluated a convolutional neural network (CNN) for identifying orthopedic implant models using radiographs. In this retrospective study, 427 knee and 922 hip unilateral AP radiographs, including 12 implant models from 650 patients, were collated from an orthopedic center between March 2015 and November 2019 to develop classification networks. A total of 198 images paired with auto-generated image masks were used to develop a U-Net segmentation network to automatically zero-mask around the implants on the radiographs. It was found that (i) classification networks processing original radiographs, and two-channel conjoined original and zero-masked radiographs, were ensembled to provide a consensus prediction; (ii) accuracies of five senior orthopedic specialists assisted by a reference radiographic gallery were compared with network accuracy using McNemar exact test and when evaluated on a balanced unseen dataset of 180 radiographs, the final network achieved a 98.9% accuracy (178 of 180) and 100% top-three accuracy (180 of 180); and (iii) the network performed superiorly to all five specialists (76.1% [137 of 180] median accuracy and 85.6% [154 of 180] best accuracy), with robustness to scan quality variation and difficult to distinguish implants, suggesting that (iv) a neural network model was developed that outperformed senior orthopedic specialists at identifying implant models on radiographs; real-world application can now be readily realized through training on a broader range of implants and joints, supported by all code and radiographs being made freely available [13]. Ren et al. [14] conducted a systematic review to summarize the scope, methodology, and performance of artificial intelligence (AI) algorithms in classifying orthopedic implant models, based on studies published up to March 2021, using search key words including AI, orthopedic implant and arthroplasty. It was reported that (i) there was a large degree of variation in methodology and reporting quality, and (ii) AI algorithms have demonstrated strong performance in classifying orthopedic implant models from radiographs; suggesting that further research is needed to compare AI alone and as an adjunct with human experts in implant identification.

13.2 AI and dental implants

AI applications are growing in dental implant procedures. In particular, AI technology has been applied for implant-type recognition, implant success prediction using patient risk factors and ontology criteria, and implant design optimization combining finite element analysis calculations to assess and identify risky portions with high stress concentration. According to systematic review [15] on 17 articles (seven investigations analyzed AI models for implant-type recognition, seven studies included AI prediction models for implant success forecast, and three studies evaluated AI models for optimization of implant designs), it was reported that (i) the developed

AI models recognized implant type by using periapical and panoramic images and obtained an overall accuracy outcome ranging from 93.8% to 98%, (ii) the AI models predicted osteointegration success or implant success by using different input data varied among the studies, ranging from 62.4% to 80.5%, and (iii) the developed AI models optimized implant designs to assess the applicability of AI models to improve the design of dental implants, which includes minimizing the stress at the implant–bone interface by 36.6% compared with the finite element model; optimizing the implant design porosity, length, and diameter to improve the finite element calculations; or accurately determining the elastic modulus of the implant–bone interface.

The reason for developing AI application for implant system (type, design, etc.) for dental implant is not necessarily same as that for the previous orthopedic implants, since revision surgery on dental implants have not been conducted, rather implant identification is principally performed to search an appropriate and the most suitable implant system for a patient. Again there are various factors influencing the implant system selection, including patient health conditions, physical parameters such as weight and gender, occlusion habits, bone quality, and quantity at the placement candidate site(s) and others. Takahashi et al. [16] identified dental implant systems using a deep learning method. A dataset of 1,282 panoramic radiograph images with implants were used for deep learning. An object detection algorithm was utilized to identify the six implant systems by three manufactures. To implement the algorithm, TensorFlow and Keras deep learning libraries were used. After training was complete, the true positive (TP) ratio and average precision (AP) of each implant system as well as the mean AP (mAP) and mean intersection over union (mIoU) were calculated to evaluate the performance of the model. It was obtained that (i) the number of each implant system varied from 240 to 1,919, (ii) the TP ratio and AP of each implant system varied from 0.50 to 0.82 and from 0.51 to 0.85, respectively, and (iii) the mAP and mIoU of this model were 0.71 and 0.72, respectively. Based on these findings, it was concluded that (iv) implants can be identified from panoramic radiographic images using deep learning–based object detection, and (v) this identification system could help dentists as well as patients suffering from implant problems [16]. Saïd et al. [17] developed a deep CNN that would identify the brand and model of a dental implant from a radiograph. A data augmentation procedure provided a total of 1,206 dental implant radiographic images of three different brands for six models (Nobel Biocare NobelActive and Brånemark System, Straumann Bone Level and Tissue Level, and Zimmer Biomet Dental Tapered Screw-Vent and SwissPlus). It was concluded that the deep CNN model had a very good performance in identifying a dental implant from a radiograph, although a huge and varied database of radiographs would have to be built up to be able to identify any dental implant. Using panoramic X-ray images, Sukegawa et al. [18] classified and evaluated the accuracy of different dental implant brands via deep CNNs with transfer-learning strategies. For objective labeling, 8,859 implant images of 11 implant systems were used from digital panoramic radiographs obtained from patients who underwent dental implant treatment between 2005 and 2019. Five deep CNN models (specifically, a basic CNN with three convolutional layers, VGG16 and VGG19 transfer-learning models, and finely tuned VGG16 and VGG19) were evaluated for implant classification. It was reported that (i) among the five models, the finely tuned VGG16 model exhibited the highest implant classification performance and (ii) the finely tuned VGG19 was second best, followed by the normal transfer-learning VGG16, suggesting that (iii) the finely tuned VGG16 and VGG19 CNNs could accurately classify dental implant systems from 11 types of panoramic X-ray images. Lee at al. [19] evaluated the efficacy of deep CNN algorithm for the identification and classification of dental implant systems. A total of 5,390 panoramic and 5,380 periapical radiographic images from three types of dental implant systems, with similar shape and internal conical connection, were randomly divided into training and validation dataset (80%) and a test dataset (20%). It was reported that the deep CNN architecture (AUC = 0.971, 95% confidence interval 0.963–0.978) and board-certified periodontist (AUC = 0.925, 95% confidence interval 0.913–0.935) showed reliable classification accuracies, indicating that deep CNN architecture is useful for the identification and classification of dental implant systems using panoramic and periapical radiographic images.

Lee et al. [20] evaluated the reliability and validity of three different DCNN architectures (VGGNet-19, GoogLeNet Inception-v3, and automated DCNN) for the detection and classification of fractured dental implants using panoramic and periapical radiographic images. A total of 21,398 dental implants were reviewed and 251 intact and 194 fractured dental implant radiographic images were identified. It was reported that (i) all three DCNN architectures achieved a fractured dental implant detection and classification accuracy of over 0.80 AUC; in particular, automated DCNN architecture using periapical images showed the highest and most reliable detection (AUC = 0.984, 95% CI = 0.900–1.000) and classification (AUC = 0.869, 95% CI = 0.778–0.929) accuracy performance compared to fine-tuned and pretrained VGGNet-19 and GoogLeNet Inception-v3 architectures, and (ii) the three DCNN architectures showed acceptable accuracy in the detection and classification of fractured dental implants, with the best accuracy performance achieved by the automated DCNN architecture using only periapical images.

For accurate implant treatment planning on mandibular jaws, the exact location of the mandibular canal is crucial, since both sides of the mandibular jaw contain the alveolar nerve. Damage to the alveolar nerve can cause the patient to suffer a tickling or burning skin sensation known as paraesthesia. Additional lesions to the inferior alveolar nerve may cause pain and alterations in the patient's sensitivity [21]. Jaskari et al. [22] employed a DLS for automatic localization of the mandibular canals by applying a CNN segmentation on clinically diverse dataset of 637 cone beam CT (CBCT) volumes, with mandibular canals being coarsely annotated by radiologists, using a dataset of 15 volumes with accurate voxel-level mandibular canal annotations for model evaluation. It was shown that the deep learning model, trained on the coarsely annotated volumes, localized mandibular canals of the voxel-level annotated set, highly accurately with the mean curve distance and average symmetric surface distance being 0.56 mm and 0.45 mm, respectively, suggesting that these unparalleled accurate results highlight that deep learning integrated into dental implantology workflow could significantly reduce manual labor in mandibular canal annotations. Lerner et al. [23] used AI to fabricate implant-supported monolithic zirconia crowns (MZCs) cemented on customized hybrid abutments. A total of 90 patients (35 males, 55 females; mean age 53.3 ± 13.7 years) restored with 106 implant-supported MZCs were examined. The follow-up varied from six months to three years. It was shown that the deep learning model when trained on the coarsely annotated volumes could locate mandibular canals of the voxel-level annotated set, while when accurately trained, DL model localized the mean curve distance and average symmetric surface distance being 0.56 mm and 0.45 mm, respectively. From these findings, it was suggested that these unparalleled accurate results highlight that deep learning integrated into dental implantology workflow could significantly reduce manual labor in mandibular canal annotations.

13.3 Telemedicine

Telemedicine allows health-care professionals to evaluate, diagnose, and treat patients from a distance using telecommunications technology and allows long-distance patient and clinician contact, care, advice, reminders, education, intervention, monitoring, and remote admissions [24, 25]. Telehealth could include two clinicians discussing a case over video conference; a robotic surgery occurring through remote access; physical therapy done via digital monitoring instruments, live feed and application combinations; tests being forwarded between facilities for interpretation by a higher specialist; home monitoring through continuous sending of patient health data; client to practitioner online conference; or even videophone interpretation during a consult [26, 27].

There are commonly recognized benefits of telemedicine [24, 28–30]. They (1) improve access to care, (2) improve cost effectiveness, (3) improve health care quality, and increased patient demand. Using telemedicine as an alternative to in-person visits has a host of benefits for patients and providers alike. For patients' side, they enjoy (i) less time away from work, (ii) no travel expenses or time, (iii) less interference with child or elder care responsibilities, (iv) privacy, and (v) no exposure to other potentially contagious patients. On the other hand, for health-care providers can also enjoy (i) increased revenue, (ii) improved office efficiency, (iii) an answer to the competitive threat of retail health clinics and online-only providers, (iv) better patient follow through and improved health outcomes, (v) fewer missed appointments and cancellations, and (vi) private payer reimbursement.

Although the terms telemedicine and telehealth are often used interchangeably, there is a distinction between the two [31, 32]. Telemedicine is the practice of medicine using technology to deliver care at a distance. A physician in one location uses a telecommunications infrastructure to deliver care to a patient at a distant site. On the other hand, telehealth refers broadly to electronic and telecommunications technologies and services used to provide care and services at-a-distance. Accordingly, telehealth is different from telemedicine in that it refers to a broader scope of remote health care services than telemedicine. Telemedicine refers specifically to remote clinical services, while telehealth can refer to remote nonclinical services. Typical telehealth applications should include (1) live (synchronous) videoconferencing: a two-way audiovisual link between a patient and a care provider, (2) store-and-forward (asynchronous) videoconferencing: transmission of a recorded health history to a health practitioner, usually a specialist, (3) remote patient monitoring (RPM): the use of connected electronic tools to record personal health and medical data in one location for review by a provider in another location, usually at a different time, and (4) mobile health (mHealth): health care and public health information provided through mobile devices. The information may include general educational information, targeted texts, and notifications about disease outbreaks [31, 32].

Telemedicine possesses a versatile procedure and it has endless possibilities of development; an application can mean accommodating more patients or discovering the best practice for a medical procedure, which will impact many lives. AI is one of the most powerful forces to enhance following tasks: (1) improvement in diagnosis, (2) convenience in patient-monitoring, and (3) aid in eldercare [33, 34]. AI in healthcare is no longer restricted to research labs alone. It has also improved many telemedicine aspects revolving around broadband technology and electronic data to assist and coordinate remote health care services. AI takes over the whole chain of clinical practice and patient-focused care by providing models of care and sustenance. AI can be effectively used in the following ways [35], including (1) analyzing medical records and other data, (2) automation of manual, repetitive tasks, (3) electronic consultation, (4) medicine management, and (5) AI-led telemedicine, which can revolutionize telehealth applications.

13.4 Teledentistry

Teledentistry (which is the dental form of telemedicine) can be defined as the remote provision of dental care, advice, or treatment through the medium of information technology, rather than through direct personal contact with any patient(s) involved, by the use of telecommunications and allied technologies [36–38]. Within dental practice, teledentistry is used extensively in disciplines like preventive dentistry, orthodontics, endodontics, oral surgery, periodontal conditions, detection of early dental caries, patient education, oral medicine, and diagnosis. Some of the key modes and methods used in teledentistry are electronic health records, electronic referral systems, digitizing images, teleconsultations, and telediagnosis. All the applications used in teledentistry aim to bring about efficiency, provide access to underserved population, improve quality of care, and reduce oral disease burden [36].

Teledentistry facilitates patient self-management and caregiver support for patients and may take several forms [37–40]. (1) Synchronous teledentistry (through teleconference) is where delivery of patient care and education is live with two-way interaction between a person or persons (e.g., patient; dental, medical, or health caregiver) at one physical location, and an overseeing supervising or consulting dentist or dental provider at another location. The communication is real-time and continuous between all participants who are working together as a group. Use of audiovisual telecommunications technology means that all involved persons are able to see what is happening and talk about it in a natural manner. (2) Asynchronous teledentistry (through chat, images, store, and forward) is different from synchronous teledentistry, in that there is no real-time, live, continuous interaction with anyone who is not at the same physical location as the patient. This is also known as "store-and-forward." Asynchronous teledentistry involves transmission of recorded health information (e.g., radiographs, photographs, video, digital impressions, and photomicrographs of patients) through a secure electronic communications system to another practitioner for use at a later time. (3) RPM is where personal health and medical information is collected from an individual in one location and then transmitted electronically to a provider in a different location for use in care. This could be used in a nursing home setting or in an educational program. (4) Mobile health involves health care and public health practice and education supported by mobile communication devices such as cell phones, tablet computers, or personal digital assistants. This could include apps that monitor patient brushing or other home care.

Moore [41] introduced various rules and regulations by the American Dental Association (ADA) governing the teledentistry activities. In 2015, the ADA adopted a policy on teledentistry that encompassed patients' rights, quality of care, licensure requirements, insurance reimbursement, the supervision of allied personnel, and technical considerations. In 2018, the ADA published its Code on Dental Procedures and Nomenclature codes for the use of synchronous (D9995) and asynchronous (D9996) teledentistry for submission to insurance companies for reimbursement for services provided. By January 2020, the coronavirus pandemic rendered most dental practices unable to provide routine services in the office. As a result, in May 2020, the ADA published an interim guidance for COVID-19 coding and billing for virtual dental visits. This guidance defined guidelines for billing and coding, as well as Health Insurance Portability and Accountability Act (HIPAA) compliance and informed consent; it also offered tips for conducting virtual evaluations. Remote dental consultations and/or evaluations can be performed via Zoom, FaceTime, Skype, Google Hangouts, Microsoft Teams, or through secure, commercially available applications that can be used with cell phones, tablets, and PDAs. However, public-facing technologies (such as Facebook Live, Twitch, and TikTok) cannot be used because they are not permitted by the federal government for this purpose.

As to benefits of the teledentistry, it can provide easier, cheaper, and less intimidating access to a licensed dentist (including specialists) and efficient means for initial case evaluation and management. At the same time, there are possible risks of causing a breach of privacy of personal medical information [42]. Patient surveys conducted by the American Teledentistry Association in 2013 found that 70% of patients were comfortable communicating with their providers via text, email, or video in lieu of meeting them in person [43, 44]. As of May 2020, 22 states had published statements promoting the use of teledentistry to expand access and maintain continuity of oral health services, with reimbursement through Medicaid programs [45, 46]. Teledentistry can benefit a broad range of populations, including Medicare and Medicaid beneficiaries, the uninsured, underserved and rural populations, people with urgent dental care needs, and people who fear going to the dentist. It also bridges critical gaps in access to care by expanding limited capacity within a practice, enabling emergency dental consults or preventive hygiene education without barriers to travel, and improving patient referrals thanks to provider-to-provider consultations. When it is more broadly adopted, teledentistry will have wider implications for both patient outcomes and care-delivery capacity [47]. As more and more states enact laws regarding teledentistry, changes will occur [41]. For example, as more insurance companies reimburse providers for teledentistry care, an increase in licensure portability across states for all providers may be necessary. The White House issued an executive order on improving rural health and telehealth access, indicating that the expansion of telehealth services is likely to be a more permanent feature of the health care delivery system [47]. In addition, the various modalities of teledentistry have applicability in dental offices, hospitals, assisted living facilities, public health settings, and educational institutions. Ultimately, teledentistry will become more common as its use is embraced worldwide [41]. The teledentistry will continue to provide an important bridge between providers, patients, and communities for many types of care and needs – from hands-on emergency care to preventive care and minimally invasive early interventions. Many providers are already embracing teledentistry in new and creative ways, from creating virtual classes on family oral hygiene for children to routinely doing teledentistry consults with new patients to understand their immediate needs [47].

AI and teledentistry are two innovations that can greatly help oral care providers in ameliorating outcomes and improving access to care as a whole [48]. There are several keys, including (1) giving individuals a first look into their oral health status, (2) mitigating the costs of care for providers and patients alike, and (3) helping make the shift to value-based care a seamless one [48]. All in all, addressing barriers of access to care primarily lies within two realms: patients' ability to access oral health care associated with socioeconomic factors, and the persisting shortage of dental care providers across the nation [48, 49]. AI and teledentistry definitely hold the potential to help the oral health fraternity address both these challenges head on [48].

References

[1]   Oshida Y. Artificial Intelligence for Medicine. De Gruyter Pub., 2021.
[2]   Wilson NA, Jehn M, York S, Davis CM. Revision total hip and knee arthroplasty implant identification: Implications for use of unique device identification 2012 AAHKS member survey results. J Arthroplast. 2014, 29, 251–55.
[3]   Wilson N, Broatch J, Jehn M, Davis C. National projections of time, cost and failure in implantable device identification: Consideration of unique device identification use. Healthcare (Amsterdam, Netherlands). 2015, 3, 196–201.
[4]   Ren M, Yi PH. Deep learning detection of subtle fractures using staged algorithms to mimic radiologist search pattern. Skelet Radiol. 2021, doi: https://doi.org/10.1007/s00256-021-03739-2.
[5]   Langerhuizen DWG, Janssen SJ, Mallee WH, Van Den Bekerom MPJ, Ring D, Kerkhoffs GMMJ, Jaarsma RL, Doomberg JN. What are the applications and limitations of artificial intelligence for fracture detection and classification in orthopaedic trauma imaging? A systematic review. Clin Orthop Relat Res. 2019, 477, 2482–91.
[6]   Yi PH, Kim TK, Wei J, Li X, Hager GD, Sair HI, Fritz J. Automated detection and classification of shoulder arthroplasty models using deep learning. Skelet Radiol. 2020, 49, 1623–32.
[7]   Urban G, Porhemmat S, Stark M, Feeley B, Okada K, Baldi P. Classifying shoulder implants in X-ray images using deep learning. Comput Struct Biotechnol J. 2020, 18, 967–72.
[8]   Murphy M, Killen C, Burnham R, Sarvari F, Wu K, Brown N. Artificial intelligence accurately identifies total hip arthroplasty implants: A tool for revision surgery. HIP Int. 2021, doi: https://doi.org/10.1177/1120700020987526.
[9]   Kang Y, Yoo J, Cha Y, Park CH, Kim J. Machine learning–based identification of hip arthroplasty designs. J Orthop Transl. 2020, 21, 13–17.
[10] Karnuta JM, with 11 co-authors. Artificial intelligence to identify arthroplasty implants from radiographs of the hip. J Arthroplast. 2020, doi: https://doi.org/10.1016/j.arth.2020.11.015.
[11] Karnuta JM, with 11 co-authors. Artificial intelligence to identify arthroplasty implants from radiographs of the knee. J Arthroplast. 2020, 36, 935–40.
[12] Yi PH, Wei J, Kim TK, Sair HI, Hui FK, Hager GD, Fritz J, Oni JK. Automated detection & classification of knee arthroplasty using deep learning. Knee. 2020, 27, 535–42.
[13] Patel R, Thong EHE, Batta V, Bharath AA, Francis D, Howard J. Automated identification of orthopedic implants on radiographs using deep learning. Home Radiol: Artif Intel. 2021, 3, doi: https://doi.org/10.1148/ryai.2021200183.
[14] Ren M, Yi PH. Artificial intelligence in orthopedic implant model classification: A systematic review. Skeletal Radiol. 2021, doi: https://doi.org/10.1007/s00256-021-03884-8.
[15] Revilla-León M, Gómez-Polo M, Vyas S, Barmak BA, Galluci GO, Att W, Krishnamurthy VR. Artificial intelligence applications in implant dentistry: A systematic review. J Prosthetic Dent. 2021, doi: https://doi.org/10.1016/j.prosdent.2021.05.008.
[16] Takahashi T, Nozaki K, Gonda T, Mameno T, Wada M, Ikebe K. Identification of dental implants using deep learning- pilot study. Int J Implant Dent. 2020, 6, 53, doi: https://doi.org/10.1186/s40729-020-00250-6.

[17] Saïd MH, Le Roux M-K, Catherine J-H, Lan R. Development of an artificial intelligence model to identify a dental implant from a radiograph. Int J Oral Maxillofac Implants. 2020, 36, 1077–82.

[18] Sukegawa S, Yoshii K, Hara T, Yamashita K, Nakano K, Yamamoto N, Nagatsuka H, Furuki Y. Deep neural networks for dental implant system classification. Biomolecules. 2020, 10, 984, doi: 10.3390/biom10070984.

[19] Lee J-H, Jeong S-N. Efficacy of deep convolutional neural network algorithm for the identification and classification of dental implant systems, using panoramic and periapical radiographs: A pilot study. Medicine (Baltimore). 2020, 99, e20787, doi: 10.1097/MD.0000000000020787.

[20] Lee D-W, Kim S-Y, Jeong S-N, Lee J-H. Artificial intelligence in fractured dental implant detection and classification: Evaluation using dataset from two dental hospitals. Diagnostics. 2021, 11, 233, doi: https://doi.org/10.3390/diagnostics11020233.

[21] Using AI to increase the efficiency of dental implant operations. 2020, https://www.innovationnewsnetwork.com/using-ai-to-increase-the-efficiency-of-dental-implant-operations/4887/.

[22] Jaskari J, Sahlsten J, Järnstedt J, Mehtonen H, Karhu K, Sundqvist O, Hietanen A, Varjonen V, Mattila V, Kaski K. Deep learning method for mandibular canal segmentation in dental cone beam computed tomography volumes. Sci Rep. 2020, 10, 5842, doi: 10.1038/s41598-020-62321-3.

[23] Lerner H, Moujhyi J, Admakin O, Mangano F. Artificial intelligence in fixed implant prosthodontics: A retrospective study of 106 implant-supported monolithic zirconia crowns inserted in the posterior jaws of 90 patients. BMC Oral Health. 2020, 20, 80, doi: https://doi.org/10.1186/s12903-020-1062-4.

[24] What is telemedicine? https://chironhealth.com/telemedicine/what-is-telemedicine/.

[25] Telehealth. https://en.wikipedia.org/wiki/Telehealth

[26] Shaw DK. Overview of telehealth and its application to cardiopulmonary physical therapy. Cardiopulm Phys Ther J. 2009, 20, 13–18.

[27] Miller EA. Solving the disjuncture between research and practice: Telehealth trends in the twenty-first century. Health Policy (New York). 2007, 82, 133–41.

[28] Thomas L. What is Telemedicine? 2021. https://www.news-medical.net/health/What-is-Telemedicine.aspx.

[29] Telemedicine Definition. https://evisit.com/resources/telemedicine-definition/.

[30] Hasselfeld BW. Benefits of Telemedicine. https://www.hopkinsmedicine.org/health/treatment-tests-and-therapies/benefits-of-telemedicine.

[31] What's the difference between telemedicine and telehealth? https://www.aafp.org/news/media-center/kits/telemedicine-and-telehealth.html.

[32] Telemedicine and Telehealth. 2020. https://www.healthit.gov/topic/health-it-health-care-settings/telemedicine-and-telehealth.

[33] Dyrda L. The role of AI in telemedicine. 2020, https://www.beckershospitalreview.com/telehealth/the-role-of-ai-in-telemedicine.html.

[34] David C. How AI is Transforming Telehealth. 2020, https://ai4.io/how-ai-is-transforming-telehealth.

[35] Artificial Intelligence and its Impact on Telemedicine. 2020, https://www.securemedical.com/telemedicine/artificial-intelligence-and-its-impact-on-telemedicine/.

[36] Khan SA, Omar H. Teledentistry in practice: Literature review. Telemed J E Health. 2013, 19, 565–67.

[37] Teledentistry. https://dentaquest.com/oral-health-resources/teledentistry/.

[38] Teledentistry. https://en.wikipedia.org/wiki/Teledentistry

[39]   ADA Policy on Teledentistry. 2020. https://www.ada.org/en/about-the-ada/ada-positions-pol icies-and-statements/statement-on-teledentistry.

[40]   Teledentistry: What it is and how it works.2020, https://www.guardianlife.com/dental-insur ance/teledentistry-what-it-is-how-it-works.

[41]   Moore TA. The Increasing Utilization of Teledentistry. 2020, https://dimensionsofdentalhy giene.com/article/increasing-utilization-teledenstiry/.

[42]   Important Information and Notice About Teledentistry. https://www.westerndental.com/en- us/teledentistry.

[43]   Lannon J. Lights, Camera, Diagnosis: Teledentistry Comes of Age. 2020, https://www.dental town.com/magazine/article/7668/lights-camera-diagnosis-teledentistry-comes-of-age.

[44]   American Teledentistry Association. Facts About Teledentistry. 2020, https://americantele dentistry.org/facts-about-teledentistry/.

[45]   Manatt P, Phillips LLP. Executive Summary: Tracking Telehealth Changes State-by-State in Response to COVID-19. 2020, https://jdsupra.com/legalnews/executive-summary-tracking- telehealth–82023/.

[46]   Oputa J, Kearly A, Leach D. Ensuring Safe Access to Oral Healthcare During COVID-19. 2020, https://astho.org/StatePublicHealth/Ensuring-Safe-Access-to-Oral-Healthcare-During-COVID- 19/06-24-20/?terms=teledentistry.

[47]   The White House. Executive Order on Improving Rural Health and Telehealth Access. 2020, https://whitehouse.gov/presidential-actions/executive-order-improving-rural-health-tele health-access/.

[48]   Kunstadter M. Artificial Intelligence and Teledentistry: Breaking Barriers to Dental Care.2021. https://thedentalgeek.com/2021/09/artificial-intelligence-and-teledentistry-breaking-bar riers-to-dental-care/.

[49]   Lim JD. AI and teledentistry. 2021, https://www.panaynews.net/ai-and-teledentistry/.

Epilogue

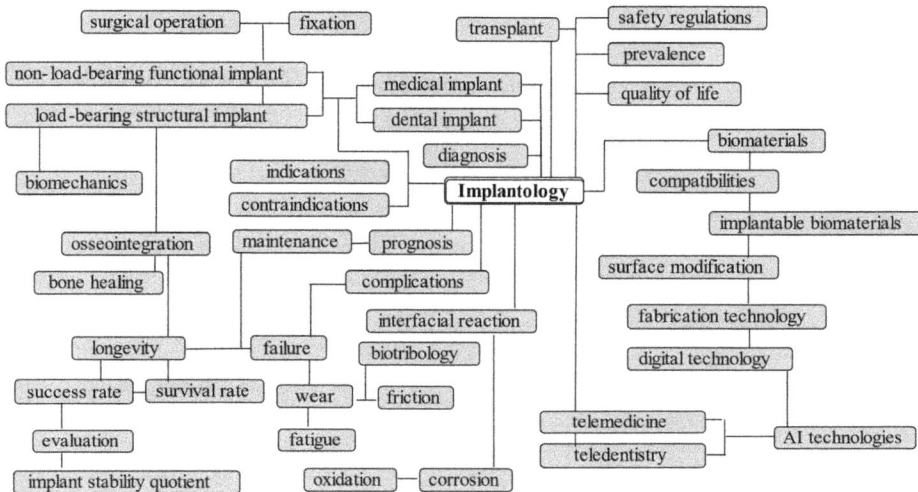

We have been covering each one of key words, shown in the above figure, which are relevant to both orthopedic implants and dental implants. Albrektsson et al. [1] pointed out that success rate and expected longevity are relied on various factors controlling the interfacial reaction to the placed implant, including (1) material's biocompatibility, (2) implant design, (3) surface of the implant, (4) the status of the host bed, (5) the surgical technique used at insertion, and (6) the loading conditions applied afterward. During the surface characterization of successfully placed implants, Oshida [2] indicated that there are three major requirements for the placed implant to exhibit successful osseointegration: biological compatibility, biomechanical compatibility, and morphological compatibility. For each of these compatibility requirements, various types of biomaterials have been developed and clinically applied, and numerous surface modifications have been advanced and proposed.

Recently developed implant surface zones are recognized with several uniqueness: (1) surface zone of successful implants possesses multiscale configurations of macroscale roughness, microscale unevenness, and nanoscale features [3–5]. To produce these surface characteristics, several technologies have been adopted and effectively employed, and can be categorized into either one of additive manufacturing or subtractive manufacturing. Additive manufacturing is often referred to as 3D printing or bioprinting. The difference between usage of the terms comes down to scale, where additive manufacturing is typically applied to industrial quantities. Additive manufacturing uses several different technologies to build parts into layers from the bottom-up. These include jet printing, deposited materials, and laser sintering. Opposite of additive manufacturing is subtractive manufacturing.

https://doi.org/10.1515/9783110740134-015

Subtractive manufacturing removes material layer by layer using lathes, mills, routers, and grinders. These tools are computer numerically controlled, and the data driving them is created by CAD/CAM systems used for mechanical design [6]. Table 1 compares these two technologies [7–11].

Table 1: Comparison between additive and subtractive manufacturing.

	Additive manufacturing	Subtractive manufacturing
Characteristics	– Adds layer of material to create an object – Layering often leaves slightly stepped or rough surface which needs to be finished post-processed by sanding or blowing – Intricate or hollow objects can be easily built up in layers – Best suited smaller items or parts, especially in plastics – Depending on the size of the object, 3D printing is a slow process	– Removes material from the object – A variety of surface finish can be machined, including smooth, stepped, mottled, etc. – Milling undercuts and intricate shapes can be difficult – Best suited for manufacturing voluminous items and parts, especially in metals – Relatively fast process
Process	– Includes 3D (bio)printing, direct digital manufacturing, rapid prototyping or additive and layered fabrication – Uses computers and specialist 3D printing equipment to create products or prototypes	– Process is either by manual removal, traditional machining, or CNC machining – Uses computers or robotics to assist standard machining processes, e.g., turning, drilling, or milling
Equipment cost	– Professional desktop printers start at $3,500 for plastics – Large-scale industrial machines for metals start from ~$400,000	– Small CNC machines for workshops start around $2,000 – More advanced workshop tools go well beyond that depending on the number of axes, features, part size, and tooling needed for specific materials
Training	– Desktop printers are practically plug and play, requiring minor training on build setup, maintenance, machine operation, and finishing – Industrial additive manufacturing systems require dedicated staff and extensive training	– Small CNC machines require moderate training for software, job setup, maintenance, machine operation, and finishing – Larger, industrial subtractive systems require dedicated staff and extensive training
Overall	– 3D printing is a fairly cheap process	– More expensive than additive method

Hybrid systems combine the versatility of additive manufacturing with some of the advantages of subtractive methods [8]. Specialist machines can perform both operations, meaning that complex parts can be made more easily. Hybrid manufacturing is particularly good for repairing worn or broken parts, as the material can be added in layers, then finished with milling tools.

Besides these additive and subtractive fabrications, there are still traditional methods to create macroscale, microscale, and nanoscale surface roughness by both additive manner and subtractive manner. Surface deposition of HA powder and fine metallic particles can create semiadditive surface features. Surface oxidation and/or internal oxidation under controlled partial pressure of atmospheric oxygen can create microscale or nanoscale irregularities. On the other hand, a chemical or electrochemical etching is a typical material removal process to create a concave surface. Shotblasting or shot-peening with fine powder can generate also surface with numerous indentations along with the beneficial negative residual stress. If extreme surface layer is manipulated to have nanoscale topology and gradually less porosity and higher strength toward a core zone of the implant body, the so-called gradually functioned structure can be established. As a result, an ideal integrated implant system can be fabricated [12]. Nanotechnology is very useful to enhance the surface functionalization. Surface layer functionalized with carbon nanomaterials in dental and orthopedic implants has emerged as a novel strategy for reinforcement and as a bioactive cue due to its potential for osseointegration, since carbon nanomaterials (including carbon nanofibers, nanocrystalline diamonds, fullerenes, carbon nanotubes, carbon nanodots, and graphene and its derivatives) have gained the attention of bioengineers and medical researchers as they possess extraordinary physicochemical, mechanical, thermal, and electrical properties [13].

Customized implant mimics anatomically to the patient's defected area. These implants can include orthopedic and dental implants. Munsch [14] reported a fabrication of prosthetics or implants by the laser additive manufacturing technology. Okazaki [15] fabricated custom-made orthopedic implants (osteosynthetic materials and prosthetic joints). Chen et al. [16, 17] designed and manufactured custom-made dental implants. Custom-made root-analogue zirconia implants were also fabricated [18].

In the advancing artificial intelligence era, implants (in particular, dental implants) can be personalized and dental treatment per se can be more digitized, as we have discussed in the previous chapter. Traditional implant treatment is based on and relied on the bone-driven implant system and concept, which is followed by the restoration-driven implant treatment. Now the paradigm of these traditional treatment concepts will be shifted to a face (appearance)-driven implant treatment, which is not necessarily limited to anterior zones, but posterior zone should also be responsible for general appearance after the completion of dental implant treatment.

References

[1] Albrektsson T, Albrektsson B. Osseointegration of bone implants: A review of an alternative mode of fixation. Acta Orthop Scand. 1987, 58, 567–77.

[2] Oshida Y. Bioscience and Bioengineering of Titanium Materials. Elsevier, 2007.

[3] Yeo I-SL. Modifications of dental implant surfaces at the micro- and nano-level for enhanced osseointegration. Materials (Basel). 2020, 13, 89, doi: 10.3390/ma13010089.

[4] Kunrath MF, Piassarollo Dos Santos R, Dias de Oliveira S, Hubler R, Sesterheim P, Teixeira ER. Osteoblastic cell behavior and early bacterial adhesion on macro-, micro-, and nanostructured titanium surfaces for biomedical implant applications. Int J Oral Maxillofac Implants. 2020, 35, 773–81.

[5] Petrini M, Pierfelice TV, D'Amico E, Di Pietro N, Pandolfi A, D'Arcangelo C, De Angelis F, Mandatori D, Schiavone V, Piattelli A, Iezzi G. Article influence of nano, micro, and macro topography of dental implant surfaces on human gingival fibroblasts. Int J Mol Sci. 2021, 22, 9871, doi: https://doi.org/10.3390/ijms22189871.

[6] What's the Difference Between Additive Manufacturing and Subtractive Manufacturing? 2020, https://www.plethora.com/insights/whats-the-difference-between-additive-manufacturing-and-subtractive-manufacturing-pt.

[7] Rathbone E. Additive vs. subtractive manufacturing – what's the difference? 2018, https://blogs.autodesk.com/advanced-manufacturing/2018/07/29/additive-vs-subtractive-manufacturing-whats-the-difference/.

[8] Rathbone E. Additive vs. subtractive manufacturing – what's the difference? 2018, https://blogs.autodesk.com/advanced-manufacturing/2018/07/29/additive-vs-subtractive-manufacturing-whats-the-difference/.

[9] Keefer N. Additive vs. Subtractive Manufacturing. 2020, https://www.perceptioneng.com/blo gold/additive-vs-subtractive-manufacturing.

[10] Additive vs. Subtractive Manufacturing. https://formlabs.com/blog/additive-manufacturing-vs-subtractive-manufacturing/.

[11] Additive Manufacturing vs Subtractive Manufacturing. 2020, https://monroeengineering.com/blog/additive-manufacturing-vs-subtractive-manufacturing/.

[12] Oshida Y, Tuna EB. Science and technology integrated titanium dental implant systems. In: Basu B, et al., ed. Advance Biomaterials. Wiley, 2009, 143–77.

[13] Kang MS, Lee JH, Hong SW, Lee JH, Han D-W. Nanocomposites for enhanced osseointegration of dental and orthopedic implants revisited: Surface functionalization by carbon nanomaterial coatings. J Compos Sci. 2021, 5, 23, doi: https://doi.org/10.3390/jcs5010023.

[14] Munsch M. Laser additive manufacturing of customized prosthetics and implants for biomedical applications. Laser Additive Manufacturing. Woodhead Publishing, 2017, 399–420; https://doi.org/10.1016/B978-0-08-100433-3.00015-4.

[15] Okazaki Y. Development trends of custom-made orthopedic implants. J Artif Organs. 2011, 15, 20–25.

[16] Chen J, Zhang Z, Chen X, Zhang X. Influence of custom-made implant designs on the biomechanical performance for the case of immediate post-extraction placement in the maxillary esthetic zone: A finite element analysis. Comput Methods Biomech Biomed Eng. 2017, 20, 636–44.

[17] Chen J, Zhang Z, Chen X, Zhang C, Zhang G, Xu Z. Design and manufacture of customized dental implants by using reverse engineering and selective laser melting technology. J Prosthet Dent. 2014, 112, 1088–95.

[18] Pessanha-Andrade M, Sordi MB, Henriques B, Silva FS, Teughels W, Souza JCM. Custom-made root-analogue zirconia implants: A scoping review on mechanical and biological benefits. J Biomed Mater Res B Appl Biomater. 2018, 106, 2888–900.

Acknowledgments

In closing, we would like to express our sincere appreciations to faculty members of Indiana University School of Dentistry and University of Gum School of Science for their involving directly and indirectly to the implantology research and lecturing. Our thanks should go to members of Japan Implant Practice Association for their support and valuable suggestions. We should not forget to thank individual authors of articles cited in this book. Last but not least important, our sincere gratitude goes to excellent editorial team of the De Gruyter Publishing. Thank you all!

https://doi.org/10.1515/9783110740134-016

Index

https://doi.org/10.1515/9783110740134-017